Human Drug Me...

Human Drug Metabolism

Third Edition

Michael D. Coleman
Birmingham, UK

WILEY Blackwell

Registered Office(s)
John Wiley & Sons, Inc., 111 River Street, Hoboken, NJ 07030, USA

Editorial Office
Boschstr. 12, 69469 Weinheim, Germany

For details of our global editorial offices, customer services, and more information about Wiley products visit us at www.wiley.com.

Wiley also publishes its books in a variety of electronic formats and by print-on-demand. Some content that appears in standard print versions of this book may not be available in other formats.

Library of Congress Cataloging-in-Publication Data

Names: Coleman, Michael D, author.
Title: Human drug metabolism / Michael D. Coleman.
Description: Third edition. | Hoboken, NJ : Wiley-Blackwell, 2020. | Includes bibliographical
   references and index.
Identifiers: LCCN 2019032038 (print) | LCCN 2019032039 (ebook) |
   ISBN 9781119458562 (paperback) | ISBN 9781119458609 (adobe pdf) |
   ISBN 9781119458616 (epub)
Subjects: LCSH: Drugs–Metabolism. | Xenobiotics–Psychological aspects.
Classification: LCC RM301.55 .C65 2020 (print) | LCC RM301.55 (ebook) | DDC 615.1–dc23
LC record available at https://lccn.loc.gov/2019032038
LC ebook record available at https://lccn.loc.gov/2019032039

Cover Design: Wiley
Cover Images: Dapsone antibacterial drug molecule © MOLEKUUL/science source
Author Photo: Courtesy of Michael D. Coleman

Set in 10/12pt Times by SPi Global, Pondicherry, India
Printed and bound in Singapore by Markono Print Media Pte Ltd

10 9 8 7 6 5 4 3 2 1

For Mark, Carol, and Devon

Contents

Preface

In the spring of 1986 in a crackly transatlantic telephone conversation, I recall discussing my impending postdoctoral work at the Walter Reed Army Institute of Research in Washington, DC, with Dr Melvin H. Heiffer, then Chief of the Pharmacology Section. I distinctly remember him saying that I could look forward to *'a beautiful adventure in Pharmacology'*. Whilst it started well, the dream of working in the United States darkened somewhat after a few months as I realised the true depth and resonance of a phrase I used to see as I passed by, which was set in stone on the front of the now demolished Stanley Hospital in Liverpool: *'I was sick and ye visited me'*.

However, a year or so after the conversation with Mel, I remember experiencing an intense epiphany of gratitude whilst cruising down the middle lane of a busy I-95 near Bethesda, Maryland, in the '69 Dodge you can see on the back cover. Like self-effacing molecular superheroes, rifampicin, isoniazid, ethambutol, and streptomycin, were in the process of returning to me the life I would have inevitably lost had I been born in another age. Indeed, our more distant ancestors would have regarded the transformative power of many of the latest therapeutic approaches as something akin to magic.

Unfortunately, we know the reality of drug therapy is not always quite as restorative, as not only does it sometimes fail, it can also cause serious harm. Pharmacology is a large subject area, and drug metabolism is just a component of it. However, I hope that this book illustrates the importance of some knowledge of drug biotransformation in the process of realising the full and dazzling potential of many modern drugs whilst minimizing adverse reactions as far as humanly possible.

This is the third edition of this book, and I have tried to improve and update the text to reflect the advances in drug metabolism in the decade since the last edition. The basic outline of the book has not changed, but the text is now referenced, using a modified Vancouver style, with each citation numbered consecutively and then listed at the end of each chapter. It is not as densely referenced as a scientific review paper might be, but it does provide sources that can act as channels to a wider understanding of each covered area. I hope that this will make the book more useful and practical as a starting point in this subject, without losing the approachability of the text. Whether I have achieved this aim remains to be seen, but the intention is to provide all those engaged in the study and use of drugs at every level, with insights into how knowledge of drug metabolism can help optimize therapeutics, whilst minimizing adverse reactions.

More than ever, therapeutics is a complex joint enterprise where many participants contribute their expertise towards the welfare of the patient. So the book is intended to be accessible to everyone who has ambitions to be involved in every stage of a drug's life, from inception to clinical use and perhaps even withdrawal. Hence, the book is intended to reach students, practitioners, health-care professionals, and scientists, providing an accessible source of information and reference material for further study.

Chapter 1 as before covers some basic concepts such as clearance and bioavailability. Chapter 2 looks at the evolution of biotransformation and how it both serves and protects us. How drugs fit into this carefully managed biological environment and what this means for drug action is examined from Chapter 3 onwards. In this edition, more is now included on how cytochrome P450s function and how they are modulated at the sub-cellular level, as well as the usual list of the basic oxidations, reductions, and hydrolyses that are necessary for any basic text in drug metabolism.

Chapter 4 opens with some concepts that I hope will help those who might struggle with the relationship of Pharmacology to Physiology. The themes briefly explored include how biological systems control themselves at the most basic level and how to map this concept onto normal and aberrant physiological processes, followed by the effect of pharmacological intervention. It also provides some very basic perspectives on why drugs sometimes fall short of expectations in terms of efficacy. This echoes a theme of Chapter 2, in relation to endocrine disruption, where living systems must battle a sea of 'hostile' chemicals to preserve their identity, purpose, and activity, and the essential prerequisite is to detect and control endogenous and exogenous chemicals, who are, respectively, our molecular servants and possible threats to our well-being. Chapter 4 also outlines the cellular systems that act on the information received in these molecules, including drugs and what this means for drug efficacy. In this edition, I have tried to reflect the recent increased understanding of biosensors such as PXR and how they influence so many endogenous functions, as well as their role in disease.

Chapter 5 focusses on inhibition of biotransformational activity, both in regard to the main mechanisms involved, classes of drug, and chemical inhibitors and finally what this means clinically. It provides an opportunity to discuss drug classes such as the azoles and antidepressants in a wider clinical, pharmacological, and sometimes even a social context.

Chapter 6 outlines what is still termed 'Phase II' conjugative and related biotransforming reactions. It reflects recent advances in knowledge in a number of enzyme systems, such as in glucuronidation, sulphation, and in some of the methylating enzyme systems and their clinical significance.

Chapter 7 develops themes from the second edition, as it attempts to chart the progress of personalised medicine from concepts towards practical inclusion in routine therapeutics. In the last decade, some extremely important advances have been made in terms of customising medicines to ethnic groups and the technology to determine genetic enzymatic identity has literally reached a retail outlet near you, in some cases. However, some of the obstacles that remain in the path towards personalized medicine are discussed. Other aspects of what the patient brings to the therapeutic situation are also covered, in relation to age, gender, and personal habits in this chapter.

Chapter 8 focusses on our understanding of how and why drugs injure us, both in predictable and unpredictable ways. The role of the immune system in drug toxicity is now better understood than a decade ago, yet still reactions linked to drug-immune problems can be as life-threatening as they are unpredictable. However, advances in pharmacogenetics have brought us to the point where we can now warn whole ethnic groups of people not to take certain drugs, which is already saving lives and preventing much suffering.

Appendix A is intended to provide a perspective on how knowledge of biotransformation is an integral part of drug development and how it is feeding into the design as well as the eventual applications of particular drugs. Some of the main experimental models are explored within the context of the business and scientific models the pharmaceutical industry currently uses. Appendix B, has been expanded to look at the less-than-beautiful adventures in Pharmacology that some drugs of abuse are bringing to users around the world over the past decade, as well as some information on their effects, disposition, and routes of metabolism as far as they are known. Appendix C as before is intended to assist students to marshal and focus their inner Einstein towards examinations, and finally, Appendix D provides a brief list of some major drugs and their assigned major biotransforming enzymes.

Once again, I would like to thank my mother, Jean, for her encouragement and my wife, Clare, for her forbearance when I spent seemingly ever-increasing amounts of time on this project. Unfortunately, no matter how much effort is devoted to a task, there are always errors and omissions. However, in memory of Mel Heiffer, whose courage matched his eloquence, I hope that this book makes your own personal adventure in Pharmacology perhaps just a shade more beautiful.

Michael D. Coleman,
Bournville, Birmingham

1 Introduction

1.1 Therapeutic window

1.1.1 Introduction

It has been said that if a drug has no side effects, then it is unlikely to work. Drug therapy labours under the fundamental problem that usually every single cell in the body has to be treated just to exert a beneficial effect on a small group of cells, perhaps in one tissue. Although drug-targeting technology is improving rapidly, most of us who take an oral dose are still faced with the problem that the vast majority of our cells are being unnecessarily exposed to an agent that at best will have no effect, but at worst will exert many unwanted effects. Essentially, all drug treatment is really a compromise between positive and negative effects in the patient. The process of drug development weeds out agents that have seriously negative actions and usually releases onto the market drugs that may have a profile of side effects, but these are relatively minor within a set concentration range where the drug's pharmacological action is most effective. This range, or *therapeutic window*, is rather variable, but it will give some indication of the most 'efficient' drug concentration. This effectively means the most beneficial pharmacodynamic effects for the minimum side effects.

The therapeutic window (Figure 1.1) may or may not correspond exactly to active tissue concentrations, but it is a useful guideline as to whether drug levels are within the appropriate range. Sometimes, a drug is given once only and it is necessary for drug levels to be within the therapeutic window for a relatively brief period, perhaps when paracetamol (acetaminophen) is taken as a mild analgesic. However, the majority of drugs require repeated dosing in time periods that range from a few days for a course of antibiotics, to many years for anti-hypertensives and antithyroid drugs. During repeated intermediate and long-term dosing, drug levels may move below or above the therapeutic window due to events such as patient illness, changes in diet, or co-administration of other drugs. Below the lowest concentration of the window, it is likely that the

Human Drug Metabolism, Third Edition. Michael D. Coleman.
© 2020 John Wiley & Sons, Inc. Published 2020 by John Wiley & Sons, Inc.

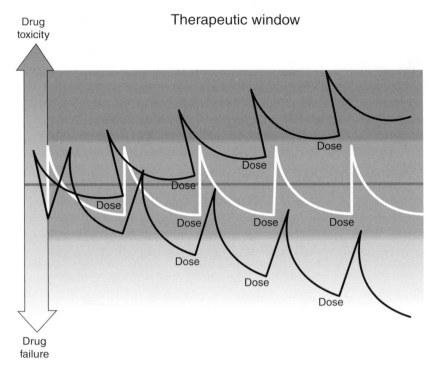

Figure 1.1   The therapeutic window, where drug concentrations should be maintained for adequate therapeutic effect, without either accumulation (drug toxicity) or disappearance (drug failure). Such is human variation that our personal therapeutic windows are effectively unique for every drug we take

drug will fail to work, as the pharmacodynamic effect will be too slight to be beneficial. If the drug concentration climbs above the therapeutic window, an intensification of the drug's intended and unintended (off-target) pharmacodynamic actions will occur. If drug levels continue to rise, significant adverse effects may ensue which can lead to distress, incapacitation or even death. To some extent, every patient has a unique therapeutic window for each drug they take, as there is such huge variation in our pharmacodynamic drug sensitivities. This book is concerned with what systems influence how long a drug stays in our bodies.

Whether drug concentrations stay in the therapeutic window is obviously related to how quickly the agent enters the blood and tissues prior to its removal. When a drug is given intravenously, there is no barrier to entry, so drug input may be easily and quickly adjusted to correspond with the rate of removal within the therapeutic window. This is known as *steady state*, which is the main objective of therapeutics. The majority of drug use is by other routes such as oral or intramuscular rather than intravenous, so there will be a considerable time lag as the drug is absorbed from either the gastrointestinal tract (GIT) or the muscle, so achieving drug levels within the therapeutic window is a slower, more 'hit and miss' process. The result from repeated oral dosing is a rather crude peak/trough pulsing, or 'sawtooth' effect, which you can see in Figure 1.1. This should be adequate, provided that the peaks and troughs remain within the confines of the therapeutic window.

1.1.2   Therapeutic index

Drugs vary enormously in their toxicity and indeed, the word *toxicity* has a number of potential meanings. Broadly, it is usually accepted that toxicity equates with harm to the individual. However, 'harm' could describe a range of impacts to the individual from mild to severe, or reversible to irreversible, in any given time frame. There is a detailed discussion on what constitutes toxicity in Chapter 8 (sections 2 and 3), but for the meantime, the broad process of harm might begin with supra-therapeutic 'pharmacological' reversible effects, progressing through to irreversible, damaging toxic effects with ascending dosage. Indeed, the concentrations at which one drug might cause potentially harmful or even lethal effects might be 10 to 1000 times lower than a much less toxic drug. A convenient measure for this is the *therapeutic index* (TI). This has been defined as the ratio between the lethal or toxic dose and the effective dose that shows the normal range of pharmacological effect.

In practice, a drug like lithium, for example, is listed as having a narrow TI if there is twofold or less difference between the lethal and effective doses, or a twofold difference in the minimum toxic and minimum effective concentrations. Back in the 1960s, many drugs in common use had narrow TIs, such as barbiturates, that could be toxic at relatively low levels. Since the 1970s, the drug industry has aimed to replace this type of drug with agents with much higher TIs. This is particularly noticeable in drugs used for depression. The risk of suicide is likely to be high in a condition that takes some time (often several weeks) to respond to therapy. Indeed, when tricyclic antidepressants (TCAs) were the main treatment option, these relatively narrow TI drugs could be used by the patient to end their lives. Fortunately, more modern drugs such as the SSRIs (selective serotonin reuptake inhibitors) have much higher TIs, so the risk of the patient using the drugs for a suicide attempt is greatly diminished. However, many drugs (including the TCAs to a limited extent) remain in use that have narrow or relatively narrow TIs (e.g. phenytoin, carbamazepine, valproate, warfarin). Therefore, the consequences of accumulation of these drugs are much worse and happen more quickly than drugs with wide TIs.

1.1.3   Changes in dosage

If the dosage exceeds the rate of the drug's removal, then clearly drug levels will accumulate and depart from the therapeutic window towards potential harm to the patient. If the drug dosage is too low, levels will fall below the lowest threshold of the window and the drug will fail to work. If a patient continues to respond well at the same oral dose, then this is effectively the oral version of steady state. So, theoretically, the drug should remain in its therapeutic window at this 'correct' dosage for as long as therapy is necessary unless other factors change this situation.

1.1.4   Changes in rate of removal

The patient may continue to take the drug at the correct dosage, but at some point drug levels may drop out of, or alternatively exceed, the therapeutic window. This could be linked with redistribution of the drug between bodily areas such as plasma and a particular organ, or protein binding might fluctuate; however, provided dosage is unchanged, significant fluctuation in drug levels within the therapeutic window will be due to change in the rate of removal and/or inactivation of the drug by active bodily processes.

1.2   Consequences of drug concentration changes

If there are large changes in the rate of removal of a drug, then this can lead *in extremis* to severe problems in the outcome of the patient's treatment: the first is drug failure, whilst the second is the drug causing harm (Figure 1.2). These extremes and indeed all drug effects are directly related to the blood concentrations of the agent in question.

1.2.1   Drug failure

Although it might take nearly a decade and huge sums of money to develop a drug that is highly effective in the vast majority of patients, the drug can only exert an effect if it reaches its intended target in sufficient concentration. Assuming that the patient has taken the drug, there may be many reasons why sufficient systemic concentrations cannot be reached. Drug absorption may have been poor, or it may have been bound to proteins or removed from the target cells so quickly it cannot work. This situation of drug 'failure' might occur after treatment has first appeared to be successful, where a patient was stabilized on a particular drug regimen, which then fails due to the addition of another drug or chemical to the regimen. The second drug or chemical causes the failure by accelerating the removal of the first from the patient's system, so drug levels are then too low to be effective. The clinical consequences of drug failure can be serious for both for the patient and the community. In the treatment of epilepsy, the loss of effective control of the patient's seizures could lead to injury to themselves or others. The failure of a contraceptive drug would lead to an unwanted

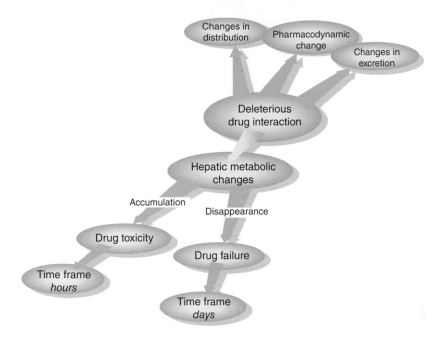

Figure 1.2 Consequences of drug interactions in terms of metabolic changes and their effects on drug failure and toxicity

pregnancy and the failure of an antipsychotic drug could mean hospitalization for a patient at the very least. For the community, when the clearance of an antibiotic or antiparasitic drug is accelerated, this causes drug levels to fall below the minimum inhibitory concentration, thus selecting drug-resistant mutants of the infection. Therapeutic drug failure is usually a gradual process, where the time frame may be many days before the problem is detected (Figure 1.2).

1.2.2 Drug toxicity

If a drug accumulates for any reason, either by overdose or by a failure of drug removal, then serious adverse reactions can potentially result. A reduction in the rate of removal of the drug from a system (often due to administration of another drug), will lead to drug accumulation. Harm to the patient may occur through the gradual intensification of a drug's therapeutic action, which progresses to off-target effects. The situation may even progress to irreversible damage to tissue or a whole organ system. For example, if the immunosuppressive cyclosporine is allowed to accumulate, severe renal toxicity can lead to organ failure. Excessive levels of anticonvulsant and antipsychotic drugs cause confusion and drowsiness, whilst the accumulation of the now-withdrawn antihistamine terfenadine can lead to lethal cardiac arrhythmias. In contrast to drug failure, drug toxicity may occur much more rapidly, often within hours rather than days.

1.3 Clearance

1.3.1 Definitions

The consequences for the patient when drug concentrations either fall below the therapeutic window or exceed it can be life threatening. The rate of removal of the drug from the body determines whether it will disappear from, or accumulate in, the patient's blood. A concept has been devised to understand and measure rate of removal; this is known as *clearance*. This term does not mean that the drug disappears or is 'cleared' instantly. The definition of clearance is an important one that should be retained:

Clearance is the removal of drug by all processes from the biological system.

A more advanced definition could be taken as:

A volume of fluid (plasma, blood or total body fluid) from which a drug is **irreversibly** *removed in unit time.*

Clearance is measured in millilitres of blood or plasma per min (or litres per hour) and is often taken to mean the 'clearance' of the drug's pharmacological effectiveness, which resides in its chemical structure. Once the drug has been metabolized, or 'biotransformed', even though only a relatively trivial change may have been made in the structure, it is no longer as it was, and products of metabolism, or *metabolites* as they are known, often exert less or even no therapeutic effect. An exception would be a pro-drug, such as the antithrombotic agent clopidogrel, which is inactive unless it is metabolised to its active form (Chapter 7.2.2).

Whether or not a metabolite retains some therapeutic effect, it will usually be removed from the cell faster than the parent drug and will eventually be excreted in urine and faeces. There are exceptions where metabolites are comparable in pharmacological effect with the parent drug (some tricyclic antidepressants, such as desipramine and morphine-6-glucuronide). In addition, there are metabolites that are strangely even less soluble in water and harder to excrete than the parent compound (acetylated sulphonamides), but in general, the main measure of clearance is known as total body clearance, or sometimes, systemic clearance:

$$Cl_{total}$$

This can be regarded as the sum of all the processes that can clear the drug. Effectively, this means the sum of the liver and kidney contributions to drug clearance, although the gut, lung and other organs can make a significant contribution. For drugs like atenolol or gabapentin, which unusually do not undergo any hepatic metabolism, or indeed metabolism by any other organ, it is possible to say that:

$$Cl_{\text{total}} = Cl_{\text{renal}}$$

So renal clearance is the only route of clearance for these drugs, in fact it is 100% of clearance. For paracetamol and for most other drugs, total body clearance is a combination of hepatic and renal clearances:

$$Cl_{\text{total}} = Cl_{\text{hepatic}} + Cl_{\text{renal}}$$

For ethanol, you may know there are several routes of clearance, including hepatic, renal, and the lung, as breath tests are a well-established indicator of blood concentrations.

$$Cl_{\text{total}} = Cl_{\text{hepatic}} + Cl_{\text{renal}} + Cl_{\text{lung}}$$

Once it is clear what clearance means, then the next step is to consider how clearance occurs.

1.3.2 Clearance and elimination

In absolute terms, to clear something away is to get rid of it, to remove it physically from a system. The kidneys are mostly responsible for this absolute removal, known as *elimination*. As we will see, the whole of the biotransforming system as it applies to xenobiotic (foreign) and endobiotic (such as our hormones) compounds could be said to have evolved around the strengths and limitations of our kidneys. Large chemical entities like proteins, cannot normally be filtered from blood by the kidneys, but they do remove the majority of smaller chemicals, depending on size, charge, and water solubility. Necessary nutrients are actively reclaimed before the soluble filtrate waste eventually reaches the collecting tubules that lead to the ureter and thence to the bladder. However, as the kidney is a lipophilic (oil-loving) organ, even if it filters lipophilic drugs or toxins, these are likely to leave the urine in the collecting tubules, enter the surrounding lipophilic tissues, and return to the blood. So the kidney is not actually capable of eliminating lipophilic chemicals or anything that is not soluble in water.

1.3.3 Biotransformation prior to elimination

It is clear that lipophilic agents *must* be made water soluble enough to be cleared by the kidney, which means they must be structurally altered, which is in turn achieved through biotransformation. The organs that have the most significant roles in this activity are the liver, gut, kidney, and lung. These organs must extract a drug from the circulation, biotransform it, then return the hopefully water-soluble metabolite to the blood for the kidney to remove. Of the biotransforming organs, the liver has the greatest role in this process and like the kidney, it can metabolize and physically remove some metabolic products from the circulation. This happens through the excretion of higher molecular weight (350–500 Daltons) drugs and metabolites into bile, where they travel through the gut to be either further metabolized by the gut microbiota, or eventually eliminated in faeces. The microbiota can have very profound effects on the disposition of a drug and its metabolites, which is discussed in Chapter 6 (section 6.2.12).

1.3.4 Intrinsic clearance

The liver and the other biotransforming organs all possess an impressive and diverse array of enzymatic systems to biotransform drugs, toxins, and other chemical entities to more water-soluble products. However, the ability of any biotransforming organ to metabolize a given drug can depend on several factors.

A key factor is termed the *intrinsic* ability of the metabolising systems in the organ to biotransform the drug. This partly depends on the structure and physicochemical characteristics of the agent and often how closely it resembles an endogenous chemical. This 'intrinsic clearance' is independent of other key factors such as blood flow and protein binding. Indeed, those two latter factors, together with its other physicochemical properties, all influence how much drug is actually presented to the biotransforming enzymes in any given timeframe. Clearly, if a drug is very tightly protein bound, it might be confined in the plasma or deep tissues. Similarly, an extremely lipophilic agent might be trapped in membranes or fatty tissues. These factors will retard the availability and presentation of free drug to the enzyme systems.

1.3.5 Clearance: influencing factors

If we can measure and account for all these factors, perhaps physically and/or theoretically, we can estimate whether a drug is relatively easy or difficult to clear. Usually, because the liver is the largest biotransforming organ and makes the greatest contribution to the clearance of most drugs, it is the main focus of attempts to categorize the degree of efficiency of biotransformation. Although much of pharmacokinetic analyses focus on the liver, terms such as extraction and intrinsic clearance can also be applied to other organs with significant biotransforming capability. So, for any given biotransforming organ, drug clearance is influenced by organ size and blood flow, the total amount of biotransforming enzymes, their intrinsic clearance, protein binding, and lipophilicity. Total body clearance, as mentioned in section 1.3.1, is the removal of drug from all tissues and is the sum total of all the individual 'clearances' of a drug, ranging from the cellular to the organ level.

1.4   First pass and drug extraction

1.4.1   First pass: gut contribution

From a therapeutic perspective, the primary goal is to use the most practical, robust, and painless method of administering a drug regularly, so that its plasma concentrations can reach the therapeutic window and hopefully stay there. Whilst oral dosage is the obvious choice here, this is rather like attempting to enter a well-garrisoned castle by politely knocking at the front gate. Indeed, in many cases (Figure 1.3), even assuming oral absorption is complete, relatively little of the drug actually enters the circulation after what is termed *first pass,* which is the result of the biotransformational processes that essentially destroy most of the pharmacological impact of many drugs, by clearing them to metabolites.

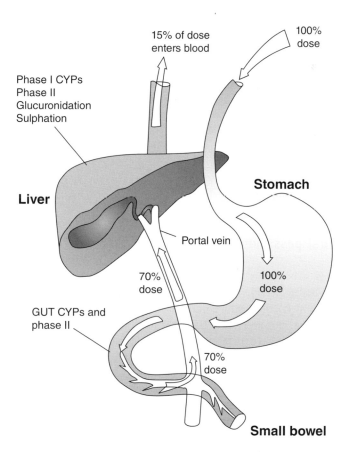

Figure 1.3   The first pass of an orally dosed highly cleared drug showing the removal of drug by the gut and liver, leading to relatively low levels of the drug actually reaching the circulation

However, the first-pass process seems to start quite promisingly, as most drugs easily diffuse through the membranes of the gut enterocytes passively, due to their relative lipophilicity or if they are more water soluble, they can enter through the spaces between the tight junctions of the enterocytes, which is known as the *paracellular pathway*. Indeed, water-soluble drugs can even be assisted by transporter systems called solute carriers (Chapter 2.6.3). These transporters normally convey vital nutrients such as amino acids as well as drugs with similar physicochemical characteristics (like some statins).

Once the drug penetrates the enterocytes, the situation changes dramatically. Until relatively recently, it was not fully understood just how great a contribution the gut biotransforming systems made to first pass. This process begins with the efflux proteins (Chapter 4.4.7), which will expel many drugs back into the gut lumen. It is thought that drug molecules can be reabsorbed and expelled repeatedly, which can retard absorption, and/or provide more opportunity for the agent to meet the biotransforming enzymes; it may even be part of a cooperative process that prevents saturation of these enzymes. Although this might occur for endogenous substrates, it is more difficult to establish with drugs[1].

It is now clear that enterocytes contain many of the most significant hepatically expressed biotransforming enzymes and they exert an impact in terms of drug metabolism which far exceeds their actual mass; this suggests that biotransformation in the gut must be exceptionally efficient. This is probably because of a combination of high local drug concentrations, high intrinsic clearance, and long gut transit times. Indeed, limiting factors, such as restricted blood flow and protein binding, are absent also. It is well established that about a third of drugs used commonly undergo significant gut metabolism and with some agents like the benzodiazepine midazolam and the immunosuppressive agent cyclosporine, up to half the dose can be biotransformed through this route before the drug even reaches the liver[2].

1.4.2   First pass: hepatic contribution

Whilst first pass is clearly an integrated enterprise between the gut and the liver, determining the respective contributions of the two organ systems to this enterprise is difficult. It could be desirable to achieve this during clinical research with existing drugs, or perhaps during the development of a new drug, when it is necessary to know where metabolism occurs in order to anticipate future drug–drug interactions, for example. In the past, isolating the contribution of the gut has been accomplished in patients by sampling directly from the hepatic portal system[3], or less invasively, through the use of inhibitors that inactivated gut metabolising capability, allowing an estimate of hepatic metabolism[4]. More recently, sophisticated pharmacokinetic models have been developed that use various estimated parameters to calculate hepatic metabolism without recourse to such problematic methods[5], and this is a key area in experimental drug development, which will be discussed in more detail in Appendix A.

At a fairly basic level, the liver's contribution to first pass is usually stated as hepatic extraction, which is the difference between the drug level in portal blood that enters the liver (100%) and the amount that escapes intact and unmetabolized (that is, 100% minus the metabolized fraction) after a single pass through the organ. Extraction is usually termed E and is defined as the extraction ratio, or

$$E = \frac{\text{Concentration leaving the liver}}{\text{Concentration entering the liver}}$$

Clinically, most drugs' hepatic extraction ratios will either be high ($E > 0.7$) or low (E $<$ 0.3), with a few agents falling into the intermediate category (E is 0.3–0.7). For high-extraction drugs, the particular enzyme system that metabolizes this drug may be present in large amounts and drug processing is very rapid. As already mentioned, a close structural resemblance to an endogenous agent, which is normally processed in great quantity on a daily basis, will probably lead to high extraction. Hence, the early anti-HIV drug AZT (zidovudine), is a close structural analogue of the DNA constituent thymidine and so possesses a half-life of an hour or less in humans. In the case of a high-extraction drug, intrinsic clearance is so high that the only limitation in the liver's ability to metabolize this type of drug is its rate of arrival, which is governed by blood flow.

So, in the case of a high-clearance drug, where the liver's intrinsic ability to clear it is very high:

$$Cl_{\text{hepatic}} = Q(\text{liver blood flow}) \times E(\text{Extraction ratio}) \text{ i.e.}$$
$$Cl_{\text{hepatic}} = QE$$

So, basically, hepatic clearance is directly proportional to blood flow:

$$Cl_{\text{hepatic}} \quad \alpha \quad Q$$

Theoretically, hepatic and intrinsic clearance would be the same value if blood flow was infinite. In the real world, there are, of course, many other factors that influence clearance, and in the early 1970s, it was realised that some form of theoretical model needed to be evolved that would guide the process of predicting hepatic clearance and that took these other factors into account, such as protein binding and intrinsic clearance, as well as blood flow. This was the 'well stirred model' of hepatic clearance and can be summarized thus:

$$Cl_{\text{h}} = \frac{Q \times f_{\text{u}} \times Cl_{\text{int}}}{(Q + f_{\text{u}}) \times Cl}$$

where Cl_{h} is hepatic clearance, f_{u} is the fraction of unbound drug, Q is liver blood flow, and Cl_{int} is hepatic intrinsic clearance.

This equation makes many assumptions, but an important one is that it concerns the clearance of drug from whole blood, rather than plasma[5]. So to employ this model, we can use data from *in vitro* methods, such as derived from human cells or tissues to calculate an intrinsic clearance. Protein binding can also be measured in the laboratory and during patient studies. Current ultrasound techniques allow for accurate and noninvasive real-time determination of liver blood flow[6]. Of course, during intensive exercise, values can fall temporarily by more than 70%, but during normal day-to-day living, blood flow through the liver does not change very much. So we can combine all this information in our well-stirred model to estimate the clearance of a high-extraction drug in a typical healthy 70 kg adult.

However, as we reach old age, hepatic blood flow can fall to approaching half what it was in our twenties (Chapter 7.3.1). In addition, liver damage from end-stage cirrhotic alcoholism (Chapter 7.7.6), or any long-term impairment in cardiac output, will also reduce liver blood flow. All these circumstances have been shown to retard the clearance of high-extraction drugs clinically and should be borne in mind during drug dosage determination in these patients.

Many drugs are bound in plasma to proteins such as human serum albumin (HSA) or alpha-1 acid glycoprotein (AGP). HSA usually transports endogenous acidic agents, such as fatty acids, bilirubin, and bile acids, although it also binds drugs such as warfarin, ibuprofen, and diazepam. AGP, which is also known as orosomucoid, is one of a number of acute phase proteins, which can increase several-fold in plasma in response to infections or increased inflammatory status and is regulated by cytokines. Elevated AGP levels can last for weeks or months and have been linked with life-threatening conditions such as stroke[7,8]. AGP will bind many basic drugs such as erythromycin, some antimalarials, and protease inhibitors; indeed, changes in AGP during infections have long been known to strongly impact drug plasma disposition[9,10].

Usually, for any given drug, there is equilibrium between protein-bound and free drug. In effect, high-extraction drugs are cleared so avidly that the free drug disappears into the metabolizing system and the bound pool of drug eventually becomes exhausted. As the protein binding of a high-extraction drug is no barrier to its removal by the liver, these drugs are sometimes described as undergoing *unrestricted* clearance. Drugs in this category include pethidine (known as meperidine or Demerol in the United States), metoprolol, propranolol, lignocaine, nifedipine, fentanyl and verapamil.

1.4.3 First pass: low-extraction drugs

On the opposite end of the scale ($E < 0.3$), low-extraction drugs are cleared slowly, as the metabolizing enzymes have some difficulty in oxidizing them, perhaps due to stability in the structure, or the low capacity and activity of the metabolizing enzymes. The metabolizing enzymes may also be present only in very low levels. These drugs are considered to be low intrinsic clearance drugs, as the inbuilt ability of the liver to remove them is relatively poor.

If a low-extraction drug is not extensively bound to protein (less than 50% bound), then how much drug is cleared is related directly to the intrinsic clearance of that drug. In the case of a low-extraction, strongly protein-bound drug, then, clearance is hampered as the affinity of the liver for the drug is lower than that of the binding protein. The anticonvulsants phenytoin and valproate are both highly protein bound ($\sim 90\%$) and low-extraction drugs, so the amount of these drugs actually cleared by the liver really depends on how much unbound or free drug there is in the blood. This means that:

$$Cl_{\text{hepatic}} \ \alpha \ Cl_{\text{intrinsic}} \times \ f_u$$

Therefore, clearance is proportional to the ability of the liver to metabolize the drug ($Cl_{\text{intrinsic}}$) as well as the amount of unbound or free drug in the plasma that is actually available for metabolism. Hepatic blood flow changes have little or no effect on low-extraction drug plasma levels, but if the intrinsic ability of the liver to clear a low-extraction drug falls even further (due to enzyme inhibition or gradual organ failure), there will be a significant increase in plasma and tissue free drug levels and dosage adjustment will be necessary. Conversely, if the intrinsic clearance increases (enzyme induction, Chapter 4.3) then free drug levels may fall and the therapeutic effects of the agent will be diminished.

It is worth noting that with drugs of low extraction and high protein binding, such as phenytoin and valproate, a reduction in total drug levels due to a fall in protein binding (perhaps due to renal problems or displacement by another, more tightly bound drug) will actually have no sustained effect on free drug plasma and tissue levels. This is because the 'extra' free drug will be cleared or enter the tissues and the bound/unbound drug ratio will quickly reassert itself. Since the free drug is pharmacologically active and potentially toxic whilst the bound drug is not, it is not usually necessary to increase the dose in these circumstances. The concentration of the free drug has the greatest bearing on dosage adjustment considerations. Laboratory assay systems are now routinely used to determine free drug levels with highly bound, low-extraction drugs that are therapeutically monitored, such as with phenytoin and valproate. Other examples of low-extraction drugs include paracetamol, mexiletine, diazepam, naproxen, and metronidazole. The term *restrictive clearance* is also used to describe these drugs, as their clearance is effectively restricted by their protein binding.

1.5   First pass and plasma drug levels

1.5.1   Introduction

As mentioned above, for many drugs, a significant amount of the oral dose is lost before it reaches the systemic circulation. To estimate a dosage regime that will place our plasma drug concentrations within our therapeutic window, we need to know how much drug is lost during the first pass process.

To accomplish this, we can measure blood or plasma concentrations after the drug is given intravenously, where we can safely assume 100% absorption. Next, we measure drug concentrations after an oral dose, which will be significantly lower. Using pharmacokinetic methods and many blood measurements, we can determine the amount of drug measured in blood after intravenous or oral dosage. We know the dose, so we can calculate the amount of drug that survives the first pass. This is termed F or the *absolute bioavailability* of the drug. It can be defined as

$$F = \frac{\text{Total amount of drug in the systemic circulation after a single oral dosage}}{\text{Total amount of drug in the systemic circulation after intravenous dose}}$$

As we have seen with clearance, F is not just one figure. It is the product of F_a (absorption into the enterocytes), F_g (amount that survives gut metabolism), and F_h (the amount that survives hepatic metabolism), assuming other organs do not contribute significantly.

Highly extracted drugs are often stated to have a poor bioavailability. This means that the oral dose required to exert a given response is much larger than the intravenous dose. If the bioavailability is 0.2 or 20%, then you might need to administer about five times the intravenous dose to see an effect orally. Drug companies work hard to try to overcome poor bioavailability with new drugs, as you can see that such a drug candidate would be vulnerable to many different and variable factors in its progress towards the systemic circulation. This, in turn, makes it more problematic to use in the real world.

1.5.2 Changes in clearance and plasma levels

Consider an extreme example. If the intravenous dose of a poorly bioavailable ($F = 0.2$), narrow TI drug X was 20 mg and the usual oral dose was 100 mg, it is clear that if the whole oral 100 mg were to reach the plasma, the patient would then have plasma levels far in excess of the normal intravenous dose, which could lead to toxicity or death. This could happen if the first-pass effect was reduced or even completely prevented by factors that changed the drug's clearance. Similarly, if the clearance of the drug was to be accelerated, then potentially none of the 100 mg would reach the plasma at all, causing lack of efficacy and subsequent drug failure.

1.6 Drug and xenobiotic metabolism

From the therapeutic point of view, it is essential to ensure that drug concentrations remain within the therapeutic window and neither drug failure nor drug toxicity occur in the patient. To understand some of the factors related to drug metabolism that can influence the achievement of these aims, there are several important points to consider over the next few chapters of this book:

- What are the metabolic or biotransformational processes that can so dramatically influence drug concentrations and therefore drug action?

- How do these processes sense the presence of the drugs and then remove these apparently chemically stable entities from the body so effectively?

- What happens when these processes are inhibited by other drugs, dietary agents, and toxins?

- What are the effects of illness, genetic profile, and other patient circumstances on the operation of these processes?

- How can changes in these processes of drug removal lead to toxicity?

- What did these processes originally evolve to achieve, and what is their endogenous function?

The next chapter considers the last point and illustrates that in a subject usually termed *drug metabolism,* modern drugs are newcomers to an ancient, complex, and highly adaptable system that has evolved to protect living organisms, to control instruction molecules, and to carry out many physiological tasks.

References

1. Shi, S, Li, Y. Interplay of drug-metabolizing enzymes and transporters in drug absorption and disposition. Curr. Drug Metab 15, 915–941, 2014.
2. Peters, SA, Jones, CR, Ungell, AL et al. Predicting drug extraction in the human gut wall: assessing contributions from drug metabolizing enzymes and transporter proteins using preclinical models. Clin. Pharmacokinet. 55, 673–696, 2016.
3. Kolars JC, Awni WM, Merion RM. First pass metabolism of cyclosporine by the gut. Lancet, 338, 1488–1490, 1991.
4. Tsunoda, SM, Velez, RL, von Moltke LL et al., Differentiation of intestinal and hepatic cytochrome P450 3A activity with use of midazolam as an *in vivo* probe: Effect of ketoconazole. Clin. Pharmacol & Ther. 66, 461–471, 1999.
5. Yang J, Jamei M, Yeo KR. Misuse of the well-stirred model of hepatic drug clearance. Drug Metab. Disp. 35, 501–502, 2007
6. Maruyama H and Yokasuka O. Ultrasonography for noninvasive assessment of portal hypertension. Gut and Liver, 4, 464–273, 2017.
7. Jorgensen JM, Yang Z, Lonnerdal B, et al., Plasma ferritin and hepcidin are lower at 4 months postpartum among women with elevated C-reactive protein or α1-acid glycoprotein. J. Nutr 147, 1194–1199, 2017.

8. Berntsson J, Östling G, Persson M. et al. Orosomucoid, carotid plaque, and incidence of stroke. Stroke. 47, 1858–1863, 2016.

9. Silamut K, Molunto M, Ho M, et al. Alpha-1-acid glycoprotein (orosomucoid) and plasma protein binding of quinine in falciparum malaria Br. J. Clin. Pharmac. 32, 311–315, 1991.

10. Lee LHN, White RF, Barr AM et al. Elevated clozapine plasma concentration secondary to a urinary tract infection : proposed mechanisms. J Psychiatry Neurosci 41(4) E67–8, 2016.

2 Drug Biotransformational Systems – Origins and Aims

2.1 Biotransforming enzymes

John Lennon once said, 'Before Elvis, there was nothing'. Biologically, this could be paraphrased along the lines of 'Before bacteria, there was nothing'. During the various moon landings of the 1970s and obviously before the release of Ridley Scott's *Alien*, I used to wonder why NASA thought it was so important to quarantine astronauts, as what could possibly survive outer space. We now know that bacterial life began on this planet perhaps only a billion years or so after its formation and that Earth then was in many ways almost as hostile an environment as Space itself. Nonetheless, bacteria managed to establish themselves above and below the surface whilst exposed to heat, cold, radiation, corrosive/reactive chemicals, and lack of oxygen. The phenomenal growth and generation rates of bacteria enabled them to evolve their enzyme systems quickly enough to not only survive but also eventually prosper in all environmental niches. It is now seriously considered that bacterial life could be found below the surface of Mars, an environment that in some ways is as challenging as the 'young' Earth.

As life evolved beyond bacteria towards the eukaryotes, cell structures for more advanced organisms eventually settled around the format we see now, a largely aqueous cytoplasm bounded by a predominantly lipophilic protective membrane. This arrangement can also demonstrate extreme robustness, as tardigrades, a species of microanimal, can survive short periods in space and near absolute zero temperatures. Although the lipophilic membrane does prevent entry and exit of many potential toxins, it is no barrier to other lipophilic molecules. If these molecules are highly lipophilic, they will passively diffuse into and become trapped in the membrane. If they are slightly less lipophilic, they will pass through it into the organism. So aside from 'housekeeping' enzyme systems, some enzymatic protection would have been needed against internal reactive species formed through radiation, as well as other invading molecules

Human Drug Metabolism, Third Edition. Michael D. Coleman.
© 2020 John Wiley & Sons, Inc. Published 2020 by John Wiley & Sons, Inc.

from the immediate environment. Among the various molecular threats to the organism would have been the waste products of other bacteria in decaying biomass, as well as various chemicals formed from incomplete combustion. These would have included aromatic hydrocarbons (multiples of the simplest aromatic, benzene) that can enter living systems and accumulate, thus deranging useful enzymatic systems and cellular structures. Enzymes that can detoxify these pollutants such as aromatics are usually termed *biotransforming enzymes*.

2.2   Threat of lipophilic hydrocarbons

Organisms that cannot rid themselves of lipophilic aromatic and nonaromatic hydrocarbons tend to accumulate these chemicals to toxic levels. One study has shown that the marine sponge *Hymeniacidon perlevis* contains 17 times more of the carcinogen benzo(a)pyrene than the Pacific Oyster (*Crassostrea gigas*)[1]. Whilst this is partly linked to the greater quantities of seawater sponges can filter, it is also clear that the biotransforming capacity of the sponges appears to have failed to adapt to the impact of that pollutant. Unfortunately, with the continued human dependence on petrochemical technology, vast amounts of lipophilic hydrocarbons are now a fixture of the air we breathe, as well as our food and drink. Dioxin (2,3,7,8-tetrachlorodibenzo -*p*- dioxin; TCDD) is part of a series of polychlorinated dibenzo derivatives and it is one the best-studied toxic lipophilic hydrocarbons. This herbicide contaminant demonstrates perhaps an extreme form of the threat of these molecules to all life on Earth. Dioxins are not only carcinogenic and teratogenic endocrine disruptors, but also their half-lives in man can be *decades*. What is particularly worrying is that we have created molecules like dioxins of such stability and toxicity that despite the fact they actually trigger a potent cellular response intended to metabolize and clear them (Chapter 4.3.1), we still cannot eliminate them quickly[2].

Although dioxins are for humans a worst-case scenario, virtually all living organisms, including ourselves, possess effective biotransformational enzyme capability, which can detoxify and eliminate most hydrocarbons and related molecules. This capability has been appropriated from either procaryotes (organisms without nuclei, such as bacteria) or eukaryotes (organisms with a nucleus, such as yeasts), or even both, over millions of years of evolution. The main biotransformational protection against aromatic hydrocarbons, for example, is a series of enzymes so named as they absorb UV light at 450 nm when reduced and bound to carbon monoxide. These specialized enzymes were termed cytochrome P450 monooxygenases or sometimes oxido-reductases. They are often referred to as CYPs or P450s. It is likely that bacteria existed on Earth when our primitive atmosphere was almost entirely carbon dioxide (as Mars is today) and CYPs may well have evolved at first to accomplish reductive reactions in the absence of oxygen. The enzymes retain this ability, although their main function now is to carry out oxidations. CYPs are expressed by a large gene family and the functional characteristics of these enzymes are reminiscent of a set of adjustable spanners in a tool kit. All the CYPs accomplish their functions using the same basic mechanism, but each enzyme is adapted to dismantle particular groups of chemical structures. It is a testament to 2–3 billion years of 'research and development' in the evolution of CYPs, that although several thousand man-made chemical entities enter the environment for the first time every year, the vast majority can probably be metabolised by at least one form of CYP.

2.3   Cell communication

2.3.1   Signal molecule evolution

At some point in evolution, single-cell life forms began to coalesce into multicell organizations, allowing advantages in influencing and controlling the cells' immediate environment. Further down this line of development, groups of cells differentiated to perform specialized functions, which other cells would not then need to carry out. At some point in evolution, a dominant cellular group will have developed methods of communicating with other cell groups to coordinate the organisms' functions. Once cellular communication was established, other cell groups could be instructed to carry out yet more specialized development. In more advanced organisms, this command-and-control chain has two main options for communication: either by direct electrical nervous impulse or instruction through a chemical. Neural impulse control is seen where the sympathetic nervous system influences the adrenal gland by direct innervation.

For an instructional chemical such as a hormone (from the Greek meaning to urge on) to operate, its unique shape must convey information to a receptor, where the receptor/molecule complex is capable of activating the receptor to engage its function. An instructional molecule must possess four features to make it a viable and reliable means of communication:

1. It must be *stable* and not spontaneously change its shape and so lose the ability to dock accurately with its receptor.

2. It must be relatively *resistant to reacting with other cell enzymes* or chemicals it might contact, such as proteolytic enzymes on the cell surface or in the cytoplasm.

3. It must be *easily manufactured* in large amounts with the components of the molecule being readily available. It is immediately obvious that the pharmaceutical industry uses the same criteria in designing its products that often mimic that of an endogenous molecule.

4. It must also be *controllable*. It is no use to an organism to issue a 'command' that continues to be slavishly obeyed long after the necessity to obey is over. This is wasteful at best, and at worst seriously damaging to the organism, which will then carry out unnecessary functions that cost it energy and raw materials, which should have been used to address a current, more pressing problem.

The production and elimination of a chemical instruction to maintain the required concentrations of the agent must be balanced during the period that is appropriate for its function. This might range from seconds to many years.

There are inherent contradictions in this approach; the formation of a *stable* molecule, which will be easily and quickly disposable. To make a stable compound will cost energy and raw materials, although to dismantle it will also cost the organism. It all hinges on what specific purpose the instruction molecule was built to achieve.

For changes that are minute by minute, second by second, then perhaps a protein or peptide would be useful. These molecules can retain information by their shape and are often chemically stable, although the large numbers of various protease and other enzymes present at or around cell membranes mean that their half-lives can be exceedingly short. This allows fine control of a function by chemical means, as the rate of manufacture can be adjusted as necessary given that the molecule is rendered nonfunctional in seconds.

2.3.2   Lipophilic hydrocarbons as signal molecules

Unlike short-term modulations of tissue function, processes like the development of sexual maturity require sustained and progressive changes in tissue structure as well as function and these cannot be achieved through direct neural instruction. Chemical instruction is necessary to control particular genes in millions of cells over many years. To induce these changes, hormone molecules need to be assembled to be stable enough to carry an instruction (the shape and properties of the molecule) and have the appropriate physicochemical features to reach nuclear receptors inside a cell to activate specific genes.

Lipophilic hydrocarbon chemicals have a number of advantages when acting as signalling molecules. First, they are usually stable and plentiful, and their solubility in oils and aqueous media can be chemically manipulated. This sounds surprising given that they are generally known to be very oil soluble and completely insoluble in water. However, those enzymes generously bequeathed to us from bacteria such as the CYPs have evolved to radically alter the shape, solubility, and stability of aromatic molecules. This is in effect a system for 'custom building' stable instructional small molecules, which are easiest to make if a modular common platform is employed, which is usually the molecule cholesterol. From Figure 2.1 you can see the position of cholesterol and steroid hormones in

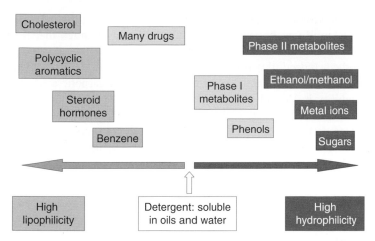

Figure 2.1 The lipophilicity (oil loving) and hydrophilicity (water loving) of various chemical entities that can be found in living organisms (not to scale)

relation to oil and water solubility, relative to a detergent, which is amphipathic, i.e. soluble in oil and water. The nearest agents with a detergent-like quality in biological systems are bile salts, which use this ability to break large fat droplets into smaller ones to aid absorption.

Cholesterol itself is very soluble in lipids and has almost zero water solubility, so it requires a sophisticated transport system to move it around the body. Although a controversial molecule for its role in cardiovascular disease, it has many vital functions, such as the formation of bile acids as well as maintenance of cell membrane fluidity. This latter function relies on the fact that cholesterol itself is so lipophilic that it is trapped in membranes. However, steroid hormones built on this hydrophobic cholesterol 'platform' also contain enough hydrophilic groups to make them much less lipophilic than their parent molecule, enabling them to cross membranes either alone or through an exocytosis/carrier molecule arrangement.

To gain a perspective on these solubilities, at room temperature ranges, oestradiol 17-β has twice the solubility of benzene in water and is 38-fold more soluble in water than cholesterol. However, oestradiol-17 β itself is around 92-fold less soluble in water than vitamin C (ascorbate)[3]. So whilst steroids are still not very water-soluble, steroid hormone synthesis has evolved to produce molecules that are able to travel through the circulation bound to the appropriate carrier molecule, and then they can leave the blood to enter cells without being confined physicochemically within membranes. Indeed, steroids can then progress through the cytoplasm, binding various sensor molecules associated with the nucleus. Ultimately, their enclosed *command* is conveyed intact to the nucleus.

Once the stable steroid platform has been built by CYPs and served its purpose, the final link in the process is the use of various other CYPs to ensure the elimination of these molecules. If the solubility of a slightly lipophilic agent such as a steroid can be significantly increased, this ensures that it will be suitable for filtration and excretion by the kidney, as it will then be too polar to be reabsorbed. Indeed, this is clear in the context of even the hydroxylation of benzene to phenol (also known as hydroxybenzene, or carbolic acid). Phenol's water solubility is 46-fold greater than benzene and 23-fold more than that of oestradiol-17 β^3. To fully satisfy the needs of the organism, the complete synthesis and degradation process is fully adjustable according to changing circumstances and can exert a remarkably fine control over steroidal pharmacology and eventual elimination. Such is the efficiency of this system that early human contraception studies showed that after an oral dose of oestradiol-17 β, systemic bioavailability was virtually zero.

2.4   False signal molecules: bioprotection

2.4.1   Endocrine disruption

As you probably are aware, drug companies work towards developing drugs one at a time, studying the effects of the new agent in various biological systems with the ultimate aim of reaching humans. Naturally, this work is intended to exert some level of control of a tissue or organ, in order to restore or improve function. At some point, they must factor into their development process the effects of perhaps several other older drugs that are likely to be prescribed to the patient population that is intended to benefit from the new drug. Indeed, *polypharmacy* is a well-known issue, especially in the elderly and chronically ill. Older drugs prescribed to these patients might affect the pharmacological or pharmacokinetic disposition of the new drug, effectively diluting or even completely eliminating the new drug's effect, thus negating the drug company's effort to regulate the process in question.

However, for biological systems, the issue of potentially losing control of a tissue or system is multiplied in importance and difficulty, perhaps thousands of times. It is now clear that oestrogen receptors will bind and function in response to a regrettably wide variety of natural and synthetic chemicals. This is mainly because large numbers of molecules have an aromatized ring in a similar orientation to a steroid. In the past, such molecules may have been polycyclic aromatic hydrocarbons from combustion, or plant sourced toxins. Such agents could enter our systems in our foodstuffs or through the air. Unfortunately, over the last half-century or so, an explosion of different manmade chemicals has been released into the environment. Some of these are prescription drugs, others are ingredients in cosmetics and toiletries like the parabens. The most significant problems include some of the agents associated with the global plastic and paint industries, such as bisphenol A, the phthalates, and various halogenated biphenyls. Many of these are now termed *endocrine disrupting chemicals* (EDCs)[4], as they can have significant oestrogenic or antioestrogenic effects, often coupled with quite long half-lives. These EDCs are now everywhere, in the water we drink, our food, and even the air. Whilst it is highly likely that EDCs do distort and even usurp hormonal regulation and effect, it is difficult to actually establish the true nature of the impact of EDCs on ecosystems, as well as human health and fertility[4,5].

2.4.2   Endocrine disruption: problems and solutions

At first glance, combatting the threat of EDCs appears to be almost impossibly difficult, as the pollutants are so similar physicochemically and structurally to our endogenous signal molecules. Indeed, to find some perspective, exogenous molecules, or *xenobiotics* (from the Greek word for 'foreign') vary enormously in their properties and water solubility. Benzo(a)pyrene, the polycyclic aromatic that the sponges in section 2.2 struggled with, is about 15-fold less water soluble than even cholesterol[3], whilst bis(2-ethylhexyl) phthalate lies between cholesterol and benzene, in terms of water solubility. At the other end of the scale, bisphenol A is nearly one and a half times more soluble in water than even phenol.[3]

Whilst this struggle for control of vital biological systems has only really become apparent to us over the last few decades, it has of course existed since life began. As early living organisms developed in complexity, they were vulnerable to xenobiotic disrupting molecules through the same sources we are today, through diet, drinking water, and inspired air. As many of the 'threat chemicals' so closely resemble endogenous signal molecules, it is entirely logical that the biotransforming enzyme systems such as the CYPs would evolve to dismantle these threat molecules as well as build and dismantle endogenous ones. Indeed, the vast duplication and multiplicity of the CYP gene families seems to have begun in earnest around 400–300 million years ago, when animals emerged from the sea and joined one of the oldest 'arms races' in history. This is the evolutionary process whereby plants and animals could make chemical agents that either poisoned their assailants or in some way influenced them to unknowingly participate in their reproductive cycles – which is again part of the battle between species to control each other. To avoid being eaten, many organisms synthesize protective toxin-like agents[6], some of which are well known to have devastating effects on both humans and animals and can be very difficult to eliminate from human diets[7,8]. On the other hand, in a harsh environment, to be able to eat such organisms safely provides an animal with a significant advantage in its survival prospects.

It is possible that the threat of xenobiotic agents was and probably remains a far more powerful driver of CYP evolution than housekeeping functions such as biosynthesising various steroid molecules. This colourful evolutionary struggle between animals, plants, and fungi is perhaps reflected in our own CYP2D6, which has a preference for substrates that resemble plant alkaloids. There are multiple copies of this gene in some ethnic groups, which has been proposed as a protective response to dietary toxins (Chapter 3, section 3.6.2, and Chapter 7, section 7.2.3). On the dark side, mycotoxins have evolved structurally to employ our own CYPs to cause lethal toxicity and carcinogenicity[8] (Chapter 8, section 8.5.6).

2.4.3   Endocrine disruption: cosmetic and nutraceutical aspects

It has long been known in folk medicine that many plants have hormone-like effects that are potent enough to influence female reproductive cycles. Over the last few decades, the production and sale of various plant extracts and the promotion of different foods rich in various phytoestrogens has become a global industry. Some of these preparations or foodstuffs have been shown to be effective in alleviating menopausal discomfort in women as an alternative to drug therapy[9]. From a cosmetic standpoint, commercial sources of plant oestrogens have long been marketed aggressively as human breast size enhancers[10], although reports of their effectiveness are largely anecdotal[9]. Interestingly, the suppliers of the various extracts of plants such as saw palmetto and fenugreek have yet, to the author's knowledge, to fund systematic investigation of these agents as viable alternatives to breast enhancement surgery, for example.

Given the potency of these chemicals and the vulnerability of breast, ovarian, and endometrial tissues to hormonally fuelled malignancy, any form of self-medication has significant risk. Indeed, long-term exposure to inappropriate hormone levels can lead to cancer in these tissues, although despite extensive investigation, it is still not clear whether the health risks of regular consumption of plant phytoestrogens outweigh their possible benefits[11]. As already mentioned, biotransforming enzymes such as the CYPs are a major defence against such unwanted molecules, and they actively protect us from exogenous hormone-like chemicals. Interestingly, the fact that plant phytoestrogens do impact human hormonal balance indicates that these agents at least partially thwart CYP systems, as they are not easy to metabolize and inactivate rapidly enough to prevent interference in human endocrine stability.

Unfortunately, it appears that our own biotransformational systems complicate the EDC issue further, as they are actually responsible for promoting the endocrine disruptive effects of some highly lipophilic toxins in our diet. In studies where benzopyrene biotransformation is prevented, much of the endocrine disrupting potency of this agent is lost[12]. Both the parent hydrocarbon and its hydroxylated metabolites have the potential to bind to oestrogen receptors, but as noted in the previous section, benzo(a)pyrene itself is so lipophilic that it is likely to be trapped in membranes or other lipid-rich areas of the cell such as the smooth endoplasmic reticulum, where it can be metabolised by the CYPs. However, the hydroxylated metabolites, such as the 7, 8 diol epoxide of benzo(a) pyrene, are more water soluble than even oestrogen itself, so their physiochemical properties may well give them greater access to cellular oestrogen receptors than the parent hydrocarbon. This situation is further complicated by the ability of polycyclic aromatic hydrocarbons to induce their own metabolism (see Chapter 4.3.1).

Whilst the future might appear bleak, in terms of the scale of environmental pollution by EDCs, the weight of evidence against agents such as bisphenol A has stung certain manufacturers into investing in alternatives, which are nonoestrogenic, effective in their role in manufacturing and economically viable to produce[13]. This suggests that it is possible for the potential damage caused by EDCs in general to be diminished, if the worldwide plastics and paint industries can be encouraged to find effective and suitable alternatives.

2.4.4   Endocrine disruption: microRNAs

If you enjoyed Francis Ford Coppola's *Godfather* films, you will be aware that the phrase *'It was Barzini all along'* could be shorthand for *'Now we know, albeit rather late in the day, who our real enemy is'*. Although we are rightly preoccupied with xenobiotic chemical threats to our homeostasis, other entities may perhaps have an altogether more profound and even potentially disturbing role to play in manipulating our endocrine and other systems through our diet. Although only discovered in the early 2000s, microRNAs (miRNAs) are noncoding single stranded molecules that regulate gene expression, including human hepatic biotransformational systems[14]. However, it also seems likely that miRNAs originating from plant species can not only survive our digestive systems but also influence human gene expression[15].

If it emerges that these miRNA are actually intended by the plant to influence protein formation in humans, it implies that this process is probably considerably more advantageous to the plant rather than to us and it raises the outlandish possibility of being effectively ordered around by our food. We know that miRNAs are already an integral part of how human biotransforming gene expression is regulated (Chapter 3.5.2) so it is likely that the full implications of such a sophisticated and high-level intervention into our biotransformational capability through dietary miRNAs will take many years to unravel.

Figure 2.2 illustrates the varying known roles of the biotransforming CYPs in living systems.

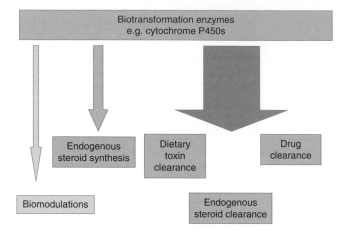

Figure 2.2 Various functions of biotransformational enzymes, from assembly of endogenous steroids, modulation of various biological processes, as well as the clearance of drugs, toxins, and endogenous steroids

2.5   Sites of biotransforming enzymes

Aside from their biotransformational roles in steroid biosynthesis and drug/toxin clearance, CYPs carry out a wide array of metabolic activities that are essential to homeostasis throughout the body. This is not surprising, as they are found in virtually every tissue. The liver and gut logically have the highest biotransformational capability. Hepatic and gut CYPs are mainly concerned with the processing and clearance of large amounts of various endogenous and xenobiotic chemicals. The CYPs and other metabolizing systems in organs such as the lung, kidney, and skin make relatively little contribution to the overall clearance of a drug, but are relevant in the formation of local toxic species from drugs and xenobiotics – of course, the mutations that eventually lead to lung cancer are linked to reactive species formed by various CYPs[16].

In the brain, again, large-scale drug clearance is not the role of the CYPs, which are often located in certain brain areas, rather than universally distributed. Overall, the net quantity of CYPs expressed is far lower than hepatic P450 levels, but in certain cellular scenarios the expression levels can actually be similar to that of the liver[17]. The scale, complexity, and range of CYP and other biotransformational functions in the brain are only beginning to emerge and are the focus of intense research. At first, it was noted that biotransformational activities by CYPs and other enzymes catalyzed specific neural functions by regulating endogenous entities such as neurosteroids, rather than larger-scale chemical processing. However, we now know that various CYPs in astrocytes and the blood-brain barrier, for example, act protectively against xenobiotic threats[18]. It is also emerging that CYPs may also impact therapeutic outcomes, despite their low net levels in the brain. Indeed, it is understood that abnormally regulated brain CYP expression may be sufficiently significant to contribute to treatment failure in epilepsy. This can occur through a combination of rapid clearance of the drug, thus preventing therapeutic concentrations being maintained in neural tissue[19], or even through the formation of an epileptogenic metabolite locally in neurones[20].

To date, nearly 60 human CYPs have been identified and, perhaps surprisingly, about half of them have highly specific biomodulatory roles that are distinct from high volume chemical oxidation. It is likely that hundreds more CYP – mediated endogenous functions remain to be discovered.

2.6   Biotransformation and xenobiotic cell entry

2.6.1   Role of the liver

Drugs, toxins, and all other chemicals can enter the body through a variety of routes. The major route is through the digestive system, but volatile and lipophilic chemicals can bypass the gut via the lungs and skin. Although the gut metabolizes many drugs, the liver is the main biotransforming organ, and the CYPs and other metabolizing enzymes reside in the hepatocytes. These cells must perform two essential tasks at the same time. They must metabolize all substances absorbed by the gut whilst also processing all agents already present (from whatever source) in the peripheral circulation. This would not be possible through the conventional way that organs are usually supplied with blood from a single arterial route carrying oxygen and nutrients, leading to a capillary bed that becomes a venous outflow back to the heart and lungs. The circulation of the liver and the gut have evolved anatomically to solve this problem by receiving a conventional arterial supply and a venous supply from the gut simultaneously (Figure 2.3); all the blood eventually leaves the organ through the hepatic vein towards the inferior vena cava.

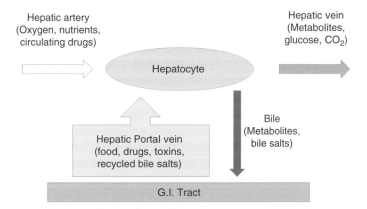

Figure 2.3   The hepatocytes can simultaneously metabolize xenobiotics in the circulation and those absorbed from the gut through their dual circulation of venous and arterial blood. Some lower-molecular-weight metabolites escape in the hepatic vein for eventual renal excretion, whilst other heavier metabolites are routed through to bile and eventually the gut

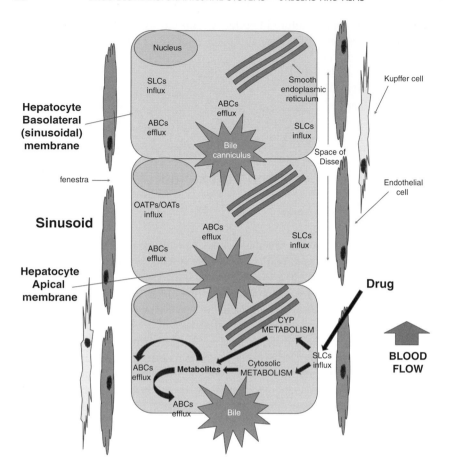

Figure 2.4 Hepatocyte transporters take up drugs from the hepatic portal and arterial blood through a system of SLC transporters (OATPs, OATs, and NTCP), which do not use ATP. After the drugs are metabolized in the smooth endoplasmic reticulum and/or by cytosolic enzymes, the polar products are either actively pumped into the bile via the ABC cassette transporters, which use ATP, or they are pumped out into the sinusoidal blood by MRP/ABC and the SLC transporters. Diagrammatically, the bile canniculi are drawn at 90 degrees to the direction of sinusoidal blood flow, which is an oversimplication but demonstrates how the blood and bile routes are anatomically separate

The hepatic arterial blood originates from the aorta and the venous arrangement is known as the hepatic portal system, which subsequently miniaturizes inside the liver into *sinusoids,* which are tiny capillary blood-filled spaces (Figure 2.4). This capillary network effectively routes everything absorbed from the gut direct to the hepatocytes, which are bathed at the same time in oxygenated arterial blood. The hepatocytes are found in large sheets known as trabeculae and the various influx transporters (Section 2.6.3) promote uptake of endogenous agents and drugs. In the case of the drugs, biotransformation may subsequently convert them into more polar metabolites, which are actively pumped out of the hepatocytes either into the sinusoidal blood, which becomes the hepatic vein, or by a separate system of *canalicali*, (Figure 2.4), which ultimately form the bile duct, leading to the gut. So, essentially, there are two blood routes into the hepatocytes and one out, which ensures that no matter how a xenobiotic enters the body, it will be presented to the hepatocytes for biotransformation. In addition, there are two routes of elimination, the venous blood and the bile.

It is usually stated that the higher the molecular weight of a drug and/or metabolite, the more likely elimination will mainly occur through bile, rather than the sinusoidal blood and thence on to the kidneys. However, this is likely to depend on several more factors than molecular weight. This area is difficult to study in humans, but it has been proposed that the approximate minimum molecular weight for biliary excretion is probably about 475 Daltons for anionic chemicals; it is not so defined for neutral and cationic agents. Among the physicochemical factors that influence biliary elimination, molecular volume and its relationship with molecular weight is probably important. These issues govern how well a given chemical fits the active sites of the ABC cassette transporters[21].

2.6.2   Drug and xenobiotic uptake: transporter systems

Although an agent might be presented to the vicinity of a hepatocyte, there is no guarantee it will enter the cell. This depends on the lipophilicity, size, charge, and other physiochemical properties of the agent. High lipophilicity in a drug as described in section 2.3 can cause an agent to become trapped in a cell membrane, preventing it from entering the cell itself. Alternatively, many charged or amphipathic drugs or toxins diffuse poorly, if at all, across lipid membranes. Their successful cellular and systemic absorption is in a large part due to their exploitation of the complex membrane transport systems that are found not only in the gut but also on the sinusoidal (sometimes called the basolateral) membranes of hepatocytes, which are bathed in blood from the portal circulation direct from the gut, as well as arterial blood. These membrane transporters regulate cellular entry of amino acids, sugars, steroids, lipids, and hormones, which are vital for homeostasis. We know this because if the hepatocyte transporters are inhibited, the bioavailability and therefore the plasma concentrations of several drugs increase because they escape hepatic clearance by the CYPs and other systems. Transporter proteins are found in all tissues and can be broadly categorized into two 'superfamilies'; those that assist the entry of drugs, toxins and nutrients into cells (uptake, or influx transporters) and those that actively pump them out using ATP in the process, usually against concentration gradients (efflux transporters: Chapters

4.4.7 and 5.6). The latter group has been much more extensively investigated, as they have such a significant role in the resistance to anticancer agents, as they pump out the drugs before they can exert any effect. Indeed, the genes that code for ATP-binding cassette efflux transporters are similar to those employed by bacteria in their resistance to antibiotics (multi-drug-resistance, or MDR genes).

2.6.3 Hepatic and gut uptake (influx) transporter systems

These influx or uptake transporters are known as the solute carriers (SLCs), and they are found in the liver, gut, brain, kidney, and the placenta. These systems are not fully understood, although intense research has been directed at understanding their roles in drug absorption and uptake into specific biotransforming organs such as the liver, as well as in clinically significant drug–drug interactions. The SLCs operate without using ATP and transport everything from small peptides, glucose, and bilirubin, as well as various anions and metabolites. To date, we know of 52 human SLC gene families that control the expression of over 400 individual carrier polypeptides[22,23]. In terms of those SLCs that are relevant to drug uptake, two subfamilies are of interest. The first is the *SLCO* subfamily, which codes for the hepatic organic anion transporting polypeptides, or OATPs, whilst the second is the *SLC22A* subfamily, which codes for the mainly renal organic anion transporters (OATs) and the organic cation transporters (OCTs)[24].

OATPs are sodium independent and effectively operate a process of facilitated diffusion, known as electroneutral exchange. For every external amphipathic molecule they pump in, they expel what was an intracellular neutralizing anion, like glutathione (GSH), bicarbonate, or even a drug metabolite. The system is rather like a revolving door and is powered by the higher concentration of the intracellular substrate. Many drugs enter gut epithelial cells and hepatocytes this way, particularly the more hydrophilic statins. The best-documented OATPs are OATP1A2, OATP1B1, and OATP1B3. These transporters are vital to the uptake of several classes of drugs (Chapter 5.6.1), and OATP1B1 can be inhibited by gemfibrozil, rifampicin, cyclosporine, and by the anti-HIV protease inhibitors such as ritonavir. Indeed, if OATP1B1 is either nonfunctional or inhibited by rifampicin, this blocks the hepatic uptake of atorvastatin[25,26] and actually has a greater impact on drug levels than if hepatic metabolism was to be inhibited[27]. Clearly, these transporters play a crucial role in determining the plasma levels of many drugs and the impact of transporter inhibition and genetic variation, will be explored in Chapters 5 (section 6) and 7 (section 2.10), respectively. In Chapter 4, it will be described how metabolizing systems respond in concert to changes in concentrations of substrates and the degree of OATP expression is modulated by the nuclear

PXR receptor system, which controls the expression of many CYPs and detoxifying enzymes.

Regarding other major transporters, bile salt re-uptake is achieved through a separate group of seven sodium-dependent transporters, mediated by a gene family known as SLC10A[28]. The two best understood include the product of *SLC10A1*, which is usually referred to as NTCP (sodium taurocholate cotransporting polypeptide) and ASBT (apical bile salt transporter), which is the product of SLC10A2[28]. NTCP is located on the basolateral membrane (blood vessel side) of hepatocytes, whilst ASBT is found in the kidney and ileum[27]. These transporters are vital for the recycling of bile salts known as enterhepatic recirculation, in which drugs can also be transported. NTCP is linked with the uptake of drugs much more than ASBT, but both are very vulnerable to inhibition by dozens of currently used drugs[29]. Indeed, NTCP is vital for the hepatic uptake of rosuvastatin, as plasma levels climb considerably if NTCP is poorly functional or inhibited[30].

The OATs are mainly found in the kidney and are responsible for removing small-molecular-weight agents, including many drugs, from the blood, into the proximal tubules. OAT1 (*SLC22A6*) and OAT3 (*SLC22A8*) are mainly renal, whilst OAT2 (*SLC22A7*) is also renal, but is found in the sinusoids of the hepatocytes[31]. OAT2 is linked with the uptake of a number of drugs, such as the nucleoside analogue entecavir, which is used in the therapy of hepatitis B infection[32].

2.6.4   Aims of biotransformation

Once drugs or toxins enter the hepatocytes, they are usually vulnerable to some form of biotransformation. Figure 2.2 presented many functions of CYPs and other biotransformational enzymes, but it is essential to be clear on what they have to achieve with a given molecule. Looking at many endogenous substances like steroids or xenobiotic agents like drugs, all these compounds are mainly lipophilic. Drugs often parallel endogenous molecules in their oil solubility, although many are considerably more lipophilic than these molecules. Generally, drugs and xenobiotic compounds must be fairly oil soluble or they would not be absorbed from the GI tract. Once absorbed, these molecules could change both the structure and function of living systems, and their oil solubility makes these molecules rather elusive in the sense that they can enter and leave cells according to their concentration and are temporarily beyond the control of the living system. As has been mentioned previously, this problem is compounded by the difficulty encountered by living systems in the removal of lipophilic molecules.

In Chapter 1, Section 1.3.2, it was outlined that even after the kidney removes them from blood by filtration, the lipophilicity of drugs, toxins, and endogenous steroids means that as soon as they enter the collecting tubules, they can immediately return to the tissue of the tubules, as this is more oil-rich than the aqueous urine. So the majority of lipophilic molecules can be filtered dozens of times and only low levels are actually excreted. In addition, very high lipophilicity molecules like some biocides, fire retardants, and plastic industry byproducts might never leave adipose tissue at all (unless moved by dieting or breastfeeding, which mobilizes fats). Potentially, these molecules could stay in our bodies for years. This means that for lipophilic agents:

- The more lipophilic they are, the more these agents are trapped in membranes, affecting fluidity and causing concentration-related disruption in membrane function.

- If they are hormones, they can exert an irreversible effect on tissues that is outside normal physiological control.

- If they are toxic, they can potentially damage endogenous structures.

- If they are drugs, they are also free to cause any pharmacological effect for a considerable period of time.

It is important to realise that biotransformational systems must assemble endogenous molecules as well as clear them, aside from their protective role concerning xenobiotics. So these systems *control* endogenous steroid hormones (assembly and elimination), as well as *protect*, in the case of highly lipophilic threats, like drugs, toxins, and hormone 'mimics' (endocrine disruptors). Metabolizing systems have developed mechanisms to control balances between hormone synthesis and clearance so the organism can finely tune the effects of potent hormones such as the various sex steroids. These systems also actually detect the presence of drugs and act to organize their elimination.

2.6.5   Task of biotransformation

Essentially, the primary function of biotransforming enzymes such as CYPs is to 'move' a drug, toxin, or hormone from the left-hand side of Figure 2.1 to the right-hand side. This means making very oil-soluble molecules highly water-soluble. This sounds impossible at first, and anyone who has tried to wash dishes without using dishwashing liquid will testify to this problem. However, if the lipophilic agents can be structurally altered, so changing their physicochemical properties, they can be made to dissolve in water. Once they are water-soluble, they can easily be cleared by the kidneys into urine and they will finally be eliminated.

2.6.6   Phase's I–III of biotransformation: descriptions and classifications

Most lipophilic agents that invade living systems, such as aromatic hydrocarbons, hormones, drugs, and various toxins, vary in their chemical stability, but many are relatively stable in physiological environments for quite long periods of time. This is particularly true of polycyclic aromatics. This means that a considerable amount of energy must be put into any process that alters their structures. This energy expenditure will be carried out pragmatically. Thus, some molecules may be subjected to several changes to attain water solubility, such as polycyclics, whilst others such as lorazepam and AZT, perhaps only one.

The stages of biotransformation are often described in scientific and medical literature as Phases I, II, and III. Phase I metabolism mainly describes oxidative CYP reactions, but non-CYP oxidations such as reductions and hydrolyses are also sometimes included in the broad term *Phase I*. The term *Phase II* describes generally conjugative processes, where water-soluble endogenous sugars, salts, or amino acids are attached to xenobiotics or endogenous chemicals. The more recent term *Phase III* describes the system of efflux pumps that excludes water-soluble products of metabolism from the cell to the interstitial fluid, blood, and finally the kidneys.

As far back as 2005 it was postulated that the 'traditional' terms of Phases I and II, had certain limitations and were even misleading[33]. Two reasons were primarily cited for this view. The first was that Phase I, for example, included many unrelated processes, and the second reason was that the use of the terms Phases I and II strongly implied that the Phase I processes must necessarily occur prior to Phase II conjugative reactions with a given molecule, and Phases I and II must have occurred prior to Phase III efflux processes. It is clear that this is not always the case, as Phase III efflux pumps can also exclude parent drugs as soon as they are absorbed from the gut, as well as metabolites.

To address the first issue of the unrelated processes, the major metabolic enzymatic reactions where molecules were altered structurally and thus physichemically, could be described more precisely through the use of two broad categories[34]. The first would be *functionalization reactions* where an existing polar group could be unmasked, or added to a molecule, which would either render the agent water soluble enough to be cleared or more easily undergo a second reaction. This might include an alcohol, amino hydroxyl, or carboxyl group. The second descriptive term would be a *conjugation reaction*, where an endogenous soluble molecule, such as glucose or a salt, will be attached covalently to a drug via an appropriate functional group, either one already present on the molecule or added/revealed by a prior functionalization reaction.

These descriptions essentially group biotransformational reactions by purpose and are more amenable to examples of drugs that may only undergo either oxidation or conjugation without any other metabolic process involved, whilst others may indeed undergo sequential metabolic processes. The main issue for those interested in drug metabolism is not to assume that metabolic processes are always sequential.

Perhaps another issue with the 'Phase' labelling of metabolism, is that whilst several so-called Phase I metabolites are associated with toxicity, such as the N-acetyl-*p*-benzoquinone imine related to paracetamol (acetaminophen), Phase II conjugative metabolites are more often seen in a 'detoxification' light. This may well not be the case, such as with some acyl (Chapter 6.2.7) and aromatic amine (Chapter 8.5.4) conjugates. Indeed, in the case of the largely obsolete sulphonamides, some conjugation products are even less soluble than the parent drug, to the point they can precipitate in the kidney (Chapter 6.7). Although the Phase I–III terminology is likely to remain popular and is sometimes used in this book, it is important to recognize the limitations of these terms in the description of many processes of biotransformation.

2.6.7   Biotransformation and drug action

Biotransformation has a secondary effect, in that there is so much structural change in these molecules that pharmacological action is often removed or greatly diminished. Exceptionally, some metabolites either retain or even exceed the parent drug's pharmacodynamic potency. Morphine-6-glucuronide, is a hundred-fold more potent in comparison with the parent drug[35]. However, increased polarity compared with the parent drug means that the efflux systems are likely to remove it relatively quickly, so diminishing the cellular residence time available for receptor interactions in the target tissue. It can be extremely difficult to really resolve the therapeutic effect of a drug in terms of the contributions of the metabolites and the parent drug.

The use of therapeutic drugs is a constant battle to pharmacologically influence a system that is actively and dynamically undermining the drugs' effects by removing them as fast as possible. The processes of oxidative and conjugative metabolism, in concert with efflux pump systems, act to clear a variety of chemicals from the body into the urine or faeces, in the most rapid and efficient manner. Logically, any system tasked with the assembly and modification of molecules to manage its activities must not only have exceptionally effective enzymatic machinery, but also sense and detect the target molecules. This allows adaptation to 'load'. The next two chapters illustrate the functionalizing biotransformational machinery in detail and its control and organization.

References

1. Gentric C, Rehel K, Dufour A. et al. Bioaccumulation of metallic trace elements and organic pollutants in marine sponges from the South Brittany Coast, France. Journal of Environmental Science and Health, Part A, 51, 213–219, 2016.
2. Abraham, K. Geusau, A, Tosun Y et al. Severe 2,3,7,8-tetrachlorodibenzo-*p*-dioxin (TCDD) intoxication: Insights into the measurement of hepatic cytochrome P450 1A2 induction. Clin. Pharm Ther 72, 163–174, 2002.

3. Pubchem Open Chemistry Database: https://pubchem.ncbi.nlm.nih.gov/compound/.

4. Tapia-Orozco N, Santiago-Toledo G, Barron V, et al. Environmental epigenomics: current approaches to assess epigenetic effects of endocrine disrupting compounds (EDCs) on human health. Env. Toxicol. Pharmacol. 51, 94–99, 2017.

5. Mallozzi M, Leone C, Manurita F, et al. Endocrine disrupting chemicals and endometrial cancer: an overview of recent laboratory evidence and epidemiological studies. Int. J. Env. Res. Pub. Health. 14, 334 (1–23), 2017.

6. Sahebi M, Hanafi, MM. van Wijnen, AJ, et al. Profiling secondary metabolites of plant defence mechanisms and oilpalm in response to *Ganoderma boninense* attack. Biodeter. & Biodegrad. 122, 151e164, 2017.

7. Runciman DJ Lee AM, Reed, KFM et al. Dicoumarol toxicity in cattle associated with ingestion of silage containing sweet vernal grass (*Anthoxanthum odoratum*) Aust. Vet. J. 80, 28–32, 2002.

8. Ketney O, Santini A, Oancea S. Recent aflatoxin survey data in milk and milk products: A review. Int. J. Dairy. Tech. 70, 320–331, 2017.

9. Begum SS, Jayalakshmi HK, Vidyavathi HG, et al. A novel extract of Fenugreek Husk (FenuSMART™) alleviates postmenopausal symptoms and helps to establish the hormonal balance: a randomized, double-blind, placebo-controlled study. Phytother. Res. 30, 1775–1784, 2016.

10. Chalfoun C, McDaniel C, Motarjem P, et al. Breast-enhancing pills: Myth and reality. Plastic & Reconstruct. Surgery. 114, 1330–1333 2004.

11. Rietjens IMCM, Louisse J Beekmann K. The potential health effects of dietary phytoestrogens. Brit. J. Pharmacol. 174, 1263–1280, 2017.

12. Vondracek J, Hyzdalova M, Pivnicka J, et al. Interference of polycyclic aromatic hydrocarbons and their complex mixtures with steroid signalling. Toxicology Letters S14–03, 2017. http://dx.doi.org/10.1016/j.toxlet.2017.07.089

13. Soto AM, Schaeberle C, Maier MS et al; Evidence of absence: estrogenicity assessment of a new food contact coating and the bisphenol used in its synthesis. Environ. Sci. Technol. 51, 1718–1726, 2017.

14. Lamba V, Ghodke Y, Guan W et al. MicroRNA-34a is associated with expression of key hepatic transcription factors and cytochromes P450. Biochem. Biophys. Res. Comm. 445, 404–411, 2014.

15. Lukasik A, Zielenkiewicz P. Plant MicroRNAs—Novel players in natural medicine? Int. J. Mol. Sci.18, 1–16, 2017.

16. Pavanello S, Fedeli U, Mastrangelo G, et al. Role of CYP1A2 polymorphisms on lung cancer risk in a prospective study. Cancer Genet. 205, 278–284, 2012.

17. Miksys S, Tyndale RF. Cytochrome P450–mediated drug metabolism in the brain. J. Psych. Neurosci 38, 152–163, 2013.

18. Tripathi VK, Kumar V, Pandey A. et al. Monocrotophos induces the expression of xenobiotic metabolizing cytochrome P450s (CYP2C8 and CYP3A4) and neurotoxicity in human brain cells. Mol. Neurobiol. 54:3633–3651, 2017.

19. Ghosh C, Hossain M, Solanki J. Overexpression of pregnaneX and glucocorticoid receptors and the regulation of cytochrome P450 in human epileptic brain endothelial cells. Epilepsia, 58, 576–585, 2017.

20. Ghosh C, Marchi N, Hossain M, et al. A pro-convulsive carbamazepine metabolite: quinolinic acid in drug resistant epileptic human brain. Neurobiol. Dis. 46:692–700, 2012.

21. Yang X, Gandhi YA, Duignan DB, et al. Prediction of biliary excretion in rats and humans using molecular weight and quantitative structure–pharmacokinetic relationships. The AAPS Journal, 11, 511–525, 2009.

22. Hediger MA, Clémençon B, Burrier RE, and Bruford EA The ABCs of membrane transporters in health and disease (SLC series): Introduction. Mol. Asp. Med. 34, 95–107, 2013.

23. Perland E, Fredriksson R, Classification Systems of Secondary Active Transporters. Trends in Pharm. Sci. 38, 305–315, 2017.

24. Zhou F, Zhub L, Wang K, et al. Recent advance in the pharmacogenomics of human solute carrier transporters (SLCs) in drug disposition. Adv. Drug Del. Rev. 116, 21–36, 2017.

25. Pasanen, M.K., Fredrikson, H., Neuvonen, P.J. and Niemi, M. Different effects of SLCO1B1 polymorphism on the pharmacokinetics of atorvastatin and rosuvastatin. Clin. Pharmacol. Ther. 82, 726–733, 2007.

26. Lau, Y.Y., Huang, Y., Frassetto, L. et al. Effect of OATP1B transporter inhibition on the pharmacokinetics of atorvastatin in healthy volunteers. Clin. Pharmacol. Ther. 81, 194–204, 2007.

27. Maeda K, Ikeda Y, Fujita T, et al. Identification of the rate determining process in the hepatic clearance of atorvastatin in a clinical cassette microdosing study. Clin Pharmacol Ther. 90, 575–81, 2011.

28. Anwer MS, Stieger B. Sodium-dependent bile salt transporters of the SLC10A transporter family: more than solute transporters. Pflugers Arch. – Eur. J. Physiol. 466, 77–89, 2014.

29. Dong Z, Ekins S, Polli JE. Structure–activity relationship for FDA approved drugs as inhibitors of the human sodium taurocholate cotransporting polypeptide (NTCP). Mol. Pharm. 10, 1008–1019, 2013.

30. Lou XY, Zhang W, Wang G, et al. The effect of Na+/taurocholate cotransporting polypeptide (NTCP) c.800C > T polymorphism on rosuvastatin pharmacokinetics in Chinese healthy males. Pharmazie 69, 775–779, 2014.

31. Burckhardt G. Drug transport by Organic Anion Transporters (OATs). Pharm. & Ther. 136, 106–130, 2012.

32. Furihata T, Morio H, Zhu M. et al. Human organic anion transporter 2 is an entecavir, but not tenofovir, transporter. Drug Metab. Pharm. 32, 116–119, 2017.

33. Josephy PD, Guengerich FP, Miners JO. "Phase I" and "Phase II" drug metabolism: terminology that we should phase out. Drug Metab. Rev. 37, 575–80, 2005.

34. Rowland A, Miners JO, Mackenzie PI. The UDP-glucuronosyltransferases: Their role in drug metabolism and detoxification. Int. J. Biochem. & Cell Biol. 45, 1121–1132, 2013.

35. Paul D, Standifer KM, Inturrisi CE, et al. Pharmacological characterization of morphine-6 beta-glucuronide, a very potent morphine metabolite. J. Pharm. Exper. Ther. 251, 477–483 1989.

3 How Oxidative Systems Metabolise Substrates

3.1 Introduction

It is essential for living systems to control lipophilic molecules, but as mentioned earlier, these molecules can be rather 'elusive' to a biological system. Their lipophilicity means that they may be poorly water-soluble and may even become trapped in the first cell membrane they encounter. To change the physicochemical structure and properties of such molecules they must be conveyed somehow through a medium that is utterly hostile to them, i.e. a water-based bloodstream, to a place where the biochemical systems of metabolism can physically attack and change the structure of these molecules.

3.2 Capture of lipophilic molecules

Virtually everything we consume, such as food, drink, and drugs that are absorbed by the gut, will proceed to the hepatic portal circulation. This will include a wide physicochemical spectrum of drugs, from water-soluble to highly lipophilic agents. Charged or water-soluble agents (if they are absorbed) may pass through the liver into the circulation, followed by filtration by the kidneys and elimination. The most extreme compounds at the end of the lipophilic spectrum will be absorbed with fats in the diet via the lymphatic system and some will be trapped in membranes of the gut. The majority of predominantly lipophilic compounds will eventually enter the liver. As mentioned in the previous chapter, the main functional cell concerned with drug metabolism in the liver is the hepatocyte. In the same way that most of us can successfully cook foodstuffs in our kitchens at high temperatures without injury, hepatocytes and other biotransforming cells such as Clara cells of the lung, are physiologically adapted to carry out millions of high-energy, potentially destructive and reactive biochemical processes every second of the day without significant cell damage occurring. Indeed, it could be argued that biotransforming cells such as hepatocytes have adapted to this

Human Drug Metabolism, Third Edition. Michael D. Coleman.
© 2020 John Wiley & Sons, Inc. Published 2020 by John Wiley & Sons, Inc.

function to the point that they are biochemically the most resistant cells to toxicity in the whole body – more of those adaptations later.

In the previous chapter it was outlined how the circulation of the liver and gut had evolved to deliver xenobiotics to the hepatocytes. The next task is 'subcellular', that is, to route these compounds to the biotransforming enzyme systems inside the hepatocytes. To attract and secure highly physicochemically 'slippery' and elusive molecules such as lipophilic drugs requires a particular adaptation inside biotransforming cells; this is the organelle that is very highly developed in such cells, that is, the smooth endplasmic reticulum (SER; Figure 3.1). You will already be aware of the rough endoplasmic reticulum (RER) from biochemistry courses, which resembles an assembly line where ribosomes 'manufacture' proteins. Regarding the SER, pictures of this organelle's structure resemble a spaghetti-like mass of tubes. The most lipophilic areas of the SER are the walls, that is, the membranes of these interconnected tubes, rather than the inside (lumen). The drugs/ toxins essentially 'flow' along inside the thickness of the walls of the SER's tubular structure (Figure 3.1) straight into the path of the CYP monooxygenase system. Indeed, this system consists of the CYPs and their 'fuel pumps', (their sources of electrons); these are cytochrome b_5 and cytochrome P450 oxidoreductase (POR), the latter requiring NADPH as a cofactor. Both the CYPs and POR are embedded in the SER through their respective membrane anchor domains, which are N-terminal ends of α helices (right-handed protein coils) rich in hydrophobic amino acids[1]. CYPs are not only believed to be anchored to the membrane, but they are also partly submerged in it[1]. Hence, once, a lipophilic chemical enters the biotransforming organ, it joins this highly lipophilic environment in a lipid-rich cell, essentially constituting a conveyor belt along which the chemical progresses through the SER and straight into the CYP monoxygenase system. This is accentuated by the agent's lipophilicity excluding it from the aqueous areas of the cell. In addition, once the CYP system metabolises the chemical into a more water-soluble agent, it is repelled from the SER walls, towards the cytoplasm, so creating a concentration gradient, which is, in turn, powered by the efflux pump systems in hepatocyte cell membranes (Chapter 4.4.7). Thus, the metabolite is driven towards the circulation and renal excretion.

3.3 Cytochrome P450s: nomenclature and methods of study

3.3.1 Classification

CYPs belong to a group of enzymes that all have similar core structures and modes of operation. Although discovered in 1958, vast amounts of research have not yet revealed all there is to know of the structure, function, and diversity of these enzymes[2]. In all living things, over 12,000 CYP gene sequences have been discovered, which are believed to have evolved from a single gene[2,3].

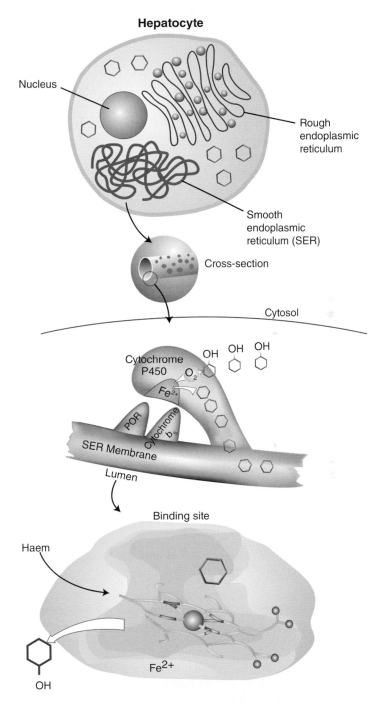

Figure 3.1 Location of CYP enzymes and their REDOX partners, cytochrome b_5 and POR (P450 oxidoreductase), in the hepatocyte and how lipophilic species are believed to approach the enzymes' active site

From a human perspective, we make do with expressing just 57[4,5] CYP proteins; of these, 7 are expressed in mitochondria, where they catalyse specific oxidations related to steroid metabolism[3]. Of the other 50 genes, only six – CYP1A2, CYP2C9, CYP2B6, CYP2C19, CYP2D6, and CYP3A4 – make significant contributions to drug clearance. To put the CYP-mediated contribution to drug clearance in context, it has been estimated that of all drug-related material administered to patients, about 25% is cleared unchanged by the kidney, 10–15% undergoes conjugation reactions of various types (Chapter 6), whilst the rest (~ 60%) is cleared by CYPs[6]. Of the CYP-cleared fraction of drugs, around half is metabolised by the CYP3A subfamily (almost entirely CYP3A4, with some CYP3A5). Other CYPs, such as CYP1A1, CYP1A2, CYP1B1 and CYP2E1 do clear some prescription drugs, but are mainly of relevance in the biotransformation of carcinogens and various toxins. The remainder carry out biosynthetic 'housekeeping' tasks in all tissues.

By the 1990s a system of classification for CYPs was established using each CYP gene's amino acid sequence homology to assign isoforms to families[4,5]. If two CYPs have 40% of the full length of their amino acid structure in common, they are assumed to belong to the same family. To date, more than a thousand CYP families have been found in nature so far, but only 18 have been identified in humans. The families are numbered, such as CYP1, CYP2, CYP3, etc. Subfamilies are identified as having 55% sequence homology; these are identified by using a letter and there are often several subfamilies in each family. So you might see CYP1A, CYP2A, CYP2B, CYP2C, etc. Regarding the individual CYP enzymes themselves, these *isoforms* originate from alleles, or slightly different versions of the same gene. They are given numbers within the subfamily, such as CYP1A1 or CYP1A2, and these isoforms have 97% of their general sequences in common. From a practical point of view, differences in the binding site amino acid sequences of the isoforms rather than the full-length structure are likely to be more relevant with regard to which specific molecules these enzymes can actually metabolise. With CYP2D6, it is known that change in just one amino acid residue in the binding site is crucial in substrate binding (Chapter 7.2.3). The amino acid sequences of many bacterial, yeast, and mammalian enzymes are now well known, and this has underlined some large differences, as well as surprising similarities between the structures of our own CYPs and those of animals, eukaryotes, and bacteria. Interestingly, the same metabolite of a given substance will be made by different CYPs across species.

3.3.2   Methods of analysis

Whilst it is obvious that a processing machine is operating optimally by examination of its products, establishing exactly how the machine is forming those products is clearly a challenge if we are unaware of the machine's full internal structure and function. Ideally, it is often easiest to understand how machines work by watching cutaway models, like the ones seen of car engines at motor shows. With enzymes in living systems, options are much more limited and the preferred method has been to 'catch it in the act', that is, to crystallize it when it is bound to a substrate. Thus, the technique of X-ray crystallography is used to explore and map the contours and features of the enzyme. The low-hanging fruit in this context was the water-soluble bacterial CYP101 (P450$_{cam}$), which was crystallized relatively early on, affording the first detailed source of information on CYP structure and function[7]. As mammalian CYPs operate in a lipid environment, this renders crystallization exceedingly difficult. As mentioned in Section 3.2, both CYPs and POR are secured to the lipid-rich SER through their membrane anchor domains. However, it was realised that if these domains were removed, along with some other structural modifications, the CYPs and POR would be water soluble enough to crystallize. This was achieved in 2000 with a rabbit CYP (2C5), followed a few years later by the crystallization of some of the most important human CYPs, CYP2C9 and CYP3A4[8]. Now, more than 100 crystal structures exist of ligands being metabolised by human drug metabolizing CYPs[9].

However, in some ways, using crystallography in this context is like studying how an animal runs or how a machine operates through a series of freeze frames, rather than being able to watch the process operating in real time in a 'natural' environment. Indeed, with any research technique, findings are influenced by the limitations of that technique and X-ray crystallography requires the subjection of the CYP isoforms to extremely unphysiological conditions. So it is hoped that each freeze frame actually does correspond to a stage in the binding and catalytic processes. In some cases, the results have been ambiguous and not easy to explain. A good example is the crystal structure of CYP2C9 binding the antihypertensive losartan; in this case, the position on the drug that is normally hydroxylated is actually oriented away from the haem iron, which does not tally with reality[9]. Indeed, this mismatch has been shown with many other CYP/substrate crystal structures. Obviously, during normal CYP operations, any given drug molecule will be aligned towards the iron at some point in the catalytic process, but it might not immediately be clear from the crystal structure how this can happen. A key source of reducing power for the CYP, that is, the POR/NADPH combination, is absent from the crystallization process and some have suggested that the reducing partners contribute to substrate alignment[9], whilst others have shown experimentally substrate binding activates the CYP whether POR is present or not[10].

Indeed, in retrospect, research carried out in the twenty years prior to the crystallization of human CYPs perhaps suggests that rather more was expected from crystallography than it could really deliver, in terms of determining function. Once it was realised that CYPs required POR's flow of electrons, efforts were made to extract POR from cells and it was established that POR would not operate unless its membrane anchor was intact and phospholipid was present[11], whilst CYPs would partially run without an intact anchor.

How CYPs and POR interact *in vivo* has been the subject of intense research over the past decades, and like many machines, when improperly assembled and run in suboptimal conditions, performance is poor. For these membrane-bound lipid environment systems, aqueous conditions cause them to distort and lose function. For full catalytic activity to occur, intact membrane anchor domains on the CYP and POR are not enough; the intimate relationship between anionic phospholipids and the CYP/POR system must be perfect also[12]. In addition, the CYP/POR complex must even be a specific height relative to the lipid bilayer they are anchored to, so as to ensure maximum coupling efficiency[10]. Once it was understood how to run the CYP/POR system in a manner approaching the physiological, it was likely that further progress in uncovering functionality secrets would involve experimental approaches that inhabited a lipid environment.

Indeed, mounting CYP/POR systems on nanodiscs, built of apoprotein scaffolding and phospholipid membranes, contributes significantly to preservation of structure and function[13]. Further studies using fluorescent analysis of metabolites have demonstrated that SER lipid material is more effective than artificial lipid mixes at promoting CYP catalysis[1]. In addition, the application of Resonance Raman Spectrometry with nanodisc-mounted CYPs has facilitated the study of substrate binding events in great detail[14]. Hence, the combination of all these methods have improved understanding of external CYP structure, including details of the access and egress pathways for substrates and products, as well as the dimensions of the inner structure such as the active site. The many basic structural similarities between the CYPs have also been revealed using crystallography. As CYPs are among the most complex enzyme systems known, many techniques have been required to understand their incredible flexibility, in terms of the range of substrates they can process, as much as their actual catalytic activity.

Detailed information from these experimental techniques has been incorporated into many *in silico*, or computer-based techniques such as molecular dynamics, which allow researchers to understand in much more detail how the protein structures of CYPs unwind and unfold to provide that remarkable degree of flexibility during substrate binding. Information on all these features is vital in areas such as drug design, where potential inhibitors as well as substrates must be modelled[7] and the powerful therapeutic and economic interest in different polymorphic CYP isoforms (Chapter 7.2.3) has driven exploration of how minor changes in CYP amino acid position have such marked impact on their metabolizing capability[8].

3.3.3   CYP key features and capabilities

So to summarize: CYPs in general are capable of metabolizing almost any chemical structure, and they have a number of features in common:

- Most mammalian drug metabolizing CYPs are partly submerged in the lipid microenvironment of the SER, as describe above. In practical terms, the chemicals access the CYP by flowing through the SER and enter the active site rather like progressing through an underwater entrance into a tropical island cave.

- All CYPs contain flexible multiple binding areas in and around their active site, which is the main source of their variation and their ability to 'capture' and metabolise a particular group of chemicals.

- CYPs feature a haem group in their active site that contains iron, which is a crucial and highly conserved part of their structures. This area is fairly rigid, but it is surrounded by the more flexible complex binding areas. It is deep inside the CYP protein structure but is above the level of the SER membrane.

- To catalyse substrate oxidations and reductions, CYPs exploit the ability of a metal, iron, to gain or lose electrons, rather like a rechargeable battery in a cordless drill (Figure 3.1).

- They all have closely associated 'REDOX partners', which are P450 oxidoreductase (POR) and cytochrome b_5 that supply them with electrons to power their catalytic activities (Figure 3.1). These associate with the proximal side of the CYP.

- They all bind and activate oxygen as part of the process of metabolism.

- They are all capable of reduction reactions that do not require oxygen.

These points give us a basic understanding of some aspects of the structure and function of these enzyme systems, but there are other more challenging aspects to CYP activities, which are more difficult to understand and investigate. Yet, they are driving research aimed at understanding how and why CYPs appear to operate as miniature self-regulating biological machine tool systems, in ways that many other enzymes do not. For instance, CYP flexibility in binding capability has been extensively investigated, but what is perhaps difficult to align with such flexibility, is the degree of stereoselectivity CYPs exhibit, where only a left-hand or right-hand version of a compound will be cleared. This level of precision is rather paradoxical in terms of such apparent fluidity of structure.

In addition, CYP catalytic activity appears to be partly controlled by substrate concentration, which would be logical, as it is advantageous to accelerate clearance of substrate when levels are high, whilst energy and reducing power are conserved when substrate levels fall. If this occurs with the same substrate, this is termed *homotropic* activity. What is also apparent is that CYPs exhibit *heterotropic* activity, where the binding of one chemical can influence the catalysis of another, structurally unrelated chemical[19-21]. This could accelerate clearance or retard it. The key questions are how CYPs can operate in these modes and, eventually, why they have evolved to do this. Such questions are discussed below in terms of what is known of CYP structure and function so far.

3.4   CYPs: main and associated structures

3.4.1   General structure

Many complex and detailed three-dimensional structures of the CYPs available online are worth viewing. However, there is perhaps a simpler way to help you visualize some idea of CYP flexibility and structure at the same time. If you place a small coin a little off centre on the palm of your hand towards your first finger, you could imagine that this is the haem iron catalytic centre, the active site of the CYP. If your hand is partially clenched, but not forming a tight fist around the coin, you might see how there are various flexible 'access channels' or entrances where a 'substrate' can enter and a 'product' leave. You can also see how flexible your fingers are in assisting the 'binding' of various substrates of different shapes to the less mobile 'catalytic centre' of the palm of your hand. In fact, to some extent, we can superimpose the actual generic structural features of real CYP isoforms onto this basic 'CYP hand' analogy. The haem – iron active site 'palm and coin' is set inside what is sometimes termed the CYP protein 'fold' (your hand). This consists of the α helices described earlier. The structural features of a CYP are often referred to as distal (far from) or proximal (close to) the haem-iron. Hence, the substrate enters the distal area of the isoform (wrist/palm edge/fingers), whilst the REDOX partners that provide the electrons to operate the enzyme are proximal to the haem iron (near the thumb and first finger). Then of course, you have to imagine your hand partly submerged in the lipid of the SER, as described previously.

3.4.2   Haem moiety

Remarkably, all the known CYP isoforms in living things have a common three-dimensional shape or protein fold, even though their amino acid sequences possess a similarly of less than 20% across the whole CYP superfamily. CYPs have twelve α helices, labelled A-L with four β sheets[14]. Among the major core α helix substructures of CYPs, the backbone of these enzymes is known as the *I helix,* which has a kink in it locating an area called the cys pocket, which, in turn, holds the haem-iron active site in place. This could be the 'palm and coin' of the CYP hand analogy. CYPs such as 3A4 and 2C9 have some flexibility in the movement of the haem, but in most CYPs this is a relatively rigid part of the protein's structure. The haem structure is also known as ferriprotoporphyrin-9 (F-9; Figure 3.2). The F-9 is the highly specialized lattice structure that supports a CYP iron molecule, which is the core of the enzyme, which catalyses the oxidation of the substrate[14-16]. This feature is basically the same for all CYP enzymes; indeed, F-9 is a convenient way of positioning and maintaining iron in several other enzymes, such as haemoglobin, myoglobin and catalase. The iron

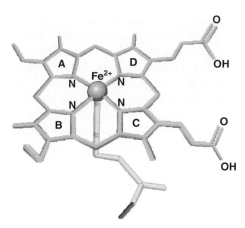

Figure 3.2 Main structural features of ferriprotoporphyrin-9, showing the iron anchored in five positions (pentacoordinate form). The cysteinyl sulphur holds the iron from below

is normally secured by attachment to five other molecules; in the horizontal plane, four of them are pyrrole nitrogens, whilst the fifth group, a sulphur atom from a cysteine amino acid residue holds the iron in a vertical plane. This is known as the pentacoordinate (five-position) state and could be described as the resting or low state, prior to interaction with any other ligand (Figure 3.2)[14-16]. The use of spectroscopy reveals that the CYP is in 'operating mode' and actively metabolising the substrate when the iron spin state is high[13]. The pentacoordinate state appears to show iron bound tightly to the sulphur and below the level of the nitrogens. When the iron binds another ligand, it is termed *hexacoordinate* and the iron appears to move upwards and draws level with the nitrogens to bind a water molecule, which is hydrogen bonded to a threonine amino acid residue that is located just above the iron, linked with proton movements during operation of the enzyme. The F-9 is held in place by hydrogen bonding and a number of amino acid residues, particularly an arginine residue, which may also stabilize the F-9 molecule. As mentioned previously, the iron is crucial to the catalytic function of CYP enzymes, and the process whereby they oxidize their substrates requires a supply of electrons, which is sourced by the dual fuel pumps of the system, POR and cytochrome b_5. These REDOX partners are sited extremely close to a relatively flat area of the proximal side of the CYP and we will look at them in more detail later on, in Section 3.4.8[14-18].

3.4.3 CYP flexible regions

Running across and over the I helix/haem active site is a series of helices that form a cover or lid on the active site. These helices are usually described as the F and G domain. This domain consists of an F helix, an F/G loop structure, and a G helix. A B/C helix loop is also part of the cover of the active site. Human CYPs contain extra F and G helices (usually termed F′ and G′), which give them added flexibility in uncovering the active site enough to accommodate large molecules[8]. To follow the CYP hand analogy, these flexible regions to the structures, the various F/G helices, their loops, and the B/C helix loop, could be regarded as the 'fingers' that are normally partially clenched, but can open out to form an access pathway to accommodate large substrates. As the human hand can grasp a pin or a beach ball, the small substrate metyrapone binds to CYP3A4 without any visible movement in the 'fingers', whilst erythromycin requires them to stretch out widely to allow binding to the active site[17]. Indeed, CYP3A4 increases its active site area by 80% to accommodate erythromycin[19]. To try to see how the whole CYP isoform is oriented in the SER membrane, then you could imagine the lipophilic substrates diffusing through the membrane and entering the isoform through the access path, which includes the highly lipophilic opening fingers. In any CYP, the access path is generally defined as the widest, shortest, and usually most lipophilic route to the haem iron active site. The active site is supplied with electrons through another access channel from the other side of the CYP by POR and cytochrome b_5. As mentioned in the CYP hand analogy, this area could be visualized as located between the thumb and first finger. These REDOX partners are also embedded in the SER membrane right next to the CYP as described earlier. Finally, there is also what is often termed a solvent or egress channel (between the 'fingers'), which is routed away from the lipid in the SER membrane into either the lumen of the SER or the aqueous cytosol, where the more hydrophilic product will naturally exit the isoform, as the other paths are so lipophilic they effectively repel the product[15-19].

3.4.4   Substrate binding in CYPs

The term *active site* of an enzyme usually means the area where structural changes in the substrate are catalysed with the help of various co-factors. This term can encompass a binding area that locates and holds the substrate in such an orientation that the appropriate moiety of the molecule is presented to the structures on the enzyme that catalyse the reactions the enzyme is intended to accelerate. In many enzymes, the dimensions and properties of the active and binding sites are quite well defined and mapped in detail. With acetylcholinesterase, for example, the anionic site is mainly responsible for attracting and locating the substrate acetylcholine through ionic forces, whilst the esteratic site is intended to catalyse the hydrolysis of the substrate. This is not the case in CYPs, as crystallographic studies have shown that what constitutes the *active* and *binding* sites of a CYP can be a very broad and flexible area indeed. To date, it has been generalized that CYP3A4, CYP2C8, and CYP2C9 have very large active sites, whilst that of CYP2D6 is intermediate and CYP2A6's site is quite small.

However, with the larger sited CYPs like CYP3A4 and CYP2C9, they can still bind very small substrates alongside the giant ones[14-16].

If we return to the CYP hand analogy, if you grasp an object like a shirt button, which is smaller than the coin at the catalytic centre, the binding site is a relatively small area on the palm and little hand/finger movement is required to grasp it. If you grasp a larger entity, which is, say, an iPhone, you can easily hold it in such a way that the edge contacts the coin, although all your fingers and thumb are now required to articulate to grasp the iPhone and the CYP binding site is pretty much your whole hand. With real CYPs, they undergo similar huge changes in movement and binding area to accommodate substrates of differing sizes like the contrasting agents metyrapone and erythromycin mentioned earlier. However, there is likely to be a size restriction to the access channels from the SER membrane.

Essentially, what constitutes the binding site of any given CYP is very difficult to define. Examination of crystallized CYPs bound to different substrates have shown that CYPs do contain small-intermediate hydrophobic pockets, as well as a capability of the rest of the F and G helices to act as extending and enclosing fingers to bind larger substrates. What is usually described as the hydrophobic pocket in a CYP comprises many amino acid residues that can bind a molecule by a number of means, including weak van der Waals forces, hydrogen bonding, as well as other interactions between electron orbitals of phenyl groups, such as pi–pi bond stacking. This provides a grip on the substrate in a number of places in the molecule, preventing excessive movement. Interestingly, when not binding substrates, CYP active site areas are full of water molecules, which are displaced upon substrate binding.

In effect, crystallographic studies have shown that the type of hydrophobic amino acid residues seen in the smaller, internal CYP hydrophobic pockets are also found on the 'fingers' of CYPs, such as CYP3A4. These are the F and F′, G and G′ helices (among others) and they are capable of binding a hydrophobic molecule by using the same pi–pi bond stacking, van der Waals forces, and/or hydrogen bonding. This effect is borne out by observations of the binding of progesterone, which appears to be held between the 'fingers' of CYP3A4, rather than in the 'palm'. Technically, progesterone actually appears to be stuck to the outside of the enzyme. So it is not unusual for many CYPs that significant areas of the interior and exterior of the isoforms are available for substrate binding.

3.4.5 Homotropic binding in CYPs

The observations that CYPs such as CYP3A4 could bind substrates in several regions of the enzyme have allowed researchers to unravel how CYPs manage their homotropic and heterotropic activity, as described in Section 3.3.3. Homotropic activity has been well illustrated by the *in vitro* study of the metabolism of testosterone by CYP3A4 using nanodisc technology[10]. When a single molecule binds, there is a small increase in reducing power consumption (NADPH) and a low-level conversion to a high spin state, but no metabolism occurs. The reducing equivalents consumed by the CYP/POR complex are essentially wasted through the formation of various reactive oxygen species and the CYP is effectively uncoupled. When a second molecule enters the active site, the enzyme elevates itself to a high spin state and testosterone hydroxylation begins and operates maximally, but wastefully; only about a twentieth of the NADPH consumed by the CYP/POR complex actually contributes to the operation of the catalytic cycle. When a third molecule binds, formation of product is no faster, but is massively more efficient, with less wastage of the NADPH (Figure 3.3).

Each successive substrate molecule binding event causes allosteric changes in the enzyme's structure and metabolic capability[10,14]. It is likely that with many drugs that are CYP substrates, their concentrations powerfully influence CYP metabolism rates, through multiple binding events in the vicinity of the CYP active site leading to better presentation of molecules to the haem iron for catalysis to occur. This process, as mentioned previously, allows a high degree of flexibility in response to immediate but transient increases in substrate load,

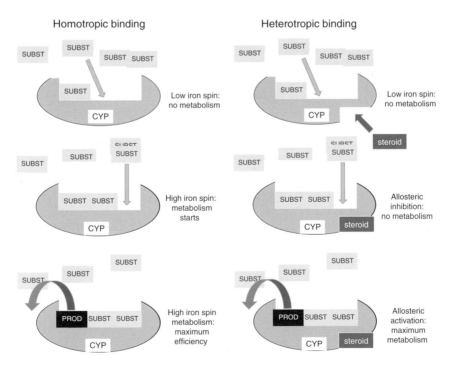

Figure 3.3   Diagram of CYP homotropic and heterotropic substrate binding. CYP activity is externally and internally modulated by substrate pressure

such as might happen during the modulation of sexual cycles by hormonal concentration pulses. Such flexibility can stave off the necessity to form more enzyme, which would not only be slow, but incur significant biosynthetic resource allocation which may then be superfluous when substrate levels decline.

3.4.6   Heterotropic binding in CYPs

It had been shown by the early 2000s in various experimental models and *in vivo*, that the presence of some steroids could directly accelerate the CYP3A4-mediated metabolism of substrates such as carbamazepine up to six times faster than when they were absent. It was postulated that this was achieved through some form of allosteric binding by the steroid[20]. The site of this allosteric binding, as mentioned previously, was revealed when CYP3A4 was first crystallized with progesterone as a substrate. The steroid was discovered to be docked outside the CYP, between the F-F' and G-G' helices[17] and was bound to an aspartate residue through hydrogen bonding, near a group of phenylalanine residues known as the *phe-cluster,* which is very clearly illustrated in the original paper[17]. Subsequent work has explored the significance of these observations and now with CYP3A4 it is postulated that a chemical can either bind productively (P-substrate), i.e., it can be metabolised at the active site near the haem iron inside the enzyme, or bind 'unproductively' outside the enzyme near the Phe cluster (Figure 3.3). This latter binding event is allosteric (A-substrate) and no metabolism occurs at this site; the docking event outside the enzyme causes a shift of a phenylalanine residue side chain (Phe-213) out of the binding site inside the enzyme, increasing the space for another substrate molecule to dock and be metabolised[21,22]. This process operates rather like a car gear lever engaging different ratios in a transmission, which is several feet from the driver. Other studies have made different versions of CYP3A4 with changes in the structure of their allosteric sites, which could be likened to jamming the enzymes in different 'gears'; these changes strongly influence the rates of metabolism and stability of the enzyme[23]. Indeed, depending on which molecule interacts with the allosteric site and in what conformation and orientation, this suggests a vast range of possible impacts on the multiple adjustable active site structure, volume, and ability to catalyse a substrate bound at the haem active site[23].

As our understanding of the processes of heterotropic and homotropic activity in CYPs improves, it is now clearer as to how CYP3A4, for example, 'squares the circle' of possessing great flexibility in accommodating different sizes of substrate whilst often paradoxically retaining regioselectivity and stereoselectivity. In addition, it is clearer now how CYP3A4 and probably most other drug metabolizing CYP isoforms respond to constant modulation, in terms of remote 'commands' issued by other tissues or organs in the form of endogenous chemicals that allosterically control acceleration or inhibition of their catalysis. Indeed, this also suggests that this considerable dynamic range of automatic capability to start, speed up, or just stop, could be achieved in time frames as short as a few minutes, as required to operate endogenous processes. It also raises the likelihood that the biosynthetic CYPs that build the myriad steroids and other biomodulatory molecules all over the body are regulated in a similar fashion. Interestingly, whilst CYP inhibition can be a negative process that leads to drug toxicity through accumulation, we should also see that some aspects of CYP inhibition must be viewed as a natural feature of their endogenous modulation, which can be exploited therapeutically (Chapter 5.8).

3.4.7   CYP complex formation

In studies from the 1970s onwards, it was clear that in rat livers there seemed to be a significant excess in CYP content, with respect to POR levels, which did not seem consistent with the required 1:1 ratio necessary for a functioning system; indeed, POR seemed to be a significant limiting factor[24]. Gradually, it was found that CYPs were internally flexible enough even when anchored within the SER to form complexes through electrostatic and/or hydrophobic interactions that either inhibited their function, through competition for POR, or the CYP complex increased their affinity for POR, or even speeded up their catalysis by a number of mechanisms[24-26]. Complexes could consist of the same CYP isoform or even different isoforms, with some isoforms, such as CYP1A2 and CYP2C9, demonstrating a particular propensity for these associations[24,25]. These observations suggest that CYPs operate in tissues as a very fluid, complex, and vast sub-cellular cooperative of different isoforms and expression levels. Such a level of cooperativity at the level of the organelle (the SER), combined with the allosteric mechanisms at the molecular level of individual control of specific CYP proteins, reveals a multilayered pyramidal adaptive system that responds to demand continually, even before the level of DNA-mediated expression changes in response to substrate load (enzyme induction) is approached (Chapter 4.4). CYP expression is known to be under the control of a hierarchy of nuclear receptors and factors and the degree of CYP complexing is likely to be no different. Clearly, the complex detail of the expression and operation of the CYP system remains to be fully understood, although what is known to date hints at the incredible flexibility and rapidity of the CYP system in the immediate, intermediate, and longer-term metabolic response to different substrates, both exogenous and endogenous.

3.4.8   CYP REDOX partners (i): P450 oxidoreductase (POR)

P450 oxidoreductase (POR), also referred to in scientific literature as CPR (NADPH cytochrome P450 reductase), is a 78 k-Dalton NADPH reductase that is a separate entity from mammalian CYPs, but it is indispensable to them

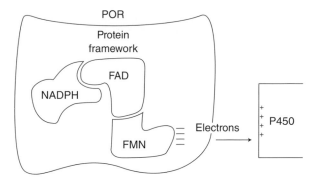

Figure 3.4 Position of CYP reductase (POR) in relation to CYP enzyme and the direction of flow of electrons necessary for CYP catalysis

(Figure 3.4). It originates from the *POR* gene, which is located on the long arm of chromosome 7 in humans[27]. Several types of NADPH reductases are found in tissues. POR and other NADPH reductases are particularly common in the liver. POR is essential to life, as removal of the gene from animal embryos is lethal[28]. A rare human condition known as Antley-Bixler Syndrome (ABS)[29] is linked with POR mutations (among others) and in severe form results in major structural malformations associated with disordered steroid metabolism[26].

Perhaps logically, the expression of POR is mainly under the control of the same nuclear receptors that control CYP expression, such as HNF4α and CAR (see Chapter 4.4.3). POR is a flavoprotein complex, which consists of a large, vaguely butterfly-shaped protein framework, which locates and binds two equal components, FAD (flavin adenine dinucleotide, an electron carrier) and FMN (flavin mononucleotide). FAD and FMN are hinged together by a linking domain of alpha-helices[27]. POR exists as locked (closed, smaller shape, no electron flow) and unlocked (open larger shape, full electron flow)[30], and the process of unlocking is crucially dependent on the ionic movements operated by the hinge section[31].

Although in tissues, NADH (used in oxidative metabolic reactions) can be plentiful, FAD has evolved to discriminate strongly in favour of NADPH, which fuels reductive reactions. NADPH is formed by the consumption of glucose by the pentose phosphate pathway in the cytoplasm. This (in part) oxidative system, which can consume up to 30% of the glucose in the liver, produces NADPH to power all reductive reactions related to CYPs, fatty acid and steroid synthesis, as well as the maintenance of the major cellular protectant thiol, glutathione (Chapter 6.4.2). Indeed, POR consumes NADPH to supply electrons to several other enzyme systems[31].

As mentioned in Section 3.3.2, POR's CYP proximity must be very close in the SER for successful electron transport to happen. POR broadly runs as follows: FAD is reduced by NADPH, which is then released as NADP+ (Figure 3.5). FAD then carries two electrons as $FADH_2$ that it passes on to FMN, forming

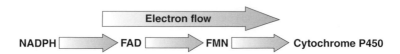

Figure 3.5 Direction of electron flow in P450 oxidoreductase (POR) supply of reducing power to CYP-mediated metabolic processes

$FMNH_2$, which in turn passes its two electrons to the CYP. The detailed operation of POR is extremely complex and not fully understood. When acting as a co-factor, normally NADPH would transfer two electrons *simultaneously* to any given enzyme. POR uses intricate single amino acid movements all the way to major alpha helix and domain shifts to supply the electrons *sequentially* at the appropriate points in the CYP catalytic cycle[30,32]. In the presence of high substrate concentrations, POR is required to provide sufficient electron flow to sustain continuous CYP catalytic activity, rather like a machine tool would need electricity in a factory (Figure 3.5). Interestingly, POR is not the only source of electrons for the CYP catalytic cycle, which is discussed in the next section.

In the lungs, POR is one of the enzyme systems that propels the toxicity of the herbicide paraquat, a weedkiller that still features in accidental and suicidal poisonings[33]. Paraquat, unfortunately, is a very good substrate for polyamine transporters and is concentrated in the lung, where it is metabolised by POR and other systems in REDOX cycles that generate vast amounts of oxidant species, that destroy lung tissue over several days, leading to death rates of more than 50%[34]. To date there is still no known antidote to paraquat poisoning, and it is an agonizing and drawn-out method of suicide. Oxidoreductases are also implicated in the reduction of nitroaromatic amines to carcinogens (Chapter 8.5.4).

3.4.9   CYP REDOX partners (ii): Cytochrome b_5

Cytochromes b_5 are electron transport haemoproteins that strongly resemble the active sites of CYP P450s, in that they also are built around a central F-9 haem group. These proteins are ubiquitous in nature, where they convert plentiful cellular supplies of NADH to NAD+, so building up proton gradients, which in turn stimulate the flow of electrons[35]. There are three main forms of cytochrome b_5; a soluble form is found in erythrocytes, where it is known as NADH-dependent methaemoglobin reductase, or sometimes NADH diaphorase[35]. This version of cytochrome b_5 converts methaemoglobin (Chapter 8.2.2), which is formed normally in small amounts and cannot carry oxygen, back into haemoglobin. Interestingly, those with the milder version of the genetic absence of NADH diaphorase spend their whole lives with blue cyanotic skin[36]. Another form is found in the outer membrane of hepatic mitochondria (known as OM cytochrome b_5), which is functionally very different to microsomal cytochrome b_5, which is of greatest interest in drug metabolism[35].

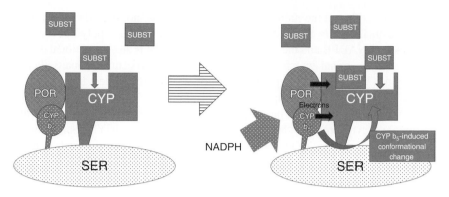

Figure 3.6 Cytochrome b_5 and POR's roles in supplying reducing power to CYP-mediated metabolic processes

Our understanding of the role of microsomal cytochrome b_5 in the operation of CYPs is far from complete but has advanced considerably over the last few years. Microsomal cytochrome b_5 is anchored by a helix which penetrates deep into the SER membrane, but its haem structure is cytosolic (it stands clear of the SER membrane) and it is physically linked with its CYP isoform and it is also closely associated with POR (Figure 3.6). For many years it was widely assumed that the flow of electrons needed to 'power' CYPs was entirely dependent on POR, which required an adequate supply of NADPH to operate. However, it appears that cytochrome b_5 also has a complex but essential role in CYP function. The standard CYP catalytic cycle (see Section 3.7 and Figure 3.8) requires two electrons to undergo a complete turn. The second electron is the *rate-limiting step* in that the speed of the CYP operation depends on how rapidly this electron can be supplied. It is now clear that cytochrome b_5 can supply this electron as rapidly and efficiently as does POR/NADPH[35]. Indeed, in steroid metabolism with CYP17A1, b_5 not only supplies electrons, but allosterically interacts with CYP/POR complexes to accelerate catalysis[37].

Although work has been carried out using antibodies to inactivate cytochrome b_5 has shown it is likely to be essential for optimal CYP activity, a key study in 2008[38] actually deleted the expression of hepatic microsomal cytochrome b_5 expression in mice. This work showed firstly that the mice suffered no ill effects and developed normally, which suggests other enzyme systems can carry out the main functions of cytochrome b_5 in cellular housekeeping. However, both *in vitro* and *in vivo*, drug clearance by CYPs was greatly reduced by up to 90% of normal with some drugs. Further studies using human CYPs expressed in mice where b_5 was knocked out, illustrated that CYP3A4 mediated (triazolam) and CYP2D6 (metoprolol) clearances *in vitro* where reduced by half, which was reflected in the *in vivo* studies, which showed 4 and 5.7 fold respective increases in area under the curve and peak plasma concentrations of triazolam and metoprolol[39].

This means that in the mouse and probably humans, not only do most CYPs rely on cytochrome b_5 to be part of the process of electron supply but also the utilization of NADPH and the efficient function of POR is also conditional on the presence of functional cytochrome b_5. If these processes occur in humans, then it is likely that the two REDOX partners are not only interdependent but their individual contributions to the supply of electrons may be exquisitely modulated. REDOX activity appears to be coupled to a combination of specific substrate binding and cytochrome b_5 binding, which then influences CYP conformation internally and externally, changing dimensions of access channels and even binding characteristics (Figure 3.6). This is complicated by the significant variation in human cytochrome b_5 expression and the ability of b_5 interactions to inhibit as well as stimulate catalysis[37-39].

The full complexity and flexibility of the modulation of CYP substrate binding and the resultant internal enzymatic conformational changes remain to be uncovered. It is likely that the CYP's response to a particular substrate is effectively customized in terms of binding, electron supply and catalytic activity. In some ways, CYPs can be regarded as part of a highly sophisticated locally and centrally controlled pump system, which draw up lipophilic agents from the SER membrane conveyor belt and expel them as more hydrophilic metabolites, thousands of times per second.

3.5   Human CYP families and their regulation

As we have seen, despite their common structure and function, CYP substrate specificity varies enormously within the main families of these enzymes. Aside from CYPs that are involved in steroid synthesis and arachidonic acid metabolism, only three CYP families are relevant to humans in terms of drug and toxin biotransformations:

CYP1 family   (CYP1A1, CYP1A2, and CYP1B1)

CYP2 family   (CYP2A6, CYP2A13, CYP2B6 CYP2C8, CYP2C9, CYP2C18, CYP2C19,CYP2D6, CYP2E1)

CYP3 family   (CYP3A4, CYP3A5, CYP3A7)

It is believed that 9 out of 10 drugs in use today are metabolised by only five of these isoforms: CYPs 1A2, 2C9, 2C19, 2D6, and 3A4/5. CYP2E1 is interesting mostly from a toxicological perspective, such as in paracetamol metabolism (Chapter 8.2.3) and the internal regulation of small hydrophilic molecules. Each CYP has its own broad substrate preferences, and in some cases they may not be expressed in some individuals at all, or in very low levels (CYP2D6 polymorphisms; Chapter 7.2.3). Using immunologically based methods that employ specific antibodies raised to bind to the various CYPs in ELISA and

Western Blotting systems, it has been a remarkable achievement that these often extremely similar isoforms have been distinguished, structurally as well as functionally, using other assay systems.

3.5.1 CYP regulation: lifespan

Whilst so far we have considered the different layers of CYP system responses to substrates, from the level of the individual isoform to the complexing of large numbers of isoforms, it is useful to consider that before all this regulation occurs, there is a basic production/destruction cycle of CYPs. Like any machine tool, the 'order' (mRNA) is sent out from the nucleus to the factory 'production line' of the rough endoplasmic reticulum where the CYP is assembled in a ribosome and self-folds to its correct shape. Near the N-terminal end (which is lipophilic and will insert into the SER), the correct 'address for delivery' is incorporated, which assists in sending the CYP to the SER, instead of, say, the mitochondria[40]. Once the N-terminal end is completed, synthesis stops and the ribosome complex transports the new CYP across the cytosol towards the SER. Once the complex finds the SER, signal recognition particles (SRPs) facilitate the precise positioning of the CYP, where its N-terminal end is located in particular domains within the SER.

For instance, very similar CYPs such as 1A1 and 1A2 have address domains that send them to different SER microdomains, which may be either *disordered* or *ordered,* depending on their composition with respect to saturated/unsaturated fatty acids, cholesterol and other components[41]. Once the installation is complete, much of the rest of the protein appears to 'float' on the lipophilic SER, as described in Section 3.2. The SER microenvironment facilitates correct delivery, folding, and functionality.

In terms of CYP end-of-life processes, to follow the 'machine tool' analogy, they have a finite lifespan and wear out and lose their functionality. CYPs are, as we have seen, extremely complex but rather fragile protein structures, whose functionality directly depends on very precise structural integrity. Wild-type CYPs are usually more stable than many polymorphic forms and, as mentioned earlier, the binding of certain substrates often enhances their stability.

It seems that CYP lifespans, given their complexity and the energy expended to build them, are surprisingly short and are usually less than two days; CYP2C8 for example, lasts about 22 hours (see Section 3.6.2). This can be shortened by some substrates that are metabolised into such reactive species that they destroy the enzyme's active site, as will be seen in Chapter 5.4.3. On the other hand, CYPs when they are 'idling', that is, without a substrate, can form reactive species in such quantity that they impact cellular REDOX potentials and are involved in several disease states, including cancer[42].

Hence, CYP control and regulation, if it were not already complex enough, also involves a system that detects and removes worn-out or defective (misfolded) CYPs, but also undamaged CYPs. This is the 'endoplasmic reticulum associated degradation-cytosolic' or ERAD-C pathway, which coordinates the recycling of SER-protein, such as CYPs, with the main cytoplasmic protein recycling Ubiquitin Proteosomal System (UPS). Part of the ERAD-C system is an essential extraction tool (P97 chaperone complex) that removes CYPs from the SER before they can be recycled[43], even though some of the CYP projects from the SER into the cytoplasm. The need for the tool underlines how firmly anchored the CYPS are.

The UPS system recycles all protein from all over the cell, including the CYPs, sometimes destroying proteins deemed surplus to requirements before they even reach their destinations, such as the SER[43]. The UPS is often likened to the cellular equivalent of a paper shredder and is a vaguely tube-shaped and contains an internal protein slice 'n' dice mechanism, which reduces various cellular proteins to peptides and amino acids for recycling. The UPS normally works in tandem with a series of ligases that attach the protein ubiquitin to any unwanted or damaged proteins or cellular structures. The UPS recognizes the ubiquitin label and destroys the protein.

The ERAD-C/UPS system operates with most CYPs, deciding their fate on the basis of their structural integrity and catalytic activity. The CYPs are then phosphorylated by cytosolic kinases, promoting P97-mediated extraction and routing to the UPS. So CYP turnover can be adjusted not only by synthesis but also by the rate of destruction. It is likely that these processes are regulated constantly at the level of nuclear factors and various sensor molecules.

3.5.2   CYP regulation: transcriptional

So far we have seen CYP *levels* controlled by synthesis and degradation, whilst CYP *activity* can be modulated at the level of the enzyme/substrate interaction, as well as through complexing of large numbers of catalytic CYP/REDOX partner units (Sections 3.4.5–3.4.7). However, even this degree of responsiveness is not sufficient to meet the demands of operating a system as complex as turning an immature human child into an adult and then maintaining their capacity for reproduction for decades. Again, if the individual encounters a high and sustained level of a potentially toxic chemical, the modest changes in CYP levels and activity gained from higher turnover and allosteric/complex-based increases in metabolic activity could not meet a sudden and sustained demand for biotransformational capacity. Such a demand, say to produce more hormones or detoxify large amounts of a xenobiotic, will need perhaps a 10- or even a 100-fold increase in CYP levels to satisfy.

For very large and sustained increases of CYP expression that are beyond a short-term response, the enzyme induction process is necessary, which is discussed in detail in Chapter 4.4. This is governed through sensor molecules in the cytoplasm or more usually the nucleus, which then bind to response elements on the genes which code for the particular CYPs. This transcriptional control thus involves the massive upregulation of the CYP expression system, which includes changes in DNA and various types of RNA activity, which results in increased ribosomal protein manufacture. This process is reversible but rather slow, taking place in humans over one to three weeks.

In a way, CYP induction as briefly described above could be likened to not only hitting the accelerator but also massively enlarging and supercharging the engine at the same time. Interestingly, at the same transcriptional level logic dictates that there must be an equally powerful and effective braking system for such a biotransformational juggernaut. An important pre-translational brake for CYP expression happens through the mRNA 3'-UTR, or three prime untranslated region[44,45]. This sequence in mRNA is one of those that does not code for a protein but is a key control panel for influencing mRNA activity. Micro RNAs (miRNAs; Chapter 2.4.4), already mentioned in relation to plants and their possible impact on our diets, regulate CYP transcription through their binding to 3'-UTR sites in CYP mRNAs. Indeed, the process is known as post-transcriptional gene silencing and miRNAs control the expression of thousands of cellular genes as well as those linked with biotransforming processes in this way.

The roles of miRNAs are still being explored but with CYP2E1, it has been shown that they can also shut off mRNA transcription through binding at the CYP gene's promotor site, as well as acting on the mRNA 3'-UTR[44]. miRNAs are of course not the only factors that can influence CYP gene expression; various signal molecules, such as hepatocyte nuclear factors, can modulate CYP2D6 expression, for example[46].

There are several reasons why it is necessary to 'slam on the brakes' of CYP expression and activity. As mentioned in Section 3.5.1, CYPs, and particularly CYP2E1, can produce significant amounts of reactive species, so if intracellular oxidative stress is detected above a certain level, it is logical to restrict sources of such stress[44]. Studies in human cellular systems have shown that the reactive species formed by paracetamol oxidation by CYP2E1 showed injury to the cells within 12 h, whilst within 24 h, the miRNA had suppressed CYP2E1, as well as 1A2 and 3A4. This was supported by the detection of higher than normal levels of the same miRNAs in children who had been overdosed with paracetamol[45]. Other studies have shown elevated levels of miRNAs in human livers during inflammatory conditions and specific miRNAs, such as miR-130b, can suppress CYP2C9 in human cellular systems. That same miRNA is upregulated in cancer and obesity[47]. CYP expression can be almost completely shut down by pro-inflammatory cytokines, and these can have very marked impacts on the clearance of drugs during acute and chronic infections (Chapter 7.9).

3.5.3   CYP regulation: post-translational

Once a CYP has been assembled, correctly folded, shipped, and then installed in the SER or the mitochondria, any other changes to the structure and function of the CYP that affect its activity can be considered as post-translational. So the partly self-regulating substrate-mediated allosteric and complexing features already described in Sections 3.4.5 and 3.4.9 are post-translational, as are the changes in CYP activity and affinities influenced by the co-factors POR and cytochrome b_5. Whilst the impact of these processes is still being understood in terms of drug clearance, perhaps better described is how CYPs are post-translationally regulated during endogenous steroid synthesis, through the processes described above and others such as phosphorylation[48]. The most dramatic post-translational modulation is of course the destruction of CYPs by the ERAD-C ubiquitin proteosomal system, as described in Section 3.5.1. Other forms of regulation may occur through changes in the stability of the CYP on substrate binding.

Overall, Figure 3.7 summarizes that CYP activity and capacity can be modulated from the molecular to the tissue level through a vast network of sensors and activating factors that allow biotransformation of endogenous and exogenous agents to be controlled over time frames ranging from minutes to years.

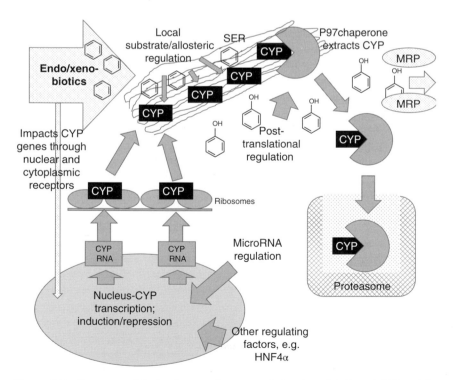

Figure 3.7   Summary of the local pre-and post-translational modulation of CYP activity as it responds to pressure from endogenous and exogenous substrates

3.6   Main human CYP families

3.6.1   CYP1A series

CYP1A1

The gene that codes for CYP1A1 is on chromosome 15, close to that of CYP1A2; both genes even share a regulatory region[49,50]. Once expressed, CYP1A1 has a molecular weight of approximately 56KDa and consists of 512 amino acids. This isoform binds and oxidizes planar aromatic, essentially flat, lipophilic molecules. The most common representatives of these compounds are multiples of benzene, such as naphthalene (two benzenes), and what are usually termed polycyclic aromatic hydrocarbons (PAHs) that are many benzene molecules in chains. There is evidence that CYP1A1 is induced by many other variants of planar aromatics, such as the dioxins (Chapter 2.2) and the polybrominated diphenylethers (PBDEs), which are carcinogens and endocrine disruptors. Interestingly, this isoform is hepatically *nonconstitutive,* i.e. it is not normally expressed or found in the liver[50]. This is probably because the accumulation of large amounts of planar aromatics in the liver should not normally occur. Hence, this CYP does not usually have a significant role in prescription drug metabolism, although the proton pump inhibitor omeprazole will induce it[51]. CYP1A1 is of interest mainly for its ability to metabolise environmental pollutants, as well as its role in the development of various cancers[52]. For instance, whilst α-napthoflavone is an experimental inhibitor of this CYP, from the cancer perspective, the inhibitory effects of various dietary agents such as the bergamottins[53] and the complex influence of various flavonoids on CYP1A1 and their impact on its ability to promote malignancy in the real world is under investigation[54].

CYP1A1 is inducible (Chapter 4.4.2) in all tissues[50-51], and this occurs in the lung in response to PAHs encountered in industrial and traffic pollution. Tobacco smokers exhibit high lung levels of this CYP due to the PAHs and other aromatics present in the smoke[55]. Interestingly, nonsmokers exposed to environmental tobacco smoke also show increased levels of CYP1A1. Metabolic products of CYP1A1, which are often epoxides, vary in their stability and the most reactive, such as those from benzpyrene derivatives, are carcinogenic. There is also evidence of higher levels of CYP1A1 in cancer sufferers[52]. Experimentally, this CYP is often studied through its ability to O-deethylate the test substrate 7-ethoxyresorufin (EROD), and it is an important biomarker in environmental contamination of PAH in many aquatic and terrestrial organisms[56].

CYP1A1 has been part of the evolutionary protection from lipophilic and planar 'threat molecules', which might mimic endogenous hormones, so the clearance of toxic agents such as the PDBEs by this CYP should be necessary and beneficial. However, the isoform is polymorphic (Chapter 7, Section 7.2.3) and its absence may predispose to toxicity with such agents. Conversely, CYP1A1-mediated production of reactive metabolites is likely in many cases to be more of a threat than a protection, as it is often overexpressed in the vicinity of carcinogenesis. One of its main endogenous roles was suggested by work with knockout mice, where animals without CYP1A1 had lower-than-normal major organ weights, alongside several other developmental problems[57]. CYP1A1 has a crucial role from embryonic stages, through to adolescence, in the metabolism of various omega-3 fatty acids, which manage the development of several organs and systems, particularly the cardiovascular system[50]. If this is its primary endogenous function, this is probably why it is not expressed significantly in adulthood, unless PAH-like agent exposure occurs. Whether CYP1A1-mediated clearance of xenobiotics is beneficial or deleterious in adulthood is probably chiefly dependent on the complex balance between the level of environmental and dietary exposure to inducers and inhibitors of this CYP[49,50,53].

CYP1A2

Whilst the CYP1A2 gene is found very close to CYP1A1 in man as described above, it has a far greater role in prescription and recreational drug clearance. It has a molecular weight of approximately 58 kDa and comprises 515 amino acids, with a sequence homology with CYP1A1 of 72%[58]. CYP1A2, along with CYP1A1, prefers planar substrates, as the F and I helices of both CYPs form an active site that promotes binding with flat molecules and pi–pi bond stacking[59]. Both CYP1A1 and 1A2 have similar active site volumes of around 375 cubic Angstroms (A^3), which are only around a quarter of the volume of CYP3A4 but are about half as large again as CYP2A6[60,61]. There is significant overlap between CYPs 1A1 and 1A2 in their endogenous substrates, although CYP1A2 also clears oestrogens, bilirubin, and hormones like melatonin[59]. Subtle differences in its active site mean that CYP1A2 metabolises planar aromatic amines, which is not seen with CYP1A1[60]. CYP1A2 has a significant impact on prescription drugs not only due to its substrate preferences, but also because it is expressed in quite high levels in the liver (13–15% of total CYP content). It also has a role in the clearance of antipsychotics, such as clozapine, as well as with tricyclic antidepressants, the muscle relaxant tizanidine, plus beta blockers such as propranolol and the airway relaxant theophylline[58,59]. CYP1A2 also clears caffeine and β-naphthylamine (a known human bladder carcinogen; Chapter 8, Section 8.5.4). Induction of CYP1A2 in smokers means that interactions with prescription drugs can occur if smoking is stopped (Chapter 7.6). CYP1A2 can be inhibited by some planar molecules that possess a small volume to surface area ratio, such as the methylxanthine derivative furafylline, as well as ciprofloxacin, enoxacin, cimetidine, mexiletine, and fluvoxamine. The CYP1A1 inducer omeprazole promotes a similar response in CYP1A2, and of course planar aromatic amines, as well as dioxins, induce it. As with CYP1A1, CYP1A2 has a role in malignancy development that is linked with the complex pattern of induction and inhibition, which results from diet and environmental factors.

CYP1B1

To provide a perspective on the vast stretches of time over which CYPs have evolved, it is thought that CYP1B1 became distinct from the CYP1A subfamily over 450 million years ago[61]. The isoform CYP1B1 has a slightly smaller active site (270 A^3) than CYPs 1A1 and 1A2,[62,63] and has a molecular weight of approximately 52 kDa and comprises 543 amino acids[62]. Interestingly, *CYP1B1* is the largest of all the CYP genes, although it is much less complex than other CYPs and has only around 40% similarity in amino acid structure with CYPs 1A1 and 1A2[62]. CYP1B1 is expressed at a low level in the liver, but is usually also found in most tissues, particularly the prostate, breast, and pancreas, but it is not normally expressed at high levels[64]. CYP1B1 is the only member of its subfamily discovered so far, and the gene that codes for it is found on chromosome 2[62]. It is inducible through the same pathway as CYP1A1/2 (Chapter 4, Section 4.4.2), and it has similar endogenous activities to its CYP1A relatives, mainly fatty acid, steroid, leukotriene and eicosanoid metabolism in embryonic development and subsequent cellular housekeeping. CYP1B1 forms a steroid vital in eye development, and it catalyses the production of an arachidonic acid metabolite, which maintains the transparency of the cornea as well as the regulation of its aqueous humour. Indeed, mutations in this CYP are linked with congenital glaucoma, where intraocular pressure is excessive and can lead to early blindness[62].

CYP1B1's role in ocstrogen metabolism is of particular interest in cancer research, because whilst CYPs 1A1 and 1A2 form 2-hydroxy derivatives of oestrogens, CYP1B1's production of 4-hydroxylated oestrogen metabolites leads eventually to mutagenic quinones, and this CYP is often overexpressed in malignant breast tissue[64]. Indeed, CYP1B1 is one of the main weapons deployed by tumours to resist chemotherapy, and in many cancers (particularly the lung and pancreas), its expression is so high it can rapidly inactivate antineoplastic agents such as docetaxel and practically eliminate their efficacy[64]. Thus, it has been a serious therapeutic goal to develop inhibitors of this CYP, possibly as adjuncts to cancer chemotherapy. Fortunately, many structures can inhibit CYP1B1, ranging from many synthetic stilbenes to dietary flavonoids; interestingly, the antineoplastic drug flutamide is also an inhibitor of CYP1B1[64].

Perhaps the most exciting prospect in this area may be with microRNAs, as they can modulate multiple systems in cells simultaneously. This has been exploited in terms of miR-27b, which can upregulate p53 to promote apoptosis in malignant cells and shut off their CYP1B1 expression at the same time, thus sensitizing them to chemotherapy[65].

3.6.2   CYP2 series

Around 18–30% of human CYPs are in this series, making it the largest single group of CYPs in man, collectively metabolizing around half of prescription drugs. They appear to have evolved to oxidize various sex hormones, so their expression levels can differ between the sexes. As with many other CYPs, they are flexible enough to recognize many potential xenobiotics and particularly toxins in the case of CYP2D6, but their endogenous functions are less well documented.

CYP2A6

This CYP originates on chromosome 19 and the protein has 494 amino acids and is really only expressed in the liver, comprising about 4% of the CYP content of human liver. Its active site is, as mentioned above, quite small (260 A^3), and it specializes in low-molecular-weight, nonplanar substrates with at least two hydrogen bond acceptors[66-68]. The endogenous roles for this isoform so far include clearance of bilirubin to biliverdin, as well as steroid, lipid, and cholesterol metabolism[66].

Whilst CYP2A6 participates in only around 3% of prescription drug clearance, the strong interest has been in its role as converting around 70–80% of absorbed nicotine to the pharmacologically less active cotinine; interestingly, CYP2A6 is 100% responsible for further cotinine clearance[67]. CYP2A6 is expressed at greater levels in women than men and is extremely polymorphic (Chapter 7.2.3), and there is a relationship between low expression and reduced smoking behaviour, easier quitting, and a lower risk of lung cancer[69-70]. This is because low CYP2A6 expression leads to slow clearance of nicotine, which reduces craving and fewer cigarettes are consumed, compared with those with high CYP2A6 expression. The relationship between nicotine levels and craving is also linked with acetylcholine receptor activity[67]. Lifelong exposure to PAHs and other mutagens in tobacco smoke is thus lower in those smokers with reduced CYP2A6 expression, and this is just as well, as this isoform can activate the many nitrosamines resulting from the pyrolysis of tobacco to mutagens[67]. Indeed, CYP2A6 activity is a good indicator of an individual's risk of developing lung cancer[70].

A key and specific marker for CYP2A6 is coumarin oxidation to 7-hydroxycoumarin, but this CYP is also involved in the clearance of a number of prescription drugs, which can lead to significant impact on efficacy and toxicity when polymorphisms are concerned (see Chapter 7.2.3.). This CYP converts the antineoplastic prodrug tegafur to its 5-fluorouracil active metabolite and can clear the anti-breast cancer drug letrozole, as well as the antiretroviral efavirenz, along with other drugs as diverse as the anticonvulsant valproic acid and the anaesthetic halothane[67]. CYP2A6 is induced by both endogenous and synthetic steroids such as oestrogen and dexamethasone, respectively, as well as moderately by the antimycobacterial rifampicin and phenobarbitone[71].

Inhibition of CYP2A6 is currently of significant interest, as a selective and safe inhibitor could not only make it easier to stop smoking, but it would also impact smoking mortality and morbidity through prevention of the formation of reactive species, theoretically even if smokers continued to smoke. Methoxsalen (an antipsoriatic agent) is a potent mechanism-based (Chapter 5, Section 5.4) inhibitor of 2A6, as is grapefruit juice, although it is also weakly inhibited by imidazoles (e.g. ketoconazole). Methoxsalen will inhibit CYP2A6 in man, and it prolongs the plasma survival of nicotine, reducing smoking[71,72].

CYP2A6's role in the activation of various carcinogens as well as those related to tobacco usage such as aflatoxins (Chapter 8.5.6) and the atmospheric pollutant 1,3 butadiene, makes the search for an inhibitor all the more urgent. However, to date, lack of selectivity has made developing effective and practical inhibitors of this CYP problematic[72].

CYP2B6

Originating on chromosome 19, the 2B series has been extensively investigated in animals, but CYP2B6 is the only 2B form found in man[73]. This isoform is found in all human livers and comprises around 1-10% of total hepatic CYPs[73,74]. CYP2B6 prefers non-planar neutral or weak bases which accept hydrogen bonding. It tends to hydroxylate at highly specific areas of molecules, particularly close to methoxy groups, which suggests that it may have a biosynthetic role in particular stages of the assembly of endogenous molecules, such as lipids and steroids[73]. Its level of expression varies by remarkable amounts – its mRNA by 40000-fold and its actual activity 600-fold[74]. Whether there is a defined sex difference in expression that is of clinical relevance is still to be decided[74]. The impact of CYP2B6 on drug efficacy and toxicity has been seriously underestimated in the past and is now better appreciated. Its importance is due to its variability, very high polymorphic expression (see Chapter 7, Section 7.2.3), and crucially it assists in the clearance of up to 10% of therapeutic drugs[75-76]. It is thought to be implicated in the metabolism of more than 70 xenobiotics, including amfebutamone (bupropion), mephenytoin, some coumarins, cyclophosphamide and its relatives and the antimalarial artemisinine. CYP2B6 accounts for more than 90% of the clearance of the anti-HIV reverse-transcriptase inhibitor efavirenz and the 8-hydroxylation of this agent is a useful CYP2B6 marker, as is bupropion hydroxylation[74,77]. Its key role in the clearance and toxicity of both methadone and ketamine is now well documented[78-79] and is detailed in Chapter 7 and in Appendix B. Inhibitors include the antiplatelet drugs clopidogrel and ticlopidine and the antineoplastic agent thiotepa[80]. It is moderately inducible by rifampicin, phenobarbitone, and the DDT substitute pesticide methoxychlor. This pesticide acts as an endocrine disruptor when it is oxidized to pro-oestrogenic metabolites by a number of human CYPs including CYP2B6 itself[76].

CYP2C8

The gene that codes for this polymorphic CYP is found in a cluster on chromosome 10, alongside those of CYP2C9, 18 and 19 and these CYPs account for about 18% of total CYP content and they share roughly 80% amino acid sequence identity[81-82]. CYP2C8 itself accounts for about 7% of human hepatic CYP content and can clear relatively large molecules, which are also weak acids[83]. The active site of this CYP is hydrophobic, large, and is sometimes described as trifurcated or shaped like a 'Y' or 'T'[84]; the haem iron resides at the bottom of the Y with the two arms as access channels. Its endogenous role appears to be linked with arachidonic acid and particularly various retinoic acids, which participate in embryonic development, so CYP2C8 foetal expression is high[83]. CYP2C8 can accommodate quite large molecules without rearranging itself as CYP3A4 is able to do. Two molecules of 9-*cis*-retinoic acid fit in the active site, whilst the leukotriene receptor antagonist montelukast fills it completely; the withdrawn hepatotoxin, troglitazone, only uses the upper part of the space[84]. Montelukast, as mentioned above, is cleared at clinically relevant levels by CYP2C8[85] and approximately 40% of the anti-cancer agent paclitaxel is cleared through 6α-hydroxylation and this reaction is a main probe for this CYP[83]. Lack of expression of this CYP can cause such accumulation-linked toxicity with this taxane derivative, that patients can no longer tolerate it[82]. CYP2C8 also clears up to 100 other drugs, including verapamil, cerivastatin, amodiaquine, rosiglitazone/pioglitazone, repaglinide, and the anti-prostate cancer drug enzalutamide[83]. CYP2C8 is vulnerable to inhibition by some glucuronides, particularly those of the lipid-lowering agent gemfibrozil (1-O-β-glucuronide) and the anti-platelet agent clopidogrel (acyl-β-D-glucuronide)[82-84]. Indeed, the interaction between cerivastatin and gemfibrozil led to such a potent inhibition of the clearance of the statin that more than 50 fatalities occurred due to rhabdomyolysis, provoking the drug's withdrawal in 2001[86,87]. CYP2C8 can also be inhibited by quercetin, and by diazepam in high concentrations, as well as the antibacterial trimethoprim. The antiretroviral agents efavirenz and saquinavir are also inhibitors of this isoform. It is moderately inducible by rifampicin and phenobarbitone[83].

CYP2C9

CYP2C9 is found on the same chromosome as CYP2C8 but it accounts for around 20% of hepatic CYP content and is linked with the clearance of around 15% of currently used clinical drugs[82,88,89]. Only CYP3A4 is expressed in greater quantities and CYP2C9 has been particularly heavily studied in terms of its structure and function, where it processes relatively small, acidic and lipophilic molecules that accept hydrogen bonds, such as the antihypertensive losartan[89,90]. Indeed, three losartan molecules can be bound simultaneously by wild-type

CYP2C9, at the active site, in the access channel and also a peripheral site respectively[91], the latter probably regulating the binding/catalysis process (see Sections 3.4.4–3.4.7). In terms of amino acid sequences and internal/external structure, CYP2C8 and CYP2C9 are very similar, but there are significant differences in their active sites and conformations which, for example, lead to CYP2C8 forming a 5-hydroxy derivative of diclofenac, whilst CYP2C9 4'-hydroxylates the same drug[91]. Indeed, some of their ability to selectively metabolise different substrates lies in the differences between their access tunnels; CYP2C8's main access is between the β sheets and the F-G α helices (see Section 3.4.2), whilst CYP2C9's access lies between the B-C and the I and G helices[91]. Hence, the same molecule can be orientated differently according to how and where it enters the CYP active site, as is the case with phenytoin with CYP2C9, which is orientated for subsequent oxidation through the phe-cluster and changes to these amino acids has a large impact on phenytoin oxidation (see Section 3.4.6)[92].

Much of the strong interest in CYP2C9 is rooted in the clinical importance of this isoform, as it tends to clear commonly used, yet relatively narrow therapeutic index drugs, such as warfarin and the anticonvulsants. In addition, the extensively polymorphic nature of this CYP[88] (see Chapter 7.2.3) means that drug interactions and marked impact on patient outcomes in terms of efficacy and toxicity are strongly associated with it. Warfarin is probably the best documented issue, as it is administered as a racemic mixture, yet the S-isomer has a five-fold higher anticoagulating action compared with the R-form; CYP2C9 has a strong role in the clearance of the S-isomer and virtually none in the R-form. As a result, lack of expression of this CYP leads to drug accumulation and potentially disastrous over-anticoagulation[90-92]. Lack of clearance of phenytoin in those with poor expression of CYP2C9 likewise leads to significant adverse effects. The CYP also clears tolbutamide, valproate, rosuvastatin, and the NSAID flurbiprofen[89]. Sulphafenazole is a potent inhibitor, as is amiodarone and fluconazole[89,90]. Naturally, co-administration of an inhibitor with a narrow therapeutic index drug such as warfarin leads to severe clinical issues and still occurs with amiodarone, for example[89-93]. CYP2C9 is inducible by the usual suspects (rifampicin and phenobarbitone) as well as the steroids prednisone and norethindrone[89].

CYP2C18

This CYP, which originates in the same chromosomal cluster as CYP2C8, 9, and 19[94], does not appear to have any real role in the clinically relevant metabolism of drugs or environmental toxins and nor has anyone found out why it is there and what it actually does. Expression is negligible in human liver and despite the high output of mRNA, no useable protein results; neither is it inducible through the usual nuclear receptor routes (Chapter 4) and it does not seem to be of much use to animals, either[95]. However, CYP2C18 is somewhat surprisingly more capable than CYP2C9 in the oxidation of phenytoin, which raises the intriguing possibility that because this CYP is expressed in the skin, it may be responsible for cutaneous reactions that are linked with phenytoin activation to reactive species[96].

CYP2C19

This fourth member of the CYP2C chromosome 10 cluster, which is closest in terms of amino acid sequence to CYP2C9, with only 43 out of 490 residues differing[91,97]. It is found in the liver, duodenum, and stomach[97], yet only accounts for around 1% of total hepatic CYP content[91]. Whilst it is very similar structurally to its sister CYPs 2C9 and 2C8, key differences between access tunnels and substrate recognition/orientation phe-clusters are reflected in the regioselective metabolism of substrates such as phenytoin and diclofenac[91,92]. CYP2C19 prefers weak bases such as amides, which have at least two hydrogen bond acceptors. This CYP has risen in prominence partly due to its role in the clearance of around 10% of prescribed drugs but particularly its role in the metabolism of several mass-market drugs such as the anti-platelet drug clopidogrel (at one time the second best-selling drug worldwide)[98] and the various proton pump inhibitors. This, combined with the discovery of more than 35 polymorphic forms to date, which range from nonfunctional to the superfast metabolising CYP2C19*17 (see Chapter 7.2.3) make this CYP potentially significant in many therapeutic outcomes. CYP2C19 also metabolises omeprazole and its related proton pump inhibiting anti-ulcer agents, as well as S-mephenytoin and diazepam. CYP2C19 is inducible through the nuclear receptor pathway (rifampicin and carbamazepine). CYP2C19 status is a strong indicator of adverse reactions and clinical efficacy with the antifungal voriconazole[99]. Interestingly, although intensive research to date shows that whilst a poorly functional variant of this CYP (CYP2C19*2) is associated with poor outcomes with clopidogrel, it is just one factor of the many that impact the clinical response of this drug and even its replacement, prasugrel[98,100,101]. However, low-functional versions of CYP2C19 can be beneficial in ulcer therapy where the proton pump inhibitor has a wide therapeutic index, so clinical response is improved (Section 7.2.3). Tranylcypromine acts as a potent inhibitor of CYP2C19, as does fluvoxamine.

CYP2D6

The gene that codes for this isoform is on chromosome 22 and expression is plentiful in the liver, duodenum and small intestine[102]; CYP2D6 amongst the various human CYPs is genuinely fascinating for a number of reasons, not least because of its catalytic and binding preferences, which means it can metabolise up to a quarter of prescription drugs[103,104]. This is because this CYP has evolved to process basic and lipophilic chemicals that feature aromatic rings and a nitrogen atom, which can be protonated in physiological conditions and as you have hopefully already learned from basic Physiology and Pharmacology, a vast number of human systems are regulated by such chemicals. Indeed, CYP2D6 is involved with more than the usual steroids and arachidonic acid derivatives; in the brain it is important in the metabolism of several biogenic amines, such as dopamine, tryptamine, and tyramine, and it is perhaps the only well-known bio-transforming CYP that influences personality[103,104]. It also is linked with the clearance of many neuronal toxins, and poorly functional expression of CYP2D6 is now strongly linked with a predisposition to the development of the neurode-generative condition, Parkinsonism[104,105]. Hence, it is logical that this CYP can function protectively to N-dealkylate and clear plant-based alkaloid toxins (Chapter 2) and will also avidly process prescription drugs, which mirror the structure and physicochemical features of the endogenous agents they are mim-icking. Indeed, these similarities are underscored in the range of drugs CYP2D6 can clear:

- TCAs, SSRI, and related antidepressants: (amitriptyline, paroxetine, venlafaxine, fluoxetine)

- antipsychotics: (chlorpromazine, haloperidol)

- antiarrhythmics: (flecainide, mexiletine)

- beta-blockers: (labetalol, timolol, propanolol, pindolol, metoprolol)

- analgesics: (codeine, tramadol, oxycodone, hydrocodone, meperidine [pethidine])

The actual CYP2D6 wild-type protein itself has some unique structural features that distinguish it from other CYPs. Crystal structures with CYP2D6 bound to the antipsychotic thioridazine show four extra amino acids lengthen its helix B'-C loop, compared with the other CYP2 series enzymes. This allows better access to a specific Asp-301 residue that binds and orientates protonated nitro-gens near the active site. CYP2D6's tertiary structure is flexible and it can also orientate the next protonated nitrogen group at the access channel as the first molecule is being processed[106]. It can be inhibited by several agents, including quinidine, fluoxetine, and paroxetine.

The most well-known feature of CYP2D6 is the sheer scale of its polymor-phisms – over a hundred alleles (versions of the gene) have been found[105] and considering the number of drugs this CYP can clear, this has major implications for efficacy and toxicity (see Chapter 7.2.3.). Another interesting feature of this CYP is that although its normal expression is controlled by nuclear factors HNF-4α (see Chapter 4), it was believed until quite recently to be noninducible, but this is actually not the case (Chapter 4.4.5). There are several means of control of its expression, and some ethnic groups express multiple copies of CYP2D6[104], which is not induction *per se*, rather just very high constitutive expression. However, it is believed that CYP2D6 is induced in areas of the brain such as the substantia nigra and that this might be post-translational through phosphorylation[104]. In addition, it was noted in the 1980s, that CYP2D6 activity virtually doubles during pregnancy and it is believed this occurs through changes in the activity of transcription by factors such as HNF-4α, KLF-9 and SHP103 and may be linked with cortisol metabolism[103]. The clinical impact of CYP2D6 polymorphisms is discussed at length in Chapter 7.2.3.

CYP2E1

In terms of drug metabolism, CYP2E1 (chromosome 10) is probably mainly of interest for its role in the formation of hepatotoxic reactive species of paracetamol (acetaminophen) in overdose, as well as of ethanol during alcohol abuse[107] (Chapter 8). Although it can process more than 70 known chemicals, perhaps only the antimycobacterial isoniazid and the antispasmodic chlorzoxazone are of clinical relevance in terms of this CYP[107-109], although the alcoholism treatment drug disulfiram is a potent inhibitor also[107,110]. Aside from its role in forming toxic species, this CYP specializes in small, often water soluble heterocyclic agents which contain some hydrophobic residues; such molecules include pyrazole, various alcohols, small ketones such as methyl ethyl ketone and methyl isobutyl ketone (MEK and MIBK) and fatty acids. Its endogenous roles are wide-ranging and complex. These include a key response to food depravation, the formation of glucose from proteins, fats, and other cellular substrates (gluconeogenesis) [111]. This process generates many different small molecule substrates for CYP2E1 and induction occurs very rapidly. CYP2E1 regulates the presence of such small, volatile, and possibly intoxicating ketone and alcohol byproducts of starvation, at least partly in order to keep one's head clear during the increasingly fraught search for food.

CYP2E's active site is very small ($190A^3$) in relation to the larger CYPs when binding low molecular weight agents and it is even smaller than CYP2A6 ($\sim280A^3$) with which is shares 40% sequence similarity[108]. However, in a manner rather like those 'Transformer' toys that were popular a few years ago, it can expand to just under 500 A^3 when binding fatty acids, which cannot be achieved by CYP2A6[108]. This internal rearrangement depends on the rotation of a Phe sidechain, which opens up the active site to accommodate larger molecules[108,109]. Indeed, it has up to three conformational changes 'in the locker' to accommodate different molecules as well as multiple binding modes, which may have allosteric relevance[108,109].

However, as with CYP3A4, the very flexibility of CYP2E1 allows it to oxidise a considerable number of chemicals that perhaps should be left alone or metabolised differently. These include nitrosamines and halogenated alkanes, various polychlorinated biphenyls, as well as several benzene derivatives, all of which can be oxidised to reactive species that are mutagenic and potentially carcinogenic[111,112]. Consequently, the polymorphic expression of this CYP can be linked to selective vulnerability to various cancers and other conditions[113].

The ongoing controversy on the real human impact of the dietary toxin acrylamide is strongly linked to CYP2E1. This agent is formed through the Maillard reaction, where amino acids (such as asparagine) and various sugars react during high temperature cooking processes and not only is acrylamide a neurotoxin, it is also oxidised to a reactive derivative, glycidamide, only by CYP2E1[114]. There may be a relationship (Chapter 8.5.5) between acrylamide exposure in food and some cancers, but it is complicated by several factors. This includes the individual's CYP2E1 expression levels, possible polymorphisms, and the propensity of the isoform to be inhibited by a number of dietary chemicals such as the flavonoids[114,115].

The general cellular mechanisms of control of CYP2E1 expression are quite unlike all the other major human CYPs (see next chapter) and is probably a reflection of two issues. First, the necessity to induce its expression to oxidise its substrates very rapidly in life-threatening conditions such as starvation as mentioned and secondly, CYP2E1's tendency to form reactive species in quantity, both the presence and absence of substrate[109,116]. CYP2E1 is more prone to forming such potentially damaging species than other CYPs probably because of its structure and relative lack of stability[109]. Its induction by ethanol abuse is thought to be a major route of hepatic damage[107] but we have mentioned that it can be reined in by several miRNAs that act to suppress its expression when the mayhem it creates becomes a major threat to cellular survival, such as in paracetamol overdose[45,116]. Indeed, synthetic versions of miRNA's are under consideration as possible therapeutic agents in the protection from CYP2E1-mediated oxidative stress[116].

3.6.3   CYP3A series

CYP3A4, 3A5, 3A7

The scale of the importance of CYP3A4 is underlined by how much it has already been discussed in detail in this book, as befitting the Cadillac of the CYPs (complicated, fast and capacious), which is known to clear more than half all prescription drugs. This includes many statins, anti-rejection drugs, opioids, antidepressants, antipsychotics, and calcium channel blockers, to name but a few. All the CYP3A protein gene loci are found on chromosome 7[117]. Aside from its hepatic expression, where it accounts for about a third of the total CYP content[118], the isoform is found in significant quantities in the duodenum and small intestine, as well as the prostate, breast, and brain[117]. It has a very large active site ($1385 A^3$) that is on a par with CYP2C8[60,61]. Some idea of the complexity of the process of CYP3A4 external and internal rearrangement that occurs during substrate and inhibitor interactions was explored in Sections 3.4.3–3.4.6. Its main endogenous functions include a wide range of the usual steroid and cholesterol suspects, but the size and flexibility of its active site make it difficult to predict where and how they might be cleared[119]. It also interacts with a large number of dietary polyphenols in the gut and the liver[118], although a disadvantage of CYP3A4's flexibility and adaptability leads it to form lethal hepatotoxic species from several plant alkaloids[120] and mould products such as the aflatoxins[121]. The significance of role of this CYP in gut enterocytes has emerged very strongly over recent years. Indeed, a combination of factors related to blood flow and cell type cause CYP3A4 gut drug clearance to be exceptionally efficient; these include high expression (80% of gut CYP material is 3A4), high local drug concentrations and residence time. These all provide greater opportunity for metabolism[118]. Hence, interactions at the level of the gut, in terms of both induction and inhibition, have great clinical relevance, especially with respect to prescription drugs and various dietary chemicals.

CYP3A4 is powerfully induced by PXR ligands (see next chapter) rifampicin, anticonvulsants, steroids, and phenobarbitone, but also the herbal remedy for depression, St John's Wort[117,118]. This can lead to effects ranging from substantial to total elimination of a drug from plasma along with its efficacy[117,122]. The most potent drug inhibitor of CYP3A4 is ritonavir[123], but there are many others, including the other protease inhibitors, as well as the azole antifungals (ketoconazole, itraconazole, fluconazole, voriconazole), macrolides (erythromycin, clarithomycin, troleandomycin), and various citrus juices, most notably grapefruit. Whilst many chemicals clearly have the potential to inhibit CYP3A4, fortunately not all are present in sufficient quantity in natural products and foods to show a clinical impact. About 300 mL of grapefruit juice easily doubles the plasma levels of felodipine, but fortunately coffee does not, although it has been estimated that 85 drugs are at risk of metabolic inhibition by the furanocoumarins in a considerable range of citrous fruits[124].

Whilst CYP3A4 is prone to mechanism-based inhibition through the formation of reactive species even this is very difficult to define conclusively[123] and inhibition can also occur through reversible allosteric means as discussed in Sections 3.4.3–3.4.6. Drugs used as 'probes' for CYP3A include the short-acting benzodiazepine midazolam, as well as erythromycin and alfentanil. The 6β-hydroxylation of endogenous cortisol can also be used reliably to measure the effects of induction and inhibition of CYP3A activity. Interestingly, to appreciate the potential scale of the impact of inhibition and induction on CYP3A4, a study once demonstrated a 400-fold difference between itraconazole and rifampicin's effects on plasma levels of midazolam[125]. Indeed, such is the quantity, flexibility, and siting of CYP3A4 that its role in drug first pass cannot be overemphasized[122].

Regarding the genetic expression of CYP3A4, this is not subject to the extremes that other CYPs can be, and slow metabolising variants are not often encountered[126], although in some Chinese populations do express CYP3A4*1G and a small proportion of individuals have impaired clearance of some CYP3A substrates such as the opiate sufentanyl[127]. A far more significant issue is with the other major member of the CYP3A family, CYP3A5. CYP3A5 exhibits only about a 17% difference in its sequence homology with CYP3A4 and the two isoforms are also difficult to distinguish catalytically, as they metabolise mostly the same substrates at similar rates and are found in similar tissues[128,129]. CYP3A5 appears to be in the minority, as only about a fifth of human livers express it. This isoform is rare in Caucasians but common in Afro-Caribbean's and is linked in some ethnic groups with hypertension and cardiovascular problems[128]. Expression of this CYP has been suggested to be advantageous for living in equatorial regions[126]. The major research focus with CYP3A5 has been its impact on immunosuppressive drugs such as tacrolimus, where high expression accelerates clearance and causes problems in achieving optimal tissue concentrations for xenograft preservation[126,130]. As with all CYPs, there are significant concerns on the issue of genetic testing for CYP3A5 in terms of optimizing therapy and these will be explored in Chapter 7.2.3. It is inducible by the same systems as CYP3A4[131] (see next chapter). Expression of CYP3A5 appears to be particularly high in a range of tumours and this CYP is emerging as a significant source of resistance to drugs in cancer, particularly with agents such as paclitaxel[132].

Of the other CYPs in the 3A subfamily, CYP3A7 has 88% sequence homology with CYP3A4 and assumes its role in the foetus and is tasked with forming estriol and other steroids during pregnancy[133], although the switching process from CYP3A7 to CYP3A4 starts within 28 days or so of birth[134]. This does not occur entirely in all individuals and CYP3A7 is expressed in adults and its expression in tumours is being heavily investigated and its polymorphisms may impact lifetime cancer risks based on their effects on oestrogen hydroxylation ratios[135]. The scenario where a CYP3A4 substrate is cleared by the mother but not by foetal CYP3A7 has already been documented for the now largely banned drug cisapride[129] and this pathway is a potential risk to foetal development of drug accumulation if it were to occur for other agents. CYP3A43 is now thought to be a variant of 3A4 and is not significant in terms of drug clearance[117]. Overall, it has been recommended that if less than a quarter of a drug's clearance is reliant on CYP3A isoforms, this is probably not clinically significant, in terms of possible CYP3A-mediated drug interactions. However, if the figure is in excess of 50%, then impact on efficacy and toxicity is strongly related to this CYP subfamily and dose adaptation may be required[122].

3.7 Cytochrome P450 catalytic cycle

Having established the phenomenal multiplicity and flexibility of these enzymes, it should be a relief to learn that all these enzymes essentially function in the same way, although again we do not fully understand the process yet. CYPs can carry out reductions (see later on), and these occur after substrate binding and before oxygen binding. However, their main function is to insert an oxygen molecule into a usually stable and hydrophobic compound. Many textbooks present this cycle, but it appears intimidating due to the many details involved in this complex process. A simplified version appears in Figure 3.8, and it is important to understand that there are only five main features of the process whereby the following equation is carried out:

$$\text{Hydrocarbon}(-\text{RH}) + O_2 + 2\,H^+ \text{ ions, gives an alcohol}$$
$$(-\text{ROH}) + H_2O$$

1. Substrate binding (reduction may happen after this stage)

2. Oxygen binding

3. Oxygen scission (splitting)

4. Insertion of oxygen into substrate

5. Release of product

3.7.1 Substrate binding

The first step, as covered in the previous section, is the binding and orientation of the molecule (Figure 3.8). The CYP's active site structure will orientate the most vulnerable part of the agent so it is presented to the haem iron, so the molecule can be processed with the minimum of energy expenditure and the maximum speed. The iron is usually (but not always) in the ferric form when the substrate is first bound:

$$Fe^{3+} - RH$$

Once the substrate has been bound, the haem iron elevates from low to high spin state and the next stage is to receive the first of two electrons from the REDOX partners, so reducing the iron:

$$Fe^{2+} - RH$$

3.7.2 Oxygen binding

The next stage involves the iron/substrate complex binding molecular oxygen sourced from the lungs. This process runs faster than the substrate binding to the

Cytochrome P-450 Catalytic Cycle

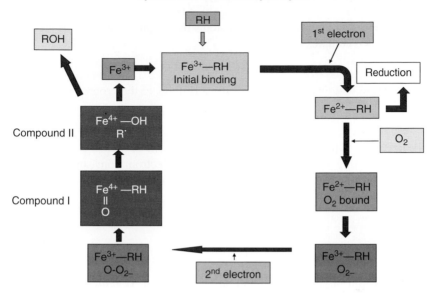

Figure 3.8 Simplified scheme of cytochrome P450 operations

iron, as there is much more oxygen present in the cell than substrate. This is known as a *ferric superoxide complex.*

$$Fe^{2+} \ RH$$
$$|$$
$$O_2$$

You will note that oxygen does not just exist as one atom. It is much more stable when it is found in a molecule of two oxygen atoms, O_2. Indeed, oxygen is almost never found in nature as a single atom as its outer electron orbitals only have six instead of the much more stable eight electrons. To attain stability, two oxygen molecules will normally covalently bond, sharing four electrons, so this gives the same effect as having the stable eight electrons. Therefore, splitting an oxygen molecule requires energy, like trying to separate two powerful electromagnets – the oxygen will tend to 'snap back' immediately to reform O_2 as soon as it is separated. Two problems thus arise: first, how to apply reducing power to split the oxygen and second, how to prevent the oxygen reforming immediately and keep the single oxygen atom separate long enough for it to react with the vulnerable hydrocarbon substrate.

3.7.3   Oxygen scission (splitting)

To split the oxygen molecule into two atoms first requires a slow rearrangement of the $Fe^{2+} O_2$ complex to form

$$Fe^{3+} \quad RH$$
$$|$$
$$O_{2-}$$

The next stage is the key to whether the substrate will be oxidized or not. This is the rate- limiting step of the cycle. A second electron from supplied by the REDOX partners feeds into the complex and forms a *ferric peroxo species.*

$$Fe^{3+} \quad RH$$
$$|$$
$$O\_O_{2-}$$

Next a proton (hydrogen without an electron, or H+) meets the oxygen atom furthest away from the haem (distal oxygen atom) and forms a *ferric hydroperoxo-complex.*

$$Fe^{3+} \quad RH$$
$$|$$
$$O\_O_{1-}$$
$$|$$
$$H$$

Next, another proton is delivered to the complex, reacting with one of the oxygens and the proton that is already there, forming water and what is known as Compound I:

$$Fe^{4+} \quad RH$$
$$||$$
$$O$$
$$H_2O$$

This solves the two problems described above; the oxygen molecule has been split, but it cannot just snap back to form an oxygen molecule again, as water is stable and takes an oxygen molecule away from the enzyme active site. Compound I is an iron IV oxo, or ferryl species, which is the high-energy form of oxygen capable of attacking the substrate-the 'lit blowtorch' essentially. For nearly half a century, its existence could not be proved with certainty, because of what was believed to be its extreme reactivity, and thus its infinitesimally small half-life, so it could not be isolated in sufficient quantity to analyse. It was eventually understood that contaminating endogenous molecules were accelerating the destruction of the Compound I and once sufficiently purified, it was possible to isolate and study it[136].

3.7.4   Insertion of oxygen into substrate

Compound I removes hydrogen from the substrate, creating Compound II and a carbon radical, which is itself extremely reactive.

$$Fe^{4+} - OH$$
$$|$$
$$R \cdot$$

The final stage is the reaction between the newly created hydroxyl group and the carbon radical, yielding the alcohol, as seen below. The entry of the oxygen atom into the substrate is sometimes called the *oxygen rebound* reaction. The CYP iron is left in the *ferric* Iron III state.

$$Fe^{3+} \text{ and ROH}$$

3.7.5   Release of product

The whole CYP catalytic process (Figure 3.8) could be described as complex yet dramatic. Although a 'pump' analogy has been used previously in this chapter, in some ways, CYPs could also be likened to the cycling of an automatic weapon, with the 'load, fire, extract, eject, reload' stages analogous to CYP substrate conversion to product. This analogy is not perfect, as the coupled CYP provides all the energy to sustain the process rather than the 'substrate' of a machine gun, but it conveys something of the rapidity and violence of the process. Once the substrate has been converted to a metabolite, it has changed both structurally and physicochemically to the point that it can no longer bind to the active site of the CYP. The metabolite is thus released and the CYP isoform is now ready for binding of another substrate molecule. It is important to break down the function of CYPs to separate stages, so it can be seen how they operate and overcome the inherent problems in their function. However, students often find the catalytic cycle rather daunting to learn and can be intimidated by it. It is much easier to learn if you try to understand the various stages, and use the logic of the enzyme's function to follow how it overcomes the stability of substrate and oxygen by using electrons it receives from the adjacent reductase systems to make the product.

3.7.6   Reductions

As mentioned earlier, CYPs probably evolved to reduce chemical agents before oxidation could occur, and once an agent is bound to the haem iron, there is an opportunity for a reduction reaction to occur prior to the oxygen binding stage. This is because a REDOX partner, usually NADPH reductase, supplies an electron that can be used to reduce the substrate. Several enzyme systems can effect reductions, such as NADPH reductases themselves, which are found in virtually all tissues and are relevant in aromatic amine-mediated carcinogenicity (Chapter 8, Figure 8.12). CYPs are just one of many systems which can reduce xenobiotics and interestingly, there are examples where drugs undergo reductions and oxidations sometimes by the same CYPs. This is thought to occur with the muscle relaxant eperisone[137] and human CYPs are capable of reducing the antibiotic chloramphenicol also (see Chapter 8).

3.8   Flavin monooxygenases (FMOs)

3.8.1   Introduction

Alongside the CYPs, other cellular systems can accomplish oxidations of endogenous molecules, and these systems are usually widespread in many tissues. Of these systems, the flavin monooxygenases (FMOs; Figure 3.9) have not received anything like the attention directed at CYPs, although realization of their significance in drug clearance is growing. Their endogenous functions are not fully documented, but include processing amino acid and nitrogen-based compounds, as well as selenium and sterol metabolism[138,139]. They are present in most tissues, usually in the smooth endoplasmic reticulum (SER), but in the brain they are found in other cellular areas[140]. Their relative obscurity with respect to CYP research is partly because it is not easy to study them in human *in vitro* systems containing standard purified SER (microsomes, see Appendix A.4.2), as conditions favour the high activity of the CYPs and various CYP inhibitors must be used to negate their influence[138,141]. Alternatively, the FMOs can be studied in isolation using human gene expression systems[142] (Appendix A.4.3).

Originally known as Ziegler's enzyme, after its discoverer, the FMOs are actually the second-most prolific oxidizers of drugs and xenobiotics after the CYPs. In humans, the five functional forms of FMO (designated 1-5) are coded for by separate genes found on chromosome 1. The nomenclature for FMOs is based around amino acid sequence commonality as with the CYPs. Individual isoforms are regarded as belonging to the same family if they possess greater than 40% amino acid sequence homology[140]. The full classification and nomenclature of the FMOs across living species is complex;[143] of the five human or *h*FMOs which have about 60% sequence similarity, FMO-3 is the most

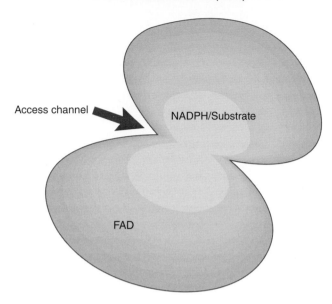

Figure 3.9   Generic structure of a flavin monooxygenase (FMO)

important in human liver[138]. This enzyme handles the clearance of trimethyl-amine (TMA), a volatile agent that has a fishy odour, which is formed from nitrogen-rich dietary components like carnitine and choline. The resulting metabolite, trimethylamine-N-oxide (TMAO), is linked with propensities towards cardiovascular damage and the polymorphic lack of FMO-3 is known to cause issues with bodily fishy odour[144] (Chapter 7, Section 7.2.4). FMO-3 is expressed around the same level as CYP1A2,[138] and forms N-oxides or hydroxy-lamines of several prescription agents, including tricyclic antidepressants such as imipramine[141] and antipsychotics, as well as several abused drugs such as amphetamines, but the actual *in vivo* contribution made by FMO-3 when CYP's are present is not easy to estimate[138,140].

FMO-5 is also found in the liver in similar proportions to FMO-3, but is rather different from the other FMOs (its gene locus is slightly detached from the other enzymes), in that it does not oxidise nucleophiles, but rather electrophiles. Its C4a-hydroperoxyflavin complex (Figure 3.10) is more stable than the other FMOs and it is able to clear drugs such as the nonsteroidal anti-inflammatory nabumetone and the anti-claudication agent pentoxifylline[142]. FMO-2 is found mainly in the lungs and the kidney, whilst FMO-1 and 4 are found in the kidney. FMO-1 is found in foetal liver, whilst FMO-3 is not expressed in neonates at all (see Chapter 7, Section 7.2.4) and it seems that FMO-1 takes its place. FMO-3 is virtually absent in the kidney. A lack of selective experimental substrates for FMO-4 and 5 have made it difficult to study them, although they are not believed to contribute to xenobiotic clearance in humans.

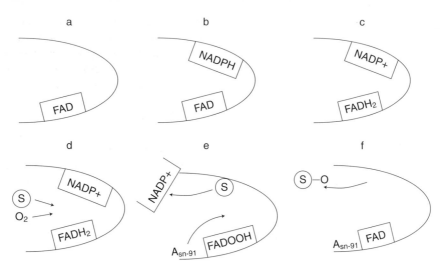

Figure 3.10 Process of flavin monooxygenase catalysis: a–c outlines the first step or priming of the enzyme. The second step (d–f) involves oxygen binding and substrate (S) displacement of NADPH, followed by reaction of oxygen and substrate and release of the oxide

3.8.2 Structure

The main features of these and variously closely related monooxygenases enzymes are pretty well conserved and are virtually 'standard issue' across the animal kingdom and mammalian amino acid sequences for the main FMOs are very similar[139,143,145]. FMOs have some structural resemblance to oxidoreductases (PORs) and bacterial (but not human to date) FMOs have been crystallized and through similar techniques used with CYPs, and the order of their catalytic operations can be worked out through a series of 'freeze-frames' where eukaryotic and bacterial FMOs can be observed binding its co-factor and substrate. The main structure of the enzyme could be described as vaguely clam-shaped, with a small 'top shell' and a larger bottom 'shell'. The access channel is obviously between the two 'shells' (Figure 3.9). On the floor of the bottom 'shell' there is a depression to which FAD is anchored firmly. On the top 'shell', there is a binding site that will fairly loosely hold NADPH, which is the main co-factor. The NADPH is bound in the orientation where the adenine portion faces the upper shell, whilst the nicotinamide portion is aimed at the flavin of FAD.

3.8.3 Mechanism of catalysis

FMOs oxidise what is termed soft nucleophiles; a nucleophile (nucleus-loving) has a surplus of electrons and is termed *electronegative,* often with a lone pair of electrons that seek electrophiles that are electron deficient. Nucleophiles can be designated 'hard' and 'soft'. A hard nucleophile would be a fluoride ion, with high electronegativity in a small area, with a sort of 'concentrated' negative charge, whereas, a 'soft' nucleophile might have the negative charge spread over a wider ionic radius, so their charge to mass ratio is relatively lower – that is, less concentrated. Good examples of 'soft' nucleophiles are sulphur, thiols (S-H groups), and various nitrogen-derivatives, as well as phosphorus and even carbon.

These enzymes use NADPH as an electron donor and FAD as the built in electron carrier, which is similar to POR. Unlike the CYPs, FMOs are found in cells in a 'primed' or 'cocked and loaded' form, so they do not need the substrate to bind to engage their operation[139]. The priming of FMO, involves NADPH transferring a hydride ion to the FAD, forming a $FADH_2$ and NADP+ complex (Figure 3.10), which is the C4a-hydroperoxyflavin[139,140]. Like many CYPs, they can leak reactive oxygen species. The presence of a soft nucleophile substrate of the appropriate dimensions to fit in the access channel initiates step two, which involves $FADH_2$ reacting with molecular oxygen (O_2) to form FADOOH and the substrate then displaces the NADP+ and is held surrounded by a sort of 'jacket' of water molecules. This means that it does not actually make contact with the enzyme. A crucial amino acid on the lower 'shell' of the FMO, called asparagine-91 (Asn-91) is instrumental in catalysing the reaction of the two oxygen molecules held as FADOOH with the nucleophilic portion of the substrate, such as a nitrogen lone pair of electrons. This leads to one oxygen molecule reacting with the substrate forming an oxide, whilst the other forms water. The enzyme is eventually regenerated to FAD and it then reloads itself with more NADPH ready to oxidize another substrate molecule[139-143].

The oxidation process is a single step transfer of two electrons, rather than the CYP system of two successive single electron oxidations. Aside from nitrogen, soft nucleophiles, favoured by FMOs, like thiols and phosphorus groups, are found in various drugs and xenobiotics and the respective oxides formed by the FMOs are more water-soluble than the parent drugs. FMOs can also metabolise tertiary amines to form N-oxides in drugs such as TCAs, as well as morphine, methadone, and pethidine (meperidine). They also can form the 1- and 3-N-oxide metabolites of the 2,4, diaminopyrimidine antiparasitics (trimethoprim and pyrimethamine)[146], as well as many other sulphoxide metabolites such as that of cimetidine. There is relatively little known as to details of substrate preferences for the different FMOs, although FMO-3 oxidizes generally smaller nucleophilic heteroatoms than FMO-1, which prefers drugs with bulkier side-chains, forming N-oxides of chlorpromazine, imipramine, and orphenadrine.

There is no evidence so far that FMOs can be inhibited as drastically as CYPs, although methimazole and indole-3-carbinol are used experimentally[147] and any relatively mild inhibition that might occur is more likely to be competitive (nitrogen/sulphur containing nucleophiles) rather than mechanistic inhibition.

If inhibition occurs *in vivo*, it is thought that it may be related to the influence of the combination of dietary factors and particular polymorphic variants of FMOs 1, 2, and 3.

3.8.4   Variation and expression

FMOs display some interesting differences and similarities in their expression and operation in comparison with CYPs. Like CYPs, their metabolites are usually more hydrophilic than the parent, although unlike CYPs, they are generally, but not always, less likely to produce reactive species and their metabolites (with some exceptions) tend to be low in toxicity, such as the various N-oxides, which CYPs can form also[147]. Our knowledge of the regulation of FMOs is much less detailed than with CYPs, but their expression can be regulated hormonally, which accounts for the marked switch from predominant expression of hepatic FMO-1 to 3, which appears to be accelerated by the birth process. Indeed, agents such as oestrogens and insulin upregulate, whilst male hormones and glucagon downregulate transcription. There is a very significant difference between the mRNA levels and actual functional protein, which suggests several layers of regulation yet to be understood, but that probably involve similar systems to the CYPs[139].

As FMOs are considered by many to have evolved at least partly as a detoxification system, it seems inexplicable that FMO expression does not appear to be capable of responding to xenobiotic challenge over, say days or weeks, which is the hallmark of the complex and highly efficient CYP enzyme induction system (Chapter 4). It is apparent that the variation in FMO expression is mostly genetic and it has been suggested that separate human populations may have evolved to express particular variants of these isoforms, which detoxify flora and fauna of specific geographical regions. FMOs are highly polymorphic and evidence is mounting of their involvement in a number of human disease states; impaired expression of FMO-3 and the failure to form TMAO has been linked to increased risk of cardiovascular disease, although this is controversial. Significant changes in FMO expression have been recorded in conditions as diverse as neurodegenerative disease, diabetes, and aging, which underlines how much more is necessary to learn about these enzyme systems[138].

3.8.5   FMOs in drug development

Clinically, FMOs do represent a potentially attractive prospect in drug development. As mentioned above, they are not easily inhibited and their general lack of induction response makes their expression stable and not subject to unexpected and potentially dangerous changes induced by diet, alcohol/drug consumption, or other small-molecule induced stimuli. From a practical standpoint, if a new drug were to be cleared by a combination of FMOs and CYPs, the introduction of a potent CYP inhibitor to a clinical regime would not completely shut down all drug clearance, so preventing accumulation-mediated toxicity, as the FMO would continue to eliminate the drug. This has been demonstrated already to some extent by the family of gastroprokinetic agents, cisapride, mosapride, and itopride. Poor gastric motility leads to accumulation of stomach acid, which causes pain and inflammation of gastro-oesphageal tissues. These drugs relieve these symptoms by improving gastric motility, although not without side effects. In the presence of a CYP3A4 inhibitor, cisapride can accumulate and extend the cardiac QT interval (Chapter 5, Section 5.7.2), potentially lethally. The drug has been withdrawn in most countries, although apparently it remains highly effective for constipation in cats. Mosapride is far safer but is a CYP inducer[148], whilst in contrast, itopride, is almost entirely cleared by FMO-3[141,149], thus potentially improving the safety of the drug in complex real-world clinical regimens.

However, you might immediately remember reading earlier about the absence of FMO-3 in fish-odour syndrome and you would feel that the polymorphism issue is clearly present with FMOs also. Well, you would be correct, as although Caucasians and Africans express defective FMO-3 perhaps at around 1% or less, with those of Chinese and Korean extraction, the frequency of defective FMO expression is up to 5%[149,150]. However, even with polymorphic less effective forms of FMO, they retain just less than half the wild-type's effectiveness and with drugs with a reasonable margin of safety, it is not currently considered that FMO polymorphisms are a significant issue[149]. This is perhaps a favourable comparison with CYP2D6, for example, where 10% of Caucasians express virtually nonfunctional enzyme (Chapter 7.2.4).

On the other hand, the polymorphisms do exist and there are also examples of FMO-mediated metabolism, which form reactive species that are toxic to cells, such as their propensity to make hydroxylamines (3.13.1; Chapter 8.2.2); indeed, the hepatic damage caused by obsolete drugs such as ketoconazole and thiacetazone is linked to FMOs[147]. However, methods to assess how FMOs contribute to drug clearance are improving, and knowledge of the preferences of these enzymes can inform drug design where groups could be inserted that are likely to be FMO substrates[140]. In addition, there is a strong interest in FMOs from the standpoint of biotechnology, and their inherent instability and fragility have been addressed to make them more accessible and practical for laboratory and industrial use[151] as well as exploring their roles in drug metabolism and health issues[142].

3.9   How CYP isoforms operate *in vivo*

The detailed processes on how living systems operate are sometimes focused on at the expense of a global understanding of how these systems might operate in the tissue. It is useful to try to visualize how CYPs and other biotransforming systems process massive numbers of molecules with a range of hydrophobicity every second. If you visualize just one hepatocyte and imagine the smooth endoplasmic reticulum, with its massive surface area, with vast numbers of CYP and REDOX partners embedded in its tubing, then you can see how the liver can sometimes metabolise the majority of drugs and endogenous substrates in a given volume of blood in just one passage through the organ. This is rather like an automotive catalytic converter that converts all the exhaust pollutants to water and CO_2 and nitrogen oxides.

3.9.1   Illustrative use of structures

Most textbooks at this point show a large number of chemical reactions that highlight how CYP isoforms metabolise specific drugs/toxins/steroids, and this one is no exception. However, it is appreciated that many students might not have studied chemistry, or struggle with it as a subject and feel intimidated by chemical structures. This is worth overcoming, as some qualitative understanding of basic organic chemistry really pays off in illustrating, understanding and appreciating how CYP enzyme systems operate at the molecular level. If you study the diagrams it should not be too difficult to eventually see a molecule in a way that approaches how a CYP enzyme might 'see' it. After all, we thrive as a species in part due to these remarkable enzymes.

3.9.2   Primary purposes of CYPs

As mentioned before, CYP isoforms have evolved to:

- make a molecule less lipophilic (and often less stable) as rapidly as possible;

- make some molecules more vulnerable to conjugation.

The first step is the binding of the substrate. As you will have seen, individual CYPs bind groups of very broadly similar chemical structures. This is partly achieved by the size and physicochemical characteristics of the molecule. For example, the entrance to CYP2C9 is not wide enough to bind part of a large molecule like cyclosporine, so this molecule is virtually excluded from all the CYPs, except the one with the largest and most flexible entrance and binding area, CYP3A4. There is a physical size restriction on FMOs also. The processes involved in the orientation of substrates to proximity with the haem iron are complex, as mentioned previously, but once an agent is presented to the iron, oxidation and occasionally reduction can then occur.

3.9.3 Role of oxidation

CYP metabolism is almost always some form of oxidation, which can achieve their main aims. Oxidizing a molecule can have three main effects on it, as follows.

Increase in hydrophilicity

Forming a simple alcohol or phenol is often carried out to make a molecule soluble in water so it can be eliminated without the need for any further metabolic input.

Reduction in stability leading to structural rearrangement

Obviously, some chemical structures are inherently less stable than others, and any prototype drugs that are unstable and have the potential to react with cellular structures are weeded out in the drug discovery process. However, the process of CYP-mediated metabolism, where a stable drug is structurally changed, can form a much more reactive and potentially toxic product (Chapter 8). A very young child hitting objects randomly with a piece of metal will not be able to discern the difference between an inert object and an extremely dangerous one (electrical equipment or an explosive device). In the same way, a molecule may be bound and metabolised by CYPs, irrespective of the impact these processes may have on the stability and potential toxicity of the product. There is a risk that the new molecule may be very reactive and dangerous indeed and may attack the CYP itself or the surrounding cellular structures. Although this does happen, evolution has retained the advantages of CYPs, such as their ability to process virtually any required molecule, through the appearance of conjugation and detoxification systems that contain and usually quench the reactivity of these agents (Chapter 6.4). This could be compared with the evolution of the Porsche 911. The weight of the engine over the driven rear wheels offers tremendous traction and thus acceleration. However, intensive modification of the car over many years has counteracted the 911's inherent tendency to carry straight on through corners and has vastly increased its safety whilst retaining its performance. With CYP oxidation processes, the evolution of attendant detoxification systems ensures that the risk to the cell of creating a reactive species usually pays off and a molecule can be quite radically changed in terms of its physicochemical properties without problems. For example, a lipophilic functional group might be oxidized to an alcohol, which may be so unstable that it breaks off. This has the dual advantage of removing a lipophilic structure that leaves the molecule more hydrophilic (see the oxidation of terfenadine). It can also pave the way for further metabolism, such as conjugation (Chapter 6).

Facilitation for conjugation

Many oxidative metabolites are much more vulnerable than their parent molecules to reaction with water-soluble groups such as glucuronic acid and sulphates. Once a conjugate is formed, this vastly improves water solubility and Phase III transport systems will generally remove it from the cell and into the blood. The conjugation and transporter-mediated clearance of polar metabolites is discussed in Chapter 6.

3.9.4   Summary of CYP operations

A sculptor was apparently once asked how he would go about sculpting an elephant from a block of stone. His response was, 'I knock off all the bits that do not look like an elephant'. Similarly, drug-metabolizing CYPs have one main imperative, to make molecules more water-soluble. Every aspect of their structure and function, their position in the liver, their initial selection of substrate, binding, substrate orientation and catalytic cycling, is intended to accomplish this deceptively simple aim.

With experience, you should be able to look at any drug or chemical and make a reasonable stab at suggesting how a CYP enzyme might metabolise it. It is important to see these enzymes not as carrying out thousands of different reactions but as basically carrying out only two or three basic operations on thousands of different molecules every second.

3.10   Aromatic ring hydroxylation

3.10.1   Nature of aromatics

Large, highly lipophilic, planar and stable molecules with few, if any, vulnerable functional groups look to be a difficult proposition to metabolise (Figure 3.11). Indeed, if there are any aliphatic groups, or nonaromatic rings associated with an aromatic molecule, these will often be attacked rather than the aromatic group. The ring hydroxylation of amphetamines is the exception to this. Polycyclic hydrocarbons are not easy to clear and they are perceived, correctly usually, by living systems as a potent threat. This is reflected in the elaborate expression system (Ah/ARNT; Chapter 4, Section 4.4.2) that modulates the nonconstitutive isoform CYP1A1, which has evolved to effectively deal with them. These include molecules such as those shown in Figure 3.11. The simplest aromatic is benzene, and this can be eventually oxidized by CYP1A1 to phenol (Chapter 2, Section 2.3.2), which is more reactive but also more water-soluble than benzene and vulnerable to sulphation and glucuronidation during conjugative metabolism.

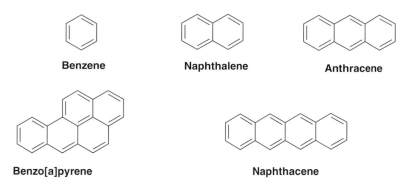

Figure 3.11 Some aromatic hydrocarbon molecules

3.10.2 Oxidation of benzene

There are several intermediates formed during the oxidation of benzene (Figure 3.12). The two main routes are the cyclohexadienone and an epoxide; in the presence of water both stages will rearrange to form the phenol. During this process, the hydrogen atom close to the oxygen will sometimes be moved around on the ring, or even lost. This is known as the NIH shift.

Epoxidation is defined chemically as a reaction where an oxygen atom is joined to an unsaturated carbon to form a cyclic, three-membered ether. Epoxides are also known as arene oxides and vary enormously in their stability, which depends on the electron density of the double bond being oxidized: the higher the density, the more stable the epoxide. This means that epoxides of varying stability can be formed on the same molecule, due to differences in electron densities, and this is most apparent in benzpyrene. The anticonvulsant carbamazepine forms a number of toxic epoxides, although the 10, 11 derivative is stable enough to be pharmacologically active. Indeed, an advantage of the latest derivative of this agent, eslicarbazepine, is that it is not metabolised to epoxides[152]. Alternatively, bromobenzene 3,4 epoxide's half-life in blood is less than 14 seconds. Generally, arene oxides form phenols or diols in the presence of water (as does carbamazepine 10, 11 epoxide), although the cytosolic enzyme epoxide hydrolase (Chapter 6.6, and Chapter 7, Section 7.2.4) is present to accelerate this process.

The phenols and diols are usually substrates for sulphation or glucuronidation. Although the process of aromatic hydroxylation is difficult to achieve and the phenolic product is more hydrophilic, the structural features of the larger polycyclics mean that this process can lead to the formation of unstable carcinogenic reactive intermediates.

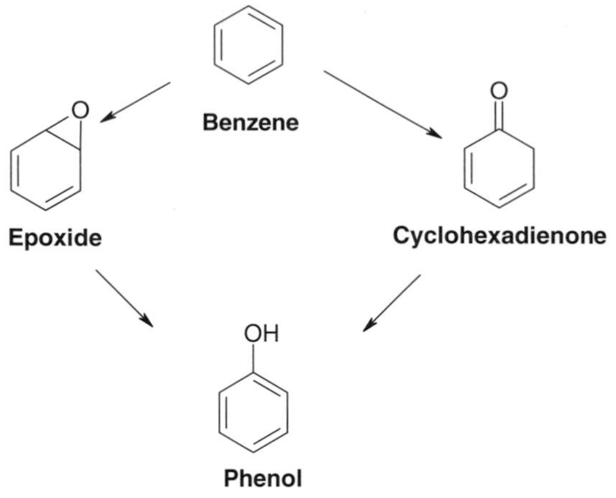

Figure 3.12   Main pathways of benzene hydroxylation

3.11   Alkyl oxidations

The saturated bonds of straight-chain aliphatic molecules are very stable; indeed, they can be even harder to break into from the thermodynamic point of view than aromatic rings, whilst molecules with unsaturated bonds are the easiest to oxidize. Straight-chain aliphatic molecules are easier to oxidize if they have an aromatic side chain. Alkyl derivatives are generally oxidized by the routes briefly described below.

3.11.1   Saturated alkyl groups

The oxidation of a saturated alkyl group can lead to the alcohol being inserted in more than one position (Figure 3.13). The 'end' carbon group of the molecule is sometimes called the *omega* group and the oxidation can result in this group being turned into an alcohol (omega oxidation) or alternatively, the penultimate group (omega minus one).

During the oxidation of saturated molecules (Figure 3.13) the CYP will operate as described in Section 3.7, abstracting hydrogen and causing the carbon molecule to form a radical. The carbon radical and the hydroxyl group then react to form the alcohol. Even though alkanes like hexane are very simple structures, they can be metabolised to a large number of derivatives (see Section 3.11.3, 'Pathways of alkyl metabolism').

As well as the formation of alcohols, CYP isoforms can desaturate carbon–carbon double bonds to single unsaturated bonds (Figure 3.14). This process can occur alongside alcohol formation and is a good example of a CYP-mediated process that leads to quite considerable rearrangement of the molecule's structure. The first hydrogen is abstracted by the CYP isoform and

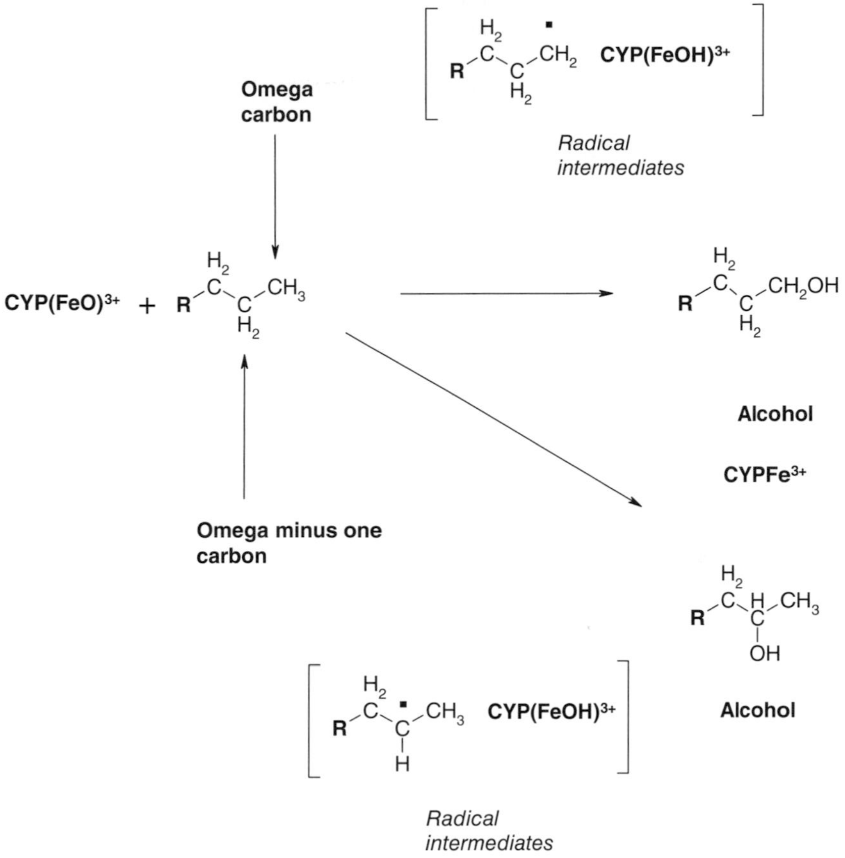

Figure 3.13 Omega and omega minus one carbon oxidation of aliphatic saturated (single bond) hydrocarbons by CYP isoforms

may leave the FeO^{3+} complex, allowing it to grab a second hydrogen atom. The highly unstable adjacent carbon radicals rearrange to form an unsaturated product. The two hydrogens and the single oxygen atom that the CYP enzyme used to accomplish this effect form water.

3.11.2 Unsaturated alkyl groups

Unsaturated or double bonds are more electron-rich than saturated bonds, and as mentioned earlier, this makes them easier to oxidize so several possible products can be formed (Figure 3.15). These include an epoxide called an *oxirane,* as well as two carbonyl derivatives, or aldehydes, which can split the molecule.

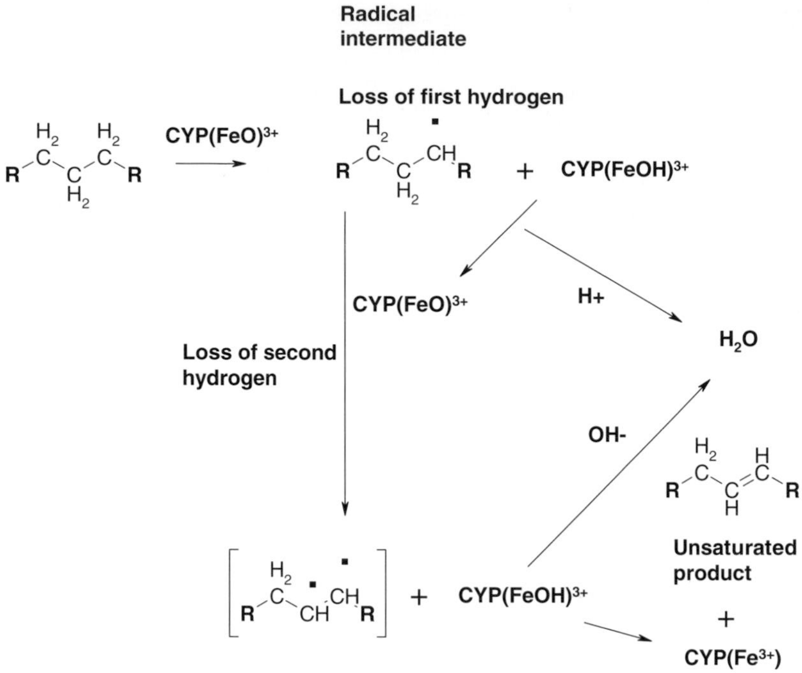

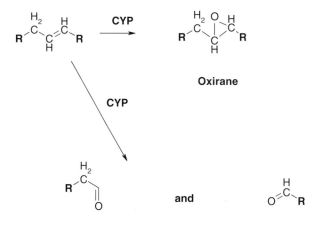

Figure 3.14   Formation of unsaturated bonds from a saturated starting point

Figure 3.15   Metabolism of unsaturated alkyl groups

3.11.3   Pathways of alkyl metabolism

A good example of where these pathways can lead is the complex metabolism of an otherwise apparently simple molecule, hexane (Figure 3.16). This hydrocarbon was once used as a volatile component of several adhesive mixtures, which were extensively applied in the leather and shoe industries. If you want an adhesive or paint to dry or cure quickly, the volatility of the carrier solvent is crucial. This makes the adhesive or paint easier to use in a mass production setting. However, it was gradually realized that many people who used hexane-based adhesives in the leather industry were suffering from damage to the peripheral nervous system, known as peripheral neuropathy. This was a progressive effect and was traced to the hexane itself. In humans, hexane is cleared at first to several hexanols, which is logical, as a volatile, water-insoluble, and highly lipophilic agent capable of causing intoxication will be a strong candidate for rapid clearance to an albeit only slightly water-soluble alcohol.

The 2-hexanol derivative undergoes further oxidative metabolism, initially to a diol, the 2,5 derivative (Figure 3.17), which can undergo further CYP isoform-mediated (probably by 2E1) oxidation to a di-ketone, which is the 2,5 hexanedione derivative. This compound is unusual in that it is a specific neurotoxin. It interferes with microtubule formation in neural fibres, causing gradual loss of neural function. It is also cytotoxic to neuronal cells and its relatives, the 2,3 and 3,4 diones, are cytotoxic to cell cultures. Consequently, n-hexane is banned from use in adhesives and should only be used where the fumes cannot be inhaled. The 2,3 and 3,4 hexanediones are used as food colourings and flavourings, although it is unlikely they can be formed in human liver. Several isomers of hexane have been used as substitutes for hexane, although the potential neurotoxicity of adhesives that use volatile alkanes should never be underestimated[153].

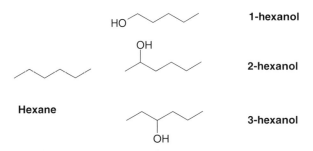

Figure 3.16   Oxidation of hexane to hexanols

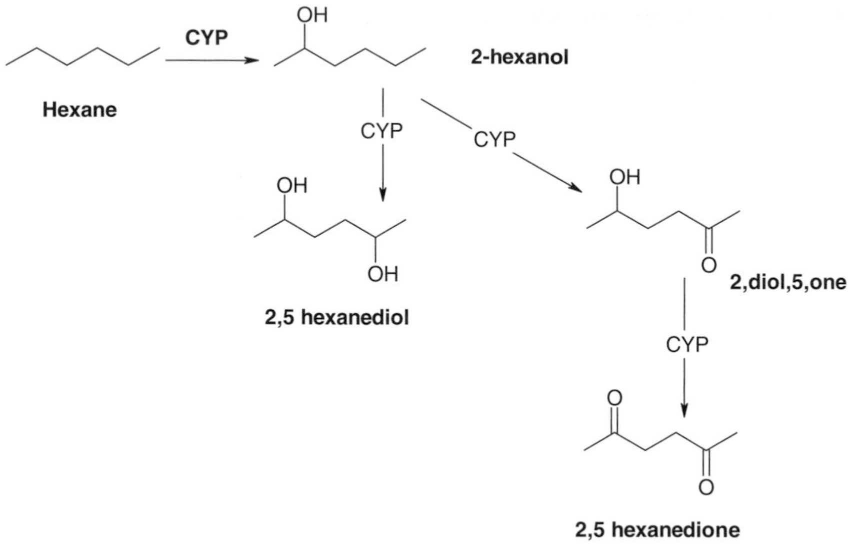

Figure 3.17   Formation of the neurotoxin 2,5 hexanedione by CYP oxidations

3.12   Rearrangement reactions

The use of oxidation as a tool to rearrange molecules to less lipophilic products
has the added benefits of unmasking other vulnerable groups and making the
products simpler to conjugate. There are several CYP-mediated oxidations that
have this effect on molecules.

3.12.1   Dealkylations

Alkyl groups, especially bulky ones, are very lipophilic and often are attached
to drugs through 'hetero' atoms, i.e. nitrogens, oxygens, and sulphurs. From the
perspective of biotransformation, it makes sense to remove the alkyl group,
leaving the hetero group vulnerable for conjugation with glucuronides or
sulphates (Figure 3.18). The quickest way to remove the alkyl group is to oxidize
it to an alcohol. This should be a win–win situation, whether the product is stable
or unstable; the alcohol (called a carbinolamine in the case of N-dealkylation) is
usually unstable and splits off, forming an aldehyde. This reveals a less lipophilic heteroatom handle for conjugation. If the alcohol is stable, then the drug
is still more hydrophilic than it was, and that might also be a target for subsequent
conjugation. With substituted aromatic compounds it is easier for the CYP to
oxidize an alkyl substituent group than the ring. Another result of dealkylation

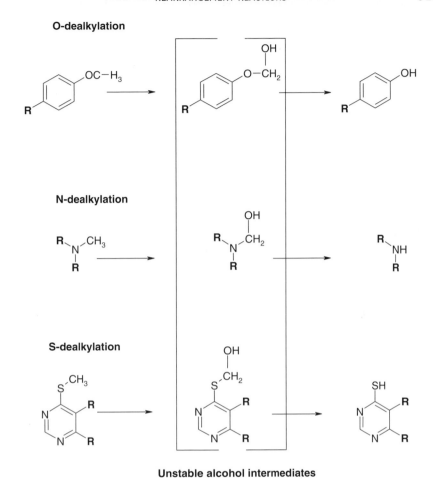

O-dealkylation

N-dealkylation

S-dealkylation

Unstable alcohol intermediates

Figure 3.18 Rearrangement reactions caused by the CYP-mediated oxidation of an alkyl group leading to the formation of a more water-soluble product, which is also more vulnerable to conjugation. The 'waste products' of the reactions are usually small aldehydes or ketones

can be the splitting of a large lipophilic molecule into two smaller, more hydrophilic ones (Figure 3.18).

Many drugs undergo this type of dealkylation. Imipramine, the TCA, is demethylated to form desmethyl imipramine, which also has pharmacological potency and is usually known as desipramine. The removal of one methyl group might not make much difference to the lipophilicity of a large molecule, although it could change its pharmacological effects. More than one alkyl group might have to be removed to make the compound appreciably less lipophilic. On the other hand, the N-dealkylation reaction of the antihistamine terfenadine has a

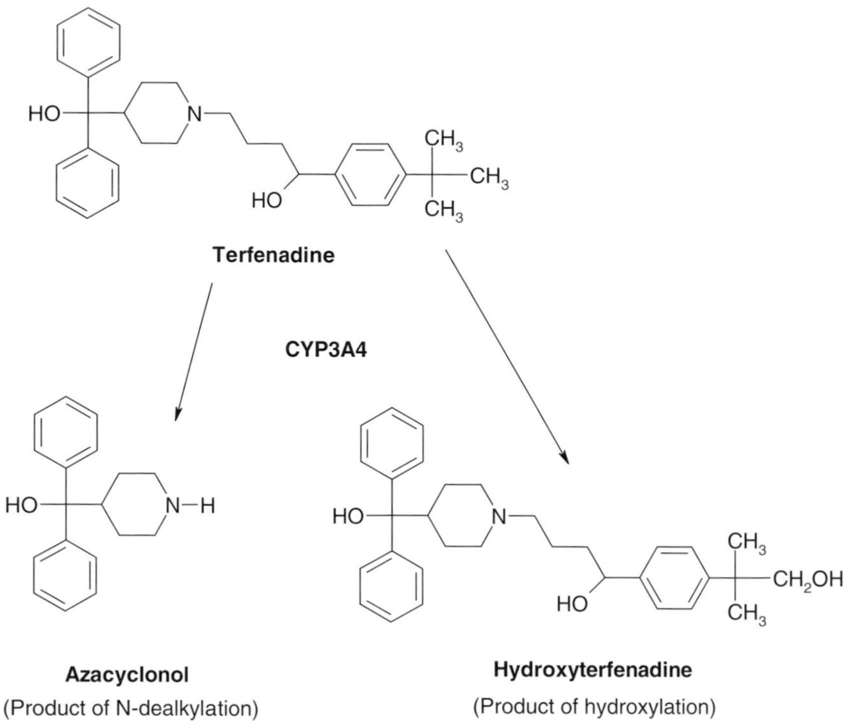

Figure 3.19   Metabolism of terfenadine: essentially the same oxidation reaction applied in two different areas of the molecule leads to vastly different effects on the structure

much more dramatic effect (Figure 3.19). The oxidation of the alkyl group adjacent to the nitrogen causes an unstable alcohol to be formed, which splits away, taking half the molecule with it. Virtually the same reaction of alkyl oxidation at the other end of the parent molecule results in a stable alcohol that is then oxidized to a carboxy derivative (fexofenadine), which is not metabolised further and is less toxic than the parent drug (Section 5.6.1).

N-dealkylation mechanisms

N-dealkylation is only part of the picture of the metabolism of how CYPs can oxidize heteroatoms. Before N-dealkylation occurs, CYPs have the option of oxidizing the substituted nitrogen itself to form an N-oxide (Figure 3.20). If N-oxide formation does not occur, then N-dealkylation can proceed. Again, this is a win–win process, as N-oxides are more water-soluble than the parent drug. As mentioned in Section 3.8, FMOs are credited with the vast majority of N-oxidations, but CYPs can also accomplish them. Whether N-oxide formation

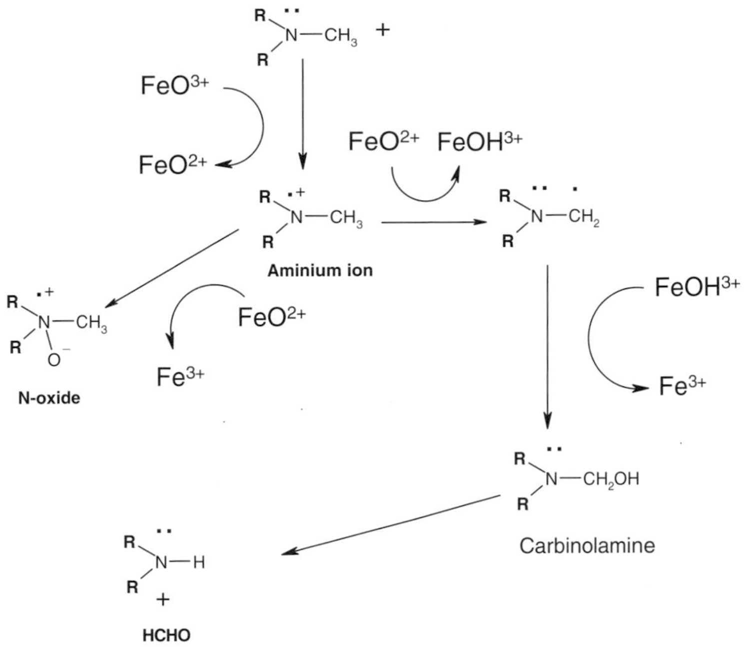

Figure 3.20 Pathways of CYP-mediated N-oxidation and N-dealkylation

or dealkylation occurs is dependent on factors such as the surrounding groups on the molecule and the CYP itself. The mechanism of N-oxidation and N-dealkylation is now believed to differ slightly from the majority of CYP-mediated hydrogen abstractions/oxygen rebound reactions. It begins with the CYP Compound I complex abstracting one of the nitrogen's lone pair of electrons (Figure 3.20), forming an aminium ion (N$^+$). Once this has been created, either the oxygen reacts with the N$^+$ giving the N-oxide, or the Compound I complex can abstract a hydrogen atom from one of the adjacent carbons, forming a carbon radical. The reaction then proceeds as with most CYP oxidations, where the hydroxyl group bounces off the haem iron to react with the carbon radical to make the (usually unstable) alcohol, or carbinolamine. Chlorpromazine and the TCAs can undergo N-oxidation or N-dealkylations, as well as sulphoxide formation (Figure 3.21).

3.12.2 Deaminations

Amine groups in drugs can be primary, secondary, or tertiary. Primary amines can be removed completely thorough conversion of the carbon—nitrogen single to a double bond, where the nitrogen loses an electron. Via a hydrogen atom

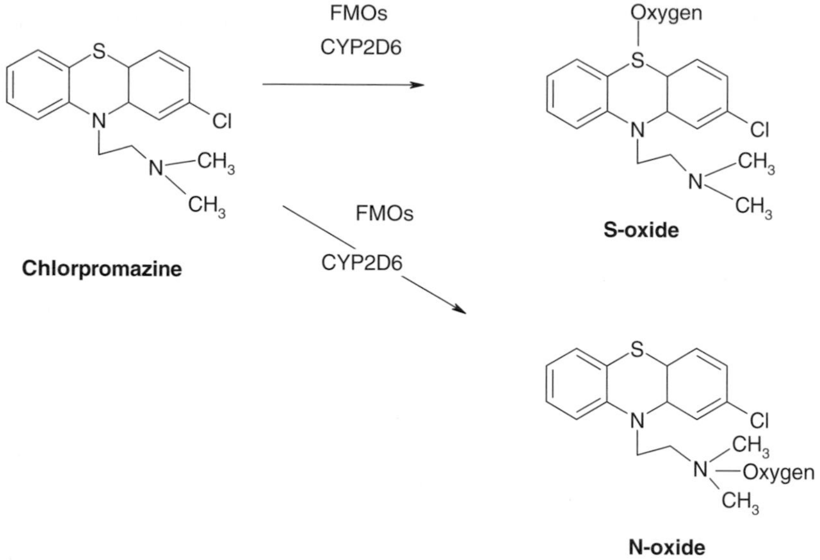

Figure 3.21   Sulphoxide and N-oxide formation with chlorpromazine, formed by either CYPs or flavin monooxygenases (FMOs, Section 3.8)

from water, ammonia is formed with a ketone product. This is one of the fates of amphetamine (Figure 3.22). More amphetamine metabolism can be found in Appendix B.5.

3.12.3   Dehalogenations

Using the same basic tool, oxidation to an alcohol, it is possible for CYPs to remove halogens (chloride, bromide, or fluoride) from molecules, forming a ketone and a halogen ion. A number of volatile general anaesthetics are subject to this route of metabolism. The adjacent carbon to the halide is oxidized to a short-lived alcohol, which causes the movement of electrons towards the halogen, which dissociates (Figure 3.23).

3.13   Other oxidation processes

3.13.1   Primary amine oxidations

Primary amines found in sulphonamides and sulphones can be metabolised to hydroxylamines, and their toxicity hinges on these pathways (Figure 3.24 and see Chapter 8.5.4). The hydroxylamines formed are often reactive, and

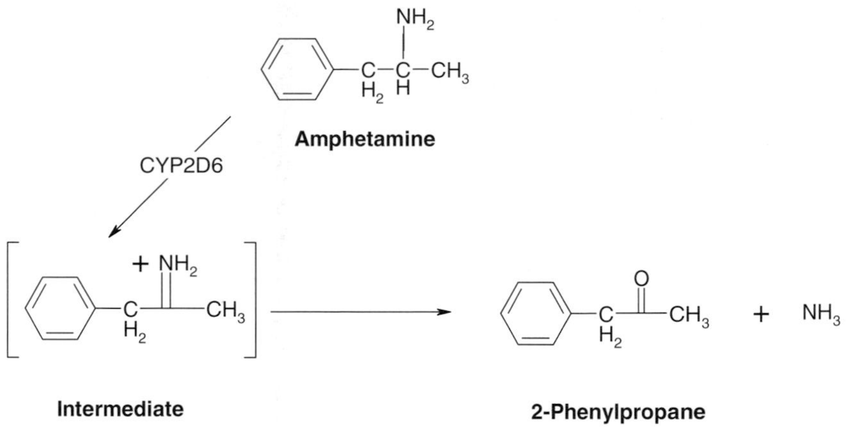

Figure 3.22 Oxidative deamination of amphetamine

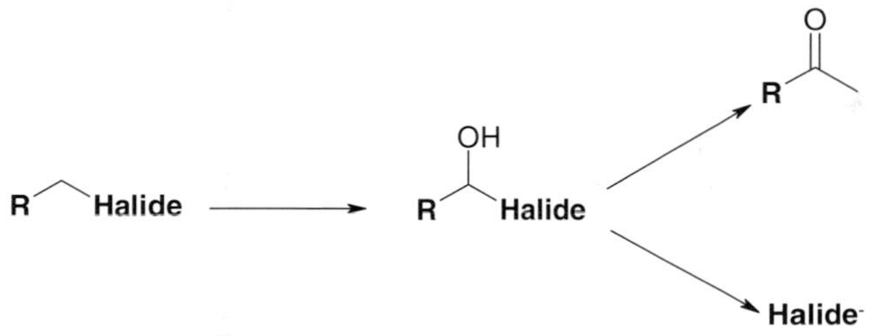

Figure 3.23 Removal of halides through an unstable alcohol intermediate

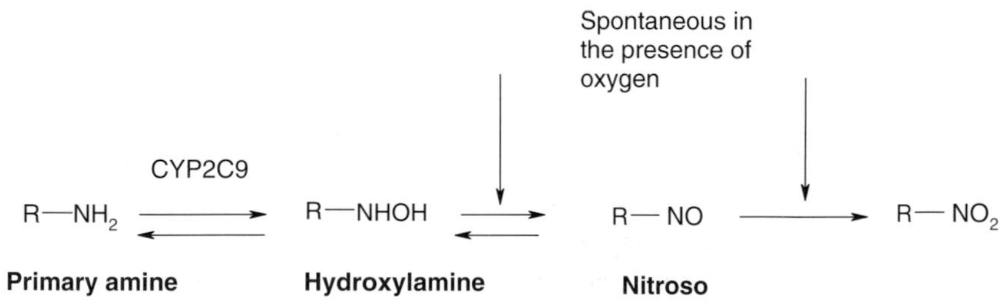

Figure 3.24 Primary amine oxidation

although they can be stabilized by glutathione (GSH) and other cellular antioxidants, they can spontaneously oxidize in the presence of oxygen to nitroso and then nitro-derivatives. The nitro forms are usually stable but are vulnerable to reductive metabolism that drives the process shown in Figure 3.24 in the opposite direction. Secondary amines can also be oxidized to hydroxylamines.

3.13.2   Oxidation of alcohol and aldehydes

Although CYP2E1 is induced by ethanol (Chapter 4.4.4), the vast majority of ethanol clearance (90%) is normally by oxidation to acetaldehyde by another group of enzymes, the alcohol dehydrogenases (ADHs). These enzymes are found in the cytoplasm and they are NAD$^+$ dependent zinc metalloenzymes. They form NADH from NAD+ in the process of alcohol oxidation. There are five classes of ADH isoforms. Class I (ADH1, ADH2, ADH3) isoforms have a high affinity for ethanol and can be blocked by pyrazoles. Classes II and III are more suited to the metabolism of longer-chain alcohols and cannot be blocked by pyrazole[147].

Aldehydes are formed from many reactions in cells, but they are oxidized to their corresponding carboxylic acid by several enzyme systems, including aldehyde dehydrogenase ALDHs, xanthine oxidase, and aldehyde oxidase. These enzymes are usually detoxifying, as many aldehydes, such as formaldehyde, are cytotoxic byproducts of CYP and other oxidative reactions. However, ADH is involved in the generation of reactive species from the nucleoside reverse transcriptase inhibitor abacavir and the anti-convulsant felbamate[147].

Of the three aldehyde dehydrogenase classes, two are relevant to alcohol metabolism. Class I ALDHs are found in the liver cytosol and specialize in acetaldehyde. Class II ALDHs are found in the liver and kidney mitochondria and metabolise acetaldehyde and several other substrates. See Section 7.7.2 for more about ADH and ALDH in alcoholism.

3.13.3   Monoamine oxidase (MAO)

Yet another important oxidative enzyme system that processes endogenous and exogenous substrates is monoamine oxidase (MAO), which exists in two isoforms, MAO-A and MAO-B. Both are found in the outer membrane of mitochondria in virtually all tissues. They have evolved to become two separate enzymes with similar functions and they originate from different genes in man. They use FAD as a cofactor and are capable of oxidizing a very wide variety of endogenous biogenic amines as well as primary, secondary, and tertiary xenobiotic amines. They accomplish their removal of amine groups through an initial reductive half-reaction, followed by an oxidation half-reaction. The reductive half oxidizes the amine and the FAD is reduced. The second half of the process involves the use of oxygen to reoxidize the FAD, leaving hydrogen peroxide and an aldehyde as products. Clorgyline blocks MAO-A, whilst deprenyl is a potent inhibitor of MAO-B. The various triptan anti-migraine drugs are extensively cleared by MAO-A and inhibition by agents such as the relatively fast-acting antidepressant moclobemide significantly prolongs their half-lives[148,154].

Perhaps the most frightening example of MAO-B toxicity is the story of the designer drug impurity MPTP, which can be found in Appendix B.2. In the

1960s, irreversible MAO inhibitors were used as antidepressants, aimed at increasing biogenic amine levels. Unfortunately, they could cause hypertensive crises (sufficient to cause a stroke) through ingestion of other amines, such as tyramine from cheese and a long list of other foods. MAO inhibitors are highly unlikely to be used now, as there are so many safer alternatives.

3.14   Control of CYP metabolic function

Although CYPs appear to be part of an impressive and flexible system for the oxidation of drugs, it is not enough just to process endogenous and xenobiotic molecules at a set rate. Endogenous and exogenous CYP substrates can vary enormously in their concentrations within the body, even on a day-to-day basis. As we have seen, steroid hormone levels must be matched to accomplish specific tasks in narrow time frames, so production and destruction must be under exceedingly fine control. This is apparent during the menstrual cycle and pregnancy. Our exposure to various exogenous chemicals, including drugs, is also variable in terms of concentration and physicochemical properties. As an advertising campaign once said, *'Power is nothing without control'*. It is essential for the CYP system to be finely controllable to respond to the often-extreme changes in the small-molecular-weight chemical presence in cells. In this chapter, we have examined various aspects of CYP control already, but the most powerful system for vast increases in CYP (and other biotransforming enzyme) capability, enzyme induction, will be discussed in detail in terms of mechanism and clinical consequences in the next chapter.

References

1. Liu KC, Hughes JMX, Hay S et al. Liver microsomal lipid enhances the activity and redox coupling of colocalized cytochrome P450 reductase and cytochrome P450 3A4 in nanodiscs. FEBS J. 284, 2302–2319, 2017.
2. Nelson DR. Progress in tracing the evolutionary paths of cytochrome P450. Biochimica et Biophysica Acta 1814. 14–18, 2011.
3. Omura T. Structural diversity of cytochrome P450 enzyme system. J. Biochem. 47, 297–306, 2010.
4. Nelson, DR. The cytochrome P450 homepage. Human Genomics. 4, 59–65, 2009. http://drnelson.uthsc.edu/CytochromeP450.html.
5. The Pharmacogene Variation (PharmVar) Consortium. https://www.pharmvar.org/.
6. Gardiner SJ and Begg EJ. Pharmacogenetics, drug-metabolizing enzymes, and clinical practice. Pharmacol Rev. 58, 521–590, 2006.
7. Asciutto EK, Madura JD, Sondej S et al. Structural and dynamic implications of an effector-induced backbone amide cis–trans isomerization in cytochrome P450cam. J. Mol. Biol. 388, 801–814, 2009.

8. Branden G, Sjogren T, SchneckeV et al. Structure-based ligand design to overcome CYP inhibition in drug discovery projects. Drug Dis. Today. 19, 906–911, 2014.

9. Maekawa K, Adachi M, Matsuzawa Y et al. Structural basis of single-nucleotide polymorphisms in cytochrome P450 2C9. Biochemistry 56, 5476–5480, 2017.

10. Denisov IG, Baas BJ, Grinkova YV et al. Cooperativity in cytochrome P450 3A4. Linkages in substrate binding spin state, uncoupling, and product formation. J. Biol. Chem. 282, 7066–7076, 2007.

11. Strobel HW, Lu AYH, Heidema J et al. Phosphatidylcholine requirement in the enzymatic reduction of hemoprotein P450 and in fatty acid, hydrocarbon, and drug hydroxylation. J Biol Chem, 245, 4851–4854 1970.

12. Jang HH, Kim DH, Ahn T et al. Functional and conformational modulation of human cytochrome P450 1B1 by anionic phospholipids. Arch Biochem Biophys, 493, 143–150, 2010.

13. Nath A. Atkins WM, Sligar SG. Applications of Phospholipid Bilayer Nanodiscs in the study of membranes and membrane proteins. Biochemistry, 46, 2059–2069, 2007.

14. Mak PJ, Denisov IG, Grinkova YV et al. Defining CYP3A4 structural responses to substrate binding. Raman spectroscopic studies of a nanodisc-incorporated mammalian cytochrome P450. J. Am. Chem. Soc. 133, 1357–1366, 2011.

15. Sridhar J, Goyal N, Liu, J. et al. Review of ligand specificity factors for CYP1A subfamily enzymes from molecular modeling studies reported to-date. Molecules, 22, 1143–1162, 2017.

16. Anzenbacher, P. et al. Active sites of cytochromes P450: What are they like? Acta Chim. Slov. 55, 63–66, 2008.

17. Williams PA, Cosme J, Vinkovic D M, et al. Crystal structures of human cytochrome P450 3A4 bound to metyrapone and progesterone. Science 305, 683–686, 2004.

18. Ortiz de Montellano PR. (editor) Cytochrome P450: Structure, Mechanism, and Biochemistry, 3e. Kluwer Academic / Plenum Publishers, New York, 2005.

19. Ekroos M & Sjogren T. Structural basis for ligand promiscuity in cytochrome P450 3A4. PNAS 103, 13682–13687, 2006.

20. Henshall J, Galetin A, Harrison A, Houston JB. Comparative analysis of CYP3A heteroactivation by steroid hormones and flavonoids in different *in vitro* systems and potential *in vivo* implications. Drug Metab Dispos. 36, 1332–1340, 2008.

21. Denisov IG, Grinkova YV, Baylon JL. Mechanism of drug-drug interactions mediated by human cytochrome P450 CYP3A4 monomer. Biochemistry. 54, 2227–2239, 2015.

22. Du H, Li J, Cai Y. et al. Computational investigation of ligand binding to the peripheral site in CYP3A4: conformational dynamics and inhibitor discovery. J. Chem. Inf. Model. 57, 616–626, 2017.

23. Polic, V and Auclair K. Allosteric activation of cytochrome P450 3A4 via progesterone bioconjugation. Bioconjugate Chem. 28, 885–889, 2017.

24. Reed JR and Backes WL. Formation of P450 · P450 complexes and their effect on P450 function. Pharmacol Ther. 133, 299–310, 2012.

25. Tan YZ, Patten CJ, Smith P, Yang CS. Competitive interactions between cytochromes P450 2A6 and 2E1 for NADPH-cytochrome P450 oxidoreductase in the microsomal membranes produced by a baculovirus expression system. Archives of Biochemistry and Biophysics. 342, 82–91, 1997.

26. Reed JR, Eyer M, Backes WL Functional interactions between cytochromes P450 1A2 and 2B4 require both enzymes to reside in the same phospholipid vesicle: -evidence for physical complex formation. J Biol Chem. 285, 8942–8952, 2010.

27. U.S. National Library of Medicine; Genetics Home Reference; https://ghr.nlm.nih.gov/gene/POR#location.

28. Shen AL, O'Leary KA, Kasper CB. Association of multiple developmental defects and embryonic lethality with loss of microsomal NADPH cytochrome P450 oxidoreductase. J Biol Chem. 277, 6536–6541, 2002.

29. Adachi M. Tachibana K, Asakura Y, et al. Compound heterozygous mutations of cytochrome P450 oxidoreductase gene (POR) in two patients with Antley–Bixler Syndrome. Am. J. Med. Genet. 128A, 333–339, 2004.

30. Aigrain L, Pompon D, Morera S et al. Structure of the open conformation of a functional chimeric NADPH cytochrome P450 reductase. EMBO reports 10, 742–747, 2009.

31. Campelo D, Lautier T, Urban P et al. The hinge segment of human NADPH-cytochrome P450 reductase in conformational switching: The critical role of ionic strength. Front. Pharmacol. 8, 755, 1–13, 2017.

32. Waskell L., Kim JJ. Electron transfer partners of cytochrome P450. In Ortiz de Montellano P. (eds.). Cytochrome P450. Springer, Cham. 2015.

33. Reczek CR, Birsoy K, Kong H, et al. A CRISPR screen identifies a pathway required for paraquat-induced cell death. Nat. Chem.Biol. 13 1274–1279, 2017.

34. Satpute RM, Pawar PP, Puttewar S, et al. Effect of resveratrol and tetracycline on the subacute paraquat toxicity in mice. Human and Experimental Toxicology 36, 1303–1314, 2017.

35. Bhatta MR, Khatrib Y, Rodger RJ, et al. Role of cytochrome b5 in the modulation of the enzymatic activities of cytochrome P450 17a-hydroxylase/17,20-lyase (P450 17A1). Journal of Steroid Biochemistry & Molecular Biology 170, 2–18, 2017.

36. Bewley MC, Ainsley Davis C, Marohnic CC et al. The structure of the S127P mutant of Cytochrome b5 reductase that causes methemoglobinemia shows the AMP moiety of the flavin occupying the substrate-binding site. Biochemistry 42, 13145–13151, 2003.

37. Simonov AN, Holien JK, Yeung JCI, et al. Mechanistic scrutiny identifies a kinetic role for cytochrome b5 regulation of human cytochrome P450c17 (CYP17A1, P450 17A1). PLoS One 10 e0141252, 2015.

38. Finn RD, McLaughlin LA, Ronseaux S. et al. Defining the *in vivo* role for cytochrome b5 in cytochrome P450 function through the conditional hepatic deletion of microsomal cytochrome b5. J Biol Chem. 283, 31385–31393, 2008.

39. Henderson CJ, McLaughlin LA, Scheer N et al. Cytochrome b5 is a major determinant of human cytochrome P450 CYP2D6 and CYP3A4 activity *in vivo*. Mol. Pharmacol. 87, 733–739, 2015.

40. Avadhani NG, Sangar MC, Bansal S et al. Bimodal targeting of cytochrome P450s to endoplasmicreticulum and mitochondria: the concept of chimeric signals. FEBS Journal 278, 4218–4229, 2011.

41. Brignac-Huber LM, Park JW, Reed JR et al. Cytochrome P450 organization and function are modulated by endoplasmic reticulum phospholipid heterogeneity. Drug Metab. Disp. 44, 1859–1866, 2016.

42. Leung T, Rajendran R, Singh S. et al. Cytochrome P450 2E1 (CYP2E1) regulates the response to oxidative stress and migration of breast cancer cells. Br. Canc. Res. 15, R107, 1–12, 2013.

43. Correia MA, Wang Y, Kim SM, et al. Hepatic cytochrome P450 ubiquitination: conformational phosphodegrons for E2/E3 recognition? Int. Union of Biochem. Mol. Biol. 66, 78–88, 2014.

44. Miao L, Yao H, Li C, et al. A dual inhibition: microRNA-552 suppresses both transcription and translation of cytochrome P450 2E1 Biochimica et Biophysica Acta 1859, 650–662, 2016.

45. Gill P, Bhattacharyya1 S, McCullough S, et al. MicroRNA regulation of CYP 1A2, CYP3A4 and CYP2E1 expression in acetaminophen toxicity. Scientific Reports 7, 12331–12342, 2017.

46. He ZX, Chen XW, Zhou ZW et al. Impact of physiological, pathological and environmental factors on the expression and activity of human cytochrome P450 2D6 and implications in precision medicine. Drug Metab. Rev. 47, 470–519, 2015.

47. Rieger JK, Reutter S, Hofmann U. Inflammation-associated microRNA-130b downregulates cytochrome P450 activities and directly targets CYP2C9. Drug Metab. Dispos. 43, 884–888, 2015.

48. Miller WL, Tee MK. The post-translational regulation of 17,20 lyase activity. Molecular and Cellular Endocrinology 408, 99–106, 2015.

49. United States National Center for Biotechnology Information. CYP1A1 cytochrome P450 family 1 subfamily A member 1 [Homo sapiens (human)] https://www.ncbi.nlm.nih.gov/gene/1543.

50. Santes-Palacios R, Ornelas-Ayala D, Cabañas N, et al. Regulation of human cytochrome P4501A1 (hCYP1A1): A plausible target for chemoprevention? BioMed. Res. Int. 5341081, 1–17, 2016.

51. Buchthal J, Grund KE, Buchmann A, et al. Induction of cytochrome P4501A by smoking or omeprazole in comparison with UDP-glucuronosyltransferase in biopsies of human duodenal mucosa. Eur J Clin Pharmacol. 47(5), 431–435, 1995.

52. Akhtar S, Mahjabeen I, Akram Z, et al. CYP1A1 and GSTP1 gene variations in breast cancer: a systematic review and case–control study. Familial Cancer 15, 201–214, 2016.

53. Olguín-Reyes S, Camacho-Carranza R, Hernández-Ojeda S. Bergamottin is a competitive inhibitor of CYP1A1 and is antimutagenic in the Ames test. Food and Chemical Toxicology 50, 3094–3099, 2012.

54. Androutsopoulos VP, Tsatsakis AM, Spandidos DA. Cytochrome P450 CYP1A1: wider roles in cancer progression and prevention. BMC Cancer 9, 187–204, 2009.

55. Anderson GD and Chan LN. Pharmacokinetic drug interactions with tobacco, cannabinoids and smoking cessation products. Clin Pharmacokinet 55, 1353–1368, 2016.

56. Nardelli-Siebert M, Mattos JJ, Toledo-Silva, G. Candidate cytochrome P450 genes for ethoxyresorufin O-deethylase activity in oyster Crassostrea gigas. Aquatic Toxicology 189, 142–149, 2017.

57. Agbor LN, Walsh MT, Boberg JR, et al. Elevated blood pressure in cytochrome P4501A1 knockout mice is associated with reduced vasodilation to omega-3 polyunsaturated fatty acids. Tox. Appl. Pharmacol. 264, 351–360, 2012.

58. United States National Center for Biotechnology Information. CYP1A2 cytochrome P450 family 1 subfamily A member 2 [Homo sapiens (human)] https://www.ncbi.nlm.nih.gov/gene/1544.

59. Sridhar J, Goyal N, Liu J. et al. Review of ligand specificity factors for CYP1A subfamily enzymes from molecular modelling; studies reported to-date. Molecules 22, 1143–1162, 2017.

60. Sansen S, Yano JK, Reynald RL. et al. Adaptations for the oxidation of polycyclic aromatic hydrocarbons exhibited by the structure of human P450 1A2. J. Biol. Chem. 282, 14348–14355, 2007.

61. Santes-Palacios R, Romo-Mancillas A, Camacho-Carranza R, et al. Inhibition of human and rat CYP1A1 enzyme by grapefruit juice compounds. Toxicology Letters 258, 267–275, 2016.

62. United States National Center for Biotechnology Information. CYP1B1 cytochrome P450 family 1 subfamily B member 1 [Homo sapiens (human)]. https://www.ncbi.nlm.nih.gov/gene/1545.

63. Achary A, Nagarajaram HA. Comparative docking studies of CYP1b1 and its PCG-associated mutant forms. J. Biosci. 33, 699–713, 2008.

64. Dutour R & Poirier D. Inhibitors of cytochrome P450 (CYP) 1B1. Eur. J Med. Chem 135, 296–306, 2017.

65. Mu W, Hu C, Zhang H, et al. miR-27b synergizes with anticancer drugs via p53 activation and CYP1B1 suppression. Cell Research 25, 477–495, 2015.

66. United States National Center for Biotechnology Information. CYP2A6 cytochrome P450 family 2 subfamily A member 6 [Homo sapiens (human)]. https://www.ncbi.nlm.nih.gov/gene/1548.

67. Zanger UM & Schwab M. Cytochrome P450 enzymes in drug metabolism: Regulation of gene expression, enzyme activities, and impact of genetic variation. Pharmacology & Therapeutics 138, 103–141, 2013.

68. Di YM, Chow VDW, Yang LP et al. Structure, Function, Regulation and Polymorphism of Human Cytochrome P450 2A6. Current Drug Metabolism 10, 754–780 2009.

69. Yuan JM, Nelson HH, Butler LM et al. Genetic determinants of cytochrome P450 2A6 activity and biomarkers of tobacco smoke exposure in relation to risk of lung cancer development in the Shanghai cohort study. Int J Cancer. 138, 2161–2171, 2016.

70. Park SL, Murphy SE, Wilkens LR et al. Association of CYP2A6 activity with lung cancer incidence in smokers: The multiethnic cohort study. PLoS One. 12(5): e0178435, 2017.

71. Tanner JA, Tyndale RF. Variation in CYP2A6 activity and personalized medicine J. Pers. Med. 7, 18, 1–29, 2017.

72. Yamaguchi Y, Akimoto I, Motegi K. Synthetic models related to methoxalen and menthofuran–cytochrome P450 (CYP) 2A6 interactions. Benzofuran and coumarin derivatives as potent and selective inhibitors of CYP2A6. Chem. Pharm. Bull. 61, 997–1001, 2013.

73. United States National Center for Biotechnology Information. CYP2B6 cytochrome P450 family 2 subfamily B member 6 [Homo sapiens (human)]. https://www.ncbi.nlm.nih.gov/gene/1555.

74. Pearce RE, Gaedigk R, Twist GP, et al. Developmental expression of CYP2B6: A comprehensive analysis of mRNA expression, protein content and bupropion hydroxylase activity and the impact of genetic variation. Drug Metab. Dispos. 44, 948–958, 2016.

75. Hammond TG & Birdsall HH. Hepatocyte CYP2B6 can be expressed in cell culture systems by exerting physiological levels of shear: implications for ADME testing. J. Toxicol. Article ID 1907952, 1–5, 2017.

76. Hedrich, WD, Hassan HE, Wang, HB. Insights into CYP2B6-mediated drug-drug interactions. Acta, Pharma. Sinica. 6, 413–425, 2016.

77. Huang SH, Lin SW, Chang SY Prediction of plasma efavirenz concentrations among HIV positive patients taking efavirenz containing combination antiretroviral therapy. Sci. Rep. 7, 16187, 1–9, 2017.

78. Ahmad T, Sabet S, Primerano DA, et al. Tell-Tale SNPs: The Role of CYP2B6 in methadone fatalities. Journal of Analytical Toxicology, 41, 325–333, 2017.

79. Rao LK, Flaker AM, Friedel CC, et al. Role of cytochrome P4502B6 polymorphisms in ketamine metabolism and clearance anesthesiology 125, 1103–1112, 2016.

80. Nishiya Y, Hagihara K, Ito T. et al. Mechanism-based inhibition of human cytochrome P450 2B6 by ticlopidine, clopidogrel, and the thiolactone metabolite of prasugrel. Drug Metab. Disp. 37, 589–593, 2009.

81. United States National Center for Biotechnology Information. CYP2C8 cytochrome P450 family 2 subfamily C member 8 [Homo sapiens (human)]. https://www.ncbi.nlm.nih.gov/gene/1558.

82. Hiratsuka, M. Genetic polymorphisms and *in vitro* functional characterization of CYP2C8, CYP2C9, and CYP2C19 allelic variants. Biol. Pharm. Bull. 39, 1748–1759, 2016.

83. Backman JT, Filppula AM, Niemi M et al. Role of cytochrome P450 2C8 in drug metabolism and interactions. Pharmacol Rev. 68, 168–241, 2016.

84. Schoch GA, Yano JK, Sansen S, et al. Determinants of cytochrome P450 2C8 substrate binding: structures of complexes with montelukast, troglitazone, felodipine, and 9-cis-retinoic acid. J. Biol. Chem. 283, 17227–17237, 2008.

85. VandenBrink BM, Foti RS, Rock DA, et al. Evaluation of CYP2C8 inhibition *in vitro*: Utility of montelukast as a selective CYP2C8 probe substrate drug. Metab. Disp. 39, 1546–1554, 2011.

86. Tornio A, Neuvonen PJ, Niemi M, et al. Role of gemfibrozil as an inhibitor of CYP2C8 and membrane transporters. Exp. Opin. Drug Metab. Toxicol. 13, 183–95, 2016.

87. Staffa JA, Chang J, and Green L. Cerivastatin and reports of fatal rhabdomyolysis. N Engl J Med 346, 539–540, 2002.

88. United States National Center for Biotechnology Information. CYP2C9 cytochrome P450 family 2 subfamily C member 9 [Homo sapiens (human)]. https://www.ncbi.nlm.nih.gov/gene/1559.

89. Van Booven D, Marsh S, McLeod H, et al. Cytochrome P450 2C9-CYP2C9 Pharmacogenet Genom. 20, 277–281, 2010.

90. Maekawa K, Adachi M, Matsuzawa Y, et al. Structural basis of single-nucleotide polymorphisms in cytochrome P450 2C9. Biochemistry 56, 5476–5480, 2017.

91. Cui YL, Xua F and Wu R. Molecular dynamics investigations of regioselectivity of anionic/aromatic substrates by a family of enzymes: a case study of diclofenac binding in CYP2C isoforms. Phys. Chem. Chem. Phys. 18, 17428, 2016.

92. Mosher Cm, Tai G, Rettie AE. CYP2C9 amino acid residues influencing phenytoin turnover and metabolite regio-and stereochemistry. JPET 329, 938–944, 2009.

93. Shaul C, Blotnick S, Muszkat M, et al. Quantitative assessment of CYP2C9 genetic polymorphisms effect on the oral clearance of S-warfarin in healthy subjects. Mol. Diagn. Ther. 21, 75–83, 2017.

94. United States National Center for Biotechnology Information. CYP2C18 cytochrome P450 family 2 subfamily C member 18 [Homo sapiens (human)]. https://www.ncbi.nlm.nih.gov/gene/1562.

95. Uno Y, Matsuno K, Nakamura C., et al. Identification and Characterization of CYP2C18 in the Cynomolgus Macaque (*Macaca fascicularis*)J. Vet. Med. Sci. 72, 225–228, 2010.

96. Kinobe RT, Parkinson OT, Mitchell DJ, et al. P450 2C18 catalyses the metabolic bioactivation of phenytoin. Chem. Res. Toxicol., 18, 1868–1875, 2005.

97. United States National Center for Biotechnology Information. CYP2C19 cytochrome P450 family 2 subfamily C member 19 [Homo sapiens (human)]. https://www.ncbi.nlm.nih.gov/gene/1557.

98. Topol EJ, Schork NJ Catapulting clopidogrel pharmacogenomics forward. Nature Med. 17, 40–41, 2011.

99. Moriyama B, Owusu Obeng A, Barbarino J. Clinical pharmacogenetics implementation consortium (CPIC) guidelines for CYP2C19 and voriconazole therapy. Clin. Pharmacol. Ther. 102, 45–51, 2017.

100. Yi X, Wang Y, Lin J, et al. Interaction of CYP2C19, P2Y12, and GPIIIa variants associates with efficacy of clopidogrel and adverse events on patients with ischemic stroke. Clin. Appl. Thromb/Hem. 23, 761–768, 2017.

101. Cuisset T, Loosveld M, Morange PE, CYP2C19*2 and *17. Alleles have a significant impact on platelet response and bleeding risk in patients treated with prasugrel after acute coronary syndrome Jacc Cardiovas. Interven. 5, 1280–1287, 2012.

102. United States National Center for Biotechnology Information. CYP2D6 cytochrome P450 family 2 subfamily D member 6 [Homo sapiens (human)]. https://www.ncbi.nlm.nih.gov/gene/1565.

103. Pan X, Ning M, Jeong H, Transcriptional regulation of CYP2D6 expression Drug Metab. Dispos. 45, 42–48, 2017.

104. Rasheed MS, Mishra AK, Singh MP. Cytochrome P450 2D6 and Parkinson's disease: Polymorphism, metabolic role, risk and protection. Neurochem. Res. 42, 3353–3361, 2017.

105. Lu Y, Peng Q, Zeng Z, et al. CYP2D6 phenotypes and Parkinson's disease risk: a meta-analysis. J. Neurol. Sci. 336, 161–168, 2014.

106. Wang A, Stout CD, Zhang Q. et al. Contributions of ionic interactions and protein dynamics to cytochrome P450 2D6 (CYP2D6) substrate and inhibitor binding. J. Biol. Chem. 290, 5092–5104, 2015.

107. United States National Center for Biotechnology Information. CYP2E1 cytochrome P450 family 2 subfamily E member 1 [Homo sapiens (human)]. https://www.ncbi.nlm.nih.gov/gene/1571.

108. DeVore NM, Meneely KM, Bart AG, et al. Structural comparison of cytochromes P450 2A6, 2A13, and 2E1 with pilocarpine FEBS J. 279 1621–1631, 2012.

109. Porubsky PR, Meneely KM, Scott EE, Structures of human cytochrome P-450 2E1: J. Biol. Chem. 283, NO. 48, 33698–33707, 2008.

110. Frye RF & Branch RA Effect of chronic disulfiram administration on the activities of CYPs 2A2/2C19/2D6/2E1 and NAcetyl transferase in healthy human subjects. Br. T. Clin. Pharmac. 53, 155–162, 2002.

111. Collom SL, Jamakhandi AP, Tackett AJ, et al. CYP2E1 active site residues in substrate recognition sequence 5 identified by photoaffinity labeling and homology modeling. Arch. Biochem. Biophys. 459, 59–69, 2007.

112. Liu Y, Han KH, Guifang J, et al. Potent mutagenicity of some non-planar tri-and tetrachlorinated biphenyls in mammalian cells, human CYP2E1 being a major activating enzyme. Arch. Toxicol. 91, 2663–2676, 2017.

113. Fang Z Wu Y, Zhang N, Association between CYP2E1 genetic polymorphisms and urinary cancer risk: a meta-analysis Oncotarget, 8 (No. 49), 86853–86864, 2017.

114. Pellé L, Cipollini M, Tremmel R, et al. Association between CYP2E1 polymorphisms and risk of differentiated thyroid carcinoma. Arch Toxicol 90, 3099–3109, 2016.

115. Östlund J & Zlabek V Zamaratskaia G, *In vitro* inhibition of human CYP2E1 and CYP3A by quercetin and myricetin in hepatic microsomes is not gender dependent. Toxicology 381, 10–18, 2017.

116. Wang Y, Yub D, Tolleson WH, et al. A systematic evaluation of microRNAs in regulating human hepatic CYP2E1 Biochem Pharmacol. 138, 174–184, 2017.

117. United States CYP3A4 cytochrome P450 family 3 subfamily A member 4 [Homo sapiens (human)], https://www.ncbi.nlm.nih.gov/gene/1576.

118. Basheer L & Kerem Z. Interactions between CYP3A4 and dietary polyphenols. Ox. Med. Cell. Long. Article ID 854015, 1–15, 2015.

119. Dai ZR, Ai CZ, Ge GB et al. A mechanism-based model for the prediction of the metabolic sites of steroids mediated by cytochrome P450 3A4. Int. J. Mol. Sci. 16, 14677–14694, 2015.

120. Fashe MM, Juvonen RO, Petsalo A et al. In silico prediction of the site of oxidation by cytochrome P450 3A4 that leads to the formation of the toxic metabolites of pyrrolizidine alkaloids. Chem. Res. Toxicol. 2015, 28, 702–710, 2015.

121. Dohnal, V, Wu, QH, Kuca, K Arch. Metabolism of aflatoxins: key enzymes and interindividual as well as interspecies differences Toxicol. 88 (9) 1635–1644, 2014.

122. Mikus G, Foerster KI. Role of CYP3A4 in kinase inhibitor metabolism and assessment of CYP3A4 activity. Transl. Cancer Res. 6(Suppl 10): S1592–S1599, 2017.

123. Samuels ER, Sevrioukova I. Inhibition of Human CYP3A4 by rationally designed ritonavir-like compounds: impact and interplay of the side group Functionalities Mol. Pharmaceutics, 15, 279–288, 2018.

124. Dresser GK, Urquhart BL, Proniuk J, et al. Coffee inhibition of CYP3A4 *in vitro* was not translated to a grapefruit-like pharmacokinetic interaction clinically. Pharma. Res. Per. 5(5), e00346, 1–9, 2017.

125. Backman JT, Kivisto KT, Olkkola KT, et al. The area under the plasma concentration-time curve for oral midazolam is 400-fold larger during treatment with itraconazole than with rifampicin. Eur. J. Clin. Pharmacol. 54, 53–58, 1998.

126. Birdwell KA, Decker B, Barbarino JM, et al. Clinical Pharmacogenetics Implementation Consortium (CPIC) Guidelines for CYP3A5 genotype and tacrolimus dosing. Clin, Pharm Ther. 98, 19–24, 2015.

127. Zhang H, Chen M, Wang X, et al. Patients with CYP3A4*1G genetic polymorphism consumed significantly lower amount of sufentanil in general anesthesia during lung resection. Medicine 96 (4) (e6013), 1–3, 2017.

128. United States CYP3A5 cytochrome P450 family 3 subfamily A member 5 [Homo sapiens (human)]. https://www.ncbi.nlm.nih.gov/gene/1577.

129. Neunzig I, Drăgan CA, Widjaja M, et al. Whole-cell biotransformation assay for investigation of the human drug metabolizing enzyme CYP3A7. Biochim. et Biophys. Acta 1814, 161–167, 2011.

130. Kim JH, Han N, Kim MG, ct al. Increased exposure of tacrolimus by co-administered mycophenolate mofetil: population pharmacokinetic analysis in healthy volunteers. Sci Rep. 8, 1687, 1–9, 2018.

131. Burk O, Koch I, Raucy J, et al. The induction of cytochrome P450 3A5 (CYP3A5) in the human liver and intestine is mediated by the xenobiotic sensors pregnane X receptor (PXR) and constitutively activated receptor (CAR). J. Biol Chem. 279, N38379–38385, 2004.

132. Noll EM, Eisen C, Stenzinger A et al. CYP3A5 mediates basal and acquired therapy resistance in different subtypes of pancreatic ductal adenocarcinoma. Nat Med. 22, 278–287, 2016.

133. United States CYP3A7 cytochrome P450 family 3 subfamily A member 7 [Homo sapiens (human)] https://www.ncbi.nlm.nih.gov/gene/1551.

134. He H, Nie YL, Li JF, et al. Developmental regulation of CYP3A4 and CYP3A7 in Chinese Han population Drug Metab. Pharmacokin. 31, 433e444, 2016.

135. Sood D, Johnson N, Jain P, et al. CYP3A7*1C allele is associated with reduced levels of 2-hydroxylation pathway oestrogen metabolites, Brit. J. Can. 116, 382–388, 2017.

136. Krest CM, Onderko EL, Yosca TH, et al. Reactive Intermediates in Cytochrome P450 Catalysis*J. Biol. Chem. 288, 17074–17081, 2013.

137. Yoo, HH, Kim NS, Kim, MJ et al. Enantioselective carbonyl reduction of eperisone in human liver microsomes. Xenobiotica 41, 758–763, 2011.

138. Wagmann L, Meyer MR, Maurer HH, What is the contribution of human FMO3 in the N-oxygenation of selected therapeutic drugs and drugs of abuse? Toxicology Letters 258, 55–70, 2016.

139. Rossner R, Kaeberlein M, Leiser SF. Flavin-containing monooxygenases in aging and disease: emerging roles for ancient enzymes. J. Biol. Chem. 292(27), 11138–11146, 2017.

140. Basaran R & Can Eke B. Flavin containing monooxygenases and metabolism of xenobiotics. Turk. J. Pharm. Sci. 14(1), 90–94, 2017.

141. Jones BC, Srivastava A, Colclough N et al. An investigation into the prediction of *in vivo* clearance for a range of flavin-containing monooxygenase substrates. Drug Metab. Disp. 45 (10), 1060–1067, 2017.

142. Fiorentini F, Romero E, Fraaije MW, et al. Baeyer–Villiger monooxygenase FMO5 as entry point in drug metabolism. ACS Chem. Biol. 12, 2379–2387, 2017.

143. Huijbers MME, Montersino S, Westphal AH et al. Flavin dependent monooxygenases. Arch. Biochem. Biophys. 544, 2–17, 2014.

144. Shih DM, Wang Z, Lee R, et al. Flavin containing monooxygenase 3 exerts broad effects on glucose and lipid metabolism and atherosclerosis. J. Lipid Res. 56, 22–37, 2015.

145. Lattard V, Longin-Sauvageon C, Lachuer J, et al. Cloning, sequencing and tissue-dependent expression of flavin containing monooxygenase (FMO) 1 and FMO3 in the dog. Drug Metab. Disp. 30, 119–128, 2002.

146. Coleman MD, Edwards IG, Mihaly GW. High performance liquid chromatographic method for the determination of pyrimethamine and its 3-N-Oxide metabolite in biological fluids. J Chromatogr. 308, 363–369, 1984.

147. Foti RS & Dalvie DK Special section on emerging novel enzyme pathways in drug metabolism—commentary cytochrome P450 and non–cytochrome P450 oxidative metabolism: contributions to the pharmacokinetics, safety, and efficacy of xenobiotics. Drug Metab. Dispos. 44, 1229–1245, 2016.

148. Kim YH, Bae YJ, Kim HS, et al. Measurement of human cytochrome P450 enzyme induction based on mesalazine and mosapride citrate treatments using a luminescent assay. Biomol Ther. 23(5), 486–492, 2015.

149. Zhou W, Humphries H, Neuhoff S, et al. Development of a physiologically based pharmacokinetic model to predict the effects of flavin-containing monooxygenase 3(FMO3) polymorphisms on itopride exposure. Biopharm Drug Dispos. 38, 389–393, 2017.

150. Phillips, IR, and Shephard EA. Flavin-containing monooxygenases: mutations, disease and drug response. Trends Pharmacol. Sci. 29, 294–301, 2008.

151. Goncalves LCP, Kracher D, MilkerS, et al. Mutagenesis-independent stabilization of Class B flavin monooxygenases in operation. Adv. Synth. Catal. 359, 2121–2131, 2017.

152. Iram F, Khan SA, b Ahmad A, et al. Eslicarbazepine acetate: A therapeutic agent of paramount importance in acute anticonvulsant therapy. J Acute Dis 6(6), 245–254, 2017.

153. Woehrling EK, Zilz TR and Coleman MD. The toxicity of hexanedione isomers in neural and astrocytic cell lines. Env. Tox. Pharmacol 22, 249–254, 2006.

154. Fleishaker JC, Ryan KK, Jansat JM, et al. Effect of MAO-A inhibition on the pharmacokinetics of almotriptan, an antimigraine agent in humans. Br J Clin Pharmacol. 51(5): 437–441, 2001.

4 Induction of Cytochrome P450 Systems

4.1 Introduction

4.1.1 How living systems self-regulate: overview

As a student, you will most likely first encounter Drug Metabolism as a subject area within the general realm of Pharmacology, which, in its turn, is usually taught after some serious grounding in Physiology and Biochemistry. This reflects the fact that Pharmacology, which of course is the impact of drugs on biological systems and vice-versa, is an attempt to modify, repair, or substitute a component or function of a patient's impaired physiological/biochemical system. Unfortunately, it is probably fair to state, that all of us – those who receive, administer, study, regulate, and manufacture drugs, are often disappointed with the patient benefit that drugs actually achieve in real life. This is against expectation, as we know that the endogenous system used to work perfectly at one time; then it failed and we believe we know why it failed. Therefore, we feel that our drug should have restored the system. That the drug often does not restore function in all individuals reveals to us rather harshly that our understanding of the physiological systems we are trying to augment or repair is short of the full truth.

So, we could perhaps simplify this extremely complex issue of poor drug performance into two aspects. The first is that our drug does not reach the target receptor or tissue, so it cannot work. Whether the drug reaches the target is governed by *pharmacokinetics*, which describes mathematically where the drug goes and how long it stays in the patient's system as a viable therapeutic entity. The second aspect is that, assuming the drug reaches the target, it does not work as well as was predicted from its developmental data. This is a *pharmacodynamic* issue, over how a drug operates at receptor and tissue level. Since we wish our drug to work and benefit the patient, we clearly need to first understand, on a strategic level, how pharmacokinetics and pharmacodynamics operate for endogenous systems – indeed, exactly how did the system run when it was

Human Drug Metabolism, Third Edition. Michael D. Coleman.
© 2020 John Wiley & Sons, Inc. Published 2020 by John Wiley & Sons, Inc.

perfect and why did it fail. Interestingly, how biological systems control cellular, tissue and organ function, in terms of the pharmacokinetics and pharmacodynamics of endogenous agents have some broad similarities which might help our understanding of how drugs work and also, why they sometimes fail to work.

Imagine a very simple scenario of activity regulation – in fact, any activity, from driving a car, to the operation of a washing machine, to any biological function. Three essential components are necessary. First, some command point or 'mission control' must exist where the 'mission instructions' for the given activity are held and this is also the place where decisions are made on regulating the activity, in the light of information received about the activity. Mission control can be a cell, a collection of cells, a tissue, or an organ. Second, there must be some form of 'sensor' that collects and sends second by second information about the activity to mission control. Biosensors range from receptors which measure a process and then translate it into a cascade of molecules (second messenger), or perhaps are connected to neurones which then send an impulse code which is translated by the central nervous system. Another word for a sensor is a *transducer,* where activity is converted from one form to another which is easily coded, then decoded by whatever system needs to know about what is happening (Figure 4.1).

Finally, there must be an activity control subsystem, which directly modulates the activity in some dynamic way and is under the direct command of mission

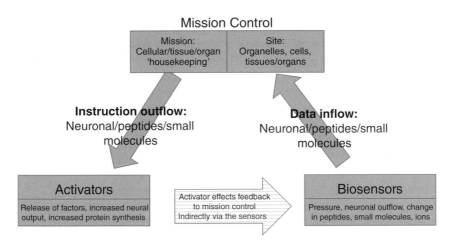

Figure 4.1   Physiological regulatory triangle. *Mission control* could be a gene, a group of genes, a cell, many cells, or a tissue. It is tasked with operating a cellular housekeeping process, within certain parameters. The mission is executed second by second, according to information gathered from the biosensors and communicated by nerves or signal molecules. The mission control sends out commands as and when required towards an activator or some kind. The biosensors and activators do not directly communicate, just indirectly through the biosensors reporting the effects of the activators to mission control

control. This could be for example, an increase or decrease in the production and release of a factor that directly changed a system or process, or perhaps even the increase/decrease in muscle tension in a tissue. Such an open 'triangular' relationship, where mission control receives information constantly from the 'sensors' and then sends instructions every second to the 'activity control' locus, is essentially how we stay alive. The sensors do not need to communicate directly with the activity control locus, but the mission control must be in second-by-second communication with the sensors and the activity control (Figure 4.1). As you can see, we have departed from drug metabolism, pharmacokinetics, and pharmacodynamics all the way back to basic physiology. Logically, mission control can operate many sensor and activator systems and maybe itself be part of many other higher mission controls.

If we return to pharmacodynamics, it is obvious that drug treatment is needed due to a fault in one or more of the members of this basic 'control triangle' in myriad systems in cells, tissues, and organs. As an exercise, you could frame this crude model on human health problems you know about already. A devastating example is type I diabetes, where the mission control, sensors, and activators all disappear with the loss if the Islets of Langehans. We know that we can remedy this with direct insulin injections. A less acute example might be blood pressure control. Hypertension is a feature of aging and is usually linked to a complex combination of factors, but part of this process is erroneous information from aging sensors, such as heart baroreceptors which tell the medulla in the brain that blood pressure is lower than it actually is (Figure 4.2). The medulla then takes measures to increase blood pressure such as raising peripheral vascular resistance incrementally; hence, we have some basis for essential hypertension. Similarly, biosensors that read different cholesterol fraction levels in blood feed incorrect information to cells which synthesis the chemical, or removal of cholesterol from the circulation is impaired or both. Either way, cholesterol fraction levels rise and if these are potentially problematic, such as VLDL cholesterol, the basis of cardiovascular damage is established. These scenarios are very simplistic, but they can emphasize the relative crudity of drug treatment, even when it is life-saving, such as in diabetes. Diabetics can continue to function whilst manually injecting insulin, but the host of issues related to complications and the loss of a decade or more of life expectancy demonstrate the chasm in effectiveness over a lifetime between the 'factory-fitted' system and their own medically supported efforts.

In type I diabetes, there is no control triangle left, but when pharmacological intervention is intended to improve less acute but nonetheless potentially damaging conditions such as high blood pressure or cholesterol levels, the 'triangular' system, although impaired, continues to operate and to some extent may well counteract the drug's pharmacodynamic efficacy (Figure 4.2). This is one reason for a lack of drug efficacy, sometimes termed *resistance* to therapy. You can imagine other scenarios where the mission control process may be faulty, the biosensors read incorrectly and the activators are impaired. The system could be

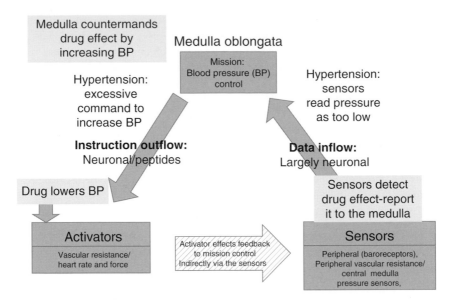

Figure 4.2   Physiological regulatory 'triangle' where 'mission control' receives faulty data and the housekeeping process, in this case, blood pressure, is increased excessively. The drug treatment effects on the circulatory system are detected by the biosensors, which communicate the change to mission control, which then tries to countermand the effect of the drug. This is one possible cause of therapeutic resistance

completely broken and nonfunctional, but if it does retain some capability, it could respond to the effects of drug treatment to try to counteract it. Sometimes dosage adjustment will improve this situation, but not always.

4.1.2   Self-regulation in drug metabolism

However, other endogenous triangular systems operate even more powerfully in the pharmacokinetics of endogenous chemicals. As described in Chapter 2.4, steroidal impact on the reproductive tract, for instance, must be modulated in terms of the synthesis and destruction of large amounts of hormones over time-frames from minutes to years. This requires a open triangular control system and drugs and other xenobiotics, through their structural similarities to endog-enous agents, activate the biosensors, which report to their mission controls. These steps can order potentially a truly vast and sustained increase in the biotransformational capabilities necessary to maintain the reproductive tissues. This process is termed *enzyme induction* and has a far greater impact on drug clearance for example, than local CYP biotransforming capacity regulation mechanisms such as the allosteric and transcriptional effects mentioned in the previous chapter. Biotransforming enzyme induction's scope and power to influence pharmacologically relevant chemical blood levels means that it can eliminate an otherwise fully absorbed drug from its therapeutic window and maintain this position even after several significant increases in the oral dose. The next few case histories illustrate how firstly the use of some 'enzyme inducer' drugs is initially 'sensed', then the information processed, which is followed by a sustained executive order to form large quantities of the appropriate biotransforming systems to control the xenobiotic threat for as long as the drug is present.

History 1

After suffering a head trauma in a motorcycle accident, a 22-year-old male was subject to recurrent grand-mal convulsions that were treated with carbamazepine. After starting on 200 mg daily, this dose had to be gradually increased stepwise over four weeks to maintain plasma levels within the therapeutic window to 1200 mg daily.

Analysis

Plasma levels were not maintained within the therapeutic window at each dose level for more than a week or so, as carbamazepine clearance appeared to gradually increase, until a dosage was reached where clearance stabilized so that drug levels remained within the therapeutic window.

History 2

A 23-year-old male epileptic was prescribed phenytoin (300 mg/day) and carbamazepine (800 mg/day). The laboratory assays showed that phenytoin was in the therapeutic range, while carbamazepine was undetectable in the plasma. A 50% reduction of the phenytoin dosage allowed the carbamazepine plasma concentrations to rise to therapeutically effective levels.

Analysis

The lack of carbamazepine in the plasma at a dosage known to exert a reasonable therapeutic effect in other patients implied that the drug's clearance was much higher than normal, to the point where bioavailability was almost zero. Cutting the phenytoin dosage slowed the high rate of clearance of carbamazepine, allowing plasma levels to ascend to the therapeutic window.

History 3

A 49-year-old male epileptic was prescribed phenytoin at 600 mg/day and carbamazepine at 2000 mg/day. The patient's condition was controlled with minimal side effects for three months. The phenytoin was then abruptly discontinued; within four days, the patient became gradually more lethargic and confused, and one week later hospitalization was necessary. The carbamazepine dosage was reduced to 1200 mg/day and the confusion and sedation gradually disappeared.

Analysis

A stable co-administration of two drugs implies that despite the high dose of carbamazepine, blood levels for both drugs were initially in the therapeutic window. The removal of the phenytoin led to gradual increase in the symptoms of carbamazepine overdose, without any change in the dose. This indicates that carbamazepine blood levels climbed way above the therapeutic window into toxicity. This was caused by a marked, but gradual, fall in carbamazepine clearance when the phenytoin was withdrawn.

History 4

A 64-year-old obese male was prescribed simvastatin 10 mg daily. Over the next three months, lack of clinical response led to a fivefold increase in dosage. He was then admitted to hospital with rhabdomyolysis. On his own initiative he had self-administered St John's Wort, which he discontinued when his mood was sufficiently elevated, around 10 days prior to the toxicity manifesting itself.

Analysis

The statin was not effective unless considerably higher doses than normal were used, indicating that the drug was being cleared at a higher rate than normal. The general practitioner was unaware that the patient was taking St John's Wort extract. The patient abruptly stopped taking the herbal extract and the clearance of the statin gradually fell while the dose did not, so the drug accumulated and exerted toxicity.

History 5

A 47-year-old female was stabilized on phenobarbitone and warfarin and her prothrombin time was optimized by substantial increase over the normal dosage of anticoagulant, although blood levels were within normal limits. Within 10 days of the abrupt withdrawal of phenobarbitone, the patient suffered a mild haemorrhage.

Analysis

As a higher than normal dosage of warfarin was necessary to maintain its plasma levels in the therapeutic window in the presence of the phenobarbitone, it suggests that the latter drug was accelerating the clearance of warfarin. Once the phenobarbitone was stopped, this accelerating effect was lost too, leading to accumulation of the warfarin to the point that blood levels rose above the therapeutic window leading to toxicity, in this case an exaggerated therapeutic effect.

History 6

A 55-year-old male being treated for tuberculosis was taking rifampicin (600 mg daily), isoniazid (400 mg daily), ethambutol (200 mg daily), and pyrazinamide (400 mg daily), was also epileptic and was taking carbamazepine (2000 mg daily). The patient decided to stop all medication over the Christmas period to enter his annual seasonal alcohol binge, where he drank heavily for several days. After approximately 13 days, he resumed his drug regimen. Before the end of the first day, he was drowsy, lethargic, confused, and eventually difficult to rouse and was hospitalized. Some of the symptoms of the tuberculosis resumed, such as fever, chills, and cough.

Analysis

The patient was suffering from carbamazepine toxicity and very high plasma levels were found on blood analysis. This indicates that the cessation of all drug intakes over the Christmas period of 13 days had led to a marked reduction in the clearance of carbamazepine, and the resumption of his previous high dosage caused drug accumulation and significant CNS toxicity. The absence of pressure of the anti-tuberculosis drugs had also allowed the disease to partially reestablish itself and may well have led to selection of partly drug-resistant forms of the bacteria.

History 7

A terminally ill cancer patient stabilized on methadone (250 mg/day) for pain relief contracted an infection soon after another patient arrived from a major hospital. The infection was determined to be MRSA (methicillin-resistant

Staphylococcus aureus) and rifampicin (600 mg/daily) along with vancomycin was started for a projected period of 14 days. At day two, the patient complained of severe breakthrough pain and the methadone dose was increased. By day 14 the infection had been cleared and the patient was taking methadone at 600 mg/day to control her pain. After rifampicin was stopped, the methadone dose was tapered to its original level over 21 days with minimal toxicity or pain breakthrough.

Analysis

The rifampicin accelerated the clearance of methadone to the point that analgesia was lost extremely rapidly and had to be restored with more than a doubling of the dose. The reversal of the induction effect appeared to take somewhat longer than the onset of the initial effect, which may have been related to the highly variable nature of methadone's half-life and the pathology of the terminal patient.

History 8

A psychiatrist prescribed the anti-epileptic drug (AED) tiagabine (16 mg daily) off-label to a patient to stabilize mood and treat severe anxiety. 'Off-label' prescribing involves the use of a drug for a condition that it is not explicitly licensed to treat, on the basis of the judgement of the health-care practitioner. The patient was already prescribed an antipsychotic drug (clozapine 600 mg/day) and an antidepressant (bupropion 200 mg/day). Two weeks later, the patient suffered an epileptic seizure for the first time in his life. Given that tiagabine is an AED, the practitioner increased the dosage to control the seizures, which made them worse. The drug was then withdrawn from the patient and a benzodiazepine substituted.

Analysis

Tiagabine is a CYP3A4 substrate, and it was evaluated during clinical trials as an adjunctive drug to be used with inducing AEDs, hence the dosage was sufficiently high to ensure therapeutic plasma levels in fully induced patients. When the drug was used in a noninduced individual (neither clozapine nor bupropion are inducers) plasma levels were significantly higher than would be experienced in combination with inducing AEDs. Both clozapine and bupropion can lower seizure thresholds and tiagabine can induce seizures in non-epileptics at higher doses or in overdose. Increasing the dose compounded the situation and caused the patient to experience convulsions, which subsided upon treatment and withdrawal of tiagabine.

4.1.3   Self-regulatory responses to drugs: summary

In all the cases, there are a number of common features:

- Some of the drugs' clearances were not stable until a relatively high dose was employed.

- One drug (or herbal preparation) was able to grossly accelerate the clearance of another agent(s).

- The changes in plasma levels were sufficiently great to either lead to toxicity or total loss of efficacy.

- The toxic effects occurred gradually over days, rather than hours.

- The increase in drug clearance caused by other drugs was fully reversible.

4.2   Causes of accelerated clearance

A number of explanations could be put forward for the effects seen above. There could be a reduction in drug absorption in the presence of another agent, possibly linked to inhibition of intestinal drug uptake transporters (Chapter 5.6), although this is relatively rare and would happen within hours rather than days of the regimen change. Perhaps the renal clearance of the drug could be accelerated in some way, although this too is unlikely. To enter the circulation from an oral dose, the drugs must pass through the gut, the portal circulation and then into the liver itself. In Histories 1 and 2, the clearance of carbamazepine was initially unstable and in the presence of phenytoin, virtually 100% cleared before it reached the circulation.

Since the liver's blood flow is not likely to undergo a sustained increase, then the only way such a large effect on drug levels can occur is that the liver is extracting much more of the drug than usual in the presence of the other drug. This acceleration of drug metabolism as a response to the presence of certain drugs is known as 'enzyme induction' and drugs that cause it are often referred to as 'inducers' of drug metabolism. The process can be defined as: *An adaptive increase in the metabolizing capacity of a tissue*; this means that a drug or chemical is capable of inducing an increase in the transcription and translation of specific CYP isoforms, which are often (although not always) the most efficient metabolisers of that chemical. Usually, hepatic induction has the most significant clinical impact, although other tissues are also involved, such as the lung, intestine and the kidneys.

4.3   Enzyme induction

4.3.1   Types of inducers

There are several broad groups of drugs and chemicals capable of inducing hepatic metabolism; these include:

- *Anticonvulsants*: the older established drugs, such as phenytoin, carbamazepine, and phenobarbitone are the most potent inducers, most strongly with CYP3A4/5 and CYP2B6, also to a lesser extent CYP2C9, CYP1A2, and CYP2A6. Topiramate and oxcarbazepine are weaker inducers and show their effects at high doses. Other weaker inducers include tiagabine, felbamate, eslicarbazepine, clobazam, and valproate[1], as well as vigabatrin.

- *Steroids*, such as dexamethasone, prednisolone, and various glucocorticoids, induce CYP3A4/5[2].

- *Anti-androgens:* These drugs are used to block androgen receptors and arrest the growth of prostate tumours. Whilst bicalutamide, which first appeared in the 1980s, is not a human enzyme inducer[3], enzalutamide, apalutamide and ONC1-13B are CYP3A4 (and CYP2C9, CYP2C19) inducers[4,5]. Enzalutamide is by far the most potent and is comparable to rifampicin[4], also inducing CYPs 1A1 and 1A2 and some glucuronidating isoforms[6] (Chapter 6.2).

- *Antibacterials/antifungals*, such as the rifamycins (rifampicin, rifabutin, rifapentin), induce most CYPs including CYP3A4, CYP1A2, CYP2C9, CYP2C19, and CYP2B6. Rifampicin has been the clinical 'benchmark' CYP inducer, against which all other inducers are compared[7]. Interestingly, rifapentine possesses about 85% and rifabutin about 40% of rifampicin's induction effect in man[8]. Rifabutin is preferred where rifampicin would unmanageably disrupt a polytherapeutic situation, such as in HIV regimes[9]. The somewhat-toxic antifungal griseofulvin is also a CYP3A4 inducer[10] and the CYP inhibitor clotrimazole is actually an inducer of CYP3A4, although this is not clinically relevant due to very low absorption from an oral dose and the usual application is topical, with virtually zero systemic absorption. The cloxacillins are also CYP inducers; dicloxacillin, albeit at a relatively high dosage (3 g daily), is a CYP3A4, CYP2C19, and CYP2C9 inducer[11].

- *Anti-virals:* the anti-HIV protease inhibitors such as ritonavir and nelfinavir are potent inducers of CYP3A4, although paradoxically, ritonavir in particular is a powerful inhibitor of this CYP. These drugs also induce

CYPB6, CYP2C8, and CYP2C9[12,13] The non-nucleotide reverse transcriptase inhibitors nevirapine and efavirenz induce CYP3A4 and CYP2B6 in a mild to moderate fashion, whilst the CYP3A4 substrate elvitegravir has a mild inductive effect on CYP2C9[9]. Other antivirals such as combinations used in hepatitis C therapy have shown mild CYP3A4 induction properties[13].

- *Herbal remedies:* St John's Wort is the most clinically relevant and investigated herbal inducer (CYP3A4/5)[14], but many more herbal preparations are being explored and marketed in a very determined manner and their CYP inducing effects remain to be explored[15].

- *Polycyclic aromatic hydrocarbons (PAHs):* these are found in atmospheric pollution, cigarette smoke, industrial solvents and barbecued meat. They also contaminate foodstuffs and watercourses (particularly dioxins and polycyclic chlorinated biphenyls). As mentioned in Chapter 3.6.1, these compounds induce the normally nonconstitutive CYP1A1 in the liver, CYP1A2, which specializes in polycyclic aromatic amines and CYP1B1. Induction of CYP1B1 is also very strong in the lungs in smokers and is a standard marker for heavy tobacco use[16].

- *Recreational agents:* smoking tobacco and related products yields a considerable dose of PAHs that are of course potent inducers of CYP1A2, CYP1A1, and CYP1B1[15-18], actually causing significant reductions in the plasma levels of several antidepressants for example[19]. Clozapine and warfarin clearance may also be accelerated. Heavy alcohol consumption will induce CYP2E1[18], which is relevant to chlorzoxazone clearance.

- *Miscellaneous inducers:* the dopamine agonist and so-called cognitive enhancer modafinil, used by narcoleptics, some helicopter pilots, and various students, is a mild CYP3A4, CYP2B6 and CYP1A2 inducer (greater than 400 mg/day), but inhibits CYP2C19[20,21]. The antiulcer drugs omeprazole and lansoprazole are inducers of CYP1A1 and CYP1A2, despite being substrates of CYP2C19[22]. The antinausea agent aprepitant is also a mild inducer and inhibitor of CYP3A4. Cruciferous vegetables have complex effects on drug clearances, but they can impact warfarin metabolism and efficacy at around 400g/day intake, most likely through CYP1A2 induction[23]. Mitotane is a relative of the organochlorine DDT, but is used to treat adrenocortical tumours; its induction of CYP3A4 is comparable to that of rifampicin[24].

- *Pregnancy:* the hundred-fold increase in oestrogens[1,25] appears to exert potent inductive effects on several drugs, including CYP3A4, CYP2B6, and CYP2C9 substrates[26]. Even CYP2D6 expression increases, probably through a post-transcriptional modulation[27].

4.3.2 Common features of inducers and clinical significance

A new drug is generally regarded as an inducer if it produces a change in drug clearance that is equal to or greater than 40% of an established potent inducer, usually taken as rifampicin. Looking at the structures of the strongest hepatic enzyme inducers, there are apparently few common features. These chemicals range in size from very small and water-soluble (ethanol) to very large and lipophilic (PAHs, rifampicin-related agents). Inducers are often not even metabolised to an appreciable extent by the CYP or biotransforming system they induce; rifampicin, for example is deacetylated. However, inducers are usually (but not always) lipophilic, contain aromatic groups and consequently, if they were not oxidized and cleared, they would be very persistent in living systems. CYP enzymes have evolved to oxidize this very type of agent and the CYP induction system organises the biotransforming power of the CYPs, other enzymes and efflux pump systems to control these agents, not just from a cellular standpoint, but systemically. At first sight, it might seem rather simplistic to regard xenobiotic chemicals that act as inducers as 'threat molecules' just because they are detected and stimulate such a powerful response. From a physicochemical standpoint, many exogenous inducers resemble key endogenous agents rather closely, so they are caught in the same physiological 'control net'. Indeed, the impact of the state of pregnancy on CYP expression demonstrates how one of the main drivers of the system is hormone management, which then impacts every other chemical subject to clearance by the induced enzymes.

However, with regard to xenobiotics, as we will see, the receptors that operate the various induction systems are often described in the scientific literature rather fruitily as 'promiscuous' as they bind so many different structures productively, that is, the binding process activates gene expression. Hence, evolutionary pressure has probably widened the 'response window' for different structures and compounds with certain physicochemical properties that favour their accumulation in particular subcellular areas, where are likely to influence endogenous functions (Chapter 2.4.1). Overall, it seems that exogenous inducers are molecular entities over which living systems vigorously exert control over through evolutionary necessity.

Indeed, the process of induction appears to have a massive dynamic range of structure detection and matched response, alongside very close control of that response. In fact, as we will see through this chapter, biotransformative enzymes are not the only systems induced. The extent and time frame of this adaptive biotransformation increase is linked more than just with the structure of the chemical. It is likely also to be matched to residence time in the cytoplasm and nucleus; for instance, ethanol is hydrophilic and distributes in total body water, moving through membranes and causing significant disruption. Logically, ethanol is a significant threat, as it can be itself a metabolic byproduct and it is structurally related to many similar molecules that also require constant modulation. On the opposite scale, several persistent lipophilic agents show the limits of the induction process. These chemicals are toxic, yet either because they cannot be cleared by CYPs due to structural issues or they are so lipophilic they are locked within membranes and do not even reach the various induction biosensors, they effectively defeat biotransformation and clearance to our long-term detriment,[28,29] as outlined in Chapter 2.4.3. Overall, clinical experience with inducers is highly variable, due to the innate variability of human drug metabolism, but some general patterns of induction have emerged.

Potent inducers: Examples are rifampicin, enzalutamide, phenytoin, carbamazepine, phenobarbitone, nevirapine, and St John's Wort. These agents stimulate a maximal response in around three weeks at clinically relevant doses, with a similar 'wear off' or de-induction period. This de-induction period is linked to half-life, with very persistent agents wearing off the most gradually.[24,30] CYP3A4 and CYP2B6 are usually induced to a greater extent than the CYP2C family. CYP1A2 induction is believed to be midway between the two groups.[1] These drugs induce all patients, and it is just a matter of how much. It is difficult to be precise, but the addition of a potent inducer to a drug regimen of CYP3A4 substrates will at the very minimum require a doubling, tripling, or even as much as a five (or more)-fold increase in dose to remain in the therapeutic window[1,31]. The effects of potent inducers can be strong enough to eventually (but not acutely) override other inhibitory effects (see Section 4.7). The US Food and Drug Administration defines a strong CYP inducer as one that decreases the AUC of a given substrate of that isoform by more than 80%.[32]

Mild inducers: Examples include the anti-HIV agents etravirine and efavirenz, the anticonvulsants eslicarbazepine and clobazam. These drugs may take much longer, perhaps months, to cause significant changes in the clearance of CYP3A4 and the CYP2C series. In comparison with strong inducers, the impact on clearance will be significantly less, although dosage adjustment will still be necessary[33]. The inductive effect is more variable and some patients may show little if any changes to clearance[1]. A mild CYP inducer is considered to impact the AUC of a substrate by 20–50%[32]. Overall, it is important to remember that inducers exert their maximal effect on orally dosed drugs, as we covered earlier, the effect is exerted on the gut as well as the liver. Indeed, with some agents such as St John's Wort, it is thought that the major inductive effect is predominantly in the gut[34] and CYP3A4 comprises around 80% of total gut CYP.[35]

4.4   Mechanisms of enzyme induction

4.4.1   Introduction

The process by which enzyme induction occurs has three main requirements:

1. The biotransforming cell must use various biosensors to detect the presence and concentrations of endogenous molecules to exert homeostatic control over their concentrations, in terms of synthesis and elimination.

2. The biosensor information is used by 'mission control' cellular DNA in the liver, gut, and lungs to command a coordinated increase in the capability of the appropriate metabolic systems within the cell, which will clear the endogenous agent/drug/toxin as efficiently as possible towards a set 'target' concentration.

3. The complete (detection and action) system should be dynamic and reversible, so it is sensitive to all potentially required tasks in controlling target molecules.

It is apparent that the main inducible CYPs, CYP1A1/1A2, CYP2C8/9, and CYP3A4, employ broadly similar systems to regulate their ability to respond to increases in drug concentration. Indeed, this commonality is borne out by observations that agents that don't induce CYP3A4, don't induce CYPs 2C8/9 and 2C19, either. The exceptions to this rule seems to be CYP2E1, which appears to have a unique system of induction and CYP2D6, which is not 'conventionally' inducible, but is nonetheless regulated in expression.

On a cellular level, we know that the type of induction mechanism involved with a given CYP is closely related to a combination of endogenous as well as xenobiotic-responsive functions. Indeed, the networks of various nuclear and cytoplasmic receptor systems described below can 'cross-talk', in that different receptors may modulate each other and even activate the same gene. This multiple means of control on the same system approach is used in aircraft to maximize safety and in the cell to reduce the possibility of potentially disruptive chemicals remaining in the body unchallenged. Indeed, the various receptor systems that govern CYP expression act in concert with other aspects of biotransformation relevant to the inducing chemical. These include stimulating the proliferation of the SER to provide space for the CYPs to be anchored, as well as the production of sufficient quantities of REDOX partners to fuel the CYPs. The expression of conjugation and Phase III transporters is also upregulated (Chapter 6.11). This all translates into relatively rapid and profound effects on drug clearance, which are often difficult to manage clinically.[1]

4.4.2   CYPs 1A1/1A2 and 1B1 induction

The AhR system-basic operation

Although enzyme induction has been known clinically since the 1960s, it was not until the 1970s that the cellular basis of the process was unravelled by studying the effects of dioxin (TCDD; Chapter 2.2). In the cytoplasm of most cells a dormant receptor complex can be found which consists of five components (Figure 4.3). These are the aryl hydrocarbon receptor, or 'AhR', two heat-shock protein molecules (Hsp90), co-chaperone p23 and an immunophilin called

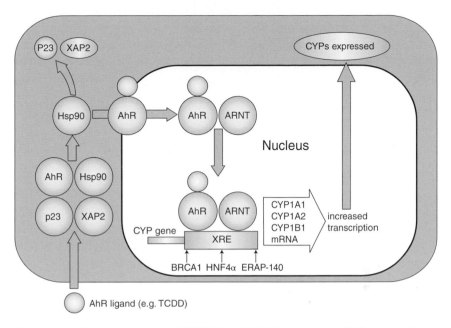

Figure 4.3 Basic mechanism of CYP1A1 and 1A2 induction: the AhR receptor binds the inducer alongside Hsp90, but only the AhR receptor and the inducer cross into the nucleus to meet ARNT and together they bind to DNA xenobiotic response elements (XRE). Co-activators that induce expression of the CYP isoforms include BRCA1, ERAP-140 and HNF4α

XAP2.[36,37] Many receptors are complexed with several chaperone-like molecules mainly to ensure they keep a certain shape so they can bind their substrate, rather like polystyrene packing is used to help fragile articles survive postal systems. The AhR itself is a ligand-activated cytosolic receptor, or biosensor if you like, which was at first thought to control the expression of some biotransforming enzymes along with a few CYPs. However, due to an explosion of recent discoveries, it is now actually rather hard to find any cellular function it does not influence. This is covered in more detail in the next section.

For now, we will just follow the AhR system as it responds to the archetypal toxic polycyclic aromatic TCDD, which binds to the previously dormant AhR/ chaperone complex, which then activates and migrates towards the nucleus. The AhR bound to TCDD then breaks away and it alone enters the nucleus and heterodimerizes (two different proteins form a complex) with the nuclear protein ARNT (aryl hydrocarbon nuclear receptor translocator), and this new complex then binds to eight specific DNA nucleotide sequences upstream of several biotransforming genes, including CYP1A1/1A2 and CYP1B1, a number of Glutathione-S-transferases, catechol-O-methyl transferase (COMT), an aldehyde reductase (ALDH3A1), and some UGT glucuronyl transferases[36,38] (Chapter 6.2).

The sequences ahead of these genes are termed xenobiotic-responsive elements (XRE), or sometimes DREs (drug or dioxin-responsive elements). There is also a competing 'off switch' called the Aryl hydrocarbon receptor repressor (AhRR), that sits in the nucleus already locked as a protein dimer with ARNT. The AhRR-ARNT complex can bind to the XRE nonproductively and thus shut down CYP gene expression[36.] Interestingly, the process of migration of the many AhR/ligand/ARNT complexes through the nucleus with resultant XRE binding not only induces the expression of AhR and its associated proteins[37] to increase system capacity for ligand detection, but also increases AhRR expression[39]. This suggests that a binding ligand accelerates the system's capability and shortly after 'applies the brakes' automatically to some extent. AhR, ARNT, and AhRR are all part of a family of transcription factors known as bHLH (basic-helix-loop-helix) proteins, whose structure allows recognition and binding to the DNA response elements of multiple genes[36.] Several other nuclear proteins are also required before CYP RNA-polymerase-mediated transcription, and then ribosomal translation is fully engaged for protein manufacture, such as the transcription regulator p300[36.] Some idea of the sensitivity and potency of this system comes from observations in human cell culture, where the presence of just 10nM of TCDD can increase CYP1A1 expression 425-fold[40.] What is also emerging, which is particularly worrying in terms of environmental pollution is that TCDD activation of AhR can also repress the expression of other CYPs in human systems, such as CYP2E1,[41] which suggests that the most powerful AhR agonists have the potential to seriously distort biotransformational enzyme expression.

The AhR system: effects of drugs

If we employed a fairly simplistic analogy, we could say that at rest, the AhR-controlled biotransforming enzyme expression systems seem to operate in an analogous way to an engine controlled by a throttle. A set 'idle speed' of expression is increased when the throttle (AhR receptor cytoplasmic complex) is 'pressed' by the sudden and perhaps sustained cellular influx of receptor agonists. The increase in biotransforming enzyme synthesis is intended to clear the offending chemical presence from the cell. Indeed, as mentioned above, the throttle itself increases in size to accommodate the demand. Release the 'throttle' and the whole system subsides to a resting level, saving energy and raw materials. Indeed, there even seems to be a 'rev-limiter' in terms of AhRR expression.

However, the finding that many other molecules as well as some prescription drugs also acted as agonists on the AhR complex, it was gradually understood that the system was much more nuanced than was at first realised. Indeed, even though TCDD was the first agent to be intensively studied in its AhR binding, it became clear that this significant toxin was not representative of how most endogenous and exogenous ligands activated this receptor system. Several well-known drugs, such as omeprazole and the anti-Parkinson's disease agent carbidopa also bind AhR and can moderately induce CYPs 1A1/2 and CYP1B1, but these drugs do not cause the carcinogenic and immunotoxic effects of TCDD[39.] Indeed, AhR agonists such as omeprazole and carbidopa are known as selective Ah receptor modulators (SAhRMs) and these agents have much lower affinities and thus bind much less tightly to the AhR than the most potent PAH-type inducer chemicals. AhR ligand binding is complex; the azole CYP3A inhibitor ketoconazole and its relative itraconazole are both weak AhR ligands, but the (+)-cis isomer of ketoconazole is 40-fold more potent than the (-)-cis isomer of the drug[42]. This essentially means that there is a spectrum of ligand-activated AhR-related effects and sensitivity on biotransforming gene expression and that there are toxic and nontoxic binding ligands for this receptor.[43,44]

The AhR system: role in carcinogenesis

Aside from its control of several biotransformational systems as described above, AhR also modulates a wide range of housekeeping genes that regulate the cell cycle and cellular activities such as proliferation and differentiation, from embryogenesis onwards all through our lives. The receptor system also has a powerful influence on the immune system and on cell death, particularly with regard to inhibition of apoptosis in tumour cells.[45] In fact, I hope you can see the picture forming already, that the AhR system is crucial to the whole carcinogenicity process, from initiation to likelihood of death.

For instance, the presence of AhR inducers such as TCDD from the environment, or complex mixtures of PAHs from smoking for instance, lead as we have seen to the induction of CYP1A1, CYP1A2, CYP1B1 and COMT. As you might recall from Chapters 2 and 3, these CYPs are strongly implicated in the metabolic activation of the inducer PAHs and also of exogenous and endogenous oestrogens. If we simply consider oestrogen metabolism, CYP1A1 promotes 2-hydroxylation that is essentially detoxification in conjunction with COMT, leading to a downregulation of breast cancer cells; however, CYP1B1/COMT induction promotes 4-hydroxylation of oestrogens leading to genotoxicity[40]. The balance between these two processes depends on which CYP combination is preferentially induced according to a variety of factors, not least the type of AhR ligand[36]. Indeed, foodstuffs and some dietary supplements such as red clover, through their mix of natural AhR-binding ligands, might tip the balance towards 4-hydroxylation and genotoxicity[40]. Whilst a toxic AhR ligand such as TCDD's effects are complex[44], it is strongly associated with breast cancer[39]. So if a tumour was to develop, AhR can then be one of the most powerful drivers of its growth, through upregulating the receptor complex itself,[45] which then provides the opportunity for various AhR agonists found in our diets to inhibit apoptosis and disable the brakes, by downregulating AHRR.[37]

However, studies with the SAhRMs have indicated that AhR could be a thera-peutic target to arrest most or even all these negative stages in the carcinogenesis process. Carbidopa, for example, in animal models, induces the same CYPs as TCDD, but carbidopa can switch off genes involved with tumour invasion. In addition, a number of drugs have been shown in cell systems to antagonize AhR activity such as the anti-arrhythmic mexiletine and to a lesser extent, the antial-lergic tranilast and the 4-hydroxy derivative of the antioestrogen tamoxifen[43,44]. Hence, there may be prospects in developing chemotherapeutic agents that can prevent the AhR system's worst excesses, despite our seeming inability to stem the flow of potent PAH AhR ligands into our environment.

4.4.3   CYP 2B6 2C8/2C9/C19 and 3A4 induction

The Nuclear Receptor System

The liver's very identity, structure, and function is regulated through a network of nuclear receptors (NRs) which like the AhR, are biosensors. These NRs reside in the nucleus in humans and bind endogenous and exogenous chemicals, in order to report their presence to DNA response elements, which will issue the appropriate protein response to regulate that chemical. The NRs themselves are commanded by a small group of master regulating nuclear receptors, known as the hepatocyte nuclear factor 4α's (HNF4α's), which exist in at least six adult forms, including HNF4α1 and HNF4α2[46]. These master regulators are not fully understood, but they are continuously active and control so many processes, (organ development as well as lipid, insulin and bile acid metabolism) that they are the key to normal liver function, although they operate in other tissues. The HNF4α group is so important that gene knockout studies have shown that animal embryos do not survive without it. The HNF4α's receive input from a large num-ber of 'executive' transcription factors that control major aspects of cell func-tion, such as the proto-oncogene C-Myc, the tumour suppressor gene p53 and immune system cytokines[46]. Sustained downregulation of HNF4α's can occur also in response to inflammation and other factors and this leads to major hepatic malfunction, ranging from cancer and cirrhosis to steatohepatitis. Upregulation promotes liver function and inhibits tumour growth, to the point that HNF4α's are now considered therapeutic targets in this area.[46] The HNF4α series directly or indirectly regulates all hepatic endobiotic and xenobiotic biotransformational processes through the next layer of command, the NRs.

Forty-eight major NRs have been discovered so far and they are fairly similar in structure, containing an N-terminal DNA binding domain (DBD) and a C-terminal hormone/chemical-binding domain, separated by a hinge section[15,47]. When the hormone or xenobiotic is absent, the C-terminal domain, usually called the ligand-binding domain (LBD), is locked in the 'off' position by a co-repressor protein complex. The binding of the appropriate ligand to the LBD causes it to rearrange and release its co-repressor. The receptor then attracts and binds a co-activator complex. There are several co-activators, part of a series of proteins known as p160s, such as SRC-1 (steroid receptor co-activator 1) and the splendidly named GRIP1 (glucocortoid receptor interacting protein 1). Once the activated complex is formed, it seeks to bind a specific DNA hormone response element (HRE), also termed XRE, or xenobiotic response element.[48] The HNF4α master regulating factor is more sophisticated in structure, with two N-terminal activation domains and the ability to silence as many genes that it can activate, as is necessary during embryogenesis. The HNF4α's also bind to DNA at Direct Repeat (DR) sites near gene promotors as a homodimer, that is, two identical HNF4α's bind, whereas the other NRs complex with RXR (the retinoic acid receptor) to bind their DNA response elements.[46]

Among all the various nuclear receptors, there is further subdivision, in that some NR receptors such as the Oestrogen Receptor (written using USA spelling as ER) thyroid and vitamin D receptors (TR & VDR), CAR (constitutive androstane receptor) and PXR (pregnane X-receptor) form complexes with RXR (the retinoic acid receptor) in order to bind HREs/XREs. The presence of HNF4α is then required to make all these binding processes productive and activate gene transcription. Of the NRs, CAR and PXR are the focus of particular attention, as they control the expression of CYPs 2B6 2C8/2C9/C19 and 3A4, which are of greatest relevance to biotransformation, as well as other systems (Figures 4.3 and 4.4a & b). The NRs are not exclusively linked with different CYPs, or even

(a)

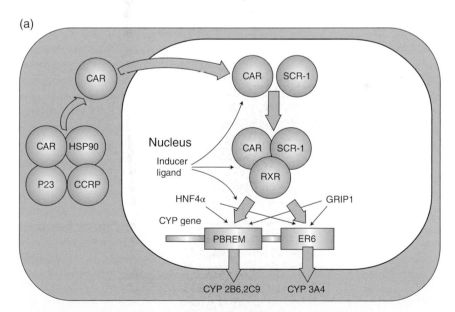

Figure 4.4 (a) Constitutive androstane receptor (CAR)-mediated control of CYP2 series and CYP3A4. CAR and SCR-1 bind the inducer ligand inside the nucleus, bind retinoic acid X receptor (RXR) and activate the CYP expression; (b) Possible mechanism for the modulation of CAR-ligand activated CYP induction: a series of endogenous deactivators cause break up of the CAR/RXR/ SCR-1 /ligand complex and induction is switched off

(b)

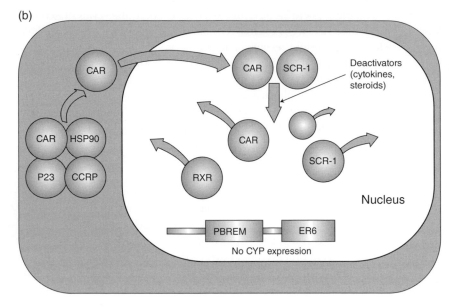

Figure 4.4 (*Continued*)

groups of CYPs. CYP expression is usually controlled by several NRs singly and in combinations and the NRs also interact with each other

CAR mediated control of CYP expression

CAR is also known under nuclear receptor classification as NR1I3 and it modulates basal biotransformational metabolism, as well as that of bilirubin, steroids, bile acids and cholesterol, alongside many aspects of energy metabolism. CAR is also under the direct control of HNF4α and is influenced by other NRs such as glucocorticoid receptor.[49,50] In common with the AhR receptor, it too is found in the cytoplasm, complexed with Hsp90, p23, an immunophilin and a protein called CCRP (CAR retention protein; Figure 4.4a). However, there is a vital difference between the AhR and CAR.

To follow the idling engine analogy used above of AhR, CAR operates like the throttle is held half way down by itself, running the engine at a decent number of revolutions. CAR does not need to be stimulated into action by agonists; it is already, as its name suggests, a 'constitutive' receptor, driving the expression and activity of CYPs and conjugation systems (Chapter 6.2) as well as transporter systems such as the OATPs and NTCP[50] (Chapter 2.6.3). The presence of inducers, such as steroid hormones and drugs (phenobarbitone, primidone, and phenytoin) merely speeds up the process, forcing the 'throttle' towards the floor, although other factors (see below) shut off CAR, sometimes as far back as 'idle speed'.

More than two decades of research have not revealed all the precise details of the cellular operations of CAR, but it is an unusual NR. Before it is activated, CAR is held in the cytoplasm locked in its complex with Hsp90 and CCRP plus the other factors, through phosphorylation at a threonine residue 38 (Figure 4.4). All CAR activators appear to trigger dephosphorylation at the thr-38 causing progression to the nucleus along with recruitment of co-activators that bind to CAR's LBD, which prepare it to interact with DNA[51,52]. CAR activation can occur in response to direct interaction, such as with a chemical known as CITCO, or indirectly with drugs such as phenobarbitone.[51] There must be an endogenous series of 'switch' molecules that untether and activate CAR either directly or indirectly, but to date they are unknown. Interestingly, even powerful inducers such as phenobarbitone do not actually bind CAR at all. It is now believed that phenobarbitone activates CAR by at least two ways; it speeds up the destruction of CCRP (unwrapping CAR) and binds the epidermal growth factor receptor (EGFR). Epidermal growth factor (EGF) itself normally acts on its receptor to keep CAR locked in its phosphorylated cytoplasm mode. Phenobarbitone essentially removes CAR's 'off' switch by competing with EGF for its receptor. Whilst other mechanisms are involved, drugs can modulate CAR migration to the nucleus where it meets its partner RXR, as well as its co-activator (SRC-1, or steroid receptor co-activator-1).

The CAR/RXR complex is a heterodimer (like AhR and ARNT), and basic CAR function comprises the continuous recruitment of RXR by CAR followed by association with SRC-1 into a complex. The CAR/RXR/SRC-1 complex binds to DNA (Figure 4.4). at the PBREM (phenobarbitone-responsive enhancer module) in the CYP2B gene (particularly CYP2B6) and to the Everted Repeat 6 (ER6) element of the CYP3A4 and CYP2C9 genes. HNF4α binding to the CYP promoter sites then ensures protein synthesis will occur. Other co-activators like GRIP1 are also involved. This is the 'half-way down throttle' stage. If inducers of CAR appear, they increase the stability of the rather 'wobbly' CAR/RXR complex making it more rigid and thus promoting a better fit with SRC-1 followed by more productive PBREM binding, pushing the throttle towards the floor-proportionally to the quantity and potency of the inducer. The more the chemical stabilizes CAR throughout its recruitment and binding processing of its co-activators, the more potent the inducer. It some ways, the system is rather like gradually engaging the clutch in a car with manual transmission. There are several mechanisms for restricting or even switching off CAR operation, and some steroids such as progesterone and various androgens inhibit CAR, so 'lifting the foot off the throttle (Figure 4.4b)[51,52]. The antidiabetic drug metformin can accentuate CAR phosphorylation, thus trapping it in the cytoplasm and this effect is powerful enough to prevent inducers from activating CAR.[51]

Variation in CAR expression is one of the main reasons why there are so many interindividual differences in biotransformation capability. Overall, CAR also regulates CYP2C8, CYP2C9, and CYP2C19 as well as CYP2A6 and UGT1A1

(Chapter 6.2.3), as well as various transporters. CAR and HNF4α control POR (Chapter 3.4.8), one of the CYP 'fuel pumps' and probably cytochrome b_5 also (Chapter 3.4.9).

PXR mediated control of CYP expression

The main NR usually associated with the control of the CYP3A series is PXR (Figure 4.5), although as mentioned above CAR is also heavily involved. PXR is also known as the SXR (steroid and xenobiotic receptor) and is classified in the nuclear receptor family as NR1I2. So far, four variant isoforms of PXR protein (between 322 and 473 amino acids in length) have been studied, although there are at least five more[53]. PXR1 and PXR2 control the expression of a portfolio of different biotransformational enzymes[53] aside from the CYP3A series, including CYP2B6 and the CYP2C isoforms, plus conjugative enzymes (Chapter 6.2) such as glucuronyl transferases and sulphotransferases, glutathione S-transferases and many drug transporter systems such as P glycoprotein, the various multidrug resistance proteins (MRP; Chapter 6.11.2) and OATP2 (Chapter 5.6.1)[14]. PXR1

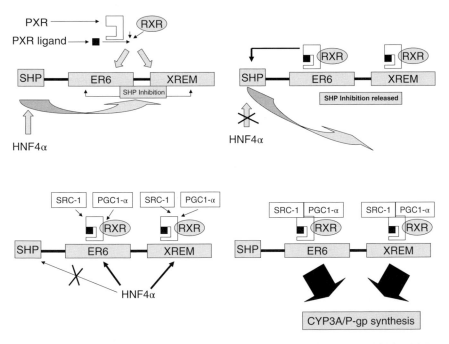

Figure 4.5   Mechanism of CYP3A induction through the pregnane X receptor (PXR) and retinoic acid receptors (RXR). HNF4α causes SHP to lock the system. The PXR/inducer ligand binding shuts off SHP and allows HNF4α to promote the binding of the co-activators, such as SRC-1 and PGC-1, so triggering full induction

is the protein expressed by the majority of us, whilst PXR2 varies from 15–60% of expression in different populations[53]. To date, it appears that PXR1 and 2 differ mainly in relation to how they are influenced by other activators and repressors, but broadly have the same activity.[53] By contrast, PXR 3 does not have a full LBD, binding DNA response elements without leading to protein expression. PXR4 seems to possess a suppressive role on many PXR functions, although full understanding of PXR activity in general remains to be understood. Henceforth, where PXR is referred to, it means PXR1/ PXR2. Although the PXRs are nuclear receptors, some aspects of their behaviour resemble AhR rather than CAR.

As with AhR and CAR, PXR appears to lie dormant in the cytoplasm of the liver, kidney gut and other tissues bound to its chaperone proteins (Hsp-90 etc.) in the absence of any binding hormones, toxins or drugs. PXR is also kept locked up by SMRT, or silencing mediator for retinoid and thyroid receptors.[48] Unlike CAR, PXR has a large LBD, or binding site, noted previously as being 'highly promiscuous', in that it will bind a very wide range of chemical structures of all shapes and sizes. Indeed, PXR's LBD has been crystallized and it is not only quite large, (1150Å^3) it can, like CYP3A4 and other CYPs, become even larger, expanding to 1540Å^{354} that you might recall from Chapter 3.6 is larger than most CYP active sites and perhaps not entirely unsurprisingly similar to the site volume of CYP3A4. PXR's LBD, like the Vitamin D receptor, possesses mostly hydrophobic, with a small number of polar amino acids, which reflects the type of chemical it has evolved to bind[48,55]. It is clear that PXR-ligand binding is much more complex than was first realized, as the LBD changes its conformation and properties in response to different chemical species, so providing a massive degree of flexibility within the binding parameters it has evolved so far[48]. Such 'intelligent' binding allows PXR to detect a vast range of possible 'threat molecules' aside from its normal endogenous trigger ligands (Figure 4.5).

Indeed, the rifamycin family of drugs illustrates both the precision and the apparently random nature of PXR ligand binding. Rifampicin, rifapentine, and rifabutin are structural analogues of each other with similar molecular weights (ranges 822–877). They differ in the side chains attached to their tricyclic naphthalene cores.[56] Rifapentine is the heaviest and the most lipophilic, with a cyclopentyl group attached to its piperazine ring, the lightest is rifampicin. All three drugs induce CYP3A and CYP2B isoforms, yet only rifampicin achieves this through PXR ligand binding.[57] Although rifampicin is the most potent inducer, it is not cleared by CYPs at all, but by deacetylation[58] to a metabolite, which rather disconcertingly colours all your secretions orange. Rifapentine is hydrolysed and deacetylated[8] and only rifabutin is actually cleared by the CYP3A isoforms. Rifabutin's clinical half-life can be up to two days, rifapentine's is around 14 h, with rifampicin around 2–3 h.[8]

Nevertheless, there is remarkable variation in the structures and sizes of the many clinical inducers of the CYP3A series. Aside from rifampicin, several steroids as well as imidazole antifungals like clotrimazole, are inducers, plus barbiturates and some organophosphate pesticides[1]. It is important to note that the induction process can be specific to species, so animal studies are not always helpful in assessing the possible human enzyme-inducing properties of a novel chemical agent. As well as with man, rifampicin is a potent inducer of pig and rabbit 3A enzymes, but it is without effect in the rat.[59]

In terms of day-to-day operations, PXR function is a multi-stage almost simultaneous process, rather like moving off in a car (Figure 4.5). You start with the car parked with the brakes engaged, then you engage the transmission, release the brake and then press the throttle and go, more or less at the same time. First, the PXR 'transmission' must be engaged: after the PXR binds a ligand, there are, depending on the ligand bound, changes in the LBD, which recruit co-activators that release one of the 'brakes' (SMRT) and help progress the PXR/ligand complex to the nucleus. Once there, RXR is recruited forming PXR-RXR heterodimers. The whole complex migrates towards the DNA and the presence of the various co-activators (probably SRC-1 and others) rearranges the chromatin to allow access to the DNA response elements in the promotor of the target gene. Binding occurs at the CYP3A gene in two separate areas. One binding site is an ER6 in the proximal promoter of the gene and at the same time, PXR/RXR complexes also bind to another ER6 and a DR3 (Direct Repeat 3) in a second area called the XREM, or xenobiotic responsive enhancer module. The system requires both the proximal promoter and the XREM to be bound by PXR/RXR heterodimers before induction can proceed.[48,55] The 'transmission' is engaged, but there is another brake to release before the 'throttle' (gene activation) can be pressed.

Normally, CYP3A transcription is kept locked up by the presence of SHP, which performs the same role in the AhR/CYP1A/1B induction story. The expression of SHP is controlled by none other than HNF4α, and SHP's 'brake' role is to prevent HNF4α from binding to the PXR/RXR bound promoter and XREM complexes, which, in turn, prevents the recruitment of the co-activators SRC-1 and PGC-1α. The presence of the inducer/PXR complex is thought to release the brake effectively.

The system will not fully launch transcription and translation unless HNF4α binds the PXR/RXR complexes already bound to both the promoter and the XREM. As SHP is no longer present, HNF4α binding then promotes recruitment of at least two co-activators, SRC-1 (also used by CAR) and PGC-1 α (peroxisome proliferator-activated receptor co-activator 1).[48,55] The throttle is being pressed, the brakes are off and we are in gear. This process is a simplified version of what is likely to happen in man, but the processes involved with PXR and indeed many NR's activities and modulation are not fully understood.

PXR system modulation

How rapidly the PXR system 'accelerates', that is, how potent an effect a particular inducer exerts on CYP expression, seems to depend on several factors. As the process outlined above has so many steps, you would expect inducers to either compete for binding to PXR, or affect the co-repressor/activator recruitment steps or all of the above. This is what seems to happen and given that the whole process is controlled by HNF4α it is clear there is almost limitless capacity for variation in terms of the basic pre-set responsiveness of the system as well as its susceptibility to different inducers and groups of inducers. There are also many layers of control of PXR and other NRs exerted through phosphorylation at specific sites on the receptor, which mostly downregulate activity, particularly related to the effect of the immune system, which can suppress PXR[48] and shut off CYP expression[60] (see next section). It has been found in some ethnic groups, but not others, that microRNA's such as miR-148a could downregulate PXR activity and PXR mRNA does not always correspond to the amount of PXR protein formed, which suggests that PXR is post-translationally regulated. However, this is not established as yet due to the complexity of human PXR activity.[55]

   Clinical induction is extremely variable in terms of patient responses in terms of timeframe and degree[1]. Some drugs have contradictory effects in terms of their induction and inhibitory effects, sometimes impacting both processes simultaneously. This has been investigated with equally contradictory results. For instance, although azoles are inhibitors of CYPs, experimental studies have shown they can nevertheless modulate CYP expression through the NR system. Clotrimazole binds to PXR and promotes SRC-1 recruitment, whilst ketoconazole, known for its potent clinical CYP inhibition, was shown in various experimental models to partly block CYP3A induction through SRC-1/PXR binding disruption[61]. However, in humans, even at the highest safe doses, ketoconazole does not have any significant impact on PXR-mediated induction[62]. This outcome highlights two issues; firstly, whilst the experimental results at the *in vitro* concentration ranges were interesting and very useful as an experimental tool,[57] it was not possible to even approach those concentrations *in vivo* through safe dosage of ketoconazole, a drug that has hepatotoxic potential. Second, the *in vitro* results suggest that perhaps there are endogenous SRC-1 inhibitors, although they are likely to be far more selective and potent than the azoles. In the long term, there may be potential in the manipulation of PXR therapeutically as has been suggested[61] and as we shall see in the next section, PXR can also cause significant therapeutic problems in drug usage. Unfortunately, the complexity and scale of PXR's other functions underlines the degree of difficulty involved in PXR modulation.

PXR system: wider therapeutic relevance

So far, we have focussed on how PXR-ligand binding stimulates large increases in body biotransformation capability, which leads to accelerated drug clearance, reduction in half-life, and a shortening in drug body residence time. Hence, as the PXR response causes drug concentrations to fall out of the therapeutic window, efficacy will be compromised. Naturally, any medical practitioner's response would be to increase the dose to the point that the concentrations are restored to their appropriate level and the issue is seen to be dealt with. The increase in dose will counteract the effects of PXR-ligand binding at the level of the major sites of biotransformation that have the greatest impact drug clearance, such as the liver and the gut.

However, it is now becoming apparent that PXR and indeed other NR activity does not just impact systemic drug levels through the major systems of drug clearance. PXR can also directly influence drug efficacy *locally* in the very cells and tissues that are drug targets. Until relatively recently, any biotransformational capability of drug target tissues, such as the brain, the skin or perhaps the heart, was only considered from the perspective of impact on total clearance, which was very little, so capability for drug metabolism in these tissues was not believed to be significant. However, *local* levels of drug in the presence of the receptor or cellular target are what counts, in terms of gaining a pharmacodynamic effect. To paraphrase the old American saying about winning; effective *local* drug levels are not everything; they are the only thing. If biotransformation in target tissues destroys or removes the drug from the immediate presence of the receptor, then efficacy will be lost, despite apparently adequate plasma levels.

Take drug resistant epilepsy, for example. Around three-quarters of patients have their condition controlled with monotherapy, with around half of the remainder perhaps requiring a change of drug at least once. However, 15–20% of patients require more than one anticonvulsant to control their condition[63] and in some cases, their epilepsy is effectively untreatable with drugs. There is certainly marked variation within patient populations in the drug levels required to control seizures. In those where the drugs fail to control the condition completely, a number of factors may be present, including brain tissue drug sensitivity and the nature of the seizure. However, in such drug-resistant patients, high levels of local CYP and multidrug transporter efflux expression has been detected that is also associated with high PXR activity[64]. In patients with intractable epilepsy, these factors unite to reduce intracellular drug levels to virtually nothing and abolish any significant pharmacodynamics effect, although the very high level of PXR-mediated induction of biotransformation may be linked to other factors besides the presence of the antiepileptic drugs acting as PXR ligands[64]. It is likely that the level of brain tissue PXR-mediated induction will certainly be a factor in all patients taking anticonvulsants and will influence their response to the drugs if they are PXR ligands. Experimentally, the potential for what could be termed *own goals* has been demonstrated, where antiepileptic drugs such as carbamazepine might be metabolised locally into a seizurogenic quinolinic derivative.[65] If this occurs *in vivo*, this is another factor that might counteract the efficacy of an anticonvulsant such as carbamazepine in neurones.

Rifampicin has been much mentioned so far with respect to its power to induce CYP responses. However, as you probably know, it is a highly effective antibacterial, usually employed clinically in drug combinations aimed at tuberculosis and leprosy. *Mycobacteria tuberculosis* is a fearsomely tough organism, whose waxy mycolic acid outer wall protects it from the immune system and even allows it to reside quite happily inside macrophages, which is for us like somehow standing nonchalantly in a vat of concentrated acid unharmed. As rifampicin and the other drugs can penetrate the infected macrophages, this trick might not seem to be so impressive, but drug response times vary widely and some infections are very hard to eliminate. Part of this problem is that the drugs can enter the infected macrophages, but often sufficiently lethal levels cannot be maintained. It appears that the drug, through its ligand binding to PXR, leads to a potent local induction of drug efflux proteins such as P-glycoprotein, which pump the drug out of the infected macrophages[57]. This effect is compounded by the ability of the organism to shed those mycolic acids, which are also potent PXR ligands, in and around infected macrophages, which again induces drug efflux capability[66,67]. These are only some of the myriad countermeasures deployed by these organisms to resist antibiotics and host defences and it underlines that full patient adherence and high drug plasma levels alone are often not enough to destroy such loathsome but determined opposition.

The PXR system has a predominantly negative impact on anticancer therapy usually through its role in the tumour's resistance effort and its immunosuppressive effects. PXR is one of the main reasons why drug treatment of solid tumours is often ineffective, even at the maximum tolerable doses for the unfortunate patients[68]. PXR is massively upregulated in tumours from whichever tissue they originated, partly in order to maximize their aggression (growth and spread), but also to ensure their invulnerability to normal cellular apoptotic control[68]. As if this propulsive effect of PXR on tumour activity was not enough, from the tumour's perspective, PXR1 & 2, are ideal, as they are so promiscuous and thus can be relied upon to detect a wide range of anticancer cytotoxics and organise a comprehensive and in-depth response. Accelerated biotransformation and detoxification of the anticancer drugs and upregulation of P-glycoprotein and other efflux transporters as described above and later (Section 4.4.7 and Chapter 5, Section 5.6.2) are all promoted powerfully by PXR in tumours. These coordinated measures unite to ensure that the cellular residence time for agents such as tamoxifen, paclitaxel, and irinotecan[53] is inadequate and often involves the induction of PXR-mediated induction of high levels of CYP3A5.[69] This can reduce the lethality of various usually fierce cytotoxics to a polite calling card that alerts the tumour to deploy yet more resistance measures. Even where PXR promotes conversion of a drug prodrug to an active agent, such as with cyclophosphamide, it is probable that the efflux transport upregulation will erode the anti-tumour effect. Indeed, tumours subvert PXR's toxicovigilance into a currently unjammable combined radar and defence system that can defeat virtually any chemical assault.

PXR system: therapeutic modulation

From the preceding sections, PXR looks to be an extremely inviting therapeutic drug target, either for antagonist or agonist development. Interestingly, PXR agonists and CAR agonists also for that matter, are already present in various popular herbal remedies[15]. Concerning St John's Wort and ginkgo biloba, for example, how much their PXR agonism impacts their CNS actions is not known, although a rather intriguing study in rats showed a role for PXR in memory restoration both after aluminium-induced toxicity[70] and even in a mouse model of Alzheimer's disease[15]. The herbally derived tanshinone IIA and notoginsenoside R1 exert their therapeutic actions in cholestatic liver and inflammatory bowel diseases respectively through PXR modulation[15]. However, for reasons not understood, PXR also has a major role in hepatic lipid and glucose metabolism and PXR agonists may potentially cause fatty liver[71]. PXR agonists are certainly hyperglycaemic; St John's Wort can impair glucose tolerance clinically[72], and the sheer number of PXR agonists in the environment may even be contributing to increased levels of diabetes.

Whilst PXR agonists are potentially problematic, antagonists are much thinner on the ground generally and the process of PXR antagonism might even approach CYP inhibition in complexity. As you will remember from previous sections, there are many opportunities to block PXR, perhaps starting with preventing the release of SMRT. Other aspects of PXR activity that could be inhibited include ligand binding, interaction with co-activators, progress to the nucleus, and the interaction with DNA response elements. Many of these events already have an endogenous ligand that modulates the process that could provide a template for a new drug. Additionally, it is highly likely that there are already herbal remedies in use that are undiscovered but effective PXR inhibitors. So far, we know of natural antagonists such as sulforaphane from broccoli and the phytoestrogen coumestrol, as well as various drugs such as ketoconazole, metformin, and leflunomide that bind either at the LBD and/or the co-activator sites, either competitively and/or allosterically[73].

The topical immunosuppressive pimecrolimus and the tyrosine kinase inhibitor pazopanib are effective and selective at the low micromolar level, in terms of PXR inhibition[73]. However, PXR is a tough target for inhibition, as not only is it species specific (narrowing studies to humans, or humanized models) in terms of the chemicals it binds[74]; its wider functions vary also – the famed ability of the liver to regrow after partial hepatectomy is controlled by PXR in rodents, but not in humans[75]. PXR is also often inhibited *in vitro* by various agents,[61] as we have seen with the azoles, but this does not translate to effective inhibition in humans[62]. Sulforaphane, for example, has long been studied as a PXR antagonist *in vitro*, but in a clinical trial, which involved its administration to the participants in a somewhat controversial cheese-based soup, sulforaphane intake for a week made no impression on rifampicin's usual CYP3A4 induction[74]. Another obvious problem with PXR inhibition is its complex impact on so many other vital homeostatic functions, ranging from the immune system, to energy metabolism and cell growth and development. This is a truly challenging therapeutic problem, and the section below underscores this complexity, as it summarizes some of the interplay known between PXR and other receptor systems.

Receptor cross-talk and CYP capabilities

Regarding CYPs and general biotransformational capability, there is a great deal of overlap in the nuclear receptor-mediated control of CYP and other metabolizing systems expression. HNF4α modulates at virtually all levels simultaneously, controlling the expression and specific activities of NRs like CAR, PXR, and AhR as well as expression of a large number of genes ranging from CYPs and their REDOX partners, through to conjugation systems and transporters (Chapter 6). Whilst CAR appears to run basal housekeeping small molecule metabolism, PXR, and AhR can respond to both 'emergencies' such as the appearance of xenobiotics in quantity, as well as operating bile salts and cholesterol processes. Indeed, PXR agonists such as rifampicin can alleviate drug-induced cholestasis (cessation of bile flow). This condition can be caused by more than 20 drugs, including oral contraceptives, anabolic steroids, and some antibiotics.[48]

The different NRs also compete with each other for coactivators such as PGC-1α or GRIP-1, and this process partly influences which enzymes are eventually expressed depending on the conditions at the time. Indeed, *CYP3A4* response elements can be engaged by CAR, PXR, and VDR (vitamin D receptor) and the GR (glucocorticoid receptor[48]). CYP3A4 induction can even occur through different NRs according to the concentration of the same inducing ligand. Low levels of dexamethasone induce via GR, whilst PXR mediates the high concentration induction[76]. However, it is not clear why PXR powerfully induces CYP3A5 rather than CYP3A4 in many tumours.[69] Prevention of that process would sensitize tumours to drug therapy, but the design of a selective agent that abolished PXR induction of CYP3A5 and not CYP3A4 is probably not an immediate prospect.

CAR predominantly regulates CYP2C9 and CYP2B6, but PXR also has a role. Drugs such as rifampicin and phenobarbitone promote the appearance of several CYPs through stimulation of several NR systems, although CITCO as mentioned previously is unusually a very specific and potent CAR stimulator. This overlap provides a 'safety net' to ensure that 'threat molecules' as well as endogenous agents that have outlived their usefulness are under control one way or another. Part of so-called NR receptor 'crosstalk' is governed in humans by exposure to different chemicals; AhR agonists can decrease PXR expression and thus rifampicin induced CYP3A4 induction[77] and CAR and PXR interact with each other and HNF4α in endogenous control of energy metabolism, as well as in that of xenobiotics.[78]

PXR and the immune system

More than two decades before PXR was even discovered, it was known that drugs such as rifampicin were immunosuppressive[78] and subsequent work indicated that other PXR agonists can show these effects also[15]. Indeed, several herbal remedies are thought to exert their anti-inflammatory actions through PXR-agonism.[15] Conversely, it has also been understood for some time that infection and inflammation can also powerfully suppress tissue CYP expression and PXR-mediated activities[78] (see Chapter 7.9). Thus, a two-way relationship between the immune system and biotransforming capability exists, partly modulated by PXR agonists and the immune system's key cytoplasmic biosensor, NF-κB. Naturally, there are several NF-κBs (at least five) and they are the immune system's first line of response to inflammatory stimuli, tuned to respond to everything from bacterial fragments, to the presence of cytokines and other inflammatory molecules. Normally, like PXR, NF-κBs sit in the cytoplasm locked to IκB and stimulation leads to release, followed by translocation to the nucleus and activation of expression of interleukins, tumour necrosis factors and other inflammatory mediators[48]. NF-κBs can also shut off PXR's interaction with RXR, preventing CYP expression, whilst PXR can prevent NF-κBs inflammatory effects[15,48,78]. What this complex cross-regulation means clinically might seem perplexing, as during tuberculosis infection and treatment with rifampicin, CYP induction is clearly the ascendant effect, in terms of its impact on CYP3A4 expression, activity and accelerated drug clearance[8]. However, it is likely that in the absence of any systemic immune factor release and effect, any significant NF-κB-mediated inhibition of PXR-agonism is probably a local issue around the site of the infection, which in most patients is the lung, rather affecting the major biotransforming organs. What would be interesting to explore is how much this PXR-NF-κB crosstalk might influence the PXR-mediated accelerated rifampicin cell efflux in infected macrophages described earlier[57]. Clearly, much remains to be learned of the practical interplay between drug modulation of the immune and biotransforming systems through NRs like PXR, CAR and the others, both systemically and locally, during both infections and inflammatory disease. With regard to PXR, subsequent sections will demonstrate that we have much more to learn about this remarkable biosensor system.

4.4.4   CYP 2E1 induction

CYP2E1 is of relatively minor interest from the standpoint of drug metabolism (it oxidizes isoniazid, paracetamol, and chlorzoxazone), but it is of serious interest in hepatotoxin (paracetamol, carbon tetrachloride, thioacetamide) and carcinogen activation (*N*-nitrosodimethylamine, benzene, vinyl chloride, and trichloroethylene). This isoform undergoes induction by apparently disparate

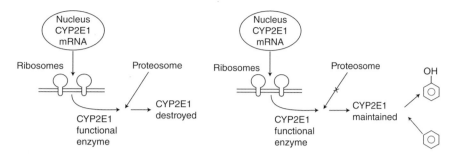

Figure 4.6 CYP2E1 induction: this CYP is not controlled by nuclear receptors and CYP enzyme is made in large quantities constantly, but in the absence of substrate, the proteosome system destroys the enzyme. The presence of the substrate effectively induces CYP2E1 by preserving it from the proteosome

factors like small hydrophilic molecules, such as ethanol, acetone and pyridine, as well as by systemic stresses, such as obesity, diabetes and starvation.[41,79] In principle, the main points of CYP2E1 induction are now understood (Figure 4.6) although the details and its main physiological purpose are still being explored. When animals are exposed to 2E1 inducers, functional CYP2E1 protein levels are increased up to eightfold, although the CYP2E1 mRNA levels remain the same, showing that 2E1 is not induced with a nuclear receptor regulated system like other CYPs. Indeed, CYP2E1 can be induced quite radically after transcription and translation are complete and the protein is fully operational in the ER or the mitochondria, which occurs with oestradiol during pregnancy, for example.[80] However, it is also regulated transcriptionally by several factors, including the AhR, and microRNAs[41,79,81,82].

In human cellular systems, the instability of this isoform and production of reactive species means that CYP2E1 functionality without substrate lasts only a couple of hours before ERAD-C degradation (see Chapter 5.1). The presence of substrate induces by increasing functional CYP2E1 survival time by up to 20-fold in comparison with substrate free cells and routing the CYP through to a different and slower degradation pathway, known as the ALD route (autophagic-lysosomal degradation), where it can remain functional for more than a day and a half[79] (Figure 4.6). Whilst other isoforms such as CYP3A4 have their half-lives extended by certain substrates, the effect seems most pronounced for CYP2E1[79].

CYP2E1 protein turnover tends to be high, yet regulation, as we have seen, both pre and post-translational, is actually very tight, despite an apparently rather wasteful amount of CYP protein made and destroyed (Figure 4.6). In some ways, the CYP2E1 system resembles a very high-pressure system, which is equally strongly dammed, rather like a fire hydrant. Why this is the case is related to endogenous functions of CYP2E1 and how these may need to be very rapidly accomplished in view of equally immediate and potentially hazardous changing cellular and systemic conditions. CYP2E1 is very sensitive to diet, even becoming induced by high-fat/low-carbohydrate intakes. Starvation and diabetes also promote CYP2E1 functionality. Insulin levels fall during diet restriction, starvation, and in diabetes and the formation of functional 2E1 is suppressed by insulin, so these conditions promote the increase of CYP2E1 metabolic capability. One of the consequences of diabetes and starvation is the major shift from glucose to fatty acid/tryglyceride oxidation, of which some of the byproducts are small, hydrophilic and potentially toxic ketone bodies[83]. These agents can cause a CNS intoxicating effect in diabetics who are very hypoglycaemic, so they may appear drunk and their breath will smell as if they had been drinking. In the nondiabetic individual who is in a state of starvation, any ketone-mediated intoxication would as mentioned earlier, hamper the search for food, so these molecules must be cleared rapidly. The key factor here is the speed at which these compounds could accumulate. The nuclear-receptor mechanism of induction, with its time frame of days, might be too slow to cope with the accumulation of ketone bodies in starvation, so the much quicker 'substrate-mediated protein preservation' system perhaps might be rapid enough to ensure that adequate levels of CYP2E1 were present to prevent intoxication of the CNS. CYP2E1 also is present in other tissues such as the brain, which would also help dispose of any volatile agents that might affect behaviour and consequent survival[41,83].

If we pursue the fire hydrant analogy, then it is clear that inappropriate use results in serious problems. Toxicologically speaking, CYP2E1-induced reactive species production leads to oxidative stress. Many CYPs also form reactive species, but they are not found in virtually all tissues as well as the liver and particularly, they are not present in mitochondria, as mtCYP2E1 is. This isoform is in close proximity to the rather delicate and vulnerable sites of cellular ATP and a key area for the regulation of life and death processes such as apoptosis. Indeed, mtCYP2E1 is linked with considerable mitochondrial dysfunction caused by the reactive species it forms[83]. With SER CYP2E1, animals where the gene is knocked out show no ill effects and are resistant to ethanol and obesity-linked liver damage[41]. The blame for much of the destruction of the human liver in alcoholism as well as ethanol-induced liver cancer has been laid at the door of CYP2E1[84]. This CYP also is linked to nonalcoholic steatohepatitis and is expressed up to three-fold higher in the obese and is implicated in many cancers[41,83]. Hence, CYP2E1 has become a key enzyme in research into cellular dysfunction and chemical-based toxicity and carcinogenesis, as well as its role in mitochondrial damage.

4.4.5 CYP2D6

CYP2D6 is important as the main source of clearance for tricyclic and SSRI antidepressants, some antipsychotics (haloperidol, risperidone) and many beta-blockers and synthetic opiates[85]. It still has a section in this chapter all to itself, partly because it was until recently believed to be unusual, in that it was not thought to be classically inducible, either by the cytoplasmic/nuclear receptor systems or the CYP2E1 model, as related above. Perhaps this was rather naïve, as there were plenty of clues in the literature, which suggested that there was some form of inductive response, just not as dramatic as the other CYPs. Given CYP2D6's rather unique substrate profile, which is oriented towards alkaloidal toxin-like structures, it is logical that this CYP should respond to demand. It has been known since the 1980s that pregnancy can increase clearance of CYP2D6 substrates by 30–100%[85,86] and as mentioned in Chapter 3, CYP2D6 expression is controlled by HNF4α, the master biotransformational regulator[85]. What is interesting is that again as mentioned in the previous chapter, CYP2D6 is polymorphic and the induction effect in pregnancy only operates if the woman is an extensive metaboliser.[86] So if no functional CYP is produced (see Chapter 7.2.3), induction does not change this. Clearly, other CYPs must take over processing of endogenous hormones and other agents in pregnant nonexpressors of CYP2D6.

In terms of CYP2D6 induction, this actually does occur in response to the classical inducers such as rifampicin,[87] and in human hepatocytes, corticosteroids can induce CYP2D6 expression tenfold,[88] which ties in well with changes seen pregnancy of these endogenous hormones[88]. In the same work, rifampicin was much less potent an inducer of CYP2D6 than the corticosteroids, although the patterns of induction were quite revealing. With the corticosteroid induction, enzyme turnover of substrate was actually greater than the induction of CYP2D6 mRNA, which suggests that some form of substrate stabilization, similar to that of CYP2E1, was occurring. By contrast, with rifampicin, CYP2D6 mRNA was induced to a seven-fold greater extent than active CYP2D6 protein, suggesting that some post-transcriptional control system was preventing full translation of the mRNA. One of the reasons this was not investigated earlier was that experimentally (see Appendix A.4.5), hepatocyte biotransformational activity was usually maintained in culture with large doses of synthetic cortiosteroids such as dexamethasone, which meant that CYP2D6 expression was already maximally induced so changes were not seen on addition of known inducers[88]. It is likely that CYP2D6 induction is controlled in a similar fashion to CYP3A4, with input from HNF4α, and various NRs, such as GR and PXR as well as microRNAs such as hsa-miR-370-3p, which can experimentally shut off transcription and translation of CYP2D6.[89]

4.4.6   Reversal of induction

Anyone who has stared in disbelief at his or her shrunken quadriceps after a full leg plaster cast has just been removed will appreciate the main imperative that drives the reversal of induction. Just as astronauts lose the ability to walk if they spend enough time in weightless conditions without exercise, the cell will usually husband valuable resources carefully and the vast increase in CYP transcription, translation and shipping to the SER is quickly curtailed in the absence of the inducer. In addition, the cell cannot afford the presence of a fully induced battery of functional CYPs without appropriate substrate, mainly because of the huge impact this biotransformational force would have on endogenous small molecules, such as hormones, seriously disrupting homeostasis. Finally, as has been mentioned previously, some CYPs, particularly CYP2E1, generate enough reactive species to cause cellular oxidative stress in the absence of substrate, which is a compelling reason to limit their unnecessary activity.

Clinically, investigation with inducers in various trials have shown that what is often termed *de-induction* (for want of a more elegant word) is usually a fairly linear process, when it is compared with the process of induction. Depending on the inducer, however long it took for the process to be fully engaged, it should take a similar timeframe to reverse,[1,90] although if even low levels of the inducer persist, so does a considerable degree of induction,[91] suggesting that there is some built-in inertia in the system. Hence, the half-life of the inducer is crucial in the process; with a relatively short half-life drug such as carbamazepine (around 15h),[92] after dosing cessation and five half-lives have passed (around 75h), the vast majority of the drug will have been cleared and the de-induction will have begun. However, it can take at least three weeks before normal CYP3A4 activity resumes.[1,90,92] At the other extreme, with mitotane, five half-lives could be several months and indeed, the induction effect on CYP3A4 does persist with this drug for that period in patients.[24] There is also gathering evidence that de-induction time frames are not the same for different CYPs. It appears that CYP3A4 is the most responsive, whilst CYP2C9 is less powerfully induced by agents such as rifampicin, but it can be significantly delayed in returning to baseline, to the point that clinically it can be difficult to ensure continuity and safety of drug treatment[93]. This underlines the complexity of the NR-modulated induction process for different CYPs, which is compounded by extremely wide human inter-individuality[93]. In practical terms, it will take about a month in a healthy individual for the inductive effect of rifampicin to subside fully on CYP3A4,[94] and mathematical models now exist for the prediction of drug–drug interactions based on the impact of induction.[95]

At the cellular level, as discussed in the sections above, the regulation of the various cytosolic and nuclear receptors all contain different braking systems, which either impede receptor movement and DNA binding, or they prevent CYP transcription, such as SHP and microRNAs. These systems act like the deadman's handle on a train and re-assert themselves as the inducer cellular concentrations fall away. The CYPs will become degraded or damaged through either substrate processing, or in the absence of substrate, production of reactive oxygen species will cause sufficient damage to attract the attention of the ERAD-C pathway (Chapter 3.5.1) and the eventual proteosomal recycling. The increased endoplasmic reticulum will also have to be degraded; the normal CYP recycling process is extremely complex and not fully investigated, but it may be accelerated once the inducer disappears from the system.[96]

4.4.7　Cell transport systems and induction: P-glycoprotein

Purpose, structure, and function

As discussed in Chapter 2.3.1, many endogenous and exogenous agents can enter and leave cells relatively easily through passive diffusion if they are reasonably lipophilic. The more charged endogenous agents need 'help' to pass across membranes and the OATPs (Chapter 2.6.3) and other transporters accomplish this. Although it is necessary for cells to pump in certain required nutrients, it is equally necessary to pump others out of one cell and into another, as part of the general circulation of endogenous agents, such as vitamins, amino acids sugars and proteins. A 'revolving door' system like the OATPs is adequate for agents moving with a concentration gradient. The movement of specific types of molecules out of a cell against a concentration gradient requires a much more specialized system. Indeed, this is rather like ejecting rowdy or unwelcome fans from a stadium into a pressing throng of people trying to enter. To stretch the analogy a little, the unwelcome visitors also need to be identified before they can be ejected. This dynamic process needs to be powered in some way. Also, molecular recognition must be reliable and rapid and finally, the process needs to fully adjustable in terms of load and velocity. Clearly, from those conditions the system required is more than just a simple pump.

The ATP-Binding Cassette transporters (ABC transporters) accomplish this and in humans there are 49 of them, now classified into seven subfamilies. The ABC transporter of greatest interest therapeutically is known as ABCB1, or MDR1, or more colloquially, Permeability-glycoprotein (P-gp). All ABC transporters use ATP to propel a vast array of endogenous and exogenous agents out of cells and P-gp is coded for by the *MDR1* gene (found on chromosome 7 in man). P-gp is a 170kDa transmembrane protein consisting of, you might say, a game of two halves. Each half has six column-like segments that span the cell membrane. This is the actual pump structure containing the binding pocket, whilst two nucleotide binding areas are embedded below the membrane in the cytoplasm and they bind the ATP, so they are the 'power pack' of the pump.[97]

The scale of substrates that P-gp can process, in terms of structure, molecular weight, function, and physicochemical properties, could be described as panoramic. Exhaustive efforts have been made to characterize the detailed properties of P-gp substrates using a variety of different experimental and biological methods,[97,98] yet it appears that if a chemical of pretty much any description partitions into the lipid bilayer of a cell, P-gp has evolved to pump it back out into the extracellular space[97]. Hence, this transporter is known as the cell's *hydrophobic vacuum cleaner*[97] and it accomplishes this function through two key facets; one of which is that the binding pocket of the pump is submerged in the membrane where its target chemicals lurk. The second facet is striking; the relationship between the pumped chemical and P-gp's binding pocket is perhaps akin to that of a receptor like PXR and the molecules it senses. P-gp's drug pocket is complex and contains 73 amino acid residues, mostly hydrophobic, except for 15 polar and two charged residues[97,99]. It has been known for several years P-gp can be inhibited to varying degrees by many drugs, either by interacting with the ATP binding process, or with the binding site.[100] However, other drugs can activate P-gp without change in mRNA or protein synthesis. It is thought that the binding of various structures to P-gp, interact with the binding pocket, resulting in a range of outcomes, from slowed or 'clogged' inhibitory effects (e.g., verapamil, paraoxetine, econazole, and the proton pump inhibitors), to a normal substrate-like effect (morphine, irinotecan, docetaxel), through to an activation (colchicine and some flavonoids), of their own or other drug's transport.[97]

Hence, when a molecule meets P-gp's binding pocket, binding occurs when the system is in what is termed an 'inward' state. The binding pocket 'reads' the agent's structure and physicochemical properties, and if it fits the 'substrate' requirements, ATP consumption is triggered and the pump cycles to rearrange itself into an outward state, thus expelling the molecule, followed by a return to the resting state. Depending on how they are 'read' by the pocket, inhibitors and activators either lock or slow the pump or speed it up, although the details are not clear[101]. ABC transporter expression is under the control of myriad different cellular systems, ranging from the immune system (NFκB) to p53, as well as PXR and CAR[97]. So the ABC transporters like P-gp are not merely pumps, but an integrated, almost infinitely responsive and selective molecular movement process system, which resembles the NRs and CYPs in terms of the multiple levels of local, translational, and transcriptional regulation.

P-gp is present in all tissues, but from the perspective of drug metabolism, its role as part of the gut's barrier function to prevent the uncontrolled entrance of xenobiotics is of key interest. The transporter effectively works in tandem with the very high levels of CYP3A that are found in the gut (three times as much as in the liver in humans). If P-gp repeatedly pumps an agent out, it has more chance of meeting a CYP on its next entry. The system also has an element of insurance, as inhibitors of CYPs such as grapefruit juice do not necessarily always inhibit P-gp, so some residual barrier function remains as seven times as much P-gp is found in the apical areas of enterocytes compared with hepatocytes. Whilst CYP3A4 and P-gp overlap in terms of their broad xenobiotic targets and are often induced by the same ligands, it is not certain whether their respective genes' expression is directly linked[97]. Overall, the presence of P-gp can certainly retard drug absorption without actually preventing it, and in animals where the ABC transporter genes have been knocked out, they are otherwise healthy, but very susceptible to accumulated toxicity from drugs.[102]

P-gp induction: mechanisms

As mentioned in the previous section, P-gp is under the control of the NRs and it is inducible with the usual suspects, such as rifampicin, St John's Wort and dexamethasone. These cause large increases in gut P-gp, plus CYP3A also, so the combined effect of reduced cellular entry and accelerated clearance of drug molecules that do enter the gut is increased by NR-mediated inducers, as you would expect. However, the response to drugs is only part of P-gp capability. Indeed, the evolutionary necessity for rapid protection against toxic insult has placed P-gp and its sister ABC transporters in a league of their own in terms of the sheer scale and velocity of their response.

It has been known for many years that P-gp and other ABC transporters can increase their capacity, in terms of expression of protein, changes in protein half-life and increased activity of existing pump molecules (like the effect of the P-gp activators described in the previous section) in response to generally noxious stimuli[103]. What is remarkable is how seriously rapid this response is. The presence of the anticancer agent cytarabine, causes P-gp mRNA's to appear within 10 *minutes* of exposure in leukaemic cell lines[104]. So as well as through increased protein expression, ABC transporter induction occurs through direct activation at the site of the pump/molecule interaction and the half-life of the P-gp pumps themselves is extended also,[103] probably through some form of repression of proteosomal degradation. These processes combine to massively increase P-gp capacity within 12–24 hours, which then falls away almost as rapidly in the absence of the stimulus.[103] This effect can happen in virtually any tissue and is stimulated by a number of cell stimuli or insults, such as immune-mediated cytokines, ionizing, and UV radiation and of course practically anything remotely toxic. This can range from pesticides, lethal herbicides such as paraquat and anticancer agents, to ethanol and of course a large number of different prescription drugs.[103]

How all this complex and coordinated response system is controlled is not entirely clear, but as so many different biosensor systems can exert transcriptional, translational and post-translational control on P-gp, it is likely that master receptors such as HNF4α and similar NRs are implicated. With regard to radiation, the cell cannot directly detect high energy electromagnetic radiation such as UV or gamma rays or ionizing α and β particles but it can detect what radiation creates in cells, that is, reactive oxygen and nitrogen species that it is well equipped to 'read' through the nrf-2 system (see Chapter 6.5.4). It is highly likely that the MDR gene is activated by nrf-2 and similar systems. From the perspective of drug treatment, it has emerged that several drugs that are not listed as CYP inducers, are detected by P-gp as threats and they trigger the above response in barrier tissues such as the gut. The antidepressant venlafaxine (Effexor, or 'side-effexor', as some pharmacists call it) is a known inhibitor of P-gp yet is a potent inducer of its expression, through PXR activation[105] and this can impact drug absorption and occurs at clinically relevant levels[105,106]. Several features of these observations are interesting; firstly, O-demethylated venlafaxine, or desvenlafaxine is not a P-gp inducer, suggesting very precise P-gp system recognition capability. Secondly, PXR helps mediate a process that not only recognizes just the parent drug, but it also induces P-gp expression in this case, but no CYPs. This underlines the complexity and flexibility of PXR molecular interactions. Finally, this suggests that agents such as venlafaxine through their effects on ABC transporter systems, can impact drug residence in many different tissues, including, of course, the target tissues of the drug. This can result in a drug accumulating towards toxicity, such as with propafenone and venlafaxine[107] or effectively shielding a therapeutic target from the drug, as with the antiglioma Wee1 inhibitors[108]. Again, apparently adequate plasma levels are not enough – the drug must not only reach the target tissue but remain in it long enough to actually achieve a pharmacodynamic effect, or not become trapped there. It is clear that in many cases, the ABC transporters are at the forefront of healthy, and as we will see later on, malignant cell determination to repel unwanted visitors.

P-gp induction: wider clinical impact

As described previously, P-gp can retard the systemic absorption of many orally dosed drugs by repeatedly expelling them from enterocytes and this is also important in drugs that are not significant CYP substrates. The cardiovascular drug digoxin is mostly eliminated unchanged, but undergoes such extensive P-gp interaction, it is considered a probe substrate for this transporter, although other systems also transport the drug[109]. Consequently there are well-documented drug interactions between P-gp inhibitors like quinidine and digoxin and such effects are problematic, as digoxin's TI is narrow (0.5–3 ng/mL) and levels that exceed 3.5 ng/mL increase the risk of patient death. The therapeutic monitoring of this valuable but toxic drug has been revised to improve its safety and it is now recommended that the therapeutic window for this agent should be between 0.5-0.8 ng/mL.[100] However, inducers of P-gp can significantly reduce digoxin bioavailability. Ten days of rifampicin induction can cut the bioavailability of digoxin by 30–50%,[110] through boosting P-gp capability by 3.5-fold. St John's Wort showed a 4-fold induction in P-gp capability after 16 days of treatment in volunteers.[111]

Anti-tissue rejection drugs such as tacrolimus, sirolimus, everolimus, and the older agent cyclosporine are problematic to manage clinically, as their narrow TI's and high levels of CYP-mediated clearance means that patient pharmacokinetic variation is considerable. Above 30 ng/mL, transplanted organ toxicity could result, below 2–3 ng/mL risks rejection. Desired trough ranges are quite precise; for example, 3–12 ng/mL for sirolimus[112] and 5–15 ng/mL for tacrolimus[113]. These drugs are P-gp substrates, but the role of the transporter in tacrolimus bioavailability is not fully resolved, with some suggesting its impact is not very significant due to the ease that drug can penetrate membranes[113]. However, there are studies where tacrolimus itself and cyclosporine bioavailability are impacted by the patient's individual genetic P-gp expression profile.[114-116] An obvious and very serious clinical issue in any xenograft recipient is infection and contracting MRSA (methicillin-resistant *Staphylococcus aureus*) would need immediate treatment. Vancomycin in combination with rifampicin is effective, yet the latter drug's impact on sirolimus bioavailability is profound. In one case study, it was necessary to increase the patient's normal dosage of 0.4 mg/day up to 16–18 mg/day to keep the tacrolimus in the therapeutic window and this was partly ascribed to the inductive effect on P-gp.[112]

We have already discussed in the PXR system section, how enzyme induction processes, through increased capacity of CYPs and transporters can act to prevent therapeutic levels of drugs building in target tissues, particularly in the brain, with respect to epilepsy treatment. Assuming first-pass and systemic metabolism leaves potentially effective levels of an anti-epileptic drug (AED) in plasma, the first hurdle to overcome is the blood–brain barrier (BBB); if you are not familiar with it, then we can start with the microcirculatory system of other organs, which are rather like the piping systems that supply our domestic water-rather leaky. Various polar agents, such as amino acids, for example, can move in and out of the organ through its microcirculation, governed by concentration gradients and transporter systems. In contrast, the endothelial cells of the brain's microcirculation have tight junctions that effectively seal off the brain as a separate entity from the rest of the body. The BBB has a complex and wide expression of transporters that then selectively take up various required polar agents. It is not a barrier to relatively lipophilic drugs, but P-gp and its sister ABC transporters are expressed very highly and expel drugs very effectively, delaying their absorption into the brain.[108] However, even if the ABC transporters in the BBB can be circumvented, the brain's astrocytes can even elevate P-gp in response to toxic challenge,[118] underlining the scale of the brain's layers of defence against toxins, right down to the level of the neurone and astrocyte. In addition, most AEDs appear to be P-gp substrates and they increase its activity, directly and through PXR, although lamotrigine is an exception.[119] As we saw earlier, the presence of other drugs, such as venlafaxine, can also promote P-gp activity to resist any therapeutic accumulation of drug in the seizurigenic brain tissues. Indeed, high P-gp expression alone correlates clinically with resistance to AED therapy in epilepsy.[120]

Whilst there are other pharmacodynamic factors in the AED drug resistance that affects up to 40% of epileptics, the ABC transporters are not only present at every stage of the AED's journey to its receptors, but they are exquisitely and ruthlessly responsive to their presence. However, preliminary beneficial clinical application of P-gp inhibitors like verapamil[121] has led to a search for practical and safe inhibitors of P-gp inhibition to overcome the resistance.[119]

P-gp and cancer

If you have read this far, you will see clearly why the P-gp system is likely to be very effective indeed in detecting and effluxing the various cytotoxins used in cancer chemotherapy. Tumours are usually very difficult to attack for a number of reasons, not least their ability to massively overexpress ABC transporters, which expel the antineoplastic agents before they can damage the cell[122]. Whilst with some cancers drug exposure is required to induce P-gp, in others, primary resistance occurs, where the tumour 'gets its retaliation in first' and already expresses high transporter levels. Either way, multi-drug resistance (MDR) to many structurally unrelated cancer agents is a characteristic of around 4 in 10 human tumours[103,122] and is associated with a poor outcome,[122] the clinical euphemism for a fairly rapid death. As mentioned previously, efforts are being made to develop inhibitors of P-gp that also downregulate its expression[122]. Whilst it is imperative to devise methods of preventing the ABC-transporters and other efflux systems from expelling the antineoplastic agents, it is just as important to prevent 'own goals' where the patient is exposed to another drug that promotes the activity and expression of the ABC-transporters. This could impact oral absorption of the antineoplastic agent and promote resistance in the tumour. As discussed later in this chapter, cancer patients take considerable numbers of supplements and drugs related to the stress of their condition and the desire to ameliorate the side effects of chemotherapy, which are frequently awful; this can have a serious impact on their therapy[123,124]. St John's Wort not only accelerates the clearance of anticancer agents such as etoposide and irinotecan plus various tyrosine kinase inhibitors such as imatinib, but its induction of P-gp retards the absorption of orally active agents. In addition, it also may promote tumour drug resistance by inducing the same transporters, which could influence treatment outcome[124]. What complicates this issue is that since the turn of the century, it has been found that St John's Wort has a spectrum of antiproliferative properties, ranging from induction of tumour apoptosis to attenuation of angiogenesis, which shows potential as an anticancer agent in its own right or in combination with other drugs[125]. Systemic application of St John's Wort and its active constituents would be complicated to use therapeutically, as you can imagine, but it is already established that the potent anti-inflammatory properties of hyperforin can be safely exploited clinically when applied topically, where its systemic absorption is likely to be minimal.[126]

4.4.8 Induction processes: summary

From Figure 4.7, you can see a broad summary of the process of enzyme induction and how it is managed. As well as the various transporters undergoing induction, several other conjugative systems, such as glucuronidation are also sensitive to induction and the processes involved are detailed in Chapter 6.2–6.5.

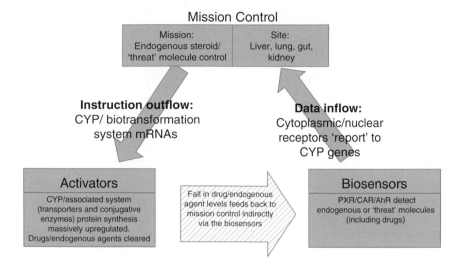

Figure 4.7 Generic summary of CYP induction. With the exception of CYP2E1 (Figure 4.6), the rest of the CYPs are induced largely through nuclear or cytoplasmic receptors, which sense endogenous molecule 'load' and report to the CYP gene response modules. Induction is responsive to local sustained concentrations of ligands, either endogenous (steroids, corticoids etc.) or drugs and other exogenous molecules

The important message is that enzyme induction is not just an increase in CYP protein, but a vast multi-process upgrade in metabolic, conjugative and transporter defence capability which can either repel (with P-gp) or transform the physicochemical properties of endogenous, drug, and threat molecules from broad lipophilicity to water solubility. The metabolites can then be vigorously expelled from the cell with other transporter systems (MRPs, Chapter 6.11).

4.5 Induction: general clinical aspects

4.5.1 Introduction

For the unwary, clinical issues with enzyme induction can be summarized in the phrase *stealth and surprise*.

- Whilst it commences within hours of inducer exposure, induction is *stealthy*, taking days or even weeks for the full effect and is compounded by remarkable patient variation.

- Drug concentrations can gradually fall out of the therapeutic window and treatment failure occurs unexpectedly - a major *surprise*.

- The induction process is usually reversible over a similar time frame to its appearance, although a minor *surprise* can occur when reversal can sometimes be slower than expected, depending on the CYP involved.

- If a treatment break of several days occurs for any reason in a patient stabilized on a high 'induced' drug dosage, drug accumulation and toxicity can occur – another potentially major *surprise*.

Perhaps it appears facetious to describe the effects of induction in such a way and health-care professionals might also resent the implication of the word *unwary*. However, consider the following factors:

- Polypharmacy in some patient groups such as HIV patients, the chronically ill and elderly

- The plethora of different prescribing professionals sometimes working with the same patient

- Time and work pressures, leading to…

- Lack of interprofessional communication and general access to patient notes

- Lack of disclosure of crucial information by the patient for whatever reason

- Patient predilections with sourcing herbal items online in quantity without full disclosure to healthcare professionals.

All these issues mean that the health-care practitioner, expert, or beginner and moreover the patient, can potentially be unwary in the clinical use of enzyme inducer drugs. The following sections look at some well-documented clinical issues in different drug classes where inductive effects on biotransformation need management.

4.5.2 Anti-epileptic agents

First-generation drugs

As already outlined in Section 4.4.3 with regard to PXR, AEDs must be given in combinations to drug resistant patients and interactions based on enzyme induction remain a significant issue with the first-generation AEDs and some of the newer agents also. As shown in the case histories, combination therapy can lead to potential problems with the induction effects of carbamazepine (CYPs 2C9, 2C19, 3A4; Histories 1 and 2), phenytoin (CYPs 1A2, 3A4) and phenobarbitone (CYPs 1A2, 2C8, 3A4). In a combination of inducing anticonvulsants, co-administered compounds metabolised by these CYP enzymes will have their plasma concentrations significantly reduced and doses must increase to compensate. Combinations of older AEDs can be difficult to manage for other reasons also; valproate is now believed to be a mild inducer in its own right, although it is better known as an inhibitor of various biotransforming enzyme systems (see next chapter). Carbamazepine will induce valproate clearance, whilst valproate can inhibit carbamazepine clearance. Add to this mix, issues with valproate and protein binding, plus pharmacodynamic interactions between the two drugs, it is recommended by some authors that such a combination should be discouraged or if deemed essential, it should be managed with therapeutic drug monitoring where blood levels of free valproate are measured[1,127]. Clearly, the first generation AEDs are problematic due also to their impact on any CYP3A4, CYP2C9, or CYP2C19 substrate, so they will accelerate the clearance and blunt the therapeutic effectiveness of the majority of prescribed drugs, ranging from many oral contraceptives, statins, to tyrosine kinase inhibitors and anticoagulants.

Second- and third-generation drugs

Felbamate appeared in 1993, and has been listed as a CYP3A4 substrate, inhibitor and mild inducer[1,63]. Felbamate can accelerate carbamazepine CYP3A4-mediated clearance to its epoxide, whilst blocking epoxide hydrolase, elevating carbamazepine epoxide levels considerably.[63] Felbamate has declined in usage due to unpredictable and potentially lethal adverse reactions, such as a relatively high frequency of bone marrow toxicity 1:4000, and a much lower, but equally potentially lethal hepatotoxicity.[63,128]

Lamotrigine was licensed in 1991 and is widely used, but is metabolised mostly by glucuronidation, where it induces its own metabolism and it is also subject to accelerated clearance caused by the inducing AEDs, which necessitates dose increase[129]. Valproate is a potent inhibitor of glucuronidation, so its impact on lamotrigine is considerable and can lead to adverse cutaneous reactions; lamotrigine can also induce valproate clearance[63]. The years 1995 and 1998 saw topiramate and tiagabine appear in clinics. About half the former drug is cleared by CYPs and the latter drug is almost entirely a CYP3A4 substrate, so both drugs are subject to induction by the first-generation AEDs. Topiramate acts as a mild inducer at higher doses[1]. Oxcarbazepine reached the European Union and US clinics over 1999–2000, after being first synthesized decades before[1,63]. Like lamotrigine, oxcarbazepine is a substrate of glucuronidation, a mild inducer itself and is induced by the first-generation AEDs and can induce their clearance also.[130] Levetiracetam was introduced in 2000 and is not subject to CYP metabolism, rather it is hydrolysed by esterases, although the inducing AEDs do accelerate its clearance,[63]

The most recently introduced AEDs include zonisamide (2005), rufinamide (2007), lacosamide and stiripentol (2008), eslicarbazepine (2009), retigabine (2011), and perampanel (2012). Whilst these drugs are not as extensively researched as the older agents, none of them is a potent inducer. Eslicarbazepine (mainly glucuronidated) and perampanel (CYP3A4 substrate) are weak CYP and glucuronidation inducers, whilst the rest are not known to be inducers. Zonisamide is cleared by CYP3A4, lacosamide by CYP2C19, whilst stiripentol is metabolised by CYP1A2, CYP2C19, and CYP3C4 and retigabine is glucuronidated so they are all liable to be cleared more rapidly in the presence of inducers. Rufinamide is metabolised mostly by carboxylesterases, but is still apparently vulnerable to induction-mediated accelerated clearance[63]. In 2016 brivaracetam was approved in the United States and the European Union and this CYP2C19 substrate appears to be well tolerated in drug combinations[131] and has not shown any significant induction or inhibitory effects to date.[132,133] As brivaracetam is extensively metabolised, carbamazepine accelerates its clearance considerably, although in common with felbamate, brivaracetam increases carbamazepine epoxide levels.[134]

Of the relatively newer drugs, gabapentin (1993) is believed not to be significantly metabolised and is cleared entirely renally as parent drug. It was also thought that vigabatrin (1989 UK, 2009 US) was cleared unchanged and along with gabapentin, any drug interactions reported would be most likely to be pharmacodynamic in mechanism. However, both drugs interact with felbamate[63] and vigabatrin interacts with carbamazepine and particularly phenytoin, although this is not fully explored.[63] It is likely that vigabatrin's impact on phenytoin clearance indicates that it is a mild CYP2C9 inducer, although its effect is somewhat delayed.[135]

Rather more seriously, vigabatrin was found by the late 1990s to induce permanent visual field destruction,[136] after which you might instinctively wish it to have been cast into the bucket marked 'toxic waste'. However, the manufacturer successfully gained US Food and Drug Administration approval for vigabatrin in 2009 with full disclosure of the retinal toxicity, which provides some perspective on the severity of the paediatric conditions where the drug was effective[137]. Indeed, with careful ophthalmological monitoring and relatively short periods of exposure, the vigabatrin is still valuable in conditions such as tuberous sclerosis complex, where the infantile seizures when uncontrolled are severe enough to cause brain damage.[137]

Drug withdrawal

There may be a variety of clinical reasons for withdrawing an AED, during mono or polytherapy, but in terms of impact on the patient, the most serious consequence of excessively rapid withdrawal can be exacerbation of the epilepsy[138]. This could be followed within a few days by the consequences of the reversal of any induction process which may have occurred, such as in History 3, where plasma drug levels began to climb and cross the threshold of toxicity. Clearly, abrupt changes in AED therapy are, except in certain circumstances,[138,139] undesirable and time and trouble must be taken to ensure a smooth transition from one regime to another. This is exemplified by a Phase III study where patients taking other AEDs were converted to monotherapy with brivaracetam and the patients existing inducing AEDs were tapered over eight weeks to zero in equal dosage reductions every second week[140]. As stated earlier, induction reversal usually occurs over a similar timeframe to the onset of induction and is very much dependent on the half-life of the inducer or inducers. Naturally, when in doubt, some form of therapeutic monitoring involving determination of plasma levels of the particular drugs may be necessary.

Epilepsy and brain tumours

Brain tumours may be benign or malignant and most of the latter metastasised from elsewhere. Of the primary tumours, around one third are gliomas, originating in the astrocytic brain tissue, which normally protects and maintains the neural cells. Naturally, depending on which area of the brain they grow, they create pressure and distortion in the neural tissue. This leads to a powerful impact on cognition and mood, as well as seizures, which with gliomas can affect almost all patients, with a lower incidence in glioblastoma of 40-60%[141,142]. It is not recommended that AEDs are used prophylactically in seizure free patients and after the removal of the tumour, the incidence of seizures in those that suffered them preoperatively can fall to single figures, so AED usage is not always necessary.[143] However, to control seizures that do arise, there are difficulties. Benign tumours burden the patient for longer, are often more drug resistant and are harder to treat with AEDs than malignant tumours[144]. However, with the advent of the new generations of AEDs, the use of the old inducing agents in this context has been left behind, as they disrupt chemotherapeutic regimes[144]. Currently, the priority is of course to control the seizures without interfering with the tumour treatment, whilst minimizing adverse reactions[144]. Hence, levetiracetam, and lacosamide, are recommended as well as lamotrigine, zonisamide, perampanel, and clobazam; of the older drugs, only valproate is recommended[141,142]. Interestingly, valproate can be a 'win–win' in that it has been shown to display a number of beneficial anti-tumour effects in brain cancers[145]. Lacosamide has been shown to be effective in brain tumour-related epilepsy, with one group reporting no side effects,[146] whilst the other mentioned unsteady gait and dizziness[144,146]. The newer AEDs have been effective in children also, although improving surgical outcomes have enabled around 80% to be seizure free and around half do not even require AEDs.[147,148]

AEDs and other drug combinations

The inducing AEDs will accelerate the clearance of any CYP3A4, or CYP2C9/ C19 substrate and the impact on the patient will depend on their age, health, and their concurrent medicine burden. Whilst polypharmacy (AEDs, antidepressants, and antipsychotics) can be an issue in younger epileptic patients, their hepatic blood flow and renal function are likely to be far superior to those of the elderly, where polypharmacy is often much more problematic for those over 60[149] (Chapter 7.3.1). Rather remarkably, given how long inducing AEDs have been in use, patients are still either not informed about the impact induction can have on concurrently used drugs, or they often do not recall being informed. For example, induction can lead to treatment failure with oral contraceptives (OCs), as the proportion of female epileptics using OCs is similar to that of the nonepileptic female population[150]. Inducing AEDs in the elderly can accelerate the clearance of several drugs, particularly cardiovascular-related agents, such as warfarin, simvastatin and to a lesser extent, metoprolol, as well as with mental health drugs, such as quetiapine, risperidone, and mirtazapine[151]. Some authors report that the increase in confidence and experience with the later generations of AEDs, has led the older inducing drugs are being prescribed in smaller quantities, so interactions are gradually falling[151]. However, others have found that inducing AEDs are more likely to be prescribed to the elderly.[149]

AEDs prescribing issues

Whilst the drug industry has not always been a 'shining light' in terms of its general conduct worldwide, it deserves credit for the second and third generation AEDs, which constitute real progress. Not only are they generally effective, but they are far simpler to add or remove from existing regimes than the inducing drugs. Crucially, there are enough of the new agents to be tried in combinations that might control a resistant patient's seizures without necessarily returning to the older problematic drugs. Indeed, there are some parallels between the SSRIs and the TCAs, in that the newer generation may not be vastly more pharmacologically effective than the old, but they are safer and easier to use. However, even after so many decades of experience with AEDs, there remains a lack of clinical understanding of their impact on regimes, significant under-reporting of the issue of induction in newer drugs and a lack of appreciation for the patient's experience of marked changes in plasma drug levels caused by interactions[63,135]. These problems are compounded by suboptimal knowledge of drug metabolism's role in potential drug interactions in some prescribers.

What is also rather surprising is how little research has been done into *rational polytherapy* in resistant epilepsy, where drug combinations are planned on the premise of complementary modes of action that improve efficacy and reduce toxicity[152,153]. Certainly, if such a promising approach becomes more widely adopted, the older inducing drugs will to some extent continue in use and knowledge of their impact on drug regimens will remain necessary to safeguard patient welfare.

AEDs are now used worldwide in a variety of conditions unrelated to epilepsy, ranging from bipolar and schizophrenic disorders to neuropathic pain, plus a wide range of anxiety-related conditions[154]. Augmentation of antipsychotics with AEDs can be useful in drug resistant schizophrenia[155]. The use of AEDs in bipolar disorder is a particularly complex issue due to the polypharmacy that these patients often experience[127]. Whilst the newer AEDs may fit into such regimes much more easily than the inducing drugs, there is a wealth of clinical experience and published dose correction tables available to counteract the impact of induction on the clearance of concurrent CYP3A4/CYP2C9/CYP2C19 substrates[1,127]. The drug industry has encouraged the 'off-label' usage of AEDs[135] and in many cases this has been beneficial. Indeed, clobazam is a benzodiazepine that started life as an anxiolytic and was found to be effective as an adjunctive AED,[135] so it has dual applicability. However, in History 8, the issue of CYP3A4 induction (indeed, the lack of it) was a key factor, when off-label use of tiagabine led to patients being inadvertently, yet persistently overdosed with the drug. This left them vulnerable to its somewhat ironic off-target seizurigenic effects in non-epileptics, which were often compounded by the concomitant presence of seizure threshold lowering drugs such as antipsychotics and antidepressants[1,156].

4.5.3   OTC (over the counter) and online herbal preparations

On the strength of reading Chapter 2, you will be thoroughly clued into the 700 million or so-year-old struggle that plants have waged to manipulate to their advantage anything and anybody that eats them. You may be aware of this issue, but the vast majority of those that use, manufacture, distribute and advertise herbal remedies, products, and dietary supplements are not. Most of us would consider departing a shop having bought all of its stock when we set out to obtain only one item rather bizarre, yet from a pharmacological perspective, this is analogous what we do when we consume a single herbal remedy, which is a mix of hundreds of potentially active chemicals. Nevertheless, although estimates of consistent or occasional herbal preparation usage around the world vary wildly, these products are a permanent part of many millions of patient's lives. It is also established that their use is intensified in those suffering from intractable conditions that are either poorly treated by conventional medicine, or their treatment is so arduous and toxic that the herbs are used to make it bearable. One study found a range of consistent herbal preparation use in a US cohort of 12–18% in healthy individuals, which rose to 16–24% in those with a cancer diagnosis,[157] although cancer patients in less developed countries with a stronger traditional use of such agents are more than twice as likely to use these products as those in the United States[158]. Some estimate that nearly all cancer patients use some form of herbal preparation at some point.[159]

Several clinical issues arise from this. First, the extent of interference exerted by herbal preparations in cancer chemotherapeutic regimes is very substantial-indeed, of 44 herbal products, 29 were considered to be either toxic, inducers or had pharmacodynamic influences on the anticancer drugs[158]. Another study emphasized the issue with oral chemotherapy drugs and found 39 interactions with the most significant problem being the PXR agonist St John's Wort (*hypericum perforatum*).[160] As mentioned in the section on P-gp and cancer, this herb not only has some efficacy in mild to moderate depression and anxiety, but has anticancer, antiviral and antiflammatory properties. Indeed, its pharmacologically active substituents, includes hypericin, pseudohypericin, and hyperforin, as well as several flavonoids, xanthones, and a considerably number of phenolics[161]. The hyperforin impacts PXR and neurotransmission systems, whilst hypericin is the antiviral and can also induce photosensitivity in high doses[161]. In cancer chemotherapy, for example, the Wort can increase the clearance of CYP3A4-metabolised tyrosine kinase inhibitors such as imatinib by more than 40%,[160] although its impact in contraception is not as pronounced, probably related to hyperforin content in the preparations studied.[14,162]

The major pharmacokinetic issues around St John's Wort, however, have been known since the turn of the century, such as its ability to accelerate cyclosporine clearance in renal transplant patients to the point that there is risk of tissue rejection unless the immunosuppressant's dosage was increased by 60%. This is also the case with the newer anti-rejection drugs tacrolimus and everolimus. Moreover, the inducing effect of the herb complicates the pharmacology of cyclosporine, as the increased production of metabolites is associated with greater nephrotoxicity, although the metabolites also exert an immunosuppressive effect.[163] St John's Wort also increases the clearances of amitriptyline and the anti-HIV agent indinavir, although it does not with carbamazepine as this drug already maximizes its own induction.

There are several key clinical problems to summarize that may be encountered by the use of herbal preparations and St John's Wort in particular:

- Many patients do not consider herbal remedies as real drugs, as they are generously provided by Mother Nature, so they do not inform their medical practitioner they are taking them (History 4). In addition, patients often fully expect these remedies to be wholly beneficial without any side effects. In short, in their self-delusion, ignorance and desperation, they can ascribe almost 'magic' properties to these products.

- The onset of the inductive effect of any PXR agonist will vary, with natural human genetic differences, but this is compounded by the variation in the hyperforin content of the extracts.

- The patient may abruptly terminate their self-medication if they feel better or encounter side effects (History 4); part of self-medication is the somewhat illusory idea of 'being in control'.

- Hyperforin, pseudohypericin, and hypericin have half-lives in the range of other PXR ligands (16–20 hours)[164] so the time frame of the diminution of the inductive effect should follow a similar pattern to other PXR agonists, but there are many other agents in the extracts, which might affect this process

- Quality, purity, and content of the active ingredients in herbal preparations can vary widely according to methods of preparation.

This final point is crucial-variation in content of the various active ingredients is well documented. A pile of freshly and carefully dried St John's Wort should contain around 5% hyperforin by weight,[165] but due to the instability of the chemical and variation in drying techniques, extracts can vary considerably (more than five-fold) in the hyperforin content[164]. One group could not even detect hypericin in some preparations at all despite their best efforts and some well-established analytical techniques.[166] The many other phenolics in the extracts varied remarkably in concentrations and proportions; in fact, in twelve preparations, there was a *29-fold* variation in one flavonoid, isoquercitrin,[166] which incidentally is an MRP2 substrate. It is potentially confusing as to how St John's Wort is discussed and dosed; this could be mg of extract or mg of hyperforin. If we consider the weight of hyperforin, it seems that dosages taken can actually vary by more than a 100-fold from 0.04 mg per day, all the way to over 50 mg per day.[164] The obvious question is how close the therapeutic (CNS) concentration might be to that which has significant PXR activity. It seems that it will manage depression in the range of 1–3 mg hyperforin daily, yet it is considered that anything above 1 mg per day activates PXR.[164] Like any drug, its CNS effects are dose dependent,[167] but the published data clearly shows that many users of St Johns Wort are overdosing themselves quite markedly.[164]

Even after comprehensive investigation, questions remain over the efficacy and impact of St John's Wort on drug regimens. However, at least it has been investigated-most other herbal preparations that are available are essentially classed as foodstuffs and have had little even in the way of preliminary evaluation for their effects on the clearance of drugs. It is now emerging that in Chinese herbal medicine, a large number of commonly used herbal preparations contain potent CAR or PXR agonists and this is actually how they exert their pharmacological effects[15] in cholestatic jaundice and inflammatory bowel disease for example, which creates a significant problem for concurrent CYP3A4/CYP2C9 cleared medication. Whilst some preparations have been investigated and are probably safe to use with medication, such as green tea,[168] ginkgo biloba extract has been shown to contain a complex mix of PXR, CAR, and AhR agonists[15] and it does induce CYP3A4 clinically,[169] but not CYP2C19[157]. The effects of Echinacea are more complex, although it has been used for centuries to combat the symptoms of colds and flu, its efficacy in the prevention of such maladies and its actions on the immune system are unproven. It has some inhibitory effects in CYPs (Chapter 5.5.4), but apparently does not induce Pg-p in humans.[170]

One point important to emphasize is that although various herbal remedies do contain active and potent substituents, there is virtually nothing known clinically about what effects mixing herbal remedies might have, in terms of dose, pharmacology and toxicity. Unfortunately, the patients will be discovering this for themselves the hard way. Kava-kava, for example, has a well-deserved reputation for liver toxicity[171] and this may have been exacerbated by St John's Wort in at least one case[171,172]. Case reports will periodically reach the literature of problematic reactions in years to come and a picture will gradually emerge in terms of which remedies interact with medicines and each other to patient detriment.

The impact of herbal remedies is also substantial with respect to CYP inhibition, which will be covered in the next chapter. What is clear is that the use of such remedies worldwide is not a passing fancy and it seems that some practitioners are either unable or unwilling to recognise the prevalence of their use. So issues such as practitioners failing to ask about and patients failing to report herbal usage are all avoidable. Hence, health-care practitioners must work with patients, partly by ensuring that they themselves understand about the major problematic herbal remedies in their respective fields, but crucially, they must explain to patients from the onset of treatment about the power of these remedies to potentially derail their therapeutic process. It is significant that if patients believe a herbal remedy will negatively impact their treatment, they will stop taking it,[173] so the situation can be approached by recognition by both sides, patient and practitioner, that all therapeutic cards must be on the table and that they are on the same side.

4.5.4   Anticoagulant drugs

Atrial fibrillation (AF) is a condition where areas of tissue in the atria trigger off random and unauthorized (so to speak) impulses that cause the muscle to twitch, promoting the formation of microemboli, which in turn increases patient risk of a stroke by five-fold[174]. AF is present in only a couple of percent of those under 60, but it is estimated to affect up to a quarter of adults over the age of 65 in the developed world[176,176] and is responsible for up to a third of all strokes,[177] so treatment is a top priority, although many individuals are asymptomatic. Treatment can be surgical, by burning away the offending area of tissue (catheter ablation), which can be effective in those resistant to antiarrthymic drugs. If the drugs are continued after that procedure, it seems half the patients will be AF free after five years[178]. Alternatively, direct current cardioversion (DCCV)[175] can be used, which is basically a hard electrical reset of the heart, although this carries a 7–10% risk of stroke due to the procedure, which is more than ten-fold less likely to happen with prior anticoagulant therapy[174]. To those old enough to remember the unreliable televisions of the 1960s, DCCV is reminiscent of thumping the set with your fist to restore the picture. Whilst a result is achieved almost immediately in 90% of cases, a year or so on, the success rate is low[179]. Both approaches carry risks and the mainstay of therapy is drug-based control of heart rate and rhythm along with anticoagulants to try to eliminate AF and its consequences[174]. However, over the last decade anticoagulant therapy has moved on, with the emergence of the direct oral anticoagulant drugs (DOACs) dabigatran, rivaroxaban, apixaban, edoxaban, and betrixaban. These are as effective as the leading Vitamin K antagonist anticoagulant, warfarin.[174,175,180] Interestingly, despite betrixaban's failure to gain European approval, these drugs have some significant practical advantages over warfarin. Indeed, if warfarin were proposed as a new drug today, it would be given short shrift for many reasons. First, because it inhibits clotting factor synthesis, coagulation continues until the existing supply of factors runs out, so it takes about five days before it will actually work; so if you need an instant effect, you must initially titrate with heparin.[181] Next, warfarin must be monitored pharmacodynamically, that is to say its effect is measured rather than its concentration. The measurement 'INR' (international normalized ratio) is used, which assumes normal blood clotting time is 1, so for warfarin to anticoagulate enough to successfully treat AF, an INR of 2.5 is required. So through frequent clinical monitoring, the dose is adjusted to achieve an INR value of approximately 2–3. At the other end of the scale, those with mechanical mitral valve replacement require maintenance of an INR of 3–3.5. This means asking the patient to commit to attend regular clinics to stabilize their dosage.[182] Warfarin is also given as a mixture of two isomers, S and R. The S isomer is up to five-fold more potent an anticoagulant than the R isomer, and the S is cleared by CYP2C9, whilst the R is metabolised by CYP1A2 and CYP2C19. CYP3A4 has a minor role for each isomer, although the role of CYP2C19 is also becoming better understood in warfarin clearance.[183,184]

From the perspective of CYP induction, plasma levels of warfarin will decline in the presence of the main CYP inducers[185] and after a lag period, its anticoagulant effects will recede (History 5). If inducers must be used, then warfarin dosage must increase and there is no substitute for monitoring INR to ensure that the drug remains within the therapeutic window. If an enzyme-inducing drug is withdrawn, there is the danger of accumulation of the increased dose anticoagulants, which will lead to haemorrhaging. It appears that most clinical problems with warfarin seem to occur as a result of the effect of inhibitors (Chapter 5.7.7) rather than inducers and some groups of patients on warfarin are more at risk than others from drug interactions, particularly those receiving cancer chemotherapy. Warfarin is also subject to very significant patient variation in terms of pharmacodynamic response, pharmacogenetic issues (Chapter 7.2.3) as well as the issues with clinical monitoring.[183,184]

By contrast, the newer drugs have a different mode of action than Vitamin K antagonism; they directly inhibit either thrombin (dabigatran) or factor Xa (rivaroxaban, betrixaban, edoxaban, and apixaban). This means anticoagulation occurs between 1-4 hours, against the best part of a week for warfarin[174]. Also, they are given as fixed doses, do not require extensive pharmacodynamic monitoring and are more predictable in clinical use.[174,175] They also have wider therapeutic indices compared with warfarin, and their short half-lives mean that rapid initiation and removal makes them easier to use. In addition, they are not subject to any dietary restrictions, unlike with warfarin,[186] although you should definitely not take betrixaban with your deep-fried meat pie, as it will not be absorbed[187]. In terms of the impact of inducers, dabigatran is administered as a prodrug, dabigatran etexilate, which is a Pg-p substrate, but it is then hydrolysed to dabigatran that is not a substrate. Dabigatran and betrixaban are not CYP metabolised, but dabigatran is mainly renally cleared, whilst betrixaban is splendidly faecally cleared.[187-189] Apixaban, betrixaban and rivaroxaban are P-gp substrates.[186] Rivaroxaban is mainly CYP3A4-cleared, with apixaban less so. Hence, inducers will accelerate the clearance of rivaroxaban and apixaban and P-gp induction will delay absorption-rifampicin cuts the rivaroxaban AUC in half[186,188]. Only about a third of apixaban is metabolised, which makes it less vulnerable to inducers than rivaroxaban, but rifampicin will significantly impact its clearance[188]. Edoxaban is a P-gp substrate, although only about a quarter is cleared by hepatic metabolism and only some of that fraction is cleared by CYP3A4, rifampicin still reduced systemic exposure of edoxaban by 34%[190]. Whilst Pg-p induction can impact these drugs, with the exception of rivaroxaban, they are less vulnerable than warfarin to induction issues, and none of them appear to be inhibitors or inducers of CYPs[190].

Unfortunately, whilst the DOACs have many positive features, the lack of antidotes until recently[191] combined with increased incidences of severe bleeding has been problematic.[192] Also, adherence is an issue with antocoagulants in general. If you are older than 65 you are more likely to be adherent, probably due to other health issues and general concerns over aging, whilst younger individuals are less adherent, partly based on issues with treatment and information overload, but also, in some countries, with the cost of the medicines.[177]

4.5.5   Oral contraceptives/steroids

Oral contraceptive steroids are combinations of different doses of ethinyloestradiol and progestogens, and the major inducers of CYP3A4 and CYP2C9 will reduce their efficacy. This has long been recognised, and there are many detailed protocols available to accommodate the inducing AEDs, such as tricycling regimens and tailored pill taking[193]. Good patient counselling is essential, over concerns about the link between increased steroid doses and hormonal side effects, which need to be explained in terms of the inductive effect of the AED and that side effects are related to blood levels rather than dose[193]. One significant interaction is caused by the contraceptive steroid induction of glucuronidation processes (Chapter 6.2.1); this can mean that a patient whose epilepsy is controlled on lamotrigine who starts taking a contraceptive steroid combination may have a seizure, as the induction of glucuronidation can cut lamotrigine plasma levels in half[193]. Just a single seizure will mean no driving for 12 months followed by an eye-watering insurance price-hike.

The rifamycins' impact on the level of unwanted pregnancies is rather poorly investigated, although it is likely they do reduce systemic exposure to oral contraceptives, and it seems that rifampicin is more likely to cause this effect than rifabutin, which is reasonable, given their respective PXR agonism differences[194]. St John's Wort would be expected to have a significant impact on oral contraceptive steroids, but again, this has not been very systematically investigated. Plenty of case reports in the United Kingdom have emerged where the use of this herb has caused unwanted pregnancies[195] and a systematic review concluded that St John's Wort caused 'weak to moderate' impact on oral contraceptive steroid levels, which is most likely to be problematic with low-dose oral contraceptive preparations. Increasing the contraceptive dose, or a recommendation to use other methods of contraception, may negate this effect[14].

Corticosteroids are both substrates and potent inducers of CYP3A4, whilst also inducing CYP2C19, CYP2C9 and CYP2B6 to a lesser extent. Even modest anti-inflammatory doses (4 mg dexamethasone daily) can cause agents such as voriconazole, a predominantly CYP2C19 substrate, to fail therapeutically[196]. Dose ranges in corticosteroid use can vary – more than a gramme daily of prednisone for a week has been used for multiple sclerosis exacerbations[197]. Hence, adding these drugs to any regime will accelerate the clearance of any substrates of these CYPs necessitating dosage adjustment unless the patient is already fully induced by another strong PXR agonist.

4.5.6   Antiviral/antibiotic drugs

Naturally, any sustained reductions in the plasma levels of an antibiotic or antiviral agent can lead to subcurative drug concentrations and a possible selection of resistant variants of the infectious agent, so plasma levels should be closely monitored to ensure minimum inhibitory concentrations (MICs) are exceeded while toxicity is minimized. Histories 6 and 7 demonstrate the impact rifampicin can have on other drugs, although as mentioned earlier, other antibiotics such as griseofulvin and dicloxacillin may also induce CYP3A4 substrate clearances.

Several of the broad group of anti-HIV drugs are also significant CYP inducers. In the context of HIV, one of the most remarkable achievements of medicine over the past 35 years has been to turn what was a certain and miserable death sentence, into a chronic but manageable condition, although to translate this feat to all those HIV sufferers outside of the developed world is another altogether more challenging issue.

The containment of HIV has been achieved through the evolution of a series of complex regimes of drug combinations aimed at simultaneously attacking various aspects of HIV's pathology, but the drugs also happen to have different biotransformationally inductive and inhibitory properties. This means that these regimes have been and continue to be extremely problematic pharmacokinetically and pharmacodynamically to evaluate and turn into practical therapeutic propositions and indeed deserve their own textbook. HIV patient polypharmacy is a very significant problem, as aside from the necessity of taking five or six drugs to contain the virus, the patient will periodically suffer from various infections, as well as other conditions that are common in their age group. This can lead to very complex drug regimes indeed, and the impact of inducers and (in the next chapter) inhibitors can be very significant.

HIV is now managed through HAART, or highly active anti-retroviral therapy, which is a combination of different classes of drugs that have different mechanisms of action. Such a combination will be assembled according to the patient's needs and is a very good example of personalized medicine, which contains the virus, but is a lifetime therapeutic adherence commitment. The main classes of the protease (such as ritonavir, nelfinavir, amprenavir), integrase (such as elvitegravir and dolutegravir), nucleoside/nucleotide reverse transcriptase (NRTIs; zidovudine) and nonnucleoside reverse transcriptase inhibitors (NNRTIs; efavirenz, etravirine), as well as the CCR5 antagonists (maraviroc) and CYP3A4 inhibitors (see Chapter 5, cobicistat and ritonavir).

Of the anti-HIV drugs, the protease inhibitors are well documented for their unpredictable effects on CYP expression and the clearances of other drugs. These drugs are PXR agonists and are mostly modest inducers (two- to five-fold) at the mRNA level, of CYPs 1A2, 2B6, 2C9, 2C19, 3A4, and glucuronyl transferases, as well as the hepatic transporter OATP1B1 and P-gp,[9,198] with amprenavir being the most potent[198.] In terms of increased functional biotransforming enzyme expression, all the isoforms are induced by these drugs, although the increased CYP3A4 protein is immediately inactivated through mechanistic inhibition (see Chapter 5.4). This usually but not always results in a net reduction in CYP3A4 activity *in vivo*.[198] The issue is whether the intensity of the mRNA induction overcomes the inhibition caused by the same drug. Ritonavir is a more effective inhibitor than inducer, whilst amprenavir is the reverse and can cause a minor net increase in CYP3A4 activity despite its inhibitory effects[198]. Overall, the protease inhibitors can increase the clearance of CYP2B6 substrates such as methadone, whilst they have less impact on CYP2C9 substrates like phenytoin[198]. The inductive effect on P-gp of the protease inhibitors has marked effects on substrates of this transporter.[198] The inhibitory effects of the protease inhibitors are used deliberately in the process of boosting of the residence times of other antiviral substrates of CYP3A4,[9,199] (Chapter 5.8).

Of the other anti-HIV drugs commonly used efavirenz is a moderately potent CYP3A4 inducer[200] and will accelerate the clearance of other antivirals and co-medications cleared by this isoform. Etravirine (a NNRTI) is also a modest CYP3A4 inducer[199]. Both drugs complicate drug regimes, as several anti-HIV agents are CYP3A4 substrates, such as maraviroc and elvitegravir, whilst others such as dolutegravir are UGT1A1 substrates with some CYP3A4 involvement,[200] so inducers do accelerate their clearance[33,200]. Powerful inducers such as rifampicin and St John's Wort will naturally strongly reduce systemic exposure of these drugs unless the dosages are increased significantly.

4.5.7 Anticancer drugs

Cancer patients as a group are not only subject to the physical and mental impact of their condition but also sometimes the unendurable toxicity and stresses of their treatment. They usually require medication to lessen this impact and also to combat other issues such as acquired infections. As if this was not enough, they are also very prone to suffering drug–drug interactions as a result of this polypharmacy[201]. Drug clearance in cancer chemotherapy is even more strongly linked to efficacy and toxicity than with other drugs, mainly due to the very narrow TI's and steep dose/toxicity curves of the antineoplastic agents[201]. Almost all cancer patients will receive drugs that are CYP cleared; inducers significantly affect therapy in three ways:

1. If the parent drug is active and the metabolites are not, an inducer may reduce efficacy and lose the patient vital time in the struggle to eliminate or contain the disease and prevent metastasis.

2. If the metabolites are active, then accelerated clearance may improve efficacy, but the lack of selectivity in these drugs means that efficacy increase usually occurs along with increased patient toxicity, which directly impacts patient tolerance and adherence through the full treatment cycles.

3. The addition of an inducer to a regime may upset the hard-won balance that has been achieved by the patient and practitioner between dose, efficacy, toxicity and tolerance.

A long list of antineoplastic drugs are cleared by the main inducible CYPs; Indeed, for example, the taxol drugs, (cabazitaxel, docetaxel, and paclitaxel), the tyrosine kinase inhibitors such as imatinib and gefitinib, the topoisomerase inhibitor irinotecan, the vinca alkaloids as well as the oxazaphosphorines (cyclophosphamide and ifosfamide) are all CYP3A4 substrates[201-203]. A moderate inducer (St John's Wort) can significantly reduce the systemic exposure of a CYP3A4 substrate such as docetaxel after 14-days exposure to a standardized version of the herb, admittedly at quite a high dose (27–57 mg hyperforin daily)[204]. As docetaxel is given intravenously, the inductive effect of the Wort on Pg-p and intestinal CYPs was bypassed, but the hepatic induction was still enough to reduce the drug's AUC and significantly, it also reduced the level of side effects in the patients. So you can see that the Wort's amelioration of the long list of unpleasant effects of docetaxel would be rather attractive to patients, given that nine out ten of them in this study suffered from the various peripheral neurological issues, nausea, fever and loss of hair that are all associated with the drug[204].

Irinotecan is a prodrug used in colon cancer and it acts through its metabolite SN-38, formed by carboxyesterases, followed by detoxification to a glucuronide (SN-38G). CYP3A4 competes for the prodrug, forming an ineffective metabolite, called APC. St John's Wort through its CYP3A4 induction can accelerate clearance to other metabolites as well as APC and thus restrict prodrug conversion to SN-38 formation by 42%,[205] reducing both the anti-neoplastic efficacy and the intensity of the side effects, as with docetaxel. A similar effect occurs with the inducing AEDs and requires dosage adjustment.[205]

Regarding the oxazaphosphorines, cyclophosphamide is activated to a 4-hydroxy metabolite mainly by CYP2B6, CYP3A4, and to a lesser extent, CYP2C9; of the other derivatives in this class, the next most commonly used is ifosphamide, which is mainly activated by CYP3A4, CYP2A6 and to a minor extent, CYP2B6[206,207]. These 4-hydroxy derivatives easily cross membranes and rearrange themselves into highly reactive phosphoramide mustard derivatives, which are similar to the blister-forming antipersonnel mustard agents used in the Great War (1914–1918). These metabolites kill malignant cells by alkylating their DNA, leading to a therapeutic effect, but their general cytotoxicity causes an appalling list of toxic effects, including hair loss, gut damage, nausea and vomiting, as well as cystitis, nephro/neurotoxicity, and immunosuppression[206]. These drugs are often given in complex combination regimes such as CPA (cyclophosphamide/methotrexate/5-fluorouracil) and the rather ominous-sounding R-CHOP (rituximab, cyclophosphamide, doxorubicin. vincristine and prednisone). Moderately successful attempts have been made to lessen the toxicity of these compounds through inserting high turnover CYPs into tumours that will form the 4-hydroxy derivatives and attempt to confine the efficacy and toxicity to the tumour (gene-directed enzyme prodrug therapy, GDEPT)[207]. However, oxazaphosphorines can induce their own metabolism (ifosphamide is the most potent PXR inducer) and are given with CYP3A substrates, such as vinca alkaloids, so other CYP inducers may accelerate the clearance to toxic metabolites making therapy intolerable[208,209].

It is emerging that drugs such as aprepitant intended to reduce CINV (chemotherapy induced nausea and vomiting) can also interfere with CYP-mediated clearances of antineoplastic agents. Whilst aprepitant and its prodrug fosaprepitant are mild CYP2C9 inducers, their more significant impacts are related to CYP3A4 inhibition but are clinically manageable[210]. Other classes of CINV drugs include the 5HT-3 antagonists, that are mainly cleared by CYP3A4 (ondansetron and granisetron) and CYP2D6 (dolasotron and palonosetron) are not known as CYP inducers, themselves, but potent inducers will accelerate their clearance and reduce their efficacy[211].

4.6 Induction: practical considerations

In changing inducing drugs within regimens, prescribers can be guided by various tables that supply approximate correction factors for particular drugs and so dosage changes in the co-administered drugs can made if necessary[127]. As we have seen in this chapter already, there is so much human variation in every stage of the induction process, it is difficult to predict what might happen. Generally, if a potent inducer is added to a regime that already has an inducer of similar magnitude, then it is unlikely that any marked changes might occur, as the CYPs will be already maximally induced. A mild inducer added to such a regime will have no effect at all[1, 127]. A mild inducer replacing a potent inducer will probably have the effect of prolonging some induction effect for several weeks. On the extreme of this scale, there are individuals who are exquisitely sensitive to inducers and might require many grammes of a drug per day just to achieve some therapeutic effect[1]. As mentioned earlier, aside from the dose, the half-life of the inducer governs the process, as it can only stimulate the nuclear receptors if it is present in sufficient concentration.

Even in well-designed clinical studies it is not easy to identify all the components of an inductive effect, say, in terms of CYPs, transporters or other systems, which may be multifactorial. Indeed, as biotransformational systems have great depth and flexibility, inducers can make unexpected impacts on drug clearance. CYP2D6 is mildly inducible, but if a patient is a poor metaboliser, they produce little or no functional enzyme, so induction will not change this. However, in some cases the presence of an inducer will cause another CYP to increase in expression, which can also clear the drug and changes in clearance will be seen where they might not be expected. This can occur with the CYP2D6 substrate risperidone, whose clearance in poor metabolisers can be increased to just short of normal by CYP3A4 inducers[212]. In the real world, the net impact of an inducer will be seen through either clinical observation and/or therapeutic monitoring, where plasma levels, pharmacodynamics, and toxicodynamic effects are assessed to ensure that the drug is tolerated and effective.

4.7 Induction vs. inhibition: which 'wins'?

Another challenging question is what happens when inducers and inhibitors are used together. Well, acutely, no matter whether the patient is taking an inducer or not, a single dose, or a series of doses over a few days of a potent inhibitor will inhibit any given CYP, irrespective of the level of induction,[62] but of course will not affect the amount of CYP protein formed. As mentioned earlier, in section 4.4.3 concerning PXR therapeutic modulation, it is difficult currently to inhibit PXR at clinically safe drug levels, so the potent CYP3A4 inhibitor keto-conazole does not impact the St John's Wort PXR-mediated induction, but does block the induced CYP3A4.[62] In 4.5.6 in the unusual case of the protease inhibi-tors, their PXR-mediated inductive effect on CYP3A4 produces more CYP3A4 protein, whose capability is promptly wiped out by their inhibitory effects. Again, the inhibitor does not affect the induction process, just its product.

However, over a longer period of time, the most potent inducers such as rifampicin can induce several different CYPs, so it is likely that at least one of them will accelerate the clearance of even very potent inhibitors. Voriconazole is second only to ketoconazole in its ability to inhibit CYPs 2C9, 2C19 and to a lesser degree, 3A4,[112,214] but it is cleared by the same CYPs. This means that in a few days, inducers can drag voriconazole plasma levels to below therapeutic levels[194] and rifampicin can make it practically disappear from the circulation, so that combination is not recommended unless significant dose increases are made[112]. So it does not matter about an inhibitor's potency – if it is not present, it cannot inhibit anything. Indeed, rifampicin PXR agonism is so strong it can defeat several inhibitors at once and still accelerate the clearance of a CYP3A4 substrate.[213] Again, there seems to be no substitute for therapeutic monitoring to be really sure about what will happen in the patient. The next chapter will look at inhibition, which has a much more immediate and in some cases dangerous effect on the patient's life and health than induction.

4.8   Induction: long-term impact

Whilst this chapter has been focussed scientifically and clinically on how biotransformational induction processes impact therapy and complicate drug regimes, it is worth considering the longer-term effects on the patient of powerful inducers. In Section 4.2, the phrase *adaptive increase* was used to describe the process of induction, which explains how we may survive periods in different environments and cope with different chemicals in our diets and of course, dif-ferent drugs in our treatment. Patients are exposed to inducers over a few days with a short course of steroids, to perhaps a couple of months with St John's Wort, or as long as six months or a year on rifampicin during anti-tuberculosis treatment. However, other patient groups are exposed to enzyme inducing drugs for several years and even decades. These include those using AEDs, but increas-ingly we must consider those enrolled on inducing drug regimens whose success has massively increased the duration of patient therapy, such as HAART in HIV and the new generations of anti-androgens used in prostate cancer therapy, such as enzalutamide.

To determine whether lifelong exposure to enzyme inducers is deleterious to health is problematic as the condition being medicated has its own impact and it is difficult to separate the two issues. Epileptic patients are the most obvious group to investigate, as whilst many have spent decades on inducing drugs, some noninducing drugs have also been in practice long enough for conclusions to be drawn on overall impact. It has been known since the 1960s that AEDs reduce bone density, but a combination of poor-quality investigations and contradictory evidence (particularly with valproate) meant that it has taken decades to establish that the inducing AEDs are more likely to cause bone problems such as fractures and reductions in density than the noninducing drugs[215-217].

The main theory that attempts to account for these observations is that inducing AEDs accelerate vitamin D metabolism, causing concomitant reduction in the active steroid's plasma and bone concentrations[215]. In support of this, it has been shown that PXR activators may stimulate the expression of genes related to bone maintenance and vitamin D metabolism that are also under the control of the vitamin D receptor[218]. Whatever the detailed mechanism, the startling reduction in life-expectancy of those over 50 that experience osteoporitic bone breaks makes the potential impact of inducing drugs in this context highly significant.[219]

What is agreed is that inducers cause a net increase in the levels of sex hormone binding globulin, which lowers free hormone levels and distorts hormone concentrations relative to each other[220]. This contributes to impacts on fertility, sexual function, and thyroid metabolism, as well as the bone density problems.[220] There may also be a link between biotransformational induction and AED modulation of the immune system which is linked to their propensity to cause severe hypersensitivity reactions in patients, which is discussed in Chapter 8.4.[221]

Overall, it is probably reasonable to suggest that in biological terms, the adaptive increase in biotransformation induction perhaps has to be viewed as relatively temporary and is not intended to be a permanent state of being. If it does become permanent, then there is a significant price to pay for this adaptation and the success achieved over the past two decades in developing alternative noninducing therapeutic agents should be encouraged and intensified.

References

1. De Leon J. The effects of antiepileptic inducers in neuropsychopharmacology, a neglected issue. Part II: Pharmacological issues and further understanding, Rev Psiquiatr Salud Ment (Barc.). 8, 167–188, 2015.
2. Niwa T, Murayama N, Imagawa Y et al. Regioselective hydroxylation of steroid hormones by human cytochromes P450, Drug Metabolism Reviews, 47(2), 89–110, 2015.
3. Cockshott ID, Bicalutamide clinical pharmacokinetics and metabolism, Clin Pharmacokinet 43, 855–878, 2004.
4. Del Re M, Fogli S, Derosa L, et al. The role of drug–drug interactions in prostate cancer treatment: Focus on abiraterone acetate/prednisone and enzalutamide, Cancer Treatment Reviews 55, 71–82, 2017.

5. Ivachtchenko AV, Mitkin OD, Kudan EV et al. Preclinical development of ONC1-13B, Novel Antiandrogen for Prostate Cancer Treatment J. Cancer 5 133–142, 2014.

6. Weiss J, Kocher J, Mueller C, et al. Impact of enzalutamide and its main metabolite N-desmethyl enzalutamide on pharmacokinetically important drug metabolizing enzymes and drug transporters. Biopharm Drug Dispos. 38, 517–525, 2017.

7. Lutz JD, Kirby BJ, Wang L, et al. Cytochrome P450 3A induction predicts P-glycoprotein induction 1: Establishing induction relationships using ascending dose rifampin. Clin Pharmacol Ther. Mar 23. doi: 10.1002/cpt.1073, 2018.

8. Regazzi M, Carvalho ACC, Villani P, et al. Treatment optimization in patients co-infected with HIV and mycobacterium tuberculosis infections: Focus on drug–drug interactions with rifamycins. Clin Pharmacokinet 53, 489–507, 2014.

9. Nguyen T, McNichol I, Custodio JM. Drug interactions with cobicistat or ritonavir-boosted elvitegravir. AIDS Rev. 18, 101–111, 2016.

10. Liu K, Yan J, Sachar M et al. A metabolomic perspective of griseofulvin-induced liver injury in Mice Biochem Pharmacol. 98, 493–501, 2015.

11. Stage TB, Graff M, Wong S, et al. Dicloxacillin induces CYP2C19, CYP2C9, and CYP3A4 *in vivo* and *in vitro*. Br J Clin Pharmacol. 84, 510–519, 510, 2018.

12. Liu L, Mugundu GM,. Kirby BJ, et al. Quantification of human hepatocyte cytochrome P450 enzymes and transporters induced by HIV protease inhibitors using newly validated LC-MS/MS cocktail assays and RT-PCR. Biopharm. Drug Dispos. 33, 207–217, 2012.

13. Garimella T, Tao X, Sims K et al. Effects of a fixed-dose co-formulation of daclatasvir, asunaprevir, and beclabuvir on the pharmacokinetics of a cocktail of cytochrome P450 and drug transporter substrates in healthy subjects drugs. R D 18, 55–65, 2018.

14. Berry-Bibee EN, Kim MJ, Tepper NK, et al. Co-administration of St John's wort and hormonal contraceptives: a systematic review contraception 94, 668–677, 2016.

15. Xu C, Huang M, Bi H. PXR-and CAR-mediated herbal effect on human diseases Biochim. et Biophys. Acta 1859, 1121–1129, 2016.

16. Li MY, Liu Y, Liu LZ, et al. Estrogen receptor alpha promotes smoking-carcinogen-induced lung carcinogenesis via cytochrome P450 1B1. J Mol Med 93, 1221–1233, 2015.

17. Hryhorowicz S, Walczak M, Zakerska-Banaszak O, et al. Pharmacogenetics of can-nabinoids. Eur J Drug Metab Pharmacokinet 43, 1–12, 2018.

18. Sweeney BP & Bromilow J. Liver enzyme induction and inhibition: implications for anaesthesia. Anaesthesia, 61, 159–177, 2006.

19. Oliveira P, Ribeiro J, Donato H et al. Smoking and antidepressants pharmacokinet-ics: a systematic review. Ann Gen Psychiatry 16:17–25, 2017.

20. Darwish M, Bond M, Yang M, et al. Evaluation of Potential Pharmacokinetic Drug–Drug Interaction Between Armodafinil and Risperidone in Healthy Adults Clin Drug Investig. 35, 725–733, 2015.

21. Robertson P Hellriegel ET. Clinical pharmacokinetic profile of modafinil. Clin Pharmacokinet. 42, 123–137, 2003.

22. Novotna A and Dvorak Z. Omeprazole and lansoprazole enantiomers induce cyp3a4 in human hepatocytes and cell lines via glucocorticoid receptor and pregnane X receptor axis. PLoS One 9(8), e105580, 1–9, 2014.
23. Scott O, Galicia-Connolly E, Adams D, et al. The Safety of cruciferous plants in humans: a systematic review. Journal of Biomedicine and Biotechnology. 503241, 1–28, 2012.
24. Van Erp NP, Guchelaar HJ, Ploeger BA, et al. Mitotane has a strong and a durable inducing effect on CYP3A4 activity. European Journal of Endocrinology 164, 621–626, 2011
25. Kolatorova L Vitku J, Hampl R. et al. Exposure to bisphenols and parabens during pregnancy and relations to steroid changes. Env. Res. 163, 115–122, 2018.
26. Ke AB, Nallani SC, Zhao P, et al. Expansion of a PBPK model to predict disposition in pregnant women of drugs cleared via multiple CYP enzymes, including CYP2B6, CYP2C9 and CYP2C19. Br J Clin Pharmacol. 77, 554–570, 2014.
27. Koh KH, Pan X, Shen HW, et al. Altered expression of small heterodimer partner governs cytochrome P450 (CYP) 2D6 induction during pregnancy in CYP2D6-humanized mice. J Biol Chem. 289, 3105–3113, 2014.
28. Buck Louis GM, Chen Z, Peterson CM, et al. Persistent lipophilic environmental chemicals and endometriosis: The ENDO Study. Env. Health Persp. 120, 811–816, 2012.
29. Zeliger HI, Lipophilic chemical exposure as a cause of cardiovascular disease. Interdiscip Toxicol. 6, 55–62, 2013.
30. Chortis V, Taylor AE, Schneider P et al. Mitotane therapy in adrenocortical cancer induces CYP3A4 and inhibits 5-Reductase, explaining the need for personalized glucocorticoid and androgen replacement. J Clin Endocrinol Metab, 98, 161–171, 2013.
31. Svensson EM, Murray S, Karlsson MO, et al. Rifampicin and rifapentine significantly reduce concentrations of bedaquiline, a new anti-TB drug. J Antimicrob Chemother 70, 1106–1114, 2015.
32. US FDA. Guidance for industry. Drug interaction studies – study design, data analysis, implications for dosing, and labeling recommendations 2012. Available at: http://www.fda.gov/downloads/Drugs/GuidanceComplianceRegulatoryInformation/Guidances/UCM292362.pdf.
33. Abel S, Jenkins TM, Whitlock LA, et al. Effects of CYP3A4 inducers with and without CYP3A4 inhibitors on the pharmacokinetics of maraviroc in healthy volunteers. Br J Clin Pharmacol. 65(1), 38–46, 2008.
34. Borrelli F, and Izzo AA. Herb–drug interactions with St John's Wort (Hypericum perforatum): an update on clinical observations. The AAPS Journal 11, 710–727, 2009.
35. Fang X, Ding XX, Zhang QY. An update on the role of intestinal cytochrome P450 enzymes in drug disposition. Acta Pharm. Sin. B 6, 374–383, 2016.
36. Androutsopoulos VP, Tsatsakis AM, Spandidos DA. Cytochrome P450 CYP1A1: wider roles in cancer progression and prevention. BMC Cancer 9, 187–204, 2009.
37. Zajda K, Ptak A, Rak A, et al. Effects of human blood levels of two PAH mixtures on the AHR signalling activation pathway and CYP1A1 and COMT target genes in granulosa nontumor and granulosa tumor cell lines. Toxicology 389, 1–12, 2017.

38. Lnenickova K, Skalova L, Stuuchlikova-Raisova L et al. Induction of xenobiotic-metabolizing enzymes in hepatocytes by beta-naphthoflavone: Time-dependent changes in activities, protein, and mRNA levelsActa. Pharm. 68, 75–85, 2018.

39. Vacher S, Castagnet P, Chemlali W. High AHR expression in breast tumors correlates with expression of genes from several signaling pathways namely inflammation and endogenous tryptophan metabolism. PLoS ONE 13(1), e0190619, 1–22, 2018.

40. Dunlap TL, Howell CE, Mukand N, et al. Red clover aryl hydrocarbon receptor (AhR) and estrogen receptor (ER) agonists enhance genotoxic estrogen metabolism. Chem. Res. Toxicol. 30, 2084–2092, 2017.

41. Attignon EA, Distel E, Le-Grand B, et al. Down-regulation of the expression of alcohol dehydrogenase 4 and CYP2E1 by the combination of α-endosulfan and dioxin in HepaRG human cellsToxicology in vitro 45, 309–317, 2017.

42. Stepankova M, Pastorkova B, Bachleda P, et al. Itraconazole cis-diastereoisomers activate aryl hydrocarbon receptor AhR and pregnane X receptor PXR and induce CYP1A1 in human cell lines and human hepatocytes Toxicology 383, 40–49, 2017.

43. Safe S. Carbidopa: a selective Ah receptor modulator (SAhRM). Biochemical Journal 474, 3763–3765, 2017.

44. Jin UH, Lee SO, Safe S. Aryl Hydrocarbon Receptor (AHR) – Active Pharmaceuticals Are Selective AHR Modulators in MDA-MB-468 and BT474 Breast Cancer Cells JPET. 343, 333–341, 2012.

45. Talari NK, Panigrahi MK, Madigubba S, et al. Overexpression of aryl hydrocarbon receptor (AHR) signalling pathway in human meningioma. Journal of Neuro-Oncology 137, 241–248, 2018.

46. Lu, H. Crosstalk of HNF4α with extracellular and intracellular signalling pathways in the regulation of hepatic metabolism of drugs and lipids Acta Pharmaceutica Sinica B 6, 393–408, 2016.

47. Hogle BC, Guan X, Folan et al. PXR as a mediator of herbedrug interaction. J. Food Drug Anal 26, S26 eS31, 2018.

48. Pavek P. Pregnane X receptor (PXR) – Mediated gene repression and cross-talk of PXR with other nuclear receptors via coactivator interactions. Front. Pharmacol. 7, 456, 1–16, 2016.

49. Yan J, Xie W, A brief history of the discovery of PXR and CAR as xenobiotic receptors. Acta Pharm.Sin.B. 6, 450–452, 2016.

50. Manikandan P and Nagini S. Cytochrome P450 structure, function and clinical significance: a review. Current Drug Targets 19, 38–54, 2018.

51. Yang H, Garzel B, Heyward S et al. Metformin represses drug-induced expression of cyp2b6 by modulating the constitutive androstane receptor signalings. Mol Pharmacol. 85, 249–260, 2014.

52. Kobayashi K, Hashimoto M, Honkakoski P, et al. Regulation of gene expression by CAR: an update. Arch Toxicol. 89, 1045–1055, 2015.

53. Brewer CT and Chen T. PXR variants: the impact on drug metabolism and therapeutic responses. Acta Pharmaceutica Sinica B. 6, 441–449, 2016.

54. Chrencik JE, Orans J, Moore LB, et al, Structural disorder in the complex of human pregnane X receptor and the macrolide antibiotic rifampicin. Mol. Endocrinol. 19, 1125–1134, 2005.

55. Smutnya, T, Manib S, Paveka P. Post-translational and post-transcriptional modifications of pregnane X receptor (PXR) in regulation of the cytochrome P450 superfamily. Curr Drug Metab. 14, 1059–1069, 2013.

56. Asif M. Rifampin and their analogs: a development of antitubercular drugs world. J. Org. Chem. 1(2), 14–19, 2013.

57. Bhagyaraj E Tiwari D, Ahuja N, et al. A human xenobiotic nuclear receptor contributes to nonresponsiveness of Mycobacterium tuberculosis to the antituberculosis drug rifampicin. J. Biol. Chem. 293, 3747–3757, 2018.

58. Nakajima A, Fukami T, Kobayashi Y, et al. Human arylacetamide deacetylase is responsible for deacetylation of rifamycins: Rifampicin, rifabutin, and rifapentine. Biochemical Pharmacology. 82, 1747–1756, 2011.

59. Nannelli A, Chirulli V, Longo V, et al. Expression and induction by rifampicin of CAR- and PXR-regulated CYP2B and CYP3A in liver, kidney, and airways of pig. Toxicology 252, 105–112, 2008.

60. Jover R, Bort R, Gomez-Lechon MJ, et al. Down-regulation of human cyp3a4 by the inflammatory signal interleukin-6: Molecular mechanism and transcription factors involved. FASEB J. 16, 1799–1801, 2002.

61. Mani S, Dou W, Redinbo MR. PXR antagonists and implication in drug metabolism. Drug Metab Rev. 45, 60–72, 2013.

62. Fuchs I, Hafner-Blumenstiel V, Markert C et al. Effect of the CYP3A inhibitor ketoconazole on the PXR-mediated induction of CYP3A activity. Eur. J. Clin. Pharmacol. 69, 507–513, 2013.

63. Patsalos PN. Drug Interactions with the newer anticpileptic drugs (AEDs)—Part 1: pharmacokinetic and pharmacodynamic interactions between AEDs. Clin Pharmacokinet. 52, 927–966, 2013.

64. Ghosh, C Hossain M, Solanki J. Overexpression of pregnane X and glucocorticoid receptors and the regulation of cytochrome P450 in human epileptic brain endothelial cells. Epilepsia, 58(4), 576–585, 2017.

65. Ghosh C, Marchi N, Hossain M, et al. A pro-convulsive carbamazepine metabolite: quinolinic acid in drug-resistant epileptic human brain. Neurobiol Dis 46, 692–700, 2012.

66. Bhagyaraj E Nanduri R, Saini A, et al. Human xenobiotic nuclear receptor PXR augments mycobacterium tuberculosis survival. J. Immunol 197, 244–255, 2016.

67. Adams KN, Takaki, K, Connolly LE, et al. Drug tolerance in replicating mycobacteria mediated by a macrophage-induced efflux mechanism. Cell 145, 39–53, 2011.

68. Wang H, Venkatesh M, Li H, et al. Pregnane X receptor activation induces FGF19-dependent tumor aggressiveness in humans and mice. J. Clin. Invest. 121(8), 3220–3232, 2011.

69. Noll EM, Eisen C, Stenzinger A et al. CYP3A5 mediates basal and acquired therapy resistance in different subtypes of pancreatic ductal adenocarcinoma. Nat Med. 22, 278–287, 2016.

70. Kaur P, Sodhi RK. Memory recuperative potential of rifampicin in aluminium chloride-induced dementia: role of pregnane × receptors. Neuroscience 288, 24–36, 2015.

71. Hakkola J, Rysä J, Hukkanen J. Regulation of hepatic energy metabolism by the nuclear receptor PXR. Biochim. Biophys. Acta. 1859, 1072–1082, 2016.

72. Stage TB, Damkier P, Christensen MM, et al. Impaired glucose tolerance in healthy men treated with St. John's Wort. Basic Clin. Pharmacol. Toxicol. 118, 219–224, 2016.

73. Burk O, Kuzikov M, Kronenberger T, et al. Identification of approved drugs as potent inhibitors of pregnane X receptor activation with differential receptor interaction profiles. Arch. Toxicol. 92, 1435–1451, 2018.

74. Poulton EJ, Levy L, Lampe JW, et al. Sulforaphane is not an effective antagonist of the human Pregnane X-Receptor *in vivo*. Toxicol Appl Pharmacol. 266(1), 122–131, 2013.

75. Shizu R, Abe T, Benoki S, et al. PXR stimulates growth factor-mediated hepatocyte proliferation by cross-talk with the FOXO transcription factor. Biochem. J. 473, 257–266, 2016.

76. Pascussi JM, Drocourt L, Gerbal-Chaloin S, et al. Dual effect of dexamethasone on CYP3A4 gene expression in human hepatocytes. Sequential role of glucocorticoid receptor and pregnane X receptor. Eur J Biochem 268, 6346–6358, 2001.

77. Rasmussen MK, Daujat-Chavanieu M, Gerbal-Chaloin S et al. Activation of the aryl hydrocarbon receptor decreases rifampicin-induced CYP3A4 expression in primary human hepatocytes and HepaRG Toxicol. Lett. 277, 1–8, 2017.

78. Oladimeji P, Cui HM, Zhang C et al. Regulation of PXR and CAR by protein–protein interaction and signaling crosstalk. Expert Opin Drug Metab Toxicol. 12, 997–1010, 2016.

79. Correia, MA, Wang YQ, Kim, SM, et al. Hepatic cytochrome P450 ubiquitination: conformational phosphodegrons for E2/E3 recognition? Int. U. Biochem. Mol. Biol. 66, 78–88, 2014.

80. Choi SY, Koh KH, Jong H. Isoform-specific regulation of cytochromes P450 Expression by estradiol and progesterone. Drug Metab. Dispos. 41, 263–269, 2013.

81. Gill P, Bhattacharyya1 S, McCullough S, et al. MicroRNA regulation of CYP 1A2, CYP3A4 and CYP2E1 expression in acetaminophen toxicity. Scientific Reports 7, 12331–12342, 2017.

82. Wang Y, Yub D, Tolleson WH, et al. A systematic evaluation of microRNAs in regulating human hepatic CYP2E1. Biochem Pharmacol. 138, 174–184, 2017.

83. Hartman JH, Miller GP, Meyer JN, Toxicological Implications of Mitochondrial Localization of CYP2E1. Toxicol Res (Camb). 6, 273–289, 2017.

84. Mueller S, Peccerella T, Qin H, et al. Carcinogenic etheno DNA adducts in alcoholic liver disease: correlation with cytochrome P-4502E1 and fibrosis alcoholism: Clin. Exp Res 42, 252–259, 2018.

85. Pan X, Ning M, Jeong H, Transcriptional Regulation of CYP2D6 Expression Drug Metab. Dispos. 45, 42–48, 2017.

86. Ververs FF, Voorbij HA, Zwarts P, et al. Effect of cytochrome P450 2D6 genotype on maternal paroxetine plasma concentrations during pregnancy. Clin Pharmacokinet 48:677–683. 2009.

87. Eichelbaum M, Mineshita S, Ohnhaus EE, et al. The influence of enzyme induction on polymorphic sparteine oxidation. Br. J. Clin. Pharmacol. 22, 49–53, 1986.

88. Farooq M, Kelly EJ, Unadkat JD. CYP2D6 is inducible by endogenous and exogenous corticosteroids. Drug Metab. Dispos. 44, 750–757, 2016.

89. Zeng L, Chen Y, Wang Y, et al. MicroRNA hsa-miR-370-3p suppresses the expression and induction of CYP2D6 by facilitating mRNA degradation. Biochem Pharmacol. 140, 139–149, 2017.

90. Magnusson MO, Dahl ML, Cederberg J, et al. Pharmacodynamics of Carbamazepine-mediated Induction of CYP3A4, CYP1A2, and P-gp as assessed by probe substrates midazolam, caffeine, and digoxin. Clin. Pharm. Ther. 84, 52–62, 2008.

91. Anderson GD, Gidal BE, Messenheimer JA, et al. Time course of lamotrigine de-induction: impact of step-wise withdrawal of carbamazepine or phenytoin. Epilepsy Research 49, 211–217, 2002.

92. Punyawudho B, Cloyd JC, Leppik IE, et al. Characterization of the Time Course of Carbamazepine Deinduction by an Enzyme Turnover Model Clin Pharmacokinet. 48, 313–320, 2009.

93. Shibata S, Takahashi H, Baba A, et al. Delayed de-induction of CYP2C9 compared to CYP3A after discontinuation of rifampicin: Report of two cases. Int. J. Clin. Pharmacol. Ther. 55, 449–452, 2017.

94. Reitman ML, Chu X, Cai, X, et al. Rifampin's acute inhibitory and chronic inductive drug interactions: experimental and model-based approaches to drug–drug interaction trial design. Clin. Pharm. Ther. 89, 234–242, 2011.

95. Yamashita F, Sasa Y, Yoshida S, et al. Modeling of rifampicin-induced cyp3a4 activation dynamics for the prediction of clinical drug–drug interactions from *in vitro* data. PLoS One 8(9), e70330, 2013.

96. Szczesna-Skorupa E and Kemper B. Proteasome inhibition compromises direct retention of cytochrome P4502C2 in the endoplasmic reticulum. Exp Cell Res. 314, 3221–3231, 2008.

97. Silva R, Vilas-Boas V, Carmo, H et al. Modulation of P-glycoprotein efflux pump: induction and activation as a therapeutic strategy. Pharmacol. Ther. 149, 1–123, 2015.

98. Dei S, Novella-Romanelli R, Manetti D, et al. Design and synthesis of aminoester heterodimers containing flavone or chromone moieties as modulators of P-glycoprotein-based multidrug resistance (MDR) Bioorg. Med. Chem. 26, 50–64, 2018.

99. Aller, SG, Yu J, Ward A, et al. Structure of P-glycoprotein reveals a molecular basis for poly-specific drug binding. Science 323(5922), 1718–1722, 2009.

100. Wessler JD, Grip LT, Mendell J, et al. The P-glycoprotein transport system and cardiovascular drugs. J. Amer. Coll. Cardiol. 61, 2495–2502, 2013.

101. Verhalen, B., Ernst, S., Borsch, M., et al. Dynamic ligand-induced conformational rearrangements in P-glycoprotein as probed by fluorescence resonance energy transfer spectroscopy. J Biol Chem 287, 1112–1127, 2012.

102. Schinkel, AH. P-glycoprotein, a gatekeeper in the blood–brain barrier. Adv Drug Deliv Rev, 36, 179–194, 1999.

103. Efferth T, and Volm M. Multiple resistance to carcinogens and xenobiotics: P-glycoproteins as universal detoxifiers. Arch. Toxicol. 91, 2515–2538, 2017.

104. Prenkert M, Uggla B, Tina E, et al. Rapid induction of P-glycoprotein mRNA and protein expression by cytarabine in HL-60 cells. Anticancer Res 29, 4071–4076, 2009.

105. Bachmeier CJ, Beaulieu-Abdelahad D,. Ganey NJ, et al. Induction of drug efflux protein expression by venlafaxine but not desvenlafaxine biopharm. Drug Dispos. 32, 233–244, 2011.

106. Levin GM, Nelson LA, DeVane CL, et al. A pharmacokinetic drug–drug interaction study of venlafaxine and indinavir. Psychopharmacol Bull 35: 62–71, 2001.

107. Gareri P, De Fazio P, Gallelli L, et al. Venlafaxine – propafenone interaction resulting in hallucinations and psychomotor agitation. Ann. Pharmacother. 42, 434–438, 2008.

108. de Gooijer MC, Buil LCM, Beijnen JH, et al. ATP-binding cassette transporters limit the brain penetration of Wee1 inhibitors. Invest New Drugs 36, 380–387, 2018.

109. Nader AM, and Foster DR. Suitability of digoxin as a P-glycoprotein probe: Implications of other transporters on sensitivity and specificity. J. Clin. Pharmacol. 4, 3–13, 2013.

110. Greiner B, et al. The role of intestinal P-glycoprotein in the interaction of digoxin and rifampin. J Clin Invest. 104, 147–153, 1999.

111. Hennessy, M, Kelleher, D, Spiers, JP, et al. St John's Wort increases expression of P-glycoprotein: implications for drug interactions. Br. J. Clin. Pharmacol. 53, 75–82, 2002.

112. Wasko JA, Westholder JS, Jacobson PA. Rifampin–sirolimus–voriconazole interaction in a hematopoietic cell transplant recipient. J Oncol Pharm Practice. 23(1), 75–79, 2017.

113. Watanabe N, Higashi H, Nakamura S, et al. The possible clinical impact of risperidone on P-glycoprotein-mediated transport of tacrolimus: A case report and *in vitro* study. Biopharm. Drug Dispos. 39, 30–37, 2018.

114. Li Y, Yan L, Shi Y, et al. CYP3A5 and ABCB1 genotype influence tacrolimus and sirolimus pharmacokinetics in renal transplant recipients. Springer Plus 4, 637, 3–6, 2015.

115. Lee J, Wang R, Yang Y, et al. The effect of ABCB1 C3435T polymorphism on cyclosporine dose requirements in kidney transplant recipients: a meta-analysis. Basic & Clin. Pharmacol. & Toxicol. 117, 117–125, 2015.

116. Vanhove, T, Annaert, P, Lambrechts, D et al. Effect of ABCB1 diplotype on tacrolimus disposition in renal recipients depends on CYP3A5 and CYP3A4 genotype. Pharmacogenom. J 17,556–562, 2017

118. Wang, XY, Huang SP, Jiang YS et al. Reactive astrocytes increase the expression of P-gp and Mrp1 via TNF-and NF-B signalling. Mol. Med. Rep. 17, 1198–1204, 2018.

119. Ferreira A, Rodrigues M, Fortuna A, et al. Flavonoid compounds as reversing agents of the P-glycoprotein-mediated multidrug resistance: An *in vitro* evaluation with focus on antiepileptic drugs. Food Research International 103, 110–120, 2018.

120. Kwan P, Li, HM, Al-Jufairi E, Abdulla R, et al. Association between temporal lobe P-glycoprotein expression and seizure recurrence after surgery for pharmacoresistant temporal lobe epilepsy. Neurobiol. Dis. 39, 192–197, 2010.

121. Iannetti, P, Spalice A, Parisi P. Calcium-channel blocker verapamil administration in prolonged and refractory status epilepticus. Epilepsia 46, 967–969, 2005.

122. Sun Y, Wang C, Meng Q, et al. Targeting P-glycoprotein and SORCIN: Dihydromyricetin strengthens anti-proliferative efficiency of adriamycin via MAPK/ERK and Ca2+-mediated apoptosis pathways in MCF-7/ADR and K562/ADR. J. Cell Physiol. 233, 3066–3079, 2018.

123. Li, C, Hansen RA, Chou C, et al. Trends in botanical dietary supplement use among US adults by cancer status: The National Health and Nutrition Examination Survey, 1999 to 2014. Cancer 124, 1207–1215, 2018.

124. Collado-Borrell R, Escudero-Vilaplana V Romero-Jiménez R, et al. Oral antineoplastic agent interactions with medicinal plants and food: an issue to take into account. J. Cancer Res. Clin. Oncol. 142, 2319–2330, 2016.

125. Khalid B, Ayman MK, Rahman H, et al. Natural products against cancer angiogenesis Tumor Biol. 37, 14513–14536, 2016.

126. Franco P, Potenza I, and Moretto F, et al. Hypericum perforatum and neem oil for the management of acute skin toxicity in head and neck cancer patients undergoing radiation or chemo-radiation: a single-arm prospective observational study. Radiation Oncology 9, 297–304, 2014.

127. de Leon J, Spina E, Possible pharmacodynamic and pharmacokinetic drug–drug interactions that are likely to be clinically relevant and/or frequent in bipolar disorder. Current Psychiatry Reports 20(17), 1–24, 2018.

128. McMillin GA, and Krasowski MD, Chapter 5-Therapeutic Drug Monitoring of Newer Antiepileptic Drugs, in Clinical Challenges in Therapeutic Drug Monitoring Special Populations, Physiological Conditions and Pharmacogenomics, eds. Clarke W & Dasgupta A 101–134, 2016, ISBN 978-0-12-802025-8.

129. Weintraub D, Buchsbaum R, Resor SR, et al. Effect of antiepileptic drug comedication on lamotrigine clearance. Arch Neurol. 62,1432–1436, 2005.

130. Patsalos PN, Zakrzewska JM, Elyas AA. Dose dependent enzyme induction by oxcarbazepine? Eur J Clin Pharmacol. 39, 187–188 1990.

131. Benbadis S. Klein P, Schiemann J et al. Efficacy, safety, and tolerability of brivaracetam with concomitant lamotrigine or concomitant topiramate in pooled Phase III randomized, double-blind trials: A post-hoc analysis. Epilepsy & Behavior 80, 129–134, 2018.

132. Stockis, A, Watanabe S, Scheen AJ. Effect of brivaracetam on CYP3A activity, measured by oral midazolam. J. Clin. Pharmacol, 55, 543–548, 2015.

133. Stockis, A, Sargentini-Maier ML, Horsmans, Y, Brivaracetam Disposition in mild to severe hepatic impairment. J. Clin. Pharmacol. 53, 633–641, 2013.

134. Stockis, A, Chanteux, H Rosa, M et al. Brivaracetam and carbamazepine interaction in healthy subjects and *in vitro*. Epilep. Res. 113, 19–27, 2015.

135. de Leon, J. False-negative studies may systematically contaminate the literature on the effects of inducers in neuropsychopharmacology. Part I: Focus on epilepsy. J. Clin. Psychopharmacol. 34(2), 177–183, 2014.

136. Westall CA, Wright T, Cortese F, et al. Vigabatrin retinal toxicity in children with infantile spasms: An observational cohort study. Neurology. 9, 2262–2268, 2014.

137. van der Poest Clement, EA, Sahin M, Peters JM. Vigabatrin for epileptic spasms and tonic seizures in tuberous sclerosis complex. Journal of Child Neurology 1–6, 2018.

138. Novitskaya, Y, Hintz, M, Schulze-Bonhage, A et al. Rapid antiepileptic drug withdrawal may obscure localizing information obtained during presurgical EEG recordings. Epilep. Dis. 20, 151–157, 2018.

139. Kumar, S, Ramanujam, B, Chandra PS, et al. Randomized controlled study comparing the efficacy of rapid and slow withdrawal of antiepileptic drugs during long-term video-EEG monitoring. Epilepsia 59, 460–467 2018.

140. Arnold S, Badalamenti V, Diaz A, et al. Conversion to brivaracetam monotherapy for the treatment of patients with focal seizures: Two double-blind, randomized, multicenter, historical control, Phase III studies. Epilepsy Research 141, 73–82, 2018.

141. Vecht, CJ, Kerkhof, M Duran-Pena, A seizure prognosis in brain tumors: new insights and evidence-based management. Oncologist 19, 751–759 2014.

142. Vecht C, Royer-Perron L, Houillier C, et al. Seizures and anticonvulsants in brain tumours: frequency, mechanisms and anti-epileptic management. Current Pharmaceutical Design 23, 42, 6464–6487, 2017.

143. Wu AS, Trinh VT, Suki D, et al. A prospective randomized trial of perioperative seizure prophylaxis in patients with intraparenchymal brain tumors. Journal of Neurosurgery 118 (4), 873–883, 2013.

144. Toledo M, Molins A, Quintana M, et al. Outcome of cancer-related seizures in patients treated with lacosamide. Acta. Neurol. Scand. 137, 67–75, 2018.

145. Hanaya R and Arita K. The New Antiepileptic Drugs: Their Neuropharmacology and Clinical Indications. Neurol. Med. Chir. (Tokyo) 56, 205–220, 2016.

146. Maschio, M, Zarabla, A, Maialetti, A, et al. Quality of life, mood, and seizure control in patients with brain tumor related epilepsy treated with lacosamide as add-on therapy: A prospective explorative study with a historical control group. Epilep & Behav. 73, 83–89, 2017.

147. Wessling C, Bartels S, Sassen, R. et al. Brain tumors in children with refractory seizures–a long-term follow-up study after epilepsy surgery. Child. Nerv. Sys. 31 1471–1477, 2015.

148. Nagarajan L, Lee M, Palumbo L. Seizure outcomes in children with epilepsy after resective brain surgery. Eur. J. Paed. Neurol, 19, 577–583, 2015.

149. Baftiu A, Feet SA, Gunnar-Larsson P, et al. Utilisation and polypharmacy aspects of antiepileptic drugs in elderly versus younger patients with epilepsy: A pharmaco-epidemiological study of CNS active drugs in Norway, 2004–2015. Epilepsy Research 139,35–42, 2018.

150. Sabers A, Pharmacokinetic interactions between contraceptives and antiepileptic drugs. Seizure 17, 141–144, 2008.

151. Faught E, Szaflarski JP, Richman J, et al. Risk of pharmacokinetic interactions between antiepileptic and other drugs in older persons and factors associated with risk. Epilepsia 59, 715–723, 2018.

152. Margolis JM, Chu BC, Wang CJ, et al. Effectiveness of antiepileptic drug combination therapy for partial-onset seizures based on mechanism of action. JAMA Neurol. 71(8), 985–993, 2014.

153. Brodie MJ, Sills GJ. Combining antiepileptic drugs—rational polytherapy? Seizure 20, 369–375, 2011.

154. Patsalos PN, Drug Interactions with the newer antiepileptic drugs (AEDs) – Part 2: pharmacokinetic and pharmacodynamic interactions between AEDs and drugs used to treat non-epilepsy disorders. Clin. Pharmacokinet. 52:1045–1061, 2013.

155. Zheng W, Xiang YT Yang XH et al. Clozapine augmentation with AEDs for treatment resistant schizophrenia. J. Clin Psychiatr. 78, e498–505, 2017.

156. Flowers CM, Racoosin JA, Kortepeter C. Seizure activity and off-label use of tiagabine. N. Eng. J. Med. 354, 773–734, 2006.

157. Li C, Hansen RA, Chou C, et al. Trends in botanical dietary supplement use among US adults by cancer status: The National Health and Nutrition Examination Survey, 1999 to 2014. Cancer 124, 1207–1215, 2018.

158. Ben-Arye, E, Samuels N, Goldstein LH, et al. Potential risks associated with traditional herbal medicine use in cancer care: a study of Middle Eastern oncology health care professionals. Cancer 122, 598–610, 2016.

159. Davis EL, Oh B, Butow PN, Mullan BA, et al. Cancer patient disclosure and patient–doctor communication of complementary and alternative medicine use: a systematic review. Oncologist 17, 1475–1481, 2012

160. Collado-Borrell R, Escudero-Vilaplana V, Romero-Jiménez R, et al. Oral antineoplastic agent interactions with medicinal plants and food: an issue to take into account. J Cancer Res. Clin. Oncol. 142, 2319–2330, 2016.

161. Vollmer JJ, Rosenson J. Chemistry of St John's Wort: Hypericin and hyperforin. J. Chem Educ. 81, 1450–1456, 2004.

162. Madabushi, R, Frank B Drewelow B, et al. Hyperforin in St John's Wort drug interactions. Eur. J, Clin Pharm. 62, 225–233, 2006.

163. Mansky PJ Straus SE. St John's Wort: More implications for cancer patients. J Nat. Canc. Inst. 94, 1187–1188, 2002.

164. Chrubasik-Hausmann S, Vlachojannis J McLachlan AJ, Understanding drug interactions with St John's Wort (Hypericum perforatum L.): impact of hyperforin content. J Pharm Pharmacol. Feb 7. doi: 10.1111/jphp.12858, 2018.

165. Zanoli P, Role of Hyperforin in the Pharmacological Activities of St. John's Wort. CNS Drug Reviews 10, 203–218. 2004.

166. Gao S, Jiang W, Yin T et al. Highly variable contents of phenolics in St John's Wort products impact their transport in the human intestinal Caco-2 Cell Model: Pharmaceutical and biopharmaceutical rationale for product standardization. J Agric Food Chem. 58, 6650–6659, 2010.

167. Shellenberg R, Sauer S, Dimpfel W. Pharmacodynamic effects of two different hypericum extracts in healthy volunteers measured by quantitative EEG. Pharmacopsychiatry 31(Suppl 1), 44–53,1998.

168. Chow, HHS, Hakim IA, Vining DR et al. Effects of repeated green tea catechin administration on human cytochrome P450 activity. Canc. Epid. Biomark. Prevent. 15, 2473–2476, 2006.

169. Robertson SM, Davey RT Voell, J. et al. Effect of ginkgo biloba extract on lopinavir, midazolarn, and fexofenadine pharmacokinetics in healthy subjects. Curr. Med. Res. Opin. 24, 591–599, 2008.

170. Gurley BJ Swain A, Williams DK et al. Gauging the clinical significance of P-glycoprotein-mediated herb–drug interactions: Comparative effects of St John's wort, echinacea, clarithromycin, and rifampin on digoxin pharmacokinetics. Mol. Nutr. Food Res. 52, 772–779, 2008.

171. Brauer RB, Stangl M, Stewart JR, et al. Acute liver failure after administration of herbal tranquilizer kava–kava (Piper methysticum). J Clin Psych. 64, 216–218, 2003.

172. Musch E, Chrissafidou A, Malek M. Acute hepatitis due to kava–kava and St John's Wort: an immune-mediated mechanism? Dtsch. Med. Wochenschr. 131, 1214–1217, 2006.

173. Cune JS, Hatfield AJ, Blackburn AA, et al. Potential of chemotherapy–herb interactions in adult cancer patients. Supp. Care Canc. 12, 454–462, 2004.

174. Gibson CM, Basto AN, Howard ML, Direct oral anticoagulants in cardioversion: a review of current evidence. Ann. Pharmacother. 52, 277–284, 2018.

175. Femia G, Fetahovic T, Shetty P, et al. Novel Oral Anticoagulants in Direct Current Cardioversion for Atrial Fibrillation. Heart, Lung and Circ. 27, 798–803, 2018.

176. Zielonka A, Tkaczyszyn M, Mende M, et al. Atrial fibrillation in outpatients with stable coronary artery disease: results from the multicenter RECENT studyPol. Arch. Med Wew 125, 162–171, 2015.

177. Obamiro KO, Chalmers L, Lee K, et al. Adherence to Oral Anticoagulants in Atrial Fibrillation: An Australian Survey. Journal of Cardiovascular Pharmacology and Therapeutics 23, 337–343, 2018.

178. Mesquita J, Cavaco D, Ferreira AM, et al. Very long-term outcomes after a single catheter ablation procedure for the treatment of atrial fibrillation – the protective role of antiarrhythmic drug therapy. J. Inter. Card. Electrophys. 52, 39–45, 2018.

179. Knoka E, Pupkevica I, Lurina B, et al. Low cardiovascular event rate and high atrial fibrillation recurrence rate one year after electrical cardioversion. Cor et Vasa 60, e246–e250, 2018.

180. Ha FJ, Barra S, Brown AJ, et al. Continuous and minimally interrupted direct oral anticoagulant are both safe compared with vitamin K antagonist for atrial fibrillation ablation: An updated meta-analysis. International. J. Cardiol. 262, 51–56, 2018.

181. Kuruvilla M, Gurk-Turner, C. A review of warfarin dosing and monitoring. Baylor U. Med Cent. Proc. 14, 305–306, 2001.

182. Tideman, PA, Tirimacco R, St John A, et al. How to manage warfarin therapy. Aust. Pres. 38, 44–48, 2015.

183. Flora DR Rettie AE, Brundage RC, et al. CYP2C9 Genotype-Dependent Warfarin Pharmacokinetics: Impact of CYP2C9 Genotype on R-and S-Warfarin and Their Oxidative Metabolites. J. Clin Pharmac. 57, 382–393, 2017.

184. Kim SY, Kang JY, Hartman JH, et al. Metabolism of R-and S-Warfarin by CYP2C19 into Four Hydroxywarfarins. Drug Metab Lett. 6, 157–164, 2012.

185. Gibbons JA de Vries M, Krauwinkel W, et al. Pharmacokinetic Drug Interaction Studies with Enzalutamide. Clin. Pharmacokinet. 54, 1057–1069, 2015.

186. Mekaj YM, Meka AY, Duci SB, et al, New oral anticoagulants: their advantages and disadvantages compared with vitamin K antagonists in the prevention and

treatment of patients with thromboembolic events. Ther. Clin. Risk Man. 11, 967–977, 2015.

187. Chan NCC, Bhagirath V, Eikelboom JW, Profile of betrixaban and its potential in the prevention and treatment of venous thromboembolism. Vascular Health and Risk Management 11, 343–351, 2015.

188. Hellwig T, and Gulseth MA. Pharmacokinetic and pharmacodynamic drug interactions with new oral anticoagulants: what do they mean for patients with atrial fibrillation? Pharmacother. 47, 1478–1487, 2013.

189. Thoenes M, Minguet J, Bramlage K, Betrixaban–the next direct factor Xa inhibitor? Expert Rev. Hematol. 9, 1111–1117, 2016.

190. Parasrampuria DA, Truitt KE, Pharmacokinetics and Pharmacodynamics of Edoxaban, a Non-Vitamin K Antagonist Oral Anticoagulant that Inhibits Clotting Factor Xa. Clin Pharmacokinet. 55, 641–655, 2016.

191. Cuker A, Burnett A, Triller D et al. Reversal of direct oral anticoagulants: Guidance from the Anticoagulation Forum. Am. J. Hematol. 94, 697–709, 2019.

192. Green L, Tan JC, Antoniou S, et al. Haematological management of major bleeding associated with direct oral anticoagulants – UK experience. Brit. J. Haematol. 185, 514–522, 2019.

193. O'Brien MD, Guillebaud J. Contraception for women taking antiepileptic drugs. J Fam Plann Reprod Health Care 36, 239–242, 2010.

194. Simmons KB, Haddad LB, Nanda K, et al. Drug interactions between rifamycin antibiotics and hormonal contraception: a systematic review. Brit. J, Obstet. Gynaecol. 125, 804–811, 2018.

195. Medicines and Healthcare Products Regulatory Agency (MHRA) of the United Kingdom. St John's Wort: interaction with hormonal contraceptives, including implants. 2014. Available at: https://www.gov.uk/drug-safety-update/st-john-s-wort-interaction-with-hormonalcontraceptives-including-implants/.

196. Wallace KL, Filipek RL, La Hoz RM, et al. Case Report: subtherapeutic voriconazole concentrations associated with concomitant dexamethasone: case report and review of the literature J. Clin. Pharm. Ther. 41, 441–443, 2016.

197. Frohman EM, Shah A, Eggenberger E, et al. Corticosteroids for Multiple Sclerosis: I. Application for Treating Exacerbations. Neurother. 4, 618–626, 2007.

198. Liu, L, Mugundu GM, Kirby BJ, et al. Quantification of human hepatocyte cytochrome P450 enzymes and transporters induced by HIV protease inhibitors using newly validated LC-MS/MS cocktail assays and RT-PCR Biopharm. Drug Dispos. 33: 207–217, 2012.

199. Song I, Borland J, Min S, et al. Effects of etravirine alone and with ritonavir-boosted protease inhibitors on the pharmacokinetics of dolutegravir. antimicrob. Agents Chemother. 55, 3517–3521, 2011.

200. Song I, Borland J, Chen S, et al. Effects of enzyme inducers efavirenz and tipranavir/ritonavir on the pharmacokinetics of the HIV integrase inhibitor dolutegravir. Eur. J. Clin. Pharmacol. 70, 1173–1179, 2014.

201. Song XM, Varker HM, Eichelbaum M, et al. Treatment of lung cancer patients and concomitant use of drugs interacting with cytochrome P450 isoenzymes. Lung Canc. 74, 103– 111, 2011.

202. Dando TM and Perry CM Aprepitant: a review of its use in the prevention of chemotherapy-induced nausea and vomiting. Drugs 64 (7), 777–794, 2004.
203. Ferron GM, Dai Y, Semion D et al. Population pharmacokinetics of cabazitaxel in patients with advanced solid tumors. Canc. Chemother. Pharmacol. 71, 681–692, 2013.
204. Goey AKL, Meijerman I, Rosing H, et al. The effect of St John's Wort on the pharmacokinetics of docetaxel. Clin. Pharmacokinet. 53,103–110, 2014.
205. Mathijssen RH, Verweij J, de Bruijn P, et al. Effects of St. John's Wort on irinotecan metabolism. J. Natl. Cancer Inst. 94, 1247–1249, 2002.
206. Wang D, and Wang H. Oxazaphosphorine bioactivation and detoxification: the role of xenobiotic receptors. Acta Pharmaceutica Sinica B2(2), 107–117, 2012.
207. Vredenburg G, den Braver-Sewradj S, van Vugt-Lussenburg BMA, et al. Activation of the anticancer drugs cyclophosphamide and ifosfamide by cytochrome P450 BM3 mutants. Tox. Lett. 232, 182–192, 2015.
208. Saba, N and Seal A. Identification of a less toxic vinca alkaloid derivative for use as a chemotherapeutic agent, based on in silico structural insights and metabolic interactions with CYP3A4 and CYP3A5. J. Mol. Mod. 24, 82, 3–14, 2018.
209. Harmsen S, Meijerman I, Beijnen JH, et al. Nuclear receptor mediated induction of cytochrome P450 3A4 by anticancer drugs: a key role for the pregnane X receptor. Canc. Chemother Pharmacol 64:35–43, 2009.
210. Patel P, Leeder S, Piquette-Miller M, et al. Aprepitant and fosaprepitant drug interactions: a systematic review. Br. J. Clin. Pharmacol. 83, 2148–2162. 2017.
211. Smith HS, Cox LR, Smith EJ et al. 5-HT3 receptor antagonists for the treatment of nausea/vomiting. Ann Palliat Med 1(2), 115–120, 2012.
212. de Leon J, Sandson NB, Cozza KL. A preliminary attempt to personalize risperidone dosing using drug–drug interactions and genetics: part II. Psychosomatics. 49, 347–361, 2008.
213. Bhaloo S and Prasad GV. Severe reduction in tacrolimus levels with rifampin despite multiple cytochrome P450 inhibitors: a case report. Transplant Proc 35, 2449–2451, 2003.
214. Li T, Liu W, Chen K, et al. The influence of combination use of CYP450 inducers on the pharmacokinetics of voriconazole: a systematic review Journal of Clinical Pharmacy and Therapeutics, 2017, 42, 135–146
215. Lee RH, Lyles KW, Colón-Emeric C. A Review of the Effect of Anticonvulsant Medications on Bone Mineral Density and Fracture Risk. Am. J. Geriatr. Pharmacother. 8, 34–46, 2010.
216. Fraser LA, Burneo JG, Alexander Fraser J. Enzyme-inducing antiepileptic drugs and fractures in people withepilepsy: A systematic review. Epilepsy Research 116, 59–66, 2015.
217. Tsiropoulos I, Andersen M, Nymark T, et al. Exposure to antiepileptic drugs and the risk of hip fracture: A case-control study. Epilepsia. 49, 2092–2099, 2008.
218. Pascussi JM, Robert A, Nguyen M, et al. Possible involvement of pregnane X receptor-enhanced CYP24 expression in drug-induced osteomalacia. J. Clin. Invest. 115:177–186, 2005.

219. Ioannidis, G., Papaioannou, A., Hopman, W., et al. Relation between fractures and mortality: results from the Canadian Multicentre Osteoporosis Study. CMAJ 181, 265–271, 2009.
220. Svalheim S, Sveberg L, Mochol M, et al. Interactions between antiepileptic drugs and hormones. Seizure 28 12–17, 2015.
221. Beghi E, and Shorvon S. Antiepileptic drugs and the immune system. Epilepsia, 52(Suppl. 3), 40–44, 2011.

5 Cytochrome P450 Inhibition

5.1 Introduction

The previous chapter focussed on problems associated with drug failure due to enzyme induction. However, when drug clearance is slowed or even stopped for any reason, the consequences are more dangerous and occur much more rapidly compared with enzyme induction. Generally, the intended pharmacological effects of the drugs will be intensified, leading to a clear manifestation of symptoms in the patient. In drugs with a wide TI, this may not be a problem and the effects of the drug accumulation will be reversible. In narrow TI drugs, the effects can be lethal in hours. In other cases, a drug may induce potentially serious unintended pharmacological effects that are only seen at high doses, considerably above the normal range. These effects, sometimes known as 'off target' pharmacological actions, may or may not have been seen in the initial pre-clinical (animal) toxicity testing of the drug. The following illustrative histories underline the effects of drug accumulation.

History 1

A previously healthy 29-year-old female used terfenadine twice daily for one year to treat allergic rhinitis. The patient drank grapefruit juice two to three times weekly. One day she consumed two glasses of juice, took her terfenadine dose, and then mowed her lawn; within one hour she became ill, collapsed, and died. Although usually undetectable, post-mortem terfenadine and metabolite plasma levels were reported as 35 and 130 ng/mL, respectively. These levels are within range of previously noted arrhythmogenic levels of terfenadine. The individual had no evidence of impaired hepatic function.

Human Drug Metabolism, Third Edition. Michael D. Coleman.
© 2020 John Wiley & Sons, Inc. Published 2020 by John Wiley & Sons, Inc.

Analysis

The presence of grapefruit juice appears to have caused unusually high levels of the parent drug to be present in the patient's plasma. Hence, some component of the juice prevented the clearance of the parent drug, leading to drug accumulation, which resulted in a fatal cardiac arrhythmia.

History 2

A 67-year-old male patient stabilised on warfarin began to drink cranberry juice twice daily in response to its reported benefits in recurrent kidney infections. Four days after starting to drink the juice, he suffered a fatal stroke. Post-mortem levels of warfarin were 40% higher than previously sampled in this patient.

Analysis

That the patient had been stabilised on warfarin indicates that his clotting time was within acceptable limits and therefore the drug was being cleared at the same rate it was entering the patient's system. The onset of the consumption of cranberry juice coincided with marked accumulation of warfarin, rendering the patient highly vulnerable to haemorrhage, which occurred within the brain and led to death. It was assumed at the time that the cranberry juice had prevented the clearance of warfarin to inactive metabolites.

History 3

A 64-year-old female with a history of depression was stabilised on amitriptyline, 150 mg/ day, but without improvement in mood. Her GP added fluoxetine, 40 mg/day, and within three weeks, the patient's symptoms subsided, although one week later, she collapsed at home and was found in a coma by a relative. The patient recovered consciousness two days later and made a full recovery.

Analysis

The addition of fluoxetine to the regime was associated with accumulation of amitriptyline, which led to unconsciousness and could have been fatal had she not been discovered. The fluoxetine must have prevented the clearance of amitriptyline.

History 4

A 44-year-old female epileptic was stabilised on carbamazepine, but on the advice of a friend started taking a liquorice preparation for stomach problems. Over a period of two days, she became gradually more sedated and confused, until she had difficulty standing up. She was admitted to hospital and recovered within three days.

Analysis

The liquorice extract was taken in large amounts and appears to have interfered with the clearance of carbamazepine, leading to drug accumulation and symptoms of toxicity.

History 5

A 55-year-old female stabilised on warfarin suffered from recurrent acid indigestion over the Christmas period and started to self-medicate with over-the-counter cimetidine on the advice of a relative. A few days later, while gardening, the patient noticed that a small cut bled profusely and did not appear to clot for a long period. The patient reported to a hospital accident and emergency room, where her INR was 3.3 compared to the usual 2.5. The hospital advised her to use an alternative anti-acid agent and her prothrombin time returned to normal over several days.

Analysis

The excessive anticoagulation was due to a reduction in the clearance of warfarin by cimetidine, which could be averted by the use of low-dose <400 mg) ranitidine or famotidine, which are not usually associated with changes in warfarin pharmacokinetics. An acceptable proton-pump inhibitor would be lansoprazole, but not omeprazole.

Overall analysis

In these cases:

- The patient was already stabilised on a particular medicine, which suggests that the dosage and clearance were approximately balanced.

- The addition to the regime prevented clearance of the first drug, leading to accumulation and toxicity.

- The toxicity manifested as an intensification of the normal pharmacological response, again indicating that drug accumulation was responsible.

- The toxic responses occurred within hours rather than days, after the addition of the inhibitor drug.

- The toxicity manifests so quickly that death can occur before even the patient realises what is happening.

- The toxic effects were rapidly reversible once the inhibiting drug was withdrawn.

- The effects can occur in response to the patient's decision to either self-medicate or change their diet routine, without consultation with medical staff, or the effect can occur after medical staff fail to be aware of the potential reaction.

5.2   Inhibition of metabolism: general aspects

In contrast to enzyme induction, drug inhibition does not at first sight,seem to be a process where a logical adaptive response can be made by the patient's metabolism. That some inhibitors can impair CYP operation for as long as they are administered indicates that the patient's homeostatic systems may not detect the inhibition effect and may fail to respond quickly, if at all, to the change in the situation within the treatment timescale. In effect, it is rather like suddenly blocking the exhaust pipe of a running engine – it will cough and then simply stop. However, you might recall from Chapter 3 that CYP activity was regulated quite intricately at the local, cellular level and inhibition was one of the modes used to temporarily modulate the activity of a particular CYP with respect to certain endogenous substrates. Likewise, in other systems such as Pg-p (Chapter 4.4.7); again, inhibition seems to be part of the modulation process of transport of substrates.

Biotransforming systems can, to a limited degree, compensate for the sustained loss of a CYP, either through inhibition or polymorphism. This can be the case with risperidone where the absence of CYP2D6 does not mean that the drug cannot be oxidised by CYP3A4 to some extent.[1] Indeed, generally, such is the overlap in substrate specificity and flexibility of CYPs, that even in the presence of potent inhibitor, some other CYPs might clear a drug, albeit much less avidly. In such a case, other routes of elimination of parent drug may become more important, such as the kidney, faeces, or the lung for more volatile agents. That the lung can clear some volatile chemicals such as alcohols is exploited in road safety in the detection of drunk drivers. Notwithstanding these points, generally, if a drug's main route of clearance is a single hepatic CYP, chances are that with respect to the quantity of drug dosed, any resultant accumulation will occur relatively rapidly, perhaps in a few hours, leading to toxicity or even death in the case of narrow TI agents.

It is clear that inhibition-based drug reactions are much more potentially clinically serious than induction effects, due to this short timescale and the speed that the patient's clinical situation can change, leading to irreversible damage (such as a stroke or heart attack) within hours of consuming the inhibitor. This is especially problematic in the light of the increasing prevalence of polypharmacy where, as mentioned already, multiple prescribers might change different medicines without full appreciation of the consequences in patient groups such as the elderly. Inhibitors may enter the regimen on the patient's initiative, through the desire to 'self-help', without informing their doctor. We visited this issue in the previous chapter with respect to herbal remedies such as St John's Wort. In addition, there are chemical inhibitors in foodstuffs and beverages that can cause lethal interactions with drugs, sometimes without the patient's or even the practitioner's prior knowledge. In the previous chapter, the clinical reality of how inducing drugs affected the efficacy of co-administered drugs was described as sometimes rather unpredictable. This issue is even more glaring with inhibitors and, as already mentioned, there can be relatively little time for the practitioner and patient to detect and circumvent an acute drug accumulation before serious problems arise.

5.3 Mechanisms of reversible inhibition

5.3.1 Introduction

Enzymes and tissue/cell receptors share similar features. A receptor binds a molecule that then acts like a switch to trigger a cascade of molecules to instruct the cell to perform a function. The molecule must fit the receptor precisely and then trigger the cascade, like a key, which first enters a lock, then turns smoothly to open it. A key that fits and enters the lock, but does not turn it, not only fails to open the door but also prevents the use of the correct key. The lock is essentially 'inhibited'.

Although they are highly specialised, CYPs are enzymes like any other in the body and they are inhibited according to the same general principles as other enzymes. How tightly a chemical interacts with a CYP isoform is based on how powerful is the mutual attraction (affinity) between the chemical and the various areas of the active site of the enzyme. In the case of CYPs and any given enzyme, affinity must be strong enough to ensure the substrate binds for sufficient time to process it to a product. The quicker this process occurs, the faster the 'turnover' of the enzyme and the more efficient it is. It is useful to try to visualise a CYP isoform, or any other human enzyme for that matter, as a three-dimensional machine tool, perhaps like a spot-welding machine. The enzyme cycles hundreds of times a second. If any single aspect of substrate binding or processing (oxidation or reduction), followed by product release is prevented, the sequential nature of these events means that the enzyme stops functioning. Another analogy might be an automatic paper stapler in a photocopier. Whatever analogy you might use, it is useful to try to visualise enzymes as dynamic nano-machines. Broadly, inhibitors of CYPs may frustrate the enzymes' operating processes reversibly or irreversibly, with varying impact on drug clearance and the individual enzyme health and survival. At high concentrations, many inhibitors might block several CYP subfamilies, but at lower concentrations, they show more selectivity and their potency in blocking individual isoforms can be measured.

Reversible inhibition processes are pharmacological/physiological and *rapid* events that do not result in any damage or change in the CYP's structure. They are essentially modulations of CYP activity, either by a drug or an endogenous molecule, like running a machine off and on and at different speeds and capacities. There are three main types of reversible inhibition; competitive, noncompetitive and uncompetitive, which usually involve parent drugs, but not always. Irreversible inhibition involves a line being crossed from a pharmacological/ physiological series of events into a toxicological, or damaging process. Irreversible inhibition is always caused by the CYP metabolising the drug into an initially reactive entity that may form a stable bond with the haem iron or some other vital structure in the CYP, which prevents the enzyme from functioning for days. This bond may progress and destroy the enzyme; this is discussed in Section 5.4.

Which type of inhibition occurs with various drugs can depend on many factors, such as drug concentration and the characteristics of a particular CYP isoform. Many drugs can act as competitive inhibitors with one CYP and non-competitive with others. There are also complex mixed forms of inhibition that can be analysed using *in vitro* studies incorporating human CYPs, either in human liver or in expressed enzyme systems (Appendix A.4.3). What is also emerging is that inhibition effects can be predicted in polymorphic CYP variants also.[2] These studies do not always reflect what will happen when the drugs are used in patients but are a reasonable starting point to predict whether a new drug might interfere with the metabolism of other agents.

5.3.2 Competitive inhibition

This is the simplest form of inhibition, where the substrate (drug) and the inhibitor are very similar in structure and have similar affinities for the same place, i.e. the CYP active site (Figure 5.1). A CYP substrate is normally processed to a different molecule, that is, a metabolite, which then has a much reduced affinity for an active site and is more water soluble, so it diffuses elsewhere. A competitive inhibitor of a CYP isoform is usually not a substrate and acts similarly to the

Main Types of Enzyme Inhibition

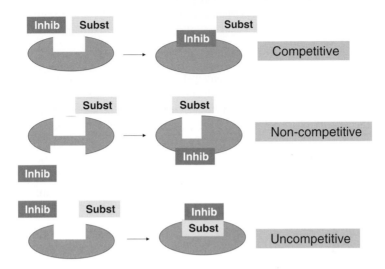

Figure 5.1 Main types of enzyme inhibition that apply to CYP isoforms

correct key for a doorlock; it may enter and leave the lock freely but does not operate it. As it cannot be metabolised, it does not leave the vicinity of the CYP, binding and detaching continually. The CYP might be unable to metabolise the inhibitor, due to particular features of the molecule that might prevent oxidation, but promote binding to the active site.

This form of inhibition is common in CYPs and is governed by the law of mass action, which states that the rate of a reaction (in this case enzyme binding) is governed by the concentration of the participants. So for CYP metabolism, whichever agent, drug, or inhibitor is in the greatest concentration, then this will occupy the active site. At low inhibitor concentrations, more of the drug can be added to overcome the inhibitory effects. However, as drug levels must be increased to overcome the inhibitor, this effectively means that the drug's affinity falls for the site (K_m increases) in the presence of the inhibitor.

Enzymes are often subject to this process of competitive inhibition because it is usually part of the endogenous feedback control mechanism on product formation. This generally involves enzymes that use cellular energy, or are at the junction of several biosynthetic pathways. When high levels of product are formed, these inhibit the substrate, so limiting the enzyme's turnover, i.e. when the desired product level is reached. This is rather like a thermostat in a heating system, which automatically maintains a preset temperature irrespective of outside temperatures. This happens in the regulation of vital endogenous molecules like NADPH and glutathione (GSH) and the process avoids unnecessary use of cellular energy. Although the enzyme is temporarily disabled, it is undamaged and has not cycled or used any reducing power. Mathematically, if a Lineweaver–Burk double reciprocal plot is made of competitive inhibition, the K_m (inverse of the affinity) changes, but the V_{max} does not; in other words, the enzyme will still run at a maximum rate if enough substrate is used, but affinity falls off.

When a new drug is evaluated as a possible inhibitor of a given CYP isoform/substrate combination and a Lineweaver–Burk plot is found as described above, then the new drug is a competitive inhibitor of that CYP and it is likely that the inhibitor is binding the CYP at its active site. There are several examples of competitive inhibitors of CYP isoforms. Indeed, if two drugs of similar affinities are cleared by the same isoform, then competitive inhibition can occur. Competitive inhibition of CYPs is, as was stated in the previous section, a pharmacological process and is part of the process of regulation of CYPs by different substrates, as outlined in Chapter 3.5. The major clinically relevant group of competitive inhibitors includes the azole antifungal agents.

Azoles: introduction

It is not surprising that these agents are potent human P450 inhibitors, as a great deal of money, time and effort went into designing them to inhibit fungal CYPs. They prevent the fungal synthesis of ergosterol, by blocking lanosterol alpha-C_{14}-demethylase, so causing the substrates (14-alpha-methylsterols) to accumulate and this disrupts fungal membranes.[2] Unfortunately, as mentioned in Chapter 2.3.2, since all living-system CYPs originate from a common bacterial source, inhibition of azole compounds also occurs in human CYPs. The azoles as a class are generally competitive inhibitors, due to their lone pair of electrons on the azole nitrogen, which temporarily binds to CYP haem groups[3]. These drugs as a class can also interact with several other enzyme systems and different nuclear receptors, such as PXR, as outlined in the previous chapter (4.4.3), as both agonists (inducers) and antagonists.[3]

In this class of drugs, miconazole and clotrimazole were introduced in the early 1970s, although their poor absorption and aqueous insolubility made them suitable mainly for topical preparations in various gels. Ketoconazole was the first orally bioavailable imidazole to be used clinically in quantity in the late 1970s. Unfortunately, it is the most potent inhibitor of the orally active azoles, with respect to CYP3A4, as well as other systems. This makes it toxic in practice, reducing testosterone levels in blood, feminizing males, leading to the appearance of breasts (gynaecomastia), loss of spermatozoa production, and impotence. The female menstrual cycle can also be disrupted. Other effects, including GI tract irritation, nausea, vomiting, and occasional severe liver toxicity, meant that ketoconazole is now only used in Europe for the treatment of Cushing's syndrome,[4] a relatively rare disorder of the adrenal glands; it also lives on in various topical antifungal preparations and anti-dandruff shampoo.[5]

The problems with ketoconazole's toxicity, as well as its potent therapeutic effects drove the continuing development of azoles to greater therapeutic potency, but with less human CYP impact. The results appeared in the late 1980s and early 1990s, in the form of fluconazole and then itraconazole, then the third-generation triazoles, such as voriconazole, posaconazole, and latterly isavuconazole, which was approved in the United States in 2015. These agents are effective antifungals, although they retain varying CYP inhibitory effects, particularly with CYP3A4. After ketoconazole, it is generally agreed that itraconazole is the most potent competitive CYP3A4 inhibitor in current usage and the only one cleared to a therapeutically active metabolite[6]. Estimates vary as to the exact rank of inhibitory effect, but posiconazole then voriconazole are the next most potent followed by isovuconazole, and then fluconazole, which is a milder CYP3A4 inhibitor[7]. In terms of the range of CYP3A4 potency of inhibition of the drugs in widespread clinical use, itraconazole is around 50-fold the potency of fluconazole[7]. Voriconazole as well as fluconazole are also potent noncompetitive/mixed inhibitors of CYP2C19 and CYP2C9[6]. Clotrimazole is a CYP3A4 inhibitor and inducer, whilst miconazole is a mild CYP3A4 inhibitor, but quite a potent CYP2C9 and CYP2D6 inhibitor[8].

Another issue that can impact the clinical management of these drugs is their relative half-lives. Whilst isovuconazole is not the most potent inhibitor, it has the longest half-life (130 h)[9], followed by itraconazole and its hydroxyl metabolite (50–60 h), then fluconazole and posiconazole (30–35 h) and the shortest is voriconazole at around 10 h[7]. The drugs also are substrates for and inhibit Pg-p to varying degrees[6]. In terms of their metabolic fates, fluconazole undergoes little oxidative clearance and posiconazole is cleared by glucuronidation (Chapter 6.2), whilst voriconazole is oxidised mainly by the CYPs it inhibits, that is CYP2C19, CYP2C9 and CYP3A4[6]. Itraconazole is cleared by CYP3A4, and isovuconazole is a CYP3A4/5 substrate[10].

The ubiquity of CYP3A4 substrates in medicine unfortunately provides a vast canvas for potential drug–drug interactions with azoles. Time frames of azole antifungal treatment can range from a few days, to many months in the case of prophylaxis in bone marrow transplants[11] and those with immune deficiencies, where very significant fungal colonisation must be tackled[9]. Such prolonged therapy is now increasingly common[12], yet even a short period of accumulation of a co-administered narrow TI drug could have significant clinical consequences. Resistance to azoles is also of increasing concern, and it is driven by both clinical and even agricultural indiscriminate use of related agents[12]. Whilst there are different modes of resistance, one of the most significant is the upregulation of ABC cassette efflux pumps[13], as described with anticancer and antibiotic drugs in Chapter 4.4.7. Whilst levels of resistance to at least one azole have reached 6% in some areas, there are enough alternative azoles to suppress the problem for now[12]. There are many groups of patients who are vulnerable to fungal infections, such as those with cancer, HIV and diabetes[13], as well as the immunocompromised, but it is sobering that those who receive azoles during hospital stays can have up to a 70% chance of a problematic drug interaction[7].

Azoles and immunosuppressants

Using azoles to combat opportunistic fungal infections in the immunocompromised and those taking immunosuppressant drugs after various organ donations or bone marrow transplants is obviously problematic as the drugs used, such as cyclosporine, sirolemus, everolemus, and tacrolimus are all CYP3A4/5 substrates. This situation can be complex, depending on whether the azoles are orally or intravenously dosed. Gut CYP3A4 comprises more than 70% of total CYPs compared with the generally lower CYP3A hepatic levels ($\sim$ 30%)[7] Interestingly, calcineurin inhibitors such as cyclosporine are actually antifungal in their own right and can act synergistically with azoles[13] to eliminate fungal infections. However, the issue of azole activity on PXR and its role in drug resistance through induction of efflux transport, such as that seen in Chapter 4.4.7. with rifampicin, needs more research. Whilst all the azoles can inhibit CYP3A4/5, any specific interaction will depend on the azole used, its half-life, the immunosuppressant used and its dosage, plus other factors related to the patient's general health and ability to clear various drugs. It is, of course, vital to eliminate fungal infections from transplant and other immunocompromised patients, but skilled clinical manipulation of the immunosuppressant dose is necessary to prevent damage to the xenograft, a real risk with these narrow TI drugs.

As mentioned at the beginning of this chapter, inhibition is unlike induction, in that it is fairly rapid in effect, so when commencing azole therapy in patients on immunosuppressants, it has been recommended that doses of tacrolimus for example, should be halved on day one, then further reduced to 70% by day 3[7]. Overall, due to the well-documented interactions between fluconazole, voriconazole and isovuconazole with tacrolimus, as well as itraconazole and cyclosporine, dose reductions of the immunosuppressants may range from 40–80% or more[7,9]. This is necessary, because the azoles can increase the AUCs of CYP3A4 substrates dramatically – posaconazole, for example, can increase sirolimus AUCs by more than eightfold[7].

Interestingly another issue is what happens after the azole is discontinued. As mentioned above, the inhibitory effect will wear off, but not immediately. With the shorter half-life azoles, 7–10 days might be enough for clearances to return to baseline, although itraconazole's effect can persist for up to a month[7]. Given that isovuconazole's half-life is more than twice that of itraconazole, it may take even longer for the newer drug's impact to diminish[9]. What is clear, is that after cessation of azole therapy, doses of drugs such as sirolemus will need to be more than doubled to maintain therapeutic concentrations; also, that this process can take several weeks to successfully manage. This requires very close clinical therapeutic monitoring to prevent immune-mediated damage to the xenograft[11] or overshooting, which would cause direct drug-mediated toxicity towards the graft.

Azoles and anticancer drugs

As outlined in Chapter 5, Section 5.2, many anticancer agents are either cleared by CYP3A4 or require metabolic activation by it in order to exert their therapeutic effect. Lack of clearance will promote drug accumulation, which becomes problematic in terms of parent drug-mediated adverse effects. If the drug is active through a metabolite, lack of clearance could seriously compromise efficacy. Either way, drug efficacy and tolerance could be radically altered in these narrow TI agents, which are, as outlined in Chapter 4 (Section 4.5.7), noted for their toxicity. Concerning the vinca alkaloids, for example, not only do azoles inhibit their CYP3A4-mediated clearance, they also block P-gp cell efflux, which promotes the retention of these drugs in tissues and particularly the CNS. This results in severe neurotoxicity, both peripheral and central with the longer half-life azoles, such as itraconazole and posaconazole, but the less potent inhibitors such as fluconazole do not have such a pronounced effect[14]. In some cases, the only way to avoid an interaction is to use antifungals of a different class[14].

With cyclophosphamide, the effect is more complex. As mentioned in Chapter 4.5.7, the first step in its activation is the CYP2B6/CYP3A4/CYP3C9-mediated oxidation to the 4-hydroxycyclophosphamide, which decomposes to the phosphoramide mustard responsible for the antitumour effect and the toxicity. About 70% of the drug is cleared through this route[15]. With fluconazole, it was suggested that its inhibition of CYP2C9 and possibly CYP2B6 inhibits 4-hydroxy derivative formation, thus diminishing efficacy and increasing parent drug levels, whilst itraconazole inhibition of CYP3A4 blocks other pathways, forcing more parent drug through to 4-hydroxy formation and increasing toxicity and probably efficacy as well[16]. Hence, even with potent CYP inhibitors, the net effect on drug clearance and efficacy/toxicity can be difficult to predict, although with ifosfamide, the impact of a potent azole inhibitor of CYP3A4 may be even more pronounced in terms of reduction of formation of the active metabolite. Azoles have a similar effect on irinotecan disposition, when CYP3A4 inhibition routes more parent drug through to SN-38 formation, promoting efficacy and toxicity at the same time[14]. The azoles retard the clearance of the taxanes (paclitaxel and docetaxel), as well as the tyrosine kinase inhibitors such as imatinib, which are cleared by CYP3A4[14].

Azoles and anticoagulants

S-warfarin is around fivefold more active than the R-version and about 85% of S-warfarin is metabolised by CYP2C9, with the R-version being handled by mostly CYP1A2 and CYP3A4[17]. It is not surprising that the azoles can significantly impact warfarin metabolism; indeed, they don't even need to be given orally to do this-clotrimazole and miconazole topical gels are sufficiently potent as inhibitors to retard the clearance of S-warfarin, promoting a serious risk of haemorrhage during candidiasis treatment[18]. However, fluconazole and voriconazole can double or even triple any given warfarin patient's INR. This is azole dose-dependent and takes about a week to fully appear and seems to be more pronounced in patients whose INRs are already quite high (2-3 plus). The same study showed that itraconazole had no INR effect, probably because CYP3A4 (and some CYP2C9) inhibition is not significant in terms of R-warfarin clinical effect. Whilst it is tempting to suggest that itraconazole might be safer than the other drugs in this context, it is recommended to take great care with all the azoles with warfarin patients[17,18].

The narrow TI, unpredictability of response, risks of under and overtreatment and sheer hassle to patients and practitioners alike in the maintenance of stable warfarin therapy has most likely contributed to the success of the DOACs (Direct oral anticoagulants), or alternatively, the NOACs (non vitamin K antagonist oral anticoagulants). However, CYPs have a role in the clearance of some of these drugs and not surprisingly, azoles can significantly affect their clinical disposition. More than half of apixaban and rivaroxaban clearance is through CYPs, mainly 3A4, 2C8/9, and 2C19 and ketoconazole can double the AUC of apixaban and strong CYP3A4 inhibitors were excluded from apixaban Phase III trials[19]. Both apixaban and rivaroxaban are, like all of this class, ABC cassette transporter substrates so the azoles are not recommended with the DOACs for their CYP and P-gp inhibitory effects[20,21], although fluconazole has a sub-clinical inhibitory effect. The combination of renal impairment and CYP3A4 inhibition, such as could be encountered in elderly patients, could cause significant accumulation, more with rivaroxaban than apixaban, as about half rivaroxaban's and about a quarter of apixaban's clearance is renal[21]. Edoxaban, dabigatran etexilate, and betrixaban are not cleared by CYPs, but they are also substrates for P-gp, so inhibition of ABC cassette transporters caused by azoles means that all the NOACs' plasma levels will significantly increase.

Azoles and other drugs

Itraconazole can increase the AUCs of intravenous steroids such as methylprednisolone, by up to threefold[7,14] and the combination is not recommended. In combination with inducing drugs such as AEDs, azole inhibition acutely can cause up to a 75% increase in plasma levels of phenytoin, which would result in marked sedation. However, over a period of a few days, the inducers will turn the tables and accelerate the clearance of the azoles to the point that they will fail therapeutically[14,22]. There is also evidence that azoles such as fluconazole and voriconazole delay the clearance of synthetic opiates such as fentanyl, potentially leading to respiratory depression[23], again through CYP3A4 inhibition. Cardiovascular-related drugs such as statins are also at risk of potent interactions caused by azoles and even fluconazole, which is considered a milder CYP3A4 inhibitor, has inhibited statin clearance to the degree the resulting accumulation led to rhabodomyolysis, which is discussed in more detail later in the chapter (Section 5.7.4)[24]. Indeed, if a statin causes this adverse reaction, it is most likely to be due to the presence of an inhibitor and the death rate is 10%[25]. The impact of miconazole gel preparation on CYP2D6 is significant enough to distort the metabolic profile of oxycodone, which normally requires activation to morphine derivatives[8]. Fungal infections in pregnancy are common, and fluconazole has been listed as an FDA Category D teratogen, although this is for chronic high doses (400–800 mg daily), whilst 400 mg/day and below showed no significant risk[26].

Miconazole is in FDA Category C; it is reasonable to suppose that the impact of these drugs on nuclear receptors and CYPs are behind these effects. It is likely that the net teratogenic effect of any given azole, will depend on many factors, including length of therapy, dose and timing (what day of pregnancy). Other key issues are the physicochemical properties of the particular azole, which will govern absorption and foetal penetration and the enzyme inhibitory system specificity and potency. Overall, any CYP3A4, CYP2C9, CYP2C19, and CYP2D6 substrate is vulnerable to inhibition by azoles and the impact of off-target and frank toxicity due to accumulation will vary according to the drug involved.

5.3.3 Noncompetitive inhibition

Noncompetitive inhibition does not involve the inhibitor and substrate competing for the same active site (Figure 5.1). In noncompetitive inhibition, there is another site involved, known as the allosteric site, which is distant from the active site. Once a ligand binds this allosteric site, the conformation of the active site is automatically changed and it becomes less likely to metabolise the substrate and product formation tails off. This process of allosteric binding is another example of the endogenous control of product formation, perhaps by another product/substrate from a related or similar pathway. The Lineweaver–Burk plot will show a fall-off in V_{max} (enzyme cannot run at maximal rate) but K_m does not change, that is, the affinity of the substrate for the active site is unchanged.

Many drugs are noncompetitive inhibitors of CYP isoforms, through allosteric effects. In Sections 3.4.4–3.4.6 it was discussed how endogenous and exogenous molecules can bind at sites distant from the main active site of CYP3A4, such as the Phe cluster and this is part of the mechanism where CYPs are locally controlled in terms of their activity. It is likely that noncompetitive interactions govern many aspects of second-by-second CYP activity, but xenobiotic agents may be able to bind much more potently than endogenous ones. There some examples of drugs that bind to different CYPs noncompetitively, such as trimethoprim and CYP2E1[27], although noncompetition is more common with dietary and medicinal phenolic agents, such as sauchinone, found in the Korean medicinal plant *Saururus chinensis*[28] and the various catechins in Japanese green tea[29].

5.3.4   Uncompetitive inhibition

This is an unusual form of inhibition, where the inhibitor binds only to the enzyme/substrate complex (Figure 5.1). This has the effect of stimulating enzyme/substrate complex formation so increasing affinity (fall in K_m), although the enzyme/substrate/inhibitor complex is nonfunctional, so the V_{max} falls. This appears to be a relatively rare form of inhibition of human CYPs by therapeutic drugs, although the NSAID meloxicam is capable of uncompetitively inhibiting quinidine *in vitro*, it is not likely to be a significant clinical interaction[30]. CYP2J2 metabolises arachidonic acid, but it can also clear the antihistamines astemizole and ebastine. Interestingly, a new experimental inhibitor of CYP2J2, known as LKY-047, is a competitive inhibitor of CYP2J2-mediated astemizole demethylation, but an uncompetitive inhibitor of ebastine hydroxylation[31].

5.4   Mechanisms of irreversible inhibition

5.4.1   Introduction

As stated in 5.3.1, irreversible inhibition is really outside the normal pharmacological classifications as outlined with competitive, noncompetitive and uncompetitive inhibitions and describes toxicological events that change the structure and even eliminate the function of the CYP, rather as nerve agents inactivate cholinesterase (Figure 5.2). To extend the 'machine gun' analogy of CYP activity mentioned in Chapter 3.7.5, irreversible inhibition is like failure to extract a spent cartridge. The nearest mechanical analogy to irreversible inhibition would be the incorrect key turning fully in the lock and not opening, followed by difficult extraction of the key, or even the key breaking off in the lock. Irreversible inhibition generally involves the same initial steps as a competitive inhibitor, which engages with the CYP active site. However, the important difference is that as the CYP catalytic cycle proceeds, reducing power is consumed, forming a metabolite that interacts with the P450 active site.

It is important to use the correct terms in this area. A time-dependent inhibitor (TDI) shows a progressive increase in its inhibiting effect before the substrate appears[32]. So to be strictly accurate, there are three types of TDI. The first is where a metabolite formed by the CYP is actually a reversible competitive inhibitor, such as with itraconazole, which is oxidised to competitively

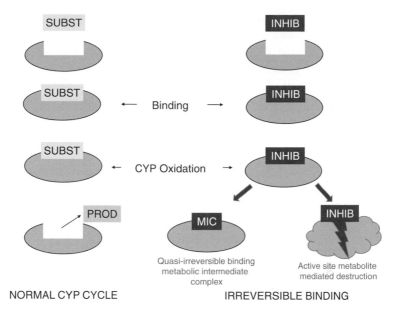

Figure 5.2 Scheme of normal substrate CYP binding (left) and mechanism-based inhibitors on the right, which result in the irreversible binding of an inhibitory complex or the destruction of the CYP active site by the inhibitor

inhibiting metabolites[33]. The second and third types of TDI are mechanism-based inactivators (MBIs), which as the name suggests, eliminate the CYP's function. Of these TDI/MBIs, the second type are termed *quasi-irreversible* inhibitors, which block the CYP with a stable metabolic intermediate complex (MIC), whilst the third type of TDI/MBI covalently binds the CYP haem or some other vital CYP structure through alkylation[32,34]. So from the three types of TDI, you could generalise that an irreversible inhibition can only begin with a metabolite, but a metabolite may not necessarily always be an irreversible inhibitor.

So to summarize, TDI/MBIs will not inhibit without CYP access to co-factors, such as reducing power (NADPH), although the presence of plenty of substrate can delay the inhibition process by protecting the CYP to some extent. This idea was exploited in the 1980s and 1990s with neostigmine to protect military personnel against nerve agents. With MBIs, V_{max} falls and affinity decreases as the inhibition becomes stronger and with the covalently binding MBIs, eventually no intact CYP is detected analytically, due to destruction of the CYP. There are other forms of TDI inhibition that are harder to classify; these involve uncoupling of the catalytic cycle and impacts on the co-factor processes that are yet to be fully understood[35]. TDI is significant, as between 25–30% of clinically relevant inhibitors cause their effect through this process[36] (Figure 5.2).

5.4.2   Mechanism-based quasi-irreversible inhibitors

Amongst clinically used drugs, the most common MBI quasi-irreversible inhibitors are alkyl amine or methylenedioxyphenyl derivatives, which comprise about half the TDI inhibitors[36]. These include the macrolides (erythromycin and clarithomycin), some antidepressants (fluoxetine and the tricyclics) and calcium channel blockers like diltiazem and verapamil,[32,37] some amphetamines, fluorquinolones and tyrosine kinase inhibitors like lapatinib. It is generally agreed that creation of the MIC requires a nitroso derivative to be formed from the drug, which then coordinately binds to the haem iron to produce the stable inhibiting complex[37]. This complex can be reversed experimentally with an oxidising agent, but obviously that can't happen *in vivo*. There has been some dispute over the sequence of metabolic events that lead to the formation of the nitroso derivative. As the tertiary amine in each alkylamine drug is metabolised to an N-dealkylated derivative, it was thought that this was necessary so that a secondary and then a primary amine would be formed, which then would be oxidised to the inhibiting nitroso group. However, studies with lapatinib and other drugs suggest that the secondary amine is hydroxylated first, forming a hydroxylamine, which is oxidized to a nitrone. This then forms a nitroso derivative, which exists in equilibrium with an oxime. The nitroso then forms the MIC[37]. Hence, it appears that these drugs can form MICs whether or not they have been N-dealkylated[36], although even within the same therapeutic drug class, such as the fluoroquinolones, some form nitroso derivatives as above and produce MICs, defining them as quasi-irreversible inhibitors, whilst others form ring-opened reactive carbon-species that attack the haem and are irreversible inhibitors[38].

5.4.3   Mechanism-based irreversible inhibitors

Several signs can reveal irreversible MBI inhibitors;

- Failure (*in vitro*) to reverse the inhibition with an oxidising agent (ferricyanide)

- Lack of effect of a trapping agent, such as glutathione (the metabolite is already reacting with the CYP active site)

- A degree of protection by a competitive inhibitor

- A lack of impact of agents such as catalase (reactive hydrogen peroxides and other oxygen species not involved)[39]

- A 10-fold dilution step will not diminish the effect of an irreversible inhibitor[40]

If none of the above measures changes the inhibition, this suggests that something very reactive (carbon-species or a nitrogen species such as an iminium ion) is involved, which covalently 'spot welds' itself to the CYP haem nitrogens, amino acids in the immediate active site, the CYP iron,[41] or all of the above. It can be detected through spectral analysis whether the haem is damaged; if this is not the case and the isoform is irreversibly inhibited, then the metabolite has bound to the CYP apoprotein i.e., one of the amino acids at the active site, so rendering it incapable of further substrate binding and catalysis[42].

This process of irreversible binding is sometimes termed 'suicide' inhibition, the enzymatic version of Private Pile's effect on Sgt. Hartmann in Stanley Kubrick's *Full Metal Jacket*. The destruction of the CYP means that only the formation of new enzyme will restore function.

There are therapeutic agents where this type of inhibition has been defined with certainty, but some drug inhibitory effects are so complex that they almost defy elucidation. For instance, ritonavir is viewed as a TDI, although some studies have suggested that it is a covalent MBI through the formation of a haem adduct, others have indicated that a very potent haem ligation occurs which gradually prevents electron transfer[43]. The effects of these drugs are very structure and CYP active site dependent; ritonavir may be a very potent CYP3A inhibitor, but it has little effect on CYP1A2[44]. There is also a very broad series of MBIs based on plant products, such as the furanocoumarins; these include psoralens, various bergamottins and many hundreds of less well-characterized compounds in fruit juices. Some of the psoralens appear to be covalent-binding MBIs, but it is not always straightforward to identify the exact nature of the inhibition[45].

Whilst we have focussed on TDI drugs with respect to CYP metabolism, there are some agents that are metabolised by different enzyme systems into chemicals that are potent MBIs. Gemfibrozil, a lipid-lowering drug, is a strong competitive inhibitor of CYP2C9 *in vitro*, but it acts as a MBI with CYP2C8 *in vivo*[46,47]. This happens as when gemfibrozil 1-O-β-glucuronide (which is very stable chemically) is formed by UDP glucuronyl transferase (see Chapter 6.2) and it then undergoes NADPH-dependent further oxidation to inactivate CYP2C8. This is partly because CYP2C8 has a much larger active site than CYP2C9 so it can accommodate the large metabolite[46]. Interestingly, gemfibrozil does not inhibit warfarin clearance but strongly inhibits repaglinide clearance as well as the glitazone drugs and any CYP2C8 substrate[46]. Indeed, gemfibrozil co-administration was a major factor in the withdrawal of cerivastatin. The statin's clearance was prevented; it accumulated and caused rhabdomyolysis, which provoked its withdrawal in 2001[47]. Another group of potent MBIs are the thienopyridine derivatives, such as ticlopidine and clopidogrel. Therapeutically, these compounds require CYP-mediated activation to a thiol that covalently blocks the P2Y$_{12}$ ADP platelet receptor[42]. Clopidogrel is the most potent mechanism-based inhibitor of CYP2B6, and it is activated into an MBI that not only destroys the haem, but promotes clopidogrel's 2-oxo metabolite attacks on the Cys475 residue on the protein, covalently binding it[48]. As with gemfibrozil, clopidogrel is cleared by UDP-glucuronyl transferaseses to an acyl-β-D-glucuronide, a potent MBI of CYP2C8. It is thought that in both cases a reactive epoxide is formed, which is a powerful electrophile that attacks the haem[48,49].

5.5   Clinical consequences of irreversible inhibition

5.5.1   Introduction

If you are reading this section from a scientific perspective, you will hopefully agree that irreversible inhibition is clearly either quasi or permanent, depending on the agent and CYP involved and the chemical processes involved. If you are reading from a clinical perspective, you might say, what difference does it make what type of irreversible inhibition occurs, and why should I care? Indeed, the MIC in quasi-irreversible inhibition blocks the CYP and cannot be reversed *in vivo,* whilst the MIB irreversible inhibitor covalently destroys the CYP.

The net short-term effect for both processes is days of inhibition, and as already mentioned, only the new CYP synthesis restores function, provided no more inhibitor is present. At this stage, how the CYP has been inhibited does not seem at first glance to matter, as the day-to-day clinical impact is the same, say, for diltiazem on CYP3A4 capability and clopidogrel metabolites' impact on CYP2C8. However, the drug companies go to great lengths to determine precisely which form of TDI inhibition occurs with a prospective therapeutic agent[32,35,41] and some of the basic methods used will be discussed in Appendix A. This happens because of the different long-term (weeks or months) consequences of the inhibition processes.

With quasi-irreversible inhibition, the CYP cannot cycle and is locked in a sort of stasis with the inhibitor. From Chapter 3.5.1, we know that CYP half-lives are short, ranging from hours to a couple of days at the most. Whether the CYP is metabolising or not, it will be replaced anyway and the lack of cycling might even preserve it from the usual 'wear and tear' of catalysis. From a hepatocyte perspective, the situation is similar to competitive inhibition, where there is 'no harm no foul' and cellular health and activity is not perturbed, but of course systemically, drug clearance is not happening. By contrast, a covalent binding event on the haem or the apoprotein changes the ultrastructure of the hepatocyte and it can trigger a series of events that attract the attention of the most ruthless, destructive and difficult to control of all our body processes – the immune system[35]. The process whereby drugs can cause immune-based adverse reactions that can end in total liver failure will be discussed in Chapter 8.4.

The next sections look in detail at some classes of TDI inhibitors of various types and focus on the impact of their inhibition from a short-term clinical perspective.

5.5.2 Quasi-irreversible inhibitors: the SSRIs

Depression: clinical context

Of those diagnosed with major depression, a quarter will attempt suicide at least once in their lives[50]. Suicide is the permanent answer to what is often a chronic but, in terms of life in general, temporary problem. Indeed, the level of irrationality that leads to the decision to end one's life is reflected in observations that those who jumped off the Golden Gate Bridge and survived (1% of participants) almost all eventually die of natural causes, and Kevin Hines' story is a good example of this[51]. Repeated suicide attempts are evidence of obvious mental anguish and diagnosing and treating the depression that usually accompanies it is vital.

The impact of depression in general is underlined by scale of the use of antidepressants in the developed world (16% of all adults)[52]. Indeed, treating depression is recognized by medical practitioners to be challenging[52]. Diagnosis can be made by the Hamilton Depression Rating Scale, which takes half an hour or so to administer, whilst the longer version contains 21 questions, known as the HAM-D21[53]. Until the 1980s, options for drug therapy were limited to TCAs and MAOIs. TCAs, for example, have narrow therapeutic indices, causing CNS and especially cardiotoxicity, alongside their unpleasant atropine-like (constipation, dry mouth) effects. Whilst the death rate from overdoses is only 3%, such episodes are difficult to treat and TCAs are a major cause of drug-related fatal overdoses[54]. MAOI use is problematic for pharmacodynamic reasons, such as the risk of serotonin syndrome when given with SSRIs and some other agents. In addition, MAOIs cause massive hypertension upon ingestion of tyramine-rich foods and are addictive with an unpleasant withdrawal process[55], and they are now unlikely to be prescribed in the United Kingdom[52]. As a response to these issues, a second generation of antidepressants appeared after the late 1980s. These broadly include the serotonin reuptake inhibitors (SSRIs), the serotonin and noadrenaline uptake inhibitors (SNRIs), as well as the noradrenergic and specific serotonergic antidepressants (NaSSAs), plus other miscellaneous drugs. Of these, the market penetration of SSRIs has been extraordinarily effective, as they comprise the bulk of prescribed antidepressants and are used in a range of other psychologically related conditions, from obsessive-compulsive syndrome, to anxiety and chronic pain[52,56].

Since the SSRIs replaced the TCAs as the first-line drugs for depression, it is clear that they are undoubtedly safer than TCAs and MAOIs in terms of the dangers of overdose, which is a major advance for vulnerable and unstable patients at risk from suicide. In 2017, a gentleman from Monza in Italy managed to walk into a hospital emergency department, four hours or so after ingesting 50 X 5 mg tablets of vortioxetine, an SSRI introduced in 2013. He was assessed and was completely fine, displaying no overt clinical symptoms[57]. This would not have been the case with a significant TCA overdose. Clinically, the second-generation drugs are no more effective than the TCAs in terms of efficacy[58], but they are much better tolerated. However, after 30 years or so of worldwide mass use, significant adverse issues have emerged. Whilst serotonin syndrome is a dose-related and understandable pharmacological consequence of elevated levels of SSRI-type agents[59], other less-predictable issues have emerged with second-generation antidepressants. These include the increased tendency to develop gastrointestinal bleeding[60] caused by their serotonin reuptake effects on platelet aggregation and brain haemorrhages, as well as the problem of dependence and withdrawal; in the polypharmaceutical elderly, extrapyramidal effects have been reported, particularly with Parkinsons' disease patients[61,62]. What is more immediately concerning is the prolongation of the QT interval by some (not paroxetine, for example) of these drugs (Section 5.7.2), with citalopram and to a lesser extent escitalopram being most associated with this effect and the subject of warnings from various world drug regulators[63].

Regarding second-generation drug efficacy, it can be difficult to be certain that a condition has been resolved with drugs where around half the patients will actually recover spontaneously within three months[52]. Indeed, SSRIs and other antidepressants are not always even as effective as placebos in studies and there has been no single stand-out drug of choice in the area[56]. This means that prescribing with the second-generation drugs has been rather idiosyncratic[52], based on different clinical experiences. Some practitioners favour fluoxetine for depression alone, whilst others try sertraline and citalopram in patients presenting with depression and anxiety co-morbidities[52]. However, the issue of QT prolongation has made a significant impact in citalopram prescribing[63].

Drug efficacy in this area has been ranked and venlafaxine and paroxetine came out most strongly, with vilazodone and bupropion showing the least effectiveness[56]. Responses to antidepressants are highly variable, with more than a third of patients showing no improvement with the first drug tried[64] and generally if there is to be a response, it will not be immediate, perhaps taking up to two or more weeks[52]. This time-lag is not always appreciated by practitioners and problematically, they can be reluctant to switch drugs for unresponsive patients, which delays recovery. Indeed, they are not always aware that SSRIs do not demonstrate greater efficacy at higher doses either; rather, it just increases the side effects without palpable gain[52]. It seems that in nonresponders to the initial SSRI, they do better if the second drug used is of another class, rather than another SSRI[65].

Another issue for the second-generation antidepressants is that dose/response relationships in patients seem poorly defined and it is rare that drug levels are assayed in patients. Indeed, it might seem rather obvious, but with escitalopram, it seems that poor responses are due to a failure of plasma drug levels to reach the therapeutic window, which with this drug is said to be greater than 20 ng/mL. This is thought to correspond with 80% occupancy of the serotonin transporter, which is believed necessary to provide a therapeutic response[64]. Overall, it is likely that the true scale of pharmacokinetically related treatment failures and adverse reactions in patients is more common than is generally realised. In the context of how they are used, it is just not practical for the lack of effect or adverse reaction to be thoroughly investigated; rather, another drug is tried until some response occurs or the patient spontaneously recovers.

The first- and second-generation antidepressants undergo oxidative metabolism and some are metabolised to potent CYP inhibitors that have been extensively studied *in vitro*. With second-generation drugs, as they are the current first-line drugs for depression, adding them to an existing regimen of several drugs requires some thought regarding the consequences of their considerable inhibitory effects, due to some significant drug–drug interactions. Of the six established SSRIs used in clinical practice: (fluoxetine, paroxetine, fluvoxamine, citalopram, escitalopram, and sertraline), citalopram and escitalopram have diminished in popularity, as although they were better tolerated, the QT interval issue has undermined this advantage. Several other drugs, including newer SSRIs (vortioxetine) and the SNRI/NaSSAs are also growing in popularity. Many of the second generation antidepressants may cause significant drug interactions with other CYP cleared drugs due to their ability to inhibit CYP2D6, CYP1A2, CYP2C19, and CYP3A4, although their *in vivo* effects are not always as potent as might be predicted.

Whilst acute medical conditions are challenging and stressful, we can all recognize that with treatment, the prospect of a relatively rapid resolution of symptoms and subsequent recovery can usually override our feelings of reactive anxiety and depression. However, long term, essentially unresolvable conditions that require frequent clinical management and impose significant restrictions on a patient's lifestyle and in many cases, prospective lifespan, are likely to cause a concomitant impact on mood and level of anxicty. As we saw in the previous chapter (Section 4.5.3), if this issue is not addressed adequately by the practitioner, then the patient may well reach for their phone or laptop and order something online to provide an answer, however inappropriate or frankly dangerous. So the second-generation antidepressants are added to many common drug regimens and their impact can be very significant on the patient's health and welfare. Some important interactions for this widely used class of TDI drugs are discussed below.

Fluoxetine and paroxetine

Dosage with fluoxetine results in the presence of four potential CYP inhibitors at once in the patient's system. This is because fluoxetine is chiral, in that it is an asymmetric molecule, so right- (Latin; Rectus) and left- (Latin; Sinister) handed versions are not superimposable. So effectively, two drugs are administered at once, (as with warfarin and many others) when the drug is given as a racemate, which is an equal mixture of R and S enantiomers. These are dealkylated by CYP2D6 and CYP2C9 to two derivatives; S-norfluoxetine is 20 times more potent therapeutically than R-norfluoxetine[66]. All four agents are strong competitive inhibitors of CYP2D6, with the S isomers being about 10-fold more potent than the R, so they cause about 90% of the inhibition *in vivo*[67]. All four agents are TDI/MIC inhibitors of CYP2C19, with S-norfluoxetine being the most potent[36,67], whilst the S-enantiomer parent and R-norfluoxetine are TDI/MIC CYP3A4 inhibitors and there is also weak inhibition of CYP2C9 by the four fluoxetine variants[66,68]. Thus, the number of different CYPs it can inhibit to varying degrees makes it the most likely SSRI to cause metabolic interactions with co-prescribed drugs[69].

Paroxetine (Paxil/Seroxat) is said to be clinically indistinguishable in terms of efficacy and adverse reactions from fluoxetine when in use[70]; however, its metabolism and its inhibitory mechanisms are different. Paroxetine is cleared mostly by CYP2D6 as well as CYP3A4, CYP1A2, and CYP2C19 to a catechol derivative, which can be further metabolised by COMT, although it also forms some reactive electrophiles[49]. It acts as an MBI of CYP2D6 through a carbene intermediate, and over time it pretty much shuts down its own clearance and virtually inactivates CYP2D6 activity[49] but has little effect on other CYPs[69].

Patients who are refractory to treatment for depression can be prescribed antipsychotics alongside SSRIs such as fluoxetine or paroxetine, and schizophrenic and bipolar patients are often prescribed second-generation antidepressants[71], so the interactions between the SSRI/SNRIs and antipsychotics have been investigated at some length as these prescribing patterns become more common. Whilst fluoxetine and paroxetine inhibit CYP2D6 through different mechanisms, the overall effect is similar, where any CYP2D6 substrate, and to a lesser extent CYP2C19 substrate, will have their respective clearances significantly retarded. Many antipsychotics are cleared by CYP2D6, so both SSRIs will markedly inhibit the clearance of older (haloperidol) and newer (risperidone and clozapine) drugs, leading to AUCs increasing several-fold. This can intensify classical antipsychotic side effects such as tardive dyskinesias, akathisias, and Parkinsonian symptoms[49,72]. The atypical antipsychotics, such as clozapine are effective in many patients who do not respond to other antipsychotics, partly through their effects on 5-HT receptors. Clozapine can cause fatal agranulocytosis and prescribing it is actually conditional on receipt by the dispenser of a normal white blood cell count (Chapter 8.4.4.). Other atypical antipsychotic drug clearances are affected by fluoxetine, causing sufficient accumulation with aripiprazole to lead to serotonin syndrome[72,73]. Ziprasodone's main route of clearance is aldehyde oxidase with less than a third of clearance via CYP3A4, making it relatively straightforward to use with SSRIs[74], although this drug has other significant problems[75]. Quetiapine is cleared mostly by CYP3A4 and to a minor degree by CYP2D6, again making it less likely to be a problem with the inhibiting SSRIs[76], but should be used with caution[65]. The combination of olanzapine and fluoxetine has become popular, most likely because olanzapine is cleared largely by glucuronidation and CYP2D6 is only a minor pathway. Fluoxetine's effect on olanzapine clearance is subclinically relevant, and this combination appears to be safe and effective particularly in bipolar depression[77-79].

Fluoxetine and paroxetine have well-documented strong impacts on the TCAs (such as desipramine, amitriptyline), increasing their AUCs by anything up to fivefold with an accompanying intensification of their atropinic side effects and increasing the risks they pose in terms of lengthening the QT interval. Indeed, the interaction can be lethal, but can also be managed successfully with dose reduction of the TCAs[74]. With nortriptyline, thought to be the best tolerated TCA, paroxetine has been used to selectively inhibit its clearance (see section 5.8.6) to reduce the side effects[72,80]. Drugs used in the management of Alzheimer's Disease such as donepezil and galantamine are also impacted by SSRIs, making their gastrointestinal off-target effects worse[72,74].

AEDs: at higher doses, fluoxetine has caused carbamazepine toxicity, probably through inhibition of CYP3A4. The drug also has a potent inhibitory effect on phenytoin clearance, leading to a several-fold increase in plasma levels of the AED and severe toxicity. It is likely that this occurs through CYP2C9 inhibition. Paroxetine has not been shown to cause this effect in a clinical trial and generally does not affect the AEDs significantly[69].

Cardiovascular agents: chronic problems in this area can promote and exacerbate clinical depression, so effective antidepressant therapy might also improve adherence to cardiovascular drug regimens and participation in clinical supervision. About 7% of warfarin patients are prescribed SSRIs[81] and unfortunately fluoxetine's CYP2C9 inhibition can impact S-warfarin clearance and also affect R-warfarin disposition due to the effect on CYP3A4 as it causes an increase in the INR. The off-target issue with bleeding caused by SSRIs, may compound this problem. Interestingly, whilst some studies show increased risk of bleeding with warfarin and SSRIs[81], others suggest that whilst there is an effect, it is not a major issue, probably because of the close clinical supervision that warfarin patients receive[82]. It is not thought to be a significant problem with acenocoumarol, paroxetine, and fluoxetine[83].

With regard to the DOACs, at the time of writing, clinical data are lacking in terms of the impact of the SSRIs in general on these drugs, beyond the obvious potential pharmacodynamic additive interaction between any anticoagulant and the ability of SSRIs to inhibit platelet aggregation through their impact on serotonin uptake[20]. From the section on the azoles, we know that apixaban and rivaroxaban clearance happens via CYPs 3A4, 2C8/9 as well as 2C19 and fluoxetine is highly likely to lead to elevated plasma levels of both drugs, potentiating their anticoagulant effect. The lack of involvement of CYP2D6 suggests that paroxetine is likely to show a less pronounced effect. Again, issues seen with the azoles, such as CYP inhibition allied to renal insufficiency are likely to increase DOAC plasma levels and pharmacodynamics effect with fluoxetine also. Regarding edoxaban, dabigatran etexilate, and betrixaban, the SSRIs are probably not likely to exert a powerful enough inhibitory effect on P-gp, to influence DOAC plasma levels at clinically relevant doses[84], so fluoxetine and paroxetine should be used with caution, but may well not cause a marked impact on those three DOACs.

Fluoxetine and paroxetine can also inhibit the clearance of some beta-blockers such as pindolol, metoprolol, propranolol, carvedilol, and the third-generation drug nebivolol, increasing AUCs in some cases by several-fold[85,86]. With metoprolol, the accumulation of the drug due to CYP2D6 inhibition with fluoxetine and paroxetine causes hypotension and bradycardia[85]. Interestingly, pindolol actually augments the efficacy of SSRIs in some clinical studies[87]. Fluoxetine, and possibly paroxetine may reduce the clearance of the antiarrhythmic propafenone[88].

- *Perhexiline*. This drug is used in Australia and New Zealand in the treatment of angina[89] but paroxetine has a potent inhibitory effect on this CYP2D6 substrate, which has quite a long half-life already[90]. Perhexiline metabolism is particularly subject to CYP2D6 polymorphisms (Chapter 7, Section 7.2.3).

- *Opiates:* Fluoxetine and paroxetine can strongly affect their efficacy by preventing the CYP2D6-mediated demethylation required to convert opiate prodrugs, such as codeine, dihydrocodeine, and tramadol to morphine or morphine-like derivatives[91]. With tramadol, the inhibition of CYP2D6 by these SSRIs can lead to CNS accumulation that has led to serotonin syndrome[92], although clinical knowledge of this problem is reportedly sketchy[93]. Interestingly, although oxycodone and hydrocodone are CYP2D6-activated to potent analgesics, the net clinical effect of paroxetine and fluoxetine is not significant as the parent drugs and other metabolites are active[91,94].

- *Other drugs*. The anti-oestrogen tamoxifen has been of immense value in reducing the risk of breast cancer long-term recurrence, although it is really a prodrug, requiring CYP2D6-mediated conversion to the active endoxifen metabolite. Indeed, levels of the active metabolite are threefold higher in the presence of the weak CYP2D6 SSRI inhibitor escitalipram, compared with that of fluoxetine and paroxetine; the latter two drugs are not recommended for use with tamoxifen[95].

Clopidogrel; this drug possesses its own inhibitory effects (already mentioned), so the impact of fluoxetine is potentially complex. There is said to be a pharmacodynamic interaction between the two drugs, due to fluoxetine's reduction in platelet aggregation and adhesion as mentioned previously, which could potentiate clopidogrel's effects[96]. In addition, it was until recently widely believed that inhibition of CYP2C19 would markedly retard the process of clopidogrel activation. In a clinical interaction study the direct impact of fluoxetine on the platelets was not significant and the pharmacokinetic interaction was only moderate (25% reduction in clopidogrel activation)[96]. This should be seen in the context of the more recent view that the contribution of CYP2C19 to clopidogrel activation is nowhere near as significant as CYP3A4 (Chapter 7.2.2). Hence, despite the potential for fluoxetine to inhibit several CYPs, the authors did not feel that specific recommendations on co-administering the two drugs were necessary and clopidogrel is likely to remain broadly effective in the presence of fluoxetine[96].

Fluvoxamine

Although clinically fluvoxamine only has minor inhibitory effects on CYP2D6, it is cleared by this isoform as well as CYP1A2. It is, however, a potent inhibitor of CYP1A2 and CYP2C19, with less effect on CYP2C8/9 and CYP3A4. Even though it is not a very potent inhibitor of CYP2C9, it is clinically relevant and fluvoxamine CYP2C9 inhibition is about twice as that of R-fluoxetine and ten times that of S-fluoxetine[68]. However, the strength of fluvoxamine's effect on CYP1A2 is such that it is an established clinical probe of this CYP[46]. Fluvoxamine can increase duloxetine and mirtazapine blood levels by several-fold, and this effect can enhance duloxetine and clozapine plasma levels[97,98]. The impact of fluvoxamine on the CYP1A2 cleared melatonin agonists is particularly potent[46,72]. Regarding AEDs, it appears not to affect the CYP3A4 substrate carbamazepine clearance, but does increase phenytoin levels by several-fold due to its CYP2C9 impact[69].

- *Antipsychotics.* Fluvoxamine can significantly increase the plasma concentrations of the first generation antipsychotics and the atypical drugs also, with olanzapine and clozapine levels increasing by more than 100%[99]. One of the issues with the atypical antipsychotics is that they can magnify the 10–25% lifetime risk of obsessive-compulsive syndrome in schizophrenics. Clozapine is most likely to cause this, with a mixed effect with olanzapine and much less of a problem with aripiprazole[100]. Adjunctive therapy with an SSRI can improve this situation, although fluvoxamine can actually make things worse as clozapine plasma levels are markedly increased[100].

- *TCAs.* Interestingly, fluvoxamine's inhibitory action on CYP2C19 means that the demethylation of some TCAs is markedly inhibited (amitriptyline, clomipramine and imipramine) and some studies have shown significant accumulation of TCAs in the presence of fluvoxamine, although this is probably dose related[101].

- *Cardiovascular agents.* In case reports, fluvoxamine can more than double INRs during concomitant warfarin therapy in a few weeks, by its inhibition of CYP1A2 (clears R-warfarin), as well as its effect on CYP2C9 (clears S-warfarin). This interaction is more significant with fluvoxamine than with paroxetine and fluoxetine, whilst sertraline is a seen as a safer alternative[102]. Although as mentioned above, warfarin's pharmacodynamic effect is usually so carefully managed clinically, that the net impact of SSRIs like fluvoxamine is minimized[82]. With the DOACs, fluvoxamine's CYP2C9 inhibition may increase plasma levels of apixaban and rivaroxaban, whilst edoxaban, dabigatran etexilate and betrixaban are probably not likely to show any major interaction with fluvoxamine, which is a substrate, but experimentally it is not a relevant inhibitor of P-gp[103].

With beta-blocking drugs, their predominant route of clearance is CYP2D6 and fluvoxamine's mild inhibitory effect on this CYP means that it is not usually thought to be a significant issue. A study in healthy volunteers with the newer agent nabivolol has indicated that concurrent fluvoxamine does not really affect the clinical performance of the beta-blocker, but it does increase the AUC by more than 1.5-fold and changes the disposition of the main hydroxylated metabolite also[86]. The authors were cautious over predicting how much effect fluvoxamine might have in patients stabilised on nabivolol.

Overall, fluvoxamine has little to recommend it, as it is no more effective and has worse gastrointestinal side effects than other antidepressant drugs of all types, and the range of its CYP inhibitory effects make it problematic in modern polypharmacy[72,104].

Citalopram, escitalopram, and sertraline

Citalopram is actually a racemic mixture of an active S-isomer and an inactive R-isomer, so the pure S-isomer was marketed cunningly in 2003 as escitalopram. Citalopram, escitalopram, and sertraline are much less potent CYP inhibitors than the other SSRIs and are consequently less likely to change the clearance of other drugs. These drugs are therefore more suitable for adding to existing multidrug regimens, particularly in the elderly and chronically ill[64]. They were also better tolerated than the other antidepressants[64,105,106]. This changed in 2011 when the USFDA warned that citalopram and escitalopram could prolong the QT interval[63] (see section 5.7.2), which could potentially lead to life-threatening cardiac emergency and that the dose should not exceed 40 mg daily in the elderly[106].

Interestingly, whilst this issue apparently appeared from nowhere, the drug's development was temporarily stopped in the 1980s when a number of dogs died through unknown causes in the high dose toxicity tests, so in retrospect, this issue should have been explored. The problem is linked to the parent drugs and the metabolites; the citalopram drugs are demethylated by CYP2C19 and CYP3A4 and partly by CYP2D6. The further CYP2D6-mediated demethylation to didesmethylcitalopram occurs at higher doses and seems to be the most potent cardiotoxic agent[105].

The citalopram drugs have multiple CYP pathways, and no single CYP inhibitor is likely to seriously retard their clearance. However, modern polypharmacy that includes drugs such as fluoxetine that inhibit multiple CYP pathways simultaneously combined with patient issues such as morbidity and old age, mean that restricting the dose of the citaloprams may not be sufficient to prevent cardiotoxicity[106].

Sertraline is probably the safest of all the SSRIs with regard to QT interval issues[63], is as effective as the TCAs and is highly ranked in terms of efficacy and tolerability, particularly with olanzapine in psychotic depression[107]. As with the citalopram drugs, its combination of weak competitive inhibition and lack of TDI activity on CYP2D6 and other CYPs, as well as its clearance by several different CYPs make it fit well into complex drug regimens without causing problematic interactions. This ease of clinical use, effectiveness, and general flexibility made it America's 'favourite' antidepressant up to 2014[108]. Indeed, any interactions it may have with other drugs such as olanzapine (increases its clearance)[107] could be related to its potent inhibitory impact (on a par with quinidine) on ABC-cassette transporters[109]. It would be reasonable to assume that sertraline would elevate the levels of drugs, which were substrates of the transporters, such as the DOAC agents. Sertraline's 'dark secret' is probably the risk it carries of hepatotoxicity, which appears to be linked to its complex and incompletely understood impact on hepatic gene expression, particularly those related to cellular stress[108]. Because idiosyncratic drug reactions in general are rare (see Chapter 8.3) there is no routine clinical monitoring for such issues, but the impact on the patient can be devastating; with sertraline, if hepatic reactions are to occur, then the time frame is quite broad (2–24 weeks)[110].

Venlafaxine

If patients do not respond to SSRIs, the next class of drugs considered is usually the SNRIs, as they are better tolerated and safer than TCAs and MAOIs. Whilst this is true, there are caveats. Whilst it is apparent that the most popular agent in this class, venlafaxine, can cause long QT syndrome, this is not a risk at therapeutic doses, but tachycardia as well as convulsions, occur in overdose. Interestingly, its major metabolite, which is also active, O-desmethylvenlafaxine, was approved to be marketed as desvenlafaxine in 2008 and it does not have such long QT issues, although its half-life is longer than the parent drug[63,76,111]. Venlafaxine's likelihood of causing sudden death due to cardiac toxicity in normal use is no greater than some SSRIs[112], however, in overdose it is twice as lethal as any of the SSRIs. In one study a comparable number of fatalities occurred with venlafaxine alone (31) than with all the SSRIs put together (39)[113]. This may partly be a reflection of the risks of treating more severely depressed and therefore more suicidal patients, but it is still true to say that venlafaxine's lethality in overdose is much closer to that of the TCAs than the SSRIs. About

70% of venlafaxine is O-demethylated to the also active O-desmethylvenlafax-ine (known as ODVEN in the literature) by CYP2D6. Whilst the inactive N-desmethyl derivative is formed mainly by CYP2C19 and CYP3A4[76,111]. There are some interactions where venlafaxine is a victim with the proton pump inhibi-tors (CYP2C19) and quetiapine and doxepin[76,114], but it is a perpetrator with other areas. Venlafaxine increases INRs with coumarin anticoagulants[81,83] due to its mild competitive inhibition of CYP2C9 although its impact on CYP2D6 is so mild that it has been recommended as a safe alternative to the SSRIs in tamoxifen therapy[115].

Duloxetine

This effective SNRI appeared in 2004 and it is easier to withdraw from when compared with venlafaxine. Duloxetine's metabolism is complex, as it is hydrox-ylated by CYP1A2 and to a lesser extent, CYP2D6, followed by sulphation and glucuronidation[40,69]. Duloxetine levels are affected by smoking-related induction, but it does not influence CYP1A2 substrate clearance. Obviously potent CYP1A2 inhibitors like fluvoxamine retard its clearance[69]. Duloxetine is thought not to be a TDI inhibitor of CYP2D6; its effect is probably via competitive inhibition by the parent drug[40]. As with sertraline, it has been linked with severe hepatotoxicity, but this is not directly related to any irreversible suicide inhibition, but a reactive epoxide may be involved[40]. Whilst they are reversible, the CYP2D6 inhibiting effects of duloxetine are considered to be more potent than citalopram, escitalo-pram, and sertaline, but less severe than the 'hardcore' inhibitors fluoxetine and paroxetine[69]. Apparently duloxetine is not contraindicated with tamoxifen[116], but there would have to be a very good reason indeed for using any agent that might compromise the effectiveness of a cancer preventing drug. The CYP2D6 inhibi-tory impact of duloxetine will affect the clearance of the major classes of drugs cleared by CYP2D6 (antipsychotics, TCAs, opiates, some beta-blockers and antiarrhythmics). Whilst duloxetine has no impact on QT interval itself[63], it can retard the clearance of drugs that do, such as quinidine, flecainide, and thiori-dazine, so promoting their QT effects[117].

Miscellaneous antidepressants

Reboxetine is cleared by CYP3A4 and is vulnerable to the usual inhibitor suspects. CYP2D6 involvement in its metabolism is not appreciable and it probably does not affect any other CYP-mediated drug clearances[69]. Other anti-depressants either antagonize serotonin's effects and/or as well as inhibiting other biogenic amines' reuptake. Bupropion is a dopamine and noradrenaline reuptake inhibitor and rejoices under the rather arch trade name of 'Wellbutrin', which is somewhat ironic given its less than stellar therapeutic effect in depression[56]. It has also been marketed successfully as 'Zyban' (which sounds altogether more dynamic) as an aid to smoking cessation, as it reduces nicotine dependence and cravings[69]. Although it is structurally dissimilar to the SSRIs and is cleared by CYP2B6, it can exert some inhibitory effects on CYP2D6 that are classed as weak to moderate, although it increases desipramine's AUC five-fold[69]. If bupropion accumulates, it can cause delirium, as in an interaction with duloxetine[118] and whilst therapeutic levels of bupropion do not affect the QT interval, it causes tachycardia in overdose[63]. Bupropion's competitive CYP2D6 inhibitory effect is enough to cause accumulation of substrates of this CYP and it is not to be used with tamoxifen[115]. Indeed, bupropion's impact on CYP2D6 is much greater *in vivo* than it should be, based on its rather weak *in vitro* activity and there may be a fascinating reason behind this. It seems that the 14-fold difference in CYP2D6 inhibitory effect in favour of the S over the R isomer of the drug is a red herring-the drug and its metabolites can actually downregulate CYP2D6 expression at the mRNA level in human cellular systems[119]. Thus, the drug effectively turns an extensive metaboliser of CYP2D6 (see Chapter 7.2.3) into a poor metaboliser in terms of restricting the functional CYP protein formed, rather than just by inhibiting expressed CYP protein. If this finding is investigated clinically in greater detail, it has considerable potential for the therapeutic manipulation of CYP expression.

Mirtazapine is another second-line antidepressant, although its side effects (causes patients to be sedated and put on weight) are used therapeutically for underweight patients and it has found a niche at low doses (~15 mg) as a nonaddictive sleeping pill[52]. Its general tendency to sedate has made it a safer alternative for severe anxiety than benzodiazepines[52]. Mirtazapine is cleared by CYP2D6, CYP3A4 and CYP1A2, so it is vulnerable to the major inhibitors, but its inhibitory CYP effects are sub-clinical[69]. Milnacipran is an SNRI introduced in the mid-1990s and more than half is renally cleared, whilst the rest of the drug is cleared by CYP3A4. *In vitro* milnacipran is said to be a a weak inhibitor CYP3A4, but this does not translate to the clinic, so the drug is not known for any serious metabolic interactions[69].

Antidepressant drug development continues, mainly driven by the poor response rates, tolerability and high propensity for drug interactions that we have seen with the existing second-generation drugs. From 2009–2013, four drugs were introduced, agomelatine (2009), vilazodone (2011), levomilnacipran (2013), and vortioxetine (2013)[120]. In terms of efficacy, clinically effective and tolerable doses are yet to be optimized[120]. However, although vilazodone is very potent in animal models, it has not been particularly startling clinically and appears to be among the least effective antidepressants[56,120]. Whilst vilazodone, levomilnacipran, and vortioxetine are superior to placebo, the usual issues with this class arise, with tolerability problems increasing with efficacy. Of the three, vortioxetine, which is a multimodal agent, was the best tolerated and possibly the safest in overdose as we have seen in Monza[57,120]. None of the four are linked with QT interval issues so far[63].

Agomelatine is one of the melatonin agonists and this and related agents such as tasimelteon can combat sleep disorders[46]. Agomelatine is almost entirely cleared by CYP1A2, which means smoking accelerates its clearance and fluvoxamine (and any potent CYP1A2 inhibitor) shuts it down, causing massive increases in plasma levels[69]. The antidepressant vilazodone is cleared by CYPs 3A4 and 2C19 and 2D6, so is subject to impact from inhibitors and inducers of those CYPs. Interestingly, it can inhibit CYP2C8 *in vitro*, so it could affect substrates of this CYP such as paclitaxel[69]. Levomilnacipran is a mirror image of milnacipran and is cleared by CYP3A4 and is subject to interactions based on that CYP, but does not appear to act as an inhibitor[121]. Vortioxetine is mainly a CYP2D6 substrate, with CYP2C19 involvement and several others, but investigations have shown that whilst bupropion can increase its plasma levels (and reduce its tolerability) through CYP2D6 inhibition, vortioxetine itself does not have any clinically relevant CYP or P-gp inhibiting properties[122]. Of the newer antidepressants, depoxetine did not make the grade due to its short half-life, but given the general propensity for many second-generation antidepressants to delay ejaculation, this drug is actually marketed as a remedy for premature ejaculation. Some of its trade names include 'Duralast' and 'Ever Long'. Dapoxetine is cleared by CYP3A4 and the usual CYP3A4 inhibitors retard its clearance, but is not known to cause inhibitory effects on CYPs itself[123].

Second-generation antidepressants: withdrawal or discontinuation?

These drugs are not advisable to abruptly terminate as they are well known for what is termed a *discontinuation syndrome*. This in many patients manifests as essentially a worse experience than the condition they started with and it includes intense emotional, neurological and other physical symptoms, lasting for weeks[124]. Short half-lives can promote problematic discontinuation (heroin and cocaine are notable examples) and the relatively short half-life of paroxetine with respect to other SSRIs exacerbates discontinuation symptoms. Fluoxetine and to a lesser extent, sertraline, have longer residence times and show far less discontinuation impact[125]. Indeed, some might suggest that *discontinuation* is just a euphemism for withdrawal, a word with rather grubby connotations of hopeless addiction to the illicit drugs mentioned above. This could be countered by the observation that withdrawal, as applied to opiates for example, is linked with intense drug craving which is absent with SSRIs; indeed, they are not illicitly abused, as the benzodiazepines still are.

The second-generation antidepressants were intended as direct and intended advances on TCAs, which, as mentioned previously, in terms of safety, they unquestionably are. However, commercially, in terms of volumes of drug sales and their marketing worldwide for anxiety as well as depression, the SSRIs effectively took the place of the vastly profitable benzodiazepines, which had, in their turn, replaced the frankly hazardous barbiturates. By the mid-1980s, the mass marketing of the benzodiazepines had to be curtailed, as they were revealed to be very addictive indeed. Fast-forward to 2012, where a systematic review found that of 42 'withdrawal' reactions reported for benzodiazepines, 37 of them were seen in SSRI *discontinuation*[125]; the authors could not see any logical distinction between benzodiazepine and SSRI dependence[125,126]. The World Health Organisation shares this view, and it could be argued strongly that it is neither necessary nor desirable that patients take second-generation antidepressants for decades of their lives.

5.5.3 Mechanism-based inhibitors: grapefruit juice

If I ruled the world, one of my immediate decrees would be to demand that a large sign, at least a foot or so in diameter should be put up behind the counter of every pharmacy. This sign would show a grapefruit with a red line through it, rather like you see in the *Ghostbusters* motion pictures. It would be even better if somebody somewhere has already invented this, but hopefully the negotiations on any image rights issues would go well. Patients would walk into the pharmacy and ask what the staff have against grapefruit, and the staff would then tell them. Whilst this would be rather tedious and repetitive for the staff, it would actually relieve suffering and save lives.

The grapefruit plant itself is actually man-made, apparently created by crossing sweet oranges and pomelos in Barbados in the eighteenth century. However, for the past century or so the vitamin C and antioxidant content of the fruit has led to the wildly successful promotion of the juice as a 'natural' beverage, whose health-giving properties can only, of course, be accentuated by the faintly self-flagellatory bitter taste. However, it was not until the late 1980s before its marked impact on drug clearance was discovered[127] and it was gradually realised that there could be a major problem with rapid and even lethal drug toxicity (History 1).

A significant victim of grapefruit juice was the first and very useful nonsedating antihistamine (terfenadine), which was removed from the list of OTC medicines in the United Kingdom in 1997 and eventually withdrawn completely[128]. What is noteworthy concerning the effect of grapefruit juice is its potency from a single 'dose', which coincides with a typical single breakfast intake of the juice, say around 200–300 ml[129,130]. It can take up to three days before the effects wear off, which is consistent with the synthesis of new enzyme.

The most interesting aspect is that grapefruit juice was thought not to inhibit hepatic CYP3A4, but gut CYP3A. It is useful to clarify this point. In the mid-1990s, it was shown that grapefruit juice did not appear to inhibit the clearance of intravenously dosed drugs, although it would inhibit when the drug was orally dosed[129,130]. More recently, it has been established that *in vitro*, there is no difference in the inhibitory effects of other agents such as azoles (fluconazole and ketoconazole) on CYP3A4 from human gut or liver. Therefore, grapefruit juice should be technically capable of blocking hepatic CYP3A4. However, when modest amounts 200–300 mls of average strength juice are consumed, the combination of its irreversible CYP binding and a high gut CYP3A expression, means that virtually none of the inhibitor physically reaches the liver. In terms of breakfast grapefruit juice consumption it is probably unlikely that anyone would be able to drink enough of the stuff to cause clinically significant hepatic CYP inhibition[129,130].

If we recall the section in Chapter 1 (5.1) on bioavailability, it becomes clear how an inhibitor of gut wall metabolism would have such a powerful effect in increasing the systemic levels of a drug. Indeed, grapefruit juice has revealed that there are many drugs that are subject to a very high gut wall component to their first-pass metabolism (pre-systemic metabolism); these include agents such as midazolam, terfenadine, lovastatin, simvastatin, and buspirone. Their gut CYP clearance is so high that if the juice inhibits it, the concentration reaching the liver can increase sixfold or more. If the liver normally only extracts a relatively minor proportion of the parent agent, then plasma levels of such drugs increase dramatically towards toxicity (see History 1). If the inhibited drug is an OATP or P-gp substrate, this toxic effect may also be influenced by grapefruit juice-mediated inhibition of these transporters (Section 5.6.1.)[129,130].

As has been mentioned, the inhibitory effect of grapefruit juice in high first-pass drugs is particularly clinically relevant as it can occur after one exposure of the juice. Obviously, the higher the pre-systemic metabolism of a drug (low bioavailability) the greater effect the juice is going to show[127,129]. So you can see that highly bioavailable drugs are not really significantly impacted by the juice. One interesting characteristic of the grapefruit juice effect is that the plasma half-lives of the drugs do not change, as the liver carries on metabolizing what drug it can take up, provided the grapefruit juice 'dose' is modest.

To summarize, drugs not to be given with grapefruit juice are ones that:

- Are orally dosed and undergo high presystemic (enteric) metabolism

- Are metabolised by CYP3A4/5

- Are narrow TI drugs

- Possess potentially lethal off-target (unexpected and not related to their main use) pharmacological effects

Here are some examples of the variable risk of grapefruit juice:

*Drugs that should **not** be taken with grapefruit juice:*
Some statins (simvastatin, atorvastatin, lovastatin), amiodarone, nisoldipine, buspirone, indinavir, sildinafil, pimozide, cilostazol, almost all the tyrosine kinase inhibitors (dasatinib, lapatinib etc.), etoposide, saquinavir, sirolimus, maraviroc, erythromycin, halofantrine, dronedarone, eplerenone, clopidogrel,

apixaban, ticagrelor, fentanyl, oxycodone, ketamine, quetiapine, ziprasidone, tacrolimus, sirolimus, everolimus, and cyclosporine.

Drugs that may be problematic with grapefruit juice:
Benzodiazepines (midazolam, triazolam, diazepam), rivaroxaban, nifedipine, felodipine, synthetic opiates (methadone, dextromethorphan), carbamazepine, quinine, sertraline, azole antifungals (itraconazole), losartan, and some steroids (prednisolone).

Alternatives in affected drug classes
There are several drugs where others in their chemical class are inhibited by grapefruit juice but they are unaffected: the statins fluvastatin, pravastatin, rosu-vastatin, are safe alternatives for hardened grapefruit consumers, whilst imatinib is almost 100% bioavailable and is not impacted by the juice[131]. The antihista-mine desloratadine is also unaffected by grapefruit juice[132].

Clinical Impact
The real-world effect of grapefruit juice on a particular drug depends on several factors as described above, but there are yet more to consider. For example, some patient groups are more at risk than others. The most fanatical consumers of the juice are those over 45 years old, who as we know, are the major recipients of prescription drugs[129,133]. Since patients often do not take seriously the idea that something you can buy in a supermarket that is 'healthy' could possibly interact with medication, means that they may not 'confess' even after a serious clinical interaction has taken place. Indeed, many quite serious interactions probably go unreported[129,133]. Older patients are also less tolerant to elevated drug levels in general and compensatory heart rate increases seen in grapefruit/felodipine inter-actions in younger individuals are not seen in the elderly as their baroreceptors are worn (Chapter 4.1.1)[134]. It appears that some individuals can reduce the impact of grapefruit with statins if they take their drugs the night before their breakfast juice[127,129,133], but this is a risky strategy – better to substitute to a non-CYP3A-cleared statin. The issue of grapefruit juice causing increased bioavail-ability of steroids (contraceptive and post-menopause) and any associated concomitant cancer risk remains to be resolved, although it does seem reasonable to avoid the interaction, if only for peace of mind.

The guilty chemical parties
As with any natural product, grapefruit and its various citrus relatives contain many hundreds of individual chemical entities. Indeed, there is now evidence that the juice contains some strong agonists of human PXR and AhR, which are quite potent in their induction of various CYPs and transporters, including CYP3A4 mRNA in the LS180 colon cancer cell line, but the CYP inhibition wipes out this effect[135]. The major flavonoid present in grapefruit juice is naringin, but the fla-vonoids are not significant in their CYP inhibition[130]. The most potent inhibitors in the juice are the furanocoumarins, which have a five-membered ring (contain-ing oxygen) attached to two aromatic rings in their 'head' and a longer aliphatic

'tail'[130]. When it is oxidised by CYP3A4/5 and CYP2B6 also, the double bond in the furan in these agents is the key to how they lead to one of the most emphatic and comprehensive demolitions of the structure and function of these isoforms. A reactive γ-ketoenal and probably the furanoepoxide formed particularly from the bergamottins (particularly 6′ 7′ dihydroxybergamottin and bergamottin itself), not only react with the apoprotein of the CYPs but also destroy the haem moiety[136]. In fact, the bergamottins can dimerize to form even more potent CYP inhibitors[127,129,133]. Interestingly, the bergamottins' interaction with other CYPs such as 1A1 is purely competitive and does not involve CYP metabolism[137].

Bergamottins are also found in many herbal remedies, as well as Seville orange juice, pomelos, and clementines, which can exert similar effects to grapefruit juice[135]. With grapefruit, however it is prepared, either as concentrate, canned, as segments, or the fruit itself, there is no escape as the inhibitory effects still occur. A further furanocoumarin, epoxybergamottin, which is present in grapefruit peel, is also a CYP3A4 inhibitor[138]. A number of other fruits and herbal remedies contain bergamottins, and it can be assumed that any foodstuff/fruit or drink that contains significant amounts of these agents will exert the similar effects to grapefruit juice. Of course, as with all natural products, the levels and proportions of the various agents differs between crops and batches. Interestingly, grapefruits in the eighteenth and nineteenth centuries were boiled to make them less bitter and more palatable. It turns out that the bergamottins are not very heat stable and could be removed from the juice, but it's likely that this would affect the taste so perhaps the heat-treated grapefruit juice might not be so much fun to drink.

5.5.4 Mechanism-based inhibitors: other juice products

In terms of the range of other juice products that are available, so far there does not appear to be a fruit juice or fruit product that can challenge grapefruit in terms of its ability to interact with drug clearance in the real world. Naturally, this does not preclude the future discovery and marketing of another 'superfood' fruit or juice product that might contain a new class of inhibitors. During the 2000s, cranberry juice was viewed in a similar light to grapefruit juice as a potentially serious drug interaction issue.

Cranberry juice gained popularity due to the perception that it is effective in preventing urinary tract infections (UTIs). The evidence suggests that this has merit, although it is probably better targeted at those at greatest risk from UTIs, such as those recovering from radiotherapy, or are immunosuppressed, or those with urinary catheters[139]. The juice has also found a recommendation to help tackle *H. pylori* infections alongside the antibiotic and proton pump inhibitor combination[140]. Cranberry juice is very rich in flavonoids known as proanthrocyanidins, which can disrupt the ability of pathogenic bacteria to coordinate their formation of biofilms, which are difficult to penetrate with antibiotics, thus making the infections more persistent and destructive[141]. The disadvantage is that the concentrated juice is really very acidic and rather indigestible – a few minutes after consuming some, you might wonder if you would have been better to top up your car battery with it. To improve palatability it is often sweetened, or mixed with other juices.

However, cranberry juice has acquired a reputation for increasing warfarin INRs, at least on one occasion (History 2) with fatal consequences[142]. Regulators issued warnings to avoid the juice when taking warfarin and given the CYPs involved, it was assumed that cranberries at least contained some fairly potent CYP2C9 inhibitors. However, a number of clinical and *in vitro* studies explored this interaction and its impact on other CYPs[143] and have not found any real issues. There may be an explanation for the original observations on INR impact, which may lie in quantities consumed, levels of flavonoids and also the ability of these compounds to displace drugs such as warfarin from human serum albumin. With such highly bound drugs, even minor displacement could increase plasma concentrations of free drug very significantly, with some impact on INR[144].

Other fruit juices

There are hundreds of fruit preparations available that have been specifically marketed for their 'superfood' antioxidant capacities, such as purple grape, pomegranate, blueberry, and acai juices. Many are not only extremely palatable, but also some studies have shown benefits in various conditions such as prostate cancer. Clearly, many people will drink them in considerable quantities and gain benefits from their consumption, although little or nothing is known of their capacities to influence drug metabolism. As they all contain large numbers of diverse phenolics and are pharmacologically active, they should probably be consumed with at least some caution during drug therapy. It appears that apple, sweet orange, and grape juice can inhibit drug transporters (Section 5.6) and distort bioavailability, but they do not appear to have any powerful effects on biotransformation[143]. Pomegranate juice can inhibit CYP2C9 *in vitro*, but is apparently not relevant *in vivo*[145] and it has little or no impact on CYP3A4, either[143]. Alternatively, chokeberry extract is quite a powerful CYP3A4 inhibitor, probably through its high flavonoid content. The authors of one case history explained that they found out that their patient was covertly using chokeberry extract *'after further questioning in a nonthreatening fashion'*,[146] which was clearly effective enough to make the patient 'spill the beans'.

Tomato juice extract showed some mechanism-based inhibition *in vitro*, using human CYP3A4[45], although no significant reports could be found whether this effect occurs *in vivo*. Concentrated tomato products like ketchups and juices are sought after due to their very high lycopene content, a chemical that is thought to protect against prostate cancer. Indeed, tomato juice itself is a popular home treatment for the consequences of overindulgence in ethanol consumption (with or without added vodka). Given the quantities of tomato products people have been known to ingest and that lycopene itself is not a significant CYP inhibitor, then if this was a relevant CYP interaction, perhaps it would have emerged as a problem before now.

5.5.5   OTC herbal remedy inhibitors

Whilst it used to be relatively straightforward to bracket OTC (over the counter) and POM (prescription only) medicines separately, online ordering means all of us can bypass regulations and obtain as much of any product, herbal or pharmaceutical, whenever we wish. In the same way, various shadowy figures design and market illicit, mind-altering substances, whilst others scour the world looking for folk remedies that might be marketable (see below).

There are thousands of herbal agents worldwide to discover and profit from, whether or not they actually benefit or even harm their consumers. As with so-called *designer drugs,* it can be difficult to evaluate the impact of a herbal product due to the issues we discussed with St John's Wort in the previous chapter (Section 4.5.3), such as how the herb was grown and processed, as well as how this impacts the amount of active ingredients. The result for each agent can often be several contradictory and inconclusive studies. However, it is probably fair to say that if an herb does contain a potent inhibitor that has real-world relevance, it will be detected and placed in context eventually.

In the previous chapter, you might recall that many of the therapeutic and potentially toxic effects of herbs are linked with the impact of their constituents on PXR and AhR (4.4.3). It is not surprising, then, given the range of chemicals in these plants, that there will be some mechanism-based inhibitors of various CYP isoforms that might approach the potency of grapefruit juice. These preparations can be spontaneously adopted by patients on the recommendation of a friend or after reading some form of publicity. As we know, patient *omerta* can apply here and informing health-care professionals about herbal usage is often not a priority.

Liquorice extract

Liquorice extracted from the roots of various *Glycyrrhiza* species has been used in Chinese and European other medicinal contexts as an anti-inflammatory, anticancer, and myriad other applications for millennia. It is also popular in foodstuffs and snacks due its flavour and sweetness. There are estimated to be 'Allsorts' of chemical entities in this root (more than 400)[147] and only a few have been evaluated for their CYP inhibitory effects. One of these entities is licochalcone A, which is found in Chinese herbal liquorice and can, as well as some acylcoumarins and flavonoids, inhibit a range of CYPs, from CYP1A2, to the 2C isoforms and also CYP3A4[147,148]; licochalcone (found only in *G. inflata*) is a mechanism-based inhibitor of CYP3A4[149]. In European varieties, glabridin is a potent mechanism-based inhibitor of CYP3A4, although it can competitively inhibit CYP2C9[149] (History 4). There is certainly potential for heavy use of some liquorice products to interact with the clearance of therapeutic drugs, but so far this has only appeared in isolated cases.

Goldenseal extract

Whilst many herbal extracts do have some positive therapeutic impact that balances their toxicity, with goldenseal (*Hydrastis canadensis*), it could be argued that the risks seem to outweigh the benefits, partly because much more has been demanded of this herb that it can deliver; indeed, it is under threat in the wild from overharvesting. Goldenseal contains two known active alkaloids, berberine and hydrastine although there are many more active agents in the extract. Goldenseal is quite a potent clinical inhibitor of CYP2D6 and CYP3A4 and thus it should not be taken with CYP2D6 substrates, such as tamoxifen, where it would reduce its clinical effectiveness. In addition, it will increase the half-life of all the usual CYP3A4 suspects such as midazolam (over 60% increase in AUC)[150]. The herb should not be used by lactating and/or pregnant women, as berberine can cause uterine contractions and may also be toxic to infants. This agent is a more potent inhibitor *in vitro* than *in vivo*[151].

Other possible herbal inhibitors

Among other popular herbal preparations, Black Cohosh has been used as a remedy for menopausal symptoms, although its CYP2D6 inhibitory effects *in vitro* suggest that its use with tamoxifen could be problematic, as with any CYP2D6 inhibitor, there is a risk of diminishing the clinical efficacy of the anti-oestrogen[152]. Black Cohosh may cause liver toxicity[153].

Kava-kava is rich in various kavalactones and is extracted from *Piper methysticum;* it has been consumed for millennia in the South Pacific Islands and has been marketed as an anti-anxiety agent. However, its propensity to cause liver failure in some individuals is linked with poor clearance of the kavalactones in CYP2D6 deficient individuals (see Chapter 7.2.3). Many more Europeans are deficient in this CYP compared with Polynesians, accounting for the toxicity, and kava-kava is on the list of banned herbs in several countries[153,154]. Incidentally, it is also a clinically significant inhibitor of CYP2E1, CYP1A2, and CYP2D6[152,153] and its risks clearly outweigh its benefits, considering how many safer alternatives there are in anxiety management. Indeed, as the cult UK TV show *League of Gentlemen* might have it, 'This is a local herb for local people; there's nothing for you here....'

Gingko biloba has been used as a memory enhancer in Alzheimer's disease, as well as for its effects on microcirculation. It has some mild but not clinically relevant inhibitory effect on CYP3A4 at the recommended dose range[151]. Milk thistle (*Silybum marianum*) and its main extract, known as silymarin, contains a number of flavonolignans and is used for liver complaints of various kinds. Several clinical studies have found no impact on the major CYPs, except for CYP2C9, where there may be a significant inhibitory effect, which requires more investigation[151]. Echinacea may have some mild inhibitory effects on CYP1A2 and CYP2C9[151], but this may not be clinically relevant[155]. Ginseng, garlic, and saw palmetto all appear to have no significant clinical impact on CYPs[151,152].

In general, there are often not enough well executed clinical trials with a particular herbal extract to make a clear judgement on real-world relevance of impact on drug clearance. This is certainly the case with the large number of Chinese herbal remedies[152]. Unfortunately, unless some catastrophic case reports emerge, such as with kava-kava, the impacts of these herbs are unlikely to be determined. Whilst it is now impossible to prevent patients obtaining herbal extracts that could potentially harm them, they can only be informed of the risks and that they should trust their healthcare providers.

5.6   Cell transport systems and inhibition

5.6.1   Uptake (Influx) transporters: OATPs

The main hepatic and gut uptake solute carriers (SLCs), or transporters, are known as organic anion transporting peptides, or OATPs. As outlined in Chapter 2.6.2, they operate a facilitated diffusion, swapping cellular molecules such as glutathione for extracellular endogenous anions and some predominantly polar drugs. In the context of this chapter, *inhibition,* particularly of CYPs, has so far translated into accumulation and subsequent drug toxicity. However, in the complex process of absorption of some polar drugs too charged to diffuse across cell membranes, inhibition of these influx transporters can cause two possible opposing outcomes, depending on whether the SLC inhibited is in the gut or the liver:

1. If OATP inhibition is in the gut, then systemic exposure can be greatly diminished, potentially causing drug levels to disappear out of the therapeutic window.

2. If the inhibition is in the liver, then the organ struggles to take the agents out of the portal circulation to metabolise them, so systemic exposure can increase dramatically.

Gut transporter inhibition

OATP1A2 and OATP2B1 are found in high levels on the luminal (inside, facing the gut contents) surface of the human small intestine and they operate the facilitated diffusion of many polar drugs such as fexofenadine, the marketed active metabolite of terfenadine[130]. Grapefruit juice is a very potent inhibitor of both OATPs. This is partly due to the furanocoumarins and the flavonoid naringenin, which can inhibit uptake of OATP2B1 substrates. Another flavonoid, naringin, is a potent inhibitor of OAT1A2, severely impacting fexofenadine uptake. The effects are concentration dependent and can cut the effectiveness of these transporters by more than 60%[130] when the drug is consumed with the juice and for a couple of hours afterwards.

Other drugs affected by grapefruit juice include the anticancer agent, etoposide, the β-blockers celiprolol, talinolol, and atenolol, plus the antihypertensive aliskiren[130]. Unlike the impact on CYPs, the effect of the juice on these transporters is quite short term, so it looks like a breakfast bout of grapefruit juice would not significantly affect drug absorption, provided the drug were to taken around lunchtime[130]. There are several known drug substrates of OATP1A2 whose absorption may be cut by grapefruit juice; these include steroids and L-thyroxin, ouabain, and the anticancer agent methotrexate[129,133,134]. Thyroxine treatment can be problematic to manage in terms of its dose/effect relationships, so any interference in absorption will have significant impact on the patient's general feeling of health and well-being, so washing L-thyroxin down with any fruit juice is a bad plan[129,133,134]. It is likely that more OATP1A2 drug substrates will be discovered in the future. The OATPs are polymorphic, so this variability in expression in different individuals has the potential for significant prescribing problems in certain patient populations (Chapter 7.2.10).

Hepatic transporter inhibition

Currently in drug development, the evaluation of drug transport in general is considered to be extremely important and specifically, the assessment of the contribution of hepatic uptake transporters to drug bioavailability is now explored extensively using human *in vitro* cellular and software-based models[156]. Rifampicin (aside from its induction effects) is a substrate for hepatic OATP1B1 and can inhibit it, to the point drugs such as atorvastatin are just not taken up by the liver, leading to several-fold increases in plasma levels. Similarly, rosuvastatin is not significantly cleared by CYPs, but is usually excreted mostly unchanged in faeces. The presence of cyclosporine increases the statin's plasma concentrations sevenfold. These two interactions are due to direct inhibition of hepatic OATP1B1 by rifampicin and cyclosporine, which both act to prevent hepatic uptake of the statins, so they cannot be metabolised. They bypass the liver and their bioavailability increases vastly, rendering the patient vulnerable to concentration-related statin toxicity, such as rhabdomyolysis[156,157]. The nature of this inhibition appears to vary according to the inhibitor, with evidence that cyclosporine, as well as several antivirals (asunaprevir, simeprevir and ritonavir) and gemfibrozil are time-dependent inhibitors of OATP1B1 and OATP1B3[157], although they do not affect transporter expression[157]. Rifampicin does not appear to show this effect. Clinically, the bioavailability of cerivastatin was increased sixfold by concurrent administration of gemfibrozil and a signifi-cant part of this effect was through inhibition of OATP1B1; more than 50 deaths from renal failure linked to rhabdomyolysis led to the withdrawal of this statin in 2001 as mentioned previously[158]. This particular interaction was due to a combination of gemfibrozil-mediated inhibition of OATP1B1 and CYP2C8 and exactly the same mechanism applies with gemfibrozil on co-administered repa-glinide[159]. Aside from the various fruit juice products, there are several other inhibitors of OATP1B1, including the food colouring and herbal agent cur-cumin, a polyphenol that increases plasma docetaxel levels through CYP2C8 and OATP inhibition, as well as possibly enhancing the pharmacodynamic effects of the anticancer agent[160].

5.6.2 Efflux transporters: P-glycoprotein (P-gp)

P-gp inhibition and clinical relevance

The previous chapter (4.4.7) discussed the depth, scale, and effectiveness of the ABC cassette efflux transporter system, as it orchestrates a finely controlled dynamic barrier to the cellular entry of a vast range of chemical entities. The nuclear receptor modulation of the best studied efflux protein, P-gp, coupled with the emerging complexity of the protein's interactions with its substrates, are now understood to the point where it is possible in drug development to realistically evaluate P-gp's role in present and future drug interactions[161], as well as design inhibitors to modulate P-gp therapeutically.

Currently, there appears to be three main ways to block P-gp; the first is through some form of direct interaction with the active site where the substrate is engaged by the pump. This is more complex than it sounds. As we saw in Chapter 4.4.7, P-gp is not just a pump but has the complexity of a fully func-tional and controllable enzyme system, which is allosterically modulated. Hence, chemical agents can bind competitively (direct competition for a com-mon site) or noncompetitively (some form of allosteric interaction) and even uncompetitively[162]. Since we know P-gp is like a CYP in its complexity and drugs bind in a mechanism-dependent way to CYPs, then it is not unreasonable to assume that other agents can do this to P-gp. So far, there does not appear to be a commonly used drug that can irreversibly bind to either the active site of P-gp or the allosteric site, but it seems plausible that it would be possible to irreversibly inhibit P-gp and the other ABC transporters in this way. The second way to inhibit P-gp is through hitting its power supply, its ATP hydrolysis pro-cess. There already is an irreversible inhibitor of this process, N-ethylmaleimide, which is a reactive experimental agent used in protein research[161]. The third method is to disrupt the surrounding cell membrane lipid structure[162], of which more will be described later.

Through clinical and *in vitro* studies, we know that there are plenty of competitive P-gp inhibitors, such as verapamil, cyclosporine, quinidine and ritonavir[162], some of which are CYP3A inhibitors, so they are capable of causing a marked impact on first pass of a P-gp substrate and especially a P-gp and CYP3A substrate. This has the effect of increasing the bioavailability of this category of drug. Of course, it is not easy to distinguish between the effects of P-gp and CYPs in the gut, so a useful P-gp substrate to study which is independent of CYP clearance is digoxin, which the USFDA recommends as a probe substrate for P-gp in humans, whilst the dye rhodamine 123 is used during *in vitro* cell studies[161,163]. Digoxin is not a perfect P-gp probe, as it is transported by other systems, is already highly bioavailable and even total blockade of P-gp will not hugely increase digoxin plasma levels, (around 1.4- to 1.8-fold[163,164].

However, any significant increase in bioavailability is clinically extremely relevant, as digoxin has a narrow TI, and a high likelihood of clinical use with several inhibitors of P-gp in older patients. The therapeutic range of the drug (revised downwards in Chapter 4.4.7) can reach 1–2 ng/mL, with toxicity ensuing around 2.8-3 ng/mL[165]. Studies have established that digoxin levels can rise by up to 60% in the presence of multiple P-gp inhibitors such as verapamil, amiodarone and atorvastatin. Digoxin is well known for its risks of adverse events[164] and elderly patients on this drug have are more likely to die when taking it[166]. Digoxin is carefully clinically monitored but there is evidence that women tend to have higher plasma levels than men at the same dosage range[165], so they are likely to be a greater risk than men from interactions due to P-gp inhibitors.

It has also become apparent that drug distribution to various tissues and excretion by the kidneys is strongly dependent on efflux and influx systems. Despite adequate absorption, drugs may still completely fail to enter target tissues as either P-gp ejects them or a drug may penetrate tissues it does not normally enter in the presence of a P-gp inhibitor. Although opiates can curtail excessive gut motility and fluid production such as in diarrhoea, their central effects are normally too severe to use in this context. Loperamide does not penetrate the brain, as P-gp ejects it very efficiently, but it is effective at the level of the gut. In the presence of P-gp inhibitors like quinidine, loperamide can reach the medulla and cause depression in respiration[167]. As discussed in Chapter 4, Section 4.4.7, the penetration of AEDs, antipsychotics, and antidepressants into the CNS is dependent on P-gp and other transporters.

The impact of various fruit juices and herbal preparations on P-gp is less well defined. The effects of grapefruit on P-gp have been difficult to ascertain, but are nowhere near as important as the CYP-inhibitory effects and are short lived[130]. Most of the other herbal remedies commonly used do not appear to strongly impact P-gp[151] at the usual dose range. Overall, drugs that act competitively or noncompetitively in the inhibition of P-gp can be seriously problematic in terms of drug–drug interactions.

P-gp and cancer: therapeutic inhibition

As mentioned in the previous chapter (Section 4.4.7), The ABC-cassette transporters are an integral feature of resistance to anticancer drugs and whilst the PXR-controlled organisation of this process has not been easy to rein in therapeutically, the search for clinically effective and low-toxicity inhibitors of P-gp has been pursued much more vigorously. Unfortunately, therapeutic P-gp inhibition has been fraught with difficulty and has even been questioned as a viable approach[168]. Whilst the reward for a successful augmentation of penetration of an anticancer agent through P-gp inhibition should be greater anticancer efficacy, it is likely to come at some cost. Since all antineoplastics are toxic to tissues, the challenge is to improve penetration in the target and not in the bystander tissues. Any inhibitor of P-gp, even assuming it did not interact in any other way with the pharmacokinetics/dynamics of an antineoplastic agent, would have to somehow be selective for P-gp in the target cancer tissue. This is not very likely, as induced P-gp is the same as constitutively expressed P-gp. What will probably happen is that the inhibitor will block P-gp wherever it distributes in terms of tissue penetration and could conceivably render that tissue vulnerable to the toxic agent, as well as any other toxic chemicals present. In addition, all the physiologically essential roles of the ABC transporters will be compromised.

There have been four generations of P-gp inhibitor, starting with the therapeutic drugs mentioned above, like cyclosporine and verapamil. Their obvious disadvantage was low selectivity coupled with their inherent potent clinical effects. Consequently, they were really only useful to prove principle, in that P-gp can be inhibited, which could yield future clinical benefit. The second and third generations were analogues of the first-generation compounds, with accentuated P-gp inhibitory effects, with greatly diminished pharmacological impact in other directions. Examples include valspodar, an analogue of cyclosporine and dexverapamil, which is the R-enantiomer of verapamil[162]. However, these agents were CYP substrates, were not specific enough for P-gp, as they interacted with other transporters and ultimately were not successful[162]. The third generation (zosuquidar, tariquidar, elacridar) did not interact with the CYPs and were potent P-gp inhibitors at the nanomolar level in some cases, but ultimately led to unacceptable toxicity[169]. The fourth generation of inhibitor is focussed partly on natural products, such as the flavonoids, which have the advantage of low toxicity. These agents are the starting point for the design of safer and more selective P-gp inhibitors[168]. This seems counter-intuitive in view of the lack of a powerful effect of these agents in juices as described in the previous section, but the basic flavonoid structures have been used to develop potent inhibitors of P-gp[168]. Some of these compounds, such as quercetin, act by disrupting ATP hydrolysis and thus shutting off the P-gp pump's power supply[162]. More fourth-generation inhibitors are various surfactants and detergents to disrupt the membrane structure surrounding the P-gp; however, it is difficult to see how this can possibly be targeted without damaging other areas of the membrane.

A more practical idea is the use of the tyrosine kinase inhibitors (TKIs) in P-gp inhibition. These drugs target the cellular signalling component of tumour growth regulation. After initial success in chronic myeloid leukaemia (CML) the latest TKIs have been approved at the time of writing for use against non-small-cell lung cancer (NSCLC). This malignancy comprises 80% of lung cancers and has a dire five-year survival rate[170]. Many older and newer TKIs, such as imatinib are potent P-gp inhibitors, with some of them acting competitively and others blocking the ATP process, such as dacomitinib, or even both at the same time[171]. Moreover, the TKIs afatinib and dasatinib, as well as the antioxidant resveratrol, can also downregulate P-gp expression through their impact on cell signalling processes, which has yielded yet another way to control P-gp activity in malignant cells[170]. This means that with several cancers, if a TKI such as afatinib is used in combination with another antiproliferative agent, perhaps paclitaxel, doxorubicin, or adriamycin, for example, the TKI can exert its therapeutic impact dually:

1. It exerts its primary therapeutic action (inhibiting epidermal growth factor receptor EGFR, for example).

2. It acts by improving the other agent and its own tumour cell penetration through downregulation and inhibition of P-gp activity[172].

The ABC cassette transporters: the future

Overall, the contribution of the multifactorial complexity of presystemic influx and efflux transport is a strong research focus and in the past, it has been difficult to separate the role of the CYPs, and the nature of the specific transport systems involved in drug bioavailability. Whilst extensive work has been done with laboratory cell lines, these do not always mirror what is happening in clinical studies[163] due to the wider expression of transporters in the living human. However, much research has gone into developing experimental clinical protocols containing probe drugs that are as selective as possible for particular transporters, so it is possible to delineate which transporters are involved with a new drug's influx and efflux in volunteers and patients. In addition, the impact of therapeutic transporter inhibitors could also be evaluated[173].

Whilst P-gp has been pursued relentlessly as a therapeutic target since the 1980s, other ABC cassette transporters such as ABCG2, or breast cancer resistance protein (BCRP) are not as well investigated. ABCG2 is a primary cause of resistance to antineoplastic agents, such as with the active metabolite (SN-38) of the topoisomerase-1 blocker, irinotecan, but the USFDA has made it a priority to increase knowledge of ABCG2's impact on drug bioavailability[174].

Likewise, other major ABC transporters, ABCC1, (MRP1) specialise in more hydrophilic drugs and metabolites as well as a well-defined role in the efflux of several anticancer drugs[175]. There is significant overlap in substrates and inhibitor selectivity across the ABC transporters[176] and much remains to learn of the roles of drugs and dietary molecules that affect the various transporter systems and drug bioavailability. Efflux transporter systems are discussed again in Chapter 6 under the heading of Phase III of metabolism, which focusses on the function of the various MRP transporters in the removal of conjugated metabolites from the cell (Chapter 6.11).

5.7 Major clinical consequences of inhibition of drug clearance

5.7.1 Introduction

Although CYP inhibition can be competitive, noncompetitive, uncompetitive, or mechanism dependent, in clinical practice the main concern is how rapidly the inhibitor causes drug levels to climb towards toxicity and whether the toxic effects can be treated before serious injury or death results. As discussed already several major clinical conditions promoted by inhibition of drug clearance can overtake even healthy individuals, in a matter of hours. These effects can be just a more intense version of the drug's usual pharmacodynamic effects that is dose related and reasonably predictable. If the drug is designed to lower blood pressure, then an accumulation will reduce it to dangerous levels. However, most drugs at higher concentrations can exert those unintended 'off-target' pharmacodynamic effects, as mentioned previously.

The speed at which these problems become manifest cannot be overemphasized. Unlike induction processes, which take days, unintended pharmacological effects caused by drug accumulation happen within hours of regime change. Clearly, the best option is prevention:

- By ensuring that health-care professionals do not make mistakes; if these do occur, someone should immediately 'pick up the ball' and ensure that the mistake is not translated to a potentially fatal prescription that could be handed to a patient. As so many different categories of health-care professional can now prescribe medicines, there are many more opportunities than in previous decades for breakdowns in communication. Alternatively, more *knowledgeable* prescribers should hopefully mean more effective vigilance.

- The patient must be informed about the dangers of some drugs in combination with inhibitors. This should prevent patient intake of both dietary inhibitors and over-the-counter/herbal preparations that could block the metabolism of prescribed drugs. Even as you read this, somebody somewhere is washing down their statins with grapefruit or other juices.

5.7.2 Torsades de pointes (TdP)

Introduction

Probably the most feared off-target drug effect is *torsade de pointes*, which literally translated means 'twisting of the points' and often abbreviated to TdP. This is the rather chaotic-looking and potentially lethal manifestation of ventricular tachycardia on an electrocardiogram (ECG), which can result spontaneously, but much more often from exposure to high concentrations of a worryingly large group of drugs. The risk of causing TdP is most common reason a candidate drug fails to reach the clinic or is withdrawn subsequent to approval[177]. It is worthwhile revisiting some basic cardiac physiology to place the risk of TdP in context.

The heart performs two vital functions simultaneously; one is to convey deoxygenated blood to the lungs and the other is to distribute the reoxygenated blood to the body. To accomplish this, the heart has evolved into an exquisitely electrically coordinated dual pump. The right atrium receives deoxygenated blood from the periphery and sends it via the right ventricle to the lungs for re-oxygenation. The left atrium receives oxygenated blood from the lungs and propels it to the left ventricle, which pumps the blood back to the periphery. The heart valves act to ensure blood flow is unidirectional. Rather like a petrol engine, the heart's pumping sequence is electrically timed precisely and reliably and ECG analysis highlights this process, the main points of which are described with the letters PQRST (Figure 5.3).

The heart's pacemaker, the sinoatrial (SA) node, begins each pumping cycle by stimulating both the atria as well as the atrioventricular (AV) node. As the right atrium is closer to the SA node it depolarizes first. On the ECG, this is the P wave (atrial depolarisation). The AV node delays firing the ventricles

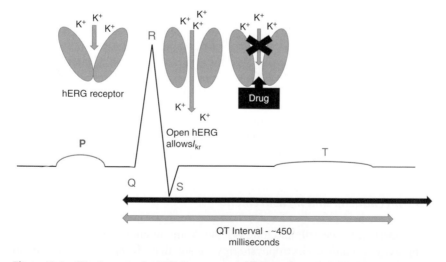

Figure 5.3   The impact of a hERG receptor-inhibiting drug on the QT interval, which prevents outflow of K^+ ions in the I_{kr} and retards repolarisation of cardiac muscle

until the atria have filled them with blood. The PR interval is the time between start of atrial depolarisation and the beginning of the next sequence, which is the large spike on the ECG called the QRS complex, which is the depolarisation of the ventricles firing. This is the real 'power stroke' of the heart, particularly the R, the largest wave, which represents the main mass of heart ventricular muscle contracting. Next, the ST segment occurs and is flat, representing the beginning of the repolarisation of the ventricles ready for the next contraction. The final T wave is the completion of ventricular repolarisation.

For the heart to maintain stability of rhythm, the QT interval, that is, the period elapsed between ventricular depolarisation and repolarisation should not exceed 0.45–0.55 of a second[178]. The QT interval will shorten at high heart rates and lengthen at slow rates, so Bazett's formula attempts to correct for heart rate, yielding a QT interval at 60 beats per min, known as the QTc. This formula has limitations and Fridericia's correction is often preferred[178], although Bazett's is easier to use from a practical perspective[177]. The QTc is useful to predict the risk of TdP, which increases dramatically beyond 0.55 of a second and indeed any rapid change of up to 60 milliseconds from baseline is also a strong risk factor for TdP[177].

The heart's tissue consists of two broad yet integrated cellular types; the autorhythmic cells, whose action potentials stimulate and coordinate the myocardial muscle cells, which actually carry out the pumping work[177]. Electrical activity during the QT interval, consists of myriad ionic movements of sodium and potassium ions associated with the action potentials, but for repolarisation, that is, the overall myocardial recovery process, the delayed rectifier potassium current, I_{kr} is crucial (Figure 5.3). This is when potassium ions flow out of the myocardial cells through channels whose entrances are coded for by a gene known as hERG or KCNH2.

Unfortunately, these hERG or I_{kr} channels contain a large entrance area, known as the vestibule, which is vulnerable to blockade by a very wide range of drugs. If the potassium current is slowed by this blockade, the QTc increases and the risk of TdP multiplies. If TdP does occur and ventricular arrhythmia/tachycardia ensues, within a few minutes of the patient becoming aware of their irregular heartbeat, they are likely to faint (syncope). If TdP is drug-induced, recovery of heart rhythm is likely to depend on the potency and concentration of the hERG-blockading agent[177]. Ideally, administration of intravenous magnesium sulphate, potassium ions, and a beta-blocker should help the heart reassert its rhythm and naturally, the causative agent should be withdrawn. If total cardiac disorganisation and no detectable QRS complex are present, rapid defibrillation (hard reset) is necessary (History 1) to prevent death.

TdP: mechanism and measurement

It is important to see TdP and long QT syndrome in clinical context. First, the complexity of cardiac electrophysiology means that all the detailed electrical events leading to TdP from long QT issues are not fully understood. Nevertheless, it is often necessary, such as in hospital intensive care units[177], to try to predict whether QT problems might lead to TdP in a given patient, so we start with gathering specific and relatively easily measured ECG-derived parameters from that patient.

Unfortunately, it is not completely agreed whether these parameters are as predictive as we need them to be. For instance, different areas of cardiac tissue, say, across the thickness of a ventricle, repolarize at a range of speeds and efficiencies from relatively slow to very rapid. The width of this range of repolarisation varies constantly and is termed *dispersion of repolarisation*. Some studies have found that the greater the dispersion of repolarisation, the more likely arrhythmia will occur[179] and this dispersion is represented by the difference between the apex and the end of the T-wave ($T_{peak} - T_{end}$ interval, or TpTe). Indeed, the TpTe and the TpTe/QT ratio have been shown in some studies to increase in patients who develop TdP[179], but not in others[180]. What is certain is that there are many risk factors to developing long QT syndrome and consequent life-threatening TdP. You can to some extent separate QT issues into genetic as well as acquired (drug or pathologically induced) long QT syndrome.

TdP: Predispositions and risk factors

The most common of the genetic abnormalities and syndromes that lead to long QT syndrome are LQT-1 and LQT-2[181], which involve abnormal hERG channels, known as channelopathies, among other pathologies[177,182]. About 20% of all deaths are sudden cardiac arrests, of which a significant proportion involves genetic long QT issues in previously healthy young people[177].

Female QT length is longer than males and subject to hormonally driven variability, although in healthy women the menstrual cycle does not significantly impact QT interval[183]. Whilst many sources state that oestrogens increase QT length and progesterone shortens it, this may be an oversimplification. Interestingly, a study found that in healthy and LQT-2 women, oestrogens can actually shorten QT intervals by accelerating repolarisation[184] and they even went as far as suggesting that oestrogens could one day be therapeutic agents in long QT syndrome. Whilst that idea will need some serious clinical investigation, it does appear that the ability of drugs to lengthen QT in women is subject to hormonal modulation, with the luteal (higher progesterone levels) phase being protective[183,185].

The hormones can interact at the molecular level to modulate myocardial cell activity; oestrogens in LQT-2 and healthy women can directly speed up the trafficking of the hERG channels to the membrane, so shortening repolarisation[184]. High doses of antioestrogens, such as tamoxifen can lengthen QT intervals, although it is not clear whether this is a direct hERG effect or linked with the ventricular tissue itself[178]. Pregnancy and its aftermath can severely aggravate LQT-2 for example, with individuals sometimes suffering multiple episodes of

TdP in the weeks after birth and therapeutic management can be challenging[186]. Obviously, drugs that block hERG channels are strongly contraindicated in these individuals. Overall, women are more prone to long QT syndrome anyway and are more vulnerable to drugs that exacerbate this problem, so TdP and ventricular tachyarrythmias are more common in females[184]. In males, testosterone shortens QT, although during prostate androgen blockade therapy, male QTcs lengthen significantly[178].

Any pathological process affecting heart physiological health will also predispose to long QT/TdP. Immune system activity, either in response to infection or autoimmune processes, can affect QT intervals through local cytokine release, such as TNFα, IL1β and IL6[185]. Rheumatoid arthritis patients, for instance, are more vulnerable to long QT syndrome and their death rates from cardiac failure reflect this[185]. Other groups are also vulnerable, such as cancer patients. Some antineoplastic agents are well known to be cardiotoxic, such as the anthracyclines (doxorubicin) and they lengthen QT partly through causing local cytokine release[178]. As if this was not enough, cancer patients have many of the risk factors for long QT, such as low potassium, calcium and magnesium, due to the diarrhoea and vomiting caused by the chemotherapy. They also receive so many drugs that long-QT inducers are likely to be among them[178]. Naturally, issues with cardiac pathology, such as myocardial ischaemia will also promote the risk of long QT.

Hence, a drug that perhaps has only a mild long QT impact in healthy individuals, in patients with multiple risk factors, both genetic and acquired, may precipitate TdP and ventricular tachyarrythmias unexpectedly. Indeed, this complex combination of causative factors is seen in intensive care patients, who are already seriously ill and may well have just received a number of different drugs elsewhere. These agents will have failed to retrieve the situation, but might still be present in significant plasma concentrations. One study found that about half of patients in an intensive care unit in Reno, Nevada, developed long QT within eight hours of admission. Around half of that number had received a QT interval-prolonging drug, but the rest had not, underlining the presence of the other factors that promote the condition in those with multiple pathologies[177]. Hence, TdP is a highly significant issue for very vulnerable patient groups, although the propensity for a drug to cause this problem in otherwise healthy individuals is another issue, which is explored in the next section.

TdP: drug-induced

If you read the 'Credible Meds' website[187] for drugs that are strongly associated with long QT syndrome, fifty-eight entries (at the time of writing) appear. If you include drugs that pose some risk of provoking the effect, the list increases to well over two hundred. Indeed, more than 70% of candidate drugs in pharmaceutical company testing 'pipelines' have been predicted to cause a problem at some concentration[188]. So on the face of it, the danger of drug-induced long-QT looks to be a significant issue in mass drug usage. Indeed, the 'Credible Meds'[187] approach, classification and regular updates make it a powerful resource to aid the rating of the risks of drugs that may cause TdP[189].

However, from a practical viewpoint, currently the main method of deciding whether to prescribe a drug that is known to be associated with TdP risk is through ECG assessment of QTc. This measurement is problematic from a number of standpoints, not least its sensitivity and selectively, regarding individual drugs and classes of drugs[190]. So it has been proposed that a more logical and systematic method of assessment would take into consideration the concentration of the drug that caused a QT prolonging effect. Since this is not practical clinically, the next best thing is to use experimental determinations in the scientific literature of the main cellular mechanism. This is routinely achieved in laboratories using patch-clamped cell lines that express recombinant human hERG channels, which can yield a half maximal inhibitory concentration for a given drug (IC_{50}). There have been concerns that these experimental determinations erred too much on the side of caution due to excessive sensitivity, so leading to the discontinuation of more candidate drugs than was strictly necessary. Part of this issue was that many hERG-determinations were carried out at ambient temperature, whilst the *in vivo* simulations were run at physiological temperature. It is apparent that due to variability in drug responses, it is critical to carry out the cellular IC_{50} determinations at 37°C.[188,190]. Hence, it has been proposed that a more logical method of decision-making with regard to drugs of QTc risk involves the ratio of the IC_{50} to the peak serum (unbound) drug concentration[190]. There is a good correlation between this ratio and drugs that are at high risk for causing TdP, such as astemizole (withdrawn for this reason) and the antipsychotics haloperidol and thioridazine. The ratio has limitations, such as with drugs that can cause TdP but through different mechanisms to the hERG receptor. The TCAs at high concentrations appear to cause TdP through sodium channel blockade[189].

Of course, this ratio might predict the background risk of TdP for a compound, but in order to make a fully informed clinical decision, issues such as patient gender (females are at greater risk), pathology (as discussed above), electrolyte concentrations, as well as the effect of inhibition on plasma drug concentrations must be considered[190]. So to summarize, the clinically relevant issues to consider whether a drug might cause long QT, which might lead to life-threatening TdP:

- The predispositions of the patient (other drugs, genetic issues, cardiac and other pathologies)

- The approximate concentration range where the drug is thought capable of lengthening QT

- Whether that concentration range might be reached, either by retarded clearance (inhibition by another drug, renal failure), or overdose

The Credible Meds site[187] underlines the major drug classes linked with TdP and these include for example, some antidepressants (mirtazapine and citalopram; higher risk) and antipsychotics, such as the older agents (thioridazine and haloperidol; higher risk). Other groups of drugs included are antiarrthymics (amiodarone, quinidine, dispyramide) macrolides, and the fluoroquinolones; several of the latter group have been withdrawn from the US market for this risk factor. Clearly any potent CYP inhibitor could be responsible for increasing a drug's risk of causing TdP from moderate to high risk, just through retarding clearance, alongside the other magnifying factors that are described above. Azoles, for example, are not only potent CYP inhibitors, but they can be cardiotoxic in their own right and promote TdP directly pharmadynamically and also pharmacokinetically through their CYP effects[191].

5.7.3   Sedative effects

The risk of sedation is obviously less of a problem in the home, rather than perhaps operating heavy machinery with razor-sharp rotating poison-tipped plutonium blades. The co-administration of inhibitors with drugs such as the benzodiazepines and others such as buspirone can potentiate their sedative effects markedly. Benzodiazepines in general have their disadvantages, as we have discussed, but midazolam is a useful short-acting sedative in hospitals for preoperative sedation, as well as in neonatal and adult intensive care units[192-194]. However, as a CYP3A4 substrate it is very vulnerable to retarded clearance due to CYP inhibition from macrolides such as erythromycin, azoles such as fluconazole, itraconazole and posaconazole and even diltiazem[193-195]. This prolongs sedation significantly and is obviously problematic when treating fungal infections in the intensive care setting. Many centres now recommend propofol rather than benzodiazepines for sedating mechanically ventilated patients, partly because of cost and recovery times but also that this agent is likely to be less vulnerable to CYP inhibitors as so much of it is glucuronidated[196,197].

Clinically, depression is often inextricably linked with anxiety disorders and although SSRIs are the first choice for depressive disorders, their characteristic delayed onset applies to anxiety as well as depression, so benzodiazepines are co-administered sometimes because they treat the anxiety immediately, although only for short time periods such as a few days. As the SSRI starts to improve the depression as well as the anxiety, the benzodiazepine is gradually withdrawn. Unfortunately, given the effects of SSRIs on CYPs, this can sometimes retard the clearance of the benzodiazepine, leading to intensification of their major side effects. Benzodiazepines are mostly cleared by CYP3A4 and CYP2C19, so SSRIs such as fluoxetine and fluvoxamine cause sedation and impairment of psychomotor activity with most of the benzodiazepines, such as diazepam (Valium) and alprazolam (Xanax), increasing their plasma levels significantly[74].

It is recommended that less than half the usual dose of these benzodiazepines should be used with fluvoxamine, or one of the benzodiazepines cleared by glucuronidation should be used (such as lorazepam). Fluoxetine retards alprazolam's clearance and intensifies its side effects, whereas venlafaxine, sertaline, citalopram, and escitalopram do not appear to be a problem[74]. Other nonbenzodiazepine sedatives such as zolpidem and buspirone are also CYP3A4 substrates, and inhibition intensifies their pharmacological effects.[74,198] Buspirone is extensively presystemically metabolised in the gut, and grapefruit juice greatly increases its systemic entry and can cause excessive sedation and confusion[130]. Administration of CYP inhibitors with AEDs such as the CYP3A4 substrate carbamazepine (History 4) will cause significant accumulation, leading to mental confusion, ataxia (staggering gait), and even unconsciousness. The macrolides verapamil and azoles are known to be able to cause this effect[74].

5.7.4   Muscle damage (rhabdomyolysis)

This is when striated muscle disintegrates and the released myoglobin and other muscle components enter the blood in quantity, causing renal damage and eventual failure. Blunt-force trauma (like a car crash), some infections, burns, electric shock, ischaemia, or severe exercise can all cause it. However, it can also occur in response to exposure of some drugs and chemicals. Classic signs are diffuse muscle pain, problems walking, and a serum creatine phosphokinase (CPK) level that is more than 10-fold higher than the upper limit of normal[146]. Why it happens is extremely complex and still under study[199]. Statins are the major drug group associated with rhabdomyolysis[200], but at least one tyrosine kinase inhibitor (sunitinib), the anaesthetic propofol, the anti-rheumatic leflunamide, the anti-sarcoma agent trabectedin[201], and daptomycin are all the subject of reports about the condition[200,201].

With regard to statins, cerivastatin was withdrawn due to rhabdomyolysis that occurred when the drug was used with gemfibrozil. It seems that with statins and the other higher risk drugs, if plasma levels are sustained at too great a level, this can trigger the condition. Whilst the risks of developing rhabdomyolysis are only around 0.1% with statins, up to quarter of patients can develop statin-induced myopathy, which in comparison with rhabdomyolsis, is clearly a much milder muscle adverse reaction[200]. Any inhibitor that increases plasma levels of the drugs listed above will increase the chances of this apparently drug-concentration dependent problem developing. Simvastatin, lovastatin, atorvastatin, and cerivastatin are CYP3A4 substrates and are vulnerable to elevation of plasma levels in the presence of potent CYP3A4 inhibitors. This is a particular concern with inhibitors such as grapefruit juice, the azoles, and erythromycin. If statin therapy must be continued in the presence of a 3A4 inhibitor, it would be wise to use those that are cleared by other CYPs, such as fluvastatin (CYP2C9) or pravastatin. Due to the multiple pathologies often suffered by patients taking statins, additions to their drug regimes, such as antiplatelet drugs like ticagrelor, a CYP3A4 substrate and mild inhibitor, can lead to interactions that result in a higher frequency of rhabdomyolysis[200]. Patients taking immunosuppressants and those with renal problems are more prone to develop rhabdomyolysis than others and are at particular risk. It is safest to avoid the interaction by substituting noninhibiting drugs or statins cleared by other CYPs.

5.7.5   Excessive hypotension

As you might recall from your basic pharmacology studies, there are several different drug options in the management of hypertension. This is fortunate, as this condition is very common, sometimes does not respond to therapy and deteriorates with age. This means that antihypertensives of various types are prescribed (like statins) in vast amounts to older patients with multiple co-morbidities, usually for many years. These include CYP3A4 substrates, such as the dihydropyridines (e.g. nifedipine, felodipine, nicardipine, amlodipine and nimodipine). There are more than 11 of these calcium channel blocking drugs in clinical use, and they vary in their physicochemical properties and degree of tissue specificity and penetration[202]. Nicardipine is more selective for heart vessels, while nimodipine is more effective in cerebral vessels. However, the most common problem with them is excessive vasodilatation that can lead to postural hypotension, dizziness, and headache. They can work too well, to the point where blood pressure is insufficient to force blood through diseased coronary arteries and they can cause reflex tachycardia; these effects can make some forms of angina worse. However, the most pressing and dangerous outcome from sustained hypotension is renal injury, as the demand in this tissue for oxygen is huge and interruptions lead to hypoxia and reperfusion injuries[203]. Naturally, this is a problem in a reasonably fit 20- to 40-year-old, but is much more significant in an elderly diabetic with already suboptimal renal function.

Obviously any marked changes in the clearance of these potent drugs could lead to potentially major deleterious changes in cardiovascular function. It is easy to see how the azoles and the macrolide inhibitors could cause severe cardiovascular problems due to nonclearance of dihydropyridines. Their high presystemic metabolism means that grapefruit juice would have a particularly potent and possibly life-threatening effect, sometimes increasing drug levels by fivefold[129]. Interestingly, the lack of gut metabolism of amlodipine makes this agent less susceptible to grapefruit juice interactions, and this is reflected in its usage, as it makes up about half of total prescriptions for this drug class[202]. Indeed, some drug additions to regimes including dihydropyridines are linked with kidney injury, such as clarithromycin, although azithromycin is not a problem, as it has long been known not to be a significant CYP3A4 inhibitor[204]. Any antihypertensive agent that is a CYP3A4 substrate will be liable to cause excessive reductions in blood pressure if they accumulate in the presence of an inhibitor. This is a particularly problematic effect in the case of potent inhibitors like grapefruit juice and norfluoxetine.

Many of the various beta-blockers still in use are predominantly CYP2D6 substrates and the various inhibitors of this isoform (SSRIs, for example) can significantly increase blood levels, causing bradycardia, which naturally leads to hypotension. This leads to dizziness and unsteadiness, but sometimes syncope on attempting to stand[205]. In elderly patients, a fall can mean a life-changing bone injury, so additions of some SSRIs (paroxetine and fluoxetine) to metoprolol regimes can lead to early discontinuation of the beta-blocker[70].

5.7.6   Ergotism

Until the advent of the highly effective group of triptan 5HT agonists, severe migraine sufferers were faced with the prospect of using ergotamine tartrate or suffering the extremely unpleasant pain that this syndrome can inflict. As a migraine sufferer myself, my one experience with ergotamine in 1980 led to a painful effect as if there were wires tightening inside my calf muscles. This took two days to wear off and I still suffered the headache.

Ergotamine can also cause severe neural derangement known as St Anthony's fire. This sometimes affected people who ate bread that had been made with mouldy flour because it contained a considerable dose of ergot alkaloids. Ergotamine is cleared by CYP3A4, so the effects of any inhibitor of this isoform on ergot clearance would lead to an extremely grim series of peripheral and CNS symptoms. Fortunately, the triptans (sumatriptan, naratriptan, zolmitriptan, etc.) are the mainstay of acute migraine treatment, and the ergotism problem with CYP3A4 inhibitors should now be of historical interest only.

5.7.7   Excessive anticoagulation

As discussed in Chapter 4.5.4, despite its many drawbacks, warfarin remains an important agent for the treatment of a number of conditions where thrombosis is at high risk, such as those with replacement heart valves, atrial fibrillation, and deep venous thrombosis. Inhibition of CYP2C9-mediated S-isomer clearance has more impact on warfarin's pharmacological effect than effects on the other CYPs, although they do show some impact. As discussed earlier, the azoles are very potent inhibitors of CYP2C9, and the most powerful clinically relevant interaction is with miconazole gel, which can increase INR threefold, whilst the next most important interaction, with the CYP2C9 inhibitor amiodarone (commonly co-prescribed) increases INR by about 1.3-fold)[206]. Amiodarone's long half-life compounds its effect, however. Other azoles such as fluconazole do not show such a potent effect, although metronizadole has caused life-threatening effects with warfarin. Cimetidine, an inhibitor of 1A2 and 2C19 (History 5), is not recommended for concurrent therapy with warfarin, although it is available OTC and thus there is potential for a moderate increase in prothrombin time.

Warfarin acts by antagonizing the effects of vitamin K, which is necessary for the formation of several clotting factors. A therapeutic dose basically knocks out around half the usual formation of these factors. It is worth noting that changes to warfarin clearance can take some time to be reflected in changes in prothrombin time. This is because the effect of the drug depends on the rate of removal of blood clotting factors that had already been formed before the drug took effect. From an initial dose, it can take up to a day and a half before any change in INR occurs. The drug already has a long half-life (one to three days) so when a drug decreases warfarin's clearance, it will take perhaps one or two days before an effect is seen in terms of prothrombin time, and this effect will persist, again for some days.

The most serious effects of excessive coagulation are GI tract bleeding and intracranial haemorrhage, both of which can be fatal. Clinically, warfarin patients are regularly checked for changes in INR and are well observed, so given that the major interacting drugs such as the azoles are well-known for their impact on CYPs, usually such interactions are well managed clinically, either by dose adjustment or use of a less potent interacting drug (fluconazole instead of miconazole)[206]. What is notable is that in a recent study of warfarin clinical interactions, the patients were taking six drugs daily on average[206].

The DOACs (Direct oral anticoagulants) as mentioned earlier, are also subject to CYP inhibitor impact, particularly from the azoles. Apixaban and rivaroxaban are probably the most susceptible due to their significant CYP clearance, but there are several drugs in the class such as edoxaban, dabigatran etexilate and betrixaban that are not cleared by CYPs and although there are P-gp issues, there are plenty of alternatives to consider clinically.

5.8 Use of inhibitors for positive clinical intervention

5.8.1 Introduction

One of the aims of Chapter 2 was to highlight and facilitate the understanding of where drug metabolism fits into the greater biological context of biotransformation. Once you have some comprehension of what a system is trying to achieve, you can to a certain extent predict how it will behave in different circumstances and use this to the advantage of the patient. Indeed, the full context of CYP function could be resolved using the often cruelly ironic LAPD motto *To protect and to serve*. CYPs protect us through their biotransformation of xenobiotics, which challenge our systems in terms of control of homeostasis.

Therapeutic drugs fall into this 'protect' category and are dealt with by the CYPs accordingly as we have seen. The 'serve' component is the vast amount of less well-understood CYP-mediated synthesis and destruction of endogenous molecules, which includes the most well-known sex hormones. As you have seen from this chapter, the effects of inhibitors on the concentrations of CYP substrates can be so dramatic and sustained that it has occurred to a number of scientists and clinicians over past decades to explore various strategies to exploit deliberate CYP inhibition to provide some form of benefit to the patient. If we pursue the 'protect and serve' idea, then a good place to start is where the 'serve' CYP processes become injurious to us and it is necessary to shut them down to prolong lives.

5.8.2   CYP inhibitors and female hormone-dependent tumours

Breast tumour analysis over the past decade has developed rapidly and understanding of the genetic signatures of the myriad different forms of the disease is impacting the specific and effective targeting of various tumour subgroups[207]. In terms of prevention of recurrence, the oestrogen receptor (ER) has been the main direct and indirect drug target so far and around 70% breast malignancies are driven by oestrogens and are regarded as hormone receptor positive (HR+), the rest are termed HR–. Tamoxifen was first used in the late 1970s to competitively inhibit the ERs and has been a mainstay of therapy ever since[207-209].

Another method to isolate the ER, is to eliminate the supply of oestrogen, which is far easier to achieve post-menopause, when ovarian production falls off and the hormone is made in many peripheral tissues, such as the adrenals and of course, in any potentially malignant tissue. The main distinguishing characteristic of an oestrogenic molecule to an oestrogen receptor is the aromatized A ring and the androgenic precursor such as 4-hydroxyandrostenedione is subject to a series of CYP19 'aromatase'-mediated reactions, rather like a spot-welding robot in car factories. This makes the androgen's 'A' ring aromatic, and then the oestrogen is completed by CYP17 that also alters the substituents on the D (far right-hand) ring of the steroid. Logically, inhibition of aromatase should shut down oestrogen production and slow tumour recurrence and growth dramatically. The first clinical attempt at this was aminoglutethimide, which blocked oestrogen formation in peripheral tissues. However, that drug was unselective[208] and therefore quite toxic and could lead to potentially lethal agranulocytosis (loss of all neutrophils, Chapter 8.4.4.).

The third-generation aromatase inhibitors include letrozole (Femara®) anastrozole (Arimidex®), which are both nonsteroidal competitive inhibitors, as well as the steroid-like exemestane (Aromasin®), which is an irreversible inhibitor. Letrozole is so potent that it can make oestrogen levels undetectable[208]. In fact, these agents are superior to tamoxifen and they reduce both cancer recurrence and mortality – the latter by 15%[209]. Indeed, compared with no treatment at all, aromatase inhibitors cut the death rate by 40%[209]. Unfortunately, the price for such efficacy can be a significantly greater risk of bone fractures[209], which is around 8% for the aromatase inhibitors compared with 5% for tamoxifen[209]. There are many other side effects related to the loss of oestrogen, such as loss of body strength, nausea, and hot flushes. However, the aromatase inhibitors do reduce the risk of an endometrial cancer compared with tamoxifen[209]. At first sight, the long-term effectiveness of these drugs seems to underline the inability of the CYP system to counteract the presence of a potent inhibitor. To a large extent this is true, because resistance to aromatase inhibitors only very rarely results from changes in the active site of the CYP and is more likely to be due to pharmacokinetic causes, as well as ER receptor upregulation[209,210]. Incidentally, the obvious idea to use tamoxifen and aromatase inhibitors together (the ATAC Trial) actually backfired and the drugs were less effective together. This was partly due to the tamoxifen showing oestrogenic effects in the near zero oestrogen environment caused by the aromatase inhibitors; tamoxifen also causes a roughly 40% fall in aromatase plasma levels[208-210].

5.8.3   CYP inhibitors and male hormone-dependent tumours

Prostate cancer is now the second most common lethal malignancy in males after lung cancer and its prevalence is increasing worldwide[210-213]. Echoing breast cancer, it is clear that the androgen receptor (AR) is the force behind prostate tumour growth and metastasis, so the condition can be contained for a while with negative feedback hormone suppressive therapy, but then it reaches a resistant stage known as CRPC or castration-resistant prostate cancer. At this point, a man might have two to four years left to live[212]. Although the process of pharmacological isolation of the AR in prostate cancer is not as advanced as ER isolation in breast cancer, there are signs the cavalry are on their way. The latest anti-androgens like enzalutamide are effective anti-androgens, but a more direct approach is to smack down androgen formation by blocking the CYP that builds the androgens in the tumour as well as peripherally[212,213]. CYP17 (also known as CYP17A1) carries out two sequential reactions. The first is a 17α hydroxylation, followed by a 17, 20 lyase reaction[212]. The first inhibitor of both reactions to reach the clinic was the pregnenolone analogue abiraterone[213,214] in 2011, where it was approved for CRPC. However, when you read the next paragraph, I hope you think back to the beginning of Chapter 4.1.1 and the triangular systems pharmacology idea.

Abiraterone must be given with prednisone because in trials, blocking the CYP17 enzymes stops cortisol formation, which is detected, and this, in turn, triggers a pulse of ACTH[215], which then massively increases mineralocorticoid levels with a host of side effects such as severe disruption of fluid balance and hypertension[212,213]. Co-administration of prednisone suppresses this effect. Abiraterone is impressively tenfold more potent than ketoconazole in its irreversible mechanism-dependent inhibition of CYP17, knocking down testosterone levels to virtually zero[216]. However, there is that ACTH issue, plus it can only be absorbed as an acetate and it inhibits several other CYPs, blocking CYP3A4 and CYP2D6, which will disrupt complex drug regimes, as we have seen[213]. Despite these issues, abiraterone was approved in 2018 for use even earlier in prostate cancer progression[214] and through its delta-4 metabolite it is as potent an AR antagonist as enzalutamide[211].

However, abiraterone is really a blunt instrument and is already obsolete. The drive in this field has been to make more selective inhibitors such as orteronel (a competitive inhibitor) galeterone, and seviteronel, which all block only the CYP17 17, 20 lyase reaction[211]; these drugs spare the 17α hydroxylation function of CYP17, so preventing the ACTH pulse and negating the need for any supplementary steroids[212]. They also have the added advantage of binding the AR receptor and essentially sabotaging its translocation and DNA interaction[212]. Unfortunately, for various reasons only seviteronel[217] is still in clinical trial to date, but it is to be hoped that an effective drug bearing this dual action concept will eventually reach the clinic.

5.8.4   CYP inhibitors and manipulation of prescription drug disposition

Some of the drug combinations in this section have been found by accident; the others have been the result of clear and methodical planning using several models to prove principle. Generally, the co-administration of a potent CYP inhibitor provides the opportunity to slow clearance of the 'object' drug significantly, resulting in its marked accumulation, as well as the diminution of the concentrations of the main metabolites. If prior to inhibition a minor proportion of the drug is excreted renally unchanged, this will significant increase in the presence of the inhibitor.

The extent to which this whole approach will be effective is naturally strongly influenced by the number and capacity of the individual CYPs involved. A potent CYP3A4 inhibitor used on a drug that is almost entirely cleared by this CYP and virtually no other will probably show the most dramatic and sustained effect. However, when the object drug is cleared by several isoforms, then unless the inhibitor is very unselective, the impact on plasma levels might only be modest and indeed rather variable from patient to patient. Other problems to be considered include polymorphisms in CYP expression (see Chapter 7.2.3), which produce high background variation in drug response. Also, different CYPs have different concentration-dependent drug affinities, again making it difficult to predict whether an inhibitor might show an effect and for how long. Since any inhibitor intervention adds to the patient's drug burden, the process needs to be at least as predictable and safe as the disposition of the object drug.

Yet another issue is that all drugs have side effects, so those of the inhibitor will be added to the 'polypharmacy' mix, not least the problems of lack of specificity of the inhibitor and the inhibitor's usual pharmacological target and off target effects. Finally, if we return to the 'protect and serve' idea, if we are essentially sabotaging the protection, then it is a near certainty that the 'serve' part of CYP activity will be disrupted also. Indeed, I hope you will immediately recall the issues with abiraterone and ACTH and you will also remember how ketoconazole appears to inhibit many more CYPs than just CYP3A4/5. So trying to specifically modify the responses of a highly evolved, complex, sensitive, and tightly regulated biotransforming system is like challenging a living supercomputer.

If we leave the mechanics of the situation aside for a moment, let us consider the strategy behind the use of CYP inhibitors. Naturally, any well-managed clinical intervention must have a clear, logical, and ethically sound motivation. Although the use of CYP inhibitors could be resolved into three major elements, there is of course blurring between them in terms of what might actually happen in the patient's system.

To increase the efficacy of the drug

Benefits: Where the parent drug is active, greater drug exposure, and tissue penetration may result, plus longer residence time and a greater likelihood of a sustained response.

Risks: Drug levels climb out of therapeutic window towards off-target and even borderline toxic effects, involving several organ systems. Pharmacological on and off target effects of the inhibitor are to be considered and therapeutic monitoring is necessary.

To reduce its side effects and/or toxicity

Benefits: Reduction of impact on patient of main metabolite-mediated toxicity, improvement in drug adherence. If the parent drug is active, greater drug exposure as above is also useful.

Risks: Parent drug accumulation and possible off-target toxicity, impact on other CYPs and endogenous systems. Pharmacological on and off-target effects of the inhibitor, monitoring again is advisable.

To reduce the cost of using the drug

Benefits: With diabolically expensive on-patent drugs, dosage can perhaps be halved, thus making a significant cost saving, which could extend treatment duration within a set budget.

Risks: Pharmacological on- and off-target effects of the inhibitor, clinical monitoring might offset cost savings.

5.8.5 Use of inhibitors to increase drug efficacy

As we have seen, for a fixed dosage, drug residence time is controlled by clearance, which is related to many factors. When a drug is in clinical trial, dosage regimes are exhaustively evaluated to provide the best effective compromise in dose/dosage interval aimed at achieving optimum efficacy and minimal toxicity. In a similar way, car engines are set to deliver power/economy/emissions performance commensurate with a decent lifespan. A modest family hatchback petrol engined car could be made to deliver 1000 horsepower, but obviously not for very long before there are injuries from flying metal. Hence, any attempts to use inhibitors to act as pharmacokinetic boosters really need a strong justification due to the risks. This is all the more important in complex multidrug 'polypharmacy' regimes. However, driven by the desire to improve clinical outcomes, clinicians sometimes 'tune' drug regimes in particular groups of patients, using one drug to increase the residence time of another to boost efficacy to perhaps the very top of the green part of the therapeutic window (Chapter 1.1). Provided there is tangible patient benefit through optimizing drug response, then most would agree that ethically the benefits outweigh the undoubted risks.

For me, the masterpiece in the gallery here is cobicistat. Going back to the mid-1990s, when HIV was well into the process of being converted from a death sentence to a chronic manageable condition, one of the antivirals used was ritonavir, which as we have seen is problematic as it is an inducer and inhibitor. As it was not very easy to use clinically, physicians started to use it to 'boost' the plasma concentrations of other HIV drugs. In a therapeutic area where drug half-lives are relatively short, it is imperative to maintain high therapeutic levels of all the drugs in the regime to ensure viral suppression. Ritonavir boosted the efficacy of co-administered drugs by blocking several CYPs in the gut and liver, as well as P-gp, so promoting bioavailability strongly. This would make regimes more effective and easier to administer[216]. Ritonavir is not the 'cleanest' drug for this purpose, as it induces many other CYPs as well as inhibiting them; it is also a PXR agonist. Hence, cobicistat was developed from ritonavir; it is similar in its inhibition of CYP3A4, does not affect other CYPs so much (apart from CYP2D6 mildly), and blocks the various transporters (OATP1B1/3 P-gp and BCRP), thus promoting bioavailability, yet is not an inducer[218,219]. It is not an antiviral either, so it can boost the efficacy of complex regimes without too much in the way of pharmacodynamic complications. It was licensed in 2012 in combination with elvitegravir, emtricitabine and disoproxil fumarate[219]. Cobicistat can overcome some of the induction effects of elvitegravir and it appears to be capable of negating even rifabutin's induction effects in the clinic[218]. Either way, cobicistat appears to be effective and is now part of several anti-HIV regimes.

Another area where inhibition has been used on a small scale and then evaluated more widely is with schizophrenia therapy. More than a third of sufferers do not respond to the first and second line drugs, leaving them vulnerable to their condition and potentially shortening their lives. The atypical antipsychotic clozapine is the alternative backed by the most evidence, although it is a narrow TI drug, which is not easy to use and has many dangerous side effects[220]. Unfortunately, half of patients will not respond well even to clozapine and some of the persistent negative effects suffered by patients such as obsessive-compulsive (OC) activity can be made worse. The SSRI fluvoxamine was used at first through a pharmacological rational with clozapine to minimise the OC effects and other negative symptoms, but also acts pharmacokinetically through its powerful CYP1A2 inhibition. This boosts parent drug levels, whilst reducing the formation of the metabolite norclozapine, so augmenting the drug's effect and make it more tolerable for patients[220,221]. However, the price can be a plasma 'pulse' of high levels clozapine, which risks a worsening of OC issues and even provoking a seizure[220]. Clearly, the desire to help the patient by suppressing the condition in the absence of any other options can drive the physician to take these risks; however, they can be mitigated by therapeutic monitoring of drug levels and symptoms. Overall, initial trials have been encouraging, if not yet definitive[221,222].

5.8.6   Use of inhibitors to reduce toxic metabolite formation

There are a number of P450-mediated metabolic reactions which result in metabolites which might vary in their stability, but they can exert a range of deleterious effects on the patient, ranging from the mild off-target, all the way to frankly toxic, which can in many cases seriously erode patient drug tolerance and adherence.

Nortriptyline toxicity and paroxetine

Whilst the TCAs are obsolete regarding front-line therapy for depression as described elsewhere, they retain their usefulness in those with severe and nonresponsive depression. Nortriptyline is considered the least toxic of these drugs, although its 10 hydroxy-metabolite is linked with many of its cardiotoxic effects[80]. Some small groups of patients, the ultra-rapid metabolisers (Chapter 7.2.3), form large amounts of the toxic metabolite and clear the parent drug very quickly, causing a bleak combination of poor therapeutic response and high toxicity. In 2001, a study was published where a small group of these patients were 'phenoconverted' using a potent CYP2D6 inhibitor (paroxetine), which effectively turned their rapid metabolism phenotype into a poor metaboliser status. This had the effect of cutting toxicity and improving therapeutic response[223]. This work has been followed up in The Netherlands over the period 2011-2015 and it reduces the level of the hydroxyl derivative by about 40%, and also the dose of paroxetine required is actually subtherapeutic in terms of its antidepressant action and also its side effects[80]. Whilst not all the patients responded as predicted, the approach shows promise with certain patient groups.

Paracetamol-mediated hepatic necrosis

Paracetamol in overdose is converted to significant levels of a reactive quinone-imine, which eventually leads to necrosis of the liver. This process is covered in more detail in Chapter 8.2.3). It was shown in the 1980s that various inhibitors could slow or prevent the metabolism of paracetamol to its reactive metabolites and several animal studies were carried out to show that this could work clinically. However, this approach never became a clinical reality. Acutely, patients presenting before liver damage was too severe could be saved with glutathione (GSH) precursor supplements, such as N-acetylcysteine. After 24–48 hr post overdose, it would be too late – with only a transplant as a last-ditch option. Including an inhibitor in paracetamol tablets could potentially prevent the formation of the toxic metabolite without really affecting clearance, as 95% of paracetamol clearance is accomplished by sulphation and glucuronidation. However, the main inhibitors of CYP2E1 are mostly sulphur-containing agents and inhibit other enzymes such as alcohol dehydrogenase and aldehyde dehydrogenase. It would not be practical to include a CYP2E1 inhibitor in paracetamol tablets because as soon as the patient drank any alcohol, they would be violently and spectacularly ill (see 'Use of inhibition in alcoholism' below). Even if CYP2E1 could be blocked, CYP2A6 is also involved in reactive species formation, so this would have to be inhibited also[224]. So the use of a CYP inhibitor to prevent paracetamol-mediated hepatic necrosis has not progressed.

Dapsone-mediated methaemoglobin formation

The sulphone drug dapsone is used in leprosy therapy in developing countries and has been effective since the late 1980s as a second line treatment of *Pneumocystis jirovecii* (formerly *carinii*) opportunistic infections as an alternative to trimethoprim and sulphamethoxazole[225]. The drug is also capable of effectively suppressing inflammatory conditions that feature the infiltration of activated neutrophils, such as dermatitis herpetiformis (DH), which affects the skin. DH is linked to glutin intolerance and glutin-free diets are more obtainable than they were, but any lapses and the condition returns. Dapsone can attenuate DH symptoms, such as the intense pruritus and skin eruptions within hours of dosage[225]. This brings rapid relief from a condition that can make patients' lives intolerable. However, the drug causes methaemoglobin formation, which is due to the CYP2C9-mediated oxidation of the drug to a hydroxylamine (Chapter 8, Figure 8.3)[226]. This particular hydroxylamine is a relatively poor substrate for glucuronyl transferase and the greater the drug dose, the more hydroxylamine escapes conjugation, enters the circulation and oxidizes haemoglobin to methaemoglobin, which cannot carry oxygen. The more methaemoglobin is formed as a percentage of total haemoglobin, the more tissue anoxia occurs; symptoms range from a headache/hangover-like effects at sub-10% levels to hospitalisation (nausea, tiredness, and breathing problems) at 20-25%[227]. The standard daily dosage in leprosy of around 100 mg of dapsone usually leads to around the 5–8% chronic level of methaemoglobin and it is just about tolerable to most patients, in the light of the alternative of the progression of the disease. However, with DH, the dosage varies wildly from patient to patient. Some can be fully controlled on 25 mg per week, whilst others must take 400 plus mg of dapsone daily and the condition is only partially suppressed. At this dosage, the patient's quality of life is much diminished by the drug and the only reason to persist with treatment might be a lack of effect of the only other drug alternative (sulphapyridine). Even moderately effective drug therapy with high side effects is better than the recurrence of the disease symptoms[228].

At this point it should be noted that I always inform students that the use of the third person is essential in scientific writing and the use of the first person is normally not acceptable. However, somehow it is acceptable for the following section. After I showed that the hydroxylamine was made by human microsomes[226], it looked plausible that if its formation could be inhibited, then this would cut the methaemoglobin formation and boost parent drug levels, thus promoting drug penetration into the skin. So I thought the first step towards a human study would be a proof of principle study in an animal model, alongside more work *in vitro* with human liver microsomes. This was before CYPs had been characterized into different families, so I did not know which CYP cleared dapsone and indeed inhibitor classifications were often very broad and were associated more with an impact on certain therapeutic drugs. I used a panel of inhibitors, ranging from the broadest CYP inhibitor, to what was perceived at the time to be more selective agents. The inhibitors were tested against a dose of dapsone that would form significant methaemoglobin levels. Piperonyl butoxide, an insecticide and broad CYP inhibitor, was very effective, as was cimetidine, although ketoconazole and methimazole were not[229]. Although it was known then that animal and human CYPs were not the same, the main families of CYPs were still being unravelled. These animal studies were reinforced by more work with human liver microsomes, which again showed that cimetidine could be effective[230]. This led to volunteer studies (including myself) that showed that cimetidine on a single dose would reduce the hydroxylamine formation[231]. Multiple dose studies in animals were also promising[232] and a clinical study in DH patients finally showed that the hydroxylamine formation could be reduced but not abolished. Methaemoglobin formation fell by nearly 30% and the drug retained its clinical effects and improved patient tolerance[228]. Subsequent studies underscored the possibilities of using cimetidine in patients who could only respond to high dapsone doses and would normally have had to endure considerable methaemoglobin formation. This method has reached clinical practice in some areas[233,234].

Interestingly, it was clear that the rat was a poor model for humans in this context, in that cimetidine was far more effective as an inhibitor in the rat. It is now established that dapsone is cleared by CYP2C9 and also CYP2C19[235]. More recent work has shown that fluconazole-mediated inhibition has an even greater impact on dapsone hydroxylamine formation, and it is likely that voriconazole would also work as it inhibits these CYPs, but posaconazole and itraconazole as CYP3A4 inhibitors will most likely have no effect[235]. Hence, the concomitant use of azoles with dapsone should not be a significant clinically relevant issue. From a longer-term toxic metabolite-suppressing perspective, it would probably be undesirable to use too potent an inhibitor, as endogenous CYP functions would have been severely affected. As a coda to this work, in 2003 the antioxidant dihydrolipoic acid (formed from lipoic acid in human erythrocytes *in vivo*) I found would partially block the reaction between the hydroxylamine and oxyhaemoglobin *in vitro*[236]. Hence, I still hold out the hope that a study could one day be designed to use lipoic acid and cimetidine in combination to make an even larger reduction in methaemoglobin formation in patients on high-dosage dapsone, without completely blocking the CYPs, with the added benefit of the lipoic acid.

5.8.7   Use of inhibitors to reduce drug costs

If you are a European reading this section, in a strange way you are rather like a scion of a mega-rich family who is lucky enough to live in Maranello, in Northern Italy. Since you were 18 you have driven around in your bespoke Ferrari, pausing only occasionally to drop it off at the factory where it is serviced, and any necessary work is done by team of skilled mechanics, who just smile and wave away your proffered credit card, while you relax with a well-earned *espresso*. However, if you drive outside Europe and you crash or break down, you suddenly discover the true nature of your car – such as the eye-watering cost of the parts and labour and its phenomenal complexity.

In countries with National Health Services, drug costs are of course a political issue, but often they only impinge on relatively small numbers of individuals, such as those suffering from the rarest cancers, and any denial of the drug to the patient still results in significant controversy on ethical and moral grounds. In the rest of the world, all medical treatment of whatever quality must be paid for, and in the United States, insurance will only cover around 80% of the costs, leaving possibly bankrupting bills to be paid. In less-developed countries, such as Nepal for example, some venomous creatures are referred to in local language as what you need to sell in order to buy the antidote to save your life. If one of those creatures is known as something like 'the farm', then you have a choice to make: your life or your family's future.

In Western medicine, the spectacular achievements in transplant therapy are founded on extremely expensive drugs indeed, and as far back as the 1990s, ketoconazole was used to try to defray some of the costs of cyclosporine therapy by effectively halving the amount of drug used in a set therapeutic time frame. Although the approach was criticized at the time, one study rather unusually in medical literature actually details the cost savings of ketoconazole's effect in US dollars in its abstract and was published in a reputable journal[237]. However, from the outset, ketoconazole was a poor choice due its toxicity, inhibitory potency, and lack of selectivity, which really made it unsuitable for long-term use[238]. The use of inhibitors solely to reduce drug costs might be necessary and desirable in some contexts, but one cannot help feeling that it is a somewhat sad reflection on humanity.

5.8.8   Use of inhibition in alcoholism

The effects of alcoholism are covered in more detail in Chapter 7.7. Among the treatments for alcoholism is the use of the potent inhibitor of aldehyde dehydrogenase and CYP2E1, disulfiram (Antabuse). This approach is based on the perceived threat of pharmacological punishment as part of the alcoholic's progress towards abstinence. In the alcoholic, ethanol is cleared mainly by alcohol dehydrogenase to acetaldehyde, which is cleared by aldehyde dehydrogenase to acetic acid and water. If alcohol is imbibed during Antabuse treatment, the clearance of ethanol to acetaldehyde occurs, but the process stops there and acetaldehyde accumulates, causing a severe effect that includes flushing, nausea, vomiting, and sweating. Even small amounts of alcohol will show this effect, such as in accidental consumption. There are many medicinal and hygiene-based products, ranging from cough mixtures to mouthwashes that can contain 5–20% ethanol, so patient awareness is valuable in this context although the whole approach is controversial. It has been difficult to design trials that would even demonstrate that disulfiram was effective, but the current view seems to be that it is[239]. Disulfiram also has been used in cocaine and even gambling addition, so it may have other CNS effects[240]. The drug has also found some interest as an anticancer agent[239].

5.9   Summary

Inhibition of drug clearance has the greatest clinical impact on patient well-being, in terms of the rapidity of the effect and its severity. This is particularly important when the patient or health-care professional is for a time unaware that a potent inhibitor has been consumed. Currently, in the light of the numbers of potent dietary and OTC inhibitors available to the patient, it is as important to educate the patient in the dangers of inhibition of narrow TI drugs as it is to educate the health-care professional.

References

1. de Leon J, Sandson NB, Cozza KL. A preliminary attempt to personalise risperidone dosing using drug-drug inter-actions and genetics: part II. Psychosomatics. 49, 347–361, 2008.

2. Niwa T, & Hata T. The effect of genetic polymorphism on the inhibition of azole antifungal agents against CYP2C9–mediated metabolism. Journal of Pharmaceutical Sciences 105, 1345–1348, 2016.

3. Mani S, Dou W, Redinbo MR. PXR antagonists and implication in drug metabolism. Drug Metab Rev. 45, 60–72, 2013.

4. European Medicines Agency: Ketoconazole http://www.ema.europa.eu/ema/index. jsp?curl=/pages/medicines/human/medicines/003906/human_med_001814.jsp

5. Niwa, T Imagawa, Y. Substrate specificity of human cytochrome P450 (CYP) 2C Subfamily and Effect of Azole Antifungal Agents on CYP2C8. J. Pharm. Pharm. Sci.19, 423–429, 2016.

6. Bruggemann RJM, Alffenaar JWC. Blijlevens NMA. Clinical Relevance of the Pharmacokinetic Interactions of Azole Antifungal Drugs with Other coadministered Agents Clinical Infectious Diseases 48, 1441–58, 2009.

7. Dodds-Ashley E, Management of Drug and Food Interactions with Azole Antifungal Agents in Transplant Recipients. Pharmacother. 30, 842–854, 2010.

8. Gronlund J, Saari T, Hagelberg N, et al. Miconazole oral gel increases exposure to oral oxycodone by inhibition of CYP2D6 and CYP3A4. Antimicrob. Agents Chemother. 55, 1063–1067, 2011.

9. Kim T, Jance T, Kumar P et al. Drug–drug interaction between isavuconazole and tacrolimus: a case report indicating the need for tacrolimus drug-level monitoring. J. Clin. Pharm. Ther. 40, 609–611, 2015.

10. Miceli MH, Kauffman CA. Isavuconazole: A new broad-spectrum triazole antifungal agent. Clin. Inf. Dis. 61, 1558–1565, 2015.

11. Nwaroh E, Jupp J, Jadusingh E, et al. Clinical impact and management of fluconazole discontinuation on sirolimus levels in bone marrow transplant patients J Oncol. Pharm. Pract. 24, 235–238, 2018.

12. Tsitsopoulou A, Posso R, Vale L, et al. Determination of the prevalence of triazole resistance in environmental aspergillus fumigatus strains isolated in South Wales, UK. Front. Microbiol. 9, 1395 1–8, 2018.

13. Tome M, Zupan J, Tomi Z, et al. Synergistic and antagonistic effects of immunomodulatory drugs on the action of antifungals against Candida glabrata and Saccharomyces cerevisiae, PEER J. 6 e4999 1–23, 2018

14. Ruggiero A, Arena R, Battista A, et al. Azole interactions with multidrug therapy in pediatric oncology Eur J Clin Pharmacol 69:1–10, 2013.

15. de Jonge ME, Huitema AD, Rodenhuis S, et al. Clinical pharmacokinetics of cyclophosphamide. Clin. Pharmacokinet. 44, 1135–1164, 2005.

16. Marr KA, Leisenring W, Crippa F, et al. Cyclophosphamide metabolism is affected by azole antifungals. Blood 103, 1557–1559, 2004.

17. Yamamoto H, Habu Y, Yano I, et al. Comparison of the effects of azole antifungal agents on the anticoagulant activity of warfarin. Biol. Pharm. Bull. 37, 1990–1993, 2014.

18. Martín-Pérez M, Gaist D, de Abajo FJ, et al. Population impact of drug interactions with warfarin: a real-world data approach. Thromb Haemost 118, 461–470, 2018.

19. Frost CE, Byon W, Song Y, et al. Effect of ketoconazole and diltiazem on the pharmacokinetics of apixaban, an oral direct factor Xa inhibitor. Br. J. Clin. Pharmacol. 79, 838–884, 2015.

20. Gelosa P, Castiglioni L, Tenconi M et al. Pharmacokinetic drug interactions of the nonvitamin K antagonist oral anticoagulants (NOACs) Pharmacological Research 135, 60–79, 2018.

21. Fitzgerald JL, Howes LG. Drug interactions of direct-acting oral anticoagulants. Drug Saf. 39, 841–845, 2016.

22. Tucker RM, Denning DW, Hanson LH, et al. Interaction of azoles with rifampin, phenytoin, and carbamazepine: *in vitro* and clinical observations. Clin. Infect. Dis. 14, 165–174, 1992.

23. Saari TI, Laine K, Neuvonen M, et al. Effect of voriconazole and fluconazole on the pharmacokinetics of intravenous fentanyl. Eur J Clin Pharmacol 64:25–30, 2008.

24. Hsiao SH, Chang HJ, Hsieh TH, et al. Rhabdomyolysis caused by the moderate CYP3A4 inhibitor fluconazole in a patient on stable atorvastatin therapy: a case report and literature review J. Clin. Pharm. Ther. 41, 575–578, 2016.

25. Law M, Ridnicka AR. Statin safety: a systematic review. Am. J. Cardiol. 97,52–60, 2006.

26. Alsaad, AMS, Kaplan YC and Koren G, Exposure to fluconazole and risk of congenital malformations in the offspring: A systematic review and meta-analysis. Reproductive Toxicology 52, 78–82, 2015.

27. Dinger J, Meyer MR, Maurer HH. Development of an *in vitro* cytochrome P450 cocktail inhibition assay for assessing the inhibition risk of drugs of abuse. Clin. Toxicol. Lett. 230, 28–35, 2014.

28. Gong, EC, Chea S, Balupuri A, et al. Enzyme kinetics and molecular docking studies on cytochrome 2B6, 2C19, 2E1, and 3A4 activities by sauchinone. Molecules 23, 555, 2018.

29. Satoh, T, Fujisawa, H, Nakamura, A et al. Inhibitory effects of eight green tea catechins on cytochrome P450 1A2, 2C9, 2D6, and 3A4 activities. J. Pharm. Pharmaceut. Sci. 19 188–197,

30. Ludwig E, Schmid J, Beschke K et al. Activation of human cytochrome P-450 3A4-catalyzed meloxicam 5 '-methylhydroxylation by quinidine and hydroquinidine *in vitro*. J. Pharm. Exper. Ther. 290, 1–8, 1999. 2016.

31. Phuc NM, Wu, Z, Yuseok O, et al. LKY-047: First selective inhibitor of cytochrome P450 2J2. Drug Metab. Disp. 45, 765–769, 2017.

32. Grimm SW, Einolf HJ, Hall SD, et al. The conduct of *in vitro* studies to address time-dependent inhibition of drug-metabolizing enzymes: a perspective of the pharmaceutical research and manufacturers of America. Drug Metab. Disp. 37, 7 1355–1370, 2009.

33. Isoherranen N, Kunze KL, Allen KE et al. Role of itraconazole metabolites in CYP3A4 inhibition. Drug Metab. Disp. 32, 1121–1131, 2004.

34. Barbara JE, Kazmi, Parkinson A. Metabolism-dependent inhibition of CYP3A4 by Lapatinib: evidence for formation of a metabolic intermediate complex with a nitroso/oxime metabolite formed via a nitrone intermediate. Drug Metab. Disp. 41 (5) 1012–1022, 2013.

35. Zimmerlin A, Trunzer M, Faller B, CYP3A time-dependent inhibition risk assessment validated with 400 reference drugs. Drug Metab. Disp. 39, 1039–1046, 2011.

36. Lutz JD, VandenBrink BM, Babu KN, et al. Stereoselective Inhibition of CYP2C19 and CYP3A4 by fluoxetine and its metabolite: implications for risk assessment of multiple time-dependent inhibitor systems. Drug Metab. Dispos. 41, 2056–2065, 2013.

37. Barbara JE, Kazmi F, Parkinson A et al. Metabolism-dependent Inhibition of CYP3A4 by Lapatinib: evidence for formation of a metabolic intermediate complex with a nitroso/oxime metabolite formed via a nitrone intermediate. Drug Metab. Disp. 41, 1012–1022, 2013.

38. Watanabe A, Takakusa H, Kimura T, et al. Analysis of mechanism-based inhibition of CYP 3A4 by a series of fluoroquinolone antibacterial agents. Drug Metab. Disp. 44, 1608–1616, 2016

39. Potega A, Fedejko-Kap, Mazerska Z. Imidazoacridinone antitumor agent C-1311 as a selective mechanism based inactivator of human cytochrome P450 1A2 and 3A4 isoenzymes. Pharm. Rep. 68, 663–670, 2016.

40. Chan CY, New LS, Ho HK, et al. Reversible time-dependent inhibition of cytochrome P450 enzymes by duloxetine and inertness of its thiophenc ring towards bioactivation. Toxicology Letters 206, 314–324, 2011.

41. Lee JY, Lee SY, Oh SJ, et al. Assessment of drug–drug interactions caused by metabolism-dependent cytochrome P450 inhibition. Chemico-Biological Interactions 198,49–56, 2012.

42. Richter T, Murdter TE, Heinkele G, et al. Potent mechanism-based inhibition of human CYP2B6 by clopidogrel and ticlopidine JPET. 308, 189 197, 2004

43. Samuels ER, Sevrioukova I. Inhibition of human CYP3A4 by rationally designed ritonavir-like compounds: impact and interplay of the side group functionalities. Mol. Pharmaceut. 15, 279–288, 2018.

44. Hossain MD, Tran T, Chena T, et al. Inhibition of human cytochromes P450 *in vitro* by ritonavir and cobicistat. J. Pharm. Pharmacol. 69, 1786–1793, 2017.

45. Sunaga K, Ohkawa K, Nakamura K, et al. Mechanism-based inhibition of recombinant human cytochrome P450 3A4 by tomato juice extract. Biol. Pharm. Bull. 35, 329–334 2012.

46. Ogilvie BW, Torres R, Dressman MA. Clinical assessment of drug–drug interactions of tasimelteon, a novel dual melatonin receptor. Agonist. J. Clin. Pharmacol. 55, 1004–1011, 2015.

47. Backman, JT, Kyrklund C, Neuvonen, M, et al. Gemfibrozil greatly increases plasma concentrations of cerivastatin. Clin. Pharmacol. Ther. 72, 685–691, 2002.

48. Zhang, H, Amunugama, H, Ney, S, Cooper, N, Hollenberg, PF. Mechanism-based inactivation of human cytochrome P450 2B6 by clopidogrel: involvement of both covalent modification of cysteinyl residue 475 and loss of heme. Mol. Pharmacol. 2011, 80, 839–847.

49. Orr STM, Ripp SL, Ballard TE, et al. Mechanism-based inactivation (MBI) of cytochrome P450 enzymes: Structure–activity relationships and discovery strategies to mitigate drug–drug interaction risks. J. Med. Chem. 55, 4896–493, 2012.

50. White NC, Litovitz T, Clancy C. Suicidal antidepressant overdoses: a comparative analysis by antidepressant type. J. med. Toxicol. 4, 238–250, 2008.

51. Kevin Hines http://www.kevinhinesstory.com/bio/

52. Johnson CF, Williams B, MacGillivray SA, et al. 'Doing the right thing': factors influencing GP prescribing of antidepressants and prescribed doses. BMC Fam. Pract. 18, 72 1–13, 2017.

53. Hamilton M. A rating scale for depression. J. Neurol. Neurosurg. Psych. 23, 56–62, 1960.

54. Kiberd MB, Minor SF. Lipid therapy for the treatment of a refractory amitriptyline overdose. Can. J. Emerg. Med. 14, 193–197, 2012.

55. Bellon A, Coverdale JH. Delirium, thrombocytopenia, insomnia, and mild liver damage associated with MAOI withdrawal. Eur. J. Clin. Pharmacol. 65, 1269–1270, 2009.

56. Monden R, Roest AM, van Ravenzwaaij D, The comparative evidence basis for the efficacy of second-generation antidepressants in the treatment of depression in the US: A Bayesian meta-analysis of Food and Drug Administration reviews. J. Affect. Dis. 235, 393–398, 2018.

57. Mazza MG, Rossetti A, Botti ER, et al. Vortioxetine overdose in a suicidal attempt. A case report. Medicine 97, 25, 2018.

58. Undurraga J, Baldessarini RJ, Direct comparison of tricyclic and serotonin reuptake inhibitor antidepressants in randomized head-to-head trials in acute major depression: Systematic review and meta-analysis J. Psychopharmacol. 31, 1184–1189, 2017.

59. Volpi-Abadie J, Kaye AM, Kaye ID. Serotonin syndrome. Ochsner J. 13, 533–540, 2013.

60. Mawardi G, Markman T, Muslem R, et al. SSRI/SNRI therapy is associated with a higher risk of GI bleed in LVAD patients. J. Heart & Lung Trans. 37, S160–S161, 2018.

61. Kim S, Park K, Kim M, et al. Data-mining for detecting signals of adverse drug reactions of fluoxetine using the Korea Adverse Event Reporting System (KAERS) database. Psych. Res. 256, 237–242, 2017.

62. Guo MY, Etminan M, Procyshyn RM et al. Association of antidepressant use with drug-related extrapyramidal symptoms: a pharmacoepidemiological study. J. Clin. Psychopharmacol. 38, 349–356, 2018.

63. Beach SR, Celano CM, Sugrue AM, QT. Prolongation, torsades de pointes, and psychotropic medications: a 5-year update. Psychosomat. 59, 105–122, 2018.

64. Florio V, Porcelli S, Saria A, et al. Escitalopram plasma levels and antidepressant response Eur. Neuropsychopharmacol. 27, 940–944 2017.

65. Pringsheim T, Gardner D, Patten SB. Adjunctive treatment with quetiapine for major depressive disorder: are the benefits of treatment worth the risks? BMJ 350, h569 1–3, 2015.

66. Wang Z, Wang S, Huang M, et al. Characterizing the effect of cytochrome P450 (CYP) 2C8, CYP2C9, and CYP2D6 genetic polymorphisms on stereoselective N-demethylation of fluoxetine. Chirality 26, 166–173, 2014.

67. Sager JE, Lutz JD, Foti RS, et al. Fluoxetine and norfluoxetine mediated complex drug–drug interactions: *in vitro* to *in vivo* correlation of effects on CYP2D6, CYP2C19 and CYP3A4. Clin. Pharmacol. Ther. 95, 653–662, 2014.

68. Schmider JL, Greenblatt DJ, von Moltke LL, et al. Inhibition of CYP2C9 by selective serotonin reuptake inhibitors *in vitro*: studies of phenytoin p-hydroxylation. Br. J. Clin. Pharmacol. 44, 495–498, 1997.

69. Spina E, Pisani F, de Leon, J. Clinically significant pharmacokinetic drug interactions of antiepileptic drugs with new antidepressants and new antipsychotics. Pharmacol. Res. 106, 72–86, 2016.

70. Bahar MA, Kamp J, Borgsteede SD, et al. The impact of CYP2D6 mediated drug–drug interaction: a systematic review on a combination of metoprolol and paroxetine/fluoxetine. Br J Clin Pharmacol. 84, 2704–2715, 2018.

71. Fjukstad KK, Engum A, Lydersen S, et al. Metabolic risk factors in schizophrenia and bipolar disorder: The effect of co-medication with selective serotonin reuptake inhibitors and antipsychotics. Eur. Psych. 48, 71–78, 2018

72. Bleakley S, Identifying and reducing the risk of antipsychotic drug interactions. Prog. Neurol. Psychiatr. 16, 20–24, 2012.

73. Bostankolu G, Ayhan Y, Cuhadaroglu F, et al. Serotonin syndrome with a combination of aripiprazole and fluoxetine: a case report. Ther. Adv. Psychopharmacol. 5, 138–140, 2015.

74. English BA, Dortch M, Ereshefsky L, et al. Clinically significant psychotropic drug–drug interactions in the primary care setting. Curr Psychiatry Rep. 14(4), 376–390, 2012.

75. FDA Drug Safety Communication: FDA reporting mental health drug ziprasidone (Geodon) associated with rare but potentially fatal skin reactions. https://www.fda.gov/Drugs/DrugSafety/ucm426391.htm

76. Paulzen M, Schoretsanitis G, Hiemke C, et al. Reduced clearance of venlafaxine in a combined treatment with quetiapine. Prog. Neuropsychopharmacol. & Biol. Psych 85, 116–121, 2018.

77. Callaghan JT, Bergstrom RF, Ptak LR, et al. Olanzapine Pharmacokinetic and Pharmacodynamic Profile. Clin. Pharmacokinet. 37, 177–193, 1999.

78. Taylor DM, Cornelius V, Smith L, et al. Comparative efficacy and acceptability of drug treatments for bipolar depression: a multiple-treatments meta-analysis. Acta. Psychiatr. Scand. 130, 452–469, 2014.

79. Walker DJ, DelBello MP, Landry J. Quality of life in children and adolescents with bipolar I depression treated with olanzapine/fluoxetine combination Child Adolesc. Psych. Ment. Health 11, 34, 1–11, 2017.

80. Jessurun N, van Puijenbroek EP, Otten LS et al. Inhibition of CYP2D6 with low dose (5 mg) paroxetine in patients with high 10-hydroxynortriptyline serum levels – a review of routine practice. Br J Clin Pharmacol (2017) 83 1149–1151 1149

81. Schelleman H, Brensinger CM, Bilker WB, et al. Antidepressant-Warfarin Interaction and Associated Gastrointestinal Bleeding Risk in a Case–Control Study. PLoS ONE 6, e21447 1–6, 2011.

82. Dong, YH, Bykov K, Choudhry NK, et al. Clinical Outcomes of Concomitant Use of Warfarin and Selective Serotonin Reuptake Inhibitors: A Multidatabase Observational Cohort Study. J. Clin. Psychopharmacol. 37, 200–209, 2017.

83. Teichert M, Visser LE, Uitterlinden AG, et al. Selective serotonin re-uptake inhibiting antidepressants and the risk of over-anticoagulation during acenocoumarol maintenance treatment. Br. J. Clin. Pharmacol. 72, 798–805, 2011.

84. Weiss J, Dormann SMG, Martin Facklam, M, et al. Inhibition of P-Glycoprotein by Newer Antidepressants. JPET 305, 197–204, 2003.

85. Mannheimer B, Wettermark B, Lundberg M et al. Nationwide drug-dispensing data reveal important differences in adherence to drug label recommendation on CYP2D6-dependent drug interactions. Br. J. Clin. Pharmacol. 69, 411–417, 2010.

86. Gheldiu AM, Vlase L, Popa A, et al. Investigation of a potential pharmacokinetic interaction between nebivolol and fluvoxamine in healthy volunteers. J. Pharm. Pharm. Sci. (www.cspsCanada.org) 20, 68–80, 2017.

87. Portella MJ, de Diego-Adelino J, Ballesteros J, et al. Can we really accelerate and enhance the selective serotonin reuptake inhibitor antidepressant effect? a randomized clinical trial and a meta-analysis of pindolol in nonresistant depression. J. Clin. Psych. 72(7), 962–969, 2011.

88. Cai WM, Chen B, Zhou Y, et al. Fluoxetine impairs the CYP2D6-mediated metabolism of propafenone enantiomers in healthy Chinese volunteers. Clin. Pharm. Ther. 66(5), 516–521, 1999.

89. Chong CR, Ong GJ, Horowitz JD. Emerging drugs for the treatment of angina pectoris. Exp. Op. Emerg. Drugs 21, Pages: 365–376, 2016.

90. Alderman CP, Hundertmark JD, Soetratma TW. Interaction of serotonin re-uptake inhibitors with perhexiline. Aust. New. Zeal. J. Psych. 31, 601–603, 1997.

91. Hersh, EV, Pinto A, Moore PA. Adverse drug interactions involving common prescription and over-the-counter analgesic agents. Clin. Therapeut. 29, 2477–2497, 2007.

92. Nelson EM, Philbrick AM, Avoiding serotonin syndrome: the nature of the interaction between tramadol and selective serotonin reuptake inhibitors. Ann. Pharmacother. 46, 1712–1716, 2012.

93. Spies, PE, Pot JLW, Willems RPJ, et al. Interaction between tramadol and selective serotonin reuptake inhibitors: are doctors aware of potential risks in their prescription practice? Eur. J Hosp. Pharm-Sci & Pract. 24, 124–127, 2017.

94. Kummer O, Hammann F, Moser C, et al. Effect of the inhibition of CYP3A4 or CYP2D6 on the pharmacokinetics and pharmacodynamics of oxycodone. Eur. J. Clin. Pharmacol. 67, 63–71, 2001.

95. Binkhorst L, Bannink M, de Bruijn P, et al. Augmentation of endoxifen exposure in tamoxifen-treated women following SSRI switch. Clin. Pharmacokinet. 55, 249–255, 2016.

96. Delavenne X, Magnin M, Basset T, et al. Investigation of drug–drug interactions between clopidogrel and fluoxetine. Fundamental & Clinical Pharmacology 27, 683–689, 2013.

97. Paulzen M, Finkelmeyer A, Grozinger M. Augmentative effects of fluvoxamine on duloxetine plasma levels in depressed patients. Pharmacopsych. 44, 317–323, 2011.

98. Polcwiartek C, Nielsen J. The clinical potentials of adjunctive fluvoxamine to clozapine treatment: a systematic review. Psychopharmacol. 233, 741–750, 2016.

99. Hiemke C, Jabarin M, Hadjez J, et al. Fluvoxamine augmentation of olanzapine in chronic schizophrenia: Pharmacokinetic interactions and clinical effects. J Clin. Psychopharmacol. 22, 502–506, 2002.

100. Gahr M, Rehbaum K, Connemann BJ. Clozapine-associated Development of Second-onset Obsessive Compulsive Symptoms in Schizophrenia: Impact of Clozapine Serum Levels and Fluvoxamine Add-on. Pharmacopsych. 47, 118–120, 2014.

101. Vezmar S, Miljkovic B, Vucicevic K, et al. Pharmacokinetics and efficacy of fluvoxamine and amitriptyline in depression. J. Pharm. Sci. 110, 98–104, 2009.

102. Nadkarni A, Oldham MA, Howard M, et al. Drug–drug interactions between warfarin and psychotropics: updated review of the literature. Pharmacother. 32, 932–942, 2012.

103. El Ela, AA, Hartter, S, Schmitt, U, et al. Identification of P-glycoprotein substrates and inhibitors among psychoactive compounds–implications for pharmacokinetics of selected substrates. J. Pharm. Pharmacol. 56, 967–975, 2004.

104. Omori IM, Watanabe N, Nakagawa A, et al. Fluvoxamine versus other anti-depressive agents for depression. Coch. Data. Syst. Rev. Art. No. CD006114, 2010

105. Patel NK, Wiśniowska B, Jamei M, et al. Real patient and its virtual twin: application of quantitative systems toxicology modelling in the cardiac safety assessment of citalopram AAPS J. 20, 1–10, 2018.

106. Wenzel-Seifert K, Brandl R, Hiemke C, Haen E. Influence of concomitant medications on the total clearance and the risk for supra-therapeutic plasma concentrations of citalopram. a population-based cohort study. Pharmacopsych. 47, 239–244, 2014.

107. Davies SJC, Mulsant BH, Flint AJ, et al. The impact of sertraline co-administration on the pharmacokinetics of olanzapine: a population pharmacokinetic analysis of the STOP-PD. Clin. Pharmacokinet. 54,1161–1168, 2015.

108. Almansour MI, Jarrar YB, Jarrar BM. *In vivo* investigation on the chronic hepatotoxicity induced by sertraline. Env. Tox. & Pharmacol 61, 107–115, 2018.

109. Kapoor A, Iqbal M, Petropoulos S, et al. Effects of sertraline and fluoxetine on P-glycoprotein at barrier sites: *in vivo* and *in vitro* approaches. PLoS ONE 8(2), e56525, 2013.

110. Conrad MA, Cui J, Lin HC. Sertraline-associated cholestasis and ductopenia consistent with vanishing bile duct syndrome. J. Pediatr. 169, 313–315, 2016.

111. Paulzen M, Haen E, Hiemke C, Fay B, et al. Antidepressant polypharmacy and the potential of pharmacokinetic interactions: Doxepin but not mirtazapine causes clinically relevant changes in venlafaxine metabolism. J. Affect. Dis. 227, 506–511, 2018.

112. Martinez C, Assimes TL, Mines D, et al. Use of venlafaxine compared with other antidepressants and the risk of sudden cardiac death or near death: a nested case-control study. BMJ, 340, c249, 2010.

113. Nelson JC, Spyker DA, Morbidity and mortality associated with medications used in the Treatment of depression: an analysis of cases reported to U.S. Poison Control Centers, 2000–2014. Am. J. Psych. 174, 438–450, 2017.

114. Kuzin M, Schoretsanitis G, Haen E. et al. Effects of the proton pump inhibitors omeprazole and pantoprazole on the cytochrome P450-mediated metabolism of venlafaxine. Clin. Pharmacokinet. 57, 729–737, 2018.

115. Desmarais, JE, Looper, KJ. Interactions between tamoxifen and antidepressants via cytochrome P450 2D6. J. Clin. Psych. 70, 1688–1697, 2009.

116. Irarrazaval MEI, Gaete GL, Antidepressants agents in breast cancer patients using tamoxifen: review of basic and clinical evidence. Revis. Med. De Chile 144, 10 1326–1335, 2016.

117. Tisdale JE, Drug-induced QT interval prolongation and torsades de pointes Role of the pharmacist in risk assessment, prevention and management. Can. Pharm. J. (Ott). 149(3), 139–152, 2016.

118. Ma SP, Tsai CJ, Chang CC, et al. Delirium associated with concomitant use of duloxetine and bupropion in an elderly patient. Psychogeriat. 17, 130–132, 2017.

119. Sager JE, Tripathy S, Price LSL, et al. *In vitro* to *in vivo* extrapolation of the complex drug–drug interaction of bupropion and its metabolites with CYP2D6; simultaneous reversible inhibition and CYP2D6 downregulation. Biochem. Pharmacol. 123, 85–96, 2017.

120. He H, Wang W, Lyu J, et al. Efficacy and tolerability of different doses of three new antidepressants for treating major depressive disorder: A PRISMA-compliant meta-analysis. J. Psych. Res. 96, 247–259, 2018.

121. Chen LS, Boinpally R, Gad N, et al. Evaluation of cytochrome P450 (CYP) 3A4–based interactions of levomilnacipran with ketoconazole, carbamazepine or alprazolam in healthy subjects. Clin. Drug. Invest. 35, 601–612, 2015.

122. Chen G, Højer AM, Areberg J, et al. Vortioxetine: clinical pharmacokinetics and drug interactions. Clin Pharmacokinet. 57(6), 673–686, 2018.

123. Abdlekawy KS, Donia AM, Elbarbry F, Chen G. Effects of grapefruit and pomegranate juices on the pharmacokinetic properties of dapoxetine and midazolam in healthy subjects. Eur. J. Drug Metab. Pharmacokinet. 42, 397–405, 2017.

124. Shelton RC. The nature of the discontinuation syndrome associated with antidepressant drugs. J. Clin. Psych. 67, 3–7, 2006.

125. Nielsen M, Hansen EH, Gøtzsche PC. What is the difference between dependence and withdrawal reactions? A comparison of benzodiazepines and selective serotonin re-uptake inhibitors. Addiction. 107(5), 900–908, 2012.

126. Nielsen M, Hansen EH, Gøtzsche PC. Dependence and withdrawal reactions to benzodiazepines and selective serotonin reuptake inhibitors. How did the health authorities react? J. Risk. Saf. Med. 25(3), 155–168, 2013.

127. Bailey DG, Spence JD, Edgar B, et al. Ethanol enhances the hemodynamic effects of felodipine. Clin. Invest. Med. 12, 357–62, 1989.

128. Bryan J. Despite its problems, terfenadine did set a new standard for hay fever treatment. Pharm. J. 287, 511, 2011.

129. Bailey DG, Dresser G, Arnold MO. Grapefruit–medication interactions: Forbidden fruit or avoidable consequences? MDCMAJ 185, 309–316, 2013.

130. Hanley, MJ, Cancalon P, Widmer WW, et al. The effect of grapefruit juice on drug disposition. Exp. Opin. Drug Metab. Toxicol. 7, 267–286, 2011.

131. Kimura S, Kako S, Wada H et al. Can grapefruit juice decrease the cost of imatinib for the treatment of chronic myelogenous leukemia? Leuk. Res. 35, e11–e12, 2011.

132. Banfield C, Gupta S, Marino M, et al. Grapefruit juice reduces the oral bioavailability of fexofenadine but not desloratadine. Clin. Pharmacokinet. 41, 311–318, 2002.

133. Bailey DG, Fruit juice inhibition of uptake transport: a new type of food–drug interaction. Br. J. Clin. Pharmacol. 70, 645–655, 2010.

134. Dresser GK, Bailey DG, Carruthers SG. Grapefruit juice–felodipine interaction in the elderly. Clin. Pharmacol. Ther. 68, 28–34, 2000.

135. Theile D, Hohmann N, Kiemel D. et al. Clementine juice has the potential for drug interactions – *In vitro* comparison with grapefruit and mandarin juice. Eur. J. Pharm. Sci. 97, 247–256, 2017.

136. Lin H, Kent UM, Hollenberg PF. The grapefruit juice effect is not limited to cytochrome P450 (P450) 3A4: evidence for bergamottin–dependent inactivation, heme destruction, and covalent binding to protein in P450s 2B6 and 3A5. JPET. 313, 154–164, 2005.

137. Santes-Palaciosa R, Romo-Mancilla A, Camacho-Carranzaa R. et al. Inhibition of human and rat CYP1A1 enzyme by grapefruit juice compounds. Tox. Lett. 258 267–275, 2016.

138. Wangensteen H, Molden E, Christensen H et al. Identification of epoxybergamottin as a CYP3A4 inhibitor in grapefruit peel. European Journal of Clin. Pharmacol. 58, 663–668, 2003.

139. Luczak, T, Swanoski M. A review of cranberry use for preventing urinary tract infections in older adults. Con. Pharm. 33, 450–453, 2018.

140. Shmuely H, Yahav J, Samra Z, et al. Effect of cranberry juice on eradication of Helicobacter pylori in patients treated with antibiotics and a proton pump inhibitor. Mol. Nutr. Food Res. 51, 746 e51, 2007.

141. Ulrey RK, Barksdale SM, Zhou W, et al. Cranberry proanthocyanidins have antibiofilm properties against Pseudomonas aeruginosa. BMC Complement Altern Med. 14, 499, 2014.

142. Griffiths AP, Beddall A, Pegler S. Fatal haemopericardium and gastrointestinal haemorrhage due to possible interaction of cranberry juice with warfarin. J. Roy. Soc. Prom. Health 128, 324–326, 2008.

143. Chen M, Zhou S, Fabriaga E et al. Food–drug interactions precipitated by fruit juices other than grapefruit juice: An update review. J. Food Analy. 26, SS61e SS 71, 2018.

144. Poór M, Li Y, Kunsági-Máté S, et al. Molecular displacement of warfarin from human serum albumin by flavonoid aglycones J. Luminesc. 142, 122–127, 2013.

145. Hanley MJ, Masse G, Harmatz JS, et al. Pomegranate juice and pomegranate extract do not impair oral clearance of flurbiprofen in human volunteers: divergence from *in vitro* results. Clin. Pharmacol. Ther. 92, 651e7, 2012.

146. Strippoli S, Lorusso V, Albano A, et al. Herbal–drug interaction induced rhabdomyolysis in a liposarcoma patient receiving trabectedin. BMC Comp. & Altern. Med. 13, 199–26, 2013.

147. Qiao X, Ji S, Yu S, et al. Identification of key licorice constituents which interact with cytochrome P450: Evaluation by LC/MS/MS cocktail assay and metabolic profiling. AAPS J. 16, 101–112, 2014.

148. He W, Wuc JJ, Ning J, et al. Inhibition of human cytochrome P450 enzymes by licochalcone A, a naturally occurring constituent of licorice. Tox. in vitro 29, 1569–1576, 2015.

149. Li G, Simmler C, Chen L, et al. Cytochrome P450 inhibition by three licorice species and fourteen licorice constituents. Eur. J. Pharm. Sci. 109, 182–190, 2017

150. Zhang XL, Chen M, Zhu LL et al. Therapeutic risk and benefits of concomitantly using herbal medicines and conventional medicines: from the perspectives of evidence based on randomized controlled trials and clinical risk management. Evid.-Based Comp. Altern. Med. 9296404, 1–17, 2017.

151. Hermann R, von Richter O. Clinical evidence of herbal drugs as perpetrators of pharmacokinetic drug interactions. Planta Med. 78, 1458–1477, 2012.

152. Wanwimolruk S, Phopin K, Prachayasittikul V. Cytochrome P450 enzyme mediated herbal drug interactions. (Part 2). EXCLI J. 13, 869–896, 2014.

153. Brown AC, Liver toxicity related to herbs and dietary supplements: Online table of case reports. Part 2 of 5 series. Food Chem. Tox. 107, 472e 501, 2017.

154. UK Medicines & Healthcare products Regulatory Agency. https://www.gov.uk/government/publications/list-of-banned-or-restricted-herbal-ingredients-for-medicinal-use/banned-and-restricted-herbal-ingredients.

155. Choi S, Oh DS, Jerng UM. A systematic review of the pharmacokinetic and pharmacodynamic interactions of herbal medicine with warfarin PLoS One 12, e0182794, 2017.

156. Mitra P, Weinheimer S, Michalewicz M, et al. Prediction and quantification of hepatic transporter-mediated uptake of pitavastatin utilizing a combination of the relative activity factor approach and mechanistic modeling. Drug. Metab. Disp. 46, 953–963, 2018.

157. Shitara Y, Sugiyama Y. Preincubation-dependent and long-lasting inhibition of organic anion transporting polypeptide (OATP) and its impact on drug–drug interactions. Pharmacol. & Ther. 177, 67–80, 2017.

158. Furberg CD, Pitt B. Withdrawal of cerivastatin from the world market. Curr. Con. Trials Cardiovasc. Med. 2, 205–207, 2001.

159. Kudo T, Hisaka A, Sugiyama Y, et al. Analysis of the repaglinide concentration increase produced by gemfibrozil and itraconazole based on the inhibition of the hepatic uptake transporter and metabolic enzymes. Drug Metab. Dispos. 41, 362–371, 2013.

160. Sun XL, Li JX, Guo CR. Pharmacokinetic effects of curcumin on docetaxel mediated by OATP1B1, OATP1B3 and CYP450s. Drug Metab. Pharmacokin. 31 269–275, 2016.

161. Netsomboon K, Laffleur F, Suchaoin W, et al. Novel in vitro transport method for screening the reversibility of P-glycoprotein inhibitors. Eur. J. Pharm. Biopharm. 100, 9–14, 2016.

162. Silva R, Carmo H, Vilas-Boas V, et al. Colchicine effect on P-glycoprotein expression and activity: in silico and in vitro studies. Chem Biol Interact. 218, 50–62, 2014.

163. Nader AM, Foster DR. Suitability of digoxin as a P-glycoprotein probe: implications of other transporters on sensitivity and specificity. J. Clin. Pharm. 54, 3–13, 2013.

164. Zhang W, McIntyre C, Kuhn M, et al. Effect of vemurafenib on the pharmacokinetics of a single dose of digoxin in patients with BRAFV600 mutation-positive metastatic malignancy. J Clin Pharmacol. 58, 1067–1072, 2018.

165. Grzesk G, Stolarek W, Kasprzak M, et al. Therapeutic drug monitoring of digoxin – 20 years of experience. Pharm. Rep. 70, 184–189, 2018.

166. Bo M, Quaranta V, Fonte G, et al. Prevalence, predictors and clinical impact of potentially inappropriate prescriptions in hospital-discharged older patients: A prospective study. Geriatr. Gerontol. Int. 18, 561–568, 2018.

167. Sadeque AJ, Wandel C, He H, et al. Increased drug delivery to the brain by P-glycoprotein inhibition. Clin. Pharmacol. Ther. 68, 231–237, 2000.

168. Dei S, Romanelli MN, Manetti D, et al. Design and synthesis of aminoester heterodimers containing flavone or chromone moieties as modulators of P-glycoprotein-based multidrug resistance (MDR). Bioorg. & Med. Chem. 26, 50–64, 2018.

169. Palmeira A, Sousa E, Vasconcelos MH, et al. Three decades of P-gp inhibitors: skimming through several generations and scaffolds. Curr. Med. Chem. 19, 1946–2025, 2012.

170. Zhang Y, Wang CY, Duan YJ, et al. Afatinib decreases P-glycoprotein expression to promote adriamycin toxicity of A549T cells. J. Cell. Biochem. 119, 414–423 2018.

171. Fan YF, Zhang W, Zeng L, et al. Dacomitinib antagonizes multidrug resistance (MDR) in cancer cells by inhibiting the efflux activity of ABCB1 and ABCG2 transporters. Cancer Letters 421, 186e–198, 2018.

172. Mao X, Chen Z, Zhao Y, et al. Novel multi-targeted ErbB family inhibitor afatinib blocks EGF-induced signaling and induces apoptosis in neuroblastoma. Oncotarget. 2017 Jan 3; 8(1), 1555–1568, 2016.

173. Stopfer P, Giessmann T, Hohl K, et al. Optimization of a drug transporter probe cocktail: potential screening tool for transporter-mediated drug–drug interactions. Br. J. Clin. Pharmac. 84, 1941–1949, 2018.

174. Nielsen DL, Palshof JA, Brünner N, et al. Implications of ABCG2 expression on irinotecan treatment of colorectal cancer patients: a review. Int. J. Mol. Sci. 18(1926), 1–16, 2017.

175. Cole SPC. Multidrug resistance protein 1 (MRP1, ABCC1), a "multitasking" ATP-binding cassette (ABC) transporter. J. Biol. Chem. 289, 30880–30888, 2014.

176. Sorf A, Hofman J, Kučera R, et al. Ribociclib shows potential for pharmacokinetic drug–drug interactions being a substrate of ABCB1 and potent inhibitor of ABCB1, ABCG2, and CYP450 isoforms in vitro. Biochem. Pharmacol. 154, 10–17, 2018.

177. Kozik TM, Wung SF. Acquired long QT syndrome; frequency, onset, and risk factors in intensive care patients. Crit. Care Nurse 32, 32–41, 2012.

178. Alexandre J, Molsehi JJ, Bersell KR, et al. Anticancer drug-induced cardiac rhythm disorders: Current knowledge and basic underlying mechanisms. Pharmacol. & Therapeut. 189, 89–103, 2018.

179. Tse G, Gong M, Meng L, et al. Predictive value of Tpeak – tend indices for adverse outcomes in acquired QT prolongation: a meta-analysis. Front. Physiol. 9:1226, 1–8, 2018.

180. Marill, KA, Dorsey P, Holmes A, et al. Is myocardial repolarization duration associated with repolarization heterogeneity? Ann. Noninvasive Electrocardiol. 23, e12519, 1–9, 2018

181. Wan E, Marx SO. Ion channels in health and disease. perspectives in translational cell biology. Chapter 5, Diagnosis, Treatment, and Mechanisms of Long QT Syndrome. Academic Press, 113–130, 2016.

182. Sicouri S, Glass A, Ferreiro M, et al. Transseptal dispersion of repolarization and its role in the development of torsade de pointes arrhythmias. J. Cardiovasc. Electrophysiol. 21, 441–447, 2010.

183. Vink AS,Clur SAB, Wilde AAM, et al, Effect of age and gender on the QTc-interval in healthy individuals and patients with long-QT. syndrome. Trends in Cardiovasc. Med. 28(1), 64–75, 2018.

184. Anneken L, Baumann S, Vigneault P. Estradiol regulates human QT-interval: acceleration of cardiac repolarisation by enhanced KCNH2 membrane trafficking. Eur. Heart J. 37, 640–650, 2016.

185. Sordillo PP, Sordillo DC, Helson L. The Prolonged QT Interval: Role of Pro-inflammatory Cytokines, Reactive Oxygen Species and the Ceramide and Sphingosine-1 Phosphate Pathways. In vivo 29, 619–636, 2015.

186. Ishibashi K, Aiba T, Kamiya C, Arrhythmia risk and beta-blocker therapy in pregnant women with long QT syndrome. Heart 103, 1374–1379, 2017.

187. Woosley, RL, Heise, CW and Romero, KA, www.Crediblemeds.org, QT drugs List, Accession Date, AZCERT, Inc. 1822 Innovation Park Dr., Oro Valley, AZ 85755.

188. Windley MJ, Lee W, Vandenberg JI, et al. The temperature dependence of kinetics associated with drug block of hERG channels are compound specific and an important factor for proarrhythmic risk prediction. Mol. Pharmacol. 94, 760–769, 2018.

189. Danielsson B, Collin J, Bergman GJ et al. Antidepressants and antipsychotics classified with torsades de pointes arrhythmia risk and mortality in older adults–a Swedish nationwide study. Br. J. Clin. Pharmacol. 81, 773–783, 2016.

190. Lehmann DF, Eggleston WD, Wang D. Validation and clinical utility of the hERG IC50: Cmax ratio to determine the risk of drug-induced torsades de pointes: a meta-analysis. Phrmacother. 38, 341–348, 2018.

191. Nix DE. Cardiotoxicity Induced by Antifungal Drugs Current Fungal Infection Reports 8, 129–138, 2014.

192. Hsieh EM, Hornik CP, Clark RH, et al. Medication use in the neonatal intensive care unit. Am. J. Perinatol. 31, 811–822, 2014.

193. Ahonen J, Olokkola KT, Takala A, et al. Interaction between fluconazole and midazolam in intensive care patients. Acta Anaesthesiol Scand 43: 509–514, 1999.

194. Panahi Y, Dehcheshmeh HS, Mojtahedzadeh M, et al. Analgesic and sedative agents used in the intensive care unit: A review. J. Cell. Biochem. 119, 8684–8693, 2018.

195. Krishna G, Moton A, Ma L, et al. Effects of oral posaconazole on the pharmacokinetic properties of oral and intravenous midazolam: a phase i, randomized, open-label, crossover study in healthy volunteers. Clin. Therapeut. 31, 286–298, 2009.

196. Pradelli L, Povero M, Burkle H, et al. Propofol or benzodiazepines for short-and long-term sedation in intensive care units? An economic evaluation based on meta-analytic results Clinicoecon. & Outcomes Res. 9, 685–698, 2017.

197. Dinis-Oliveira RJ. Metabolic profiles of propofol and fospropofol: clinical and forensic interpretative aspects. Biomed. Res. Int. 6852857, 1–16, 2018.

198. Hanley MJ, Masse G, Harmatz JS, et al. Effect of blueberry juice on clearance of buspirone and flurbiprofen in human volunteers. Br. J. Clin. Pharmacol. 75, 1041–1052, 2013.

199. Hohenegger M, Drug induced rhabdomyolysis. Curr. Opin. Pharmacol. 12, 335–339, 2012.

200. Danielak D, Karaźniewicz-Łada M, Główka F. Assessment of the risk of rhabdomyolysis and myopathy during concomitant treatment with ticagrelor and statins. Drugs 78, 1105–1112, 2018.

201. Grosso F, D'Incalci M, Cartoafa M, et al. A comprehensive safety analysis confirms rhabdomyolysis as an uncommon adverse reaction in patients treated with trabectedin. Canc. Chemother. Pharmacol. 69, 1557–1565, 2012.

202. Uesawa Y, Mohri K. Quantitative structure-interaction relationship analysis of 1,4-dihydropyridine drugs in concomitant administration with grapefruit juice. Pharmazie 67, 195–201, 2012.

203. Mishima E, Maruyama K, Nakazawa T et al. Acute kidney injury from excessive potentiation of calcium–channel blocker via synergistic CYP3A4 inhibition by clarithromycin plus voriconazole. Intern Med. 56(13): 1687–1690, 2017.

204. Westphal JF. Macrolide – induced clinically relevant drug interactions with cytochrome P-450A (CYP) 3A4: an update focused on clarithromycin, azithromycin and dirithromycin. Br. J. Clin. Pharmacol. 50(4), 285–295, 2000.

205. Gupta R, Rahman MA, Saeed M, et al. Symptomatic bradycardia and postural hypotension. Postgrad. Med. J. 80, 679–680, 2004.

206. Martín-Pérez M, Gaist D, de Abajo FJ et al. Population impact of drug interactions with warfarin: a real-world data approach. Thromb Haemost. 118(3), 461–470, 2018.

207. Godon RLN, Leitã GM Araújo NB et al. Clinical and molecular aspects of breast cancer: Targets and therapies a molecular prospection and bioinformatics. Biomed. & Pharmacother. 106, 14–34, 2018.

208. Bhatnagar AS, The discovery and mechanism of action of letrozole. Breast Canc. Res. Treat. 105, 7–17, 2007.

209. Early breast cancer trialists' collaborative group (EBCTCG)* Aromatase inhibitors versus tamoxifen in early breast cancer: patient-level meta-analysis of the randomised trials. Lancet 386, 1341–1352, 2015.

210. Miller WR, Larionov AA. Understanding the mechanisms of aromatase inhibitor resistance Breast Canc. Res. 14, 201, 1–11, 2012.

211. Alex AB, Pal SK, Agarwal N. CYP17 inhibitors in prostate cancer: latest evidence and clinical potential. Ther. Adv. Med. Oncol. 8, 267–275, 2016.

212. Norris JD, Figg WD, McDonnell DP. Androgen receptor antagonism drives cytochrome P450 17A1 inhibitor efficacy in prostate cancer. J. Clin. Invest. 127, 2326–2338, 2017.

213. Bonomo S, Hansen CH, Petrunak EM, et al. Promising tools in prostate cancer research: selective nonsteroidal cytochrome P450 17A1 inhibitors. Scient. Rep. 6, 29468 1–11, 2016.
214. FDA approves abiraterone acetate in combination with prednisone for high-risk metastatic castration-sensitive prostate cancer. https://www.fda.gov/drugs/informationondrugs/approveddrugs/ucm596015.htm
215. Rehman Y, Rosenberg JE. Abiraterone acetate: oral androgen biosynthesis inhibitor for treatment of castration-resistant prostate cancer. Drug Des. Dev. Ther. 6, 13–18, 2012.
216. Vasaitis TS, Bruno RD, Njar VCO. CYP17 inhibitors for prostate cancer therapy J. Ster. Biochem. Mol. Biol. 125, 23–31, 2011.
217. Gupta S, Nordquist LT, Fleming MT, et al. Phase I study of seviteronel, a selective CYP17 lyase and androgen receptor inhibitor, in men with castration-resistant prostate cancer. Clin Canc. Res. 24, 5225–5232, 2018.
218. Tseng A, Hughes CA, Wu J, et al. Cobicistat versus ritonavir: similar pharmacokinetic enhancers but some important differences. Ann.Pharmacother. 51, 1008–1022, 2017.
219. Nguyen T, McNicholl I, Custodio JM, et al. Drug interactions with cobicistat-or ritonavir-boosted elvitegravir. AIDS Rev. 18(2), 101–11, 2016.
220. Gee S, Howes O. Optimising treatment of schizophrenia: the role of adjunctive fluvoxamine. Psychopharmacol. 233, 739–740, 2016.
221. Gahr M, Rehbaum K, Connemann BJ. Clozapine-associated development of second-onset obsessive-compulsive symptoms in schizophrenia: impact of clozapine serum levels and fluvoxamine add-on. Pharmacopsychiat. 47, 118–120, 2014.
222. Kishi T, Hirota T, Iwata N. Add-on fluvoxamine treatment for schizophrenia: an updated meta-analysis of randomized controlled trials. Eur. Arch. Psychiat. Clin. Neurosci. 263, 633–641, 2013.
223. Laine K, Tybring G, Hartter S, et al. Inhibition of cytochrome P4502D6 activity with paroxetine normalizes the ultrarapid metabolizer phenotype as measured by nortriptyline pharmacokinetics and the debrisoquin test. Clin. Pharmacol. Ther. 70, 327–335, 2001.
224. Mazaleuskaya LL, Sangkuhl K, Thorn CF, et al. Pharm GKB summary: Pathways of acetaminophen metabolism at the therapeutic versus toxic doses. Pharmacogenet Genomics. 25, 416–426, 2015.
225. Coleman MD. Dapsone: modes of action, toxicity, and possible strategies for increasing patient tolerance. Brit. J. Dermatol. 129, 507–513, 1993.
226. Coleman MD, Breckenridge AM, Park BK. Bioactivation of dapsone to a cytotoxic metabolite by human hepatic microsomal enzymes. Brit. J. Clin. Pharm. 28, 389–395, 1989.
227. Coleman MD, Coleman NA. Drug-induced methaemoglobinaemia. Drug Saf. 14, 394–405, 1996.
228. Coleman MD, Rhodes LA, Scott AK, et al. The use of cimetidine to reduce dapsone-dependent methaemoglobinaemia in dermatitis herpetiformis patients. Brit. J. Clin Pharmac. 34, 244–249, 1992.

229. Coleman MD Winn MJ, Breckenridge AM, et al. Inhibition of dapsone-induced methaemoglobinaemia in the rat. Biochem. Pharmac. 39, 802–805, 1990.

230. Tingle MD, Coleman MD, Park BK, Effects of pre-incubation with cimetidine on the N-hydroxylation of dapsone by human liver microsomes. Brit. J. Clin Pharmac. 32, 120–124, 1991.

231. Coleman MD, Scott AK, Breckenridge AM, et al. The use of cimetidine as a selective inhibitor of dapsone N-hydroxylation in man. Brit. J. Clin. Pharmac. 30, 761–767, 1990.

232. Coleman MD, Tingle MD, Park BK. Inhibition of dapsone-induced methaemoglobinaemia by cimetidine in the rat during chronic dapsone administration. J. Pharm. Pharmacol 43, 186–190, 1991.

233. Scheinfeld N. Cimetidine: A review of the recent developments and reports in cutaneous medicine. Dermatol Online J. 9(2), 4, 2003.

234. Goolamali SI, Macfarlane CS. The use of cimetidine to reduce dapsone-dependent haematological side effects in a patient with mucous membrane pemphigoid. Clin. Exper. Dermatol. 34, E1025–E1026, 2009.

235. Corallo C, Coutsouvelis J, Avery S, et al. Dapsone and azole interactions: A clinical perspective. J. Oncol. Pharm. Practice. 24(8), 1–4, 2017.

236. Coleman MD, Taylor CT. Effects of dihydrolipoic acid (DHLA), α-Lipoic acid. N-Acetyl cysteine and ascorbate on xenobiotic-mediated methaemoglobin formation in human erythrocytes in-vitro. Env. Tox. Pharmacol. 14, 121–127, 2003.

237. Keogh A, Spratt P, McCosker C, et al. Ketoconazole to reduce the need for cyclosporine after cardiac transplantation. New Eng. J. Med. 333, 628–633, 1995.

238. Greenblatt HK, Greenblatt DJ. Liver injury associated with ketoconazole: review of the published evidence. Journal Clin. Pharmacol. 54, 1321–1329, 2014.

239. Skinner MD, Lahmek P, Pham H, et al. Disulfiram efficacy in the treatment of alcohol dependence: a meta-analysis. PLoS ONE 9(2), e87366, 2014.

240. Skrott Z. Mistrik M, Andersen KK. Alcohol-abuse drug disulfiram targets cancer via p97 segregase adaptor NPL4. Nature. 552, 7684, 194–199, 2017.

6 Conjugation and Transport Processes

6.1 Introduction

You might recall from Chapters 1 and 2 that drugs and many endogenous chemicals, such as steroids, are essentially oil-soluble (lipophilic) agents that exploit their lipophilicity and stability to carry out their biological function, crossing membranes, binding specific carrier molecules and finally entering cells to bind to the appropriate receptors. This lipophilicity and stability also means that they are very hard to control. In this context, control entails the termination of biological function and subsequent removal from the body. With steroid hormones, for example, the ability to modulate synthesis and destruction allows the exertion of exceedingly fine control over the structural and biochemical changes they promote. As radical chemistry was involved in their assembly, body clearance of such stable molecules means that equally radical chemistry must also be applied to change these oil-soluble agents to water solubility.

Hence, the objectives of metabolizing systems could be summed up thus:

- Terminate the pharmacological effect of the molecule.

- Make the molecule so water-soluble that it cannot escape clearance, preferably by more than one route, to absolutely guarantee its removal.

These objectives could be accomplished by taking these steps:

- Change the molecular shape so it no longer binds to its receptors.

- Change the molecular lipophilicity to hydrophilicity to ensure high water solubility.

- Make the molecule larger and heavier, so it can be eliminated in bile as well as urine.

Human Drug Metabolism, Third Edition. Michael D. Coleman.
© 2020 John Wiley & Sons, Inc. Published 2020 by John Wiley & Sons, Inc.

- Deploy efflux pump systems, which ensure that a highly water-soluble metabolite actually leaves the cell to enter the bloodstream before it is excreted in bile and urine.

The CYP system ensures that virtually all lipophilic (and many hydrophilic) molecules can be oxidized and made at least slightly more water-soluble. However, many hydroxylated metabolites are not water-soluble enough to ensure that they remain in urine when filtered by the kidney and not be reabsorbed into the surrounding lipophilic tissue of the collecting tubes. So CYP-mediated metabolism can increase hydrophilicity, but it does not always increase it enough and it certainly does not make the molecule any bigger and heavier; indeed, sometimes the molecule becomes lighter as alkyl groups are removed during O, N, and S-dealkylations. As we have noted previously, CYP-mediated metabolism does not always alter the net pharmacological effects of a drug. In the case of the benzodiazepines, and some opiates, for example, metabolites do exert significant pharmacological effect.

However, CYPs do perform two essential tasks: the initial destabilization of the molecule, creating a 'handle' on it. A crude analogy would be to liken a stable lipophilic molecule to a solid block of steel, and the CYP would be the high-speed drill that bores a hole in it, so that a hook or bolt could be attached. CYPs also 'unmask' groups that could be more reactive for further metabolism. The best examples of this would be the various dealkylation reactions. These reveal amine, hydroxyl, and sulphide groups that can undergo more metabolism to make the molecule heavier, a different shape, and more water-soluble.

This CYP-mediated preparation can make the molecule vulnerable to the attachment of a very water-soluble and plentiful agent to the drug or steroid, which accomplishes the objectives of metabolism. This is achieved through the attachment of a modified glucose molecule (glucuronidation), or a soluble salt such as a sulphate (sulphation) to the prepared site. Both adducts usually make the drug into a stable, heavier, and water-soluble ex-drug.

It is important not to think of oxidative and conjugation reactions as part of an inevitable progression that a drug must follow, such as a 'Phase I' metabolite must enter 'Phase II' before elimination. Some oxidative metabolites are more than soluble enough to be found in urine and some conjugated metabolites do not require any prior oxidation 'preparation' before they are conjugated. However, many drugs are so stable and lipophilic that they do need several metabolic operations to make them water-soluble.

A final problem is created by the formation of highly water-soluble metabolites, in that they can be too hydrophilic to easily leave the cell. The control of this process is sometimes termed Phase III of metabolism. In this case, a series of molecule pump systems have evolved, or efflux transporters, which provide a powered gradient to encourage the egress of these molecules into the interstitial cell fluid. Once out of the cell, there is no way back for such hydrophilic molecules and from the blood they are filtered by the kidneys into urine.

6.2   Glucuronidation

6.2.1   UGTs

Glucuronidation is the largest capacity conjugative clearance system in humans and is accomplished by a set of enzymes known as UDP (uridine diphosphate) glucuronosyl transferases, or UGTs. These are found in many tissues, but the greatest concentration is in the liver, followed by the gut. Their main homeostatic roles are in the control of bilirubin, bile acid and steroid hormone conjugation, which usually eliminates their pharmacological effects by changing their physicochemical properties and structures, thus facilitating clearance into bile and urine. These functions are particularly important, as bilirubin and bile acids are toxic if they accumulate and of course failure to clear steroid hormones results in loss of homeostatic control of several processes[1]. Whilst many gains have been made in the understanding of UGTs in terms of their structure and function, our knowledge of CYPs is much more extensive. Indeed, a complete UGT has not been crystallized. This is changing as the roles of UGTs in areas such as anticancer drug resistance are gradually unravelled[2]. However, the study of UGTs *in vitro* has also been hampered by the standard experimental conditions of isolation of the SER (microsomes, see Chapter 3.3.2), which appears to render their activity rather unphysiological, which then means that *in vivo* activity is underpredicted, a problem known experimentally as *latency*[3]. It is likely that the application of similar techniques used to explore CYP activity more realistically (nanodiscs, Chapter 3.3.2) might be applied to UGTs will make estimates of *in vivo* capacity and impact on clearance more realistic.

Virtually all of a UGT, including its active site, is exposed to the lumen of the SER, rather like seaweed attached to underwater rocks. They are thought to be able to waft around and interact with the ER lipids whilst remaining anchored in the ER (Figure 6.1). UGTs are like several other enzymes (it seems to me, anyway) and are vaguely butterfly-shaped, with a large standard (highly conserved) C-terminal 'wing' that binds the co-factor (UDPGA, see below) which is also attached to the SER membrane. The other 'wing' of the butterfly is the N-terminal section, which varies structurally to bind different substrates[3]. Depending on their lipophilicities, the connection to the SER might act like a 'scoop' or access channel (rather like the CYPs) to facilitate the entry of some substrates, although more polar substrates will approach the N-terminal site luminally also. The luminal positions of the UGTs places them in close proximity to the CYP isoforms and it is thought that the two enzyme systems interact and can regulate each other[4], so the oxidation and conjugation activities can be coordinated to respond to the presence of certain molecules. Complex interactions between the lipid ultrastructure of the SER and the UGTs also regulate their enzymatic activity, although this is far from being understood[3]. Although it is not always necessary for the CYPs to oxidize the drug prior to glucuronidation (Figure 6.1), it is logical that the two systems should co-regulate at the enzymatic level. UGTs use

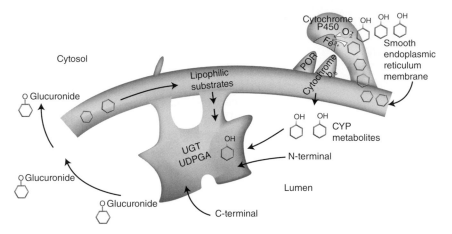

Figure 6.1 Scheme showing the approximate position of glucuronyl transferases (UGTs) in the smooth endoplasmic reticulum in relation to that of CYPs

an activated glucose derivative (UDPGA) as a co-factor to convert a huge number of chemicals to beta-D-glucopyranosiduronic acids, or glucuronides, by a nucleophilic substitution reaction[3].

6.2.2   UGT mode of operation

All UGT systems process their substrates in the same basic way (Figure 6.2). The enzyme functions by attaching a polar glucose derivative in the most favourable position on the substrate, as glucose itself is not reactive enough to achieve this unless it is activated in some way. Preparation of the glucose thus makes it thermodynamically easy for the enzyme to employ during catalysis and UDPGA formation takes place in the cytoplasm away from the SER.

Glucose-1-phosphate is reacted with uridine triphosphate (UTP) to eventually form uridine diphosphate-glucuronic acid (UDPGA). The problem of how to force a very polar co-factor through the SER wall and into its lumen is solved by a series of transmembrane pumps that include the nucleotide sugar transporters (NSTs), such as UDP-galactose transporter-related protein 7 (UGTrel7). These pumps are similar to OATPs, in that they are an automatic 'revolving door' exchange system that does not require ATP. The UDPGA enters and exchanges with UDP-N-acetylglucosamine, which is then forced out into the cytoplasm. The rate of UGT isoform cycling and product formation is closely linked to the supply of its co-factor[3] – in enzyme parlance, UDPGA is rate-limiting for UGT[3].

As mentioned earlier, UGTs use nucleophilic attack to glucuronidate substrates, attacking hydroxyl groups attached to alcohols, or phenols, as well as sulfuryl, carbonyl, carboxyl, and a range of various aromatic amino groups. This gives them

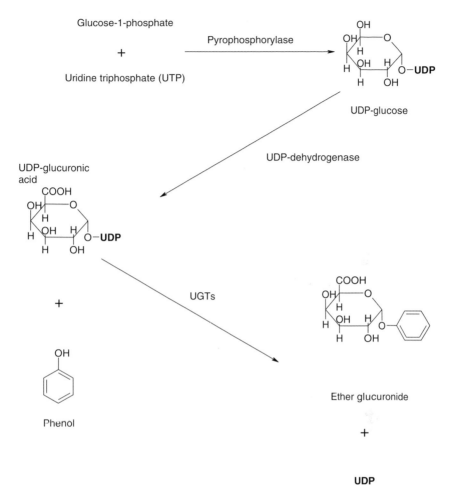

Figure 6.2   Process scheme for the preparation of glucose to UDP glucuronic acid, the co-factor for UGT-mediated formation of an ether glucuronide of phenol

a very wide range of target molecules so that they form O-, N-, or S-glucuronides, which might vary in stability, but they usually lead to detoxification and successful elimination in urine[2]. The basic reaction with phenol is shown below:

$$\text{Phenol} + \text{UDPGA} \rightarrow \text{Phenolic glucuronide} + \text{UDP}$$

This reaction is actually reversible, but the SER system has evolved to ensure that it cannot reverse, as although they have not been conclusively identified, OATP-like anion transporters sweep the glucuronides out of the SER into the cytoplasm extremely rapidly[3], whilst the UDP is also pumped out and recycled. Some β-glucuronidases are found in the ER and they convert hormone glucuronides back to the parent drug as part of their shipment around cells. It is thought that loss of the main production of SER glucuronides to β-glucuronidases is likely to be trivial given the amounts of glucuronides, which are formed and how quickly they are ejected from the SER.

6.2.3   UGT isoforms

Although these enzymes had been documented for many years, little detailed information about them was available until cDNA cloning technology allowed the expression of large quantities of the enzymes, so enabling the study of the different UGT isoforms. It emerged that in the same way different families of CYPs had evolved to metabolize broad groups of substrates; the same system applies to UGTs. Of the human UGT superfamily, there are four families and two are relevant to drug metabolism, UGT1 and UGT2. Of these, three subfamilies UGT1A, UGT2A, and UGT2B[3,5] are responsible for the clearance of endogenous and xenobiotic agents. As mentioned earlier, the UGTs have a fairly standard C-terminal co-factor binding area whilst the N-amino termini confer their substrate specificity. They have been extensively studied by expressing the human enzymes in various cellular heterologous systems. These are systems where human genes have been inserted into different bacterial and eukaryotic cell systems and the enzymes expressed in large enough amounts to study specificity and catalytic activity (Appendix A.4.3). Regarding the UGT1A subfamily, this originates with a single gene that is located on chromosome 2 that can give rise to nine functional variants, UGT1A1, and UGT1A3–10. The UGT1A7–10s are much closer in structural similarity to each other than the others in this subfamily[6]. Of the UGT2 families (originating on chromosome 4), there are three relevant UGT2A variants (UGT2A1–3), and seven UGT2B isoforms (UGT2B4, 7, 10, 11, 15, 17, and 28). Whilst the liver and gut are rich in these enzymes, they are found elsewhere also, such as the brain and the lung. The UGTs share so many structural and catalytic similarities that one substrate might be metabolized by many isoforms, to the point that it is very difficult to find out which one is predominant. However, whilst paracetamol is cleared by at least four isoforms (UGTIA1, 6 and 9, as well as UGT2B15), other substrates are cleared by only one isoform, such as UGT1A1 and bilirubin metabolism[2], so some UGTs have quite specific preferences, as we will see. Some important UGT1A and UGT2B isoforms and some of their substrates are summarized below.

UGT 1A family

UGT 1A1 (hepatic/lung)
This heavily studied and vital isoform is a very efficient metabolizer of bilirubin, oestrogenic endogenous steroids, thyroxine (T4), as well as ethinyloestradiol (not androgens or other steroid molecules). Other xenobiotics processed include

paracetamol, gemfibrozil, some statins (atorvastatin), some metabolites of buprenorphine and statins, plus large aromatic carcinogenic hydrocarbons such as benzopyrenes, particularly in the lung. Anticancer agents such as etoposide, belinostat, and SN-38 (the active metabolite of irinotecan) are also cleared by this variant of UGT[3,5]. UGT1A1 has a major role in the clearance of several antiretrovirals, such as the newer agents, dolutegravir and atanazanvir, as well as older agents like indinavir[1]. Overexpression of UGT1A1 and other isoforms is now recognized as a significant contributor to anticancer drug resistance, such as with irinotecan[1].

UGT1A3-10

UGT1A3, processes carboxylic acids, bile acids, thyroxine (T4), some tertiary amines and flavonoids, as well as some statins and phenolics. UGT1A4 does not clear small phenolics[2] but prefers bile acids and aromatic amines, trifluoperazine, trycyclic antidepressants, such as imipramine, as well as midazolam (N-glucuronidated) and around 70% of lamotrigine conjugation[1]. UGT1A3 & 4 convert tamoxifen into an active 4-OH N'-glucuronide[1]. UGT1A5 is extrahepatic, but has not been well investigated. UGT1A6 (found also in the stomach, kidney, and brain) clears small planar, phenolics like naphthols and paracetamol, as well as the anti-cancer agent methotrexate[1] and the antihypertensive telmisartan. This isoform also inactivates morphine to its 3-glucuronide. UGT1A7 is extrahepatic and is located mostly in the stomach and oesphagus[6] and, like UGT1A6, is also found in the blood–brain barrier. Reflecting the emergence of our understanding of UGT expression in malignancy, UGT1A7 overexpression is now recognised as a prognostic indicator of pancreatic cancer[7]. UGT1A8 is expressed in the gut, but at a lower level than other UGTs. It is known to clear raloxifene, along with a minor fraction of etoposide[1] and it inactivates both morphine (to the 3-glucuronide) and tamoxifen (to a trans-4-OH glucuronide). UGT1A9 is also found in the kidney and clears various phenolics and coumarins, paracetamol, flavones, amines, SN-38, the diuretic furosemide, telmisartan and raloxifene. UGT1A9 was linked to irinotecan resistance due to its overexpression and clearance of SN-38, although UGT1A1 is more important in this context[1]. Vasculature regulating eicosanoid concentrations are controlled partly by UGT1A9. UGT1A10 is extrahepatic and is highly expressed in the intestines. This isoform clears oestrogens, as well as raloxifene and SN-38. UGT1A8-10 together make a significant contribution to the clearance of many drugs in the gut, including naloxone and some β-blockers like propranolol, as well as dietary antioxidants such as resveratrol and quercetin[6].

UGT 2 family

UGT2A1-3

The UGT2A series was believed to be linked to the glucuronidation of small aromatics in nasal tissue, but UGT2A1-3 were found to be widely distributed in many tissues within the respiratory tract, as well as the liver, colon, lung, and kidney. UGT2A1 & 2 can metabolise PAHs and steroids, whilst UGT2A3 can glucuronidate smaller aromatics. Whilst UGT2A1 particularly is thought to be important for the detoxification of PAHs from tobacco and atmospheric pollution[8], the main endogenous function of the three UGT2As is to detoxify bile acids.

UGT 2B7

This isoform is the best-documented UGT in the 2B family and was the first (and so far the only one) to have its C-terminal crystallized, which confirmed the UDPGA binding site structure[3]. To give some idea of its capacity, when AZT was first introduced it had to be taken several times a day as it was glucuronidated by UGT2B7. AZT is now used as a probe for this isoform[9]. UGT2B7 is extensively expressed in the liver, and also it comprises about 20% of the expression of gut UGTs, although it is found elsewhere such as the kidney and blood-brain-barrier[3]. UGT2B7 clears many bile acids and oestrogenic steroids, as well as agents such as naproxen, hydroxymidazolam, gemfibrozil, and anticonvulsants such as carbamazepine. UGT2B7 also metabolises anticancer agents such as epirubicin[3] and also anti-HIV drugs. Indeed, UGT2B7 has such a strong impact on efavirenz disposition and efficacy that its activity governs clinical outcomes[10]. UGT2B7 is well documented in its clearance of opioids; whilst like other UGTs it clears morphine to its inactive 3-glucuronide, it alone forms the super-potent analgesic morphine-6-glucuronide[1], which is actually responsible for morphine's analgesia, to the point that morphine is now regarded as a prodrug[11]. Whilst knowledge of inhibitors of UGTs is not extensive, UGT2B7 is inhibited by fluconazole[12] and is competitively blocked by long-chain fatty acids[9], which might provide some clues as to how it is regulated metabolically. UGT2B7 has been investigated to provide more information on its expression and capability in the human liver, with a view to predicting its participation in the clearance of newer and older drugs[9].

UGT2B10

UGT2B10 has attracted much attention in recent years, as it clears about 90% of cotinine, the main metabolite of nicotine, to an N-glucuronide[12]. Indeed, UGT2B10 is, like UGT1A4, rubbish at forming O-glucuronides of the usual UGT substrates such as steroids and various hydroxylated drugs. However, if a specific leucine residue is changed for a histidine in UGT2B10's N-terminal binding site, it can suddenly produce those O-glucuronides from the phenolics[12]. It seems UGT2B10 is very similar to UGT1A4, in that it prefers aliphatic tertiary amines, as well as aromatic N-heterocyclics, such as the TCAs and desloratidine, as well as nicotine and cotinine[12]. UGT2B10 is also competitively inhibited by many of its substrates, particularly desloratidine and amitriptyline at therapeutic levels[12]. The role of UGT2B10 is likely to attract a great deal more attention in the future, given the greater understanding of the differing pharmacodynamic properties of cotinine (better CNS anxiolytic properties) with respect to nicotine[13]. Indeed, changes in the expression or activity (such as inhibition) of UGT2B10 and their impact on cotinine plasma levels[14] are likely to affect smoking behaviour quite strongly, especially in some patient groups such as those suffering from psychiatric conditions[13]. The pharmacology of tobacco will be covered in more detail in the next chapter (7.6).

UGT2B4, 11, 15, 17, and 28

UGT2B4 is almost exclusively hepatic and handles bile acids and several xenobiotics, such as catalysing the formation of O-glucuronides from the CYP3A4 product hydroxymidazolam[15]. UGT2B11 and 28 are very similar structurally[16], with both being found in the gall bladder in quite high levels, but UGT2B11 is found in the liver whilst UGT2B28 is extra-hepatic[17]. UGT2B15 and 17 gluronidate androgens, as well as phenols, flavonoids, and oxazepam; UGT2B15 activity is linked with initiation processes in prostate cancer[1].

6.2.4   UGTs and bilirubin

As with all biotransforming enzymes, the endogenous functions of UGTs are part of cellular 'housekeeping'. Of the 19 human UGT relevant to drug metabolism, probably the most important is UGT1A1, which plays a key homeostatic role in bilirubin processing. Once erythrocytes have reached the end of their lifespan, they are dismantled and the major waste product is the iron-containing haem from haemoglobin. The haem is converted to biliverdin by haem oxidase, followed by conversion to bilirubin by a reductase. Bilirubin is very lipophilic and can enter the brain to cause neurotoxicity, so it cannot be allowed to accumulate. But it does react rapidly with reactive oxygen species, forming biliverdin, which is then re-reduced to bilirubin[5]. Indeed, bilirubin is such a good antioxidant that those who have relatively high (but non-neurotoxic) plasma levels have a lower risk of developing heart disease and even cancer; indeed, some authors have highlighted that for a 'waste' product, a suspiciously large amount of cellular energy goes into making it. This supports the contention that bilirubin is a seriously effective physiological asset. If you would like to increase your own bilirubin levels, get into the habit of sustained exercise, as this causes higher bilirubin levels that, in turn, promote longevity[18]. If you want to know for some reason how many mg of bilirubin your system forms every day, just multiply your weight in kilograms by four[19].

Bilirubin is taken up by the liver from its sinusoidal blood flow through OATP1B1 and 3,[20]. Overall, the balance of bilirubin plasma levels and excretion into bile is crucial, and the main function of UGT1A1 and the transporters is to manage bilirubin plasma levels by converting it to water soluble conjugates. UGTs actually reside in the SER rather like torch (flashlight) batteries in packs of two (dimers) and four (tetramers). The advantage in this is that UGT dimers tend to form bilirubin monogluconides, but the UGT tetramers can concentrate the bilirubin monoglucuronide and form bilirubin diglucuronides. These are hydrophilic enough to be cleared in bile after they are pumped out of the SER and eventually eliminated into the sinusoids in the bile by a type of transporter known as multi-drug-resistance proteins (MRP1 and 3)[20]. If there is a problem with bile flow, the MRPs can pump the bilirubin glucuronides back into the blood[19].

As it is the only UGT capable of handling bilirubin, impairment of UGT1A1 expression can have life-threatening consequences. The most severe condition is Crigler–Najjar syndrome, classed as CN-1 and CN-2. CN-1 is fatal (without a liver transplant) due to non-expression of UGT1A1, because of a fault in the UGT1A1 promoter system. CN-2 is a less dangerous version with reduced UGT1A1 expression. A milder polymorphism of UGT1A1 is Gilbert's syndrome, which is another promoter defect (UGT1A1*28) that restricts expression of UGT1A1 to less than 30% of normal. The condition is highlighted by lower clearance of major drug glucuronidation substrates like paracetamol, ethinyloestradiol and gemfibrozil[5,18]. This is predominantly found in those of African ancestry, as well as other populations to varying degrees (Chapter 7.2.7). Any polymorphic impairment in UGT1A1 (Chapter 7.2.7) predisposes a patient to gut toxicity from the main metabolite of irinotecan, SN-38, which is normally cleared to a glucuronide by this isoform. Even a partial inhibitor of UGT1A1 such as atazanavir in Gilbert's syndrome patients under treatment for HIV, will cause jaundice[18]. UGT expression levels differ widely, so conferring different degrees of bilirubin-mediated antioxidant protection, which appear to outweigh any possible negative issues (e.g. gastrointestinal issues) related to hyperbilirubinaemia and any vulnerability to xenobiotic-mediated toxicity. Indeed, Gilbert's syndrome halves the risk of cardiovascular-linked mortality[18].

Around 20–30% of neonates can be jaundiced due to bilirubin accumulation, which usually clears in a few days, unless the baby has one of the conditions above involving impaired UGT1A1 expression[21]. However, if the condition persists for 14 days or more, clinical investigations are necessary[22]. Indeed, if bilirubin levels keep rising, then potentially brain damage (kernicterus) may result[5]. In the past, UGT inducers like phenobarbitone were used to accelerate bilirubin clearance, but currently phototherapy is more usual with blood transfusions *in extremis*.

6.2.5   UGTs and bile acids

Several UGTs (UGT1A3, UGT1A4, UGT2A1-3, UGT2B4, and UGT2B7) are responsible for the clearance of around a third of bile acids. These amphipathic (detergent-like) molecules facilitate the absorption of fats, various fat-soluble vitamins and cholesterol. They are made from cholesterol, and their conservation is normally extremely efficient, as extensive recycling means daily losses are less than 5%. However, due to their hydrophobicity and detergent effects they can be hepatotoxic, probably by damaging membrane integrity. UGTs are part of the process to control their concentrations[23]. Bile acids partly regulate bile flow, so if routes of clearance such as glucuronidation were to be impaired, accumulation would occur, which can lead to cholestasis, or shutdown of bile flow. Although there are many different routes to cholestasis, ranging from liver malignancies to biliary cirrhosis, it can cause serious liver damage and organ failure. In addition, cholestasis is often associated with severe chronic pruritis (itching), which can drive patients almost to suicide and impacts quality of life as much as intractable pain[24]. This is believed to be linked with bile salts as well as bilirubin. The itching is difficult to control and patients are not always responsive; cholestyramine can be effective as it sequesters the bile salts but is associated with severe constipation, whilst sertraline has also been used. Rifampicin induces the bile salts CYP3A4-mediated 6α-hydroxylation and the UGTs simultaneously, so promoting systemic clearance of the salts. There is less than a one in twenty chance of hepatotoxicity with rifampicin in this context, but it resolves on cessation of the drug, so it remains a good second-line choice to control hepatic itch[25].

6.2.6 Role of glucuronidation in drug clearance

With many xenobiotics, glucuronidation appears to happen after some form of oxidative metabolism; however, there are a number of drugs that are cleared virtually entirely by direct glucuronidation, without any prior oxidative metabolism. These include lorazapam, oxazepam, temazepam, and morphine (Appendix B); as stated earlier, the latter narcotic is cleared to a 3 and a 6-glucuronide. Glucuronidation is the major route of clearance for a vast array of chemicals: these include endogenous substances such as steroids, bilirubin, bile acids, fatty acids, retinoids and prostaglandins, as well as environmental pollutants, dietary constituents, and, of course, drugs. Chemically, phenolics, carboxylic acids, hydroxylamines, amines, opioids, and exogenous steroids can all be conjugated with glucuronic acid by these enzymes[3]. The products are extremely water-soluble and, provided that they are stable, they ensure that the agent is cleared into urine or bile. In general, glucuronides are so radically different in shape and water solubility from their parent drugs that they do not exert any pharmacological effect. The most often quoted exception as mentioned above is morphine-6-glucuronide. UGTs have a seriously high capacity and are practically unsaturable; the first-generation anti-HIV agent AZT is a good example of this. With the exception of UGT1A1 and bilirubin, there appears to be a much larger overlap in many of the UGT specificities compared with those of the CYPs, so even if several UGTs were poorly expressed, the glucuronidation of a drug could still occur in quantities sufficient to clear the agent from the body.

6.2.7 Types of glucuronides formed

There are several options for UGTs to insert a glucuronic acid group into a molecule. The ether glucuronide as formed from simple phenols (Figure 6.2) is one option, as is an ester (acyl) glucuronide, which forms when chemicals similar to benzoic acid are glucuronidated. With such chemicals that possess a carboxylic acid group, the UGTs will attack the hydroxyl and form the acyl glucuronide, such as in some statins and diclofenac[26] (Figure 6.3). Acyl glucuronides appear with many drugs and do not lead to significant problems. However, it is suspected that in some cases, these glucuronides are unstable and reactive (Figure 6.3), leading to covalent binding to tissue macromolecules. This in turn can cause idiosyncratic toxic reactions, either through cytotoxicity pathways or possibly even through immune toxicity (Chapter 8.3.4).[26] The process of acyl migration has been linked with this toxicological pathway, although it has been suggested that toxicity with drugs that undergo acyl glucuronidation may still be predominantly dependent on oxidative metabolism.[26] S-glucuronides are also formed, whilst glucuronidation of amines is more complex and interesting from a toxicological perspective. Aromatic amines can be glucuronidated directly on the nitrogen of the amine to form an N-glucuronide (Figure 6.4), or the amine can first be oxidized to form a hydroxylamine, where the glucuronide can be attached to either the nitrogen or the oxygen of the hydroxylamine to form an N-glucuronide or an O-glucuronide.

With most aromatic amines, the N-glucuronide is formed directly from the amine and oxidative hydroxylation is unnecessary, although the N-hydroxy (hydroxylamine) metabolites can also be glucuronidated. The question is whether this would occur on the nitrogen (N-glucuronide) or the oxygen of the hydroxylamine (O-glucuronide).

The sulphone drug dapsone, is an interesting example of the conjugative fate of an aromatic amine. In volunteers, the parent drug could be monoacetylated, which could then be N-hydroxylated along with the parent drug (CYP2C9) to form acetylated and non-acetylated hydroxylamines[27]. These hydroxylamines were then predominantly glucuronidated, as around a third of the dose is recovered in 48 hours as glucuronides[28]. Acid hydrolysis with β-glucuronidase (which splits the glucuronic acid off the drug or metabolite at acid pH) left intact dapsone hydroxylamine as the major metabolite. Irrespective of the position of the glucuronide, the hydroxylamine was just stable enough to survive after the glucuronic acid was hydrolysed away[28]. With dapsone, oxidation to the hydroxylamine is so rapid that there is probably little opportunity for the parent amine to be directly conjugated to form an N-glucuronide.

Whilst dapsone is not carcinogenic, in terms of either its oxidative or conjugative metabolites, the stability and position of glucuronides on other amines are crucial issues with regard to the potential carcinogenicity. Many N-glucuronides, unlike O-glucuronides, can be very susceptible to acid hydrolysis, which can occur in urine made acid by a diet rich in meat and dairy products. Human urine also contains β-glucuronidases, which operate most efficiently at acid pH. Those who

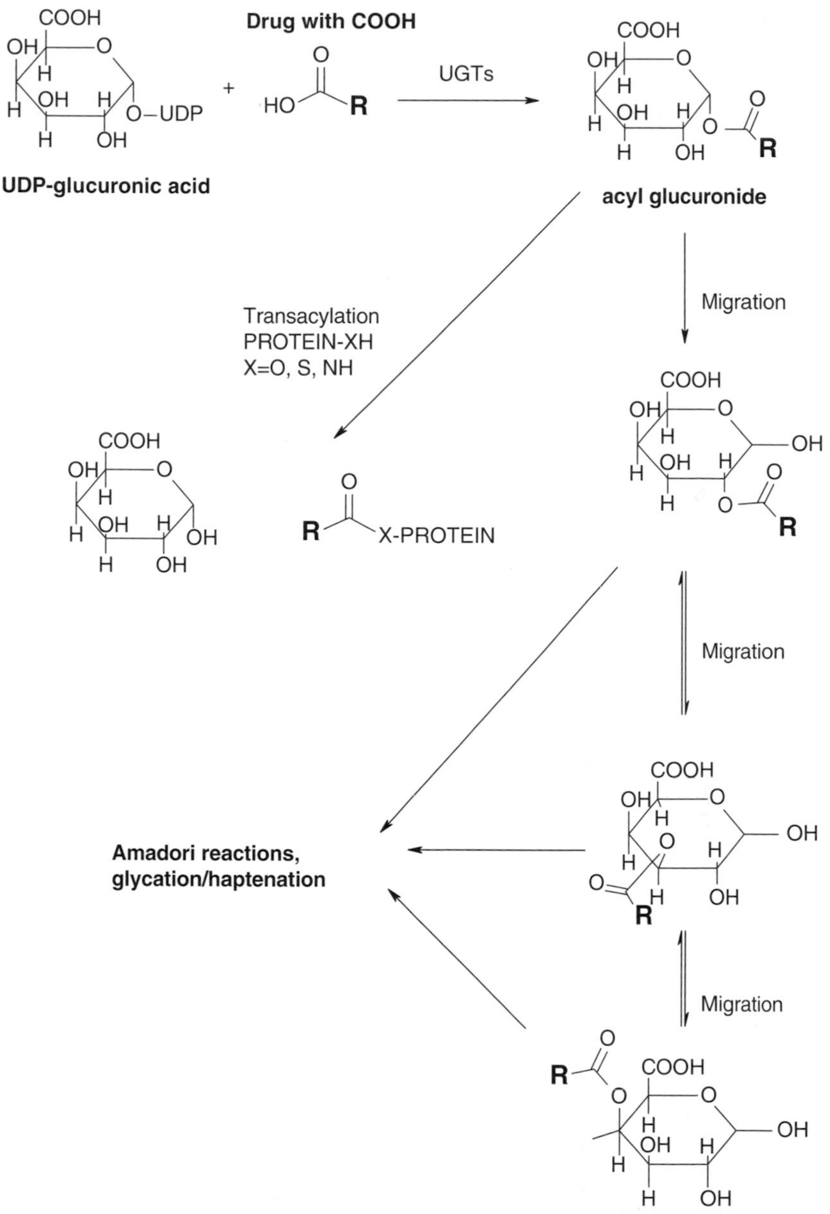

Figure 6.3 Formation of acyl glucuronides from drugs containing a carboxylic acid group. These glucuronides can be unstable, rearranging themselves to form cytotoxic reactive species

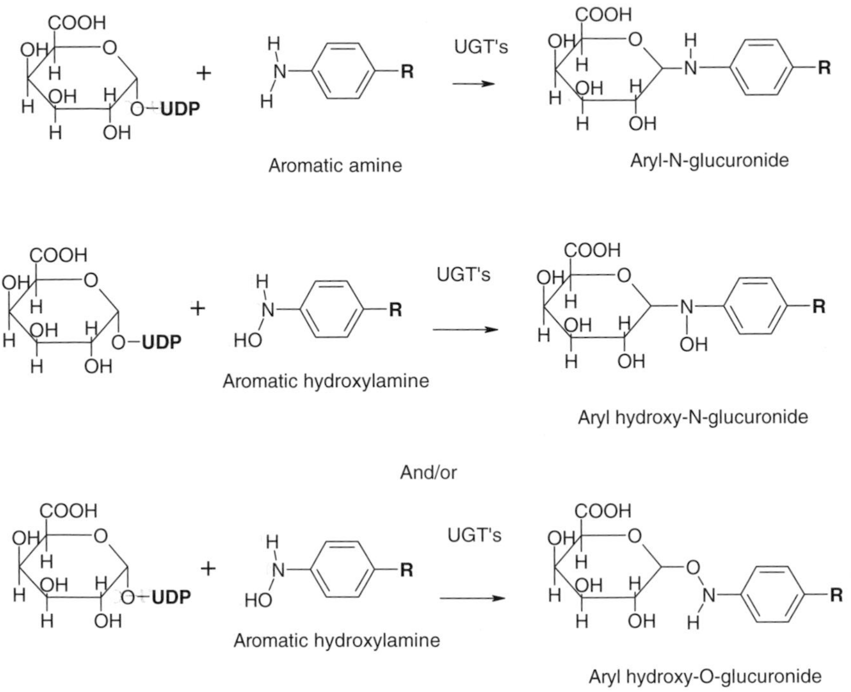

Figure 6.4 Scheme for glucuronidation of aromatic amines and hydroxylamines

have a predominantly vegetarian diet have more alkaline urine and less hydrolysis will occur. It appears that the acid hydrolysis of N-glucuronides of aromatic amines like benzidine, as well as their acetylated derivatives leads to liberation of the parent amine, hydroxylamines and the acetylated derivatives in the urine. What occurs next is complex, but one or other of the released parent compound, hydroxylamines and/or acetylated metabolites may be further oxidised by bladder epithelial enzyme systems to form reactive species[29-31] which lead to benzidine-derivative DNA adducts that can eventually lead to bladder cancer. This malignancy occurs in many workers exposed to benzidine-like compounds in various dye industries, now sited in Third World countries[29]. The UGT1A family is most involved in the glucuronidation of amine derivatives, particularly UGT1A9, as well as UGT1A4, 6 and 1, although one member of the UGT2B sub-family UGT2B7 is also implicated[29]. This is discussed in more detail in Chapter 8 (Figure 8.11).

6.2.8 Control of UGTs

As you might recall from Chapters 3 and 4, a multilayered control system has evolved to both coarse and fine-tune CYP isoform expression according to substrate 'load'. The cooperation between CYP and UGT activity has already been alluded to, so naturally glucuronidation is controlled through the same broad hepatic nuclear transcription factor and ligand binding sensor system, which is reasonable, given the capacity, importance and flexibility of this conjugative system. If a student were to be asked which nuclear receptor (NR) systems controlled glucuronidation and they were to reply 'all that we know about plus a few we don't', this answer might sound rather facetious, but it would not be far from the truth. Regarding the liver expression of the UGTs, the master transcription regulators like HNF1α and HNF4α exert overall control, with the molecular sensors like AhR, CAR, PXR and nrf2 (see later in this chapter) acting more immediately on a daily/hourly basis to respond to specific substrate/xenobiotic demand[1,32]. With extra-hepatic UGTs, expression is regulated by tissue-specific transcription factors such as CDX2 (Caudal related homeobox 2), which is the intestinal equivalent of the HNFs, in that CDX2 controls many aspects of the expression of organ identity and activity[6]. Other more exotic control factors involved with the UGTs include the somewhat elliptically named Forkhead Box A1, a transcription factor that is the link between various hormone systems and glucuronidation[1]. With regard to particular substrates, we have already established the importance of UGT1A1 in bilirubin homeostasis and this is believed to be operated by the sensing of bilirubin concentrations by AhR, as well as nrf2. Now CYPs and other biotransforming genes contain areas where various NRs and other factors can bind as they respond to a drug such as phenobarbitone, known as a PBREM (phenobarbitone responsive enhancer module). UGT1A1 is responsive to so many different NRs, that it has a large 290 base-pair cluster of transcription binding sites known as the gtPBREM (Figure 6.5).

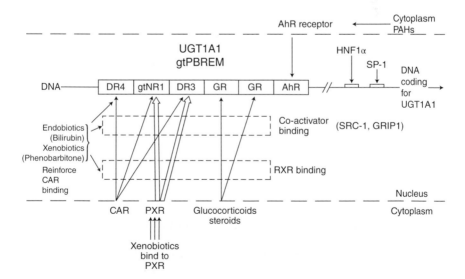

Figure 6.5   Control of UGT1A1 expression: the multi-site gtPBREM allows the isoform to respond to many receptor systems and a variety of endogenous, dietary and therapeutic stimuli

The gtPBREM (glucuronyl transferase phenobarbitone-responsive enhancer module; Figure 6.5) contains six receptor binding sites: a DR4 site that responds to CAR; a gtNR1 and a DR3 site, both of which bind CAR and PXR; two glucocorticoid receptor GR binding sites and finally a site which binds AhR[32,33]. Upstream from gtPBREM, there are two more promoter sites that bind the transcription factors HNF1 α and SP-1. CAR and PXR appear to operate in the same way that they do with the CYPs; PXR binds the ligand in the cytoplasm, such as rifampicin, and migrates to the nucleus to seek the heterodimer partner RXR. CAR operates continuously and its complexing with RXR is tightened by the presence of the ligand. In general, after the NRs recruit RXR they lock onto the DNA response elements with their co-activators such as SRC-1 and GRIP1. With gtPBREM, HNF1α must also bind at the upstream site before maximal induction can occur[33]. The gtPBREM is activated by other NRs such as GR in response to dexamethasone and other steroids, as well as AhR binding PAH's. In the gut, a combination of HNF4α and CDX2 control the expression of UGT1A8-10[6]. Bile acid elimination by UGTs is controlled by CAR and PXR, as well as by the farnesoid X-receptor and the vitamin D receptor. UGT1A3's bile acid regulation is controlled by other NRs such as the liver X-receptor (LXR) and peroxisome proliferator-activated receptor α (PPARα). The fibrate class of drugs act as agonists on this latter receptor and that is part of how they reduce plasma lipid levels. Again, these NRs seek the heterodimer partner RXR to bind DNA response elements[33]. There is evidence that the tumour suppression gene p53 can also modulate UGT activity. The UGT2B7 gene has a p53 response element in its promotor site and the anticancer agent epirubicin upregulates p53, which, in turn, induces UGT2B7, unfortunately protecting the malignant cells from the drug[34].

UGTs are also very responsive to immune system modulation through various cytokines, such as interleukins 1α and 6, as well as tumour necrosis factor α, although the expression of the UGTs can be upregulated or downregulated, according to the inflammatory process and the specific UGT isoform[2]. The impact of oxidative stress on UGT expression as regulated through such pathways as nrf2 is also unpredictable, sometimes suppressing expression, sometimes enhancing it. The full spectrum of UGT modulation through multiple systems is far from understood, but as the influence of these isoforms on drug efficacy and toxicity is studied further, UGTs' responses to xenobiotic and pathological condition exposure will be better understood.

6.2.9　Induction of UGTs: clinical consequences

Over the past decade, a great deal of research has revealed that the variety and depth of UGT responsiveness to a range of xenobiotic molecules has significant impact on clinical outcomes in a number of different therapeutic arenas. This is not just a systemic reduction in plasma concentrations due to an induction-style accelerated clearance. UGT induction has very local consequences in terms of shielding tissues and organs from drug action, as well as protecting tumours from anticancer drugs. Because polar metabolites such as conjugates cannot diffuse across membranes, UGTs are regulated in tandem with various efflux systems (Section 6.11.2), so the combination of the rapid formation of conjugates followed by cellular efflux can be extremely effective in actively preventing local drug efficacy in spite of evidence of apparently therapeutic plasma levels.

UGTs and cancer chemotherapy

Some tumour populations show much reduced UGT expression, and this lack of detoxification may be why the malignancy occurred in the first place[1]. However, in general, resistance to anti-cancer drugs can be intrinsic (the tumour gets its retaliation in first, so to speak) and/or extrinsic, where the tumour senses the agent and upregulates the expression of appropriate detoxification systems, as part of its overall resistance portfolio. If resistance is intrinsic, then prior to medication, the tumour may already overexpress batteries of biotransforming drugs and transporters, so drugs such as etoposide, irinotecan, and epirubicin may be less effective than expected. With tamoxifen, this is more complex, as whilst the O-glucuronide is inactive, the N-glucuronide contributes to the therapeutic effect on the oestrogen receptor[1].

When resistance is acquired through the tumour's detection of the agent, then either specific or broad sweeps of UGTs and efflux systems may be overexpressed as part of this process. Resistance to the dihydrofolate reductase inhibitor methotrexate is probably through AhR-induced UGT1A6 expression, and if other drugs are given in combination with this agent and they are UGT substrates, their effectiveness will be blunted[1]. With irinotecan, the levels of the active SN-38 dictate efficacy and toxicity and UGT1A1 expression has a major impact, as has been outlined. SN-38 is a PXR agonist, so not only will this metabolite induce its own accelerated clearance, it will upregulate a host of CYPs and UGTs as well as UGT1A1, which will impact the clearance and efficacy of other co-administered drugs[1]. In Chapter 4.4.3, upregulation of PXR in tumours was outlined as a potent tumour cellular defence, and it will certainly promote resistance to irinotecan. Other forms of resistance include transcription factors produced by tumour cells, such as glioma associated protein (GLI1), which can cause a post-translational stabilization of UGT1A1, increasing its activity against drugs like cytarabine[35].

UGT induction in other therapeutic areas

HIV exposure itself appears to upregulate efflux systems and several antiretroviral drugs undergo glucuronidation, such as the older agents AZT and indinavir, as well as newer drugs such as atazanavir, raltegravir and abacavir. Efavirenz not only induces multiple UGTs through CAR, it also lowers bilirubin levels and accelerates its own metabolism by UGT2B7, negatively impacting efficacy[1]. All the usual PXR/AhR inducers will accelerate the UGT-mediated clearance of these antivirals.

There is intensive interest in drug resistance in mental health, particularly with antiepileptic drugs (AEDs). As discussed in Chapter 4.4.7, there are a number of mechanisms as to how areas of the brain can prevent therapeutically effective drug residence times and MDR proteins such as P-gp is one, whilst UGTs may be another. Lamotrigine is effectively resisted in some patients through upregulated blood-brain barrier and neuronal/astrocytic expression of UGT1A4[1]. Whilst other AEDs and/or their metabolites are cleared by UGTs, such as valproate (an inhibitor, see later), phenytoin and carbamazepine, there is not so much evidence that UGTs make a significant contribution to drug resistance. There is also ongoing research to determine if UGTs are involved in resistance to antipsychotics and antidepressant drugs[1].

Glucuronidation is a major pathway for statin clearance, and other inducers will accelerate their clearance as expected. Atorvastatin is a PXR agonist and can induce its own metabolism (UGT1A1), whilst other UGTs are involved in statin clearance include UGT1A3 and not surprisingly, UGT2B7[1]. Several anti-hypertensives are UGT substrates, and clearance by these isoforms has been linked with loss of efficacy with telmisartan and fimasartan, for example. Variations in patient responses are also linked to expression of different UGTs and this will be covered in the next chapter (Section 7.2.7).

6.2.10   UGT inhibition: bilirubin metabolism

In terms of impact on the patient, a major concern is the level of inhibition of UGT1A1's handling of bilirubin. Whilst UGT inhibition is important, it has to be remembered that bilirubin has to be taken up from the blood by transporters (OATPs), before processing by UGT1A1 and then elimination into bile must occur via other transporters (MRPs). The clinical impact of an inhibitor depends on its effects on part or all of this system. Whilst we have discussed the positives of mild hyperbilirubinaemia, clearly excessive levels affect the patient through yellow skin coloration, the itching as described previously, as well as a range of other possible symptoms such as fever, diarrhoea and nausea.

Protease inhibitors

Major UGT inhibitors include some protease inhibitors, which are, with the exception of indinavir, not metabolized by UGTs[19]. This class of drugs are mixed (non-competitive/competitive) inhibitors of UGT1A1, UGT1A3 and UGT1A4, whilst they have little or no effect on UGT1A6/9 and UGT2B7[19]. Based on *in vitro* studies with UGT1A1, atazanavir is about three- to four-fold more potent than lopinavir, and more than five-fold more potent than nelfinavir and saquinavir. However, just because the drugs show inhibitory effects *in vitro* using expressed human UGT isoforms and microsomes, this does not necessarily translate into a clinical effect. Indeed, of the protease inhibitors, atazanavir and indinavir are the ones that are associated with significant clinical hyperbilirubinaemia. Of these two, atazanavir *in vitro* is about 40-50 fold more potent in UGT1A1 inhibition than indinavir[19] and this does translate into the clinic, where atazanavir is associated with higher levels of bilirubin than indinavir[36]. However, many of the protease inhibitors are not associated with significant hyperbilirubinaemia (saquinavir, ritonavir, nelfinavir) and this is partly because like saquinavir, they most likely do not inhibit the OATPs. Indinavir's inhibition of OATP1B1 means it prevents bilirubin uptake as well as its UGT1A1-mediated clearance[19], thus compounding its impact. Both atazanavir and indinavir are less protein bound than the other protease inhibitors, so their greater proportion of free drug available for inhibition of both the UGT and the transporters is also a factor in their ability to cause hyperbilirubinaemia[19]. Atazanavir and indinavir may affect the pharmacokinetics of any drug or metabolite that is cleared through UGTs 1A1/3 and 4. Also, lopinavir and ritonavir in combination can increase SN-38 levels due in part to UGT1A1 inhibition; this caused neutropenia in the patients[36]. Additionally, patients with impaired UGT1A1 expression such as in Gilbert's syndrome will be much more sensitive to hyperbilirubinaemia in the presence of some protease inhibitors[19,20,36].

Tyrosine kinase inhibitors

As mentioned previously, these drugs have been very successful in treating several cancers and many derivatives are now available for clinical use. However, they do cause significant liver issues, particularly hyperbilirubinaemia[20]. These effects are serious, leading to levels in around 5% of patients treated, which can be tenfold higher than the top end of normal[20]. Derivatives such as nilotinib, sorafinib, erlotinib, and lapatinib are the most potent inhibitors of UGT1A1, whilst imatinib is less potent. The inhibition is thought to be noncompetitive, affects the OATP/MRPs involved with bilirubin uptake and clearance and occurs at clinically relevant drug concentrations[20,37]. Again, those with genetically impaired bilirubin metabolism, like CN-2 and Gilbert's syndrome will be vulnerable to high bilirubinaemia.

6.2.11   UGT inhibition: drug clearance

The pharmaceutical industry is paying increasing attention to the impact of UGT inhibitors on both the bioavailability and systemic clearance of drugs that are mainly, or even entirely, cleared by UGTs. Systemically, the AED valproate can inhibit UGT1A4, increasing the lamotrigine AUC by two and a half fold. It also can retard lorazepam and AZT clearance[38]. Fluconazole and probenecid are UGTB7 inhibitors, and fluconazole can retard AZT clearance, whilst the anti-asthmatic zafirlukast is a broad UGT inhibitor, blocking UGT1A1, as well as UGT1A8 and 10[38]. In terms of impact on bioavailability, inhibition in the gut exerted on high capacity UGTs like UGT2B7 can cause a marked fall in pre-systemic clearance, causing very significant increases in plasma levels in the presence of the inhibitor. For example, in normal usage the oestrogen receptor modulator raloxifene is almost entirely wiped out by gut UGTs to the point that it is only 2% bioavailable. Inhibition of UGT1A1, 8 and 10 would very significantly increase plasma levels[36,38] accentuating its side effects (hot flushes and leg cramps). UGT inhibition will also impact furosemide and diclofenac bioavailability. Other UGT inhibitors include mefenamic (UGT1A9), ketoconazole, and the lipid-lowering agent gemfibrozil (UGT1A1)[2]. This lipid-lowering fibrate has gone out of favour due to various adverse reactions. Gemfibrozil was used with statins in some hyperlipidaemias, but the risks of rhabdomyolysis are fifteen-fold higher than with a statin/fenofibrate combination. This is linked with gemfibrozil's inhibition of OATP1B1 and UGT1A1. Hence, hepatic uptake of statins, as well as a main route of clearance (in the gut and liver), are both inhibited at the same time, resulting in a doubling of plasma statin levels[39]. Indeed, it has been recommended that gemfibrozil should be contraindicated with lovastatin, simvastatin and pravastatin and only used with reduced statin dosage with other drugs, such as a 10mg per day maximum with rosuvastatin[39]. Overall, the most potent UGT inhibitors (zafirlukast, the tyrosine kinase and protease inhibitors and gemfibrozil) will significantly increase the bioavailability of drugs like raloxifene, diclofenac, lorazepam, naproxen, and many statins, whilst increasing the toxicity of metabolites such as SN-38, which are cleared by UGTs.

It is not surprising that given the substrate preferences of the UGTs, dietary or nutraceutical phytochemicals will be major substrates and inhibitors. Many of these agents, either pure or in plant extracts, can inhibit human UGTs at the micromolar level. Indeed, curcuminoids are particularly potent, but many other extracts have UGT inhibitory effects *in vitro* which are likely to translate to *in vivo* if consumed in a high enough quantity[40]. Unfortunately, these agents are sold without any guidance as to their possible impact on prescription drugs. Hence, many patients assume that if something is effective at 1 g it will be 10 times better at 10 g, so as discussed in the previous chapter (Section 5.5.5) concerning other herbal preparations, it is likely that serious interactions are likely to be discovered when unfortunate patients are immortalized in case reports.

6.2.12   Microbiome and drug metabolism: passengers or crew?

In earlier versions of this book, there was a short section on enterohepatic recirculation that discussed gut bacterial action on drug conjugates that led to significant reabsorption of the parent drug. However, we now know that so-called enterohepatic recirculation is actually the tip of a very large, vitally important and beneficial bacterial iceberg indeed. The population of bacterial species that humans carry on skin, guts, mouths, nasal cavities, and vaginas is termed our microbiome. This constitutes as many cells as in our entire bodies and about a hundred times more DNA[41]; together we are actually a *supra-organism* and our microbiome is essential for our health and resistance to infection. Indeed, many phytochemicals in our diets, such as flavonoids, are actively metabolized by the microbiome to highly beneficial compounds that we immediately absorb and it is likely that we would not have been able to form these agents with our own systems[42]. On the other hand, damage to the microbiome with indiscriminate use of antibiotics, or in the case of the vagina, with cynically marketed and harmful 'hygiene' products[43], leads to significant and long-term health detriment. About 99% of these organisms reside in our guts, and this has the greatest influence on orally administered drugs.

Hence, the key issues with drug interactions with the microbiome are metabolism *prior* to absorption and further metabolism of drug-related material *after* it has been eliminated into bile.

Prior to drug absorption

Drug is metabolised by microbiome, which will potentially:

- liberate a pharmacologically active agent

- restrict parent drug entry

- form metabolites which may/may not be absorbed (which may or may not be toxic or pharmacologically active)

After absorption/metabolism

Drug metabolite enters gut via bile and is further metabolised by microbiome:

- parent drug released

- potential toxic metabolite formed/released

Microbiome drug metabolism

The various bacterial species present in the gut have as wide, if not actually wider, biotransformational capability than we have. Indeed, many microbial species can accomplish reactions that are beyond mammalian systems, such as breaking into aromatic rings. From the research perspective, the most obvious way to determine the contribution of the microbiome in drug metabolism in animal and human studies is to determine drug disposition before and then after, wiping out the bacteria with antibiotics. From animal studies, it is also likely that human gut and liver biotransforming enzyme profiles and background levels of expression (e.g. modulated by CAR, PXR, and AhR) is significantly influenced by the gut microbiome[44,45]. To date, we know that the human gut microbiome can accomplish a wide range of drug oxidative and reactions as described in Chapter 3.10–13, but the key reactions affecting prescription drugs that we understand currently are based around reductive and hydrolytic reactions. The latter group are directly relevant to enterhepatic recirculation[45,46].

Microbiome drug metabolism-clinical relevance

In terms of reductive metabolism a good example is the treatment of ulcerative colitis, where various agents such as sulphasalazine are prodrugs that rely on bacterial reduction to release the anti-inflammatory 5-aminosalicylic acid. Nitroreductase capacity is also significant, enabling the reduction of nitrogroups on drugs to amines in benzodiazepines such as nitrazepam. The frankly disturbingly named antithrombocytopenic drug eltrombopag, is pretty much cut in half by bacterial cleavage of its hydrazone linkage and in some individuals as much as 45% of the drug meets this fate, although there is marked variability in the ability of gut populations to manage this[44]. Human biotransforming enzyme systems cannot accomplish this reaction. Digoxin is not extensively metabolised by biotransforming systems in man, but at least 16% of the dose is reduced to a dyhydro derivative that human systems cannot make. Indeed, it was discovered as far back as the late 1970s that an antibiotic could actually significantly increase digoxin plasma levels, as it prevented the bacterially mediated metabolism[47]. The anti-Parkinson's drug L-dopa is decarboxylated by gut bacteria and infection with *Helicobacter pylori* can clear so much L-dopa that it actually impairs its clinical effect[48].

Of major interest is the ability of gut bacteria to hydrolyse glucuronides, often through β-glucuronidase activity[46]. Whilst this might liberate parent drug from a glucuronide, which could be reabsorbed to show a classic later 'pulse' of parent drug appearing in the plasma, it is problematic with a toxic metabolite. NSAIDs are thought to exert their well-known gut toxicity due to hydrolysis of glucuronide metabolites. As mentioned previously, irinotecan is cleared to SN-38, which is, in turn, conjugated into an inactive glucuronde. Liberation of SN-38 through hydrolysis of the glucuronide leads to serious gut toxicit[46,49] and later onset diarrhoea that is so bad some patients must be hospitalized[49]. This problem has been tackled with a number of different strategies, including the use of antibiotics to deplete the gut bacteria, to a more refined approach involving specific inhibitors of bacterial β-glucuronidase with some success in reducing the diarrhoea burden with irinotecan. However, there are drawbacks to these approaches, including the side effects of the antibiotics, damage to the microbiota, and fungal infections[49].

Enterohepatic recirculation can occur with many drugs, such as opiates, antiparasitics such as ivermectin, many dietary antioxidant flavones, and other phenolics. It can occur in steroids, although it is not thought to be clinically relevant in the cases of norethisterone and gestodene. Of course, antibiotic administration can accelerate the clearance of enterohepatically recirculated drugs by killing the gut flora responsible for the hydrolysis of the conjugates. Whilst the microbiota can influence drug disposition, it is itself very vulnerable to damage by antibiotics and other agents, such as the proton pump inhibitors (PPIs). Indeed, long-term use of PPIs predisposes patients to *Clostridium difficile* infections[46] and different drugs, particularly NSAIDs, can cause drug-specific changes in the combination of bacterial species in an individual's gut microbe population. Overall, the microbiome influences our health in terms that are far beyond the scope of this book, and as our understanding of our relationship with these organisms has grown, we should no longer see them as merely passengers, but essential members of our 'crew'[46].

6.3   Sulphonation

6.3.1   Introduction

Sulphonation is accomplished by a set of enzyme systems known as sulphotransferases (SULTs) and they are found in most tissues to varying degrees of activity. The major sites of activity are the liver, small intestine, main intestine, and colon, although they are also found in the brain and the placenta. The main detoxifying isoforms are cytosolic enzymes, but there are membrane bound forms located in the cytosolic Golgi bodies, and these forms are engaged in biosynthetic cellular housekeeping tasks rather than biotransformation. Rather than directly attach the sulphate molecule to the xenobiotic, they require sulphurylase + APS phosphokinase to manipulate the sulphate into a sulphonate (SO_{-3}) form that is thermodynamically easiest for the enzyme to attach to the xenobiotic (Figure 6.6). The first step is to 'load' the sulphate into adenosine 5′ phosphosulphate.

The general aim of sulphonation is to make the substrate more water-soluble and usually less active pharmacologically. Sulphonated molecules are more readily eliminated in bile and urine. All the SULTs are virtually identical structurally in the area of the enzyme where they bind the co-factor, although there are obviously considerable differences in the substrate binding sites, which confer on the different enzymes the ability to bind groups of similar substrates in the same manner as CYP and UGT enzymes[50].

Although sulphotransferases do not have the vast capacity that the UGT systems possess, they are very selective for certain classes of molecules and they can form large quantities of sulphate metabolites in a relatively short timescale[50]. However, the full range of the endogenous and xenobiotic-metabolizing roles of SULTs remains to be uncovered, as these enzymes are extremely important in many arenas, often because SULTs regulate many endogenous receptor coupled processes, by sulphating their agonists and changing their affinities, such as with steroids, thus 'recoding' metabolic instructions[51].

Hence, SULTs are part of the complex molecular switching systems that regulate many hormonally dependent cellular housekeeping processes[50, 51]. SULTs

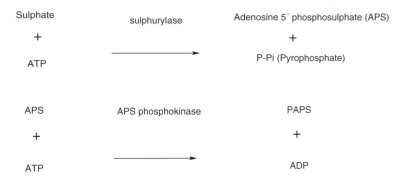

The co-factor known as PAPS (3′-phosphoadenosine-5′-phosphosulphate) acts as the final sulphate carrier.

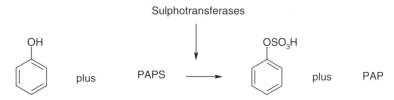

Figure 6.6 Basic reactions of sulphotransferases

are also important in bile salt processing, with the majority of sulphonated bile acids (more than 70%) cleared in urine. They are especially active at the beginning of life as they are found in high levels during foetal development in many species[50]. This is closely linked with the SULTs' role in thyroxine metabolism and is in strong contrast to UGT expression, which is low early in life and does not mature for several years. Consequently, SULTs are a very significant part of foetal and neonatal defence from xenobiotics[50]. However, SULT roles regarding xenobiotic metabolism in adults seem to be contradictory, with many SULTs involved in the activation of carcinogens, as well as the detoxication of other reactive species. All SULTs are subject to genetic polymorphisms, with a high degree of individual variation in their expression and catalytic activities (Section 7.2.8); this is currently under investigation, particularly from the standpoint of individual risk factors for carcinogenesis. Indeed, high SULT expression is a liability in the prostate, as it massively increases the risk of developing a malignancy[52].

Regarding classification of the superfamily of SULTs, it is assumed that 47% amino acid sequence homology is indicative of same family members and 60% homology for subfamily members. There are 13 SULT isoforms in four families in humans[53], with SULT1A1 being the most studied and is found in the largest quantities in the liver[54], with the gut expressing significant quantities of this and other variants.

SULT 1 family

This is the major group of human sulphotransferases, which contains four sub-families (1A, 1B, 1C, and 1E) and a considerable number of isoforms, including 1A3, 1B1, 1C1, 1C2, and the oestrogen sulpho-transferase 1E1. 1A3 (not found in the liver) as well as 1E1 are potent sulphators of xenobiotic oestrogens and are inhibited by oestrone and quercetin. However, the most important isoform from the xenobiotic metabolism point of view is the phenol and aryl amine sulphating sulphotransferase SULT1A1. The genes that code for SULTs 1A1, 1A2, and 1A3 are found very close together on chromosome 16. SULT1A1 is found in large quantities (grammes of it, apparently) in the liver[54], whilst SULT1A3 is found mainly in the foetal liver, gut and lung where it forms sulphonates from catecholamines. Toxicologically and from the perspective of drug clearance, SULT1A1 has become more and more important as its role in the clearance and activation of xenobiotic molecules has been unravelled. The polymorphic variants of this isoform (designated SULT1A1* 1-3; Chapter 7, Section 7.2.8) are found mainly in the liver and gut, but also the lung, skin, and brain[54].

SULT 2 family

These enzymes are hydroxysteroid sulphotransferases, and there are only two subfamilies, SULT2A and SULT2B, with SULT2A1 and SULT2B1 being the best understood. There are a large number of individual enzymes that sulphonate a variety of sex hormones and hydroxysteroids. SULT2A1 metabolizes sex steroids and many drugs; it is very selective for dehydroepiandrosterone, the steroid found in the highest plasma levels in humans[55].

6.3.2   SULT structure related to catalytic operation

SULT isoforms are capable of metabolizing a wide variety of hydrophobic molecules, and a great deal of progress has been made as to how SULT1A1 and SULT2A1 can be so selective for certain substrates. The activity of these isoforms is important because together they make up around 90% of liver sulphonation activity[51]. The SULTs are quite similar in basic structure and from crystallized (SULT1A1 * 2) versions, it is apparent that they are generally vaguely clam-shaped, possessing a highly conserved binding site for the co-factor PAPS, with a much more variable substrate binding area. What is particularly interesting is that the SULTs have a 30-amino-acid residue shutter or lid, that when completely closed, covers PAPS and almost, but not all, of the substrate binding site.

The shutter has two sections; one completely covers bound PAPS and the other covers the vast majority of the substrate-binding site. When the enzyme is resting, then both shutters are open. What is believed to happen is that once the

PAPS cofactor binds to the 'open' enzyme, it triggers the closing of this two-section shutter on itself and most of the substrate-binding site, like somebody disappearing into a submarine and closing the hatch behind them. Hence, the large substrate-binding site is almost entirely restricted to only allow a small 'pore' or channel for small molecules in via their particular geometry and properties. This has been termed a molecular 'sieve' that confers very precise small substrate selectivity[50,51,55,56]. SULTs can metabolize larger substrates, provided the enzyme is fully open, or if it is closed, then the part of the shutter which covers the substrate binding site can open, which it can independently of the PAPS nucleotide shutter. Although the enzyme can sulfonate the substrate only if PAPS is fully locked in by the shutter, affinity for large substrates is controlled by the local concentration of PAPS; very high levels discourage the processing of large substrates[55,56].

Indeed, the rate-limiting step for the enzyme is movement of cofactor in and out of the PAPS binding site, with the shutter opening and closing as required. This PAPS movement is controlled allosterically by two separate-binding sites. One site is part of the enzyme structure designated 'α-helix 6' and this site, once a ligand binds to it, causes the stabilization of the PAPS shutter in the closed position, thus slowing down the enzyme activity. This site is bound by NSAIDs such as mefenamic acid and obviously explains why these drugs are so effective (at the nanomolar level) in blocking SULT activity[57,58].

Another allosteric site is operated by catechins (found in green teas), which is completely separate from the NSAID site[56], so it is likely that SULTs are subject, like CYPs, to very precise local allosteric control by an array of different endogenous molecules, which probably bind in a range of different affinities and potencies. Hence, very precise, minute-by-minute control of the velocity of sulfonation is possible for the requirements of the tissue and the type and concentration of substrates it is necessary to modify.

Whilst the two-section shutter arrangement, rate limitation of PAPS, and allosteric control of SULTs is common across all these isoforms, they do differ in their substrate binding sites, as you might expect, so that they can process specific groups of substrates. In the case of SULT1A1, this is an L-shaped and hydrophobic binding pocket which can accommodate two substrate molecules[59]. To illustrate the subtle differences in these isoforms in terms of structure that lead to huge differences in function, SULT1A3 shares 7 out of 10 aromatic residues in its substrate binding site with SULT1A1, but it has much narrower specificity, only really metabolising catecholamines like dopamine to sulphonates. The ten aromatic residues in SULT1A1's substrate binding pocket allow it to metabolise extremely hydrophobic substrates[59].

In summary, SULTs differ radically from CYPs in terms of their structures, functions, and metabolic roles. However, through features such as their two-section shutters and allosteric sites, SULTs can emulate CYPS in displaying extraordinary flexibility in substrate processing, and they can be modulated locally through allosteric manipulation. SULTs have essentially evolved to solve similar problems to CYPs, but using entirely different approaches. Proof of principle has already been found that molecules such as mefenamic acid can be used to exploit the SULT allosteric control apparatus, and in time custom designed inhibitors will appear that will be able to improve the residence time of substrates such as the thyroid drugs and prevent activation of potential carcinogens[50].

6.3.3   Control of SULT enzymes

It has now been established that SULT expression at the cellular and tissue level is controlled by all the same nuclear/cytoplasmic receptors and factors that are involved with CYPs and the rest of the biotransformation library of enzymes and transporters. These include PXR, CAR, AhR, the liver X receptor (LXR), the farnesoid X receptor, vitamin D receptors, and many more[53].

However, SULTS respond to different receptors in ways that are not necessarily predictable. The AhR, for instance, suppresses SULT expression, whilst PXR (through rifampicin) can increase some SULT isoforms and not others. The SULTs are responsive to oxidative stress through the nrf2 pathway (see 6.5.6) and several antioxidants such as caffeic acid may provide a beneficial effect through this route with SULT1A1[60]. On the other hand, some cytokines and inflammatory processes suppress SULT expression, and these isoforms can be induced by their substrates and other endogenous chemicals[60]. The steroid oestradiol-17β is sulfonated by SULT1A1 and SULT1E1, but can also induce the expression of a major androgen sulfonating isoform, SULT2A1, which demonstrates the complex interplay in steroid regulation[60].

6.3.4   SULTs and cancer

The role of SULTs in cancer has been extensively investigated, and abnormally high expression increases the risks of developing malignancies in male (prostate) and female (breast) cancers, mainly linked to steroid metabolism, as well as other agents[61]. Overexpression of SULT1E1 is particularly strongly linked with breast cancer[62]. The SULTS can either accelerate the clearance of one form of a steroid, causing a distortion in the levels of other variants, or they can form reactive species through the formation of unstable sulfonated metabolites that disproportionate into highly reactive carbocations that attack DNA[57]. It is thought that some dietary agents such as resveratrol may translationally suppress SULT expression, which could be protective, but their other effects may cancel this out[61]. So of the thousands of potential dietary phytocompounds, many will be SULT substrates and inhibitors, so it is very difficult to predict whether they will have a beneficial effect in protecting high-risk (high-SULT-expressing) individuals from malignancy. Part of tamoxifen's benefit is its conversion to 4-OH tamoxifen, which is catalysed by SULT1A1, among other enzyme systems. The 4-OH derivative of tamoxifen is a potent inducer of malignant cell apoptosis, and poor expression of SULTs is correlated with lower survival rates[62]. It seems that in some cases, expression of defective SULTs (SULT1A1*2) can either protect patients from some cancers (smokers from bladder cancer) or increase the risk of others (lung cancer in smokers). So the SULTs may, depending on the tissue, detoxify or activate potential carcinogens. Given the potency of some allosteric inhibitors of SULTs such as the NSAIDs, this provides some hope for the development of more selective inhibitors of SULTs, which might make therapeutically successful interventions in patients who may be vulnerable to SULT-related malignancies.

6.4   The GSH system

6.4.1   Introduction

One of the main problems with the oxidation of various molecules by CYP enzymes is that they are often destabilized and sometimes form highly reactive products. The analogy used previously in this book was that of a child given a hammer and told to hit anything metallic hard. CYPs occasionally form metabolites so reactive that they immediately destroy the enzyme by reacting with it, changing its structure and, therefore, its function. Sometimes, this kind of damage can be self-limiting because the reactive species formed destroys the CYPs, so no more reactive species are formed until several days later, while more enzyme is assembled. Meanwhile, the chemical substrate itself is not a problem unless it is oxidized and it might well be cleared through other routes in the meantime.

   The most dangerous forms of reactive species, either formed by an enzyme or a result of a further series of chemical reactions, are those that evade detoxifying enzymes, or are inadvertently created by conjugation processes. These can be reactive oxygen species (ROS), or the complex reactive versions of nitrogen, such as peroxynitrite (ONOO−), as well as reactive carbonyl groups[63]. These electron-rich unstable species (nucleophiles) will attack positively charged areas of cellular structural proteins and change their shape and function. On the other hand, very positively charged unstable species, such as nitrenium ions N^+, will attack negatively charged areas of the cell, such as DNA, and naturally this form of damage is just as destructive as the electrophilic toxins and may cause mutagenesis. These species may have different half-lives, and how far they travel and retain their reactivity is dependent on their physicochemical properties. Critical damage caused by these species may lead the cell into apoptosis, or if the damage is too rapid and widespread, then cellular destruction and necrosis may result. (Chapter 8.2.3).

CYPs are not the only source of reactive species generated within cells. Around 75% of our food intake is directed at maintaining our body temperature and a great deal of energy must be liberated from the food to accomplish this. Cells derive the vast majority of their energy through oxidative phosphorylation, and this takes place in the 'engine rooms' of the cell, the mitochondria. With a car or any fuel-burning machine, aside from the achievement of the desired function, energy is also wasted in the form of heat and toxic exhaust gases. In cells almost all the oxygen we breathe is consumed in oxidative phosphorylation, forming ATP, heat and reactive oxidant species in the mitochondria that could cause severe damage to the structure and function of the cell if they were allowed to escape.

So all cells, particularly hepatocytes, have evolved a separate system to accommodate such reactive toxic products and this is mostly based on exploiting the redox properties of the thiol group of the essential amino acid cysteine, although methionine can be converted to cysteine if necessary. Cellular protein folding and stability depends on the precise positions of cysteine residues, as well as the functionality of their thiol groups and the bonds formed from the reactions of two cysteine groups, known as disulphide bridges[63]. Cells have evolved a low-molecular-weight 'portable' form of thiol, a three-amino-acid (cysteine, glycine and glutamate) thiol known as glutathione, or GSH. Thiols in general are extremely effective at reducing and thus 'quenching' highly reactive, electrophilic species. GSH has a very high redox potential of −0.33, and donates electrons to reactive species. During this process, it loses a proton and forms a glutathionyl radical (GS•), which is capable of causing oxidant damage itself, but more usually undergoes a complex series of reactions, resulting in the disulphide bridged GSSG. This is now stable and is spent, and by itself it is useless in controlling reactive species. However, GSSG can be regenerated to GSH by GSSG reductase to resume its antioxidant role.

Hence, the thiol/S–S relationship is of incredible importance to cells, as not only can they protect themselves from unstable species with the thiol but the covalent S–S bond is a cornerstone of cellular structure. Cells can 'unpick' this strong bond whenever it is necessary, with reductase enzymes (ubiquitous in cells), so conferring great flexibility in cellular structural and functional modifications. Indeed, GSH has an essential maintenance role in repairing damaged proteins, through a thiol disulphide exchange, where the GSH effectively repairs a protein-SSG group and restores it to a protein-SH, via a thiol transferase[64]. This process is reversible, but GSH levels are so high in the cell that there is a constant law of mass action mediated 'positive pressure' on protein repair by GSH. You can also see that a thiol molecule in such quantity that has a system of reductases and transferases behind it, can react and change protein structures, so behaving as a molecular switch, running large numbers of cellular processes off and on as required. There is evidence that we have only scratched the surface of the complexity of thiol and sulphide metabolism, with many more derivatives operating key cellular functions[64,65].

To say that GSH can act as a cellular 'fire extinguisher' is an imperfect analogy, as it implies that GSH is only useful in emergencies. In fact, if cells are depleted of GSH by blocking its synthesis (by using buthionine sulphoxime), cell death follows and the organism itself will die in a few days, due to uncontrolled activity of endogenous radicals. A better analogy for GSH would be the oil in a car engine. Without oil, the engine will essentially weld itself together and seize. Proteins, for example, are partly held together by disulphide bridges, but also contain crucial thiol groups that must not be oxidised; otherwise, function is impaired or shut down. To stretch the engine analogy somewhat, a combination of a good oil quality, pressure, and supply to high-stress components means that a well-designed car engine will last almost indefinitely. If GSH levels are not maintained in the cell over a long period of time, the cell wears out more quickly; for example, diabetic complications and HIV infection are linked with poor GSH maintenance. In fact, among its many functions, the GSH system acts like a cellular 'battery charger', recharging the oxidized spent versions of other antioxidants such as ascorbic acid and vitamin E.

6.4.2   GSH system maintenance

In the same way a central heating system maintains a set temperature, GSH levels are thermostatically maintained in different subcellular 'pools' at preset levels. The ratio between GSH and GSSG will be very high in mitochondria, to ensure protection against reactive species, whilst the ratio is lower in other less radical-threatened areas of the cell such as the rough and smooth endoplasmic reticula. Overall, the hepatocyte maintains GSH at a very high (8–10 mM) intracellular level, whilst normal erythrocytes hold it at around 1–2 mM. Plasma levels are usually only in the micromolar range.

Intracellular GSH concentrations are maintained by two methods: firstly, by a two-stage ATP-dependent direct synthesis involving γ-glutamylcysteinyl ligase (formerly synthetase) and glutathione synthetase, and secondly, by recycling the GSSG to GSH by the operation of GSSG reductase, which consumes cellular reducing power (NADPH; Figure 6.7). This system is so efficient that at any one time, 98% is GSH and less than 2% will be GSSG. The cell can completely restock its GSH level from nothing in less than 10 minutes. GSH levels are maintained through allosteric negative feedback mechanisms, where high GSH concentrations inhibit GSSG-reductase and both the GSH synthetic enzymes. GSH maintenance can only be frustrated by a lack of raw materials (a sulphur containing amino acid, methionine or cysteine) or lack of reducing power. The cell will often succeed in maintaining high GSH levels even under severe attack by oxidative species. It is not usually necessary for GSH to leave a cell, and it cannot cross membranes without a specific transporter such as MRP-1 or SLCs like the OATPs. Also, GSH is resistant to degradation because the peptide bond that links the cysteine to the glutamate can only be broken by an enzyme that is extracellular[66].

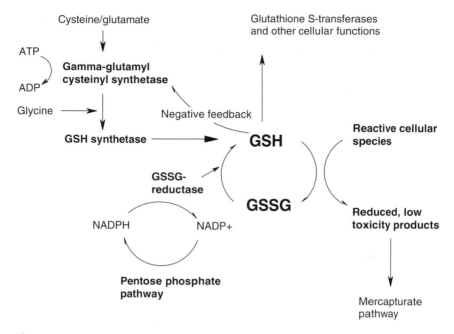

Cysteine/glutamate

Glutathione S-transferases
and other cellular functions

ATP

Gamma-glutamyl
cysteinyl synthetase

ADP

Glycine

Negative feedback

GSH synthetase

GSH

Reactive cellular
species

GSSG-
reductase

GSSG

NADPH   NADP+

Reduced, low
toxicity products

Pentose phosphate
pathway

Mercapturate
pathway

Figure 6.7   The GSH maintenance system in man

Transporters actively pump GSSG out of cells in times of oxidative stress to prevent it reacting with other cellular thiols. GSSG egress is actually a good indicator
of oxidative stress in a cell. When GSH quenches a reactive species, it is sometimes written as a GS-conjugate, which undergoes further processing to emerge
as a mercapturate, which is excreted in urine (Figure 6.8)[64,66].

Currently, over 30 essential cellular functions have been found for GSH, and
the GSH system is so highly evolved that it does not solely rely on GSH spontaneously reacting with dangerous species. There are several enzymes that promote and catalyse the reaction of GSH with potential toxins to ensure that
reactive species are actively dealt with, rather than just passive GSH-mediated
reduction. Probably the most important from the standpoint of drug metabolism
are the GSH-S-transferases.

6.5   Glutathione S-transferases

6.5.1   Structure and location

Forming around 5% of total cellular protein, the glutathione transferases, or
S-transferases as they are usually known (GSTs) are found in the smooth endoplasmic reticulum (SER), mitochondria and the cytosol; the latter group have
been the most intensively investigated[67]. GSTs are primarily of interest as key

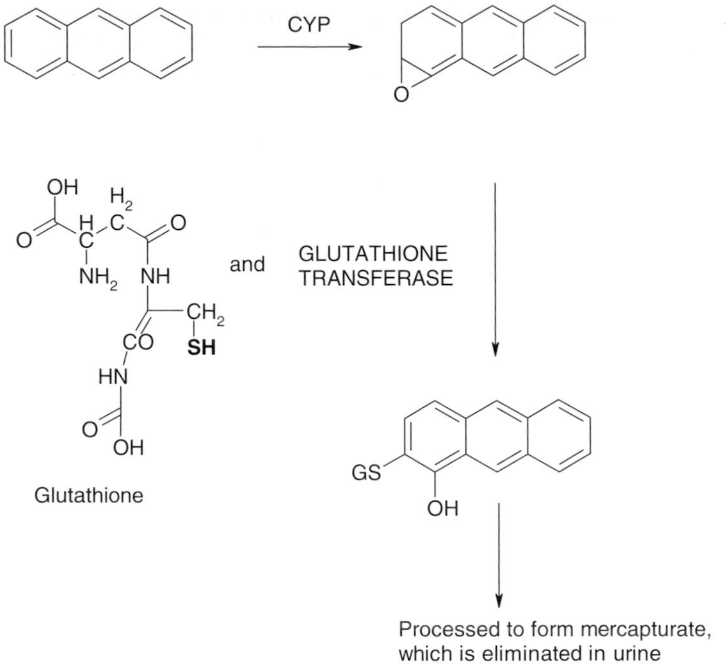

Figure 6.8 A typical GST catalysed reaction of GSH with a potentially reactive xenobiotic. The hydrophilic GSH molecule substantially increases the water solubility and molecular weight of the aromatic and will also detoxify it and ensure it will be transported out of the cell. It is unlikely anyone would make you learn the structure of GSH, but it is useful to look at it and appreciate that it is highly water-soluble

cellular defences against electrophilic agents formed from endogenous or xenobiotic oxidative metabolism. However, we now know that they are involved with many other processes, ranging from cell signalling to inflammation, as well as their role in protection from oxidative stress[67-69]. Although GSH is capable of reacting directly with reactive species, the GSTs could be likened to using a fire hose instead of a hand extinguisher.

The basic GST monomer weighs about 30 KDa and rather than butterflies or clams, they seem to resemble a paper stapler. The top half (known as domain I) binds GSH and is known as the G site (the amino-terminal end). As with the SULTs, the G site binds the thiol cofactor and so does not need to vary. The substrate binds to the hydrophobic H site, which is on the carboxy-terminal and can be up to three times larger than the G site[67]. The H sites differ according to the GST isoform and confer their abilities to bind their particular range of substrates. There is a short hinge or linker amino acid sequence that holds them together and allows them to articulate to accommodate substrate and cofactor[68]. In some GSTs, such as the α series, there is a hydrophobic lock and key style arrangement of protein substructure that stabilizes the enzyme as it opens and shuts to accommodate substrate and co-factor[67]. Both the substrate and cofactor binding sites undergo considerable conformational change to accommodate their respective molecules and the H site can accommodate many different substrate structures, which echoes CYP and SULT flexibility.

GSTs are believed to operate mainly as dimers, but also possibly as monomers, although this may depend on the type and location of the isoforms and their substrates[67]. The monomers are believed to form dimers by the G site of one monomer interacting with the H site of another, so to visualize this, you bring your hands together as if to pray (as you would perhaps, just before an examination) and rotate one hand through 180 degrees. This dimer arrangement is thought to promote enzyme stability and efficiency of catalysis[70].

The hydrophobic SER GSTs are structurally different from the cytosolic GSTs and they metabolize leukotrienes and prostaglandins and are known by the acronym MAPEG. Aside from the detoxification roles of cytosolic GSTs, they carry out those repairs of damaged proteins (described earlier), which have been S-thiolated. The seven major classes of mammalian enzymes are Alpha, Mu, Pi, Theta, Sigma Zeta, and Omega, although these enzymes are found right across the animal and plant kingdoms. In humans, the presence of mercapturates in urine (Figure 6.8) is usually a reasonably good indication of the formation of a reactive species somewhere in the hepatic handling of a xenobiotic agent. It has been noted that the term *glutathione-S-transferase* is not very precise, because these enzymes do not always act as transferases and have a much broader capability than the standard GST designation implies. They can be better classified as RX:glutathione R-transferases, as the X is the leaving group (halide or sulphate ion) and the R is the electrophilic substrate[71], although it is likely that the GST designation will persist from sheer habit.

6.5.2   Mode of operation

The GSTs firstly bind GSH at the G site and the pKa in the immediate environment is lowered from 9.2 to 6.2–6.6, which promotes the removal of the proton (the hydrogen ion) forming the reactive anionic thiolate radical (GS •). This is a strong nucleophile that attacks the electrophilic substrate immediately, and the resultant thioether can be rearranged through further metabolism (mercapturate pathway) to form mercapturates, which are stable and nontoxic. It is still not exactly known how the enzyme manages this process and some very significant structural rearrangement must be necessary to manage so many different reactions with different aliphatic and aromatic substrates. Some theories suggest that a tyrosine residue holds the GSH and draws the proton away (tyrosine assisted proton transfer)[71]. The tyrosine (a serine in the theta class)[72] may have a crucial role to play, because if it is substituted or removed, catalysis is lost. Once the thiolate radical has been formed, it seems that this probably triggers the H site to manipulate the substrate in such a way as to promote the nucleophilic attack and the conjugate is formed. The departure of the conjugate may be the rate-limiting step of some GSTs[71]. What complicates things is that it may be that all the GSTs may not actually operate in the same way, depending on their favoured substrates. Either way, GST catalytic operation is spectacularly complex and currently unresolved and will require detailed studies involving building mutants with minor structural changes and studying how these impact catalysis.

6.5.3   GST classes

The GSTs are found in humans in several major classes. Generally, 70% amino acid sequence homology is required for an isoform to be assigned to a particular class. The classes contain several subfamilies, with around 90% common sequence homology. These enzymes are polymorphic (Chapter 7, Section 7.2.9) and their individual expression ranges from complete absence in some isoforms to over-abundance as a response to anticancer therapy.

GST Alpha class

GST Alpha class originates on chromosome 6 and codes for five proteins. GSTA1–5 (1-1 is an important representative of the A class, which is found in high quantity in the liver and kidney and to a lesser extent other tissues (intestine, lung, testis). Hepatic GSTA1-1 accounts for 1% of cytosolic protein, providing considerable protection against electrophiles. Finding GSTA1-1 in the blood is a clear sign of liver damage and is a more sensitive marker for monitoring the progress of liver toxicity. This is because it is more closely associated with the liver than aspartate aminotransferase (AST) or alanine aminotransferase (ALT) and is more rapidly eliminated, so a more up-to-date picture of liver pathology is available[73], although practicality and cost implications may have prevented the widespread uptake of this idea clinically.

The substrate-binding site of the Alpha GSTs is most efficient at processing small hydrophobic molecules, such as diol epoxides, various lipids and smaller PAHs and it displays great structural flexibility to accomplish this. Indeed, it has been suggested that GSTA1-1 has a gate-like α-helix-9 that modulates substrate access and release of product from the active site, perhaps a little like that shutter arrangement of the SULTs described in the previous section[74]. The different alpha class GSTs display a range of affinities for various groups of small hydrophobic molecules. Of the other human GSTA isoforms, GSTA2-2 is also abundant in the liver, whilst GSTA3-3 is found in the adrenal glands, placenta, as well as the testes and ovaries. GSTA1-4 can be inhibited by ethacrynic acid, lipid hydroperoxides, as well as 4-hydroxyalkenals (products of lipid breakdown). These enzymes are also capable of carrying out GSH peroxidase activities.

GST Mu class

There are five members of this class, which originates on chromosome 1[72]; representative Mu class variant GSTM1-1 lacks an α-helix-9 so has no barrier to its larger more open active site than the alpha GSTs and it contains a deeper and very hydrophobic binding cleft, compared with the GSTP variants[70]. GSTM1-1 is found in high levels in the liver, as well as the brain, testis, kidney, and lung and will oxidize many bulkier electrophilic agents, such as 1-chloro-2; 4-dinitrobenzene (CDNB), trans-4-phenyl-3 buten-2-one and benzpyrene diols. It is also an important detoxifier for aflatoxin B1-8, 9-epoxide, as well as the myriad large electrophilic polycyclics found in tobacco smoke[75]. Lack of this GST predisposes to a higher risk of developing malignancies in barrier tissues such as the oesophagus[76].

GST Pi class

GSTP1-1 (chromosome 11) is not found in the liver but is the most highly expressed GST pretty much everywhere else; it is found particularly in the oesophagus, thyroid, and erythrocytes[67,77]. Its H binding area is half hydrophobic and half hydrophilic[70]. It is overexpressed in tumour cells by as much as 200%. It will process a variety of toxicologically dangerous agents as well as endogenous species; these include CDNB, acrolein, adenine, propenal, benzyl isothiocyanate, and 4-vinylpyridine. Aside from their functions in xenobiotic metabolism, GST Pis and the Mu class GSTs can regulate many cell signally kinase systems, such as the c-Jun NH2-terminal kinase (JNK) pathway that modulates cellular apoptosis. The role of GSTP1 is especially troublesome in the induction of resistance to alkylating agents in cancer chemotherapy. Part of this process involves tumour upregulation of GSH formation, but GSTP1 and other isoforms in the series defend the tumour cells by direct detoxification as well as by blocking apoptosis through their effects on cell signalling kinases[70]. There is a version of GST Pi where a valine residue is substituted for an isoleucine residue at codon 105 and this GST Pi Ile105Val variant is linked with reduced defence against reactive species and a greater risk of malignancy as its catalytic activity is much less than the wild-type isoforms[76,78].

GST Theta class

Originating on chromosome 22, these isoforms (GST T-1 and 2) are found in the liver and other tissues in lower levels than the other GSTs differ from the other GSTs as they do not use the tyrosine residue to catalyse the reaction between the substrate and GSH[67,70]. Serine accomplishes this activity in the GST-T isoforms, and it is likely that the site is capable of some structural rearrangement that assists in the catalytic process. This group of GSTs is associated with the metabolism of environmental and industrial carcinogens, including planar polycyclic aromatic hydrocarbons, halomethanes, dihalomethanes, and ethylene oxide. GST T-2 has a hydrophilic binding pocket that allows it to hydrolyse sulphate esters, although GST T-1 cannot do this, as it does not have a suitable binding area[70]. Interestingly, GST-T in erythrocytes is identical to the hepatic version, so methyl bromide or ethylene oxide turnover by the enzyme in sampled erythrocytes is used to determine if an individual expresses this isoform.

GST Omega and Zeta class

This class of two isoforms (chromosome 10) can produce ascorbate from dehydroascorbate, as well as conjugating some bulky hydrophobic molecules[70]. They are found in most tissues. Instead of tyrosine, they use cysteine as the key active site residue[67]. These isoforms are thought to be responsible for protein repair, where thiol adducts are trimmed off cytosolic structures as the enzyme acts as a thiol transferase. It has a very large and open hydrophobic binding site, which allows it to bind polypeptide chains. This isoform is also involved in preventing cellular apoptosis by blocking calcium ion mobilization from intracellular stores. There is a single GST Zeta isoform in humans and it relies on a serine in its active site. This GST carries out various endogenous isomerization and peroxidase reactions.

GSTs and drug action

GSTs tend to be viewed exclusively as detoxifying enzymes, although they have a key role in the activation of azathioprine, an immunosuppressive drug that is capable of causing malignancies with prolonged use[79]. Azathioprine is a pro-drug, as it is converted to 6-mercaptopurine (6-MP), which is processed further to active (and toxic) 6-thioguanine nucleotides (6-TGN) that block purine synthesis. Whilst several GSTs can accomplish azathioprine activation, GSTM1-1 is the major isoform involved and lack of its expression leads to lower efficacy and toxicity with this drug[77]. The balance between 6-MP formation and its destruction by TMPT (Chapter 7.2.6) is also a major influence on how successful therapy is with azathioprine[80]. The GSTs and TMPT are polymorphic and the impact of polymorphisms on therapeutic outcomes will be discussed in the next chapter.

GST therapeutic inhibition

As mentioned above, overexpression of GSTP-1-1 in tumour cells makes a powerful contribution to cancer chemotherapy resistance. If an anticancer drug kills tumour cells through reactive species generation, upregulated GSTs can curtail efficacy either directly by quenching the species and/or by conjugating the drug and forming a polar metabolite, which is then pumped out of the cell by the various efflux pump systems that are upregulated alongside the GST. Cisplatin, melphalan, and doxorubicin can all be detoxified by GSTs such as GSTP1-1[81], which as mentioned previously can block tumour cell apoptosis. However, the high expression of GSTP1-1 in tumour cells, for example, actually provides two therapeutic targets. The isoform could be blocked, or its activity exploited to deliver pro-drugs directly to the tumour cells.

If we consider efforts to block GSTP1-1, this would appear to be a very attractive concept, although initial studies with ethacrynic acid were problematic due to the diuretic effect of this drug[81]. A more promising approach was with ezatiostat (TLK-199), which is a GSH analogue that is metabolised to TLK-117. That derivative binds to the G site of GSTP1-1 and is very selective[81]. Although ezatiostat caused gastrointestinal problems in clinical trials, it has shown promise in the treatment of myelodisplastic syndrome and has also been applied to lung cancer.

In terms of exploiting high GSTP1-1 levels to release prodrugs, a number of potential drugs have been synthesized, such as doxorubicin analogues[81], but canfosfamide (Telcyta) reached clinical trials. This drug used GSTP1-1 to release a phosphorodiaminate alkylating agent that would alkylate the tumour DNA. It got through Phase II, but in Phase III trials involving doxorubicin and topotecan, median advanced ovarian cancer patient survival was worse with Telcyta, and the manufacturer (Telik Inc.) committed a serious ethical breach during this process by not disclosing these data to patients in other ongoing Telcyta trials for several months. Once the data were released, the other Phase III trials were halted[82] and Telik's reputation never recovered.

Finally, both the blocking and prodrug concepts were explored simultaneously with ethacraplatin, which combines cisplatin with ethacrynic acid, as the somewhat unfortunate name suggests. This agent has shown itself effective in studies with cell lines, as it releases both the GST inhibitor and the cytotoxin cisplatin into the tumour cell at the same time[81,83].

Whilst the various GST inhibition concepts are ingenious, the various approaches have been hampered by a lack of specificity for GSTs in the inhibitors and the broad specificity overlap between the various individual isoforms and even families of these enzymes[81]. At this point, it is probably a good plan to look at what the 'manufacturer' uses to control GSTs and consider some variation on that approach. This is already in train, and GSTP1-1 expression is known to be suppressed by microRNA-133b (miR-133b) as well as miR-513a-3p and studies with prostate cancer cell lines have shown potential in terms of downregulating GSTP1-1, although there is much more research to be done[67]. Whilst GST regulation, as we will see in the next section, is complex, the best hope of specifically modulating these enzyme systems in the distant future probably lies in some form of interaction with their gene promotors, and perhaps exploiting endogenous 'switches' could address the perennial problem of resistance.

6.5.4   Control of GSTs: overview

Over the past decade or so, something of the scale of the complexity of the structure and functions of the GSTs has been discovered, and this far exceeds xenobiotic metabolism and detoxification that are the narrow foci of this book. However, one of the positive outcomes of researching the impact of aggressive toxins such as anticancer drugs is that they can often unmask the full machinery of a living system's defensive response. As mentioned earlier, GSTP1-1 expression in cancer research has been particularly revealing in how cells sense and defend themselves in terms of redox and cell signalling. Exploring and defining pathways such as these are the future in terms of evolving new therapeutic concepts, as relying on screening thousands of potential structures, or repurposing existing drugs, or just blind luck, all have their limitations.

Clues to the value of the GSTs lie in their sheer quantity of expression all over the body, as well as their ability to carry out many more reactions than just conjugation of electrophiles with GSH. They can catalyse isomerizations, disulphide exchanges, peroxidase reactions, and cellular thiol repairs[84]. They are linked with many different disease states, and they are part of what has been termed the *GST interactome,* which is a convergeance of many signalling and response systems coordinated by the GSTs[84]. Indeed, with just one of these isoforms, it has been found that GSTP1-1, interacts with at least 35 proteins, either physically or chemically. Some of the physical interactions, where the GST interlinks with another protein, are with MAP kinases, such as the JNK apoptosis system. GSTs also interact physically with many other key regulators of the cell cycle and apoptosis such as the tumour suppressor p53, as well as factors that modulate cellular protein structure and apoptosis, such as tissue transglutaminase-2 (TGM2). GSTs are involved in all these systems that can either be part of the response to redox challenges or part of general top-down cellular housekeeping processes[84].

In terms of their responses to other cellular factors, GST genes possess a complex series of response elements. Whilst in Chapter 4, the CYP/conjugative biotransformation system 'eyes and ears' were exemplified by the nuclear and cytosolic receptors, PXR, CAR and AhR, the GSTs are operated by different factors, which reflect their multiple roles in cellular defence and housekeeping. On the GST gene promotor and enhancer sequences there are binding sites for factors such as activator protein-1 (AP-1), which controls many cell cycle functions; NF-κB, which coordinates cytokine production and cell survival activities; and c-Jun, which also coordinates the cell cycle[67,84]. However, from the standpoint of xenobiotic threats such as toxins, drugs, and of course, anti-cancer drugs, GSTs are a focus of response to oxidative stress.

6.5.5   Control of GSTs and reactive species

Detecting and coordinating the detoxification of reactive species poses different challenges with respect to other forms of biosensor-mediated detection, such as the PXR/CAR/AhR systems. As we saw in Chapter 4.4.3, PXR, for instance, is a very 'plastic' sensor that detects a variety of potential environmental threat molecules. Part of why these molecules are a threat is that their very stability allows them to wander through cells and tissues, binding endogenous receptors and disrupting housekeeping. PXR has evolved to detect such well-defined and stable structures. Also, PXR and the other biosensors, with the exception of CAR, are 'demand led', in the sense that level of expression rises and falls in accordance with the presence of the activating ligands. The ligand sensing and upregulation of the sensitivity of the system (number of receptors available for binding) is a relatively slow and conservatively managed system, in terms of cellular effort expended.

Such a leisurely PXR/AhR style response system is entirely inappropriate for detecting reactive species, as perhaps only few microseconds are available to report their existence before they attack something and are 'off the radar'. We need a swifter CYP2E1-style response, as unless the increasing presence of reactive species is recognised and a response mounted, then (for the cell) potentially fatal events such as apoptosis could be set in train.

Let's say we wanted to design an effective reactive species detection system. It must possess at least three key features. The first is excellent cellular coverage, so threats could be detected anywhere in the cytoplasm. The second would be exquisite sensitivity, which could record the existence of reactive molecules and even grade them in terms of degree of hazard. The third would be a very rapid reporting/response time, as the cell's fate depends on the expression of the appropriate enzymes to avert triggering apoptosis.

If make the basis of our detection system a sensor protein, then the coverage issue we could solve by producing and distributing the protein in large quantities. This would ensure that it would be pretty much everywhere in the cell all the time in considerable concentration, but it would have to be shielded in some way from being triggered by normal cell processes. Regarding the sensitivity issue, the assumption must be made that the chemistry of the reactive species cannot be prevented from proceeding towards attacking a macromolecule and attaining stability. So if part of the sensor protein is adapted to be very reactive towards electrophiles, then the ubiquity of the sensor means that the species cannot avoid reacting with it. Ideally, some means of discriminating between reactive molecules would be advantageous, to sort them in terms of potential for cellular damage. Hence, we have coverage, sensitivity, and selectivity. Next, the result of the reaction between sensor and species should release a signal molecule that heads to the nucleus quickly and binds to the antioxidant response elements in various genes, including GSTs. Such a system will be expensive in terms of cellular investment in energy and materials, but worth it, in terms of the capacity for rapid and effective response.

6.5.6   Control of GSTs: the nrf2 system

Of course, something far more advanced than our basic ideas already exists in our cells, and like the GSTs, we are only beginning to fully appreciate the spectrum of its functions. The system is based on nuclear factor erythroid-2-related factor-2, known as nrf2. At first it seems counterintuitive that there is actually very little functional nrf2 detectable in a normal cell. However, as this is actually the signal molecule that proceeds to the nucleus to bind DNA response elements, then in the absence of threat, it is logical from our concept in the last section that nrf2 should be thin on the ground when it is not needed.

What happens in 'peacetime', so to speak, is that as soon as nrf2 is produced, it is immediately bound (through its N-terminal Neh2 domain) to two monomers of the protein Kelch-like ECH-associated protein 1, usually abbreviated to KEAP1, which is linked to a Cullin 3(CUL3)-based E3 ubiquitin ligase complex[85]. This process effectively labels (ubiquinates) nrf2 for immediate destruction by the proteasome (Chapter 3.5.1). This apparently wasteful process is reminiscent of CYP2E1 (Chapter 4.4.4). At this point, our signal molecule is being made in quantity but is essentially shielded from the cell and the safety catch is on. So as things stand in an unstressed cell, streams of nrf2 proteins are intercepted by an excess of KEAP1 proteins. Once they are bound together, nothing will save the nrf2, which is headed for the proteasome. However, if something was to inactivate KEAP1 and prevent it from binding to nrf2, then the signal molecule would be free to escape and head towards the nucleus.

Now KEAP1 has a considerable number of cysteine thiols exposed, which as we know, avidly react with electrophiles. If the reactive species appear in the cytoplasm, they hit the KEAP1 cysteine thiols in such a way as to generate a molecular signature[85,86] that corresponds to a particular 'threat' structure, which influences the change in structure of KEAP1, preventing it from scooping up the nrf2s as they come off the ribosome assembly line[85,86]. Such interactions with the KEAP1/nrf2 complex do not release nrf2[85]. It might be logical to expect that KEAP1 is assembled so that the signature of the species controls the rate of degradation of the KEAP1, acting as a volume dial to reduce its ability to bind nrf2 with the most dangerous signature molecules and shutting down KEAP1 most quickly (Figure 6.9).

However, once nrf2 enters the nucleus, it is phosphorylated and then recognised by β-TrCP (β-transducin repeat-containing protein), which is also linked with a ubiquinating system and marks the now-nuclear phosphorylated nrf2 for destruction. This second stage of nrf2 control is clearly intended to fine-tune the cytoplasmic wave of nrf2 generated by the reactive species. So whilst CYPs are induced through PXR/CAR etc., nrf2 is actually 'derepressed'[85], which is essentially the same type of process than governs CYP2E1 expression and response to xeno and endobiotics.

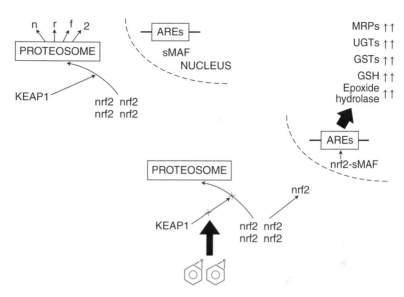

Figure 6.9 The nrf2 system is intended to respond to electrophiles, which prevent KEAP1 from guiding nrf2 towards proteosomal destruction and allow surplus nrf2 to bind Maf transcription activators. The complex binds DNA antioxidant response elements (AREs), expressing a range of detoxification enzymes such as the GSTs, epoxide hydrolases, UGTs, and GSH synthesis

If nrf2 escapes phosphorylation, it meets and forms a heterodimer with sMAF (small (musculoaponeurotic fibrosarcoma), which is a transcription factor that allows the nrf2 to bind to DNA response elements not only in GST genes but also in a host of systems, particularly involving 'defensive' genes. These include various NADPH reductases, glucuronyl transferases, epoxide hydrolases (next section), and various ABC cassette pump genes[85,86]. Nrf2 also upregulates GSH synthesis by increasing the expression of γ-glutamyl cysteinyl ligase to keep pace with increased demand for detoxification by overexpressed GSTs[87]. It has been known for some time that dietary phytochemicals of various types can activate nrf2, so possibly conferring some protection from reactive species generated from other dietary horrors, such as the aflatoxins (see Chapter 8.5.6). Curcumin (from turmeric) and sulforaphane (Chapter 4.4.3) act as nrf2 expression agonists[85], which contributes to their protective effects against DNA damage. Interestingly, clinical trials with broccoli showed some protective effects in terms of the impact on toxic metabolite excretion, but the variable bioavailability of the sulforaphane meant that there was no statistically significant protective effect[88].

If you have read this far, then you can work out what malignant cells would do to ensure they unlocked the full nrf2-mediated protective armoury and ran it at full capacity so they could evade the impact of anticancer drugs. This they achieve through a variety of means, mainly aimed at preventing KEAP1 from locking up nrf2, either by downregulating KEAP1 or jamming it with different metabolites and activating factors that inhibit it[85]. Through these methods, malignant cells can divert a large amount of cellular resources to reactive species defence. Indeed, this had led to various attempts to inhibit nrf2 to try to bypass malignant cell defences. The experimental agent ML385 binds to Neh1, so preventing nrf2 from interacting with DNA. A derivative of a Chinese herb halofuginone has been evaluated clinically, but to date nothing effective has reached the mainstream[85].

6.6   Epoxide hydrolases

6.6.1   Nature of epoxides

Epoxides (sometimes termed oxiranes) are strained three-membered rings with 60 degree bond angles (109 is preferable) sometimes formed by CYPs, even though a phenol would be a safer alternative. They are also consumed in our food, or we inhale them. Depending on their surrounding groups, epoxides vary in their stability and reactivity. They can be formed from alkenes (double carbon bonds), leading to either *cis* (functional groups are on one side of the double bond) or *trans* (groups on either side of the bond). With aromatic hydrocarbons *cis* epoxides are formed. The reason they can be so reactive is that the oxygen draws electrons away from the two carbons in the ring, which is already unstable, so it makes them positive and thus highly electron-seeking (electrophilic)[89]. Many highly reactive electrophilic epoxides are formed in the smooth endoplasmic reticulum that could threaten electron dense DNA and other critical cell macromolecules, so it is necessary to detoxify them.

6.6.2   Epoxide hydrolases

Whilst the GSTs can clear epoxides, epoxide hydrolases (EHs) have also evolved to detoxify them and in endogenous signalling, they regulate these molecules[90]. In the same way as CYPs, higher organisms have 'inherited' epoxide hydrolases from bacteria, with little real change in structure and function. Indeed, all the EH isoforms are very similar, with 80% similarity across various mammals as well as us in the amino acid composition of these enzymes. From the drug/toxin metabolism standpoint, their main function is to convert any endogenous or exogenous epoxide, to a diol (from an alkene oxide) or a dihydrodiol (from an epoxide on an aromatic ring), which is usually less reactive, more water-soluble, and more likely to be cleared by other pathways. There are two main variants of

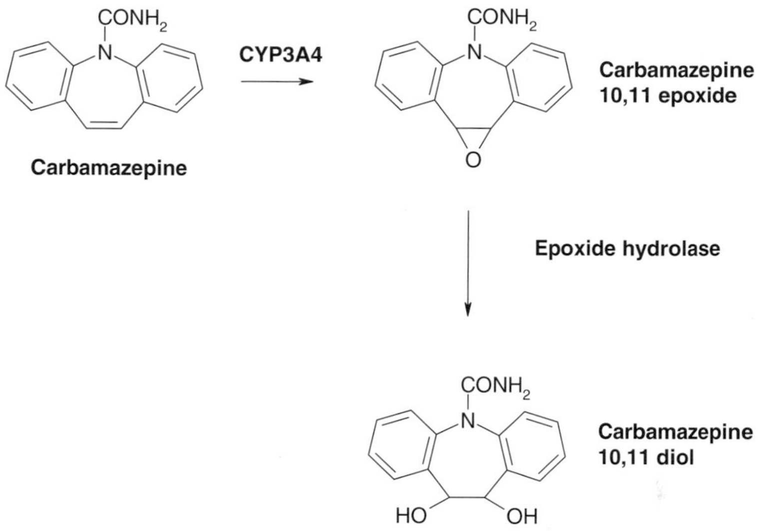

Figure 6.10 A typical microsomal epoxide hydrolase reaction: formation of carbamazepine 10,11 diol, which is thought to have anticonvulsant activity

epoxide hydrolases that have different amino acid sequences but have the same 3-D structure and catalytic sequences[89].

mEH

Microsomal epoxide hydrolase (mEH, from the *EPHX1* gene) is ideally positioned in the hepatic endoplasmic reticulum to 'intercept' epoxides of alkenes (1,3 butadiene), polycyclic aromatics (benzpyrene), or drugs (carbamazepine 10,11 epoxide; Figure 6.10) formed by CYPs and make them into either diols/dihydrodiols.

sEH

Soluble epoxide hydrolase (sEH, from the *EPHX2* gene) also forms diols from many endogenous and exogenous epoxides. Unlike mEH, it has been found in most tissues as well as the liver, lung, and kidney, as it is often closely associated with CYP enzymes such as CYP2C8/9. Soluble EH appears to regulate many pathways related to endogenous systems such as fatty acid and leukotriene epoxide metabolism and is involved in blood pressure regulation and inflammatory responses. Therapeutic inhibition of sEHs may promote the concentrations and benefits of their substrates, many of which are derivatives of arachidonic acid known as epoxyeicosatrienoic acids. These chemicals have potent anti-inflammatory and vasodilatory properties that prevent atherosclerotic disease and hypertension[92].

6.6.3   Epoxide hydrolases: structure, mechanisms of action, and regulation

Microsomal EH is anchored in the SER through its N-terminal end, and the C-terminal end that faces the cytosol carries out the catalytic functions. There is also a version of mEH that is attached to the cellular plasma membrane that faces into the extracellular space. The EHs have access tunnels that shield their active sites, to allow some substrate specificity. With mEH, cis-epoxides will fit down this tunnel, but trans ones will not. If a chemical is chiral, there are nonsuperimposable left- (S, after the Latin, 'sinister') and right- (R) handed versions; mEH will attack the S carbon[90]. The mechanisms of action of these enzymes have been described (Figure 6.11), partly on an experimental basis and partly on the crystallization of sEH.

Both mEH and sEH have a five-pronged formation in their active sites. This consists of two tyrosines, an aspartate residue (common to sEH and mEH), and finally a histidine and glutamate (mEH) or histidine and aspartate (sEH). The substrate wanders down the tunnel and is immediately locked into place by the two tyrosines using hydrogen bonding. Next, the aspartate residue acts as a nucleophile and attacks the epoxide S carbon whilst one of the tyrosines donates a proton to the activated oxygen of the former epoxide, which

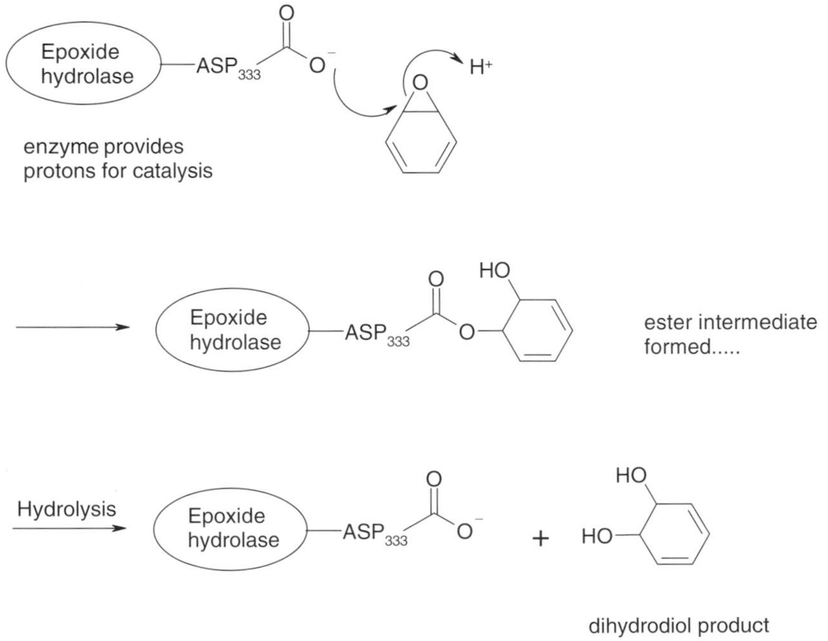

enzyme provides
protons for catalysis

ester intermediate
formed.....

Hydrolysis

dihydrodiol product

Figure 6.11   Mechanism of action of epoxide hydrolase in the conversion of a typical CYP-formed epoxide to a dihydrodiol

immediately becomes an ester. So far, three 'prongs' have been employed in the fastest part of the enzyme's activity. The ester is a covalent bond that would normally be quite stable, but the other two prongs, the histidine and its partner, manage to relatively slowly hydrolyse the ester, forming the diol or dihydrodiol as appropriate[89,90].

What is particularly effective about this process is that once the ester has been formed, the epoxide has been detoxified. Indeed, provided the enzyme is very highly expressed (which it usually is), even though its hydrolysis step is slow, it will effectively 'vacuum up' any epoxides as soon as they are formed by the CYPs before they have the chance to damage anything valuable[89,90].

Epoxide hydrolases form dihydrodiols from epoxides from a number of well-documented and formidable carcinogens, including many PAHs, aflatoxins, 1,3 butadiene and insecticides such as dieldrin. They also form dihydrodiols from drugs such as phenytoin and carbamazepine[90]. Generally, EH isoforms are an effective defence against reactive epoxides, although there are instances with some PAHs where dihydrodiols are formed by mEH, which are then reoxidised by CYPs to dihydrodiol epoxides, which mEH cannot clear. In that case, only GSTs can save the day[90]. There are slow and fast versions of these enzymes, and there is considerable human variation in their phenotypes (Chapter 7.2.4). Hence, defence against reactive epoxides varies considerably between individuals.

Although the full range of the cellular receptors, transcription factors, and other biosensors that can modulate the EHs is not clear, nrf2 inducers such as various reactive species, as well as dietary agents like sulforaphane, can induce mEH in humans. It is believed that nrf2 acts through an antioxidant response element (ARE) in a region of the mEH gene known as HS-2[91]. The other major isoform, sEH can be induced by clofibrate through the peroxisome proliferator-activator receptor α (PPARα) and suppressed through glucocorticoids[90], and it is likely that many more hormones can induce or suppress these isoforms.

6.7   Acetylation

Nature of acetylated derivatives

By now, you will hopefully be familiar with the central idea that biotransformation increases water solubility at the expense of lipid solubility. Acetylation is generally classified as a conjugative or 'Phase II' process, although it appears to be rather contradictory as acetylated metabolites are less water-soluble than the parent drug. Indeed, you might know that diacetylmorphine (heroin) is much less water soluble than morphine. Excreting acetylated metabolites in urine is correspondingly difficult and *in extremis*, acetylated metabolites of some old sulphonamides were so poorly water soluble that they crystallized (painfully) in the patient's kidneys. In this basis, acetyltransferases do not appear to be very effective at contributing positively to the clearance of their substrates. However, these cytosolic enzymes are present in virtually every tissue, which highlights their importance in cell homeostasis. DNA transcription is regulated through histone acetyl transferases (HATs) acetylating histones, allowing DNA to be released to be read, whilst histone deacetylases (HDACs) remove acetyl groups so DNA binds more tightly to histones. Although most metabolic reactions are, to some extent, reversible, acetylation is much more reversible than most. As with DNA, deacetylase and acetylase activity means that many acetylated molecules can act as 'on' and 'off' switches in the regulation of the functions of nuclear receptor systems. Many signalling cascades and hormones such as melatonin are regulated through acetyltransferases[93].

There are several drugs that NATs acetylate, but they also have a very significant role in the acetylation of potentially carcinogenic acyl amines that can either detoxify these agents to some extent or in some cases lead them towards activation to reactive species such as nitrenium ions that attack DNA (Chapter 8.2.3).

N-acetyl transferases; structure and functions

The acetyl transferases relevant to human xenobiotic metabolism include two families of genes on chromosome 8. They are expressed as N-acetyl transferase 1 (NAT-1), which is found mostly all over the body and is crucial in folate metabolism; and N-acetyltransferase 2 (NAT-2) which is most relevant to xenobiotics and is found in quantity in the liver and gut. These were the first enzymes in biotransformation where genetic polymorphisms were documented with the anti-tuberculosis drug isoniazid[94]. NAT isoforms are made up of two main protein sections or α and β domains, which contain the active site. There is a third section (the C-terminal), which (like several other enzymes we have looked at) fits over the first two sections like a lid and regulates the substrate specificity. If the lid is absent, the enzyme just uses up its co-factor and does not form any metabolites[94]. So, like GSTs and the SULTs, the lid or enzyme covering not only allows the enzyme to seize and locate its substrate, but it is essential for the conversion to product. The active site of NAT2 is a triad, which possesses a key cysteine residue that is in close proximity to histidine and aspartate residues[94]. These enzymes bind acetyl Co enzyme A as a cofactor, although not necessarily before binding the substrate. The acetylation process appears to conform to a double displacement (ping-pong) mechanism. There are two sequential steps to the reaction: firstly, the acetyl group from acetyl CoA reacts with the cysteine to form an acetylated enzyme intermediate, then the histidine and aspartate probably promote the acetylation of the substrate and CoA is released[94]. Iodacetate and N-ethylmaleide are irreversible inhibitors of the process, whilst reversible inhibitors (salicylamide) are similar in structure to the substrates.

NAT-1

NAT-1 is found in many tissues, particularly in the colon, but also in erythrocytes. NAT-1 expression is polymorphic (Chapter 7.2.5) and its role in the susceptibility to many cancers related to polymorphic expression has emerged over the last decade[94]. NAT-1 can metabolize aromatic amines and this can lead to carcinogenesis (Chapter 8.5.4). In human cell systems, NAT expression is regulated by hormones and the androgen receptor is involved, but the receptors do not directly bind the NAT-1 gene. Rather, the process of hormonal regulation occurs through another transcription factor[95]. NAT-1 is strongly upregulated in breast cancers[94] and may be inducible in response to certain xenobiotic and endogenous substrates. It prefers to process substrates such as para-aminobenzoic and para-aminosalicyclic acids and is not generally associated with the acetylation of drugs. The consequences of genetic variation in acetyltransferases are discussed in Chapter 7.2.5).

NAT-2

This isoform is very polymorphic (Chapter 7.2.5) and more of its endogenous functions are becoming better understood, not least its role in insulin function and the development of type-2 diabetes[96]. NAT-2 has a role in the clearance of several drugs that tend to possess an amine or hydrazine group. The degree of expression of NAT-2 thus has implications for drug clearance as well as susceptibility to various potential carcinogens[93,94]. Drugs acetylated by NAT-2 include the anti-leprosy and anti-inflammatory dapsone, the largely outmoded sulphonamides, as well as procainamide and the anti-tuberculosis agent isoniazid. Although NAT-2 is found in a narrower range of tissues compared with NAT-1, it is found in various human white cell populations and it is regulated post-translationally by a group of proteins known as sirtuins[93].

6.8   Methylation

Methylated derivatives

Methyl groups are, as you probably know, the smallest alkyl entity that can be added or removed from a molecule. Methylation makes little difference to the molecular weight of a chemical and does not make much impact on its lipophilicity either. Indeed, you might recall from Chapters 3 and 4 that much of the CYP effort towards demethylation is aimed at providing a 'handle' for a conjugation reaction. However, methyl groups can sometimes make massive contributions to pharmacological potency by inducing a much better fit to a target molecule, such as a receptor or enzyme active site[97]. So from the cellular housekeeping perspective, you can see that manipulation of methyl groups can make a large impact on the pharmacological potency of regulatory small molecules such as hormones and especially neurotransmitters.

The many cytosolic- and membrane-bound methylases are vital for DNA regulation and many other cellular housekeeping tasks. S-adenosyl methionine (SAM) is used as a carrier for the methyl group. SAM is made from ATP and L-methionine. SAM-dependent methyltransferases methylate RNA and DNA, many proteins, polysaccharides, lipids and many other molecules. N, O, and S-methyltransferases are also widely expressed, and N-methyltransferases can methylate various histamine-related compounds and amines. Indeed, the key role of the methyltransferase TPMT in the clinical outcomes of purine antimetabolites will be outlined in Chapter 7.2.6. However, from a pharmacological perspective, the best-known methylating enzyme systems are probably the catechol-O-methyl transferases (COMT).

Catechol-O-methyl transferase (COMT)

There are two versions of this enzyme in humans, a membrane-bound one (MB-COMT) and a cytosolic or soluble version (S-COMT). Both types are found in the liver and many other tissues, but in the CNS MB-COMT predominates[98]. COMT attacks catechols, which are dihydroxybenzene derivatives. COMT operates in a manner that is termed 'ordered sequential'; it first binds the SAM. Next, a magnesium ion, which is not only divalent but of the correct size and properties, binds and allows the enzyme to coordinate the position of SAM on the enzyme. The substrate is bound to the magnesium ion and the substrate is deprotonated, facilitating the transfer of the methyl group to the substrate to form the methylated product[99]. Understanding why the magnesium ion is so vital to the functionality of COMT was greatly promoted from trying different divalent and trivalent metals experimentally and noting how they deranged the enzymes functionality[99].

COMT processes

These enzymes act as key regulators of neurotransmitter effect through their processing of the catecholamine neurotansmitters, adrenaline, noradrenaline, and dopamine[9,98]. Whilst catechols need to be controlled partly for pharmacological reasons, they are also reactive and can generate damaging species in redox cycles[98]. Hence, COMT's role is crucial in CNS neurotransmitter disposition and therefore the enzyme's expression and activity (it is also polymorphic)[100] impacts human behaviour, ranging from personality and degree of impulsivity, through to the likelihood of falling into substance abuse and even one's sensitivity to pain[99,100].

COMT has long been a target in the control of the symptoms of Parkinson's disease, as the enzyme would normally metabolize levodopa, the standard treatment for the condition. Whilst levodopa is effective, it is vulnerable to clearance by COMT as soon as it is absorbed, so relatively little might reach the brain. This, combined with its short half-life of an hour or so, means that as its effects subside, the patient suffers distressing and troublesome motor fluctuations. COMT inhibitors prolong the effect of levodopa and are intended to minimize the end- of- dose movement problems[101]. Whilst inhibitors of COMT, such as tolcapone and entacapone have some effect, they are hepatotoxic and in the case of tolcapone, fatally so in some cases. Opicapone is a new COMT inhibitor[102] that is more potent and long lasting in its inhibition than the other drugs and to date has not shown any hepatotoxicity, although it was only licensed in 2016. It is cleared mainly by conjugative processes such as sulphation and is believed to have minimal CYP inhibitory interactions[101].

6.9   Esterases/amidases

Ester and Amides

Many drugs possess ester or amide linkages, and these are vulnerable to the hydrolysis activity of esterases and amidases. Indeed, with some well-known drugs, these enzymes provide a major pathway of clearance, such as with the anti-platelet agent clopidogrel, whilst other drugs rely on esterases to activate them to their pharmacologically effective derivatives (angiotensin converting enzyme inhibitors). The esterases share a common catalytic mechanism, based around a key serine group, but there is a wide range of different isoforms expressed in all tissues. Some esterases are circulating in the plasma (acetyl, benzoyl, and butyrylcholinesterases), which are known for the rapid clearance of heroin and cocaine, whilst others are key specialized tissue components such as neuronal acetylcholinesterase, the target for so-called nerve agents. The esterases are now in the front line of research as they are overexpressed in the obese and those with type-2 diabetes. The carboxyesterases have complex and wide-ranging roles in controlling fatty acid, tryglyceride, and cholesterol ester metabolism and they are currently considered as strong targets for inhibition in tryglyceride metabolism[103]. In terms of the impact on drug clearance, the most relevant isoforms are the liver and gut carboxyesterases (CE, or sometimes CES), which are not found in plasma[104].

Carboxyesterases (hCEs) structure and function

There are five mammalian families of CESs, with two isoforms expressed in humans; CES-1A (hCE1), which is found in high amounts in the liver and not the gut, whilst CES-2A1 (hCE2) is mainly expressed in the small intestine[104]. Whilst both hCE1 and hCE2 are highly flexible in terms of the sheer scale of the number of drugs they can hydrolyse, it seems that they do differ significantly from a structural perspective (only 47% amino acid homology)[104]. hCE1 prefers substrates with a small alcohol and large acyl (double bonded oxygen attached to an alkyl moiety) group, whilst hCE2 prefers large alcohol/small acyl combinations[105,106]. The hCEs locate themselves in cytosol, but also extensively in the smooth endoplasmic reticulum where they operate catalytically in the lumen.

The hCEs have a common structure that comprises a large β-sheet with various α-helices[104] and a very basic way to visualize it is if you join hands as if in prayer again, and keep your hands joined at their base and open out your fingers. As mentioned above, all esterases employ a catalytic triad, serine residue, in proximity with histidine and glutamate residues. The triad, which is located near two glycine residues, is low down on one of your palms at the join between your hands, so to speak. There is one binding site near the active site where smaller ester substrates bind close by (sort of low down on the opposite palm). hCES1 has another flexible (hydrophobic) binding site that accommodates and orientates larger molecules which is just below the fingertips of one of your hands, just near the 'entrance'. Very large molecules can trail out of the top of the enzyme between your fingers[103]. Whatever its size, the substrate is aligned using either or both binding sites as necessary and the ester carboxyl group is manipulated towards the active site. The serine hydroxyl group uses nucleophilic attack to break into the carboxyl group, as the glycines stabilize the briefly acylated enzyme. Then the bond between the serine and the substrate is hydrolysed and the ester is split (Figure 6.12)[104]. With esterases in general, whether an agent is a substrate or inhibitor is related to how and where it binds, as well as how rapidly the acylated enzyme can be hydrolysed.

hCEs regulation

Although the hCEs clear many commonly used drugs and their endogenous roles in lipid regulation is becoming better understood, how they are controlled

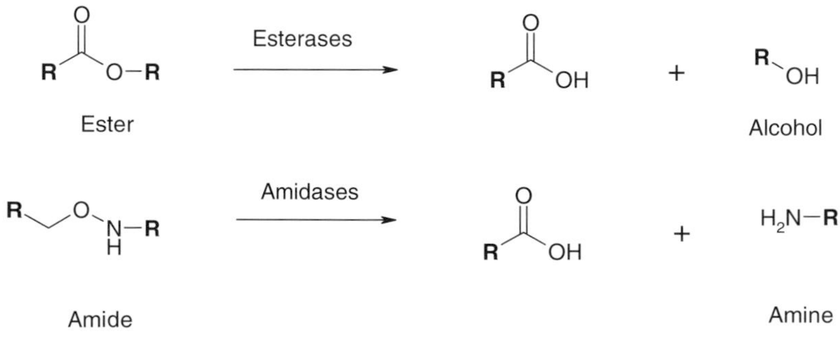

Figure 6.12   Esterase and amidase reaction sequences

and induced is not that well defined to date. Work in human cell systems suggests that they are not regulated by PXR or AhR, as they do not become induced in response to agonists which stimulate those receptors (rifampicin and omeprazole respectively). However, hCE1 does respond to oxidative stress, as well as to some drugs (diclofenac and ticlopidine) as well as some antioxidant/preservatives (butylated hydroxyl anisole), as used in foodstuffs. It appears that there is an ARE that responds to nrf2 on hCE1 and in some cell models, this factor regulates basal expression, along with other transcription factors such as Sp1 and C/EBP. So biological systems seem to partly 'classify' hCESs as a response to oxidant stress and the challenge of reactive species[107]. That hCES1 is highly expressed in the lung already, means that it is of value in supporting other biotransformation systems in tackling polycyclics and other lipophilic atmospheric threats[108].

hCEs in drug metabolism

hCE1's endogenous tasks include processing cholesterol and fatty acid esters, but it clears around 80–85% of clopidogrel to an inactive carboxy derivative, as well as the anti-influenza drug oseltamivir to its active carboxylate[109]. hCES1 also converts the directly acting anticoagulant dabigatran etexilate to dabigatran, as well as several angiotensin converting enzyme inhibitors (benazepril, imidapril, enalapril) to their active versions[108,109]. Other well-known substrates include methylphenidate (better known as Ritalin) and the opiates pethidine and heroin(Appendix B.2).

hCES2 expression in the small intestine makes it well positioned to process drugs such as the anti-androgen flutamide, the anti-cancer drug capecitabine and of course irinotecan. The latter drug is essentially a prodrug, as hCES2 attacks the ester link, which cuts the bipiperidine group off irinotecan converting it into SN-38, a very toxic topoisomerase I inhibitor. This is problematic as SN-38 is so toxic to normal gut tissue and it causes severe diarrhoea in patients[49]. However, whilst inhibiting the formation of this active metabolite would certainly reduce the gut problems, it might also affect efficacy. When the hCEs breaki into amide bonds, they can liberate toxic amines that can undergo further oxidative metabolism. This is thought to be possible with flutamide[110].

There is significant pharmaceutical interest in developing inhibitors of the hCEs, due to their endogenous roles in obesity pathology, but also in terms of their ability to clear drugs. A considerable number of dietary phytochemicals, such as flavonoids, tanshinones, and various triterpenoids can inhibit hCEs, although the challenge is building selective and practical therapeutic agents[103]. For instance, quercetin inhibits CYP3A4 and hCEs, but is only 1% bioavailable in man[111]. There is also a great deal of interest in the drug discovery world in understanding what features of molecules are particularly attractive to hCEs in terms of future drug design and how this relates to further metabolism of hCE-formed products[108], although no agent has reached the mainstream as yet[103]. This is partly as hCEs are not easy to work with as animal models are not very similar to human systems in this regard[112] and the level of understanding of the control of esterases also needs advancement. Other drugs cleared by various esterases include procaine, acetyl salicylate, and chloramphenicol, although genetic polymorphisms can prolong the clinical effects of a number of neuromuscular blocking drugs (succinylcholine, atracurium, mivacurium) that are hydrolysed by butyrylcholinesterase[113] (BChe; Chapter 7.2.4). The liver and kidney enzymes can hydrolyse the drugs listed above as well as several organophosphate and carbamate insecticides, as well as some herbicides.

6.10   Amino acid conjugation (mainly glycine)

The process of conjugation of glycine and other amino acids with small organic acids occurs in many organs, but mainly the liver and kidney, mostly in the mitochondrial matrix. The organic acid, which is usually benzoate, is activated to benzyl-CoA by HXM-A, an ATP-dependent CoA ligase. An enzyme known as GLYAT (glycine N-acyltransferase) catalyses the formation of hippurate (the first urinary metabolite I ever measured in laboratory work), with CoA thiol (CoASH) formed at the same time. Hippuric acid is much less lipophilic than benzoate acid and is excreted in urine. Until recently, this was regarded as a minor detoxification pathway and not considered to be particularly exciting. This was due mainly to a lack of attention paid to the process over the years and a sketchy understanding of its true endogenous purpose.

This has changed significantly with new thinking on this pathway, not least the 'glycine deportation hypothesis'[114]. This idea departs from the organic acid detoxification view, towards considering the pathway as a method of regulating amino acid levels that could become neurotoxic. However, the purpose of the pathway could well be three-fold; first, amino acid regulation, second, the control of potentially toxic organic acids, through increasing their polarity, so it is easier to pump them out of the mitochondria into the cytoplasm. Thirdly, CoASH is a vital cofactor for many reactions and the pathway maintains cellular levels[115].

There are many sources of organic acids from our diets, as well as those formed by the microbiome. Drugs such as salicylate and some carboxylic acids (some bile acids also) are caught in this pathway. There are concerns that excessive benzoate consumption in a diet heavy in preserved foods might deplete glycine levels, which could impact GSH maintenance[115]. The organic solvent toluene is cleared by CYP2E1 to benzyl alcohol[116], which then is conjugated to hippuric acid.

Little is known of the regulation of the glycine pathway, although it is downregulated in some cancers. Given its necessity in controlling organic acids that can form many reactive species[115], it is highly likely that it is operated at least in part by the nrf2 through response elements on both HXM-A and GLYAT genes.

6.11   Phase III transport processes

6.11.1   Introduction

Obviously, once xenobiotics have been converted into low-toxicity, higher-molecular-weight and high-water-solubility metabolites by the combination of CYPs, UGTs, SULTs and GSTs, this appears at first sight to be 'mission accomplished'. However, these conjugates must be transported against a concentration gradient out of the cell into the interstitial space between cells. Then they will enter the capillary system and thence to the main bloodstream and filtration by the kidneys. The biggest hurdle is the transport out of the cell, which is a tall order, as once a highly water-soluble entity has been created, it will effectively be 'ion-trapped' in the cell, as the cell membrane is highly lipophilic and is an effective barrier to the exit as well as entry of most hydrophilic molecules. In addition, failure to remove the hydrophilic products of conjugation reactions can lead to:

- toxicity of conjugates to various cell components;

- hydrolysis of conjugates back to the original reactive species;

- inhibition of conjugating enzymes.

If the cell can manage to transport them out, then they should be excreted in urine or bile and detoxification can proceed at a maximal rate, although enterohepatic recirculation can of course complicate this situation.

Consequently, an impressive array of multipurpose membrane bound transport carrier systems has evolved that can actively remove hydrophilic metabolites and many other low molecular weight drugs and toxins from cells. The relatively recent (1990s) term of Phase III metabolism has been applied to the study of this essential arm of the detoxification process.

6.11.2   ABC Efflux transporters

In the context of the elimination of biotransformed and now mainly hydrophilic substrates, transporters must address several tasks. Substrate identification, binding, and then efflux against a concentration gradient. The process must be responsive to local load and also subject to precise regulation and coordination with the other biotransformational systems. Such detailed demands can only be met at the level of the highly sophisticated ABC-type ATP-powered transporters, which as we discussed with P-gp, these are really closer to enzymes than just pumps.

Whilst the ABC transporters P-glycoprotein (P-gp, Chapters 4.4.7 and 5.6.2) BCRP (breast cancer resistance protein, Chapter 5.6.2) and related systems are usually associated with effluxing mainly lipophilic molecules from cells, the task of removing charged and mainly hydrophilic molecules falls to the multi-drug resistance associated proteins, known as MRP1-9, or MDRs or ABCCs. These transporters are part of the human 12 member 'C' family of ABC trans-porters and are thus designated ABCC1-6 (MRP1-6), and then later on MRPs7-9 are termed ABCC10-12[117]. The gene whose defect causes cystic fibrosis is part of the ABCC/MRP family (cystic fibrosis transmembrane conductance regula-tor; CFTR/ABCC7) and is a chloride ion transporter[117]. The terms ABCC and MRP are used interchangeably, but ABCC links them to their genes and their wider family, such as P-gp and BCRP.

As outlined in Chapter 4.4.7, the ABC transporters have a common structure, with a set of transmembrane hydrophobic domains (TMDs), which as the name suggests, span the plasma (or endoplasmic reticulum) membrane. These bind/recognise substrate and transport it. The TMDs are linked to two nucleotide-binding domains (NBDs), which are the power packs that process ATP[117]. Concerning the ABCC transporters, some have two TMDs and two NBDs and this basic structure is seen in ABCC4, 5, 11, and 12, which are known as the 'short' ABCCs. However ABCC1, 2, 3, 6, and 10 have an extra TMD, which makes them much larger ('long' ABCCs)[118]. The differing substrate profiles of the ABCC (MRPs) and the lipophilic substrate pumping P-gp is reflected in many structural differences, particularly in the nucleotide binding areas and in their substrate binding sites[119].

The ABCCs have been intensively investigated for their roles in drug resistance during cancer chemotherapy, but they are not as well understood as P-gp. How the ABCCs recognize and process their substrates is emerging through studies that substitute particular amino acids at key sites and then measure the functional impact on the transporters. The internal allosteric mechanisms of regulation for ABCC1 can now be studied in this fashion, as well as how they recognize their substrates[120]. Whilst the ABCCs are similar in their amino acids and general struc-ture, they tend to have very distinct substrate preferences and together, they pro-vide a formidable and exquisitely controllable cellular efflux system, which is coordinated with the main biotransformational processes within cells.

ABCC1 (MRP-1)

This 190 kDa glycoprotein is found in all tissues, particularly the liver, kidney, and blood–brain barrier. ABCC1 specializes in organic and some inorganic ani-ons (arsenic derivatives), so it can process a wide variety of conjugated drug metabolites, such as mercapturates (GSH-derived), glucuronides and sulphates, as well as many anticancer drugs, such as doxorubicin, vincristine (pumped out with GSH), and the taxanes[120,121]. ABCC1 will also transport metabolites of significant toxins, such as those of the aflatoxins[122]. ABCC1 operates in concert with high levels of GSH in its environment both for the transportation process itself and the conjugation of molecules for it to transport[122]. ABCC1's affinity for other drugs varies; it is not very efficient at transporting protease inhibitors[122], but better at clearing irinotecan and SN-38[117]. Endogenously, ABCC1 is thought to handle the steroid conjugates formed during normal metabolism as well as unconjugated bilirubin[117].

ABCC2 (MRP2)

This transporter has a strong overlap with ABCC1, although it is actually the largest of this class and is most commonly found in the liver, gut, and kidney[117]. It processes many sulphated bile acids and bilirubin conjugates and a condition known as Dubin-Johnson syndrome is associated with lack of expression of ABCC2 and patients cannot clear bilirubin conjugates efficiently. That condition can cause jaundice, but as we know, moderate accumulation of bilirubin is not necessarily a problem and the condition is benign. ABCC2 transports many of the anti-cancer drugs ABCC1 can move, but has a greater affinity for other agents such as cisplatin and the protease inhibitors[122]. ABCC2 overexpression is seen in cisplatin-resistant cells[117]. ABCC2 also extensively transports flavonoids and their metabolites and can export so much GSH it can promote oxidative stress in cells[117].

ABCC3 (MRP3)

Expression of this glycoprotein is very high in the liver and it can, to some extent, compensate for low expression of ABCC2[122]. ABCC3 has a major role in handling bile acids and can transport them with and without conjugation, which ABCC1 and 2 cannot. Aside from that, it is very similar structurally and functionally to ABCC1 (58% homology)[122]. ABCC3 transports anionic conjugates of GSH and glucuronides, but does not transport GSH by itself[117]. It is also less efficient at promoting resistance to anticancer agents in comparison with the previous two transporters. ABCC3 is believed to be a backup system that can compensate when bile acid levels rise too high in cells (such as in cholestasis).

ABCC4-6 (MRP4-6)

The focus of ABCC4 is in extra-hepatic tissues such as the kidney and lung, as well as the prostate. It appears to specialize in transporting signal molecules like cAMP, cGMP and eicosenoids, as well as the usual steroid conjugates[122,123]. It seems to have a major role in controlling subcellular levels of cyclic nucleotides as well as a strong part in the clearance of urate[123]. ABCC4 has a significant task in the integrity of the blood–brain barrier, behaving in a similar way to Pg-p to push drugs and toxins out into the circulation. It also clears diuretics such as furosemide and a wide variety of drugs, ranging from antibiotics to various anticancer drugs. Indeed, it is linked with resistance to irinotecan. ABCC4 acts with ABCC3 to protect the liver from excessive levels of bile acids and can be expressed in basolateral and apical areas, which allow it to move substrates in different directions in different cell types, such as in the liver and the kidney[123].

ABCC5 seems to have a very similar profile of activity to ABCC4 and is found in all tissues.

ABCC6 again shares functionality profiles with the other transporters and is also associated with resistance to anticancer drugs[117,123]. ABCC6 is very close in structure and function to ABCC1, although it is not yet known what it actually does in terms of endogenous substrates[124]. ABCC6 mutations are strongly linked with a genetic disorder of connective tissue, called PXE or *pseudoxanthoma elasticum*, although the nature of the relationship is not yet known[124]. ABCC6 is heavily involved in the transport of tyrosine kinase inhibitors, as are several others of its family, and it is linked with resistance to nilotinib and dasatinib[124].

ABCC10-12 (MRP 7-9)

Much less is known about the functions and endogenous substrates of ABCC10-12. ABCC10 is very similar in functionality to ABCC1-4, clearing lipophilic anions from most tissues. ABCC10 is upregulated by vinca alkaloids and the taxane derivatives, thus conferring resistance to these agents[118]. ABCC11 is found in several tissues including the liver and is part of the process of ear-wax production. ABCC11 is also linked to anticancer drug resistance and clears cyclic nucleotides and various purines and pyrimidines. Its expression in patients with breast cancer appears to change, but some studies have shown increases and others decreases. To date, very little is known about ABCC12[118,124].

ABCC regulation

The same battery of nuclear receptors (such as PXR, CAR, and FXR) and transcription factors (nrf2) that regulate the CYPs, UGTs, SULTs, GSTs, and EHs control the ABCCs. Logically, the induction of the ABCCs is part of the general coordination of the response to both endo and xenobiotics, in terms of their biotransformation and the channelling of the metabolites towards the circulation and the gut for renal and faecal elimination. Whilst during normal metabolism this coordinated response is highly desirable and efficient, it is extremely problematic in cancer chemotherapy. In that case, upregulation of the ABCCs is one of the very effective weapons malignant cells deploy to ensure little or no effective residence time for cytotoxic drugs[117,118]. This has led to interest in trying to inhibit the ABCCs to improve chemotherapeutic outcomes.

ABCCs: inhibition

As these transporters can seriously erode anticancer drug penetration into malignant tissue, there has been strong interest in co-administering inhibitors that might diminish this problem. As with P-gp in Chapter 5.6.2, transporters are the same in all tissues, even after a selective inhibitor for ABCC1 for example, has been developed, there is nothing to prevent it from blocking transporter function in the healthy tissue in the organ in which the malignant cells reside and indeed everywhere else in the body. Even worse, blocking the transporters in healthy tissues locally and systemically would prevent them from excluding the anticancer drug, so promoting its local bystander and systemic toxicity. So ideally, an inhibitor would somehow need to be selective for malignant tissue transporters only and this seems somewhat unrealistic for a small molecule.

Nevertheless, of the ABCC inhibitors under consideration, the tyrosine kinase (TK) inhibitors such as ibrutanib have shown experimental promise in preventing ABCC1 from expelling vinca alkaloids from cell models. Ibrutanib appears to operate pharmacologically on the transporter and does not affect its expression[121]. However, it also overcomes taxane resistance with P-gp and ABCC10 inhibition[121], so whilst it is effective experimentally, it is not exactly selective. ABCC10 is blocked by many TK inhibitors, and this raises the possibility of these drugs having a dual impact in chemotherapy and potentiating the effects of co-administered anti-neoplastic drugs[118]. Indeed, the third-generation P-gp inhibitor tariquidar has shown efficacy in inhibiting ABCC10, but does not impact BRCP, or ABCCs 1-3[117,118]. Several other derivatives of GSH conjugates and triterpenes can inhibit the ABCCs that are under consideration.

If a tumour or malignant tissue was identified to overexpress a specific ABCC that was protecting it from chemotherapy, then a selective inhibitor of that transporter could perhaps improve the chemotherapeutic response. As mentioned above, the bystander tissues would be unable to expel the cytotoxin, but other ABC transporters might be adequate to protect them. Systemically, other ABC transporters might protect tissues around the body from the drug. It is likely that the cytotoxin will induce other transporters in the tumour and this will diminish response.

If we consider other therapeutic approaches to prevent ABCC overexpression, since they respond to so many nuclear receptor and transcription factor systems, it would be very difficult to prevent their overexpression as we saw with efforts to control PXR-mediated tumour defence in Chapter 4.4.3. One possible route would be the use of post-transcriptional methods, and progress has been made in this area. The micro-RNA miR-326 expression seems to control ABCC1 expression and miR-297 can rein in ABCC2 and manipulating these miRs can downregulate the transporters and increase the potency of cytotoxics[115]. Whilst this exciting prospect suggests that specific ABCCs can be switched off using miRNAs, it still requires some means of restricting the radius of the miRNA efficacy, or transporters in the nearby healthy cells and other tissues will also be downregulated.

6.11.3   RLIP76

When I was writing this section I had spent a great deal of time looking at the ABC cassette transporters, particularly their role in cancer chemotherapy resistance. It seemed that the non-ABC transporter RLIP76[125] could be the Keyser Söze[126] of transporters. RLIP76 has many aliases, ranging from RalA Binding Protein 1, to Dinitrophenyl S-Glutathione ATPase (DSG-ATPase), with RLIP76 (Ral-interacting protein of 76 kDa)[127,128] being the most common title to date. RLIP76 originates from the gene *RALBP1* found on chromosome 18 and it makes the ABC cassette transporters look rather primitive. RLIP76 can accomplish the remarkable feat of not only acting as the major efflux system for glutathione-related electrophile conjugates, but it also coordinates a vast array of cellular functions. RLIP76 organizes its complex multi-signalling function through its internal binding sites for Ral GTPases and Rho G-proteins[125], which allow it to participate in myriad key processes such as stress defence, apoptosis, cell migration, and mitochondrial division[126,127]. RLIP76 even coordinates the response to oxidative stress in terms of glucose and lipid metabolism[127].

In healthy tissue, this ubiquitously expressed protein is not actually essential (knockout animals are fine without it) and interestingly, RLIP76 has not been shown to be involved with resistance to AED drugs clinically[128], for example. However, it is essential for a rapid and effective response to protect us from elec-trophilic toxins by accelerating the efflux of their metabolites. So it is also emerging as a major issue in resistance to cancer chemotherapy and is upregulated in malignant cells[127,129]. RLIP76 can rapidly upregulate transporter-mediated efflux of anticancer drugs, whilst suppressing apoptosis and marshalling the gen-eral aggression and resistance to chemotherapeutic attack of malignant cells[127,130]. Experimentally, shutting off RLIP76 with antibodies is very effective at driving malignant cells towards apoptosis. RLIP76 complements other parallel defence systems such as those coordinated by AhR, PXR, and most closely, nrf2. Understanding of the internal modulation of RLIP76 activity is growing, with the discovery of internal transcription factors such as POB1 (partner of Ralbp1), which can inhibit its transport function, for example[127]. Indeed, given the scale of the processes the RLIP76 system controls, it looks in the future to be a far more productive therapeutic target in malignant cells than the ABC transporters.

6.12   Biotransformation: integration of processes

As has been mentioned already, it is clear that the whole process of detection, metabolism, and elimination of endobiotic and xenobiotic agents is minutely coordinated and is responsive to changes in load in individual tissues. The CYPs, UGTs, ABCs, and RLIP76 are all tightly regulated through the NR system of PXR, CAR, FXE, PPARα, LXR, plus the cytoplasmic systems of AhR and nrf2.

Some enzyme/pump processes are closely linked, such as CYP3A4 and P-gp, as inducers powerfully increase both systems capacity. The reactive species protection 'arm' of biotransformation (particularly of reactive electrophiles) is also controlled through multiple systems such as RLIP76 and nrf2, which coordinate not only the interception of reactive species by GSTs and EHs, but also the supply of their GSH substrate. This latter coordination is particularly relevant in resistance to cancer chemotherapy and happens because overexpression of any one entity alone cannot rid the cell of the toxin. If the GSTs or the UGTs alone were upregulated, their products would soon negatively feedback on their activity and they would also run out of cofactors. This is particularly true of GSH, as not only is it the only GST cofactor/substrate, it is also necessary for the activity of the ABCCs (but not ABCC3) as they use it to extrude the conjugates from the cell. The upregulation of GSH synthesis usually by nrf2 requires more ATP, which must come from upregulation in mitochondrial production. The ABCCs, GSH production, GST/UGT, and other stress responses are induced in concert. Whilst much of the integration and coordination of detoxification processes remains to be uncovered, the mere fact that Keith Richards has survived three editions of this book and continues to thrive testifies to their resilience and effectiveness.

Figure 6.13 shows a scheme that tries to sum up the main aspects of how biotransformation is coordinated alongside the Phase III systems of efflux proteins such as the ABCCs. This system operates to some extent in virtually every tissue in the body, as well as the main areas of metabolism, such as the liver, gut, lung, kidney, and gonads.

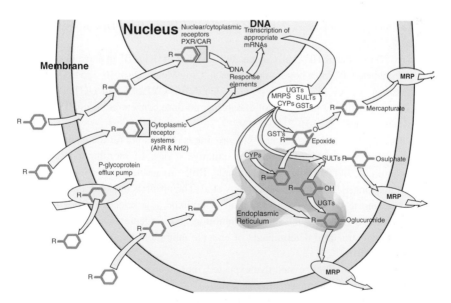

Figure 6.13 A scheme showing some major biotransformational systems operating in concert for the detection and elimination of xenobiotic molecules such as aromatic hydrocarbons

References

1. Mazerska Z, Mróz A, Pawłowska M, et al. The role of glucuronidation in drug resistance. Pharmacol. & Ther. 159, 35–55, 2016.

2. Yang N, Sun R, Liao X, et al. UDP-glucuronosyltransferases (UGTs) and their related metabolic cross-talk with internal homeostasis: A systematic review of UGT isoforms for precision medicine. Pharmacol. Res. 121, 169–183, 2017.

3. Liu Y, Coughtrie MWH. Revisiting the latency of uridine diphosphate-glucuronosyl transferases (UGTs)—How does the endoplasmic reticulum membrane influence their function? Pharmaceut. 9, 32, 1–16, 2017.

4. Ishi Y, Iwanaga M, Nishimura Y, et al. Protein–Protein Interactions between Rat Hepatic Cytochromes P450 (P450s) and UDP-Glucuronosyltransferases (UGTs): evidence for the functionally active UGT in P450–UGT complex. Drug Metab. Pharmacokinet. 22, 367–376, 2007.

5. Sanchez-Dominguez CN, Gallardo-Blanco HI, Salinas-Santander MA, et al. Uridine 5'-diphospho-glucronosyltrasferase: its role in pharmacogenomics and human disease. Exper. & Ther. Med. 16, 3–11, 2018.

6. Mubarokah N, Hulin JA, Mackenzie PI et al. Cooperative regulation of intestinal UDP-glucuronosyltransferases 1A8,–1A9 and 1A10 by CDX2 and HNF4⊠ is mediated by a novel composite regulatory element Mol. Pharm. 93, 541–552, 2018.

7. Yilmaz L, Borazan E, Aytekin T, et al. Increased UGT1A3 and UGT1A7 expression is associated with pancreatic cancer. Asian Pacif. J. Canc. Prev. 16, 1651–1655, 2015.

8. Bushey RT, Dluzen DF, Lazarus P. Importance of UDP-glucuronosyltransferases 2A2 and 2A3 in tobacco carcinogen metabolism. Drug Metab. Dispos. 41, 170–179, 2013.

9. Xu C, Gao J, Zhang HF, et al. Content and activities of UGT2B7 in human liver *in vitro* and predicted *in vivo*: a bottom-up approach. Drug Metab. Disp. 46, 1351–1359, 2018.

10. Kwara A, Lartey M, Sagoe KW, et al. CYP2B6, CYP2A6 and UGT2B7 genetic polymorphisms are predictors of efavirenz mid-dose concentration in HIV-infected patients. AIDS 23, 2101–2106, 2009.

11. Klimas R, Mikus G. Morphine-6-glucuronide is responsible for the analgesic effect after morphine administration: a quantitative review of morphine, morphine-6-glucuronide, and morphine-3-glucuronide. Brit. J Anaesth. 113, 935–944, 2014.

12. Pattanawongsa A, Nair PC, Rowland A et al. Therapeutics human UDP-glucurono-syltransferase (UGT) 2B10: validation of cotinine as a selective probe substrate, inhibition by UGT enzyme-selective inhibitors and antidepressant and antipsychotic drugs, and structural determinants of enzyme inhibition. Drug Metab. Dispos. 44, 378–388 2016.

13. Moran VE. Cotinine: beyond that expected, more than a biomarker of tobacco consumption. Front. Pharmacol. 3, 173, 2012.

14. Murphy SE, Sipe CJ, Cho K et al. Low cotinine glucuronidation results in higher serum and saliva cotinine in African American compared to White smokers. Canc. Epidemiol. Bio. Prev. 26, 1093–1099, 2017.

15. Hyland R, Osborne T, Payne A, et al. *In vitro* and *in vivo* glucuronidation of midazolam in humans. Br. J. Clin. Pharmacol. 67, 445–454, 2009.
16. Ohno S, Nakajin S. Quantitative analysis of UGT2B28 mRNA expression by real-time RT-PCR and application to human tissue distribution study. Drug Metab Lett. 5, 202–208, 2011.
17. UGT2B11 UDP glucuronosyltransferase family 2 member B11 [Homo sapiens (human)] https://www.ncbi.nlm.nih.gov/gene/10720…UGT2B11.
18. Bulmer AC, Bakrania B, Du Toit EF, et al. Bilirubin acts as a multipotent guardian of cardiovascular integrity: more than just a radical idea. Amer. J. Physiol. – Heart & Circ. Physiol. 315, H429–H447, 2018.
19. Zhang DL, Chando TJ, Everett DW, et al. *In vitro* inhibition of UDP glucuronosyltransferases by atazanavir and other HIV protease inhibitors and the relationship of this property to *in vivo* bilirubin glucuronidation. Drug Metab. Disp. 33, 1729–1739, 2005.
20. Qosa H, Avaritt BR, Hartman NR et al. *In vitro* UGT1A1 inhibition by tyrosine kinase inhibitors and association with drug-induced hyperbilirubinemia. Canc. Chemother. Pharmacol. 82, 795–802, 2018.
21. Weng YH, Cheng SW, Yang CY et al. Risk assessment of prolonged jaundice in infants at one month of age: a prospective cohort study. Scient. Rep. 8, 14824, 2018.
22. McKiernan P. Neonatal jaundice Clin. Res. Hepatol. Gastroenterol. 36, 253–256, 2012.
23. Perreault M, Gauthier-Landry L, Trottier J, et al. The Human UDP-glucuronosyltransferase UGT2A1 and UGT2A2 enzymes are highly active in bile acid glucuronidation. Drug Metab Dispos. 41, 1616–1620, 2013.
24. Dull MM, Kremer AE. Management of chronic hepatic itch. Dermatol. Clin. 36, 293–300, 2018.
25. Gordon-Walker TT. Editorial: alleviating the itch–the safety of rifampicin in the real world. Aliment. Pharmacol. & Ther. 47, 1332–1333, 2018.
26. Van Vleet TR, Liu H, Leeb A et al. Acyl glucuronide metabolites: implications for drug safety assessment. Toxicol. Lett. 272, 1–7, 2017.
27. Coleman MD, Holden LJ. The Methaemoglobin Forming and GSH depleting effects of dapsone and monoacetyl dapsone hydroxylamines in human diabetic and non-diabetic erythrocytes *in vitro*. Env. Tox. Pharmacol 17, 55–59 (2004).
28. Coleman MD, Scott AK, Breckenridge AM et al. The use of cimetidine as a selective inhibitor of dapsone N-hydroxylation in man. Brit. J. Clin. Pharmac. 30, 761–767 (1990).
29. Zenser TV, Lakshmi VM, Hsu FF et al. Metabolism of N-acetylbenzidine and initiation of bladder cancer. Mutat. Res. 506–507, 29–40, 2002.
30. Wang CY, King CM. Metabolic activation of benzidine. Int. J. Cancer. 121, 1640–1641, 2007.
31. Carreon T, Kadlubar FF, Ruder AM, et al. Reply to the letter to the editor: "N-Acetyltransferases and the susceptibility to benzidine-induced bladder carcinogenesis". Int. J. Canc. 121, 1637–1639 2007.
32. Bock KW. Homeostatic control of xeno-and endobiotics in the drug-metabolizing enzyme system. Biochem. Pharmacol 90, 1–6, 2014.

33. Sugatani J, Mizushima K, Osabe M et al. Transcriptional regulation of human UGT1A1 gene expression through distal and proximal promoter motifs: implication of defects in the UGT1A1 gene promoter. Naunyn-Schmied. Arch. Pharmacol. 377, 597–605, 2008.

34. Hu DG, Rogers A, Mackenzie PI. Epirubicin upregulates UDP glucuronosyltransferase 2B7 expression in liver cancer cells via the p53 pathway. Mol. Pharmacol. 85, 887–897, 2014.

35. Zahreddine HA, Culjkovic-Kraljacic B, Assouline S, et al. The sonic hedgehog factor GLI1 imparts drug resistance through inducible glucuronidation. Nature 511, 90–93, 2014.

36. Marques SC, Ikediobi ON, The clinical application of UGT1A1 pharmacogenetic testing: gene–environment interactions. Human Genomics. 4(4), 238–249, 2010.

37. Fujita K, Sugiyama M, Akiyama Y, et al. The small-molecule tyrosine kinase inhibitor nilotinib is a potent noncompetitive inhibitor of the SN-38 glucuronidation by human UGT1A1. Canc. Chemother Pharmacol 67, 237–24, 2011.

38. Oda S, Fujiwara R, Kutsuno Y, et al. Targeted screen for human UDP-glucuronosyltransferases inhibitors and the evaluation of potential drug–drug interactions with Zafirlukast. Drug Metab. Disp. 43, 812–818, 2015.

39. Wiggins BS, Saseen JJ, Morris PB. Gemfibrozil in combination with statins—is it really contraindicated? Curr. Atheroscler. Rep. 18, 18, 2016.

40. Husni Z, Ismail S, Zulkiffli MH, et al. *In vitro* Inhibitory effects of *Andrographis paniculata, Gynura procumbens, Ficus deltoidea, and Curcuma xanthorrhiza* extracts and constituents on human liver glucuronidation activity. Pharmacog. Magaz. 13, S236–S243, 2017.

41. Rajpoot M, Sharma AK, Sharma A, et al. Understanding the microbiome: Emerging biomarkers for exploiting the microbiota for personalized medicine against cancer. Sem. Canc. Biol. 52, 1–8, 2018.

42. Cassidy A Minihane AM. The role of metabolism (and the microbiome) in defining the clinical efficacy of dietary flavonoids. Am J. Clin. Nutr. 105, 10–22, 2017.

43. Jenkins AL, Crann SE, Money DM, et al. "Clean and Fresh": understanding women's use of vaginal hygiene products. Sex Rol. 78, 697–709, 2018.

44. Deng Y, Rogers M, Sychterz C, et al. Investigations of hydrazine cleavage of eltrombopag in humans. Drug Metab. Disp. 39, 1747–1754, 2011.

45. Wilkinson EM, Ilhan ZE, Herbst-Kralovet MM. Microbiota–drug interactions: Impact on metabolism and efficacy of therapeutics. Maturitas 112, 53–63 2018.

46. Wilson ID, Nicholson JK. Gut microbiome interactions with drug metabolism, efficacy and toxicity. Transl Res. 179, 204–222, 2017.

47. Lindenbaum J, Rund DG, Butler VP Jr, et al. Inactivation of digoxin by the gut flora: reversal by antibiotic therapy. N. Engl. J. Med. 305, 789–794, 1981.

48. Hashim H, Azmin S, Razlan H, et al. Eradication of Helicobacter pylori infection improves levodopa action, clinical symptoms and quality of life in patients with Parkinson's disease. PLoS One 9, e112330, 2014.

49. Swami U, Goel S, Mani S. Therapeutic Targeting of CPT-11 Induced diarrhea: a case for prophylaxis. Curr. Drug Targ. 14(7), 777–797, 2013.

50. Cook I, Wang T, Leyh TS. Isoform-specific therapeutic control of sulfonation in humans. Biochem. Pharmac. 159, 25–31, 2019.
51. Cook I, Wang T, Falany CN, et al. A nucleotide-gated molecular pore selects sulfotransferase. Biochemistry 51, 5674–5683, 2012.
52. Nowell S, Ratnasinghe DL, Ambrosone CB, et al. Association of SULT1A1 phenotype and genotype with prostate cancer risk in African-Americans and Caucasians. Canc. Epidemiol. Biomark. Prev. 13, 270–276, 2004.
53. Dubaisi S, Barrett KG, Fang H, et al. Regulation of cytosolic sulfotransferases in models of human hepatocyte development. Drug Metab. Disp. Drug 46, 1146–1156, 2018.
54. Riches Z, Stanley EL, Bloomer JC, et al. Quantitative evaluation of the expression and activity of five major sulfotransferases (SULTs) in human tissues: The SULT "pie". Drug Metab Dispos. 37, 2255–2261, 2009.
55. Cook I, Wang T, Almo SC, et al. Testing the sulfotransferase molecular pore hypothesis. J. Biol. Chem. 288, 8619–8626, 2013.
56. Cook I, Wang T, Falany CN, et al. The allosteric binding sites of sulfotransferase 1A1. Drug Metab. Dispos. 43, 418–423, 2015.
57. Wang T, Cook I, Leyh TS. The NSAID allosteric site of human cytosolic sulfotransferases. J Biol Chem. 292, 20305–20312, 2017.
58. Wang T, Cook I, Falany CN, et al. Paradigms of sulfotransferase catalysis: the mechanism of SULT2A1, J. Biol. Chem. 289, 26474–26480, 2014.
59. Gamage N, Barnett A, Hempel N, et al. Human sulfotransferases and their role in chemical metabolism. Tox. Sci. 90, 5–22, 2006.
60. Suiko M, Kurogi K, Hashiguchi T, et al. Updated perspectives on the cytosolic sulfotransferases (SULTs) and SULT-mediated sulfation. Biosci. Biotechnol. Biochem. 81, 1, 63–72, 2017.
61. Poschner S, Maier-Salamon A, Zehl M, et al. Resveratrol inhibits key steps of steroid metabolism in a human estrogen-receptor positive breast cancer model: impact on cellular proliferation front. Pharmacol. 9(742), 1–16, 2018.
62. Leung AWY, Backstrom I, Bally MB. Sulfonation, an underexploited area: from skeletal development to infectious diseases and cancer. Oncotarget 23, 55811–55827, 2016.
63. Benedikter BJ, Weseler AR, Wouters EFM et al. Redox-dependent thiol modifications: implications for the release of extracellular vesicles. Cell. Molec. Life Sci. 75, 2321–2337, 2018.
64. Lu SC. Regulation of glutathione synthesis. Mol. Aspects Med. 30, 42–59, 2009.
65. Fukuto JM, Ignarro LJ, Nagy P. Biological hydropersulfides and related polysulfides–a new concept and perspective in redox biology Febs. Lett. 592, 2140–2152, 2018.
66. Lu SC. Glutathione synthesis. Biochim. Biophys. Acta. 1830, 3143–3153, 2013.
67. Chatterjee A, Gupta S. The multifaceted role of glutathione S-transferases in cancer. Cancer Letters 433, 33–42, 2018.
68. Dasaria S, Ganjayia MS, Yellanurkondab P, et al. Role of glutathione S-transferases in detoxification of a polycyclic aromatic hydrocarbon, methylcholanthrene. Chem. Biol. Interact. 294, 81–90, 2018.

69. Zhang J, Yea ZW, Singh S, et al. An evolving understanding of the S-glutathionyla-tion cycle in pathways of redox regulation. Free Rad. Biol. Med. 120, 204–216, 2018.

70. Wu B, Dong D. Human cytosolic glutathione transferases: structure, function, and drug discovery. Trends Pharmacol. Sci. 33, 656–668, 2012.

71. Deponte M. Glutathione catalysis and the reaction mechanisms of glutathione-dependent enzymes. Biochim. Biophys. Acta 1830, 3217–3266, 2013.

72. Pathania S, Bhatia R, Baldi A, et al. Drug metabolizing enzymes and their inhibitors' role in cancer resistance. Biomed. & Pharmacother. 105, 53–65, 2018.

73. Liu F, Lin Y, Li Z, et al. Glutathione S-transferase A1 (GSTA1) release, an early indicator of acute hepatic injury in mice. Food Chem. Tox. 71, 225–230, 2014.

74. Le Trong, I, Stenkamp RE, Ibarra C, et al. 1.3-A° resolution structure of human glutathione S-transferase with S-hexyl glutathione bound reveals possible extended ligandin binding site. Proteins 48, 618–627, 2002.

75. Suthar, PC. Glutathione-S-transferases: a brief on classification and GSTm1-t1 activity. Int. J. Pharm. Sci. Res. 8, 1023–1027, 2017.

76. Zheng Y, Ni X, Jiao Y, et al. Genetic polymorphisms of glutathione S-transferase (GSTM1, GSTT1 and GSTP1) with esophageal cancer risk: a meta-analysis. Int. J. Clin. Exp. Med. 9, 13268–13280, 2016.

77. Stocco G, Pelin M, Franca R, et al. Pharmacogenetics of azathioprine in inflamma-tory bowel disease: a role for glutathione-S-transferase? World J. Gastroenterol. 20, 3534–3541, 2014.

78. Sailaja K, Surekha D, Rao DN, et al. Association of the GSTP1 gene (Ile105Val) polymorphism with chronic myeloid leukemia. Asian Pac. J. Canc. Prev. 11, 461–464, 2010.

79. Kwong YL, Au WI, Liang RHS. Acute Myeloid Leukemia after Azathioprine Treatment for Autoimmune Diseases: association with −7/7q−. Canc. Gen. Cytogenet. 104, 94–97, 1998.

80. Relling MV, Gardner EE, Sandborn WJ, et al. Clinical pharmacogenetics implementation consortium guidelines for thiopurine methyltransferase genotype and thiopurine dosing: 2013 update. Clin. Pharmacol. Ther. 93, 324–325, 2013.

81. Allocati N, Masulli M, Di Ilio C, et al. Glutathione transferases: substrates, inhibitors and pro-drugs in cancer and neurodegenerative diseases. Oncogen. 7(8), 1–15, 2018.

82. Vergote NJ, Finkler JB, Hall O, et al. Randomized phase III study of canfosfamide (C, TLK286) plus pegylated liposomal doxorubicin (PLD) versus PLD as second-line therapy in platinum (P) refractory or resistant ovarian cancer (OC). J. Clin. Oncol. 27, 15S, 5552–5552, 2009.

83. Kenny RG, Chuah SW, Crawford A, et al. Platinum (IV) prodrugs – a step closer to Ehrlich's vision? Eur. J. Inorg. Chem. 1596–1612, 2017.

84. Bartolini D, Galli F. The functional interactome of GSTP: A regulatory biomolecular network at the interface with the Nrf2 adaption response to oxidative stress. J. Chromatogr. B, 1019, 29–44, 2016.

85. Taguchi K, Yamamoto M. The KEAP1–NRF2 Systemin Cancer. Front. Oncol. 7, 85. 1–11, 2017.

86. Suzuki T, Yamamoto M. Molecular basis of the Keap1–Nrf2 system. Free Rad. Biol. & Med. 88 93–100, 2015.

87. Steele ML, Fuller S, Patel M, et al. Effect of Nrf2 activators on release of glutathione, cysteinylglycine and homocysteine by human U373 astroglial cells. Redox Biology 1, 441–445, 2013.

88. Kensler TW, Chen JG, Egner PA, et al. Effects of glucosinolate-rich broccoli sprouts on urinary levels of aflatoxin-DNA adducts and phenanthrene tetraols in a randomized clinical trial in He Zuo Township, Qidong, People's Republic of China. Canc. Epidemiol. Biomark. Prev. 14, 2605–2613, 2005.

89. Oesch-Bartolomowicz, B, Oesch F. Vol 5 'ADME Tox Approaches' in Editors-in-Chief: Taylor JB, Triggle DJ. Comprehensive Medicinal Chemistry II 2nd Edition, 2007, ISBN 978-0-08-045044-5.

90. El-Sherbeni AA, El-Kadi AOS. The role of epoxide hydrolases in health and disease. Arch. Toxicol. 88, 2013–2032, 2014.

91. Su S, Yang X, Omiecinsk CJ. Intronic DNA elements regulate nrf2 chemical responsiveness of the human microsomal epoxide hydrolase gene (EPHX1) through a far-upstream alternative promoter. Biochim Biophys Acta. 1839, 493–505, 2014

92. Wu HF, Chen YJ, Wu SZ, et al. Soluble epoxide hydrolase inhibitor and 14,15-epoxyeicosatrienoic acid-facilitated long-term potentiation through cAMP and CaMKII in the hippocampus. Neural Plast. 3467805, 2017.

93. Salazar-González RA, Turiján-Espinoza E, Hein DW, et al. Expression and genotype-dependent catalytic activity of N-acetyltransferase 2 (NAT2) in human peripheral blood mononuclear cells and its modulation by Sirtuin 1. Biochem. Pharmac. 156, 340–347 2018.

94. Sim E, Abuhammad A, Ryan A. Arylamine N-acetyltransferases: from drug metabolism and pharmacogenetics to drug discovery. Brit. J. Pharmacol. 171, 2705–2725, 2014.

95. Butcher NJ, Tetlow NL, Cheung C, et al. Induction of human arylamine Nacetyltransferase type I by androgens in human prostate cancer cells. Cancer Res 67: 85–92, 2007.

96. Marzuillo P, Di Sessa A,. Umano GR, et al. Novel association between the nonsynonymous A803G polymorphism of the N-acetyltransferase 2 gene and impaired glucose homeostasis in obese children and adolescents. Pediatr. Diab. 18, 478–484, 2017.

97. Sun S, Fu J. Methyl-containing pharmaceuticals: methylation in drug design. Bioorg. & Med. Chem. Lett. 28, 3283–3289, 2018.

98. Maser T, Rich M, Hayes D, et al. Tolcapone induces oxidative stress leading to apoptosis and inhibition of tumor growth in Neuroblastoma. Canc. Med. 6, 1341–1352, 2017.

99. Sparta M, Alexandrova AN. How Metal Substitution Affects the Enzymatic Activity of Catechol-O-Methyltransferase. PLoS ONE 7(10), e47172, 2012.

100. Cope LM, Hardee JE, Soules, ME, et al. Reduced brain activation during inhibitory control in children with COMT Val/Val genotype. Brain and Behav. 6, e00577, 2016.

101. Fabbri M, Ferreira JJ, Lees A. Opicapone for the treatment of Parkinson's disease: A Review of a New Licensed Medicine. Movement Dis. 33, 1528–1539, 2018.

102. Rodrigues FB, Ferreira, JJ. Opicapone for the treatment of Parkinson's disease. Expert Opin. Pharmacother. 18, 445–453, 2017.

103. Zou LW, Li YG, Wang P, et al. Design, synthesis, and structure-activity relationship study of glycyrrhetinic aid derivatives as potent and selective inhibitors against human carboxylesterase 2. Eur. J. Med. Chem. 112, 280–288, 2016.

104. Wang D, Zou L, Jin Q, et al. Human carboxylesterases: a comprehensive review. Acta Pharm Sin B. 8, 699–712, 2018.

105. Kabeya T, Matsumura W, Iwao T, et al. Functional analysis of carboxylesterase in human induced pluripotent stem cell-derived enterocytes. Biochem. Biophys. Res Comm 486, 143–148, 2017.

106. Gabriel M, Puccini P, Lucchi M, et al. Presence and inter-individual variability of carboxylesterases (CES1 and CES2) in human lung. Biochem. Pharmacol. 150, 64–71, 2018.

107. Maruichi T, Fukami T, Nakajima M, et al. Transcriptional regulation of human carboxylesterase 1A1 by nuclear factor-erythroid 2 related factor 2 (nrf2). Biochem. Pharmacol. 79, 288–295, 2009.

108. Yao J, Chen X, Zheng F et al. Catalytic Reaction Mechanism for Drug Metabolism in Human Carboxylesterase-1: Cocaine Hydrolysis Pathway. Mol. Pharmac. 15, 3871–3880, 2018.

109. Oh J, Lee S, Lee H, et al. The novel carboxylesterase 1 variant c.662A>Gmay decrease the bioactivation of oseltamivir in humans. PLoS ONE 12(4): e0176320, 2017.

110. Kang SP, Dalvie D, Smith E, et al. Bioactivation of flutamide metabolites by human liver microsomes. Drug Metab. Disp. 36, 1425–1437, 2008.

111. Palle S, Neerati P, Enhancement of oral bioavailability of rivastigmine with quercetin nanoparticles by inhibiting CYP3A4 and esterase. Pharmacol. Rep. 69, 365–37, 2017.

112. Potter PM. Evaluating esterase-mediated drug metabolism should be easy, why do we fail? Drug Metab. Pharmacokinet. 33 S1eS15 2018.

113. Lee LA, Athanassoglou V, Pandit JJ. Neuromuscular blockade in the elderly patient. J. Pain Res. 9 437–444, 2016.

114. Beyoglu, D, Idle JR. The glycine deportation system and its pharmacological consequences. Pharmacol. Ther. 135, 151–167, 2012.

115. Badenhorst CPS, Erasmus E, van der Sluis R, et al. A new perspective on the importance of glycine conjugation in the metabolism of aromatic acids. Drug Metab Rev. 46, 343–361, 2014.

116. Nong A, McCarver DG, Hines RN, et al. Modeling interchild differences in pharmacokinetics on the basis of subject-specific data on physiology and hepatic CYP2E1 levels: A case study with toluene. Tox. Appl. Pharmacol 214, 78–87, 2006.

117. Choudhuri S, Klaassen CD. Structure, function, expression, genomic organization, and single nucleotide polymorphisms of human ABCB1 (MDR1), ABCC (MRP), and ABCG2 (BCRP) efflux transporters. Int. J. Toxicol. 25, 231–259, 2006.

118. Zhang YK, Wang YJ, Gupta P, et al. Multidrug Resistance Proteins (MRPs) and Cancer Therapy. The AAPS Journal, 17, 802–812, 2015.

119. Lee JY, Rosenbaum DM. Transporters Revealed. Cell 168, 951–953, 2017.

120. Weigl KE Conseil G, Rothnie AJ, et al. An outward-facing aromatic amino acid is crucial for signalling between the membrane spanning and nucleotide binding domains of multidrug resistance protein 1 (MRP1; ABCC1). Mol. Pharmacol. 94, 1069–1078, 2018.

121. Zhang H, Patel A, Wang YJ, et al. The BTK inhibitor ibrutinib (PCI-32765) overcomes paclitaxel resistance in ABCB1-and ABCC10-overexpressing cells and tumors. Mol. Canc. Ther. 16, 1021–1030, 2017.

122. Roy U, Barber P, Tse-Dinh YC, et al. Role of MRP transporters in regulating antimicrobial drug inefficacy and oxidative stress-induced pathogenesis during HIV-1 and TB infections. Front. Microbiol. 6, 948, 1–13, 2015.

123. Russel FGM, Koenderink JB, Masereeuw R. Multidrug resistance protein 4 (MRP4/ ABCC4): a versatile efflux transporter for drugs and signalling molecules Trends Pharm. Sci. 29, 200–207, 2008.

124. Eadie LN, Dang P, Goyne JM, et al. ABCC6 plays a significant role in the transport of nilotinib and dasatinib, and contributes to TKI resistance *in vitro*, in both cell lines and primary patient mononuclear cells. PLoS One. 13, e0192180, 2018.

125. Mott HR, Owen D. Structure and function of RLIP76 (RalBP1): an intersection point between Ras and Rho signalling. Biochem. Soc. Trans. 42, 52–58, 2014.

126. Baldwin S, Byrne G, del Toro B, et al. 'The Usual Suspects' dir. Singer B. 106 min, Bad Hat Harry Productions, released 25 January 1995.

127. Singhal J, Nagaprashantha L, Vatsyayan R, et al. RLIP76, a glutathione-conjugate transporter, plays a major role in the pathogenesis of metabolic syndrome. PLoS ONE 6(9), e24688, 2011.

128. Manguoğlu El, Akdeniz S, Dündar N, et al. RLIP76 Gene variants are not associated with drug response in Turkish epilepsy patients. Med. Genet. 14, 25–30, 2011.

129. Wang CZ, Yuan P, Xu B, et al. RLIP76 expression as a prognostic marker of breast cancer. Eur Rev Med Pharmacol Sci. 19, 2105–2111, 2015.

130. Yogesh CA, Rajendra S, Sushma Y, et al. The non-ABC drug transporter RLIP76 (RALBP-1) plays a major role in the mechanisms of drug resistance. Curr. Drug Metabol. 8, 4, 2007.

7 Factors Affecting Drug Metabolism

7.1 Introduction

As you are may be aware, drugs are tested at the preclinical stage in animal populations that are usually inbred and display little variation from animal to animal. Data from animal studies in one country are usually comparable with that of another, provided the animal species and strain are the same. This provides a consistent picture of the basic pharmacological and toxicological actions of a candidate drug in a living organism, although controversy still rages over the value of this picture. Currently, animal data, combined with human tissue studies, are intended to give some approximation as to how the drug might affect humans. Unfortunately, you might also be aware that some drugs reach the clinic level only to be withdrawn or have severe strictures placed on their usage, based on some form of toxicity or unintended pharmacological effect. Although it has been obvious since animal testing began that there would be large differences in the way a drug might perform in humans compared with animal species, perhaps in the last 15–20 years it is clear just how vast and ultimately costly these differences can be. Unfortunately, there is no experimental model yet designed that can not only consider human biochemistry and physiology, but also the effects of age, smoking, legal and illegal drug usage, gender, diet, environment, disease and finally genetic variation. Indeed, many clinical studies have revealed enormous differences in drug clearance and pharmacological effect even in age, sex, and ethnically matched individuals. In effect, this means that the first year or so of a drug's clinical life is a vast, but monitored experiment, involving hundreds of thousands of patients, and there is no guarantee of success. This chapter discusses some of the major factors related to patient genetic identity and how they live their lives, on drug metabolism and how these factors affect the goal of maintaining drug levels within the patient's therapeutic window.

Human Drug Metabolism, Third Edition. Michael D. Coleman.
© 2020 John Wiley & Sons, Inc. Published 2020 by John Wiley & Sons, Inc.

7.2 Genetic polymorphisms

7.2.1 Introduction

Apparently, cheetahs are all genetically identical and can receive skin grafts from any other cheetah, as an identical twin could in humans. The same goes for elephant seals. Environmental and hunting pressures shrank the species to such small numbers that only one genetic variant now exists. The long-term survival of this type of population is unlikely, as genetic diversity is vital to ensure that no one toxin, bacteria, or virus can eliminate all the population.

In practice, genetic diversity manifests in differences in single DNA nucleotides and/or whole genes that code for particular proteins. This results in a proportion of a population that expresses a protein that is different in structure and function to the majority. These differences may manifest as a substitution of a single amino acid with another, or a whole amino acid sequence may be different. These are termed *polymorphisms* and their study is termed *pharmacogenetics* or *pharmacogenomics*. A polymorphism can be defined as a genetic variant that appears in at least 1% or more of a population. If you recall earlier Chapters where enzyme structures were described, you might remember (hopefully) that a single amino acid in a precise position is often vital for the enzyme to function. The most common polymorphisms can be due to a change in one nucleotide, which means that the gene now specifies a different amino acid in a critical position. These are called SNPs or single nucleotide polymorphisms and the resultant enzyme is structurally and often functionally different from the wild type. Although occasionally, polymorphisms can mean increased enzyme activity, they usually have a deleterious impact on enzyme performance, ranging from no enzyme appearing at all, all the way to the formation of incomplete isoforms due to arcane faults in mRNA expression. In Chapter 4.4, you saw that control of gene expression itself required a complex series of promotors and co-activators, all of which have to possess the exact structural and physicochemical properties to fulfil their function. Even if the area of the gene that codes for an enzyme itself is potentially functional, defects in expression regulation will impair functional translation and transcription. If you take into account the myriad systems that provide 'quality control', chaperone and position various enzymes, it is little short of miraculous that there are relatively so few serious genetic failures. Again, this is the result of two billion years of evolutionary 'research and development'.

Polymorphisms arise due to mutations, but persist in human populations due to factors that may involve some advantage of the milder forms of the polymorphism (heterozygotes). As we saw in Chapter 6 (Section 6.11.2) with some transporters, if a faulty enzyme's endogenous function can be shouldered by other systems, then even homozygotes for the polymorphism may never be aware of their genetic makeup, as they suffer no significant problems. Glucose-6-phosphate dehydrogenase deficiency (G-6-PD) in Afro-Caribbeans is only a problem if the individual takes certain antimalarials (primaquine or dapsone, Chapter 8.2.7), which generate toxic species that overwhelm the G-6-PD individual's compromised oxidant defence mechanisms. Other defects may not affect some tissues, but may affect risk elsewhere. In Sections 6.2.4, and 7.2.7, Gilbert's syndrome does not lead to hepatic injury, but actually improves resistance to cardiovascular disease and carcinogenicity.

However, if only one enzyme can clear a particular therapeutic drug, a polymorphism in that enzyme may cause extremes of either drug failure or drug toxicity. Of course any given drug response is based partly on the sensitivity of the appropriate receptor as well as the amount of drug in the system. Although both these factors are subject to polymorphisms, the following sections concentrate on polymorphic effects on the mechanisms that govern drug clearance. Most biotransformational polymorphisms that might potentially cause a problem clinically are due to an inability of those with defective enzymes to remove the drug from the system. Or, treatment failure occurs if the agent is administered as a prodrug and there is no metabolic conversion to an active metabolite. Either way, drug accumulation can lead to side effects and loss of patient tolerance for the agent.

Polymorphisms: terminology

A polymorphic form of a CYP is usually written with a * and a number for each allelic variant, or translated version of the gene. The wild-type allele, or the major form expressed in the population is often termed *1, as in *CYP2B6*1*, for example. Whatever version of the protein is being discussed is written in uppercase (CYP2B6) and the gene is written in upper case italics (*CYP2B6*). The animal versions are written in a similar way in lower case.

Regarding the polymorphic forms, they might contain one or more SNPs in the same allele. For example, *CYP2B6* has eight other significant allelic variants besides the wild-type; among these variants, the allele *CYP2B6*4* has just one SNP, whilst *CYP2B6*6* possesses two SNPs.

The scientific literature often refers to SNPs either by the particular codon and nucleotide substitution and/or by the actual amino acid that is substituted in the finished enzyme. So for the *CYP2B6 *4* variant the SNP will be termed A785G or A>G 785. This means that a wild-type adenine base has been replaced by a guanine in this SNP in codon 785 on the gene. On the enzyme, at amino acid 262, a wild-type lysine (DNA codon AAA or AAG) is now an arginine (codon AGA or AGG). The respective single letter codes for lysine and arginine are K and R, so this SNP is also known as K262R. Polymorphisms will be rather confusingly referred to by either of these identifications. So for the dual SNP *CYP2B6*6* variant, it has a G516T (or Q172H; glutamine replaced by histidine) as well as the A785G (K262R) described above[1]. There are also variants of CYP2B6 that have three SNPs, although these are extremely rare. Overall, there are over fifty allelic versions of *CYP2B6* alone catalogued to date. More are being discovered and this is not even the most polymorphic CYP[2-4].

Over the last decade, polymorphic forms are increasingly referred to as a reference SNP identification number, which is termed the lower case letters 'rs' and a number, which is given by the US National Centre for Biotechnology Information[5]. This number maps the precise location of the gene on the chromosome. So, for instance, CYP2B6*4 (Lys262arg) is termed rs2279343 and you can see from the NCBI website (dbSNP) exactly where it is on chromosome 19[5]. It is also useful to note that polymorphic forms of an isoform might describe a specific change in the catalytic or binding sites, but it can also describe a defective or more efficient regulatory process for the isoform. So whilst CYP1A2*1F (see later in Section 7.2.3) is as far as we know catalytically competent as the wild type, it is induced considerably more rapidly as there are polymorphic changes in how various factors bind to its DNA response elements. Sometimes in the description of a polymorphism, you might also see a minus sign (−1639G>A) in front of the SNP, such as with the vitamin K oxidase gene *VKORC1*2*. This means that the SNP is actually in a promotor upstream of the main gene, which in this case restricts mRNA formation.

Most population studies of human polymorphisms list the allelic frequency, that is, how many of an ethnic group contain the alleles in question. For CYP2B6, about 50% of Caucasians have the wild-type allele, but only 28% have the *CYP2B6*6* allele. This does not mean that 50% of Caucasians actually express wild-type enzyme. The actual diplotypes in the population, that is, which individuals express which combinations of alleles, are not the same as the population allelic frequency. In a real population, a considerably smaller number, in the case of CYP2B6, about 30% of individuals are homozygous for the wild type (CYP2B6 *1*1). These individuals are usually classed as extensive metabolisers (EMs). There will be various combinations of the wild-type and other defective alleles, such as *1/ *6, (about 30% of the population) and *1/ *5 (about 10%). These heterozygotes might be classed as intermediate metabolisers or IMs, as the enzyme is slightly impaired but serviceable. Then there are various combinations of defective versions of the gene, such as *5/ *6 or homozygous *6/ *6 which might be only 7–8% of the population. These individuals are usually termed poor metabolisers (PMs)[1].

If an SNP or a combination of SNPs is a fairly mild defect in the enzyme when it is homozygously expressed, then the heterozygotes will show little impairment and the polymorphism may be clinically irrelevant. With other SNPs, the enzyme produced may be completely nonfunctional. Homozygotes will be virtually unable to clear the drug and heterozygotes will show impairment also. There are also smaller populations of UMs, or ultra-rapid metabolisers, which may have a feature of their enzyme which either makes it super-efficient or expressed in abnormally high amounts. There are also versions of genes where they are much more efficiently induced, such as with *CYP2B6*22* allele, that has enhanced sensitivity to PXR-agonists[1].

Sometimes SNPs in different enzyme families are inherited together, where the individuals might express two different defective CYPs (or other enzyme) at once. This combination may only come to light with a particular drug that relies on both CYPs for clearance and is also seen with UGT1A1 and 1A7 (Section 7.2.7). Essentially, the variation appears to be almost for its own sake- anything rather than just have a homogeneous expression for any given enzyme.

Genotype and phenotype

In practical terms, it is important to distinguish between genotyping and pheno- typing patients for a particular enzyme isoform. Genotyping uses the patient's DNA to determine whether they are homozygous or heterozygous for particular alleles. This will show that for only a few alleles, there are likely to be dozens of haplotypes (group of alleles all expressed together) combinations contributing to the full spectrum of drugs clearance. Phenotyping usually involves administer- ing a single probe drug for a particular enzyme and measuring clearance and comparing it with data from other patients.

As you can imagine, what should be predicted from the genotypic tests is not necessarily true phenotypically for all substrates of that enzyme and *vice versa*, as there are so many confounding factors, such as age, gender, weight, alternative routes of clearance, and other variables, known and unknown. Phenotyping will group patients in very broad EMs, IMs or PM categories, but will be unable to distinguish between heterozygous and homozygous EMs. Although genotyping may be very helpful in dosage estimation in the initiation of therapy, there is no substitute for the normal process of therapeutic monitoring, which is effectively phenotyping the individual in the real world in terms of maximizing response and minimizing toxicity.

Indeed, you may see the terms *phenocopying* or *phenoconversion*, which essentially means that if an inhibitor blocks the clearance of another drug it effectively reduces a genotyped EM to the status of a phenotypic PM for that drug[4]. If the drug inhibits its own metabolism, it is sometimes called *autopheno- copying*. In effect, a whole patient population, including the EMs and UMs can be 'converted' to phenotypic PMs by phenocopying or autophenocopying. This can occur during real-world chronic administration, as many drugs inhibit their own clearance after a few days, such as SSRIs and some protease inhibitors.

Another issue when discussing polymorphisms is the description of the cat- egories of ethnic groups in humanity. We can use the exceedingly broad terms Caucasian, African, Afro-Caribbean and African American without a problem. However, in the United Kingdom, the term Asian generally refers to people from the Indian subcontinent and Pakistan, whilst in the United States, it refers to people from countries such as China, Japan, and Korea, what we would term in the United Kingdom as the Far East. However, if I was from China, I might find being referred to as being from the 'Far East' in publications as annoying as Latin Americans might feel about being designated South Americans. Hence, even if the term 'Asian' when used to describe those from China or Japan grates on the UK ear, I feel that using this American expression is a good plan in this context.

7.2.2   Clinical implications

Personalised medicine

Logically, the vast amount of genetic variation across humanity in terms of biotransformational capability seems to suggest that in therapeutics, 'one size fits all' should seem not only outdated but fabulously naïve. Already, in certain areas of medicine drug administration is personalised, such as with warfarin and in cancer chemotherapy. However, these personalisation processes involve time-consuming basic therapeutic monitoring combined with clinical experience, perhaps alongside some trial and error. It could be argued that determination of a patient's genotype or phenotype for a key biotransforming enzyme would be of great assistance in accelerating therapy optimisation and also reducing the inherent risks involved. Indeed, to take this further, personalised medicine has been promoted by the slogan 'the right drug at the right dose *the first time*'.[4] Those latter three words are striking – they exclude slow therapeutic monitoring-style processes and demand a solid basis for the accurate prediction of drug and dose. Such a basis naturally can only be derived from determination of genotype and/or phenotype through some form of testing. Hence, all the necessary information for the 'one-shot deal' of personalised medicine would be determined from a genetic or phenotypic test. The result would be analysed using an extensive database and presumably various algorithms.

However, these test scenarios have not yet gained general acceptance even in areas such as anticoagulant or cancer therapy. Also, when any of us might visit a general practitioner, the conversation is still highly likely to include a phrase to the effect of 'I think we will start you off on this drug/dose and see how things go', rather than a blood test for a specific group of biotransforming or pharmacodynamically relevant enzyme expression levels. Hence, the majority of clinical decision making over drug choices is not yet informed by such detailed prior information. This situation might seem very much at odds with the scale of the scientific and medical efforts currently and over the past thirty years or so, to explore, predict, and catalogue biotransforming enzyme polymorphic expression in just about every ethnic group under the Sun. To date, such knowledge in an individual patient seems really only of interest to account for some idiosyncratic lack of, or excessive response to, a drug regime that cannot be explained in any other way. Therefore, the following chapter will try to summarize what we know of polymorphic impact on biotransformation, its limitations, and its possible future.

Polymorphic realities

Indeed, determining the practical therapeutic relevance of human biotransformational polymorphisms has proved to be extremely problematic. In terms of polymorphism detection, this area is a classic illustration of how the exploration of the human genome with powerful molecular biological tools as applied to blood samples from a group of individuals may unearth many apparently marked polymorphic defects that may not necessarily translate into a measurable clinical impact in terms of efficacy and toxicity. In reality, it is much easier to use molecular biology with a single blood sample to discover and publish such polymorphisms *in vitro*, than there are clinical scientists, resources and patients in sufficient quantity to run full trials to determine practical clinical relevance, in terms of loss of or gain of function and its impact on drug efficacy and toxicity.

Clinical studies of polymorphism relevance are expensive, very onerous and perhaps need three essential components

1. Sufficient numbers are required to convince in terms of statistical relevance.
2. Pharmacokinetic changes must be adequately detected.
3. Establish how those changes affect drug pharmaco/toxicodynamics.

Many studies cover one or other of these three essentials, but not always all three. Many fall at the first hurdle, as the costs of clinical trials are so high that resourcing a high-quality trial is highly problematic.

Even if an extensive trial is mounted and adequately financed, in order to research the relationship between a drug and a known polymorphism relevant to that drug, we will base the study on certain key assumptions, which may not necessarily be the whole truth. Clopidogrel and tamoxifen are two instructive examples in this context. Both drugs are aimed at long-term use – that is, many years of protection for the patient from their condition – and both drugs are capable of significantly extending lifespan. However, they are prodrugs and require metabolic activation, and it is logical to determine which patients would potentially not benefit from the drugs, as they could not activate them in sufficient quantity for optimum efficacy. This assumes that it is not always possible to use conventional therapeutic monitoring with blood analysis of metabolites, and it also assumes that the relationship between the levels of active metabolites and efficacy are known. Therefore, it could be argued that to commit the patient to such long-term risk of poor efficacy is unethical, so there is a strong motivation to link the respective polymorphisms that are associated with the drugs to treatment outcomes.

Polymorphic realities: clopidogrel

With clopidogrel, the role of various esterases was initially described, but a significant role for CYP3A4 was also noticed. However, despite this, it gradually became accepted that its efficacy rested mainly on CYP2C19-mediated conversion of the 2-oxo-metabolite to the active thiol[6]. Subsequent work with inhibitors of CYP2C19 such as omeprazole seemed at first to support the role of this CYP. However, it transpired that CYP3A4 inhibitors such as grapefruit juice had a several-fold larger impact on cutting conversion to the active metabolite than any effects on CYP2C19[7], which is now known to be a minor pathway, so it is no wonder that repeated investigations on the impact of polymorphic expression of CYP2C19 and clopidogrel efficacy were inconclusive[6]. The investigations with CYP2C19 were based on an erroneous picture of clopidogrel activation and that CYP2C19 and CYP3A4 were not part of the same pathways, but they were competing and CYP3A4 was the 'winner'[6], although there are other polymorphic issues, as will be discussed in a later section.

Polymorphic realities: tamoxifen

It is fair to say that tamoxifen is a much more complex issue than clopidogrel, but it is a very instructive example of the struggles involved in the process of the effort to link polymorphic isoform status to treatment outcome. Tamoxifen's clinical effect is that of oestrogen antagonism, which denies breast cancerous cells the hormonal driving force necessary for their growth. This is undeniably effective, and it cuts mortality by 30–50%[8]. How it achieves this is through metabolic activation. More than 90% of the drug is N-demethylated by CYP3A4 to N-desmethyl tamoxifen and the rest is cleared by CYP2D6 to 4-hydroxy tamoxifen. CYP2D6 hydroxylates the major metabolite N-desmethyl tamoxifen to endoxifen, and 4-hydroxy tamoxifen is converted to endoxifen by CYP3A4. So now we have two active metabolites, endoxifen and 4-hydroxy tamoxifen. In a patient with fully functional wild-type CYP2D6, they will form around 10 times as much endoxifen as 4-hydroxy tamoxifen, although this varies massively[8]. Hence, pharmacokinetics and biotransformation take us to the point where endoxifen appears to be the key metabolite for efficacy. Indeed, it has been reported that even when CYP2D6 allele status is not consistently linked with efficacy, the level of endoxifen (>15 nM) is a crucial measure of positive therapeutic outcome[9].

From a pharmacodynamics perspective, endoxifen not only can block the oestrogen receptor better than any other tamoxifen derivative, it also accelerates the proteosomal destruction of the oestrogen receptors, which gives it a potent dual mechanism of antagonism[8]. At this point, whilst it is clear that CYP2D6 is important in the formation of active metabolites of tamoxifen, there are other processes occurring. From Chapter 5.8.2, you might recall how effective the aromatase inhibitors such as letrozole were in shutting down oestrogen production. Interestingly, tamoxifen is also converted to norendoxifen that is actually as potent as letrozole in aromatase inhibition. CYP2C19 catalyses this process, but those with a low-function allele CYP2C19*2 form less norendoxifen[10]. Conversely, CYP2C19*17, an ultrafast version of the isoform, is associated with better treatment outcomes with tamoxifen and even reduced susceptibility to developing malignancy anyway[8].

If these issues were not complex enough, there are several other CYPs involved in tamoxifen biotransformation, and the conjugative clearance of endoxifen, norendoxifen and 4-hydroxy tamoxifen depends in part on UGTs and SULT isoforms. Their expression also can dictate the half-lives of these metabolites as outlined in Chapter 6.3.4. Additionally, whilst rapid conjugative clearance of the active metabolites of the drug might diminish efficacy, equally rapid clearance of endogenous oestrogens would actually improve outcomes. Given that so many enzymes and events are involved with the activity of tamoxifen, reliable tests would be essential so that decisions could be made on the basis of the information produced. However, it was also found that some of the testing processes that assigned CYP2D6 status to patients were flawed due to inaccurate determination of particular alleles for various technical reasons[8].

Hence, given tamoxifen's labyrinthine clearance pathways, its route of efficacy, and the length of time it has taken to unravel all these processes, plus the many practical difficulties, all underline that it is not surprising that it has taken decades to establish the current consensus that an association between CYP2D6 and tamoxifen efficacy[11] does exist. Such an agreement does support some form of genetic testing[11-13] and advice not to co-administer CYP2D6 inhibitors with tamoxifen[11] has been reinforced. Indeed, any CYP inhibitor is contraindicated with tamoxifen therapy.

If determination of CYP2D6 status did become routine prior to initiation of tamoxifen, whether or not it addresses other important clinical issues with the drug remains to be determined. Tamoxifen therapy is essentially life-long and can be very difficult for women to tolerate. It can cause hot flushes, potential thromboembolisms, and can lead to significant depression[8]. There is also evidence that extensive metaboliser status, so desirable for efficacy, is associated with even more side effects and so all this effort and technology eventually does not avert the depressing situation that only half of women to whom the drug is prescribed actually take it[14].

Polymorphic realities: the future

In the longer term, if you consider a recent publication that highlights issues related to the real-world impact of polymorphisms, the authors cite four drugs that treat life-threatening conditions where genotype may play a significant role in outcomes. These agents were warfarin, tamoxifen, irinotecan, and drugs that are converted to thiopurines[4]. With warfarin and tamoxifen, there are now well-established alternatives (directly acting anticoagulants and the aromatase inhibitors) that may conceivably in the future replace them in clinical practice. Replacement of the other drugs might be more distant.

If we return to clopidogrel, as we discussed in the previous chapter, Section 6.9, this was one of the world's best-selling drugs, then competition from agents such as prasugrel, gradually overtook it in practice. A new drug may not be particularly more effective, but if it is safer and easier to use in clinical practice, then it may well be more successful. Hence, there may be limited enthusiasm for engaging in decades-long efforts to research the real-world impact of biotransforming enzyme genotypes if a particular drug's commercial and clinical cycle is actually too short to justify the effort and resources involved. Drug companies now do try to determine the impact of genotype on the disposition of a new drug, and appropriate tests are becoming available, although the vastness of the human genome makes this a far-from-precise exercise.

Determining real-world polymorphism impact

Whilst the difficulties with the drugs we have discussed so far were very significant, fortunately, in clinical practice many drugs are effective despite polymorphic pressures on their clearance and efficacy. Naturally, it will help if a drug has a wide therapeutic index, like many modern drugs, so the higher plasma levels seen in PMs will rarely be enough to cause serious off-target effects and frank toxicity. In the case of prodrugs, it is often apparent that even though production of the most potent metabolite is lacking in PMs, the drug retains enough clinical effect for the polymorphism to have only academic interest and a minor change in dosage optimizes treatment. However, if a drug is not likely to be replaced or outmoded in the intermediate or distant future, it may be worthwhile to invest in optimizing therapeutic performance and thus improving tolerance and adherence through some form of personalised drug therapy.

Motivations for pharmacogenetic testing

With some drug categories, certain combinations of factors can make a particular polymorphism exert strong impact on treatment outcome and drug tolerance in some groups of individuals. Some of these factors are as follows:

- *Narrow therapeutic index (TI)*. Aside from the greater likelihood of a toxic event occurring in a narrow TI drug, the nature of the drug's toxicity and its target organ are crucial. For instance, off-target cardiotoxicity such as TdP (Chapter 5.7.2) is far more dangerous over a short timescale than most other organ-directed toxicities.

- *Single pathway mediated clearance*. If only one enzyme is responsible for the clearance of an agent, inhibition of this pathway has much more impact than a drug with multiple pathways of elimination.

- *Background variability of pharmacodynamic response*. In the case of off-target responses, in effect, the TI in one patient may be considerably narrower than in another and it may not be possible to predict or anticipate this.

- *Delay in toxic response to high drug levels*. This is most marked in a drug that may have a complex mode of action, and the onset of deleterious effects is hard to predict, such as with warfarin.

- *Drug levels equate directly with toxicity*. If the relationship between concentration and effect are consistent and predictable, then once a polymorphism is identified in an individual, dose escalation can be better informed.

- *Long-term therapy*. Protecting a patient from potentially damaging toxicity or ensuring that efficacious drug levels are maintained over many months or years more than justifies the considerable effort and expense of polymorphism determination. This may be particularly applicable to drugs used in mental health, cardiovascular disease and cancer, where regimes last many years.

The following sections discuss the detection of major polymorphisms in human biotransformation and reviews current appraisals of their clinical relevance. As there are now so many polymorphic variants of biotransforming enzymes, it is probably most practical to try to draw attention to the impact of fast (gain of function) and slow (little or no function) versions. Although polymorphic changes in the catalytic and binding sites of different biotransforming enzymes are now well documented, more has emerged over the impact of polymorphisms on the regulation of the various enzymes. This is particularly the case with their various induction pathways. In practical terms, this can mean that patient consumption of inducers of various forms can have a disproportionately high impact on drug efficacy. Overall, the complexity of polymorphic relationships with drug disposition underlines some of the obstacles in the path of 'the right drug at the right dose the first time'.

7.2.3   Genetic polymorphisms in CYP systems

CYP1A1

CYP1A1 is not a constitutively expressed hepatic isoform and it is not really relevant to normal drug clearance, but its greatest significance is as the inducible producer of potentially carcinogenic species from PAHs, particularly in smokers' lungs[15]. Induction of CYP1A1 can contribute to anti-cancer drug resistance with flutamide, tamoxifen, and ifosfamide[16]. Indeed, as we have discussed in Chapter 3.6.1, high levels of this isoform are often seen in many tissues affected by cancers, ranging from the breast to the cervix. The enzyme has evolved as a protection against PAH-like molecules, activating them into epoxides or other reactive species, which are then conjugated by adjacent conjugative systems to nonreactive excretable products[17].

This system might be linked to many cancers for a variety of reasons, such as the large inter-individual variation in the degree of induction (Chapter 4.4.2) of CYP1A1, as well as polymorphisms in the isoform itself. From Chapter 4, it can be recalled that there are several steps in the AhR induction of this CYP (4.4.2) and there are several mutations in this pathway, not least in the xenobiotic response elements on the *CYP1A1* gene itself, which can impact the responsiveness of the induction process[17].

Of the currently described polymorphic forms of CYP1A1, CYP1A1*2A (3798T>C) can be found in 5% of Caucasians. CYP1A1*2C (2454A>G) is not common on its own in Caucasians but can expressed in up to 19% of Japanese[15]. CYP1A1*2C can be found with CYP1A1*2A in a joint variant, which is termed CYP1A1*2B[15]. There is also CYP1A1*3, (T3205C), which is very rare in Caucasians. The last variant is termed CYP1A1*4 and is found at a 3% level in Caucasians. These variants mostly show gain of function, in that they have higher activity than the wild type, and although it is controversial[16], some studies show that the faster versions of CYP1A1 (particularly CYP1A1*2A) do increase the risk of developing a lung malignancy[15], presumably because more reactive species are made in the lung over the patient's smoking career. Faster and more easily induced CYP1A1 variants thus promote the risk of developing a cancer and probably hinder the efficacy of the anticancer chemotherapy as well[18].

CYP1A2

CYP1A2 specializes in xenobiotic polycyclic aromatic amines, as well as oestrogen metabolism. A considerable amount of liver CYP content has been identified as this isoform (approximately 10%), and several drug classes are cleared at least partly by this CYP, such as the 'R' isomer of warfarin, the antipsychotic clozapine, as well as caffeine and theophylline. In looking at the various polymorphic forms of CYP1A2, CYP1A2*1F has been extensively investigated. It is also known as rs762551, −164A>C, or −163C>A. CYP1A2*1F in homozygotes (known as 'AA') provides significant gain of function, in that it is much more readily inducible than the wild type (CYP1A2*1)[19,20]. Clozapine has a strong evidence base for its efficacy in treatment resistant schizophrenia, where patients have not responded to two trials of other antipsychotics.

There are those who are also resistant to clozapine, and this has been linked to AA CYP1A2*1F individuals. Since so many schizophrenics smoke (Section 7.6), it is logical to suggest that AA patients that are smokers might struggle to attain the minimum required clozapine plasma concentration for efficacy (~350 ng/mL). However, conclusive evidence for CYP1A2*1F as a major factor in clozapine treatment failure has not appeared. Interestingly, this may be linked to differences between racial groups. In one study in the Netherlands, the *1F allele appeared in more than 66% of the subjects, so although the link between this allele and rapid clozapine clearance may be strong in Caucasians, it may not be the case in other groups[21]. Currently, personalising clozapine dosage based on CYP1A2 genotype does not appear to be justified[19].

CYP1A2*1C is also known as rs2069514, as well as –3860G>A. With this allele, the binding of the AhR complex to the response elements on the gene is thought to be defective, so induction is not as efficient as in the wild type alleles and is much slower than the gain-of-function *1F allele[22]. This less-sensitive version of the CYP1A2 is rare in Caucasians, at around 1%[21]. However, its relevance to health appears to be with its rather contradictory association with some cancers. A less-responsive version of CYP1A2 could be beneficial, in terms of producing fewer reactive species from environmental toxins such as PAHs and aromatic amines to attack DNA. However, in terms of endogenous oestrogen metabolism, poorly responsive CYP1A2 leads to increased levels of 16α hydroxyloestrone, which can also attack DNA[22]. Therefore, in the latter case, a fast CYP1A2 is beneficial. Hence, it is easy to see that assigning cancer risks to expression of individual isoforms is problematic, as CYP1A2 is a modest fragment of a very complex picture of biotransformation, detoxification, DNA repair, and environmental pressures.

CYP1B1

This CYP is strongly inducible and oxidizes many precarcinogens, such as PAHs, heterocyclic amines, aflatoxins and other environmental toxins. It is also the most efficient oxidizer of oestrogens and is linked with hormone-dependent cancers. Overall, interest has centred on its role in predisposition to carcinogenesis, particularly in tobacco users and its toxicological significance has not been easy to elucidate. For instance, CYP1B1*3 (rs1056836) has the wild-type leucine changed to a valine at 432. Since the mid-2000s, studies have focussed on the risks posed by this CYP variant[23]. Over 70% of Japanese and Chinese are homozygotic for CYP1B1*3, whilst a third of Caucasians and less than a tenth of African Americans share this situation[24]. CYP1B1*3 was linked with responses to taxanes, as well as breast cancer, as it is thought be catalytically two to three times faster than the wild type with leucine. This suggested that more 4-OH oestradiol would be formed that might be carcinogenic, although recent studies have not been conclusive. This CYP does seem to protect against prostate cancer[25], although it is linked with dose delay and reduction of taxanes due to neutropenia[26], which is not encouraging for those patients.

There are a number of other CYP1B1 variants, such as CYP1B1*4 (rs1800440; Asn453Ser), but it is often very difficult to assign precise risks in terms of impact on various cancers on particular isoforms[27]. However, the role of CYP1B1 in resistance to anticancer drugs is probably the most fruitful to pursue in terms of the possibility of making practical progress clinically. It is now becoming clear that micro RNAs crucially regulate CYP1B1 activity during antineoplastic drug pressure. In experimental renal tumours, miR-200c is effectively switched off, allowing CYP1B1 to upregulate and defend the tumour[28]. Indeed, it is not a great leap to assume that there are polymorphic aspects to micro-RNA control of CYPs, and this may be an added dimension of complexity in tumour susceptibility and responsiveness to treatment, which may even be why it has been so difficult to tie different CYP1B1 variants to particular conditions.

CYP2A6

The major interest in this CYP is its role in the metabolism of more than 80% of nicotine to cotinine and thence to 3'hydroxycotinine (3HC)[29]. Whilst CYP2A6*1A is considered the reference wild type, there are many other similar wild-type versions with adequate activity. However, CYP2A6*2 (rs1801272), as well as *3, *4, and *5 are all nonfunctional, whilst CYP2A6*6 (Arg128Gln) is slower than wild type in activity, as is *7, although *8 is similar to the wild types[29]. There are at least 45 other variants, mostly, not always decreasing in activity, although there are some ultrafast duplicated versions (CYP2A6*1X2A & B).

Experimentally (such as in clinical trials with intravenous nicotine) and whilst smoking, smokers are very sensitive to nicotine dose and regulate it quite well. This seems to be gender specific and is linked to their systemic nicotine amounts, which is around 1–1.5 mg at any one time[30]. Many studies have tried to equate smoking behaviour with approximate rate of nicotine clearance, as related to CYP2A6 genotype. In this regard, CYP2A6*1A, with its highly efficient clearance of nicotine, is held 'responsible' for the somewhat fearsome addiction homozygous (*1A*1A) smokers have to their tobacco.

From a pharmacological perspective, rapid elimination of a compound with such complex and soothing CNS actions as nicotine is likely to promote tolerance, dependence, and eventually craving. Clearly, the grip tobacco still has on many is most likely linked to the very widespread presence of wild-type fast CYP2A6 alleles in humanity. CYP2A6*1A is expressed in Caucasians, African Americans, and Brazilians at between the 50–70% level, whilst Chinese and Japanese were lower, at 27–52%[29]. This can be contrasted with the frequency of the nonfunctional CYP2A6*2 that is found at less than 5% in most races. Some of the other low or modest functional alleles are more common in Asian races, particularly *7 and *9[29]. Japanese smokers have about half the CYP2A6 capacity of Caucasians[31]. Overall, the variants of CYP2A6 that are most linked with reduced nicotine clearance are *2, *4, *9, and *12[32]. Effectively, homozygote CYP2A6*1's show maximum clearance and often addiction, with other combinations, such as CYP2A6*1/*9, or CYP2A6*1/*12 leading to around 70% of maximum, with homozygotic *9's, or *1/*2 variants for example, showing around 50% of maximal nicotine clearance. It is much rarer to find people with two nonfunctional genes. However, from a practical perspective, because there are many other influences on smoking behaviour, the ratio between 3HC and cotinine (slightly confusingly known as the nicotine ratio) is taken as a more reliable indicator than genotype determinations. This is because the two metabolites have a much longer half-life than nicotine so the ratio is much less dependent on how smokers smoke and when they last smoked[32].

Unfortunately, high nicotine intake means high risk of exposure to PAHs (but not necessarily in *vaping;* see Section 7.6) and of course the risk of cancer. Naturally, the high level of dependence a smoker exhibits due to their rapid clearance genotype can make it much more difficult to stop using tobacco and any nicotine-based product. Indeed, around 8 out of 10 heavy smokers fail to maintain abstinence over 12 months and any lapses over the first 14 days of supposed abstinence are strongly predictive of failure to give up[33]. It is still quite sobering that about a fifth of all preventable deaths are due to smoking[32]. As an aid to giving up, bupropion (cleared by CYP2B6) and varenicline can be more effective than nicotine replacement therapy (NRT) in fast CYP2A6 individuals, partly through their actions on nicotinic receptors[32]. Bupropion generally shows success in fast and slow CYP2A6 individuals, although slow metabolisers often can stop smoking with NRT alone[31]. Overall, possession of poorly or nonfunctioning CYP2A6 variants substantially reduces the risks of addiction to tobacco (and the risk of lung cancer) and allows the individual to essentially 'take it or leave it' throughout their lives. My grandmother called this smoking OP's or 'other people's' cigarettes.

CYP2A6 clears several other significant compounds, and its polymorphisms can impact the antimetabolite tegafur, the aromatase inhibitor letrozole and efavirenz, the anti-HIV agent, more usually associated with CYP2B6. Tegafur is a prodrug that must be converted to 5-fluorouracil (5-FU) by CYP2A6. Whilst slower versions of this gene in Japanese cancer patients showed less conversion to the active drug, there was usually still sufficient 5-FU to provide some efficacy. Probably only in homozygous *2 individuals would there be hardly any 5-FU conversion[31] and another drug warranted. Letrozole's residence time is significantly greater in those with slow CYP2A6 variants, but this does not seem to be therapeutically problematic. Efavirenz is interesting, as around 80% of clearance is through CYP2B6 to an 8-OH metabolite. However, when patients possess a slow diplotype of CYP2B6, then how CYP2A6 converts the remaining 20% of the drug to a 7-OH derivative becomes relevant. So a homozygote for CYP2A6*1A will compensate to some degree in forming plenty of the 7-OH. Slow versions of CYP2A6 will not do this and are associated with loss of adherence due to toxicity with efavirenz patients[31]. The antimalarial artesunate is converted to its active dihydroartemisinine by CYP2A6, and slow versions will reduce therapeutic efficacy, whilst *1 forms will form high levels of the metabolite, leading to side effects and issues with adherence[31].

CYP2B6

CYP2B6 comprises a lower proportion of total hepatic CYPs than CYP1A (approximately 6%) and it is only responsible for around 7% of drug metabolism, but it has gained importance in view of its high inducibility and very high baseline variability in expression[34]. Although relatively few drugs are cleared by this CYP, several are quite toxic in high concentrations and many are inducers of this isoform. Cyclophosphamide, efavirenz, rifampicin, nevirapine, and carbamazepine are substrates and inducers, whilst methadone, ketamine, MDMA (ecstasy), pethidine, and propofol are also substrates. The numbers of new polymorphisms of CYP2B6 keep appearing[34], but there remain only a few really significant variants in terms of their allelic frequency and influence on drug disposition.

The most important variant as mentioned earlier in section 7.2.1, is CYP2B6*6, which is found in up to 60% of some human populations. It contains two SNPs, meaning that two incorrect amino acids are coded for; these are designated G516T (Q172H) and A785G (K262R). CYP2B6*6 is strongly associated with toxicity with efavirenz; compared with *1/*1 individuals, clearance is 40 and 70% reduced in *1/*6 and *6/*6 patients, respectively[35]. Also, whilst efavirenz is a CYP inducer, this effect is strong in *1, but weak in *6 alleles[35]. The population distribution of *6 is much greater in Africans than Europeans, with up to half of West Africans possessing this allele and only 8% of Finns[35]. In addition, 60% of Papua New Guineans, 40% of African Americans, Chinese, Hispanics, and South Indians, as well as about 28% of Caucasians, and 17% of Japanese have this allele. Hence, issues with efavirenz toxicity and adherence will be more common in Africans than Caucasians and some authors have recommended that with this drug some form of testing should determine CYP2B6 status, alongside therapeutic monitoring[35].

Paradoxically and typical of the variability of CYP2B6, some *in vitro* studies have shown that CYP2B6*6 is actually up to three-fold more efficient at metabolizing efavirenz[34], whilst others disagree[36]. However, *6 is more effective at oxidising cyclophosphamide, compared with the *1 variant[34,36]. What is agreed is that *in vivo*, of the two SNPs in *6, G516T causes a variable and aberrant gene-splicing process that leads to less than a quarter of the required amount of mRNA formed, so relatively little of the CYP protein is actually made[34]. Hence, *6/*6 homozygosity in HIV+ Africans is a liability with efavirenz in the clinic, but for Japanese cancer patients it may be a qualified advantage, where higher levels of 4-hydroxylation of cyclophosphamide compared with *1/*1 patients increase clinical efficacy, but also toxicity[37].

Among the other variants of this CYP, G516T alone (no A785G as in *6) is termed CYP2B6*9 (rs3745274) and again has shown rapid efavirenz metabolism *in vitro*[34] but is strongly associated with up to five-fold higher efavirenz levels in patients due to low expression[38]. CYP2B6*9 is found in up to 38% of human populations, which makes it probably the next most common variant after *6 and is also associated with skin reactions to the anti-HIV drug nevirapine[38,39]. CYP2B6*4 just contains A785G, and unlike *6 (with the two SNPs) this variant is less efficient at hydroxylating cyclophosphamide[34] but faster with efavirenz 8-hydroxylation[36]. CYP2B6*5 allele (C1459T) shows approximately equivalent 8-OH efavirenz metabolism to wild type[34] *in vitro*. The variant CYP2B6*18 (T983C) is found only in Africans at a level of 5–10% and has also become associated with accumulation of nevirapine. The accumulation is serious enough to predispose around 1–2% of those individuals to the catastrophic hypersensitivity reaction Stevens-Johnson syndrome (Chapter 8.4.3)[39].

CYP2B6 has a significant role with the opiate methadone (Appendix B.2). The pharmacokinetics of this opiate have been described as extremely variable, with half-lives varying from 5–130 h[40]. Methadone is used as an analgesic as well as a means to withdraw from heroin addiction, and it is given as a racemic mixture. The R-isomer is up to fifty-fold more potent as an opiate and less dependent on CYP2B6 clearance than the S, which is three to four times more likely to block the hERG channels and cause TdP (Chapter 5.7.2) than the R-isomer[41]. What seems to happen in CYP2B6 *6, *9, and *11 patients is that they have much higher plasma levels of both isomers, but particularly the S isomer[42]. Indeed, *in vitro* studies have shown *6 to be six- and three-fold, respectively, less capable than *1, in the N-demethylation of the R and S isomers[43]. Fatalities with methadone in Caucasians have also been linked with slower versions of CYP2B6, such as *9 (rs3745274) and *5 (rs3211371)[42]. Hence, although other CYPs clear the drug, defective CYP2B6-mediated clearance contributes to fatal outcomes with methadone probably from both long QT syndrome, leading to TdP (Chapter 5.7.2) as well as respiratory depression[41].

CYP2C8

Approximately 6% of commonly prescribed drugs are cleared by this CYP in some form, although only a few agents rely on it as their main route of clearance. These are usually stated to be the antineoplastic agent taxol (paclitaxel), and two antidiabetic agents, rosiglitazone and repaglinide. Aside from the wild type (*1), 17 polymorphic forms of CYP2C8 have been described so far, although the most documented includes *2–5[44]. In those of African ancestry, the *3 variant is very low in frequency, but the *2 allele is found at levels of up to 20%. The *5 is found at a very low frequency in Japanese and the isoform is useless as most of the enzyme's structure is missing[44].

The *3 and *4 variants are found in Caucasians (the *2 is absent in this group) whilst *3 contains two SNPs, G416A and A1196G. The *3 variant is linked with CYP2C9*2 and CYP2C9 *3 polymorphisms (see next section) and studies with the R isomer of ibuprofen, a safe and useful probe for CYP2C8, have shown a four-fold increase in drug half-life in homozygous *3 variant individuals[45]. Given that the *3 allele frequency is 13–23% in Caucasians, as well as the high frequency of the *2 variant in Africans, it was not surprising that those with the *2, 3, and 4 variants are at much higher risk of taxane-induced neuropathy than those with wild-type alleles[46]. Highly toxic drugs such as the taxols are used under close supervision in a hospital or clinic, so even if the CYP2C8 *3/*3 (homozygous) variant was to show marked impact on a Caucasian patient in the initial stages of a chemotherapeutic cycle, dose modulation would swiftly follow resulting in efficient customizing of treatment to that individual. Alternatively, any impact of CYP2C8 polymorphisms on a relatively casually used over-the-counter (OTC) drug such as ibuprofen's clinical effects are probably not really likely to be relevant clinically, given the good safety profile of the drug.

The CYP2C8*3 variant has been associated with a dose-dependent increase of efficiency of repaglinide clearance, leading to reduced concentrations, although this could be partly due to OATP1B1 transport rather than a CYP effect. The same allele also is more efficient than the *1 version in clearing rosiglitazone and pioglitazone, again lowering plasma concentrations by 20–40%[45].

CYP2C9

Around 18–20% of hepatic CYP content is accounted for by this isoform. Indeed, polymorphisms in this CYP are likely to be very significant in terms of clinical impact, due to the considerable number of commonly prescribed drugs it clears (about 15%), that also may have narrow TIs and can be lethal in high concentrations[47]. Worldwide, there are two major polymorphic defective variants, each with their unique SNP; the CYP2C9 *2 (C432T; rs1799853) shows a mild reduction in activity and the allele is found in about 10–25% of Caucasians, 11–20% of Middle Eastern people, 2–16% Central and South Asians, 1–2% of African Americans, and is absent in Japanese people[48,49].

The second more severely compromised variant (CYP2C9*3; A1077T; rs1057910) has a catalytic capability that is less than 10% of the wild type and is present in up to 20% of Caucasians and about 3–5% of Japanese, but is rare (~1%) in other ethnic groups such as Chinese/Koreans and virtually absent in Africans[47,48]. Usually, most literature focusses on the CYP2C9 issues with *2 and *3, although until relatively recently, when it emerged that African Americans, who have very low levels of *2/*3 expression, have a much higher expression of CYP2C9*8, with up to 12% expressing this poorly catalytic variant[49].

Clinically, homozygous individuals (CYP2C9*3/*3, or *8/*8), will require half the standard dose of phenytoin, for example, with intermediate metaboliser (*1/*2, *1/*3, 1*8) combinations of CYP2C9 requiring dosages to be reduced by a quarter[47]. Phenytoin is a narrow TI drug that exhibits nonlinear pharmacokinetics, which means that there is little direct relationship between dose and plasma levels. Also, 70–90% of the clearance of this drug is due to CYP2C9[50]. The toxicity of phenytoin includes many CNS-based and cutaneous reactions (Chapter 8.4), and these are much worse in polymorphic individuals due to poor drug clearance. Indeed, as with slow versions of CYP2B6, the CYP2C9*3 allele is also significantly associated with phenytoin-induced Stevens-Johnson syndrome[51] (Chapter 8.4.3).

However, possession of a slow version of CYP2C9*3 (and possibly *8) is actually a significant advantage with sulphonylurea antidiabetic drugs. These agents are mostly cleared by CYP2C9, and often their glycaemic control efficacy is suboptimal (~66% over 5 years)[52]. Heterozygotes and homozygotes with *3 show significantly better glycaemic control compared with the wild type with glibenclamide, simply due to slower clearance and longer drug residence time[53].

Several NSAIDs are cleared by CYP2C9 and retarded clearance of these drugs does promote the risk of upper gastrointestinal bleeding with the *3 variant isoform, but not with *2[54]. However, clinical studies are not always consistent in this area and may not necessarily show significant impact of genotype; indeed, a report with diclofenac and celecoxib did not reveal any differences between fast and slow versions of CYP2C9[55]. However, subsequent work has suggested that those with *3/*3 diplotype can still clear celecoxib through CYP2D6, as well as CYP3A4[56]. Hence, clinically, the impact of genotype with these drugs might not always be apparent, unless an individual expresses both CYP2C9 and CYP2D6 slow metabolising alleles and even with those individuals, CYP3A4 may provide some clearance. Other drugs affected by slower forms of CYP2C9 include the angiotensin II blockers losartan and irbesartan, along with dapsone, some sulphonamides, amitriptyline, and the leukotriene receptor antagonist, zafirlukast[57].

CYP2C9 and anticoagulants

A major class of drugs cleared by CYP2C9 where toxicity is particularly important is the coumarin anticoagulants. However, as we will mention through this section, it is important to recognise that this CYP is a significant part, but not the whole part of the polymorphic issues with coumarins. Whilst there are many disadvantages with these drugs, they will remain part of the clinical picture for generations to come. This is partly because the latest drugs, such as the directly acting anticoagulants (DOACs), until recently lacked antidotes, and their costs are such that for millions, they or their insurance companies are unwilling or unable to pay the costs, which are much higher than for warfarin[58].

As you might remember from earlier in this book (Chapter 5.3.2), about 8/10ths of the S-isomer of warfarin is cleared by CYP2C9, and this isomer supplies about 70% of the drug's clinical effect. CYP3A4 and CYP1A2 clear the R-isomer. In wild-type individuals, R-isomer plasma levels are about twice those of the S-isomer. The *2 and *3 variants cause significant change in this ratio, with S-isomer plasma levels rising according to the severity of the polymorphism[48]. At the extreme end of the spectrum, CYP2C9 *3/*3 individuals can be fully anticoagulated on 10 times less warfarin than wild-type patients. Regarding other coumarin drugs, acenocoumarol is active through its R-isomer rather than its rapidly CYP2C9-cleared S-isomer. It too is markedly affected by CYP2C9 *3 variants, with a 20–30% reduced dose needed for even heterozygotes and down to 60–70% reduction for homozygotes. A less marked dosage might be required for CYP2C9 *3 /*3 individuals with phenprocoumon, as CYP2C9 has a much smaller role with this anticoagulant as CYP3A4 is significant[59].

However, there are other factors involved in coumarin clinical response, such as age, weight, sex, and particularly the status of the Vitamin K epoxide reductase subunit 1 *VKORC1* gene, which codes for the coumarin therapeutic target. In fact, the largest variance (nearly a third) in the coumarin patient experience is actually due to *VKORC1,* which is a greater predictor for response than CYP2C9 status[48]. The key polymorphism for this enzyme is rs9923231, −1639G>A, or VKORC1*2 and because it is a fault in an upstream promotor, it causes a 70% fall in mRNA, leading to a much-reduced expression of the protein. This means that much less coumarin is necessary to anticoagulate VKORC1*2 individuals[48]. Africans show low expression (<10%) of VKORC1*2, but about half of Europeans and virtually all of some Chinese and East Asian races express it. There is yet another influence on warfarin response, in terms of a polymorphism in a vitamin K_1 oxidase (*CYP4F2*; rs2108622). Carriers of this variant need higher doses of the drug[48]. Africans tend not to express this form of *CYP4F2*, but it is much more common in Europeans, Middle Easterners, and Central/South Asians (up to 20–57%).

Regarding warfarin and acenocoumarol, four polymorphisms can influence warfarin tolerance and efficacy[59]. You can see that if you are an African, you may well have a wild type of CYP4F2 (lower dose required), high VKORC1 expression (higher dose needed) plus CYP2C9*1/*1 diplotype (higher dose needed). However if you are African American you might have the other variants, but be CYP2C9 *8 heterozygous or even homozygous, which leads to a reduction in S-warfarin clearance of around a third[49] so you can see the complexity involved in trying to accommodate all the possible permutations. Indeed, warfarin use clinically is already quite complex. Once the drug starts to block the formation of the clotting factors, factor II's half-life is around three days, so it can take up to five days or more before a patient started on 5 mg warfarin daily for, say, atrial fibrillation, reaches their INR target of between 2–3, which should be steady for two consecutive days[58].

Some authorities have already recommended that warfarin should now be dosed using at least the VKORC1 and CYP2C9 status of the patient, although this does not always prevent overdosage[58]. In the absence of this information, other authorities have developed dosage advice recommendation web platforms, which are backed by algorithms that take into account population genetics[60]. Hence, in the absence of genetic information, the site will make a reasonable estimate, which of course becomes more precise as more information is available to submit[60]. Whilst this approach was successful with Caucasians, it was less so with African Americans, leading to overdosing, which emphasised that insufficient relevant information was incorporated into the algorithms[58]. Indeed, at the time of writing, the site did not accept genetic data on *8 status. Platforms for the estimation of warfarin dosage are improving all the time as more population genetic information becomes available, but this cannot happen fast enough, as anticoagulation-linked patient fatalities are unacceptably common[48]. Overall, it appears that with warfarin and the other coumarins, genetic information and algorithm-driven dosage determination platforms will always require an element of therapeutic monitoring to improve the safety of these drugs into the future.

CYP2C19

First detected as S-mephenytoin hydroxylase, CYP2C19 has emerged as important in the clearance of several classes of exceptionally widely used drugs, such as the proton pump inhibitors (PPIs), phenytoin and S-mephenytoin, as well as several barbiturates, benzodiazepines, SSRIs, and the antimalarial proguanil. The isoform has 90% of its amino acid structure in common with CYP2C9, but it has very different properties. Of the more than 30 variants found so far, three are of significant relevance. Of these, two code for completely nonfunctional enzyme (CYP2C19*2, rs4244285, and CYP2C19*3, rs4986893), which leads to virtually zero activity in homozygotes, whilst one variant (CYP2C19*17; rs12248560) causes what is termed 'ultra-rapid' metabolism[50]. This latter version's promoter is thought to be faulty, causing very high levels of isoform production, rather like a throttle stuck on full[61]. The allele for *2 is the commonest, found in 15% of Caucasians, 17% African Americans, and 30% of Chinese. The *3 variant is rare in most ethnic groups, with frequencies in Caucasians of 2-6%, and only around 1% in African-Americans, by contrast, as many as 25% of Asian races such as Koreans have this allele[62]. Indeed, up to 14% of Koreans are CYP2C19 poor metaboliser status compared with around 3.8% of Swedes[63]. The ultra-fast metabolizing *17 allele is found in about 18–26% of Caucasians, but only 1–2% of Asian races[64]. Hence, in terms of population impact, CYP2C19 slow metabolisers are clearly much more common in Asian races, although the faster version (*17) is rare in Asians[65].

CYP2C19 and PPIs

A major concern in CYP2C19 polymorphisms is in the therapy of peptic ulcers. Until the discovery that most ulcers were actually caused by the bacterium *Helicobactor pylori*, this condition was treated by years of the extremely commercially lucrative use of gastric acid suppressives like cimetidine and ranitidine. By the late 1980s, it was established that using any two of amoxicillin, clarithromycin, or metronidazole with a PPI agent would eliminate the bacteria. The PPI would not only suppress the acid formation to let the ulcer heal but would improve the stability and efficacy of the antibiotics in the stomach[66]. However, failure rates began to exceed 30% in some areas, partly due to predictable antibiotic resistance and adherence issues, but also due to insufficient maintenance of effective PPI levels. This is partly due to the relatively short half-lives of these drugs that do not always protect the stomach from the acid overnight[67]. In addition, the short residence time is obviously exacerbated by rapid clearance by avid wild-type and *17 homozygotic CYP2C19[67].

Unfortunately, around 7 out of 10 Caucasians have *1/*1 CYP2C19 capability, which contributes to treatment failure. Only 30–40% of Asian races are wild-type CYP2C19 homozygotes, so the issue is less acute for them. Of course, *17 rapid clearance capability is even worse for maintaining adequate PPI concentrations. So conversely, CYP2C19 deficient individuals have better treatment outcomes than EMs as the drugs (omeprazole, lansoprazole, and pantoprazole) are cleared four to six times more slowly and they build more pharmacodynamically effective concentrations. Fortunately, the high levels in the slow metabolisers are not a significant issue, as the PPIs in general are safe and well tolerated drugs.

With omeprazole, drug persistence in EMs is improved by its tendency to inhibit its own CYP2C19-mediated metabolism, which does not happen with the other drugs. Studies have shown that virtually all PMs are cured in a week of therapy, whilst cure rates range from 60% to 70% with EMs. The ultra-rapid CYP2C19*17 homozygotes are recommended to require higher doses, as well as supplementation of the regimes with other antibacterials such as bismuth compounds[66]. Other means of improving the response in wild-type (*1) and fast (*17) individuals is to increase the dose, or use rabeprazole, which is less vulnerable to CYP2C19 status as it is cleared by other CYPs. Omeprazole is given as a racemic mixture and it was found that the S-isomer is more resistant to CYP2C19 clearance (especially at first pass) than the R-isomer. It was marketed as esomeprazole and maintains better plasma levels for a given dose compared with the racemic mixture[66].

Because they are less expensive, the older PPIs will continue for some time, but the future of PPIs may well be with newer agents such as tenatoprazole, which has six times the half-life of omeprazole and addresses the issues with acid suppression overnight[67]. In addition, there is also ilaprazole, which is as effective as omeprazole but does not rely on CYP2C19 for its clearance and is much less subject to variation in response[68].

CYP2C19 and clopidogrel

In Section 7.2.2, clopidogrel was stated to be more reliant on CYP3A4 than CYP2C19 for activation and that CYP2C19 status was much less relevant than was previously believed. Studies are still published that show a link between CYP2C19 status and resistance to clopidogrel, which is termed HTPR (high on-treatment platelet reactivity). The studies can show increased resistance, or excessive bleeding, depending on whether a slow (*2 or *3) or fast (*17) variant of CYP2C19 is concerned[69,70]. However, what has also emerged is the large number of other polymorphisms that directly impact the pharmacological effects of clopidogrel in all races. There are several platelet glycoprotein and receptor polymorphisms, plus some relevant proteases[71] as well as the direct ADP drug receptor target itself, P2Y12, which exists in multiple alleles whose impact has not been resolved[72]. Indeed, a key study found that not only did a third of rapid CYP2C19 metabolisers respond poorly to clopidogrel, but slow metabolisers had no worse outcome than others[73]. Indeed, many other studies revealed that even the known inhibition of CYP2C19 by omeprazole and esomeprazole (but not pantoprazole) did not affect clinical outcomes[72]. The US FDA still at the time of writing warns of the risk of slow CYP2C19 alleles on clopidoregl outcomes, whilst other authorities have recommended that CYP2C19 testing could be carried out in patients undergoing specific clinical procedures only[74]. Interestingly, others have rejected genotyping completely[73]. Indeed, there are so many factors operating in the patient, in terms of their health, habits, and environment, plus any phenoconverting drugs or dietary agents they may have ingested, that genotyping in many contexts is without practical meaning. However, one effective compromise that would inform clinical decisions with respect to CYP2C19-cleared drugs is the pantoprazole ^{13}C breath test (Ptz-BT), which can be carried out in a clinical setting within half an hour and provides an accurate reading of CYP2C19 phenotypic patient metabolizing status[75].

Other platelet-modulating drugs have been developed, such as the thienopyridine, prasugrel, which shares clopidogrel's mode of action, but is activated through several CYPs and is not reliant on CYP2C19 status. Whilst prasugrel is safer and more effective than clopidogrel clinically[76,77], there is still resistance to it in some patient groups[72]. In that case, there is always ticagrelor, which targets the same receptor as clopidogrel but is reversible, does not need activation, is faster acting[72], and is probably superior to clopidogrel in terms of safety, ease of use, and efficacy[69].

CYP2C19 and other drugs

There are several other drugs whose clearance depends on CYP2C19, although the loss of function alleles does not necessarily impact clinical outcomes. The antimalarial prodrugs chlorproguanil and proguanil must be cleared to their active metabolites cycloguanil and it would be expected that PMs would form insufficient levels of the active agent to destroy the *Plasmodia* parasites, or fast versions might cause adverse reactions. However, this issue does not have clinical impact[78,79]. This might be linked with the biguanides use in combination with other drugs with different modes of action. On the other hand, despite most of phenytoin's clearance being linked with CYP2C9, slow versions of CYP2C19 (*2/*3) do require reduced doses in those of Chinese ancestry[80].

With respect to the antifungal voriconazole, this drug is well known for variability of clinical response, particularly with respect to CNS and visual issues that can occur in up to half of patients[81]. Whilst it is not always possible to demonstrate a strong relationship between CYP2C19 status and outcomes with voriconazole[81], the general feeling is that poor CYP2C19 metabolisers status is linked with adverse reactions severe enough to discontinue therapy, whilst treatment failure has occurred in ultrafast *17 individuals[82,83]. It is recommended that some form of phenotypic analysis take place with this drug, such as the pantoprazole breath test, which would save time in terms of establishing curative drug levels whilst minimising potential toxicity that could impact adherence[82]. Although TCAs are usually considered to be cleared by CYP2D6, the role of CYP2C19 in some individuals with these drugs has been sufficiently researched for the Clinical Pharmacogenetic Implementation Consortium (CPIC) to recommend that fast and slow CYP2C19 homozygotes should not be exposed to tertiary amine TCAs such as amitriptyline[74].

The SSRIs citalopram, escitalopram, and sertraline all have significant CYP2C19 clearance, which is not considered an issue for wild-type and intermediate CYP2C19 metabolisers; for rapid phenotypes, the citaloprams are contraindicated, whilst with sertraline it is 'proceed with caution' and if necessary, seek an alternative[84]. For all three drugs, reduction of the dose by half is recommended for slow CYP2C19 phenotypes, not least to minimize the risk of TdP (Chapter 5.7.2.) with the citaloprams and possible hepatic issues with sertraline (Chapter 5.5.2).

CYP2D6

Although only 2% or so of hepatic CYP content, CYP2D6 is responsible for the clearance of 25–30% of prescription drugs and it was noticed in the 1970s that the clearance of some (now obsolete) drugs (sparteine and debrisoquine) was greatly retarded in a small proportion of Caucasian populations. The enzyme is still occasionally termed *debrisoquine hydroxylase*, or *sparteine/debrisoquine hydroxylase*. There are over 100 different haplotypes described so far, and the clinical impact of the polymorphic forms of this CYP is still the subject of intense research and powerful debate. CYP2D6 has a role in the clearance of several drug classes, which are important in the management of many CNS, cardiovascular, and anti-nociceptive clinical conditions.

The full picture of CYP2D6 variation is somewhere between fiendishly and sumptuously complicated. From a practical perspective, perhaps the easiest way to view this CYP is to look at the alleles in terms of decreasing functionality, as well as frequency. The wild-type allele, which is fully functional (CYP2D6*1), is present in roughly a third of humanity. CYP2D6*2, is close to *1 in capability, but apparently has many minor subversions[85]. The variant *2 is found in about a quarter of Caucasians, but is rarer in Asian races (16%) and Africans (11%)[85]. In terms of alleles that are defective, the most common defective allele is CYP2D6*10 (P34S; rs1065852). The protein formed lacks stability and has lower affinity for substrates compared with *1 and is found in up to half of Chinese, Korean, Japanese, and other Asian peoples, but only 1–2% or so of Caucasians and about 4–6% of other races[86,87]. The next most common version is CYP2D6*17 (T107I, R296C, S486T; rs28371706), which is again partly functional due to changes in the active site that reduce drug affinity, although this can vary according to substrate[86-88]. The allele *17 is found in 35% of Africans, but not in other races. CYP2D6*41 (G2988A, rs28371725) causes a splicing fault that leads to less-functional protein produced, but there is still activity. It is found in 7–10% of Caucasians, but below 5% in other races[87].

The most relevant alleles from the therapeutic angle are the nonfunctional ones, where no usable protein is produced. The most common null allele is found in about a quarter of Caucasians (CYP2D6*4, 1846G>A, rs3892097), which is known as a splice site mutation, which mean the mRNA for the protein is not assembled properly, causing the absence of functional protein[86,87]. About 12–15% of Hispanics have *4, but it is rarer (<5%) in Africans and Asian races. Of the other null alleles, the gene is completely absent with *5, which is found in 2–5% of Caucasians, and *36 is nonfunctional also, but is found in 4% of Asian races[85]. The other null alleles (*3, *6, and others) are vanishingly rare.

In terms of gain of function, for many years it was assumed that this CYP was not inducible, partly because it was thought that the ability *CYP2D6* has to repeatedly copy itself in some individuals in certain ethnic groups provided a sort of permanent induction status. This is well documented in 16% of those of Ethiopian and Saudi-Arabian heritage[86]. It is now known that duplicated copies of nonfunctional CYP genes occur also and that the CYP2D6 is actually inducible in certain circumstances (Chapter 4.4.5). Ultra-rapid metabolisers of CYP2D6 can be termed xN after their diplotype, such as 'CYP2D6*1/*1xN', and an individual might have a dozen functional copies of the gene such as *2 or *1, providing spectacular metabolic activity. Ultra-rapid metabolisers of various types comprise around 5–10% of most ethnic groups[85]. Multiple copies of a less-functional allele such as *10 or *17 would probably confer similar or slightly greater than normal capacity.

In practical terms, around 70% of humanity has what could be termed 'normal' CYP2D6 capacity[85], which, given the scale of the variation, is actually remarkable in itself. The spectrum of intermediate metabolism will be, of course, very wide. On the high end, you might see those with a normal and a slightly defective allele, such as *1/*10, or *2/*17, for example, whilst those with a normal and nonfunctional allele (*1/*4) might be anywhere around the mid to lower end of intermediate. Those with combinations of defective/null allele combinations, such as *10/*4 or *17/*3 for example, would be at the lowest area of intermediate to slow and in practical terms they would not test very differently from those with two null alleles in terms of drug clearance[87].

What is interesting is that ethnically, CYP2D6 *10 and *17 are so common in Asian and African races that their clearance of substrates of this CYP will be significantly lower than that of other groups. It is also remarkable that there are such large numbers of Caucasians with no functional CYP2D6, which may be up to 40 million people in Europe alone. The frequency of ultra-metabolisers may also be higher in Caucasians than in other races, perhaps 10% or more[87]. Outside of drug exposure, *4 carriers are predisposed to developing Parkinson's disease, but only if they are exposed to pesticides[89].

CYP2D6 and antipsychotics

Schizophrenia is generally poorly controlled, to the extent that some reports suggest that only a quarter of patients are stably and successfully treated to the point of meeting remission criteria[90]. This is partly due to the complexity of the condition and the high levels of adverse reactions associated with antipsychotic drugs. In addition, antipsychotic drugs cause cumulative structural and functional damage to the brain, leading to a range of permanent Parkinsonian symptoms and the well-researched extrapyramidal movement disorders[91,92]. These latter distressing effects include tardive dyskinesias (TDs), which are involuntary movements of the lips, tongue, face, arms and legs, which begin to occur during the first year or so of treatment with some of these drugs. It is little wonder that these issues contribute to poor compliance that feeds the high lifetime suicide risk amongst schizophrenics. Clearly, any measure that could ameliorate this desperate situation in any meaningful way would have the wholehearted support of patients and practitioners. Given that it might be expected that the high level of null drug metabolising alleles present in Caucasians might increase residence time and toxicity in many patients, it appears at first sight surprising that genotyping of CYP2D6 is not part of clinical procedures at the time of writing. Although the majority of old and newer antipsychotic drugs are cleared or partially cleared by CYP2D6 and despite decades of clinical studies, it is still controversial whether slow *4-related clearance really significantly impacts treatment to the point that it is necessary to test for it and act on the results in terms of dosage modulation. Some studies associate higher extrapyramidal symptoms (risperidone)[93] or high drug plasma levels and nausea and vomiting (aripiprazole)[94] with slow CYP2D6 variants. Indeed, the US FDA recommends that doses of the aripiprazole successor brexpiprazole (CYP2D6 and CYP3A4 substrate) should be halved in CYP2D6 poor metabolisers. However, others require more data[95] and yet more show no significant relationship between genotype, plasma levels, and impact on the patient[90,91]. What is particularly problematic is that some studies show no benefit from genotype-informed dosage adjustments[91].

Why it has been so difficult to form a consensus is obviously multifactorial. The combination of the impact of many years of drug therapy on already disordered dopaminergic pathways in the schizophrenic brain may make it less responsive to dosage changes in terms of efficacy and adverse reactions[91]. Many other issues impact drug clearance. These can range from other unknown pathways, through to the effects of smoking and phenoconversion of fast to slow metabolisers caused by co-administered drugs. In addition, as with CYP2C19 and clopidogrel, the influence of other polymorphisms that impact antipsychotic drug response are only now becoming better understood, such as the issues with polymorphic type 2 dopamine receptors[90]. The polymorphic DRD2/ANKK1 Taq1A, (rs1800497) found in 22% of Caucasians, may decrease dopaminergic transmission characteristics[90]. This raises the possibility of complex combinations of homozygotic ultra-fast CYP2D6/rs1800497 individuals being inadequately treated, through a combination of low-plasma drug levels and low receptor sensitivity. On the other hand, homozygotic CYP2D6*4/rs1800497 patients may actually be adequately treated, as the higher plasma drug levels were more effective on the poor-response receptor. However, this also does not take into account other polymorphic issues with catechol-O-methyl transferase (COMT, section 7.2.6) and other enzymes that modulate dopamine levels in the schizophrenic brain.

Whilst ultra-fast CYP2D6 individuals might at first not be responsive to the usual doses, practitioner experience would trigger cautious increased dosage until either a response occurred or adverse reactions were intolerable to the patient and another agent employed that did not rely on CYP2D6 (quetiapine, for example). However, with drugs such as risperidone, the situation is complicated by the clearance to 9-hydroxy risperidone (marketed now as paliperidone), which has the efficacy of the parent drug, but with different physicochemical properties. As the 9-hydroxy derivative is more polar, it is likely to penetrate the CNS less and perhaps cause more peripheral side effects, although this requires more investigation[95]. Certainly more of this metabolite is formed in ultra-rapid CYP2D6 individuals so their response to this drug will be different from other patients[95]. Risperidone and paliperidone appear to show similar side-effect profiles and have been the subject of lawsuits related to gynaecomastia caused by hyperprolactaemia. Whether ultra-rapid metabolisers are more at risk for this issue with either drug is suspected, but not proven[95].

Overall, it is likely that whilst CYP2D6 genotyping with antipsychotics might not necessarily be as helpful as was once envisaged, phenotyping may be of some value in adjusting doses as part of efficient therapeutic monitoring, as the therapeutic ranges of all the major drugs are well established[96]. However, there are fundamental underlying issues with antipsychotic drugs aside from genotyping. In the long term, marketing the metabolites of existing antipsychotics as new drugs could be seen at best as a somewhat intellectually bankrupt wheeze from the distant past. Nothing less than a radical and revolutionary change must occur in antipsychotic drug design to improve efficacy and eliminate their cumulative and devastating neurotoxicity.

CYP2D6 and tricyclic antidepressants (TCAs)

These drugs were the mainstay of treatment for depression until the advent of the SSRIs and newer mixed-function agents discussed in Chapter 5.5.2. TCAs are essentially superseded drugs, although they are still used in a number of complaints aside from depression, such as intractable pain. They were and remain difficult to use and dangerous from a number of perspectives, not least because with antidepressive agents in general, the onset of therapeutic effect is so slow (six weeks or more) and their atropine-like side effects (dry mouth, constipation, etc.) make them poorly tolerated. If the patient did not respond to these drugs, then the patient might choose to overdose (Chapter 5.5.2). This, combined with a narrow TI, led to many TCA fatalities. Inability to clear these drugs due to a patient's status as a 'poor metaboliser' would provide the twin problem of a high level of the atropinic side effects, combined with an even narrower TI in that patient and a risk of death from a modest overdose.

Over the past decade, it has been realised the role of CYP2C19 is also significant with many TCAs and must be considered as well as CYP2D6 status. Evidence has mounted that responses and toxicity are strongly linked to CYP status with these drugs. Indeed, it is now clear that ultra-rapid CYP2D6 homozygotes are more likely to be unresponsive to the drugs than other individuals and at first doubling the doses of drugs like desipramine was recommended. However, the CPIC at the time of writing took this further and now recommends that TCAs are contraindicated with these individuals as well as those who are homozygous for CYP2D6*4[74]. CPIC also suggests a 25% reduction in dose for intermediate metabolisers[74].

CYP2D6 and beta-blockers

Although atenolol is entirely renally cleared, the beta-blocker with the largest CYP2D6 component to its clearance is metoprolol (about 80%), whilst timolol, propranolol, pindolol, and nebivolol are cleared by other CYPs[97]. Carvedilol is cleared by CYP2D6 as well as CYP2C9 and various UGTs[98]. Their metabolism is complex as many are chiral, such as with carvedilol for example. This drug is highly effective and has a wide therapeutic index, but it is given as a chiral mixture of R and S enantiomers. This means that its pharmacodynamics and pharmacokinetics are strongly influenced by this stereoselectivity. The R-isomer is predominantly vasidilatory, whilst the

S-isomer binds cardiac β1 receptors. The R-isomer is more bioavailable than the S, and CYP2D6 is said to clear the S more than the R, although CYPs 1A2, 2C9 and 3A4 are also involved. The drug is also cleared by UGTs and is transported by P-gp[98]. Already, you can see that polymorphisms may intervene at several pharmacodynamic (β-receptors) and pharmacokinetic stages and this is just in healthy individuals.

These drugs are not prescribed in the quantities they were in the 1970s and 1980s, as there are many more alternatives, but they continue to be used for atrial fibrillation, angina and hypertension as well as other conditions. They also prolong the lives of patients with heart failure by protecting them from excessive sympathetic stimulation[99]. When beta-blockers are used for hypertension they are dosed towards a target area response, although in the United States in heart failure, the patient dosage is directed towards reaching an established target for effectiveness in clinical trials[99]. Interestingly, less than half of heart failure patients reach these targets in clinical practice. This is a partly linked with the issues described in the previous paragraph, as well as the almost 50% reduction in drug clearance seen in heart failure patients[99]. In addition, slow CYP2D6 status will also contribute to those factors and will prevent dose escalation towards the targets. This form of target-driven dosage escalation has been criticised as the antithesis of personalised medicine. However, with heart failure patients, it is clear that there are many factors exerting greater influence on clinical outcomes than CYP2D6 status, although some studies show that slow allele patients do not reach the target drug doses due to poor tolerance of the drugs, such as with metoprolol[99].

It is clear that *4/*4 CYP2D6 status does impact beta-blocker pharmacokinetics, and with metoprolol this can show as a significantly greater pharmacological effect[97]. However, in the context of how the drugs are used clinically, when dose is titrated towards effect, it is not surprising that CYP2D6 status, perhaps even the ultra-rapid genotype, is just one more factor in a complex process. Indeed, several other studies suggest that the *4 allele seems to make the patients more sensitive to metoprolol earlier in the titration process, causing issues such as dizziness, but ultimately it does not have any decisive impact[100]. With carvedilol, due to the role of other CYPs and the other systems (P-gp and UGTs) that participate in its clearance, the relationship between CYP2D6 status and clinical outcome is less than with metoprolol. Indeed, carvedilol's pharmacology is complex enough even without other drugs in a typical polypharmacy regime affecting the CYPs, P-gp, and UGTs that are involved in its clearance. CYP2D6 activates carvedilol to a metabolite, which is 13-fold more active than the parent[99], but CYP2C9 has a major role in carvedilol's clearance also. Hence, CYP2D6 status in Caucasians may have some impact on carvedilol response[99].

As with other beta-blockers, in Koreans the pharmacokinetic impact of the *10 alleles was clear in terms of slower carvedilol clearance, but this was not translated into a significant pharmacodynamic effect[101]. Whilst knowledge of patient CYP2D6 status may be helpful in the process of establishing a patient on a new regime with a beta-blocker, in the real world of polypharmacy and significant pathology, only therapeutic monitoring based around drug levels and measurement of efficacy are practical in clinical usage.

CYP2D6 and SSRIs

With the SSRI and SNRI drugs, an observer could be forgiven for thinking that the use of these drugs in depression is similar to the saying about second marriage (the triumph of hope over experience). The failure rate as a first line in depressive illness is over 50%[84] and as discussed in Chapter 5.5.2, whilst these drugs are much safer than their predecessors, they are no more effective. In addition, they are normally prescribed by general practitioners rather than specialists, without any therapeutic monitoring save, 'Do you feel any better this week?' Indeed, considering the low bar of expectation, their very slow onset of action (if any), plus the plethora of similar alternatives, a health-care professional might well switch to another related drug, or after some fruitless months move onto a TCA or even an antipsychotic such as olanzapine rather than investigate in depth as to why the SSRI did not work.

It is also the case that SSRI clearance is complex, as they are extensively oxidatively metabolised by CYP2D6 as well as CYP2C19 as mentioned in a previous section, leading to saturation and inhibition on chronic dosage. With fluoxetine, the complexity of its clearance (both CYP isoforms, plus the impact of the different enantiomers (Chapter 5.5.2) means that there is insufficient evidence that dosage guidance based on genotype is useful[84]. However, it can cause long QT syndrome and the risk might increase with null CYP2D6 alleles. For paroxetine, it is recommended that in ultra-rapid metabolisers plasma levels may not be therapeutic, so it is contraindicated. Although paroxetine can inhibit its own clearance, it is not certain whether this would lead to useful plasma levels. At the other end of the scale, the dose should be halved for poor metabolisers[84]. Whilst there is insufficient evidence with fluvoxamine and ultra-rapid metabolisers to make any recommendations, it is suggested that its dose should be halved also with poor CYP2D6 metabolisers.

As with other CYPs, there is growing evidence that a whole range of other pharmacodynamically relevant polymorphisms may have at least as marked impact on efficacy in depression as the metabolic issues. These include the *COMT* gene as mentioned earlier in this section, plus the serotonin receptor genes *HTR1A* and *HTR2A*, as well as *BDNF* (brain-derived neurotrophic factor) and the various ATP cassette transporters and many other neurotransmitter-related proteins[102]. The different combinations of polymorphisms as they impact pharmacokinetic and pharmacodynamically related drug responses in depression are yet to be unravelled.

CYP2D6 and antiarrhythmics

Flecainide, mexiletine, and propafenone have potent effects on cardiac electro-physiology and these are sensitive to dosage, so high concentrations of these agents may promote arrhythmias rather than reducing them. Flecainide is mainly renally cleared and does not seem to be strongly influenced by genotype, although AUCs are larger with slower CYP2D6 alleles[88]. To date, any adverse effects in CYP2D6 PMs have mainly been seen in mexiletine-treated rather than flecainide-treated patients, with light-headedness and nausea seen in CYP2D6 PMs, although there is less variability in the clearance of the drug with the slow metabolisers[88]. With propafenone, the elimination kinetics of the drug are more linear in poor compared with extensive metabolisers, due to its inhibition of CYP2D6, but it takes up to three times longer to reach steady state with poor metabolisers, so gene identification would be helpful in this context[88]. The newer agent vernakalant (Brinavess) does not appear to be very significantly affected by CYP2D6 genotype[103], as the intersubject variability is high and there is only around 15% difference in AUCs between the slow and fast metabolisers[103]. Vernakalant was in the rather unusual situation of gaining European Union approval in 2010, but not US FDA approval, apparently on safety grounds[104]. This has been perplexing to many[104], although at the time of writing the manu-facturer intends to make another application to the FDA in late 2019.

CYP2D6 and opiate prodrugs

These include agents such as codeine, dihydrocodeine, oxycodone, hydroco-done, and tramadol (Appendix B.2). These drugs are activated to morphine or one of its derivatives, and CYP2D6 status is an important influence on their potency as analgesics. Codeine and dihydrocodeine are prodrugs, as they are methylated versions of morphine. Codeine is less potent but more commonly used than the dihydro derivative and usually only about a tenth of the codeine dose needs to be O-demethylated by CYP2D6 to liberate morphine to act anal-gesically. In PMs, codeine and to some extent dihydrocodeine both show much less efficacy for a given dose, although some efficacy comes from combination of the quantity of codeine-6–glucuronide formed (about 80% of the dose) and its weak analgesic effects[105]. Such is the clinical context of opiates and the perceived dangers of toxicity and addiction that when individuals who are PMs complain of poor analgesia they can be unjustly accused of 'drug-seeking' behaviour[106]. Indeed, there can be as much as a 30-fold difference between poor and ultra-fast metabolisers of codeine in terms of morphine AUC[107].

However, it is easy to perhaps underestimate the potency of codeine, as it is found in a variety of POM and OTC preparations and it is perhaps seen as a rather minor opiate compared with heroin and morphine. Theoretically, most of the dose is available for conversion to morphine by CYP2D6 and there have been several reports of fatalities in children with ultra-fast CYP2D6 genotypes linked with codeine. This has been exacerbated by the high numbers of codeine prescriptions issued to young children who in any case are more sensitive to opiates compared with adults. Post-mortem analysis in the children revealed very high plasma levels of morphine, with almost no codeine present[105]. The danger of ultra-rapid CYP2D6 codeine toxicity has caused its use in children to be more carefully monitored or even discouraged[105]. Another problem with ultra-rapid clearance of codeine is that it explains why some individuals become so tolerant and dependent to what is apparently a minor opiate[106].

CYP2D6 also converts oxycodone (OxyContin) to oxymorphone, as well as hydrocodone (marketed as Vicodin, with paracetamol), to hydromorphone. About half of hydrocodone conversion is CYP2D6 and the rest is CYP3A4 (forms norhydrocodone). This drug is very sensitive to CYP2D6 status, as it works poorly with slow metabolisers and large amounts of hydromorphone are formed with fast metabolisers[105]. The clearance of oxycodone is more complex. About 80% is cleared to noroxycodone by CYP3A4 which is shows little analgesia, but CYP2D6 forms oxymorphone (stronger than morphine) and this CYP also forms noroxymorphone out of the noroxycodone that was made by CYP3A4. Noroxymorphone is weaker than morphine, but stronger than noroxycodone. Whilst all this is slightly confusing, CYP2D6 status has a strong bearing on the net analgesic effect. The faster CYP2D6 version the patient has, the more powerful the oxycodone efficacy[105,106], although ultra-rapid patients are also subject to more psychiatrically disturbing effects with all the activated synthetic opiates such as hallucinations and confusion[108].

Tramadol is employed in many contexts as it is much less toxic than morphine, although it should be used with caution with SSRIs and other seratoninergic agents, as its mild inhibition of biogenic amine reuptake can cause serotonin syndrome with these drugs. CYP2D6 forms O-desmethyl tramadol (often called M1) that has a 200-fold stronger opioid action compared with the parent drug. Tramadol is also N–demethylated to the inactive M2 by CYP3A4 and CYP2B6 may also be involved[105]. M1 is mainly glucuronidated and to some extent sulphated. In homozygous PMs, around 75% less M1 is formed compared with the EMs and there is of course less opioid effect. However, there are other factors that to some extent preserve tramadol's efficacy in PMs. The formation of M1 is rather drawn out even in heterozygotes and in PMs its half-life is more than three-fold longer than in the EMs. In addition, PM parent drug levels are nearly double those of EMs and tramadol's amine re-uptake inhibition (noradrenaline and serotonin) contributes to analgesia[105,109,110]. In neonates also, M1 clearance is delayed by a lack of glucuronidation, so UGT status will also impact M1's residence time. Whilst in some PMs very little if any analgesia occurs, in many slow metabolisers some analgesic effect will occur with the bonus of almost no side effects as these are directly related to M1 levels[109]. However, as these levels increase in extensive metabolisers, the problems increase as with the other opiate prodrugs; ultra-fast CYP2D6 metabolisers can experience severe reactions, such as cardiac, renal and respiratory failure[109].

As with other drug classes, the impact of opiate receptor polymorphism has come under increasing focus. Opiates interact with the μ receptor subtype OPRM1, and some studies suggest that almost half of some human populations possess the *OPRM1* 118A/G variant, which is less sensitive than the wild type[106]. Of course, there are the usual suspects again, such as COMT and other enzyme systems that impact biogenic amines, plus the various transporter systems that modulate CNS drug entry. As you can see, this can compensate some individuals or leave them very vulnerable, depending on their combinations of relevant genotypes.

CYP2D6 and antiemetics

The 5-HT-3-receptor antagonist antiemetics are also CYP2D6 substrates. These include ondansetron, dolasetron, tropisetron, palonosetron, and andramosetron. Although it has been controversial, it is now accepted that reliance of these drug's clinical efficacy does rest on CYP2D6 status[111]. For instance, in a study with an 8 mg preoperative dose of ondansetron a third of wild-type metabolisers suffered vomiting, whilst the poor metabolisers experienced 100% efficacy[112]. CYP2D6 UMs will have little if any efficacy from a standard dose. Currently, it is recommended that there is a strong case for CYP2D6 status testing with these drugs[112,113] and ultra-rapid metabolisers should be given granisetron[113], which is cleared by CYPs 3A4 and 1A1, although it too can be subject to marked variation in efficacy[114]. However, if testing does occur, it does look as if the 8 mg dose may not have sufficient efficacy in wild-type homozygotic CYP2D6 patients[112].

CYP2D6 status and TdP-mediated sudden death

As detailed in Chapter 5, the list of drugs that lengthen the QT interval (Chapter 5.7.2) is ever increasing, and many are CYP2D6 substrates. Although some groups of patients, such as females, those with heart disease, hypokalaemia, aged over 65 years old, those on polypharmacy regimes in intensive care, plus certain genetic variants of hERG channels, are much more susceptible to TdP than the rest of the population[115], there are even more factors to consider, which complicates the issue. Whilst many agents can cause long QT syndrome, this does not necessarily lead to TdP. As with any pharmacological effect, susceptibility to long-QT syndrome is linked to the potency of the drug's blockade of the potassium channels, as well as to local drug concentration. The danger of sustained hERG inhibitory levels could be accentuated by CYP inhibition as well as CYP status or a combination of the two. Outside of those individuals with combinations of risk factors for long QT, it is also problematic to identify whether a sudden death is linked to drug-induced TdP, as undetected heart disease is already extremely prevalent in Western countries, and very large numbers die from heart failure every day.

The role of CYP2D6 status in the risks of sudden death linked with TdP is, of course, difficult to predict. Of the drugs where CYP2D6 has a significant role in their clearance, older antipsychotics such as thioridazine (the worst offender, now mostly withdrawn), haloperidol, pimozide, and chlorpromazine are well known to put patients at risk of TdP-induced cardiac failure compared with non-drug users (with the exception of loxapine[115,116]. Iloperidone has also shown a significant risk, as has risperidone[115,117], whilst quetiapine is of moderate risk[118]. Of other psychoactive CYP2D6 substrates linked with long QT, the SNRI venlafaxine and some TCAs have also been recorded as problematic[115].

Several studies point to the role of slow CYP2D6 alleles in promoting the risk of long QT with all these drugs[117-119], although again CYP2D6 status is just one of many other factors subject to polymorphisms aside from those mentioned above. These include variants of the hERG channel gene (*KCNH2*), *MDR1*, and *NOS1AP* (codes for nitric oxide synthase that regulates heart contractility) and also combinations of these alleles found in specific ethnic groups; Caucasians are at greater risk of drug-linked long QT syndrome than other groups[119].

CYP2D6 and other drugs

Venlafaxine is a noradrenaline and serotonin reuptake inhibitor and is often a second-line choice after SSRIs have failed to help, although patients are often then locked into this agent for many years. It is intermediate in safety between the TCAs and the SSRIs (Chapter 5.5.2) and the anticholinergic effects seen with the TCAs are not seen with this drug. Venlafaxine is oxidised by CYP2D6 to two major active metabolites, O-desmethyl venlafaxine (ODV) and N-desmethyl venlafaxine (NDV). With this drug, opinion seems divided. It has been reported that slow CYP2D6 alleles are associated with more side effects, including TdP as mentioned in the previous section, plus others such as gut problems[120,121]. In addition, the drug seems to work much better in faster metabolisers with fewer side effects than slow metabolisers[122]. Taken together, this suggests that the smaller the ratio of parent to ODV (and perhaps to NDV also), the more effective and tolerable the drug becomes. However, other studies did not report a strong association with CYP2D6 status[123] and currently the US FDA does not recommend genotyping CYP2D6 status. Others recognise the association between efficacy and parent metabolite ratio and suggest that with slow and intermediate metabolisers, alternatives should be used[124].

The antianginal agent perhexiline (Chapter 5.5.2) is a known CYP2D6 substrate, and its issues with neuropathy and hepatotoxicity led to its withdrawal in many countries, with the exception of Australia and New Zealand. It is now recognised that CYP2D6 status is usually predictive of perhexiline toxicity and that this CYP is virtually the only route of clearance of perhexiline. It has been suggested that this drug is a worthwhile candidate for CYP2D6 genotype testing alongside careful therapeutic monitoring to exploit its high clinical value in angina therapy[125,126].

CYP2E1

As mentioned in Chapter 3.6.2, the major interest in this CYP is toxicological rather than pharmaceutical, as apart from chlorzoxazone, a minor fraction of paracetamol (Chapter 8, section 8.2.3.) and theophylline clearance, this CYP2E1 has little role in drug metabolism[127,128]. It is of intense interest due to the various environmental toxins it clears, such as PAHs, nitrosamines, organic solvents, and small carcinogenic heterocyclics, as well as its role in the metabolism and toxicity of ethanol. Its toxic substrates and its high expression in many 'barrier' tissues, such as the lung, gut, and oesophagus, make it very influential in the production of potentially carcinogenic reactive species. In addition, it is rapidly inducible (Chapter 4.4.4) and forms many reactive species when not processing substrate, so the combination of its polymorphic expression and activity means it exerts a significant influence on life expectancy in all human ethnic groups. With many variants of CYP2E1, expression is directly proportional to reactive species formation, although with some toxins failure to clear the agent can be injurious, so low expression may not necessarily be protective.

CYP2E1 nomenclature can be confusing, but using the *allele convention, the wild-type is CYP2E1*1A and in some populations such as in Brazil, Caucasian and non-Caucasian populations, 85–96% are homozygous for the wild-type[127]. There are many other variants, including *1D, *1B (rs2070676), *5B (rs2031920), *6(rs6413432), and *7B[128,129]. CYP2E1*5B (rs2031920), found in around half of Asian races but only 5–10% of Caucasians, has impaired transcription, possibly lowering risks for some cancers[129,130], but there is no consistent picture. *7B has a faster promotor, nearly doubling its expression, and may be linked with cancer risks, whilst the very rare *2 version has lower enzyme activity and the equally rare *3 is similar to the wild type[129].

As with CYP1B1, due to large variation and all the complicating factors seen in human studies, many reports trying to link variants of CYP2E to diseases have produced conflicting data. With drugs, it appears that slower versions of CYP2E1 can make a small impact on theophylline clearance, as a minor fraction of this drug is cleared by CYP2E1[127]. Many studies have tried to link different CYP2E1 alleles with alcohol damage and tendencies to abuse, although this has been problematic, mainly because even when fully induced, it does not clear anything like as much ethanol as the alcohol/aldehyde dehydrogenase systems. Also, many previous studies have looked at individual SNPs when it seems that combinations of different SNPs in various haplotypes are more likely to be linked with ethanol consumption, although this is less likely in Caucasians and some

Asian races than in other ethnic groups[128]. There are many other metabolic and also myriad CNS predispositions that are likely to be more predictive of ethanol addiction propensity than CYP2E1 alleles.

CYP3A

Whilst the adult CYP3A (CYP3A4 and 3A5) isoforms metabolise up to 60% of all drugs and handle a large proportion of endogenous steroid metabolism, the patterns of their polymorphisms are rather different to that seen in the other CYPs, which are usually fairly distinct in terms of their substrate preferences and modes of expression modulation. For instance, if a drug is confined to CYP2D6 as a clearance pathway, possession of two null alleles for this CYP makes a marked clinical impact, as we have seen. However, such is the substrate overlap and general variation in CYP3A4 expression and sensitivity to induction, that even in the absence of CYP3A5 expression it might be quite difficult to distinguish the clinical impact on drug clearance and thus efficacy and side effects. What is also apparent is that despite the CYP3A family being so important for drug clearance, the scientific literature regarding its polymorphic expression can be rather contradictory and the clinical relevance is still emerging, with the focus being rather narrow on drugs such as the immunosuppressants in transplant therapy.

CYP3A4

Compared with other CYP genes, *CYP3A4* appears to be remarkably well conserved across humanity. The main form is CYP3A4*1, sometimes written as *1A. Perhaps the most significant polymorphism in terms of population numbers, health risks, and impact on drug clearance is seen mostly in Africans/African Americans, which is CYP3A4*1B (rs2740574), which is an adenine to guanine substitution (c.-392A>G) in the 5' promotor region of the gene. What this actually means in terms of CYP3A4 capability is not always clear in the literature, but several studies have linked *1B with patients requiring significantly larger drug doses to maintain efficacy, particularly with immunosuppressants (tacrolimus, cyclosporine)[131], but not sirolimus[132]. With AEDs, drug-resistant epilepsy was strongly associated with *1B[133]. These studies point to the issue with the 5' promotor in this gene increasing the basal expression of the protein that makes clinically relevant and sustained reductions in drug plasma levels, necessitating dose increases to maintain appropriate efficacy. The *1B allele is expressed in less than 3% of Caucasians and less than 10% of Mexicans, but studies suggest that 25–50% of those of African ancestry possess it[131-133]. This variant is linked with the high levels of prostate cancer seen in those of African ancestry[134]. CYP3A4*1G is found in around 25% of Asian races (not in

Caucasians), and is by far the commonest CYP3A4 variant in Chinese and Japanese people[74,135]. It is associated with the need for around two-thirds of the usual doses of several drugs, such as analgesics[74,135-137] and antiarrythmics like amlodipine[138]. Other forms, such as *16, are slower catalytically, but are only found at a low level (1–2%) in Korean and Japanese[139,140].

CYP3A5

CYP3A5*1 is the fully functional wild type. In contrast to CYP3A4, there is one major null allele, *3, or rs776746, (6986A>G), which causes a change in RNA splicing where an early stop codon is inserted, which leads to rapid destruction of the mRNA, so no usable protein appears and homozygotes are termed CYP3A5 non-expressors[141,142]. So far, it is apparent that around 45% of those of African ancestry, under 10% of Indians, and around 7% of Chinese are CYP3A5*1/*1, but only 1% of Caucasians are homozygous. Around half of Chinese, Black, and Indian groups have intermediate metabolizing capacity (*1/*3), whilst just under 90% of Caucasians, about 60% of Indians, and 47% of Chinese *3/*3 are nonexpressors of CYP3A5[141]. Nonexpression of CYP3A5 is rarer in Africans, with a level of around 15%. Variants *2 and *4–*7 are rarer and mostly of reduced function or nonfunctional.

The success of anti-rejection drugs and tacrolimus in particular has been a strong driver of research into the clinical implications of CYP3A5 polymorphism, chiefly because of the narrow therapeutic indices of these agents and the desire to maximize their safety and efficacy in preserving the massive personal, clinical, and financial investment in a successful organ transplant. It could be said that the patients live on an anti-rejection drug 'knife-edge' where excessive plasma levels damage the xenograft and insufficient levels risk rejection. Hence, the trough concentrations of tacrolimus must be monitored carefully to ensure graft preservation. So pretty much any means of improving clinical outcome related to modulating drug dosage is worth exploring. The question is, what effect will CYP3A5 polymorphism have on clinical immunosuppressant efficacy? Some like cyclosporine are cleared predominantly by CYP3A4, whilst with others, particularly tacrolimus, CYP3A5 clearance is about double that of CYP3A4[143]. The exploration of polymorphisms in both isoforms has been pursued experimentally.

CYP3A4 and CYP3A5

Let us say for argument's sake that an individual (Mr Wild-Type) who is homozygous for wild-type full function CYP3A4 (*1/*1) and also for CYP3A5 (*1/*1), should have his CYP3A capability provided roughly 50/50 by the two isoforms[141]. From what we have seen, Mr Wild-Type is unlikely to be Caucasian, but far more likely to be African. Now, if we tested some individuals and looked for the highest theoretical metabolizing capacity combination of CYP3A and we found Mr Super (CYP3A4*1B/*1B) and (CYP3A5*1/*1), he would probably be African also. If we consider the other extreme, testing to find low-capacity isoform combinations, then Mr Slow (CYP3A4*1G/*1G with CYP3A5*3/*3) perhaps could be found, who is highly likely to be Chinese.

Some of these variants (CYP3A*1B and CYP3A5*1) are already well known in Caucasians and Africans to be in linkage disequilibrium (found in the same haplotype) in the majority of Caucasians and virtually all Africans[144]. If we look at these combinations, immediately, it is clear that there is no large-scale 100% null CYP3A capability, but that also with Mr Super, there is some very high capacity available, which on full induction would be considerably higher. However, there is a very large (mainly Caucasian) 'middle ground', as several studies have found that almost all CYP4A5 non-expressors (*3/*3) also have CYP3A4 *1/*1, as these alleles are also in linkage disequilibrium[144,145]. There is so much more in terms of complex allelic linkages to be discovered across our genomes – for instance, CYP3A5 *1/*1 is linked to elevated blood pressure in many studies[141]. To summarize, if we consider all the other allelic combinations possible across the spectrum from Mr Slow to Mr Super, it is possible to see how robust CYP3A capacity is across humanity. The obvious question is, what effect does this have in the clinic?

We have already seen that the CYP3A4*1B allele causes tacrolimus trough concentrations to be anything up to a third lower than those with wild-type CYP3A4[141]. With CYP3A5, expressors (homozygous and heterozygous) show lower trough tacrolimus levels, whilst conversely, *3/*3 nonexpressors exhibit higher drug concentrations, although evidence conflicts for cyclosporine, probably as its CYP3A4 metabolism is greater[132,141]. With nonexpressors, there is greater variability in tacrolimus levels, probably because the CYP3A4 component is greater and this is very sensitive to inducers and inhibitors, as mentioned earlier[141].

Hence, CYP3A5 expression can have around a 35% impact on the dose of tacrolimus. However, the impact of CYP3A5 polymorphism on patient outcomes is not currently considered to be significant[131,132,141] and does not significantly affect outcomes with tacrolimus[141,146]. Indeed, the CPIC advice published in 2015[147] explicitly did not recommend using genotype in dosing and remains unchanged at the time of writing this book. It really echoes basic therapeutic monitoring, with the regular dose recommended for nonexpressors and 1.5- to 2-fold increase for hetero and homozygous expressors, but not to exceed 300µG/Kg/day[147]. What does appear to be the case is that nonexpressors reach the correct plasma range for tacrolimus a few days before the expressors, so genotyping would benefit the initial stabilisation of the patients on the drug[141]. With the immunosuppressants, the lack of impact on outcomes is probably a reflection on the intensive therapeutic monitoring process, which has been built on 30 years of clinical experience, so genotype testing in this context would not necessarily be expected to have a radical impact; it might be just another useful tool. Indeed, it is now established that African Americans, for example, will require higher doses of tacrolimus to stay in the therapeutic window[143]. Of the many other CYP3A substrates, those in clinical contexts with therapeutic monitoring will probably also not see marked benefit from genotypic testing, whilst other drugs will be cleared by one isoform or another, except in rare cases of very unusual allele combinations.

7.2.4   Genetic polymorphisms in nonconjugative systems

Flavin mono-oxygenases (FMOs)

As outlined in Chapter 3.8, the FMOs role in drug clearance has only relatively recently been appreciated and some of the features of these enzymes, such as their lack of induction, still holds the attention of drug designers who wish to exploit their more predictable clearance of certain drugs is part of novel clinical regimens. As alluded to in Chapter 3.8.1, it is thought that FMOs are not inducible because, perhaps like CYP2D6, there is so much genetic variation in their expression 'built in' to humans. In fact, FMOs are fabulously polymorphic, with 300 different SNPs in FMO-3 alone, which is the major hepatic form[148]. Of the five main isoforms, the least variable (FMO-4) has over 30 variants and the most variable possesses over 50 (FMO-2). Although attempts have been made to look at various drug substrates as markers for expression (sulindac, ranitidine and S-nicotine), there are many other enzymes (such as CYPs) that clear FMO substrates often with greater capacity and efficiency[149].

Ironically, in a field that requires complex DNA array technology to detect elusive polymorphisms, all that is necessary to detect some clinically relevant FMO-3 deficiencies is a sense of smell. Although all of FMO-3's main homeostatic tasks have not been elucidated, from a personal perspective, its most important function is to clear trimethylamine (TMA: which gives off a fishy smell) to the odour-free TMA N-oxide. In wild-type individuals (about 90–98% of Caucasians), FMO-3 clears around 90–95% of the TMA. A clearance level of 60–90% is classed as mild fish-odour syndrome (trimethylaminuria, TMAU) and clearance levels of below 60% are associated with the most severe syndrome[149].

Forty SNPs are associated with FMO-3 fish-odour syndrome[148], but there are three alleles (two are linked) of strong interest and two relatively recently discovered that are also significant. E158K (rs2266782. G>A) and E308G (rs2266780) are found in 42% and 20% of Caucasians and are in linkage in 36% of Japanese, as well as 20–25% of Caucasians. Because of the nature of the gene linkage, homozygotes of these alleles are rarest in Africans, (<1%), about 5% in Asian populations, and up to 10% of some Caucasian populations[148]. V257M (rs1736557) is found in about 20% of Chinese and Japanese, and 8% of African

Americans[148,149,150]. The two recently found alleles rs1800822 and rs909530 have also been linked with fish-odour syndrome[148].

In comparison with the wild type, E308G causes a change to the enzyme surface which affects NADP+ binding, E158K changes the structure of the access channel and V237M is thought to cause changes in both FAD+ and NADP+ binding[151,152] (Chapter 3.8.3). Normally, it is assumed that such considerable polymorphic changes would show some marked and probably negative impact on the enzyme's performance, and this was the case with itopride, which is about 70% cleared by FMO-3[150]. The impact of homozygotic polymorphic alleles roughly doubled plasma levels, although the authors of the study did not feel that this was significant enough to warrant genetic testing. However, surprisingly, the effect of these polymorphisms can be substrate specific and not necessarily detrimental. V257M during *in vitro* work showed half and double the wild-type clearance of tamoxifen and clomiphene, respectively[151,152]; E158K is 1.5-fold better at clearing tamoxifen than the wild-type enzyme. With two anticancer aurora kinase inhibitors, V257M shows similar to wild-type performance with the aurora kinase anti-proliferative tozasertib, but 3.5-fold less turnover than wild type with tozasertib's relative danusertib[152].

If we return briefly to tacrolimus, it has been suggested that neither CYP3A5 polymorphisms nor presumably CYP3A4, account for all the marked variation seen in its pharmacokinetics[151]. At least two studies have shown that rs1800822 and rs909530 are surprisingly frequent (10-20%) and were linked with significantly faster than normal tacrolimus elimination[143,153], in both African Americans and Chinese. What is interesting is that the most investigated alleles (E158K, V257M, or E308G) did not affect tacrolimus clearance[151]. However, not all studies searching for impact on tacrolimus kinetics found FMO-3 to be significant[154] and other reports have shown that rs909530 has impaired clearance with sulindac[155].

To summarize, whilst several *FMO-3* alleles have impaired endogenous function with TMA, whether they will clear any given drug seems completely substrate specific and unpredictable. This introduces yet another element of clinical uncertainty that would be of relevance with narrow therapeutic index drugs. If we take into consideration the health implications of FMO alleles (E158K and E308G) that may protect against obesity[156], these enzymes are worthy of much more investigation to determine their true clinical impact.

Dihydropyrimidine dehydrogenase (DPD) and 5-fluorouracil

Antineoplastic agents based on 5-fluorouracil (5-FU) (capecitabine & tegafur) are still used for various solid tumours (particularly colorectal), usually in combination with agents such as cyclophosphamide and methotrexate. However, 5-FU is more than capable of seriously damaging the patient on its own, as it can cause severe gut and neurotoxicity, as well as bone marrow damage that together can force patient withdrawal from chemotherapy and prolonged hospitalisation[157]. The reason it can be so toxic and difficult to use is the striking effect of the polymorphism that affects *DPYD*, the gene that codes for DPD, which is 5-FU's main route of clearance. Those with wild-type DPD show an average 15-minute half-life, with 80% cleared to inactive metabolites and rest is renally eliminated unchanged. Conversion of only 1–2% of the parent to the active anti-metabolites is necessary to inhibit tumour growth[158].

*DPYD*2A* (rs3918290), and *DPYD*13* (rs55886062) form nonfunctional proteins, and homozygotes will suffer the greatest 5-FU toxicity. An alternative drug is recommended. Heterozygotes have less than half the wild-type metabolising capacity so a commensurate reduction in dose (half) is suggested. Whilst rs67376798 and rs75017182, (HapB3) have some function[158] and heterozygotes have about two-thirds and homozogotes around half of wild-type capability in Caucasians, HapB3 is the most common allele (<5%), followed by *2A (1.6%), whilst rs67376798 is less than 1%. Africans have an allele (rs115232898), which is found at less than 5%[158]. The true capability of the various alleles is still under investigation, as *DPYD*4* functionality, for example, has been shown to be of much less significance by some clinicians[157] more than others[158].

So far we have seen many instances where intermediate metabolism in heterozygotes is adequate for clinicians to 'work with', in terms of therapy optimisation. However, with 5-FU, the situation seems to be much less forgiving. You might expect that with homozygotic nonfunctional DPD individuals, the most severe toxicity will result, but given that the frequency of that diplotype is only around 1% in Caucasians, it seems surprising that in one study in Caucasians, 23% of the patients suffered severe toxicity[157]. This suggests that the combination of the high cytotoxicity of the drug and the steep drop in enzyme performance seen even in heterozygotes with partly functional haplotypes (HapB3) lead to what is essentially a clinically quite narrow therapeutic index situation. Indeed, genetic testing prior to therapy is desirable in the context of *DPYD* polymorphisms to protect patients from the horrors of potentially lethal 5-FU toxicity[158,159]. The evidence suggests that it is more than just financially cost-effective[160]; it is actually a no-brainer[157].

Butyrylcholinesterase polymorphisms

This polymorphism was first seen around 60 years ago, when patients under anaesthesia were treated with nondepolarizing neuromuscular blocker suxamethonium. This type of drug is still used to paralyze skeletal muscle during surgical procedures, and it usually works in seconds and is cleared in a few minutes, making its pharmacological effect extremely controllable. After drug clearance, the patient should resume a normal breathing pattern in 5 minutes or so. In some patients, this did not happen and their intercostal muscle paralysis might take more than half an hour to resolve. The condition was called Scoline (the tradename of suxamethonium) apnoea, and it was found that drugs such as suxamethonium were cleared by butyrylcholinesterase (BChE, also known as pseudocholinesterase) that was subject to polymorphisms. BChE is a 340 KiloDalton tetramer that is similar to acetylcholinesterase that has the crucial task of hydrolyzing acetylcholine in the neuromuscular junction and in autonomic nervous system ganglia. BChE is also responsible for the metabolism of cocaine and heroin (Appendix B.2/3). The wild-type BChE allele is known as BChE U, but the most common impaired allele is usually termed the K-variant, after Professor Werner Kalow, a pioneer of pharmacogenetics. Now also known as rs1803274, this allele is found in up to a third of Caucasians and results in loss of activity of more than more 30% in homozygotes[161,162]. The atypical allele (A) or rs1799807 is found in around 2% of Caucasians and has more than 70% less activity and less protein is produced than the wild-type[163]. However, there is a strong linkage between the alleles, so almost all the A alleles have the K allele too[162]. Hence, diplotypes with AK/K are more likely to be found than A/K, for example.

The clinical impact with neuromuscular blockers such as mivacurium extends the recovery time according to the allele combination, as you would expect. Those who are heterozygotes (U/K, or U/AK) have their recovery time increased by about 30% and U/A heterozyogosity also causes impairment. K/K doubled recovery time, but A/A patients take up to 8 hours on a full dose of mivacurium to recover, when a half-hour would suffice for for a U/U patient[161]. Since around a quarter of Caucasians have a defective allele, then variability with drugs such as mivacurium can be left behind by agents that spontaneously degrade, such as atracurium or cisatracurium, which have their issues, but are far more predictable in clinical practice.

In Alzheimer's disease, BChE expression is upregulated as acetylecholinesterase activity falls and there is evidence that cholinesterase inhibitors such as donepezil lose efficacy in those with the K allele, although the disease progression is not apparently linked with the allele[162]. Indeed, donepezil actually accelerates the condition in those with the K-variant and this probably is the case for other cholinesterase inhibitors such as rivastigmine. It is thought that this occurs as the K-variant is much more susceptible to drug inhibition than the wild type, causing excessive and deleterious CNS acetylcholine concentrations in Alzheimer's disease, so genetic testing has been recommended to prevent this problem[162].

BChE also clears cocaine and heroin, and some studies have suggested that homozygotic K-variant individuals have a preference for the use of more intensely addictive forms of cocaine, such as smoking 'crack'[163]. As the K-variant prolongs the half-life of cocaine, it would be likely that more of the drug reaches the brain and its residence time is increased. The authors in this study suggested that these higher concentrations reinforced addition, increasing the compulsion of drug use[163]. However, this appears to be the opposite effect to that seen in nicotine, where slower elimination significantly lessens addiction.

As BChE is one of the targets for nerve agents such as sarin, soman, VX, and of course, the 'novichok' compounds, it is under investigation as a possible antidote.

The idea is that large amounts of transgenically produced human BChE are infused into the poisoned individual and the BChE binds and removes the nerve agent from the plasma. Whilst the idea has been around for several years, there are issues with the half-life of the injected BChE and making it in sufficient quantity[164]. At the time of writing there are still research programmes aimed at making BChE into a practical injectable defence against nerve agents[165], which perhaps could augment pharmacological antidotes such as the oximes and atropine.

Epoxide hydrolases

Most of the research into polymorphisms of EPHX1 (microsomal epoxide hydrolase) and EPHX2 (soluble epoxide hydrolase) has centred on the link between these enzymes and various disease states. Several studies have tried to link the risks of carcinogenicity EPHX1, usually in relation to the activation of various aromatic hydrocarbons, although with mixed results. Wild-type EPHX1 has a tyrosine at position 113 and a histidine at 139. There are two main substitution polymorphisms. The first is a slow variant, which operates at 40–50% of wild-type capacity, cuts the risk of lung cancer, and is an exon 3 EPHX1 Tyr to His at amino acid 113 (usually called EPHX1 Y113H, or 337T >C on the gene; rs1051740). There is a 25% faster than wild-type version, known as EPHX1 His to Arg H139R (416 A>G; rs2234922), which increases the risk of lung cancer[166]. There are also individuals with both alleles and more than 29 other SNPs have been found. About half of Japanese populations have the slow variant, and just under 15% have the fast version. The enzyme variants can be inconsistent in their clearance of different epoxide substrates, which may explain the variability in studies aimed at linking variants to disease predisposition, although EPHX1 is also linked with prostate cancer risks in African Americans and other races[167].

EPHX1 can metabolise vitamin K like VKORC1, and hence rs1051740 is one of the many factors that can affect warfarin sensitivity[168]. Carbamazepine is cleared by CYP3A4 to its pharmacologically active 10,11 epoxide, which is also responsible for some of its side effects, which are related to concentration and lead to vomiting and ataxia. It seems well established that both the two EPHX1 variants rs1051740 (Y113H) and rs2234922 (H139R) have, respectively, significant impacts on either retarding or accelerating carbamazepine 10,11-trans dihydrodiol formation[169,170], although this does not necessarily have any predictive bearing on the development of resistance to the drug in epileptics[169].

Paraoxonases

These enzymes were first described in the 1940s and have been known far longer than many other biotransforming systems, yet their significance in drug, potential toxin, and susceptibility to disease remains to be fully elucidated. There are three human paraoxonases (PON1–3), all named after their ability to hydrolyse the organophosphate insecticide paraoxon[171]. The paraoxonases are found on chromosome 7 and are around 70% similar in terms of their gene sequences[171]. These enzymes can act as esterases or lactonases. They are made in the liver and can operate there or be released into plasma[172]. PON1 is associated with high-density lipoprotein (HDL) in the plasma and appears to assist in the antioxidant effects of HDL that protects against atherosclerotic disease[172]. Hence, it has been put forward that any process that affects PON1 activity will have a direct bearing on cardiovascular health. Indeed, PON1 expression is very sensitive to a very wide range of factors. Its expression is controlled by a transcription factor known as Sp1 (specificity protein 1), as well as by the AhR[171] (Chapter 4.4.2). Some studies have shown that part of the beneficial pharmacological effects of statins is their mild stimulation of paraoxonase activity[173], which is mediated through Sp1. Other drugs may increase PON1 activity, such as the fibrates and aspirin, but this is controversial[173]. High-sugar diets, smoking, and drinking suppresses PON1 activity[173].

In terms of the impact on drug clearance of the paraoxonases, they were suggested to be part of the activation of clopidogrel, but this was found to be due to flawed study methods[173] and, if paraoxonases are involved with clopidogrel metabolism at all[173], it is only through formation of a very minor and inactive endothiol derivative[174]. PON1 and PON 3 can protect against the toxicity of the pharmacologically inactive metabolite atorvastatin δ-lactone, by hydrolysing it[173].

The paraoxonases are highly polymorphic; there is a variation in the active site at amino acid 192 (Q192R), which has a very strong bearing on the catalytic activity with different substrates. The Q version is a better antioxidant than the R version. The expression of PON1 is governed strongly by PON1-L55M, with the L version showing much higher tissue expression. In addition, there is a promotor polymorphism (T108C), with the 108C version showing greater tissue activity[172]. Regarding Q192R, there are said to be three main genotypes, Q/Q, Q/R as well as R/R[173] and the R allele is more common in Asian populations (up to 90%) and rarer in Caucasians (15–30%)[172]. Because the paraoxonases are subject to so many influences regarding their expression, it has been very difficult to show consistent relationships between different forms of the enzymes and predispositions to pathological conditions[172,173]. Indeed, lack of success in pharmacologically modulating HDL suggests that its role and that of the paraoxonases, is more complex in cardiovascular disease than was first thought[173]. In addition, although rapid PON-mediated clearance of organophosphate insecticides and nerve agents should be protective against their toxicity, again, associating specific polymorphic forms with protection or susceptibility has been problematic[173].

7.2.5   Conjugative polymorphisms: acetylation

Acetylation was the first polymorphism to be investigated and was based on the observation that the antitubercular drug isoniazid caused a different level of neural toxicity in Japanese compared with US patients. There are two

N-acetyltransferase (NAT) cytosolic isoforms, NAT-1 and 2, and they share 81% amino acid sequence identity. NAT-1 is found in most tissues and can metabolise drugs, but is often linked with risks of developing various cancers. NAT-1 metabolises a whole raft of toxic chemicals, which perhaps should be left to other systems to dispose of. Aside from the various aromatic amines, it also attacks arylhydrazines and hydroxamates, which can result in either activation or attenuation of reactive species[175]. There is a great deal of variability built into *NAT-1* expression, as it has three possible promotors. Two gain-of-function alleles of *NAT-1* (*4 is wild-type) are of interest, NAT-1*10 (rs1057126) and NAT-1*11, which is a haplotype that contains no less than five different SNPs plus a nine-base deletion[175]. This results in faster enzymes expressed in greater quantity. These alleles may protect individuals with slow NAT-2 variants from the risks of hypersensitivity with sulphonamides[175].

NAT-2 (found mostly in the liver and gut) is relevant to the metabolism of vintage drugs such as the sulphonamides, the sulphone dapsone, and of course the antimycobacterial isoniazid. NAT-2 polymorphisms are much better understood than those of NAT-1 and for *NAT-2* (which unlike *NAT-1* has a single promotor), three distinct groups have been described – that is, fast, intermediate, and slow acetylators[176]. The wild type is *NAT-2*4*, but as detailed above with *NAT-2*11*, *NAT* genes have a marked tendency to exhibit several SNPs in one allele, which effectively makes them haplotypes. Of the fast alleles of NAT-2 (which has a single promotor) aside from *4, there is also *11, *12A-C, and *13. The fast alleles *12C and *12B are haplotypes, which contain two SNPs. These alleles are as fast as *4, or may have some gain of function[177].

The scale of the variation in NAT-2 is seen in the slow alleles that are mainly haplotypes. There are over twenty variants described so far, ranging from NAT*5A-*14G (SNP) and only three contain a single SNP, whilst *5G and *14C contain four SNPs each. To make sense of this, currently seven SNPs are searched for in clinical studies. These are rs1799929 (C481T, found on *5A), rs1799930 (G590A, only SNP on *6B), rs1799931 (G857A, only SNP on *7A), rs1801279 (C282T, found on all *14 alleles, A-G), rs1801280 (G191A, found on all the *5's A-J), rs1041983 (T341C, found on most *6 alleles), and rs1208 (A803G, found on *5B-D & *14C, E-G).

The highest frequencies of NAT-2 fast acetylators are in Asian countries, particularly Japan, with over 90% of the population. Frequencies in China are lower, around 60–70%, whilst those in the Indian subcontinent head towards 30–40%. European and African-ancestry populations are mostly slow 60–80% slow acetylators around 20–40% are fast/intermediate acetylators. In Caucasians, Indians, and Moroccans, the frequency of *4 is about 13%. The *4/*4 diplotype is not common in these groups[178]. Indeed, looking at typical population of mainly slow acetylators, the fast allele *4 has a frequency of around 30%, whilst the most common slow alleles are the *5 and *6 haplotypes that account for 30% of the population each. Other alleles such as *7 and *14 series together comprise less than 10%[179].

From a drug metabolism perspective, as these isoforms process aryl amines, the width and scale of the potential toxicity of such groups (Chapter 8.5.4) mean that it has been decades since new drugs have been developed that are likely to be acetylated in quantity. Also, in any case, it is an inefficient form of clearance due to the lipophilicity of the acetylated derivatives (Chapter 6.7). However, a narrow spectrum of niche or very old drugs is still vulnerable to NAT polymorphisms, which have dramatic effects on the plasma levels of the parent form of acetylated drugs. With fast acetylators, perhaps only 20% of drug-related material in the plasma will be the parent drug and therefore clinically effective (assuming metabolites are not active). This situation could lead to potential treatment failure. In contrast, in the slow acetylators, more than 80% of drug-related material in the blood will be the parent drug and these levels may be so high as to approach toxicity in some individuals.

Acetylation and sulphonamides

In a study of patients taking sulphasalazine (sulphapyridine and 5-aminosal-icylate linked by an azo bond), for ulcerative colitis, parent drug (sulphapyridine) plasma levels were between three-and fourfold higher in the slow acetylators compared with the fast acetylators, whilst the acetylated drug was found in approximately three-fold higher levels in the fast acetylators compared with the slow[180].

Despite the potential problems caused by wide disparities in plasma concentrations between populations and individuals, sulphonamides were cheap, effective and were used in vast amounts as broad-spectrum antibacterials for over 50 years. In fast acetylators, providing there was enough drug in the plasma to suppress bacterial growth or exert an anti-inflammatory effect, efficacy would be adequate. However, in some individuals, it is likely that sustained low plasma levels of the drug contributed to the selection of resistant bacterial populations that reduced drug effectiveness. Resistance was offset by the use of sulphonamides in combination with 2,4 diamino-pyrimidines (pyrimethamine and trimethoprim), which are synergistic in effect, as the two drug types attack the bacterial/protozoan DNA synthesis process in two places (the sulphonamide blocks dihydropteroate synthetase, whilst the diaminopyrimidine blocks dihydrofolate reductase). Thus, much lower plasma concentrations are effective of both drugs together, rather than either singly. Currently, a combination of sulphamethoxazole with trimethoprim (SMX and TMP)[181] remains effective in treating *Pneumocystis jirovecii*-induced pneumonia in HIV-positive individuals who have T-cell counts below 200.

Sulphonamides and their combinations also fell out of favour for broad-spectrum antibacterial usage, because of a relatively high rate of adverse reactions, which included some gruesome conditions such as Stevens–Johnson syndrome (Chapter 8.4.3.). Indeed, more than half of HIV patients taking the

SMX/TMP combination suffer from milder adverse cutaneous reactions such as various rashes, which can diminish patient tolerance of the drugs. These reactions are not connected with the parent drug or the acetylated metabolites, which are not cytotoxic on their own. Unfortunately, all aromatic amine-based drugs such as sulphonamides and sulphones and other substrates of NAT-2 are also subject to extensive CYP-mediated (usually CYP2C9) oxidative metabolism. These metabolites are the main route of clearance of these agents, and they are mostly hydroxylamines, which are usually eliminated as conjugated sulphates or glucuronides. However, enough of the cytotoxic hydroxylamines escape conjugation to be the cause of the adverse reactions associated with these drugs.

So you can see that NAT enzymes could be seen as detoxification pathways for aromatic amine-related drugs; this is because once an amine has been N-acetylated, then there should be less opportunity for it to be oxidised by the CYPs to hydroxylamines (Figure 7.1). This would only apply if the acetylated derivative was excreted into urine and did not undergo further

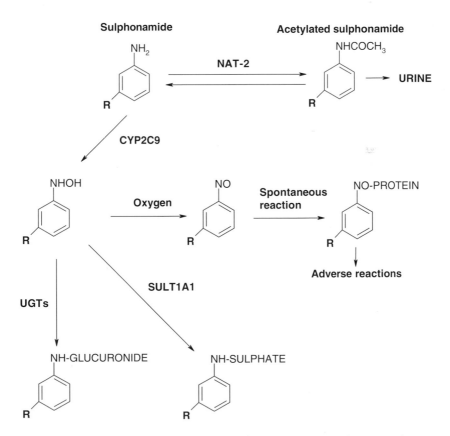

Figure 7.1   Basic scheme for sulphonamide metabolism and how adverse reactions are related to metabolism

oxidative metabolism. With sulphapyridine, a modest amount (approximately 20%) of the dose was found in urine as the acetylated derivative, so suggesting in single aromatic amine agents (sulphonamides), the frequencies of adverse reactions related to the oxidative metabolites should theoretically be lower in fast acetylators and correspondingly higher in slow acetylators. However, whilst many studies have shown that common slow acetylator phenotypes (*5–*7), are linked to hypersensitivity reactions, others have not been able to show this[181,182]. Frequencies of sulphonamide reactions of 3–5% in the otherwise healthy and perhaps 30–40% in the immunocompromised, in groups with similar slow NAT-2 allelic frequencies, suggests that immune status and other factors are more predictive of the risk of reaction to sulphonamides than NAT-2 status alone.

Sulphones

However, in drugs with two aromatic amine moieties, such as dapsone, this protective effect does not apply, as diacetylation is a minor pathway in dapsone metabolism and only a few per cent of the dose is found in urine as the monoacetylated derivative and trace amounts of the highly lipophilic diacetylated sulphone (Figure 7.2). Monoacetylated dapsone is much more lipophilic than the parent drug, so essentially, acetylation can slow dapsone clearance, by delaying it in the acetylation/deacetylation pathway. N-hydroxylation is the only effective way of making dapsone less lipophilic and even the free amine group of acetylated dapsone is N-hydroxylated. The hydroxylamines are eliminated as N-glucuronides and sulphates of monoacetyl and dapsone hydroxylamines. So hydroxylamines are formed regardless of the acetylator phenotype of the individual, even though plasma parent drug levels are higher in slow compared with fast acetylators[183,184]. The toxic oxidative metabolism of aromatic amines like dapsone will be discussed in Chapter 8.5.4).

Isoniazid metabolism

Isoniazid (INH) has been the cornerstone of the therapy of tuberculosis for over sixty years. Resistance has emerged relatively slowly due to the deployment of other drugs with INH. This drug has a unique mechanism of action that only operates in *Mycobacteria tuberculosis* and its closely related pathogens. INH remains effective in many areas, but it is toxic, and if tuberculosis were a major disease of the developed world, it certainly would have been superseded. Unfortunately, no new antituberculosis drug has been introduced since the 1960s.

INH's tendency to cause peripheral neurotoxicity can be pretty much removed by taking pyridoxine (vitamin B_6). However, its major clinical drawback is

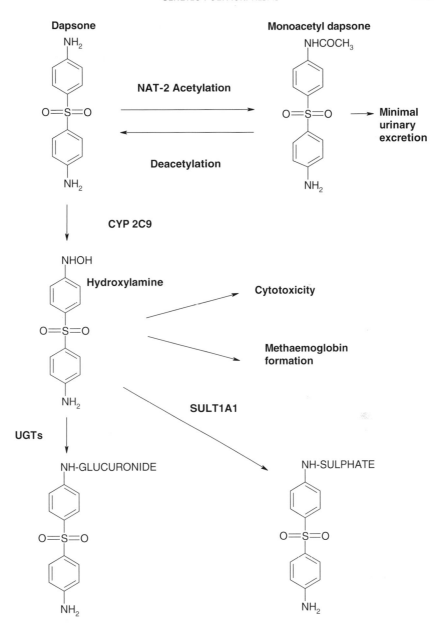

Figure 7.2 The role of acetylation in dapsone metabolism

significant hepatotoxicity, which reveals itself as elevation of liver transaminase enzymes (AST, ALT) in many patients. Around a fifth of patients show perhaps a three- to five-fold increases in AST and ALT over the first 4–8 weeks of treatment, which often settles as the liver adjusts and probably upregulates GSTs and GSH levels to compensate. This and other issues (gut and hypersensitivity problems) mean that adverse reactions occur in up to a third of patients and in 10% discontinuation of therapy is necessary[176]. Once the enzymes rise to the 300–500 range (around 10-fold normal), more worrying hepatotoxicity may be developing, which could progress to liver failure in 1–2% of individuals if the drug is not stopped immediately. Liver problems are sometimes more associated with females than males, whilst older patients and heavy drinkers are more susceptible also.

NAT-2 forms N-acetylisoniazid that can hydrolyze to form acetylhydrazine, which is a potent nucleophile in its own right and does not need any more metabolism to be cytotoxic. It is thought that fast acetylators can eventually form greater quantities of diacetylhydrazine, which is not toxic, whilst the slow acetylator livers must contend with acetylhydrazine for much longer periods of time. It might be expected that fast acetylators would initially form more acetylhydrazine and then be more susceptible to toxicity, than the slow acetylators, but this is not borne out by clinical studies (Figure 7.3). Slow acetylation appears to be

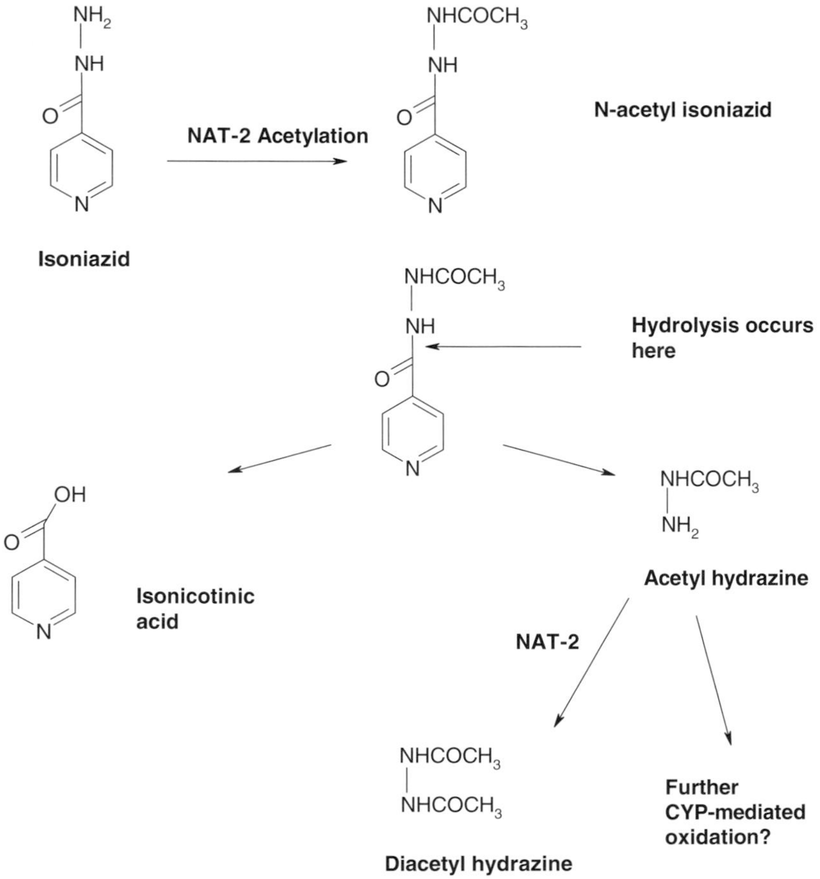

Figure 7.3 Acetylation and isoniazid metabolism

associated with greater risk of liver toxicity and up to three-fold higher plasma INH levels, although parent drug levels are not associated with the risk of liver damage. Several studies with INH have shown that slow NAT-2 alleles are associated with INH mediated liver damage, although they differ over which of the slow alleles exhibit the greatest risk[176,185].

However, any concurrent therapy such as rifampicin (or heavy ethanol use) predisposes to hepatotoxicity through a number of mechanisms, not least CYP induction, particularly CYP2E1[176,178]. Interestingly, with some reports there was no impact of CYP2E1 polymorphism on the risk of INH hepatotoxicity. More than one CYP is probably be involved with conversion of either the parent drug or acetylhydrazine derivatives to more reactive necrotic species. This would account for the protective effect of acetylation, which essentially promotes the formation of a relatively reactive but containable metabolite (acetylhydrazine) on the way to a safe one (diacetylhydrazine), so restricting entry of the parent drug and acetylhydrazine itself into oxidative metabolism. Interestingly, rifampicin is extensively deacetylated (that wonderful orange metabolite), and it may increase the risk of liver damage by competing for the NAT-2 and leaving more drug to be oxidised by other pathways, as well as its other inductive and disruptive effects on liver function. There is no evidence that fast acetylators are disadvantaged in terms of INH's efficacy.

Toxicological significance of acetylation

To complicate this situation further, some acetylated metabolites can undergo further oxidation themselves to form highly reactive cytotoxic and carcinogenic species. Indeed, in recent years, acetylation has become intensively studied almost entirely due to its role in the carcinogenic activation of aromatic amines and this is discussed briefly in Chapter 8.5.4. Perhaps the best context to see acetylation is a homeostatic process that unfortunately xenobiotics do enter and are metabolised, often leading to outcomes that are at best rather equivocal.

7.2.6   Conjugative polymorphisms: methylation

TMPT

A particularly dangerous polymorphism clinically was identified in the 1980s for one of the methyltransferases. The endogenous role of *S*-methylating thiopurine *S*-methyltransferase (TPMT) is not that clear, but it is capable of *S*-methylating thiopurine derivatives like 6-mercaptopurine (6-MP) and the 6-MP pro-drug azathioprine. These drugs are antimetabolites that mimic purine to disorder DNA metabolism and can act as powerful immunosuppressants (azathioprine) or anticancer drugs. They are effective through activation to 6-thioguanine nucleotides by an enzyme called HGPRT (Figure 7.4). TPMT is the main route

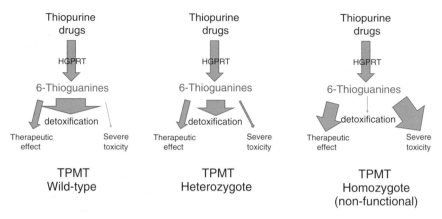

Figure 7.4 Scheme of the role of TPMT polymorphisms in the metabolism of S-methylating thiopurine derivatives and clinical outcome

of clearance of these toxic 6-thioguanines and this polymorphism can greatly distort the clinical behaviour of this drug in the patients[186]. Those who are poor metabolisers and have low TPMT expression can suffer from lethal myelosuppression and abnormally high levels of the active 6-thioguanine nucleotides[186,187].

As with CYPs and acetylation, in Caucasians, there are PMs (<1%) IMs (10%) and the rest (EMs) are wild type. The most common allele variant is *TPMT*3A* (5% Caucasians) which contains two SNPs 460 G>A (rs1800460) and 719A>G (rs1142345) and this haplotype codes for unstable non-functional TPMT[186]. TPMT*3C (2% of Asian races and 2-3% of African Americans; rs1142345) is the next most prevalent, whilst *TPMT*3B* contains only rs1800460. Other rarer alleles include TPMT*2 (rs1800462), which is found in about 0.5% of Caucasians[186]. *3A, *3C, *2 account for virtually all the non-functional alleles in humans and generally, the chances of a single individual expressing non-functional TPMT are low, around 0.3%. Clearly, rate of clearance of the toxic thiopurines directly influences treatment success and failure. Too-rapid clearance causes insufficient residence time and reduces efficacy. Wild-type and intermediate clearance lead to more than sufficient residence time of the 6-thioguanines to secure a good therapeutic outcome, although poor clearance leads to neutropenia and interruptions in treatment[187]. Drug failure, either through toxicity necessitating cessation of treatment or lack of efficacy in the context of a progressive and life-threatening disease, is obviously extremely serious.

Genotypic and phenotypic tests were developed from the late 1980s onwards for TPMT expression based on the fortunate fact (in common with GST-theta and NAT-1) that enzyme expression in red cells is linked with systemic expression[187]. This means that the patients can be tested for TPMT expression prior to therapy, which should reduce the risk of life-threatening toxicity.

TPMT: Genotyping or activity measurement

TPMT highlights the genotyping/phenotyping issue mentioned earlier in the management of patients with polymorphisms. Genotyping will reveal the level of TPMT expression that should be expected in the otherwise healthy patient and it is virtually 100% accurate for wild-type individuals. However, it can much be less precise in identifying intermediate allele combinations. Phenotypic testing, which is essentially a functional assay of TPMT activity, seems at first to have many advantages over genotype testing. Issues such as the presence of enzyme inhibitors such as benzoic acid derivatives and specifically salicylates (used with azathioprine in inflammatory bowel disease) can be allowed for, and the result should then reflect the true nature of the patient's TPMT capability, provided of course, that the salicylates are continued by the patient. The disadvantage to phenotypic testing is ethnic group variation and lack of international standardisation and again, intermediate activity can be misclassified.

Whilst genotypic testing is more reliable than phenotypic testing, the agreement between them varies also, particularly in the intermediate range. So overall, the answer seems to be that both need to be deployed. Worldwide, TPMT testing is not universally accepted, although nearly 70–90% of UK clinicians use it depending on their specialty[187]. Currently, many take the view that with azathioprine, heterozygotes should be able to tolerate the drug at standard doses, but it is beyond dispute that homozygotes with nonfunctional TPMT are in serious danger from such doses. The financial bottom line is, however, very persuasive; the cost of a *3A/*3A individual being hospitalised by a 'wild-type' dose of azathioprine is 400 times the cost of the tests[187].

COMT: Warrior or worrier?

Catechol-O-methyl-transferase (COMT) methylates a catechol hydroxyl group on dopamine, noradrenaline, and other biogenic amines and inactivates them[188]. Hence, this enzyme's performance would be expected to have a very significant day-to-day impact on catecholamine synaptic concentrations. Since so many theories of human CNS pathology, personality, and mood rest on synaptic neurotransmitter concentrations, any polymorphism in COMT activity you might expect would be influential in many aspects of behaviour and personality, both normal and pathological. This has already been alluded to in Chapter 6.8. The idea of COMT's role in all these areas rests on several polymorphisms, and there are at least 23 with this enzyme[189], of which rs4680 (427G>A) is the most prominent. So in COMT, in the membrane version of the enzyme, in codon 158, if you are val/val, COMT activity is up to four times faster than met/met, or val/met combinations[188]. Hence, the val/val version means rapid inactivation of catecholamines, which is said to protect the individual from vulnerability to stress. This apparently makes val/val/ individuals 'warriors' and val/met and met/met individuals, 'nonwarriors', or 'worriers'[190]. Given how potentially subjective, simplistic, but newsworthy these ideas could be, much research has been completed in this area. Some suggest that warrior status provides greater brain reserve compared with worriers in dealing with stress such as encountered after surgery[190].

COMT and depression

To address the generally poor patient responses to anti-depressant drugs (Chapter 5.5.2), many studies have tried to relate COMT status with depression susceptibility and therapeutic drug responses. Regarding susceptibility, in Caucasians and Asians the differences between depressed and nondepressed val carriers are not very marked (52.6 vs. 49.6% and 60.5 vs. 65.9%) and there are no differences at all in Africans (68.69%) and in some other studies on Caucasians[188,191]. However, it is believed that val/val Caucasians are more likely to be severely depressed, although the met allele is associated with early depression in Asians[188]. With Caucasians and val, it is assumed that their tendency to depression is linked with low catecholamine levels due to their fast version of COMT. Unfortunately, in terms of drug responses with venlafaxine, for example, COMT status does not affect drug efficacy and is not a useful biomarker for a potential response[188]. However, other studies with fluoxetine, paroxetine, and milnacipran have shown poorer responses in val homozygotes, and generally the met/met combination promotes response to antidepressants[188,192].

With respect to the biogenic amine theory, this is logical in that the slower version of COMT will allow the drugs to increase the persistence of various neurotransmitters in the locality of the synapses. It does appear that in terms of depression, other factors influence predisposition to depression as well as drug response, both in terms of ethnic group and expression of CNS and peripheral enzyme systems.

COMT and other CNS conditions

As already mentioned, given the fundamental link between local CNS neurotransmitter disposition and brain function, COMT status is likely to influence a variety of different conditions and drug responses, and the current evidence supports this idea. With drug-resistant schizophrenia, clozapine responses were better with those who were hetero or homozygotic for the met variant, compared with val homozygotes[193]. Responses to the attention deficit disorder drug methylphenidate may also be modulated by val158met status[194].

More worryingly, the relationship between COMT and illicit drugs such as cannabis is also emerging. A study in 2002 showed a relationship between early (15 years or below) cannabis use and the risks of schizophrenia by the age of 26[195], and it is now apparent that COMT genotype may influence these outcomes. Essentially, the val/val combination is the risk genotype for cannabis-linked psychosis, with decreasing risk with hetero and then homozygotic (met/met) combinations[196].

This brings us to the interesting perspective that a 'worrier' dope smoker is less likely to turn psychotic than a 'warrior'. Those of you who are keen amphetamine or amphetamine derivative users (Appendix B.5) might also be interested in COMT status, but it does not strongly impact dopamine levels in the presence of amphetamine[197]. In fact, val/val status was associated with poorer visual-motor processing, which amphetamine use actually improved[198], but there were no effects on the mood-altering aspects of the drug. This is more a reflection of the tissue localisation of COMT, which is less in the limbic (emotional) areas and more in the cortical (analytical/processing) areas of the brain[198]. What COMT status means in terms of the effects of plethora of other amphetamine and cannabis ('Spice') related agents detailed in Appendix B is anybody's guess. Hence, our understanding of the influence of COMT polymorphisms on CNS pathology and drug responses is potentially as exciting as it is difficult to study in the future.

7.2.7   Conjugative polymorphisms: UGT 1A1

UGT1A1 variation spectrum

As *UGT1A1* performs such a key function, that is, the regulation of bilirubin elimination, it is not surprising that it is actually quite a resilient gene, usually maintaining some activity in the face of some serious evolutionary pressures[199]. The list of polymorphic alleles in UGT1A1 is growing[200] and the impact of these variants on drug clearance in different ethnic groups is a growing area of research.

The worst and probably rarest (1 in 10^6 children) UGT polymorphism is Crigler-Najjar syndrome variant CN-1, with complete deletion of UGT1A1, resulting in no enzyme and death from bilirubin damage a couple of years after birth without liver transplantation. CN-2 expresses some enzyme and phenobarbitone induction will elevate UGT1A1 levels to around 20% of normal[201]. The most common UGT polymorphisms in the general population come under the general area of Gilbert's syndrome, which appears to cover a range of UGT1A1 polymorphisms, including the TATA box promotor faults (see next section). Indeed, some variants of Gilbert's syndrome are severe enough to overlap with CN-2, in terms of impact on bilirubin levels[199]. So the full spectrum of impairment of UGT1A1 runs from nonviable CN-1 through to CN-2 (5% of wild type), then homozygous (up to 30% of wild-type capacity) and then myriad heterozygous variants of Gilbert's syndrome. What this means from the perspective of human populations is that 5–15% of humanity have a homozygotic significant impairment of UGT1A1, and heterozygotes possess a range of milder impairments, which means that in many populations there is a considerable range of bilirubin plasma levels which are above what is considered normal (0.3–1mg/dL). It is now argued persuasively that this is something of an evolutionary masterstroke, where the waste product is actually highly protective, significantly reducing mortality, whilst extracting a relatively minor cost in terms of some gastrointestinal issues[202]. Of course, it may not be entirely as simple as that, and sustained elevated bilirubin levels and impaired UGT1A1 exposes an individual to other risks, most obviously in drug therapy as detailed in Chapter 6.2.10. The major Gilbert's syndrome variants are described below.

Gilbert's syndrome

This condition was first identified in France and was named after Professeur Nicolas Augustin Gilbert (1858–1927), so as a tribute, the French pronunciation of 'Gilbert' is most appropriate. When otherwise healthy patients presented with high unconjugated bilirubin levels but normal ALT and other liver enzymes, the condition would be suspected, but it was not until the 1990s that the genetic basis was understood. Normally, around half of all our gene expressions are promoted by a fairly conserved final 'switch' which is termed a TATA box, a series of precisely coded repeating thymine/adenines which in UGT1A1*1/*1 (wild-type) is actually six TAs followed by a TAA sequence. This switch is thrown by the binding of a TATA box binding protein (TBBP) that is a key step in the initiation of transcription. In the UGT1A1 (e.g. rs34815109 and rs8175347) seen in Gilbert's syndrome, there are can be five, seven, or even eight TAs instead of six, which are known as deletion (TA5) or insertion (TA7 and 8) polymorphisms[203]. This makes a large dent in the efficiency of the basal promotion of the expression of the gene in TA7/7 (even worse in TA8/8) individuals, as it runs at only 30% of wild-type (*1/*1), probably due to inefficient binding of the TBBP[200,204]. The vast majority of individuals (over 90%) who have Gilbert's syndrome have the TA/7/7 diplotype, with the TA5 variants being much rarer. The TA8 variant is more common in Africa[205].

Several other alleles are also associated with Gilbert's syndrome diplotype, such as the lower activity UGT1A1*6 allele (211G>A; rs4148323). UGT1A1*6 can be rare in Caucasians but is found in 20% or so of Asian races[206]. However, in Caucasians, this is a cause of hyperbilirubinaemia in neonates and does not appear to impact adult bilirubin levels[204], which does not seem to be the case in Han Chinese[206,207].

There is much still to research about Gilbert's syndrome, such as how some individuals with combinations of apparently abnormal TA repeats have normal bilirubin levels[204]. Indeed, in India a subset of 10% of Gilbert's syndrome patients were discovered to have a trinucleotide CAT insertion, which cut their UGT1A1 expression down to 5%, which is on a par with CN-2[199].

However, it is emerging that homozygosity in Gilbert's syndrome also is associated with other UGT loss of function alleles, such as UGT1A7*3 and UGT1A6*2 (both at 77% of Gilbert's patients), and over 90% have UGT1A3 -66 T>C[208]. Interestingly, the latter SNP is part of the UGT1A3*2 haplotype, which is associated with a gain of function[209], so this is effectively a compensatory measure for the Gilbert diplotype. Homozygotic Gilbert's syndrome is most common in Africans (about 20%), followed by Caucasians (up to 16%) then Indians (12%), but only around 6% in Han Chinese[206,208]. In general, Gilbert's syndrome alleles have quite high frequencies in human populations, ranging from 30–40% in Caucasians alone[206,208,210]. Not only is expression low with this syndrome, the linked alleles *UGT1A1*60* (3279T>G) and UGT1A1*93 (3156G>A) impair PBREM-mediated xenobiotic induction. Half of Caucasians and about 85% of Africans have *60[200] and around 30% have *93[211].

The narrow 'headroom' Gilbert's patients have regarding UGT1A1 capacity is reflected in their susceptibility to several severe drug-adverse reaction issues. HAART therapy contains potent inhibitors of UGT1A1, such as atazanavir, which have a disproportional impact on Gilbert's patients[212]. The strong bearing that Gilbert's syndrome has on SN-38 toxicity (gut toxicity and neutropenia) during irinotecan therapy (Chapter 6.2.10) has long been recognised in the literature and is included with the drug packaging. Whilst the TA7/7 diplotype is rarer in Asian races, a strong association has emerged with UGT1A1*6 and SN-38 toxicity, particularly those with the TA7/7 and *6/*6 combination[213]. Although it has also been well documented that the higher levels of SN-38 seen in Gilbert's syndrome individuals can boost therapeutic response[213], the disadvantage is the increased risk of SN-38 mediated neutropenia and, to a lesser extent, diarrhoea.

Many authorities now recommend genotype testing for Gilbert's syndrome followed by reductions in irinotecan doses by a third in confirmed cases. Genotypic testing is also recommended for other anticancer drugs with high UGT1A1 clearance like belinostat[211]. Other isoforms can clear SN-38 such as UGT1A7, and the impaired allele UGT1A7*3 is linked with a high risk of neutropenia[214]. Overall, in the future, it may be necessary to evolve a protocol that tests for all the major alleles involved in the spectrum of Gilbert's syndrome, including UGT1A1 TA5, 7, and 8, as well as UGT1A1*6, *60, and *93, as well as the UGT1A7 low-function alleles, plus the transporters such as the MRPs. Indeed, if irinotecan is administered with 5-FU, then DYPD[203] would also have to be evaluated. However, there is no reason why a standardised panel of tests could not be introduced, alongside readily available online advice such as that seen with warfarin. Such measures would go some way to alleviating the suffering of patients with Gilbert's syndrome undergoing therapy with drugs such as irinotecan, as adverse reactions with this drug already can impact half of those who have to take it for bowel cancer[214].

Other UGTs

UGT2B7 is a high-capacity pathway that clears a variety of different drugs, particularly opiates. This isoform has attracted interest due to its expression in the gut as well as the liver, but also for its impact on antineoplastic drugs, particularly anthracyclines. As we have seen, changes in clearance of these agents can have a significant bearing on toxicity as well as efficacy. *UGT2B7* –161 C>T (rs7668258) is a promotor SNP, which increases UGT2B7 metabolising capacity and is found in nearly half of Caucasians[215]. This SNP can protect against myelosuppression and cardiotoxicity with epirubicin, although it is not clear whether it reduces efficacy as residence time is shorter[216]. Paradoxically, with the AED lamotrigine, the same SNP was reported to reduce clearance by 20%[215]. Another UGT variant, *UGT2B7* 802 C>T, is a slower version of the isoform, retarding tamoxifen clearance as well as epirubicin, thus promoting toxicity. Other variants of interest include UGT2B7 372 A>G, which is fairly rare in homozygotic form (2% of Caucasians) but shows rapid clearance of lamotrigine, which although is metabolized by UGT1A4, it also has a significant component of UGT2B7 to its clearance[215].

UGT2B15 is highly expressed and clears a variety of endogenous and environmental toxins, such as the endocrine disruptor bisphenol A, as well as the drugs lorazepam, oxazepam and paracetamol[217]. About half of Caucasians express a UGT2B15*2 which contains the SNP 253 G>T (rs1902023) variant, although its activity and expression may be greater than wild-type. Other alleles of this UGT do not appear to differ significantly from the wild-type[217].

7.2.8   Conjugative polymorphisms: sulphonation

Understanding of sulphonation and its roles in endogenous as well as xenobiotic metabolism is steadily improving. The role of SULTs in the activation of carcinogens has been studied, but often with conflicting results. One of the major influences on SULT activity is their polymorphic nature; in the case of one of the most important toxicologically relevant SULTs, SULT1A1, this isoform exists as three variants, SULT1A1*1 (wild-type), SULT1A1*2, and SULT1A1*3. The *1 variant allele is found in the majority of Caucasians (around 65%), whilst the *2 variant (638G>A, rs9282861) lacks thermal stability, although much of its activity with various substrates is similar to the wild-type[218]. The *2 variant is common in Caucasians (33%) and Africans (29%), but much less in Asians (<10%). SULT1A1*3 (667 A>G, rs1801030) is found mostly in Africans (22%), but is virtually absent in Chinese and is rare (1%) in Caucasians and is thought to be a gain-of-function allele, with more affinity for substrates[218]. There is another SNP in SULT1A1 found in the promotor area (rs750155) in over 60% of Africans and 48% of Caucasians and 42% of Chinese, that is linked with SULT1A1*2, which interestingly retards expression in Africans and accelerates it in Caucasians[218]. The *2 haplotype including rs750155 is common in platelets[218].

What is perhaps underexplored is that some SULT genes possess a similar feature to CYP2D6, that of repeatedly copying themselves. This is termed CNV, or copy number variant, and in large-scale studies it has been shown in Caucasians that not only did some individuals not express SULT1A1 protein at all (<1%), but that only 4% had a single copy of the gene, whilst the majority (64%) had two copies, 26% had three, and 4% had four. There were even individuals with six copies, but this was less than 1% of the population[219,220]. What was also surprising is that the more copies of SULT1A1 possessed, the lower the concentrations of oestrone sulphates, which may have implications of long-term cancer risks[220].

SULT1A1 clears several drugs such as those related to steroids and hormone metabolism and the efficacy of tamoxifen also is linked to SULT1A1 metabolism, but this does not seem to have been extensively investigated in terms of CNV. The number of copies of SULT1A1 surely must affect the availability of the 4-OH derivative of tamoxifen for conversion to the active endoxifen and influence efficacy quite strongly in those with three or more copies of the gene[219]. Clearly, analysis of gene copy of SULT1A1 as well as CYP2D6 would be extremely advantageous in customizing tamoxifen therapy to the individual. Other SULT isoforms can clear steroids and steroid-like compounds, although some have considerable variation (SULT2A1 and SULT2B) and others such as SULT1E1 do not[220].

From the cancer risk standpoint concerning SULT1A1, a highly active SULT1A1*1 may be advantageous if it removes reactive species rapidly as stable sulphates, but a problem if it activates others (acetylfluorene) to reactive species. SULT1A1*2, for instance, is less efficient at activating some carcinogens[218]. The activity of this variant differs between ethnic groups due to the promotor SNP variant, as well as the differences between the same allele activities in different groups, so overall, it is not surprising that the balance between SULT handling of protective dietary flavonoids like quercetin, chrysin and genistein and the activation or attenuation of carcinogens is complex. It will differ according to individual, isoform, copy numbers, levels of expression, diet, and environment. The combinations are endless, so it is often extremely difficult to predict risks of carcinogenicity for individuals and toxin exposures.

7.2.9 Other conjugative polymorphisms: Glutathione S-transferases

GSTs are polymorphic, and much research has been directed at linking increased predisposition to cytotoxicity and carcinogenicity with defective GST phenotypes. With GST-M and GST-T genes, the main polymorphisms involve complete deletion of the gene. Hence, active wild-type GSTM-1 is found in around

50-60% of Caucasians, but the remainder don't express it at all (null diplotype). GST-Theta null allele homozygotes are less common (15–20%) than null GSTM-1 individuals[210]. The GST-Pi class is also important in protecting cells from oxidative stress and its polymorphisms are perhaps more similar to other metabolizing enzymes. The major polymorphic variant of GST-Pi is the missense substitution 313A>G, (rs1695). The valine (instead of the isoleucine) in position 105 of the isoform makes the variant unstable, but can also massively increase its ability to detoxify some epoxides, but reduce its ability to clear other aromatics[222]. About 40% of Caucasians have GST-Pi wild type, whilst 48% are heterozygous for rs1695/wild-type and just under 10% are homozygous for rs1695[221]. Individuals with null GST-M1, GST-Theta and homozygous rs1695 should theoretically be more at risk from oxidative stress and cancers, but this not always the case[221].

GST-M1 null (nonfunctional alleles) can predispose to risks of prostate abnormalities, and GST Pi is also subject to several SNPs. Many attempts have been made to link these SNPs with the consequences of failure to detoxify reactive species, such as in various cancers[221]. Regarding the effects of dietary and smoking habits, linking GST expression and increased DNA damage is problematic partly because other detoxifying enzymes such as epoxide hydrolase (mEH; also polymorphic) might be operating and masking the effects on the test cell system. Carcinogenesis may be due to a complex mix of factors, where different enzyme expression and activities might combine with particular reactive species from specific parent xenobiotics that lead to DNA damage only in certain individuals. Resolving specific risk factors may be extremely difficult in such circumstances.

However, in cancer chemotherapy, there is evidence that the presence of GST-M1 and GST-Theta1 null (nonfunctional) alleles actually improve treatment outcomes, possibly because the tumours cannot overexpress a functional isoform to defend themselves against the anticancer drugs[223]. Other studies with breast cancer treated by anthracyclines in GSTM1/GST-Theta null and GSTPi1 rs1695 individuals showed a combination of good response, but greater susceptibility to toxicity and the likelihood of breaks in treatment cycles[224]. Hence, whilst some studies suggest that the risks of developing some cancers are greater with the null and poor function GST alleles, the responses to therapy are better, albeit at a cost in adverse reactions. This does suggest that reduced dosage in the light of genotypic testing to detect null allele combinations of GSTs could lessen the toxicity and reduce treatment breaks, all without affecting efficacy, as the null status leaves the tumour vulnerable to attack from the antineoplastic drugs.

7.2.10   Transporter polymorphisms

The various transporters, such as the solute carriers and the ATP-binding cassette transporters, are pivotal in the battle waged by biological systems to defend themselves from small molecule threats both systemically, such as in the gut, and locally, in organs such as the liver and the brain. Hence, differential expression of these systems in various populations is likely to be a significant part of the variation in drug absorption and clearance seen in human populations, as well as resistance to drug therapy. With the solute carriers such as OATP1B1, there are at least 14 SNPs in 15 haplotypes that usually, but not always, result in less-efficient drug transport. OATP1B1 *1B (388A>G (rs2306283) is found in over 70% of those of African ancestry, 30% of Caucasians and 30-60% of Asians and is actually faster than the wild type. Whilst 521T>C (rs4149056) is found in Caucasians (15–20%) but is rarer (<2%) in Africans and only has 15% of wild-type activity. There are various haplotypes with combinations of these two alleles, such as *5 and *15, which are impaired, such that the liver struggles to extract several drugs (various statins, rifampicin, methotrexate, and docetaxel) from the portal blood, leading to significantly increased systemic exposure[225]. The haplotype OATP1B1 *1B (388G/521T), has been suggested to be faster than wild type, leading to reductions in systemic exposure due to excessive hepatic uptake.

As OATP1B1 is responsible for the vast majority of uptake of statins, it is not surprising that the impaired allele (521T>C) is directly linked with myopathy risk. This allele is carried by significant numbers of Caucasians as mentioned above, but is rare in Africans[225]. Indeed, the slow allele also impairs statin clinical effect, as they have to get into the liver to actually block HMG-CoA reductase. Conversely, the faster allele 388A>G has been linked with increased clearance of statins, but so far, has shown no significant impact on efficacy. Rosuvastatin is cleared half as fast in Japanese individuals compared with Caucasians and the dose is correspondingly lower, although it is not clear whether this is linked with OATP1B1 status[225]. The null alleles of OATP1B1/3 are the main cause of Rotor Syndrome, which is another (rare) cause of hyperlipidaemia[226].

Not surprisingly, *ABCB1* (P-gp) is also highly polymorphic, with more than 60 SNPs determined already. The most studied includes the triallelic *ABCB1*10* (2677G>T or A, rs2032582), *ABCB1*6* (3435C>T, rs1045642) and *ABCB1*8* (1236C>T, rs1128503). The alleles *10 and *6 are linked and are found together in 90% of Japanese and 60-80% of Caucasians[227]. In cancer chemotherapy, it would be expected that if the P-gp variants were less efficient than the wild type, then both clinical response and toxicity would be increased as more of the antineoplastic drug would reach the tumour. Several, but not all studies, did show improved clinical response and issues with neutropenia and fever with anthracyclines and docetaxel, as well as improved survival with paclitaxel, with *6 and *10 homozygotes[227]. Interestingly, the three variants did not significantly differ from the wild-type ABCB1 proteins in a cell system in their transport of the DOAC rivaroxaban[228]. Overall, the different haplotype and diplotypes of P-gp seem to overlap across many ethnic groups and are relatively common, Also, there do not seem to be many null or very poor function alleles and much needs to be discovered in relation to the effects on the promotors and other areas of these complex genes. So overall, studies on the impact of P-gp polymorphisms, particularly those focussing on the three main alleles of interest (*10, *6, and 8*) struggle to see clear trends in terms of population and drug ABCB1 variant relationships[229].

7.2.11   Polymorphism detection: clinical and practical issues

Lack of progress

In the decade since the last edition of this book, the use of pharmacogenomic testing to make drug selection and dosage decisions has perhaps not become as widespread as might have been anticipated. Even the 'Jewel in the Crown' of genotypic testing, TPMT status assessment, is not yet routine practice every-where. However, there are several organisations around the world that now make widely disseminated, clearly argued and frequently updated recommendations on specific drugs, such as the CPIC and the Dutch Pharmacogenetics Working Group (DPWG). There is clearly willingness on the side of patients and clinicians to engage with pharmacogenomics testing financially and therapeutically[241] so it is useful to consider some of the issues that are retarding the routine adoption of pharmacogenomics testing in general clinical practice.

Lack of evidence

At the moment, with several drugs and drug classes, there is simply not enough convincing evidence that a particular genetic variant or combination of variants is/are sufficiently strongly linked with either impaired efficacy, or increased toxicity, to justify the deployment of the pharmacogenetic test that is currently available. Such a situation should evolve as more data are published and the quality, reliability and cost of the tests become acceptable and can be considered by the various authorities concerned. However, this evolution can be hampered by myriad practical difficulties as discussed earlier, as well as shortcomings in the design and scope of the investigations planned. As we saw with clopidogrel, investigations can be deflected or sidetracked, or in the case of tamoxifen, bogged down by the sheer complexity of the relationship of its metabolism and its efficacy.

Also, some of the assays used to assign genotypes and phenotypes are some way from perfect. Some groups can be significantly more accurately identified than others, such as with intermediate TMPT variants, whilst with tamoxifen issues have occurred where tumour tissue provides misleading data on CYP2D6 status. In addition, many prospective studies have focussed perhaps on one or two SNP variants, without always addressing the linkage with other genes, particularly in the framework of pharmodynamics as well as pharmacokinetics.

By the time the full implications of the pharmacogenetic impact of a drug on human populations has been elucidated, as noted in Section 7.2.2, the drug may well have been superseded. Hence, there is the question how such work will be funded and by whom.

Other more practical issues involve convincing clinicians of the value of pharmacogenetic testing as part of their daily decision-making. In many areas, such as cancer chemotherapy, or xenograft immunosuppression, clinical supervision is already very intense and backed by evidence and clinical experience. Clinicians will be rightly concerned about the risks of under- or overdosing a patient on the basis of a pharmacogenetic test, unless the supporting evidence convinces them. In addition, the issues related to interpreting the tests clinically and explaining these interpretations to the patient in nonspecialist, nonfrightening but accurate language is also a significant and recognised set of challenges[230]. Who should pay for the tests is, of course, a key issue that also requires justification to whoever funds the health care in the particular country.

Current developments: evidence

Whilst the personalised medicine ideal of 'the right drug at the right dose the first time' is clearly a long way off, there are many signs that pharmacogenetic testing is finding favour in terms of benefit and cost. Patient benefit of course arises from the wider and successful use of testing, which is in turn, directly related to the accumulation of supporting evidence in greater quantity and quality. One very important development in this area is the movement away from doggedly pursing particular SNPs to the adoption of GWAS (Genome-Wide Association Studies) where the appropriate statistical power is built into the experimental design to adequately determine the impact of combinations of linked variants that may require complex analytical simulation and processing[231].

Hence, as we have seen initially with warfarin, but now with several other drug groups, the most informative testing concept aimed at minimizing toxicity and maximize efficacy, must have an approach that is more complete. Pharmacokinetically, the full range of possible oxidising and conjugating biotransforming isoform variants should be examined, as well as the various transporters. The pharmacodynamic aspects of the drug target, receptor systems and tissues, need also to be investigated. In addition, all the markers involved should be approved by organisations such as the CPIC, USFDA, and the DPWG. In addition to gathering the information, accurate guidance should also be supplied to the clinician and patient to inform final decision-making. To date, such a complete picture has not and may not always in the future necessarily be possible to assemble. With warfarin, for example, what information the clinician may be able to determine in terms of specific genomic tests can be inserted into the warfarindosing.org website, and guidance is formulated according to the level of patient data inputted[60].

Currently, the most complete pharmacogenomics testing approach has appeared with the advent of Commercial Pharmacogenetic Decision Supporting Kits (CPDSKs). These are backed by laboratories that will analyse patient saliva for all the appropriate markers, validated by the major regulatory and advisory groups. They will also supply the appropriate clinician and patient friendly reports, detailing the appropriate recommendations on drug and dosage based on interpretation of the genetic markers. Some of these CPDSKs are even available through US food and drug retailers. This awakens concerns over access and interpretation of pharmacogenomics data by the general public who may obtain such information with or without the consent or knowledge of healthcare practitioners. Whilst it is probably appropriate to be cautious, in the sphere of depressive illness, several reports have indicated that the use of these CPDSKs has been beneficial in reducing side effects and improving outcomes[232,233].

Practical applications

Already there is evidence that whilst pharmacogenetic testing is not yet transformative, it is used increasingly reactively as part of retrospective investigation processes in drug toxicity or efficacy failure. The prospective use of such tests is also gaining ground as it is realised that significant toxicity and hospitalisation caused by narrow TI drugs can be prevented, such as described with 5-FU[157]. Moreover, such toxicity leads to treatment breaks and /or dose reductions, which are problematic, as whilst the patient is recovering the drug's toxicity, the underlying disease is progressing unchecked, shortening survival time. Most of us would willingly pay for a pharmacogenomic test that would effectively prevent such a scenario and perhaps prolong our lives. Again, with narrow TI drugs in very stressful clinical situations, such as in both cancer chemotherapy and xenograft maintenance, pharmacogenetic tests can speed up drug optimisation significantly, as seen with tacrolimus. This might not necessarily markedly affect outcomes, but a smooth progression into therapy can have very significant impact on patient morale and well-being, as well as increasing their faith in the therapeutic process. This should help address the often-dire drug adherence seen even in serious clinical conditions, such as encountered with tamoxifen. There is also enormous unmet need in terms of poor drug performance in mass-drug use areas such as mental health, where it is estimated that over 40% of first line drug failures in depression are linked with pharmacogenetic issues[232]. In the longer term, pharmacogenetic testing will continue to widen into more mainstream areas such as cardiovascular disease and obesity prevention.

Costs and the future

From a purely financial perspective, convincing analyses are emerging where the cost savings of pharmacogenetic testing are very significant indeed[160]. The report on 5-FU mentioned earlier, showed that to test a whole group of patients costs around a tenth of the hospital stays caused by toxicity in the same group[157]. Cost benefits have also been shown in other spheres such as with antipsychotics[234]. The issues of evolving evidence strength and the cost/benefit of pharmacogenetic testing will remain, but it would make sense to build at least the foundations of such determinations into the new drug development process, unless a drug can be designed that bypasses much of pharmacogenetics, such as atracurium, for example. Perhaps pharmacogenetics is now at the stage of 'the right drug at the right dose, *more efficiently with the test than without*'.

7.3   Effects of age on drug metabolism

The effects of age on drug clearance and metabolism have been known since the 1950s, but they have been extensively investigated in the last 20 or so years. It is now widely accepted that at the extremes of life, neonatal and geriatric, drug clearance can be significantly different from the rest of humanity. Usually, neonates, i.e. those less than four weeks old, cannot clear certain agents due to immaturity of drug metabolizing systems. Those over retirement age cannot clear the drugs due to loss of efficiency in their metabolizing systems. Either way, the net result can be toxicity due to drug accumulation.

7.3.1   The elderly

Significant features

It has been estimated that just under a sixth of the world's population is over 60, and this figure has increased by nearly half in 20 years[235]. The elderly population (usually taken as 65 or over) are obviously heterogeneous, with a spectrum ranging from the fit to the frail, alongside increasing health issues, often associated with chronic pain[235]. In the fit elderly, their mortality and morbidity after elective surgery is often similar to the general population, which signifies a high degree of robustness; however, several studies have shown the frail elderly to have much more impaired drug clearance than the fit elderly, usually because of multiple pathologies such as hip fractures and organ failure[235].

Aside from concerns over polypharmacy, memory and information processing speeds decline with age, but fortunately intelligence apparently does not. Hence, elderly patients may not always remember instructions related to their medicines or conditions, but it should be noted that they are likely to understand the explanation when they are reminded. Unfortunately, many other aspects of the human body decline fairly precipitously beyond the age of around 70, and the ability to clear drugs is no exception. In general, there is a decline in the inducibility in most CYPs, although the main conjugative reactions like sulphation and glucuronidation are really only impaired in those on polypharmacy[235,236].

It has been suggested that elderly livers do not supply enough oxygen to the CYPs *in vivo*, which was known as the 'oxygen limitation theory'. This, however, is rather academic, as the major limitations on drug clearance in the elderly are due to other more pressing physiological factors. There are significant changes in the liver itself, as it decreases in mass and blood flow is reduced as we age. This occurs at the rate of around 0.5–1.5% per year, so by the time we hit 60–70, we may have up to a 40–50% decline in liver volume and blood flow compared with a 30-year-old[237]. In fact, as aging progresses, the long-term ultrastructural changes in the liver are not dissimilar to those of cirrhosis[238]. Other factors include gradual decline in renal function, increased fat deposits, and reduction in gut blood flow, which can affect absorption. Other issues include significant changes in distribution of drugs, leading to reductions in volume of distribution in agents such as paracetamol[237].

Impact on high-clearance drugs

Clearance of drugs that are metabolised in proportion to blood flow such as propranolol, are most noticeably impaired in the elderly (high intrinsic clearance; $E_H > 0.7$, Chapter 1.4.2). These include beta-blockers, TCAs, fentanyl, propofol, and verapamil. Bioavailability with some drugs such as paracetamol declines with age, but increases with others[237]. In general, clearance declines with increasing frailty, with significant differences (>30%) between robust and frail elderly people[235]. This can lead to considerably higher blood levels than would be expected in a 40-year-old. This can be a serious problem in drugs with a narrow TI, such as antiarrhythmics[238].

Impact on low-clearance drugs

This is more complex and concerns drugs such as ibuprofen, benzodiazepines, and the AEDs valproate and phenytoin, where clearance is capacity limited and linked to both the fraction unbound and the ability of the biotransforming pathways to metabolise them (intrinsic clearance, Chapter 1.3.4). Hence, with low clearance drugs, the key variables are the activity of the CYPs/conjugative pathways and the amount of unbound drug available, which is in turn linked to protein binding. The impact of age on CYPs is variable, with many showing little decline in activity in the elderly[236]. Although with flecainide, which is mainly cleared by CYP2D6 with some CYP1A2 component, a steep age-related decline in clearance is seen in CYP2D6 PMs[239]. Another issue is the decline in the formation of plasma proteins, such as albumin, α_1 acid glycoprotein and various lipoproteins, which has the effect of increasing the unbound fraction of many drugs, which in turn increases the availability of the agent for clearance. This can be a significant bearing on drug toxicity with some agents, such as phenytoin. So this combination of factors has the effect of leaving some low clearance drugs unchanged in the elderly (diazepam, temazepam, and valproate), slightly increasing some (phenytoin) and reducing others (lorazepam and warfarin)[238]. Average doses of warfarin required to provide therapeutic anticoagulation in the elderly are less than half those required for younger people. Sometimes, even if the fraction unbound has increased due to age, the CYPs/conjugation systems have declined so they cannot take advantage of the increased availability of the drug, so clearance is around 30–50% lower with drugs such as ibuprofen and naproxen with respect to the young. This can be an issue with the elderly and chronic pain, where they can overdose themselves over a significant period of time, causing gut problems and tinnitus.

Summary

Life-long habits such as drinking and smoking, combined with increasingly erratic diets, can influence liver and kidney health in the elderly, and of course very frail elderly individuals with multiple pathologies may have significant hepatic inflammation, which can downregulate drug metabolism. The ever-present polypharmaceutical pressures can also lead to marked clearance effects when drugs are added or (much more rarely) removed from regimes. Generally, liver metabolising powers are remarkably tough, although tests on CYPs such as CYP3A4 or CYP2C19 show variable resistance to decline in the elderly[238]. Nearly 20% of elderly people are believed to be clinically depressed and SSRIs dosages should be reduced in older patients. Fluoxetine and paroxetine therapy inhibit various CYPs so sustained reductions in the clearance of co-administered agents will occur just as they do in the young. Obviously, the inhibitory effects of SSRIs on CYP2D6 and to a lesser extent on the CYP3A series suggest that caution should be exercised when prescribing these drugs to the 'polypharmaceutical' elderly. A major issue with the accumulation of CNS-active drugs including the SSRIs, due to age-impaired metabolism is the effects on balance and alertness, which can lead to falls. These can be life-changing, particularly in the female elderly[240] due to the risks of osteoporosis-linked fractures which will mean hospitalisation and permanent loss of independence. SSRIs use in the elderly can alleviate depression, but the price can be increasing the risk of fracture in older people[241], and accelerating the process of cognitive impairment in the elderly[242].

Overall, the population of fit elderly has been increasing and life expectancies continue to rise in the developed world, possibly for the next couple of decades. Although drug clearance is impaired in this group generally, once therapy is optimised in the fit elderly, the situation should be stable in many cases into their 80s provided no major pathologies develop.

7.3.2   Drug clearance in neonates and children

Neonates, older babies, toddlers, and young and older children (aged 10 and beyond) are structurally the most heterogeneous stages of human life, due to vast and rapid changes that occur over a relatively short time in size, blood volume, organ development, and biotransforming enzyme capability. Perhaps neonates and children have the greatest need of all patients for personalised medicine, yet our understanding of the changes in drug clearance and their relationships with efficacy do not compare with those of adult populations. Part of this is the scale of the changes, from tiny premature neonates with skin permeability can be up to 100 times that of adults, to robust school-age children[243]. There is a clear and ongoing need to formulate general dosing guidelines in neonatal and older children for the very high proportion of drugs that are used off-label (essentially unapproved) in this context[243]. However, there are many obvious obstacles to assessing drug clearance in neonates, babies and young children experimentally, such as parental permission, ethical considerations and limitations on sample volume and frequency. Hence, many published studies produce data from babies who might be under intensive treatment, could be seriously ill and thus not necessarily representative. This situation is changing and more studies are being designed using specialised techniques that provide the basis for dosing recommendations[243]. A neonate of 1 Kg in weight only has 90 mL of blood, so no more than 3 mL can be taken in a drug study in 24 h. In addition, sparse and scavenged (vital lab test) sampling or even dried blood spots of a few microliters, combined with sensitive analytical techniques, are used to conduct trials in neonates[243]. Pharmacokinetic models are now more advanced in their ability to predict the clearance of drugs in neonates and children, although predictions in pre-term neonates remains problematic[244].

Neonates

Babies are classed as neonates when they are less than four weeks old, although in some reports, older babies are sometimes included as neonates. The major organs are underdeveloped, with lower liver blood flow and low renal function, although GFR increases rapidly as the neonates grow. There are other issues which influence drug effect, such as low protein binding, higher fluid volumes with respect to muscle mass and their blood brain barrier is immature also. Hence, drug distribution and penetration into areas such as the brain are different in neonates with respect to adults[243]. All these issues are accentuated in premature neonates compared with full-term babies. Interestingly, recent research has showed that premature and term neonates can actually clear some drugs faster than older children, despite their lower CYP activities. CYPs 1A1/2 and CYP2C9/19 show low activity[243], with neonates clearing caffeine and theophylline slowly. Some studies have shown CYP2D6 to be the highest expressed

CYP[245], or the lowest in the first weeks of life[243]. However, CYP3A7 is very active pre-birth and peaks in the first week or so after birth and then declines away by about a month. CYP3A4 expression then accelerates to the point of reaching 30–40% of adult levels within a month. Likewise, CYP2E1 is absent in the foetus but is around 10% activity at birth and then increases[246]. FMO-1 tails off rapidly after birth while FMO-3 increases in expression. There is little glucuronidation expression in neonates, but SULT activity is very high, so with paracetamol its clearance is SULT-mediated. Hence, whilst CYP levels can be hugely variable (1000-fold in some studies[245], pre- and term neonates are capable of surprisingly rapid drug clearance in some cases.

Regarding drugs used on neonates, analgesia, and antipyrexia can be achieved with paracetamol, which is used more frequently in neonatal intensive care as it avoids opiate use and intravenous formulations are now more widely available[247]. Whilst in pre-term and term neonates paracetamol is unlikely to be significantly oxidatively converted to reactive species, as age increases, CYP2E1 capacity rises and SULTs can be saturated, so inadvertent toxic overdoses can become problematic. Neonates can be vulnerable to infections due to immaturity of their immune systems, so treatment with antiviral/bacterial/fungals can be necessary. With micafungin, clearance is nearly twice as fast in premature neonates compared with older children, whilst metronidazole, meropenem and clindamycin are cleared more slowly in neonates[243]. Other drugs such as fluconazole's clearance is closely matched to renal capability. However, there are still difficulties using drugs such as sedatives in neonatal intensive care, such as midazolam, whose clearance can increase by 50% in critically ill neonates[244]. In addition, the drug increases the risks of death or brain injury as well as length of stay in intensive care[248]. Other drugs which rely on CYP activity to be cleared have prolonged half-lives in neonates, such as diazepam and phenytoin[249].

Conjugative metabolism in neonates

In contrast to the high-birth SULT activity, glucuronidation, in neonates and term babies is low, due to impaired UGT expression. Indeed, around 6 out of 10 newborns have jaundice, which usually fades within 10–14 days[250]. This was thought to be due to immaturity of UGT-mediated bilirubin clearance (Chapter 6.2.4). However, it is now believed to be a specific PXR-mediated repression event that, as well as breast milk (suppresses gut UGT1A1), deliberately elevates bilirubin levels, presumably linked with its antioxidant, or some other beneficial effect[250]. A number of processes, including de-repression and induction of UGT1A1 by bilirubin[250], gradually reins in the high bilirubin levels. In babies who are not jaundiced, their low UGT capability can be revealed by drug administration.

In the 1950s, the use of chloramphenicol for infections in babies led to *grey baby syndrome* where the drug was cleared to a hydroxylamine (probably by a

CYP2C variant), which would be glucuronidated in adults by UGT2B7, but in children the poor UGT capability led to the metabolite escaping the liver and causing methaemoglobin formation[217] (Chapter 8.2.2). Chloral hydrate was once used as a sedative in neonates; however, it is cleared to trichloroethanol, which is not only poorly glucuronidated by neonates but also competes with bilirubin for this pathway, thus exacerbating jaundice. In neonates with Gilbert's syndrome (Caucasians and Africans, TA7/7; Asians, UGT1A1*6), jaundice is likely to be more pronounced than unaffected individuals, and glucose-6-phosphate dehydrogenase deficiency exacerbates this (G-6-PD). This is most likely to occur in those of African ancestry and is due to increased haem processing caused by G-6-PD that overloads a very low capacity system. Neonates and young children with Gilberts and G-6-PD will exhibit long half-lives in drugs cleared partly or wholly by glucuronidation such as lorazepam, morphine, and AZT. Indeed, normal premature neonates show greater than four-fold slower glucuronidation of such drugs compared with adults.

In general, whilst GSTs rapidly attain adult levels not long after birth[249] other systems, such as NATs are poorly developed at birth. Neonates and babies are slow acetylators until their genotype is fully expressed by the age of 2–4 years[251]. The various esterases are also at low capacity, which is thought to be why babies are very susceptible to the effects of anaesthetics[249]. The maturation process of glucuronidation systems is mainly much slower than other biotransforming systems according to current knowledge. In one study, UGT1A6 and UGT2B7 were undetectable in newborns[245]. Indeed, codeine is cleared partly through UGT2B7 and neonates can be at risk from CNS depression as they struggle to clear codeine and morphine from their mothers who are CYP2D6 UM's[252]. UGTs are generally not expressed foetally until the third trimester, but UGT1A1, 1A3, 1A6 and 1A9 were found at birth[245], with UGT1A3 expression about 30% of adult levels[245]. This pattern of rapid UGT development soon after birth has been followed with other UGTs. UGT1B15[217] levels were less than 20% before birth and reached maturity after 15 weeks. UGT2B7 and UGT1A1 reach adult levels by 6 months, although some UGTs, such as 1A6, do not reach adult levels until puberty. So overall, conjugative capacity in very young children appears to be extremely variable, but all the main systems, UGTs, SULTs, GSTs, NATs and esterases are either at adult capacity or well on the way by the time the child is beginning to walk[217].

Drug clearance in older children

By the ages of 1–2 years, CYPs 1A1/2, CYP2C9/19, CYP3A4/5 and UGT2B7 have reached adult levels, as have the various ABC cassette and solute transporters, with CYPs 2D6 and 2E1 and FMO-3 taking one to three years longer[243,249]. In addition, the organs progress towards maturity; renal capability has reached adult values and liver blood flow increases. However, their metabolic rates and renal clearances can be quite different to adults, resulting in marked differences in drug disposition. With antifungals, fluconazole, voriconazole, and micafungin show significant increases in clearance in children with respect to adults[253]. Several AEDs such as lamotrigine, phenytoin, and carbamazepine are also cleared faster in young children. Indeed, with zonisamide, clearance in children up to 4 years old can be almost twice that of adults[254]. With some antibiotics such as levofloxacin, children younger than 5 years old cleared them twice as rapidly as adults. However, clearance of other drugs are within the range of adults, such as many opiates, proton pump inhibitors, fluoxetine, methylphenidate and risperidone[251,255].

SSRIs are used in children aged 6–12 for the treatment of obsessive-compulsive disorder and emotional/depressive conditions. There is evidence that fluvoxamine is cleared more slowly than in adults and it is recommended that doses should be reduced, especially in girls[256]. SSRIs in general have a negative effect on bone density and this is the case with children also, especially the spine and the long bones[257]. Many clinical researchers emphasize that children are not just miniature adults, but have many pharmacodynamic and pharmacokinetic differences with respect to adults. Indeed, there are still difficulties in first-time dosing in children with many drugs and although many pharmacokinetic models have been developed and they may contain enough relevant data to be useful with children from 2–12 years, it remains difficult to model neonatal and preterm drug disposition[244,249].

7.4 Effects of diet on drug metabolism

The old saying 'you are what you eat' still has resonance, and diet can influence the clearance of drugs as much as general health. Some of the possible diets that are available in Western countries can be more toxic than many drugs, and Morgan Spurlock's documentary *Super-Size Me* did induce some well-known purveyors of fast food to pay some attention to the composition of their products. Food contains a vast array of chemicals, such as antioxidants, plant toxins, preservatives, polyphenols, polycyclic aromatics, and various other environmental pollutants. It is clear from the preceding chapters that the profile of our biotransformational systems has been designed to respond to the pattern of our xenobiotic intake, and since most of our dietary habits vary relatively little over our lives, our metabolizing systems adapt to the xenobiotic stimuli in our food intake also. Chapter 5's descriptions (Section 5.5.4) of the effects of various fruit juices notes that many of the compounds absorbed from our diet are as potent as drugs in terms of their impact on CYP expression and catalytic activity.

7.4.1 Polyphenols

Thousands of polyphenols are found in plants, vegetables, fruit, as well as tea, coffee, wine, and fruit juices. These molecules have specific uses in the plants, either as antioxidants to extend the 'shelf-life' of fruit on the plant to maximize the attraction to insects and birds to spread seeds, or as toxins to ensure a hideous lingering death for any animal that dares to eat the plant. These compounds are a broad classification and can include phenolic acids, stilbenes, flavonoids, and lignans. Each subclass, such as the flavonoids, has a wide array of related agents (flavonols, isoflavones, etc.). The drastic effects of the polyphenols such as naringenin and bergamottin from grapefruit juice on biotransformational activity have already been discussed in Chapter 5 (Sections 5.5.3–5.5.4). Flavonoids such as quercetin and fisetin are excellent substrates for COMT, so competitively inhibiting the metabolism of endogenous catecholamine and catechol oestrogens. Quercetin and other polyphenols are found in various foods such as soy (genestein), and they are potent inhibitors of CYPS and SULT1A1 which sulphate endogenous oestrogens, so potentiating the effects of oestrogens in the body. Many of these flavonoids and isoflavonoids not only inhibit or modulate oestrogen metabolism, but they also act as phytoestrogens on the ER receptor[258]. As discussed in Sections 5.5.3–5.5.5, many of these agents are found in foodstuffs as complex mixtures of many highly active chemicals, and our understanding of how these agents behave in mixtures as opposed to singly is not well developed. If you recall the complexity of evaluating the effects of St John's Wort on various medicines (Section 4.5.3), then it is clear that we will probably never know what effects all these agents will do in relation to prescription drug efficacy and clearance. The marketing of these agents for such a variety of anti-inflammatory, cardiovascular, antibacterial, and antidepressant-linked conditions strongly suggests that the intended buyers are already suffering from these conditions and are taking myriad prescription drugs already[258].

It is also likely that various polyphenols influence other endogenous substrates of sulphotransferases, such as thyroid hormones and various catecholamines. It is gradually becoming apparent that polyphenols can induce UGTs, indeed; it would be surprising if they did not. Certainly drug-efflux transporters are modulated by polyphenolic isoflavones. This effect has been seen with P-glycoprotein and MRP1–3. The flavonoid chrysin, found in honey, is a potent inducer of UGT1A1. Overall, as we saw in Chapter 4.4.3, many plant remedies actually exert their therapeutic effects through nuclear receptors such as PXR and CAR and these polyphenols will thus potently modulate CYPs and conjugative enzymes in exactly the same way[258]. As discussed in Chapters 4 and 5, there is probably no way to prevent patients seeking, buying, and consuming online supplements, but all we can do is encourage them to be honest with their health-care professionals so at least some investigation can be carried out to determine whether a particular agent is a risk. It is probably also sobering to consider that if prescription drugs worked optimally, then perhaps fewer patients would feel driven to consider herbal supplements.

7.4.2 Barbecued meat

In the United Kingdom, the weather effectively prevents the excessive consumption of so-called charbroiled or barbecued meat. However, many fast-food outlets claim that their burgers are 'flame-broiled', so it is possible through some modest effort to maintain a high consumption of barbecued meat in the United Kingdom. This exposes the consumer to a combination of polycyclic aromatics, nitrosoamines, and hetercyclic aromatic amines that all induce CYP1A1, CYP1A1, and CYP1B1. This has no measurable impact on CYP3A4 or P-gp, but it has a potent effect on CYP1A2 expression, both in the gut and liver, pretty much doubling capability from only a week of heavily cooked meat intake[259]. This can accelerate the clearance of several drugs that have a CYP1A2 component to their metabolism, including amitriptyline, clozapine, fluvoxamine, mexiletine, haloperidol, naproxen, olanzapine, and paracetamol. GSTs and UGTs will also be induced, so that the carcinogenic effects of the induction of these CYPs can be offset to some extent. However, meat eaters who want to avoid any risk of CYP1A2/CYP2E1 induction, yet cannot face a life eating nothing but vegetables can switch to eating the traditional South African marinated and uncooked dried meat, biltong[260]. The carcinogenic effects of the heterocyclic amines and polycyclic aromatics in cooked meat can also be to some extent fended off by consumption of considerable amounts of vegetables with cooked meats. They contain a variety of inhibitors of these CYPs, as well as inducers of GSTs and epoxide hydrolases so reducing the risk of activating the polycyclics. Interestingly, apiaceous vegetables, such as carrots, contain furanocoumarins that are quite potent CYP1A2 inhibitors[261], although stacking your flame-broiled burger with carrots may take some time to catch on.

7.4.3   Cruciferous vegetables

Most of us now accept that a diet containing plenty of vegetables can stave off many cancers. However, it seems that the cruciferous vegetables may the most effective anti-cancer agents in our diets. These include cabbages, cauliflower, watercress and of course, broccoli. They are not to everyone's taste; a former US President (George H.W. Bush) was once heard to refuse to eat broccoli. Whilst he lived to be 94 years old, who knows how long he would have lasted had he enjoyed the benefits of these vegetables. As detailed in Chapter 4.4.3, they contain a broad array of beneficial polyphenols, glucosinolates, and isothiocyanates, especially the young shoots. If you would like to avoid malignant disease in later life, it is apparent that you need to consume cruciferous vegetables regularly from an early age. The most studied is probably broccoli, which has been bred selectively since antiquity. Its major glucosinolate is glucuraphanin, which is converted enzymatically to sulforaphane when broccoli products are crushed, chopped, or otherwise manhandled[262]. It seems that preparation of cruciferous vegetables is important. Boiling is rather destructive, but steaming is better and microwaving accelerates the formation of sulforaphane by several-fold[251].

Many of the benefits of sulforaphane were detailed in Chapter 4.4.3 and its activation of the nrf2 system (Chapter 6.5.6), thus inducing detoxification enzymes such as the GSTs and epoxide hydrolases, as well as its boosting of

GSH formation. Sulforaphane has a range of anti-inflammatory and anticancer properties plus cardiovascular benefits[263]. To obtain this apparently impressive package, you just need to eat around 200–300 g of broccoli daily, and within one to two weeks the benefits kick in. It is, of course, not quite that simple. Although many studies have shown that around five servings of cruciferous vegetables weekly can reduce the risk of bladder, prostate, and colon cancers, not all reports show this. There is evidence that cruciferous vegetables induce CYP1A2 like cooked meats, as well as GSTs, but if you eat them with carrots as mentioned in section 4.2 above, the CYP1A2 effect is inhibited, although the effects of the carrots last only about a day or so[261]. Several clinical trials have been mounted to explore the benefits of sulforaphane and trials continue to be organised with the various brassicas, extracts and other cruciferous preparations. On balance, they have shown benefits[264], but the US FDA has already had to warn against inflated and distorted claims made for some of the marketed products[264]. On balance, it is probably a good long-term plan to eat some broccoli, watercress, or sprouts regularly, maybe around the recommended five servings per week level and no more.

7.4.4   Other vegetable effects on metabolism

Several studies have been carried out with different garlic preparations and extracts, as well as with diallyl sulphide, one of the active agents in garlic. The main thrust of such work has been to inhibit CYP2E1 so that it lessens its contribution to the formation of carcinogenic species. The inhibitory effect of diallyl sulphide on CYP2E1 is considerable and can be demonstrated on the CYP2E1 probe substrate chlorzoxazone, and this effect may be linked to the healthy reputation of garlic in preventing various diseases[265]. Overall, consumption of vegetables leads to an alkaline, rather than acid urine, and this is beneficial in maintaining the stability of a number of conjugated metabolites of aromatic amines and preventing their degradation to the parent metabolite in the bladder, an organ not exactly renowned for its detoxification capability.

7.4.5   Caffeine

Moderate caffeine (1,3,7,trimethylxanthine) consumption is a good idea that also makes getting out of bed worthwhile, particularly if you grind your own fair-trade preferably organic beans and brew good-quality coffee, rather than the instant version, which is a crude form of coffee flavouring. Caffeine is a mild stimulant that enhances alertness and performance, and it is found in a vast array of products. These range from various teas and carbonated cola drinks through to some cold cures, chocolate, and specialised vile-tasting stimulant drinks aimed at young people. It is also available pharmaceutically OTC in tablet form.

In order of caffeine dosage, the stimulant tablets can be 200 mg or more, whilst a strong brewed cup (about 200 mL) of good coffee is around 100 mg, and an equivalent volume of tea, made from black loose tea leaves, will contain perhaps around 50 mg, slightly less than that of disgusting instant coffee. Many caffeinated drinks yield around 35–40 mg per can and are unsuitable for young children, as the drug affects appetite as well as acting as a stimulant. Almost caffeine's entire metabolism is by CYP1A2, and it appears to act as a competitive inhibitor of substrates that are cleared by this CYP. An average consumption of caffeine could be taken to be around 3–4 cups of tea or coffee per day, which is in the range of 250–400 mg per day, or about 6mg/kg body weight. Current recommendations for a caffeine intake not associated with adverse reactions, for healthy adults, are 400 mg/day, whilst for children and adolescents, it is 2.5 mg/Kg/body weight[266]. The intake for pregnant women has been the subject of controversy, linked with low birthweights and other issues. However, recent studies have not supported these findings[267] and the most recent assessment of the recommendation of 300 mg/day for pregnant women has not been challenged[266].

The intake of 300–400mg daily is not considered performance enhancing and results in around less than half the International Olympic Committee limit of about 12 mg/litre excreted in urine. If a patient is stabilised on a CYP1A2 substrate (fluvoxamine, warfarin, and clozapine), then this habitual level of consumption of caffeine should not make much practical difference to other CYP1A2 drug clearances, as most of us are creatures of dietary habit. However, if the patient binges very seriously on caffeine or stops it altogether, changes in the CYP1A2 substrate clearance will occur. This is more likely to register in terms of adverse reactions in a narrow TI drug such as clozapine, and patients should be made aware of this possible effect. A dosage of around 400–1000 mg of caffeine can make an impact on clozapine metabolism by around 20%, although this level of consumption seems rather high in normal diets[268] and caffeine intake is not normally considered to make an impact on clozapine efficacy in the real world. The situation is more complex in smokers, where CYP1A2 is strongly induced, and they will require perhaps two to three times more caffeine than nonsmokers. Of course there is huge variation in caffeine metabolism, and women clear caffeine more slowly than men due to lower CYP1A2 activity (see Section 7.5).

7.4.6   Diet: general effects

It is clear that diet can substantially modulate biotransformation – the consumption of high levels of cooked meat induces CYPs 1A1/2, but prior established consumption of cruciferous vegetables and watercress, for example, can negate the carcinogenic effects of the meats. This is why in the Mediterranean, the local diet is sufficiently rich in stimulators of detoxification pathways and inhibitors of oxidative metabolism, that it is possible to eat what you like, drink everything in sight, and smoke pounds of shockingly pungent tobacco and live to be 116 years old. As to the effects on prescription drugs, this is more difficult to measure reliably, although abrupt changes in a person's diet may significantly alter the clearance of drugs and lead to loss of efficacy or toxicity.

7.5 Gender effects

Women vs. men in drug clearance

For many years, there was a very large scientific literature on the sex differences in animal models in their clearance of drugs. In humans, this has not been as heavily investigated. Some differences have been found, although they are not often taken into account clinically. In hindsight, it should be glaringly obvious that the highly sophisticated control system for menstrual steroid metabolism, one of the *raisons d'etre* of human biotransformation, should indicate that women are likely to clear drugs differently from men and that drugs which modulate metabolism are more likely to lead to adverse effects on female metabolism than male. Indeed, drugs that have been withdrawn in the past caused greater adverse issues in women than in men[269].

Aside from size issues, there are many obvious differences in male and female basic physiology that impact drug disposition, ranging from higher fat proportions and lower total body water in women, to reduced lean muscle mass. Women also show blood flow differences, with lower major organ and muscle, but higher adipose tissue blood flow[270]. Another intriguing effect is that oestrogens can decrease hepatic blood flow, to the point that it reduces imipramine clearance[271]. All of these factors influence drug distribution, but there are clear differences in drug biotransformation that are now well documented, The major difference is in CYP3A4, which is expressed in double the quantity and also operates at a higher activity in women[272]. Hence, they can potentially clear around half of prescription drugs more rapidly than men, although the hormonal influence on blood flow may moderate this effect. Female levels of PXR are higher than men also. Other biotransforming enzymes are expressed at up to twice the level of men in women, include SULT1E1 and BChE. These effects are gradually lost in post-menopausal women, although it is thought that the regulation of the CYPs involves more than just direct hormonal effects and may involve other factors such as growth hormone[272]. Whether CYP2B6 and CYP2A6 are expressed at higher levels in women is not confirmed.

Apart from the higher capability of several biotransformational systems in women, the menstrual cycle exerts a significant effect on drug clearance and efficacy. This was studied in the 10–12% of epileptic women who suffer from catamenial (menstrual cycle-influenced) epilepsy. As the cycle progresses to the luteal phase in the 14 days or so after ovulation, phenytoin concentrations can fall by as much as 40% in some women[273] promoting the possibility of seizures.

This effect is also seen in carbamazepine, but does not seem as pronounced[274]. Hormonal cyclic differences influence drug metabolism, but also the use of oral contraceptives (OCs) have more sustained effects, such as the reduction in imipramine clearance mentioned in the previous paragraph. The impact of OCs is often not considered clinically, but there is evidence that they are mild inhibitors of CYP1A2, CYP2C19, and to a lesser extent, CYP2B6 and CYP3A4. OCs slow the clearance of bupropion, which is a CYP2B6 substrate[271].

Overall, it has been difficult to separate gender differences in medicine responses in terms of pharmacodynamics and pharmacokinetics. Women are twice as likely to be diagnosed with depression than men, but young women appear to respond better to SSRIs than TCAs, and this distinction is not seen in males. This is linked to slower female clearance of TCAs, leading to more adverse reactions[271]. Other studies have shown some CYPs, such as CYP2A6 and possibly CYP2B6, are directly induced by oestrogens, which may be relevant to nicotine clearance and effects in women. Other studies have drawn attention to the effects of inhibitors on CYP3A4 and other CYPs in women and what effects these may have on hormone clearances. There are other marked differences; fluoroquinolones such as ciprofloxacin and ofloxacin are cleared significantly more slowly in women, so that peak drug levels are more than a third higher than those of males and females suffer more adverse effects from these drugs compared with men. It is also apparent that females in general are more susceptible to drug-adverse reactions than males, especially hepatotoxic effects. This may be a reflection of their increased formation of reactive or toxic metabolites and less robust detoxification, although there are also pharmacodynamic differences in drug sensitivity. Women are much more likely to develop *torsades des points*, or dangerous lengthening of the QT interval (Chapter 5, section 5.7.2). This is because endogenous oestrogens lengthen the interval anyway and so QT-lengthening drugs can exert an effect at lower concentrations than in men. Many more females died from terfenadine-induced *torsades des pointes* than males before the drug was withdrawn.

There are some contradictions, such as with CYP1A2, as males express about twice the level seen in women, so women demethylate caffeine more slowly than men. Indeed, studies with human liver microsomes and hepatocytes have shown that CYP1A2 and CYP2E1 are markedly more active in men[272]. Once the menopause occurs, many of the differences in CYP expression disappear[272] and this is reflected in improvements in responses to TCAs in post-menopausal compared with premenopausal women[271].

Drug metabolism in pregnancy

Whilst casual therapeutic drug use in pregnancy is discouraged, around half of women take at least one medication during this period in their lives and in the USA three to five prescribed medicines are used[275]. There are several conditions where continuity in therapy is essential, or the benefits of drug treatment outweigh the disadvantages. This involves certain drug classes, such as AEDs, antidepressants, antipsychotics, and anti-HIV agents. There are some obvious physical changes in pregnancy, such as the gradual increase in plasma and organ volumes, cardiac output, and glomerular filtration rates[275]. Hepatic blood flow can increase by 60%, whilst the concentration of plasma proteins falls due to the plasma volume increases. This has the effect of increasing the clearance of high extraction drugs (such as many beta-blockers and opiates) and the free concentrations of low extraction drugs (phenytoin) also increases. This latter effect, plus an increase in biotransforming enzyme activity, together has the effect of reducing bioavailability of high pre-systemically cleared drugs and accelerating the clearance of low extraction drugs as well. Of course, these changes are very gradual from conception and not only the growing foetus can clear limited drug amounts through CYP3A7, but the placenta has a full complement of CYPs, conjugation and transporter systems[276]. All these factors have incremental but increasing impact on drug clearance and thus drug levels must be added to the usual monitoring over the normal progress of the pregnancy.

Since the main female hormones can interact with the nuclear receptors and modulate biotransforming enzymes, the 100-fold increases seen in pregnancy[277] are certain to have a significant effect. That CYP2A6 is known to be induced by oestrogens explains the acceleration in nicotine metabolism, which is wholly negative, as it will induce the mother to smoke more to maintain the effect. The net increase in carbon monoxide and PAHs are harmful to foetal development. UGT1A4 is also induced by ostrogens, which accounts for the massive (up to eightfold) increases in clearance of the AED lamotrigine[278].

Other CYPs, such as CYP2D6, CYP3A4/5, CYP2C9 activities all increase, gradually causing AEDs, antidepressants, antipsychotics, and many other drugs to fall out of the therapeutic window[275] – potentially with significant consequences to the mother and pregnancy. Alternatively, CYP1A2 and CYP2C19 decrease in activity in pregnancy. Indeed, the decrease in activity in CYP1A2, in concert with the changes in blood flow cancel each other out, so there is no net change in olanzapine and clozapine clearance during pregnancy[279,280], even if some of the mothers were smokers[280]. However, quetiapine (CYP3A4) plasma levels fell by more than 75% by the third trimester and aripiprazole (CYP2D6) more than 50%, whilst antipsychotics such as haloperidol and perphenazine also showed declines in plasma levels[279]. Hence, drugs involving mental health and epilepsy, which are necessary to treat conditions that may be of risk to the mother and baby, need to be monitored in pregnancy, particularly if they are CYP3A4 and CYP2D6 substrates and particularly towards and during the third trimester.

The old 'state of immunosuppression' idea in pregnancy has developed into an understanding of the hugely complex mutual cooperation between the maternal and developing foetal immune systems[281]. However, drug therapy during pregnancy may be essential to suppress existing infections. With HIV, the goal is to prevent transmission of the virus to the child. Whilst antiretroviral therapy with drugs such as AZT, lamivudine, and nevirapine can be successful in viral suppression in pregnancy, the use of mefloquine in preventing malaria can through its CYP3A4 inducing carboxy derivative[282] cause HIV drugs to fall out of the therapeutic window and expose the offspring to the risk of viral transmission[283].

Pregnancy represents a uniquely complex and challenging pharmacokinetic environment, but for obvious reasons it is difficult to gather data on drug clearances from large-scale clinical trials. However, several pharmacokinetic models have been designed[278] that are part of the process of optimizing drug dosages according to the dynamics of the process of pregnancy and its impact on drug disposition and efficacy.

7.6 Smoking

General context

Tobacco still has a very strong hold on humanity, although its well-documented tendency to end the lives of its adherents prematurely, as well as smoking bans in public and workplaces have facilitated its decline in most sections of Western societies except perhaps cautiously rebellious teenagers. Not surprisingly, the tobacco manufacturers have exploited the general aspirational tendencies of Third World economies to market smoking as a habit of raffish affluence and this has resulted in expanding tobacco consumption worldwide. As far back as the 1860s, it has been known that cigarettes are toxic (there is a tirade against the dangers of cigarettes in Dostoevsky's *Crime and Punishment*), so it is frankly bizarre for anyone today to be unaware of tobacco's toxicity. Although incomprehensible to nonsmokers, it has been said to be easier to give up heroin than cigarettes. I remember reading in the Guinness Book of Records as a teenager that the Sheriff of Turku (Johan Katsu) was the first to stop smoking successfully in 1679, which presumably allowed him to feel rather smug for the final month of his life.

To stave off the horror of smokers quitting and affecting their share prices, cigarette manufacturers have long manipulated cigarette design to improve the volatility of nicotine during pyrolysis and in some cases frustrate accurate regulatory analysis of their product to make it appear safer than it actually is when smoked[284]. This, among other features, effectively makes a cigarette as far from a natural product as a sophisticated, consistent and efficient nicotine delivery system can be. Therefore, as a pharmaceutical nicotine delivery system, cigarettes should always have been subject to the rigorous testing a new drug must complete, but they are somehow exempt, as are riot control and restraint chemicals.

Perhaps you might wonder how the tobacco industry is still in existence at all. Almost every year, it appears to be sued successfully for vast punitive damages through the US court system by the disgruntled family of a deceased chain-smoker involving sums of money that comfortably exceed the gross national product of several entire countries. However, tobacco products world-wide make money on a par with that of the pharmaceutical and armaments industries and the tobacco companies are not only thriving[285], but are slowly and almost imperceptibly turning themselves into single product pharmaceutical companies through the marketing of nicotine delivery systems of various types[286]. Of the many organisations that control the vast worldwide business of purveying pleasure chemicals, it is perhaps ironic that even the most seriously ruthless, vicious and viscerally savage ones do not cost us all as much as the tobacco industry does.

The relatively new electronic cigarette/vaping craze has been a heaven-sent opportunity for this industry to 'pharmaceuticalize' and legitimize their nicotine delivery process, but it also allows them to luxuriate in the righteousness of the 'let us help you stop this terrible habit' concept. Hence, they now market even more blatant and only more marginally regulated nicotine delivery systems than a conventional cigarette. With such new devices, we now should consider not only the impact of the nicotine but also the chemical contents of the aerosol formed from vaping. Thankfully, this has been the subject of European Union legislation since 2014[287,288].

Traditional cigarettes contain around 4000 chemicals, of which around 45–50 are carcinogens. Among the few proven human bladder carcinogens, β-naphthylamine and 4-aminobiphenyl are present in tobacco smoke, which keep smoking ahead of occupational amine exposure as the primary cause of bladder cancer worldwide especially in males. Most of the aromatic hydrocar-bon carcinogens induce CYP1A1 and CYP1A2 as well as GSTs Mu and Alpha particularly in the lungs. CYP1A1 contributes little to the clearance of drugs as it is extrahepatic and is subject to polymorphisms, and its inducibility is variable in Caucasians. The key issue for smoking and drug metabolism is CYP1A2.

Smoking and drug clearance

Nearly half of Caucasians are homozygous for CYP1A2*1F, (rs762551, -163C>A)[289], which is highly sensitive to inducers, chiefly polycyclic aromatics (PAHs), although rifampicin and carbamazepine also have an effect. Smoking is the largest source of PAHs, but *1F has a significant effect on accelerating olan-zapine clearance even in non-smokers[290] as mentioned in the previous section on barbecued meat. Potentially, smoking can induce CYP1A2 to the extent that drug clearance may double with respect to nonsmokers for a number of drugs that are wholly or partly cleared by this CYP, which includes tacrine, flutamide, mexiletine, mirtazapine, ramelteon, and of course, several antipsychotics, including olanzapine and clozapine. This becomes an issue if an individual smoker becomes established on a CYP1A2 substrate regime and then abruptly stops smoking. Over two to four weeks, the drug plasma levels will climb, causing major adverse effects[291]. Naturally, patients should be exhorted to inform their health-care practitioners if they do stop smoking and moreover if they switch to vaping[291]. The sudden loss of the PAHs will remove the CYP1A2 induction stimulus and the various additives and flavourings do not appear to have the same inductive effects[291].

Electronic cigarettes: 'vaping'

The various manifestations of electronic (e-) cigarettes have gained popularity extremely rapidly since the last edition of this book, partly in response to their powerful marketing as aids to stop smoking 'proper' cigarettes and partly because vaping is seen as safer than conventional tobacco use. Whilst e-cigarettes are based on solvent evaporation/cooling to form a nicotine aerosol, manufacturers immediately set about adding a plethora of attractive flavourings from the food industry to market their product. One of the additives was diacetyl, the buttery power behind yummy microwave popcorn, although huge amounts of salt also help. Of course, since diacetyl is so wonderful in popcorn, many were surprised to learn that when inhaled, it can cause a devastating lung condition, known as *bronchiolitis obliterans,* first seen in workers in the food industry, where it was duly dubbed 'popcorn lung'[292]. Fortunately, the anticipated flood of popcorn lung victims from vaping did not appear, but the European Union wisely stepped in and banned its inclusion in e-cigarette products[288], alongside the long-standing tobacco additive 'cool as a mountain stream' menthol, dearly beloved of 1970s disaffected teenagers and heroin addicts.

Clearly, the jury is out on whether vaping in all is manifestations is safer than combustible tobacco products, and proponents will sidestep the horrors of oral cancers caused by other noncombustion forms of tobacco[293] by insisting that vaping just contains nicotine, some harmless flavours, and none of the carcinogenic nitrosoamines found in raw and processed tobacco. However, the emerging 'sleeper' role of nicotine as a promotor of carcinogenicity[294] means that many vaping enthusiasts are likely to suffer significant pathology in the years to come.

Smoking and schizophrenia

The schizophrenic's relationship with tobacco may begin much earlier than was first envisaged; mothers who smoke while pregnant are more likely to produce schizophrenic offspring[295]. Indeed, schizophrenia is very strongly associated with cigarette consumption, with 50–60% of sufferers using tobacco, often in serious quantities. Why this happens has been the subject of significant research. At first glance, it appears similar to the situation seen in combatants under fire, that is, a combination of anti-stress and self-medication. Some studies have suggested that not only nicotine, but also cotinine, which has a longer half-life and better brain penetration, may be a strong driver of this process[296]. Others have shown that tobacco use does affect the pharmacology of schizophrenia, with several studies indicating that it actually makes the condition worse. This can involve more aggression and disorders such as akathisia (restlessness), as well as deterioration in cognition, although the impact of smoking remains controversial[297,298].

One thing that is beyond dispute is the doubling of the clearance of clozapine and olanzapine in smokers and any changes in smoking habits will have a commensurate impact on plasma levels[297,298]. With clozapine, if a patient is struggling to stop, or decides to start vaping[299], this can cause their 'induced' dosage (often around 500 mg per day)[297] to be an effective overdose, with the side effects with clozapine for example, appearing as eating massive amounts of food, blurred vision and drowsiness[301]. Schizophrenics are difficult to persuade to stop smoking[299] and bans around hospitals make little impression[299]. Gene testing to determine CYP2A6 status might assist with helping schizophrenics to stop smoking, alongside therapeutic monitoring to minimize the toxicity of the drugs, improve their efficacy and reduce the general stress of the condition.

7.7   Effects of ethanol on drug metabolism

7.7.1   Context of ethanol usage

Around 10% of men in the United Kingdom regularly exceed the healthy limits of ethanol intake, and probably half of those are fully dependent on ethanol. Ethanol is basically a sedative and its ability at modest doses to relax inhibitions in a generally pleasant manner and its easy availability has made it humanity's primary legal source of chemical succour. Of course, it is famously addictive and increasing proportions of women and adolescents are also becoming dependent on a high ethanol intake. It has an unfortunate tendency to promote risk-taking whilst impairing reflexes and physical awareness. This translates to ethanol's central role in more than half of all violent crime and fatal road accidents; the drug also makes many city and town centres in the United Kingdom increasingly hazardous places to venture at night. It is not often appreciated that children are very sensitive to ethanol, as it causes hypoglycaemia particularly in toddlers and intoxication in young children at very small doses. There is more than enough ethanol in a modest helping of homemade sherry trifle to cause ataxia and vomiting in a toddler. Nevertheless, as ethanol is the major legalised mind-altering substance, the vast majority of the population drinks regularly and the ethanol industry is exceedingly keen to bring to our attention more imaginative, palatable, and profitable ways to consume it, although less effort has gone into an effective hangover cure. Since ethanol use is all pervading in most societies, whether it is legal or not, it is important to consider its effects on real-world prescription drug usage.

The question as to whether ethanol intake can affect the clearance and efficacy of prescribed drugs does rather depend on patient honesty – the difference between what a patient tells a doctor he drinks and what he actually consumes can be considerable. The range of ethanol consumption and the accompanied self-delusion is correspondingly wide, from a couple of glasses of beer per week, to bottles of spirits per day. Since drinking habits build gradually, many patients are genuinely unaware that they are exceeding the limit for what is generally considered healthy. Since 2016, in the United Kingdom this has been taken to be approximately two spirit measures ('units' of alcohol in the UK) per day[300]. Around 10 spirit measures per day, or 5 pints of reasonable-strength beer, is well into dependence. The current UK limits for women are now the same as for men but this is more a reflection of finding a safe level for both women and men, rather than trying to emphasize any gender disparities in sensitivity to alcohol-mediated damage. What is clear is that the medical profession, as of late 2018, think that neither the UK government nor the alcohol industry is taking the threat to public health caused by alcohol consumption sufficiently seriously. This ranges from failure to deploy measures to discourage consumption to a lack of understanding of the consequences the combination of alcohol abuse and other health problems, such as obesity[300].

7.7.2 Ethanol metabolism

The main route of metabolism is cytosolic alcohol dehydrogenase (ADH), which is found in many other tissues such as the stomach and gut, but mainly the liver. ADH oxidizes ethanol to acetaldehyde, which is extremely toxic. ADH is a zinc-containing dimer that is under the control of the AhR[301], although it is not inducible. ADH uses NAD as a co-factor to remove the hydrogen atoms that are abstracted from the ethanol to form NADH, which then dissociates from the enzyme. Around 80–90% of ethanol is cleared by the ADH/ALDH system, with the remainder oxidised by CYP2E1, which has a 10-fold lower affinity for the drug, whilst around 10% is eliminated unchanged. The genes that code for ADH are found on chromosome 21 and there are five classes (ADH1-5) in humans. Each class has separate allelic variants: ADH1A ADH1B, and ADH1C are found in the liver, whilst ADH3, ADH4, and ADH1C are found in the gut[302]. ADH is actually saturated at quite low concentrations of ethanol as it has a high affinity for it, although CYP2E1 contributes to the clearance at higher alcohol levels. Overall, ethanol's clearance is not considered as zero order as it was in the past, as it does have a concentration-related component[302].

The acetaldehyde formed by ADH is dealt with by aldehyde dehydrogenase ALDH, which is sourced on a variety of chromosomes. The two most relevant forms are cytosolic ALDH1A1 and mitochondrial ALDH2, which actually function as tetramers, unlike ADH. Rather cunningly, both ALDH variants are fearsomely efficient at processing acetaldehyde-in the case of ALDH2, it is several-hundred-fold more efficient at clearing acetaldehyde than ADH1A*1 is at producing it. You might say that this adaptation signals a healthy evolutionary respect for the systemic toxicity of acetaldehyde. ALDH and ADH are polymorphic and as we all know through experience, the variation in the ability to clear ethanol in humans is enormous, of which, more later.

Ethanol is primarily an inducer of CYP2E1, although CYP1A1 and CYP3A are also affected. The degree of induction will of course be dependent on the patient's usual consumption and it is now believed that CYP2E1 contributes more to ethanol clearance at higher doses and indeed may be of significance in the massively increased capacity for ethanol seen in the heavily dependent[302]. The toxicological consequences of its induction are also linked to chronic hepatotoxicity, as CYP2E1 forms acetaldehyde. Interestingly, although ethanol can be so destructive to the liver, this organ is not the only route of elimination; some investigators have reported substantial extrahepatic clearance of ethanol even in alcoholics.

7.7.3   Ethanol and inhibitors of ALDH

Patients often believe that they should not drink when given antibiotics. This is only true for antibiotics that block ALDH, which as mentioned above, normally clears acetaldehyde formed by ADH. Inhibition of acetaldehyde clearance causes a severe flushing/vomiting/ sweating and nausea effect that is exceedingly unpleasant. There is a surprising list of ALDH inhibiting drugs such as metronidazole, cefoperazone, cefamandole, griseofulvin, chloramphenicol, nitrofurantoin and sulphamethoxazole. Other agents, which can be inhibitory, include isoniazid and sulphonyl ureas. Antabuse, or disulfiram, is intended to block ALDH, so exploiting acetaldehyde toxicity to help the alcoholic stop drinking (Chapter 5.8.8). Given that ethanol clearance is polymorphic, if you have been paying attention you will see that this is a form of phenocopying (Section 7.2.1).

Tuberculosis patients should definitely not drink when on isoniazid, as induced CYP2E1 converts this drug to several reactive species. Isoniazid (INH) is not that well tolerated in otherwise healthy individuals, and it causes high liver enzymes in heavy drinkers and can lead to severe liver damage in those cases[303]. If the patient can stop drinking, then their CYP2E1 induction will fall and their problem with INH should diminish. Of course, many who contract tuberculosis in Western countries are at the margins of society and are highly likely to be very serious drinkers indeed. Hence, they are more likely to sustain liver damage due to the INH/ethanol problem, which means they stop the INH, and the disease reignites, this time resistant to the drug.

7.7.4   Mild ethanol usage and drug clearance

ADH has no bearing on drug clearances, although CYP2E1 is responsible for the clearance of a number of anaesthetics, paracetamol, isoniazid and chlorzoxazone. In non-drinkers, in the absence of ethanol, or in occasional drinkers, ADH metabolises most of the ethanol and CYP2E1 induction is likely to be modest. However, a dose of 0.8 g of ethanol per kilo will inhibit CYP2E1[254] that in the mythical 70 kg man equates to 56 mL. Given that the average single measure of spirits (25 mL) contains about 10 mL of ethanol, then it will take more than five and half measures of spirits to block CYP2E1 consumed all at once, which seems a rather impulsive and expensive gesture. Since this is hopefully not usual behaviour in most of us, we can infer that the clearance of CYP2E1 substrates is not likely to be significantly affected by social ethanol intake. When ethanol is consumed in considerable quantities (a good night, but not to the point of being taken home in a supermarket trolley), then some inhibition might occur and half-lives may be lengthened and their CNS effects pronounced. This effect will be combined with the standard intoxicating effects of ethanol. Therefore, a single moderate-heavy drinking session is likely to change the pharmacokinetics of prescribed CYP2E1 substrates. Ethanol in higher doses can start to inhibit CYP2C9 and it can have a pharmacodynamic interaction with tricyclic antidepressants and benzodiazepines (increased drowsiness) but whilst binge drinking can cause serious anticoagulation issues as it can partially block warfarin clearance[304], current advice suggests that there is no problem at or below the current recommended maximum of two units per day.

7.7.5   Heavy ethanol usage and paracetamol

For those chronically dependent on ethanol their CYP2E1 levels can be three to ten-fold higher than nondrinkers and they would clear CYP2E1 substrates extremely quickly, if they chose to be sober for a while. This may lead to the accumulation of metabolites of the substrates. Regarding the impact of paracetamol, a significant overdose would have little or no effect in a heavy alcohol user, as their CYP2E1 would be fully inhibited and it would be unable to form the hepatotoxic reactive species responsible for liver failure in untreated individuals (Chapter 8.2.3). However, there is a considerable medical literature on newly abstinent alcoholics who may be at risk of suffering paracetamol (acetaminophen)-induced liver toxicity at doses which are considerably lower than that of non-drinkers, which has been styled 'alcohol-paracetamol' (or alcohol-acetaminophen) syndrome[305]. This was believed to be an increased vulnerability to paracetamol-mediated toxicity due to a combination of persisting CYP2E1 induction, low-protective GSH and absence of ethanol, so there is nothing to stop the CYP from forming the toxic reactive species and little to stop its cytotoxicity (Chapter 8.2.3). Indeed, such a view was reinforced by a study that even in healthy individuals liver enzyme levels were elevated after two weeks of 4 g of the drug per day, which is the dose considered safe by most authorities[306].

Several investigations then found that even with this narrow TI drug, abstinent cirrhotics could tolerate 2–3 g of paracetamol daily without significant problems[307,308]. Subsequently, the view that alcoholics were at special risk from paracetamol overdose was not seen to be supported by the evidence and if they overdose on the drug in the United Kingdom, they are treated the same way as everyone else, concerning the antidote dosage criteria[309]. However, some maintain that paracetamol toxicity in newly abstinent alcoholics is nevertheless a significant issue and that alcoholics are vulnerable to staggered overdoses and they could suffer significant liver damage after modest paracetamol intake[309]. Certainly, as alcoholism is primarily a mental illness, alcoholics are extremely unpredictable, so it is highly likely that many would engage in repeated bouts of drinking relapses, followed by staggered overdoses and then sobriety behaviour, which would cause significant hepatic stress. On balance, the argument that the combination of high CYP2E1 induction, general loss of hepatic resilience due to years of chronic abuse and poor health leads to significantly increased vulnerability to paracetamol in comparison with the healthy is persuasive. However, in the light of the current evidence, it seems that whilst it is not unreasonable that physicians will be reluctant to allow paracetamol in abstinent cirrhotic patients, provided the 2–3 g/day levels are not exceeded and the patients are monitored, then sufficient analgesia should be achievable without toxicity for as long as is necessary[307].

7.7.6   Alcoholic liver disease

Predisposition to alcoholism: ADH / ALDH polymorphisms

As mentioned above, alcoholism is a mental illness, although there are many predispositions that can help or hinder the desire to become dependent on the drug. It is generally true that those who pride themselves on their high tolerance to ethanol are more likely to develop alcoholic liver damage than those with little or no tolerance. This is partly because there is more opportunity to build high tolerance and dependence at the level of the brain. Although ethanol is cleared at a leisurely pace compared with nicotine, cocaine, and heroin, craving can nevertheless be promoted by relatively rapid removal of any additive drug. On a practical level, the longer you can stay conscious and able to drink, the addiction can be serviced and the more ethanol-related toxic species will damage you over your drinking lifetime. If you can detoxify the acetaldehyde rapidly and efficiently, you will recover quickly from the previous night's excesses and you are ready to drink more. These factors facilitate dependence and eventually the tediously well-trodden path to health destruction. In contrast, those who become hilariously incoherent and/or spitefully abusive after one or two drinks are less likely to develop such heavy dependence on alcohol. Indeed, those who cannot detoxify acetaldehyde and thus suffer its systemic toxicity after exposure to ethanol are virtually immune to alcoholism, although incredibly, there are exceptions. These general observations are supported by the polymorphic clearance of ethanol by both ADH and ALDH.

In terms of capability, the vast variation in ADH catalytic activity across the human race is mainly due to just a few SNPs that profoundly change the efficiency of the isoforms. ADH1B/*1 is the ADH wild type, with an arginine residue at amino acid position 48. This variant is by far the most common in 80–90% of Caucasians, Afro-Caribbeans, and American Indians, but only 15% of Japanese. ADH1B/ *2 (rs1229984 Arg48his) is common in >85% of Japanese, Chinese, and Koreans, but found in only 5% of Caucasians and has a histidine at position 48. This means lower affinity than wild type, but up to a 40-fold increase in maximal activity, which means it produces large amounts of acetaldehyde[296]. ALD1B/ *3 is found in some African races (~15%) and has a cysteine at position 369 and exhibits low capability[288]. ALD1C/*1 and ALD1C/*2 are found in the Chinese/Japanese/Koreans, as well as Caucasian and black populations. The high acetaldehyde output of ADH1B*2 can exceed the ability of ALDH to process it all, so *2 homozygotic individuals can drink heavily, but they can suffer acetaldehyde flushing and sickness also, which protects them from alcoholism[296]. So it seems that the risk of becoming a full-fledged alcoholic is up to four-fold more likely if you are homozygotic for wild-type ADH1B/*1[311]. The other defective isoforms are found in lower frequencies in alcoholics and cirrhotics.

As mentioned above, most Caucasians and Africans, whatever their variant of ADH, can process acetaldehyde to acetate with the gold standard aldehyde dehydrogenase ALDH2*1, which has the highest activity of all these isoforms and ALDH2*1/*1 is the second essential component for an alcoholic career. The majority of cirrhotics have the maximally efficient combination of ADH1B/*1*1 and ALDH2*1/*1[311]. The variant ALDH2*2 (rs671, Glu504Lys) is virtually nonfunctional and is found in 50–60% of those of Asian extraction[312]. Usually with polymorphic isoforms, the heterozygous version will have some activity, but in this case, both ALDH2*2/*1 and ALDH2*2/*2 are useless[296] and render more than half of those of Asian extraction very sensitive to acetaldehyde poisoning and vulnerable to truly shocking hangovers. So statistically and not surprisingly, possession of this ALDH2*2 variant and even better, ADH1B*2, are both highly protective against the chances of developing alcoholism. However, the combination of a fast ADH1B1 (*2/*2) and null ALDH2*2/*2 can predispose towards the risk of oesophageal cancer when the individual does drink alcohol, due to the very high acetaldehyde exposure[313,314].

Alcoholic liver disease: cirrhosis

Social drinking is of little direct effect on hepatic metabolism, but in chronic heavy drinking, long-term disruption of the liver results. As you probably know, the liver is the toughest organ in the body in terms of its formidable detoxification systems, and very high (6–10 mM) levels of GSH are maintained intracellularly. Consequently, it takes enormous quantities of ethanol to damage the liver beyond repair. An estimate of how much ethanol is needed to kill the liver is interesting.

Assuming the average alcoholic drinks heavily for 20 years prior to liver failure using about a bottle and a half of spirits per day, this amounts to about 2000 litres of pure ethanol in many thousands of drinks ranged against a 1.5 kg liver. That the majority of extremely heavy drinkers die of something other than liver failure (about a third die of cirrhosis) is further testament to hepatic resilience.

In chronic drinkers, ethanol and its major metabolite, acetaldehyde, impair the complex control of hepatic lipids, partly by inhibiting the production of the major lipid regulator adiponectin and other pathways such as sirtuin 1, AMP-activated kinase, and PPARα, as well as the ADH-mediated excess of NADH formation[302]. This causes alcoholic 'fatty liver' or steatosis, which is a well-known staging post to more severe liver disease. Glucuronidation can be disrupted, and sustained high NADH levels affect UDP-glucuronic acid formation. At this stage, there are few symptoms beyond low blood sugar, although hepatic weight can more than double. The condition will progress under pressure from ethanol consumption gradually to alcoholic hepatitis (severe liver inflammation). The effects of the condition become more visible, such as jaundice due to impaired bilirubin clearance[315]. At this stage, if patients stop drinking and reform their lifestyle, eating properly and taking care of themselves, they can walk away intact with a large fund of appalling 'back when I was drinking' stories.

Regarding those who continue to drink, one of the major toxic pressures in alcoholic liver disease at the cellular level appears to be CYP2E1 induction that of course churns out acetaldehyde and large amounts of reactive species (Chapter 3.6.2). Therefore, whilst the alcoholic is asleep, gradually CYP2E1 may overcome its ethanol inhibition as levels of the agent fall. The combination of the formation of various toxic hydroxyethyl radicals from ethanol and the stream of reactive oxygen radicals from the CYP itself, all must be detoxified by the hepatocyte. CYP2E1 is also induced in mitochondria as well as the SER and excess acetaldehyde formed in the cellular 'engine rooms' can cause mitochondria to release apoptotic factors that can induce cell death[315]. In addition, damage to mitochondrial ATP production can cause cell death through cellular necrosis. Ethanol abuse and CYP2E1 induction together lead to long-term oxidative stress. It is not yet clear whether CYP2E1 variant and level of induction have a significant bearing on the gradual oxidative destruction of the liver in alcoholism as the evidence is not consistent across ethnic groups[316]. However, it does seem that there is a relationship, as this could explain why some individuals can survive decades of very heavy drinking, whilst others die within a few years. The oxidative stress caused by alcohol metabolism also gradually erodes GSH maintenance, as thiols are required to detoxify reactive species formed by acetaldehyde. As alcoholic liver disease progresses, liver enzymes, such as serum ALT and AST, climb steadily from 10–50 times the normal limits towards the thousands[308].

Chronic acetaldehyde toxicity leads to hepatocyte death and a sustained inflammation phase that drives the progressive replacement of hepatocytes with fibrotic connective tissue. This is cirrhosis, where nodules and lumps of this fibrous tissue appear all over the liver and disrupt its blood circulation and the removal of bile[315]. The liver hardens to the touch and shrinks, developing varicose veins known as vascular 'spiders'. At this stage, provided they stop drinking, there is still a chance of survival with some residual liver function. Although there are other causes of cirrhosis (hepatitis, drug therapy, genetic conditions), over 90% of cases are sustained by ethanol consumption. Biochemically, the liver fights a losing battle as production of essential proteins gradually falls as the drinking progresses the disease and cholesterol, sugar, and triglyceride metabolism is compromised. Elsewhere, the gut is damaged by ethanol intake to the point that it becomes excessively porous and even undigested food particles enter the blood, which are recognised by the immune system. This leads to virtually permanent gut inflammation and variable drug absorption. Heavy alcohol use also is a very strong risk indicator for hepatocarcinoma[315].

As cirrhosis progresses, hepatic back pressure becomes so high that the blood from the portal vein has difficulty in entering the liver to the point that blood starts to leave the liver and run back down the portal vein. This causes swollen varices, or varicose veins in the oesophagus and stomach, which can be at high risk of bleeding and may require treatment[317]. Fluid is forced out into the tissues, causing abdominal ascites. The circulatory system adapts to the problem of the liver back pressure through a gradual process known as portosystemic shunting, where the vasculature actually bypasses the liver. Approaching end-stage liver failure, serum ALT and AST levels are well in the thousands and bilirubin appears in the blood in quantity leading to severe jaundice as the liver effectively waves the physiological white flag in desperation. Cirrhotic liver damage is permanent and is graded clinically though the Child-Pugh score[318], which predicts the individual's outcome, but the five-year death rates even if the drinker stops can be more than 50% from cardiac arrest, coma, and malnutrition (alcoholics never eat properly). Alcohol abusers in end-stage liver failure have particularly poor outcomes in hospital intensive care units, especially in hepatorenal syndrome, where kidney failure occurs due to the liver disease. Liver cirrhosis is actually responsible for 1 in 50 of all global deaths[319]. At this stage, the only therapy is a new liver, provided the alcoholic has been 'dry' for six months or more, or is exceedingly rich and famous. The decision whether to offer a liver transplant to an alcoholic is as complex as it is controversial[320]. It could be argued that alcoholism is unlike many other addictions, as there are more than an average number of warning signs and chances to 'get off the bus' as the liver and the body in general both put up an exceedingly stout long-term defence against ethanol's cytotoxicity and solvent-like effects. Tragically, with many confirmed alcoholics it is probably easier to rescue a fish from the sea than to stop them drinking.

Women and alcoholic liver disease

Women are much more vulnerable to ethanol damage and on average die in half the time it generally takes for a male alcoholic to drink himself to death. The numbers of reformed, suitably grateful, and elderly alcoholics in the entertainment industry seem to be mostly male. Indeed, one of the first to admit his alcoholism publically, Dick Van Dyke, was born in 1925 and is still with us at the time of writing. Women consume much less alcohol than men do, yet they show a faster progression through the various stages of alcohol dependency and eventually cirrhosis[321]. One study showed women consumed about 14,000 drinks to induce cirrhosis, whilst men required more than 45,000 to achieve the same effect[322]. Another showed more than double the amounts of ethanol were needed to cause cirrhosis in men, compared with women[321]. There are some distributional differences in women, such as the distribution of ethanol in total body water only, so it is suggested that the greater fat content of females means that blood ethanol levels are higher than men of similar weight and age. In addition, female stomach ADH is less effective than that of men, which also promotes the entry of more ethanol into the blood[302].

The causes of this gender disparity are not exclusive to ethanol toxicity. Women are more vulnerable to acute liver failure and drug-induced liver damage in general, so their susceptibility to alcohol is not surprising in this context[321]. Certainly female susceptibility to alcoholic liver failure is probably multifactorial, mainly linked to the immune system as controlled by female hormones, alongside a combination of excess reactive species formation, possibly linked to CYP2E1 induction, and compromised detoxification in terms of GSTs and thiols. Women do less well after transplants than men also[321] and they are less likely to seek help than men to control their drinking[323].

7.7.7   Effects of cirrhosis on drug clearance

Main features

The global effects of cirrhosis on the liver include the dramatic reduction in blood flow (indeed, the total derangement of the portal system), the loss of functional hepatocytes, and changes in biotransforming enzyme expression. CYP1A2 activity drops steeply, with less than 20% of normal levels, whilst CYP2C9 and CYP2C19 decay to about 60% and CYP2E1, about 40% of normal. CYP3A4/5 holds up to about 70–80% of normal, whilst CYP2D6 was unchanged[324]. With conjugation systems, SULTs are diminished in expression, whilst UGTs can be more robust, with oxazepam clearance unchanged, but AZT, morphine, and lamotrigine conjugation decline in cirrhosis[325]. What is also interesting is that not only are the nuclear/cytoplasmic receptor expressions unchanged in cirrhosis (PXR, AhR, and CAR), but the variation in different cirrhotic livers can be often 10–20 times greater than that of normal livers[324]. Indeed, such unpredictability and the other factors unite to account for the high level of adverse reactions suffered by cirrhotic patients during drug therapy[319,326]. Other liver housekeeping functions, such as drug-binding protein production are seriously compromised, which can lead to increases in available free drug, which will contribute to pushing drug levels out of the therapeutic range towards toxicity. This situation is compounded by up to a fifth of cirrhotic patients during hospital stays being dosed inappropriately or incorrectly.

Overall, it is believed that many of these issues can be addressed, although often there is a lack of access to appropriate clinical advice or information outside specialised hospital units. Several groups have attempted to address these this knowledge gap and to improve the safety and efficacy of drugs used to treat cirrhosis and its consequences[319,326], as well as other more widely used agents. Whilst intravenous drug administration is more predictable in cirrhosis, the impact of the condition on oral drug effect is considerably harder to predict, and this can be addressed by considering drugs in terms of their extraction and thus their presystemic metabolism.

High-extraction drugs

Regarding the clearance of high-extraction drugs, it was established even before the 1980s that as cirrhosis compromises blood flow, the clearance of such drugs will be severely reduced. The portosystemic shunt effect can contribute significantly in denying the cirrhotic liver exposure to the drug. The beta-blockers (propranolol, metroprolol, labetalol), as well as pentazocine, lignocaine, and opiates such as pethidine and morphine hepatic extraction can be reduced by 50% or more. The key issue here is that this results in major increases in bioavailability after an oral dose, which in turn means plasma concentrations increase by up to several-fold. In the case of the sedative chlormethiazole, a study from the 1970s clearly shows that cirrhosis increases its bioavailability by 10-fold by wiping out the first pass of the drug[327]. The dangers of this drug were highlighted by its role in the death of Keith Moon, one of rock's truly great drummers. So for alcoholics, it is recommended to reduce oral dosing by 5–10 times and halve the dose intravenously.

Low-extraction drugs

The clearance of low-extraction drugs (phenytoin, carbamazepine, warfarin) is more about the intrinsic capability of the biotransforming enzymes with particular drugs and the fraction unbound available for metabolism. Such drugs are usually quite high in bioavailability, and this will not necessarily change markedly during cirrhosis[325]. As mentioned already, protein binding falls due to the failure of the liver to make enough proteins, and this increases the free fraction of the drug, which theoretically could increase the clearance. However, the general biotransforming enzyme deterioration will be very significant and, depending on the Child-Pugh score, the liver may not be able to take advantage of the increase in free drug. Indeed, in cirrhotic patients, the most potentially hazardous drugs are those with low hepatic extraction and narrow therapeutic indices, such as warfarin[325]. Maintenance doses should be halved, as clearances may fall by more than 50–70%[325].

Current recommendations in cirrhosis

Some authorities have classified various drugs in terms of their safety according to different Child-Pugh scores, alongside suitable recommendations of appropriate doses based on published literature[307]. Some drugs have been contraindicated completely, such as NSAIDs[307]. These often low-extraction drugs are well known for the gastrointestinal and renal toxicity even in the healthy, and they are poorly cleared in cirrhotics who often have significant renal and gut damage, so they are potentially very hazardous[326]. Other agents were contraindicated on the grounds of high extraction status, such as many calcium antagonists, specifically isradipine, and barnidipine, as well as drugs in other classes, such as nebivolol, pantoprazole, and atorvastatin[319,326]. There is also evidence that opiates are problematic in cirrhotics, either because they are not activated to morphine (codeine), or are very slowly cleared, or that they cause hepatic encephalopathy[307]. Overall, there is now much more information available to manage cirrhotic patients so as to optimize drug therapy and prevent drug toxicity[328].

7.8 Artificial livers

Liver failure

Liver failure can be broadly split into two groups. The first, acute liver failure (ALF) patients, will have experienced some major issue that has caused the collapse of liver function, such as a paracetamol overdose (about half the cases), DILI (drug-induced liver injury), or a fulminant infection or immune-mediated damage. With ALF, in many cases, the rest of the patient might be otherwise healthy, although obviously this will only be a few days without a functioning liver. The second group is classed as acute on-chronic liver failure (ALCF), where the problem is linked to a chronic condition, such as alcoholism or the consequences of a previous infection. When ALCF is diagnosed, only a third of patients will last a month[329]. Liver transplant is still the only long-term option for both categories of patients, but naturally, livers are scarce. In the short term, extracorporeal liver support (ECLS) can theoretically give the ALF patient's organ enough time to demonstrate those genuinely awe-inspiring 'Christine'-like[330] powers of hepatic recovery. ECLS should be able to carry both ALF and ALCF patients for such time as a transplant liver does eventually appear in a helicopter. Ideally, such systems need to detoxify, synthesize proteins, and help maintain blood chemistry.

Artificial systems

On a basic level, the systems use microfilters and exploit the ability of albumin to bind various endogenous and exogenous toxins as required in renal dialysis machines. SPAD (single pass albumin dialysis) operates by routing blood past an albumin-rich fluid that binds the toxins and/or drugs and then it is discarded, like in total-loss lubrication systems in very early cars and, indeed, like the Rover SD1 V8 I owned in the late 1980s.

A more sophisticated version was invented in 1993 called MARS (molecular adsorbent recirculating system) that uses a combination of albumin and activated charcoal to bind and remove drugs, hepatotoxins, and accumulating metabolic waste (bilirubin and ammonia). In MARS, the albumin system is regenerated by a system of adsorbents that detach the toxins from the blood to the albumin and out to the adsorbents. The SPAD system can be a useful temporary support, although some studies have mentioned that it does not affect clinical outcome in liver failure patients. The Prometheus system, or FPSA (fractional plasma separation adsorption), is a further innovation. This is a complete unit that separates the albumin-mediated removal of toxins from the clearance of water-soluble toxins. It uses the patients' own albumin to achieve this, whilst sparing fibrinogen for adequate clotting. The Prometheus unit uses two high-affinity adsorbers to clear off the toxins, and then the albumin is returned to the patient's circulation. The unit then removes water-soluble toxins using high-flux haemodialysis[329].

Bioartifical systems

These contain a live cellular matrix that is theoretically intended to perform serious biotransformational activity as well as all the basic detox and housekeeping activities performed by the other devices. Human hepatocytes are, of course, ideal, but for obvious reasons they are scarce and expensive, difficult to acquire in sufficient quantity, and even harder to culture in a practical clinical environment. Instead, transformed cell systems (C3A hepatoblastoma), or porcine hepatocytes are used, which naturally have their own risks and drawbacks. In all attempts to reproduce functioning tissues from cells, recapitulating the relationship between the functioning cells and the intercellular matrix is one of the most challenging aspects of bioengineering, and there is a long way to go to make these systems as practical, efficient, and affordable. These systems include ELAD (extracorporeal liver assist system) and HepatAssist[329].

Effectiveness

Human trials with both the artificial and bioartificial systems on the whole have not been very impressive. Although early studies with MARS were promising and it does regulate the major toxins, a large trial did not show increased survival or benefit beyond existing therapy[331]. Prometheus has shown some promise, but the bioartificial systems are yet to dramatically impress[329]. Overall, cruder and more basic measures such as high-volume plasma exchange (8–12 L) seem to be more successful, particularly in immune-mediated conditions. Overall, these systems require major investment and some significant basic science advance before they will supersede conventional therapy for liver failure patients.

7.9 Effects of disease on drug metabolism

As seen with cirrhosis, any disease state, caused either by infection, malignancy or autoimmune factors exerts very potent effects on the architecture, ultrastructure and cellular health of the liver, as well as other metabolising organs like the gut. Aside from the physical impact on the liver and its blood flow, the full extent of the power of the immune system to modulate biotransformation is gradually emerging and is even more clinically relevant with the approval of more biological (cytokine/antibody) drugs.

The immune system exerts its control on cellular activity through a host of different cytokines and factors, including tumour necrosis factors (TNF-α and the various interferons and interleukins (IL-1β, IL-2, and IL-6)[332]. These factors downregulate not only the CYPs but all aspects of the biotransformational system, including the transporters and conjugation apparatus[333]. From a patient's perspective, many inflammatory and malignancy conditions are very gradual in onset and it may have taken many years for their drug metabolising systems to depart from the broad norm. Hence, when they are given a standard drug course, there is a significant risk of accumulation and adverse reactions in the absence of therapeutic monitoring. It is now understood that the immune system's response is usually, but as we shall see in Chapter 8 (Section 8.4), not always, commensurate to the challenge. Inflammatory conditions such as psoriasis show mild changes in CYP suppression-mediated drug clearance, whilst the impact of surgery, liver cancers, HIV infection, influenza, rheumatoid arthritis, and flaring episodes of Crohn's disease can cause drug AUCs to increase very significantly (2- to 5-fold)[332].

Rheumatoid arthritis (RA), as well as many cancers are accompanied by high expression of IL-6. This cytokine can suppress the expression of most major CYPs (1A2, 2C9/19, 2B6, 3A4/5), although not CYP2D6, which is controlled by other factors such as interferons. In RA IL-6 can cause 30–50% reductions in drug clearances, which can be rather impressively restored by the inhibition of IL-6 with the monoclonal antibody sirukumab[334]. Hepatitis, either from the B or C viruses, can result in chronic liver disease, such as cirrhosis or even liver cancer. Hepatitis causes significant changes to the liver's general health as well as the suppression of the CYPs. Also, treatment involving injection of interferons can in itself cause 20–40% reductions in drug clearances[332]. In hepatocellular cancer, there is a more severe collapse in CYP expression even compared with cirrhosis[324].

There are also significant deficits in the GSH detoxification system in alcoholics, diabetics and those suffering from HIV, suggesting that these patients are much more at risk than healthy individuals of hepatic oxidative stress and vulnerability to reactive species formation by drugs. Effectively, the 'window' where reactive species generated by oxidative metabolism of drugs or environmental toxins have the opportunity to bind irreversibly within cells before they can be detoxified may well be considerably longer in these patients, thus predisposing them to greater hepatotoxic risk compared with healthy patients. This oxidative stress process is probably linked with the development of hepatic cancers in the first place.

Indeed, in diabetes, oxidative stress is complicated by CYP2E1 induction, but this only seems to occur in obese type II diabetics, who can clear the CYP2E1 marker chlorzoxazone more than 2.5 times faster than healthy individuals can. In animal studies, CYP2E1 induction in untreated diabetic individuals disappears once they receive insulin therapy, and it has been suggested that insulin is part of the control of CYP2E1 expression. It is plausible that type II diabetics might be more at risk from paracetamol overdose and carcinogenicity linked with CYP2E1 metabolism. There does not seem to be a CYP2E1 inductive effect in well- to moderately controlled type I diabetics.

7.10 Summary

Overall, there are a large number of factors that can influence drug metabolism, ranging from those that increase clearance to cause drug failure, as opposed to those that prevent clearance thus leading to toxicity. In the real world, it is often impossible to delineate the different conflicting factors that result in net changes in drug clearance that cause a drug to fall out of, or climb above, the therapeutic window. It may only be possible clinically in many cases to try to change what appears to be the major cause to bring about a resolution of the situation to optimize drug levels. Figure 7.5 tries to form a summary picture of the major influences on drug clearance.

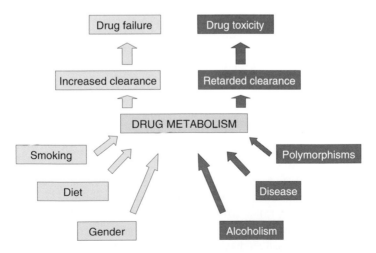

Figure 7.5 Generalised scheme of factors that influence drug metabolism, leading to changes in clearance, which are responsible for extremes of drug response

References

1. Zanger UM, Klein K. Pharmacogenetics of cytochrome P450 2B6(CYP2B6): advances on polymorphisms, mechanisms, and clinical relevance Front. Genet. 4, 24, 1–12, 2013.
2. Gaedigk A, Ingelman-Sundberg M, Miller NA, et al. PharmVar steering committee. Clin. Pharm & Ther. doi:10.1002/cpt.910Full Text, 2017.
3. The Pharmacogene Variation (PharmVar) Consortium: Incorporation of the Human Cytochrome P450 (CYP) Allele Nomenclature Database. www.pharmvar.org/.
4. Shah RR, Smith RL. Addressing phenoconversion: the Achilles' heel of personalised medicine. Br. J. Clin. Pharmacol. 79, 222–240, 2014.
5. US NCBI Clustered RefSNPs (rs) and Other Data Computed in House. https://www.ncbi.nlm.nih.gov/books/NBK44417/pdf/Bookshelf_NBK44417.pdf dbSNP Short Genetic Variations. Https://www.ncbi.nlm.nih.gov/projects/SNP/snp_ref.cgi?rs=2279343
6. Ford NF. The metabolism of Clopidogrel: CYP2C19 Is a minor pathway. J. Clin. Pharm 56, 1474–1483, 2016.
7. Holmberg MT, Tornio A, Neuvonen M, et al. Grapefruit juice inhibits the metabolic activation of clopidogrel. Clin. Pharmacol.Ther. 95, 307–313. 2014
8. Klein DJ, Thorn CF, Desta Z, et al. PharmGKB summary: tamoxifen pathway, pharmacokinetics. Pharmacogenetics and genomics. Pharmacogenet. Genom. 23, 643–647, 2013.
9. Lee CL, Low SK, Fox P, et al. Simplified cyp2d6 metaboliser phenotype categorisation of patients treated with tamoxifen: role for endoxifen level monitoring Asia-Pacific Journal of Clinical Oncology. 12, 59, 1743–7555, 2016.
10. Lim JSL, Sutiman N, Muerdter TE, et al. Association of CYP2C19*2 and associated haplotypes with lower norendoxifen concentrations in tamoxifen-treated Asian breast cancer patients. Br. J. Clin. Pharmacol. 81, 1142–1152, 2016.
11. Brauch H, Schwab M. Prediction of tamoxifen outcome by genetic variation of CYP2D6 in post-menopausal women with early breast cancer. Br. J. Clin. Pharmacol. 77, 695–703, 2014.
12. Brauch H, Schroth W, Schroth W, et al. Clinical relevance of CYP2D6 genetics for tamoxifen response in breast cancer. Breast Care 3, 43–50, 2008.
13. Zembutsu H, Nakamura S, Akashi-Tanaka S, et al. Significant effect of polymorphisms in CYP2D6 on response to tamoxifen therapy for breast cancer: a prospective multi-center study. Clin Canc. Res. 23, 2020–2026, 2017.
14. Goetz MP. Update on CYP2D6 and Its Impact on tamoxifen therapy. Clin. Adv. Hematol. & Oncol. 8, 536–538, 2010.
15. Ezzeldin N, El-Lebedy D, Darwish A, et al. Genetic polymorphisms of human cytochrome P450 CYP1A1 in an Egyptian population and tobacco-induced lung cancer Gen. Env. 39, 7, 1–8, 2017.
16. Iscan M, Ada AO et al. Cytochrome P-450 polymorphisms and clinical outcome in patients with non-small cell lung cancer. Turk. J. Pharm. Sci. 14, 319–323, 2017.
17. Liu D, Qin S, Ray B. et al. Single nucleotide polymorphisms (SNPs) distant from xenobiotic response elements can modulate aryl hydrocarbon receptor function: SNP-dependent CYP1A1 induction. Drug Metab. Disp. 46, 1372–1381, 2018.

18. Ding B, Sun W, Han S, Cytochrome P450 1A1 gene polymorphisms and cervical cancer risk A systematic review and meta-analysis. Medicine 97 (e0210), 1–8, 2018.

19. Lally J, Gaughran F, Timms P, et al. Treatment-resistant schizophrenia: current insights on the pharmacogenomics of antipsychotics. Pharmacogen. Personaliz. Med. 9, 117–129 2016.

20. Fulton JL, Dinas PC, Carrillo AE. et al. Impact of genetic variability on physiological responses to caffeine in humans: a systematic review. Nutrients 10, 1373, 2018.

21. Kootstra-Ros JE, Smallegoor W, van der Weide J et al. The cytochrome P450 CYP1A2 genetic polymorphisms 1F and 1D do not affect clozapine clearance in a group of schizophrenic patients. Ann Clin Biochem. 42, 216–219, 2005.

22. Ayari I, Fedeli U, Saguem S, et al. Role of CYP1A2 polymorphisms in breast cancer risk in women. Mol. Med. Rep. 7, 280–286, 2013.

23. Gehrmann M, Schmidt M, Brase JC, et al. Prediction of paclitaxel resistance in breast cancer: is CYP1B1*3 a new factor of influence? Pharmacogen. 9(7), 969–974, 2008.

24. Sowers MF, McConnell D, Mary L. Jannausch ML, et al. Estrogen metabolites and their relation to isoprostanes as a measure of oxidative stress. Clin Endocrinol (Oxf). 68, 806–813, 2008.

25. Beuten J1, Gelfond JA, Byrne JJ, et al. CYP1B1 variants are associated with prostate cancer in non-Hispanic and Hispanic Caucasians. Carcinogen. 29(9), 1751–1757, 2008.

26. Tulsyan S, Chaturvedi P, Singh AK, et al.,Assessment of clinical outcomes in breast cancer patients treated with taxanes: multi-analytical approach. Gene 543, 69–75, 2014.

27. Zhang HT, Li L, Xu Y. CYP1B1 Polymorphisms and susceptibility to prostate cancer: a meta-analysis. PLoS One. 8, e68634, 2013.

28. Chang I, Mitsui Y1, Fukuhara S, Gill A, et al. Loss of miR-200c up-regulates CYP1B1 and confers docetaxel resistance in renal cell carcinoma. Oncotarget 2015.

29. López-Flores LA, Pérez-Rubio G, Falfán-Valencia R. Distribution of polymorphic variants of CYP2A6 and their involvement in nicotine addiction EXCLI J. 16, 174–196, 2017.

30. Jensen KP, DeVito EE, Valentine G, et al. Intravenous Nicotine Self-Administration in Smokers: Dose–Response Function and Sex Differences. Neuropsychopharm. 41(8), 2034–2040, 2016.

31. Tanner JA, Tyndal RF. Variation in CYP2A6 Activity and Personalized Medicine. J. Pers. Med. 7, 18, 1–29, 2017.

32. Gold AB, Lerman C. Pharmacogenetics of smoking cessation: role of nicotine target and metabolism genes. Hum. Genet. 131, 857–876, 2012.

33. Chivers LL, Higgins ST, Heil SH, et al. Effects of initial abstinence and programmed lapses on the relative reinforcing effects of cigarette smoking. J. Appl. Behav. Anal. 41(4), 481–497, 2008.

34. Watanabe T, Saito T, Gutiérrez Ricoa EM, et al. Functional characteriszation of 40 CYP2B6 allelic variants by assessing efavirenz 8-hydroxylation. Biochem. Pharmacol. 156, 420–430, 2018.

35. Kitabi EN, Minzi OMS, Mugusi S et al. Long-term efavirenz pharmacokinetics is comparable between Tanzanian HIV and HIV/ Tuberculosis patients with the same CYP2B6*6 genotypeSci. Rep. 8, 16316, 1–12, 2018.

36. Ariyoshi N, Ohara M, Kaneko M, et al. Q172H Replacement Overcomes Effects on the Metabolism of Cyclophosphamide and Efavirenz Caused by CYP2B6 Variant with Arg262. Drug Metab. Disp. 39, 2045–2048, 2011.

37. Nakajima M, Komagata S, Fujiki Y, et al. Genetic polymorphisms of CYP2B6 affect the pharmacokinetics/pharmacodynamics of cyclophosphamide in Japanese cancer patients. Pharmacogenet Genom. 17, 431–445, 2007.

38. Kwara A, Lartey M, Sagoe KW, et al. CYP2B6 (c.516G→T) and CYP2A6 (*9B and/or *17) polymorphisms are independent predictors of efavirenz plasma concentrations in HIV-infected patients. Br. J. Clin. Pharmacol. 67(4), 427–436, 2009.

39. Carr DF, Chaponda M,Cornejo Castro EM, et al., CYP2B6 c.983T>C polymorphism is associated with nevirapine hypersensitivity in Malawian and Ugandan HIV populations. J. Antimicrob. Chemother. 69, 3329–3334, 2014.

40. Eap, C.B., Buclin, T., Baumann, P. Interindividual variability of the clinical pharmacokinetics of methadone: implications for the treatment of opioid dependence. Clinical Pharmacokinetics, 41, 1153–1193, 2002.

41. Ahmad T, Sabet S, Primerano DA, et al. Tell-tale SNPs: The role of CYP2B6 in methadone fatalities. J. Anal. Toxicol. 41(4): 325–333, 2017.

42. Dobrinas M1, Crettol S, Oneda B, et al. Contribution of CYP2B6 alleles in explaining extreme (S)-methadone plasma levels: a CYP2B6 gene resequencing study. Pharmacogenet. Genom. 23, 84–93, 2013.

43. Gadel S, Crafford A, Regina K, et al. Methadone N-demethylation by the common CYP2B6 allelic variant CYP2B6.6. Drug Metab. Disp. 41, 709–713, 2013.

44. Kaspera R, Naraharisetti SB, Tamraz B, et al. Cerivastatin *in vitro* metabolism by CYP2C8 variants found in patients experiencing rhabdomyolysis. Pharmacogenet. Genom. 20, 619–629, 2010.

45. Backman JT, Filppula AM, Niemi M, et al. Role of cytochrome P450 2C8 in drug metabolism and interactions. Pharmacol. Rev. 68, 168–241, 2015.

46. Cliff J, Jorgensen AL, Lord R, et al. The molecular genetics of chemotherapy-induced peripheral neuropathy: A systematic review and meta-analysis. Crit. Rev. Oncol. Hematol. 120, 127–140, 2017.

47. Silvado CE, Terra VC, Twardowschy CA. CYP2C9 polymorphisms in epilepsy: influence on phenytoin treatment Pharmacogenom. Pers. Med. 11, 51–58, 2018.

48. Ross KA, Bigham AW, Edwards M, et al. Worldwide allele frequency distribution of four polymorphisms associated with warfarin dose requirements. J. Hum. Genet. 55, 582–589, 2010.

49. Liu Y, Jeong H, TakahashiH, et al. Decreased warfarin clearance associated with the CYP2C9 R150H (*8) polymorphism. Nature 91(4), 660–665, 2012.

50. Dagenais R, Wilby KJ, Elewa H et al. Impact of genetic polymorphisms on phenytoin pharmacokinetics and clinical outcomes in the Middle East and North Africa region. Drugs R. D. 17(3), 341–361, 2017.

51. Wu X, Liu W, Zhou W, et al. Association of CYP2C9*3 with phenytoin-induced Stevens-Johnson syndrome and toxic epidermal necrolysis: A systematic review and meta-analysis. J. Clin. Pharm. Ther. 43, 408–413, 2018.

52. Kahn SE, Haffner SM, Heise MA, et al. Glycemic durability of rosiglitazone, metformin, or glyburide monotherapy. N. Engl. J. Med. 355, 2427–2443, 2006.

53. Castelan-Martinez OD, Hoyo-Vadillo C, Bazan-Soto TB et al. CYP2C9*3 gene variant contributes independently to glycaemic control in patients with type 2 diabetes treated with glibenclamide. J. Clin. Pharm. Ther. 43, 768–774, 2018.

54. Figueiras A, Estany-Gestal A; Aguirre C, et al. CYP2C9 variants as a risk modifier of NSAID-related gastrointestinal bleeding: a case-control study. Pharmacogenet. Genom. 26, 66–73, 2016.

55. Brenner, SS; Herrlinger, C; Dilger, K; et al. Influence of age and cytochrome P4502C9 genotype on the steady-state disposition of diclofenac and celecoxib. Clin. Pharmacokinet. 42, 283–292, 2003.

56. Siu YA, Hao MH, Dixit V, et al. Celecoxib is a substrate of CYP2D6: Impact on celecoxib metabolism in individuals with CYP2C9*3 variants. Drug Metab Pharmacokinet. 33, 5, 219–227, 2018.

57. Lee HJ, Kim YH, Kim SH, et al. Effects of CYP2C9 genetic polymorphisms on the pharmacokinetics of zafirlukast. Arch. Pharm. Res. 39, 1013–1019, 2016.

58. Deng J, Vozmediano V, Rodriguez M. Review genotype-guided dosing of warfarin through modeling and simulation. Eur. J. Pharm. Sci. 109, S9–S14, 2017.

59. Teichert M, Eijgelsheim M, Uitterlinden AG, et al. Dependency of phenprocoumon dosage on polymorphisms in the VKORC1, CYP2C9, and CYP4F2 genes. Pharmacogenet. Genom. 21, 26–34, 2011.

60. www.warfarindosing.org

61. Huang B, Cui DJ, Ren Y, Han B, Yang DP, Zhao X. Effect of cytochrome P450 2C19*17 allelic variant on cardiovascular and cerebrovascular outcomes in clopidogrel-treated patients: A systematic review and meta-analysis. J. Res. Med. Sci. 22, 109, 2017.

62. Park S, Hyun YJ, Kim YR, et al. Effects of CYP2C19 Genetic Polymorphisms on PK/PD responses of omeprazole in Korean healthy volunteers. J. Kor. Med. Sci. 32, 729–736, 2017.

63. Ramsjö M, Aklillu E, Bohman L, et al. CYP2C19 activity comparison between Swedes and Koreans: effect of genotype, sex, oral contraceptive use, and smoking. Eur. J. Clin. Pharm. 66, 871–877, 2010.

64. Hashemizadeh Z, Malek-Hosseini SA, Badiee P. Prevalence of CYP2C19 genetic polymorphism among normal people and patients with hepatic diseases. Int. J. Org. Trans. Med. 9, 27–33, 2018.

65. Justenhoven C1, Hamann U, Pierl CB, et al. CYP2C19*17 is associated with decreased breast cancer risk. Breast Canc. Res. Treat. 115, 391–396, 2009.

66. Kuo CH, Lu CY, Shih HY, et al. CYP2C19 polymorphism influences Helicobacter pylori eradication Wor. J. Gastroenterol. 20, 16029–16036, 2014.

67. Hunt, R. H.; Armstrong, D.; Yaghoobi, M.; et al. The pharmacodynamics and pharmacokinetics of S-tenatoprazole-Na 30 mg, 60 mg and 90 mg vs. esomeprazole 40 mg in healthy male subjects. Aliment. Pharm. & Ther. 31, 648–657, 2010.

68. Savarino E, Ottonello A, Martinucci I, et al. Ilaprazole for the treatment of gastroesophageal reflux. Exp. Opin. Pharmacother. 17, 2107–2113, 2016.

69. Wang D, Yang XH, Zhang JD, et al. Compared efficacy of clopidogrel and ticagrelor in treating acute coronary syndrome: a meta-analysis. BMC Cardiovasc. Dis. 18, 217, 2018.

70. Chouchene S, Dabboubi R, Raddaoui H, et al. Clopidogrel utilisation in patients with coronary artery disease and diabetes mellitus: should we determine CYP2C19*2 genotype? Eur. J. Clin. Pharmacol. 74, 1567–1574, 2018.

71. Wang JY, Zhang YJ, Li H, et al. CRISPLD1 rs12115090 polymorphisms alters anti-platelet potency of clopidogrel in coronary artery disease patients in Chinese Han. Gene 678, 226–232, 2018.

72. Strisciuglio T, Franco D, Di Gioia G, et al. Impact of genetic polymorphisms on platelet function and response to anti platelet drugs. Cardiovasc. Diagn. Ther. 8, 610–620, 2018.

73. Collet JP, Hulot JS, Cuisset T,. et al. Genetic and platelet function testing of anti-platelet therapy for percutaneous coronary intervention: the ARCTIC-GENE study. Eur J Clin Pharmacol. 71, 1315–1324, 2015.

74. Clinical Pharmacogenetics Implementation Consortium (CPIC). https://cpicpgx.org/

75. Klieber M, Oberacher H, Hofstaetter S, et al. CYP2C19 Phenoconversion by routinely prescribed proton pump inhibitors omeprazole and esomeprazole: clinical implications for personalized medicine. J. P.E.T. 354, 426–430, 2015.

76. Faggioni M, Baber U, Chandrasekhar J, et al. Use of prasugrel vs clopidogrel and outcomes in patients with and without diabetes mellitus presenting with acute coronary syndrome undergoing percutaneous coronary intervention. Int. J. Cardiol. 275, 31–35, 2019.

77. Mangiacapra F, Pellicano M, Di Serafino L, et al. Platelet reactivity and coronary microvascular impairment after percutaneous revascularization in stable patients receiving clopidogrel or prasugrel. Ateroscler. 278, 23–28, 2018.

78. Elewa H, Wilby KJ, A Review of Pharmacogenetics of Antimalarials and Associated Clinical Implications. Eur. J. Drug Metab. Pharmacokinet. 42, 745–756, 2017.

79. Helsby NA, CYP2C19 and CYP2D6 genotypes in Pacific peoples. Br J Clin Pharmacol 82 1303–1307, 2016.

80. Liao K, Liu Y, Ai CZ, et al. The association between CYP2C9/2C19 polymorphisms and phenytoin maintenance doses in Asian epileptic patients: A systematic review and meta-analysis. Int. J, Clin. Pharm. Ther. 56, 337–346, 2018.

81. Sienkiewicz B, Urbaniak-Kujda D, Dybko J, Influence of CYP2C19 Genotypes on the Occurrence of Adverse Drug Reactions of Voriconazole among Hematological Patients after Allo-HSCT. Pathol. Oncol. Res. 24, 541–545, 2018.

82. Moriyama B, Obeng AO, Barbarino J, et al. Clin. Pharmacogenet. Implementation Consortium (CPIC) Guidelines for CYP2C19 and Voriconazole Therapy. Clin Pharmacol. & Ther. 102, 45–51, 2017.

83. Blanco-Dorado S, Amigo OM, Latorre-Pellicer A, et al. Impact of CYP2C19 genetic polymorphisms on voriconazole exposure in patients with invasive fungal infections. Bas. Clin. Pharm. & Tox. 123, 93–94, 2018.

84. Hicks JK, Bishop JR, Sangkuhl K, et al. Clinical pharmacogenetics implementation consortium (CPIC) guideline for CYP2D6 and CYP2C19 genotypes and dosing of selective serotonin reuptake Inhibitors. Clin. Pharm. Ther. 98, 127–134, 2015.

85. Del Tredici AL, Malhotra A, Dedek M et al. Frequency of CYP2D6 alleles including structural variants in the United States. Front. Pharmacol. 9, 305, 1–13, 2018.

86. Don CG, Smieško M, Out-compute drug side effects: Focus on cytochrome P450 2D6 modeling. Comput. Mol. Sci. 8, e1366, 1–16, 2018.

87. Kiss AF, Tóth K, Juhász C, et al. Is CYP2D6 phenotype predictable from CYP2D6 genotype? Microchem. J. 136, 209–214, 2018.

88. Zhou, SF. Polymorphism of human cytochrome P450 2D6 and its clinical significance. Clin. Pharmacokinet. 48, 11, 689–723, 2009.

89. Elbaz A, Levecque C, Clavel J, et al. CYP2D6 polymorphism, pesticide exposure, and Parkinson's disease. Ann Neurol. 55, 430–434, 2004.

90. Kurylev AA, Brodyansky VM, Andreev BV, et al. The combined effect of CYP2D6 and DRD2 taq1a polymorphisms on the antipsychotics daily doses and hospital stay duration in schizophrenia inpatients (observational naturalistic study). Psychiatria Danubina 30, 2, 157–163, 2018.

91. Koopmans AB, Vinkers DJ, Poulina IT, et al. No effect of dose adjustment to the CYP2D6 genotype in patients with severe mental illness. Front. Psychiatr. 9, 349, 2018.

92. Moncrieff J, Leo J. A systematic review of the effects of antipsychotic drugs on brain volume. Psychol. Med. 40, 1409–1422, 2010.

93. Ito T, Yamamoto K, Ohsawa F, et al. Association of CYP2D6 polymorphisms and extrapyramidal symptoms in schizophrenia patients receiving risperidone: a retrospective study. J. Pharm Health Care Sci. 4, 28, 1–6, 2018.

94. Belmonte C, Ochoa D, Roman M, et al. Influence of CYP2D6, CYP3A4, CYP3A5, and ABCB1 Polymorphisms on pharmacokinetics and safety of aripiprazole in healthy volunteers. Bas. Clin. Pharmacol. & Toxicol. 122, 596–605, 2018.

95. Dodsworth T, Kim DD, Procyshyn RM, et al. A systematic review of the effects of CYP2D6 phenotypes on risperidone treatment in children and adolescents. Child Adolesc Psych. Ment. Health 12(37), 1–10, 2018.

96. Urban AE, Cubała WJ, Therapeutic drug monitoring of atypical antipsychotics. Psychiatr. Pol. 51, 1059–1077, 2017.

97. Bijl MJ, Visser LE, van Schaik RH, et al. Genetic variation in the CYP2D6 gene is associated with a lower heart rate and blood pressure in beta-blocker users. Clin. Pharmacol. Ther. 85, 45–50, 2009.

98. Parker BM, Rogers SL, Lymperopoulo A. Clinical pharmacogenomics of carvedilol: the stereo-selective metabolism angle. Pharmacogen. 19, 1089–109, 2018.

99. Luzum JA, Sweet KM, Binkley PF, et al. CYP2D6 Genetic variation and beta-blocker maintenance dose in patients with heart failure. Pharm. Res, 34, 1615–1625, 2017.

100. Batty JA, Hall AS, White HL, et al. An investigation of CYP2D6 genotype and response to metoprolol CR/XL during dose titration in patients with heart failure: a MERIT-HF substudy. Clin Pharmacol Ther. 95(3), 321–30, 2014.

101. Jung E, Ryu S, Park Z, et al. Influence of CYP2D6 Polymorphism on the pharmacokinetic/pharmacodynamic characteristics of carvedilol in healthy Korean volunteers. J. Kor. Med. Sci. 2(33), 27, e182, 2018.

102. Amare AT, Schubert KO, Baune BT. Pharmacogenomics in the treatment of mood disorders: strategies and opportunities for personalized psychiatry. EPMA J. 8, 211–227, 2017.

103. Mao ZL, Wheeler JJ, Clohs L et al. Pharmacokinetics of novel atrial-selective anti-arrhythmic agent vernakalant hydrochloride injection (RSD1235): influence of CYP2D6 expression and other factors. J Clin Pharmacol. 49(1), 17–29, 2009.

104. Camm J. The Vernakalant story: how did it come to approval in Europe and what is the delay in the U.S.A? Curr Cardiol Rev. 10(4), 309–314, 2014.

105. Chidambaran V, Sadhasivam S, Mahmoud M. Codeine and opioid metabolism – implications and alternatives for pediatric pain management. Curr. Opin. Anaesthesiol. 30, 349–356, 2017.

106. Agarwal D, Udoji MA, Trescot A. Genetic testing for opioid pain management: a primer. Pain Ther. 6, 93–105, 2017.

107. Kirchcheiner J, Schmidt H, Tzvetkov M, Keulen JT, Lotsch J, Roots I, Brockmoller J. Pharmacokinetics of codeine and its metabolite morphine in ultra rapid metabolisers due to CYP2D6 duplication. Pharmacogenom J. 7, 257–65, 2007.

108. Ruaño G, Kost JA. Fundamental considerations for genetically guided pain management with opioids based on CYP2D6 and OPRM1 polymorphisms. Pain Physician 21, E611–E621, 2018.

109. Lassen D, Damkier P, Brøsen K. The pharmacogenetics of tramadol. Clin. Pharmacokinet. 54, 825–836. 2015.

110. Slanar O, Nobilis M. Kvetina J. Miotic action of tramadol is determined by CYP2D6 genotype. Physiol. Res. 56. 129–136, 2007.

111. Wesmiller SW, Henker RA, Sereika SM, et al. Association of CYP2D6 Genotype and postoperative nausea and vomiting in orthopaedic trauma patients. Biol. Res. Nurs. 15, 382–389, 2013.

112. Niewinski PA, Wojciechowski R, Sliwinski M, et al. CYP2D6 basic genotyping as a-potential tool to-improve the antiemetic efficacy of-ondansetron in-prophylaxis of-postoperative nausea and vomiting. Adv. Clin. Exp. Med. 27, 1499–1503, 2018.

113. Bell GC, Caudle KE, Whirl-Carrillo M, et al. Personalized clinical pharmacogenetics implementation consortium (CPIC) guideline for CYP2D6 genotype and use of ondansetron and tropisetron. Clin. Pharmacol. Ther. 102, 213–218, 2017.

114. Bustos ML, Zhao Y, Chen H, et al. Polymorphisms in CYP1A1 and CYP3A5 Genes contribute to the variability in granisetron clearance and exposure in pregnant women with nausea and vomiting. Pharmacother. 36, 1238–1244, 2016.

115. Dietle A. QTc Prolongation with antidepressants and antipsychotics. US Pharm. 40, HS34–HS40, 2015.

116. Popovic D, Nuss P, Vieta E. Revisiting loxapine: a systematic review. Ann. Gen. Psychiatr. 14, 15, 2015.

117. Llerena A, Berecz R, Dorado P, et al. QTc interval, CYP2D6 and CYP2C9 genotypes and risperidone plasma concentrations. J. of Pharm. 18, 2, 189–193, 2004.

118. Wenzel-Seifert K, Wittmann M, Haen E: QTc prolongation by psychotropic drugs and the risk of torsade de pointes. Deutsch. Arztebl. Int. 108, 687–693. 2011.

119. Niemeijer MN, van den Berg ME, Eijgelsheim M, et al. Pharmacogenetics of drug-induced QT interval prolongation: an update. Drug Saf. 38, 855–867, 2015.

120. Shams ME, Arneth B, Hiemke C, et al. CYP2D6 polymorphism and clinical effect of the antidepressant venlafaxine. J. Clin. Pharm. Ther. 31, 493–502, 2006.

121. Johnson EM, Whyte E, Mulsant BH. Cardiovascular changes associated with venlafaxine in the treatment of late-life depression. Am. J. Geriatr. Psychiatr. 14, 796–802, 2006.

122. Ahmed AT, Biernacka JM, Jenkins GD, et al. Pharmacogenomics of serotonin noradrenergic reuptake inhibitors (SNRIs) antidepressant response after selective serotonin reuptake inhibitors (SSRIs) treatment failure in major depressive disorder. Biol. Psychiatr. 83(9), Suppl. S307–S307, F177, 2018.

123. Taranu, A, Colle R, Gressier F, et al. Should a routine genotyping of CYP2D6 and CYP2C19 genetic polymorphisms be recommended to predict venlafaxine efficacy in depressed patients treated in psychiatric settings? Pharmacogenom. 18, 639–650, 2017.

124. Swen JJ, Nijenhuis M, de Boer A, et al. Pharmacogenetics from bench to byte – an update of guidelines. Clin. Pharmacol. Ther, 89, 662–673, 2011.

125. Inglis SC, Herbert MK, Davies BJL, et al. Effect of CYP2D6 metaboliser status on the disposition of the (+) and (-) enantiomers of perhexiline in patients with myocardial ischaemia. Pharmacogenet. Genom. 17, 305–312, 2007.

126. Davies BJ, Coller JK, James HM, et al. The influence of CYP2D6 genotype on trough plasma perhexiline and cis-OH-perhexiline concentrations following a standard loading regimen in patients with myocardial ischaemia. Brit. J. Clin. Pharmacol. 61, 321–325, 2006.

127. Sutrisna EM. The Impact of CYP1A2 and CYP2E1 Genes Polymorphism on Theophylline Response.J. Clin, Diag. Res. 10, FE1–FE3, 2016

128. Huang CY, Kao CF, Chen CC et al. No association of CYP2E1 genetic polymorphisms with alcohol dependence in Han Taiwanese population. J. Formos. Med. Assoc. 117, 646–649, 2018.

129. Shahriary MG, Galehdari H, Jalali A, et al. CYP2E1*5B, CYP2E1*6, CYP2E1*7B, CYP2E1*2, and CYP2E1*3. Allele frequencies in Iranian populations. Asian Pacif. J. Canc. Prev. 13, 6505–6651, 2012.

130. Boccia S, De Lauretis A, Gianfagna F, et al. CYP2E1PstI/RsaI polymorphism and interaction with tobacco, alcohol, and GSTs in gastric cancer susceptibility: a meta-analysis of the literature. Carcinogen. 28, 101–106, 2007.

131. Shi W-L, Tang H-L, Zhai S-D. Effects of the CYP3A4 1B genetic polymorphism on the pharmacokinetics of tacrolimus in adult renal transplant recipients: a meta-analysis. PLoS ONE 10(6), e0127995, 2015.

132. Żochowska D, Wyzgał J, Pączek L. Impact of CYP3A4*1B and CYP3A5*3 polymorphisms on the pharmacokinetics of cyclosporine and sirolimus in renal transplant recipients. Ann. Transplant. 17(3), 36–44, 2012.

133. López-García MA, Feria-Romero IA, Serrano H, et al. Pietro Fagiolinod, Marta Vázquezd, Consuelo Escamilla Influence of genetic variants of CYP2D6, CYP2C9, CYP2C19, and CYP3A4 on antiepileptic drug metabolism in pediatric patients with refractory epilepsy. Pharmacol. Reps 69, 504–511, 2017.

134. Fernandez P, de Beer PM, van der Merwe L et al. Genetic variations in androgen metabolism genes and associations with prostate cancer in South African men. S. Afr. Med. J. 100, 741–745, 2010.

135. Dong AN, Tan BH, Pan Y, et al. Cytochrome P450 genotype-guided drug therapies: An update on current states Clin Exp Pharmacol Physiol. 45, 991–1001. 2018.

136. Zhang W, Chang YZ, Kan QC, et al. CYP3A4*1G genetic polymorphism influences CYP3A activity and response to fentanyl in Chinese gynecologic patients. Eur. J. Clin. Pharmacol. 66, 61, 2010.

137. Lv J, Liu F, Feng N, et al. CYP3A4 gene polymorphism is correlated with individual consumption of sufentanil. Acta Anaesthesiol Scand. 62, 1367–1373, 2018.

138. Huang Y, Wen G, Lu Y, et al. CYP3A4*1G and CYP3A5*3 genetic polymorphisms alter the antihypertensive efficacy of amlodipine in patients with hypertension following renal transplantation. Int J Clin Pharmacol Ther. 55, 109–118, 2017.

139. Maekawa K, Harakawa N, Yoshimura T, et al. CYP3A4*16 and CYP3A4*18 Alleles Found in East Asians Exhibit Differential Catalytic Activities for Seven CYP3A4 Substrate Drugs. Drug Metab. Disp. 38, 2100–2104, 2010.

140. Lee JS, Cheong HS, Kim LH, et al. Screening of genetic polymorphisms of CYP3A4 and CYP3A5 genesKor. J. Physiol. Pharmacol. 17, 479–484. 2013.

141. Chen L, Prasad GV. CYP3A5 polymorphisms in renal transplant recipients: influence on tacrolimus treatment. Pharmacogen. & Pers. Med. 11, 23–33, 2018.

142. Uno T, Wada K, Matsuda S, et al. Impact of the CYP3A5*1 Allele on the Pharmacokinetics of Tacrolimus in Japanese Heart Transplant Patients. Eur. J. Drug Met. Pharmacokinet. 43(6), 665–673, 2018.

143. Jacobson PA, Oetting WS, Brearley AM, et al. Novel polymorphisms associated with tacrolimus trough concentrations: results from a multicenter kidney transplant consortium. Transplantation 91, 300–308, 2011.

144. Hesselink DA, van Gelder T, van Schaik RH, et al. Population pharmacokinetics of cyclosporine in kidney and heart transplant recipients and the influence of ethnicity and genetic polymorphisms in the MDR-1, CYP3A4, and CYP3A5 genes. Clin Pharmacol Ther. 76, 545–556, 2004.

145. Anglicheau D, Le Corre D, Lechaton S et al: Consequences of genetic polymorphisms for sirolimus requirements after renal transplant patients on primary sirolimus therapy. Am. J. Transpl. 5, 595–603, 2005

146. Takaya U, Kyoichi W, Sachi M, et al. Impact of the CYP3A5*1 Allele on the Pharmacokinetics of Tacrolimus in Japanese Heart Transplant Patients. Eur. J. Drug Metab. Pharmacokinet. 43, 665–673, 2018.

147. Birdwell KA, Decker B, Barbarino JM, et al. Clinical Pharmacogenetics Implementation Consortium (CPIC) guidelines for CYP3A5 genotype and tacrolimus dosing. Clin. Pharmacol. & Ther 98, 19–24.

148. Ferreira F, Esteves S, Almeida LS, et al. Trimethylaminuria (fish odor syndrome): Genotype characterisation among Portuguese patients. Gene 527, 366–370, 2013.

149. Hisamuddin IM, Yan VW. Genetic polymorphisms of human flavin-containing monooxygenase 3: implications for drug metabolism and clinical perspectives. Pharmacogenomics. 8, 635–643, 2007.

150. Zhou W, Humphries H, Neuhoff S, et al. Development of a physiologically based pharmacokinetic model to predict the effects of flavin-containing monooxygenase 3 (FMO3) polymorphisms on itopride exposure. Biopharm Drug Dispos. 38, 389–393, 2017.

151. Catucci G, Occhipinti A, Maffei M, et al. Effect of human flavin-containing monooxygenase 3 polymorphism on the metabolism of aurora kinase inhibitors. Int. J. Mol. Sci. 14, 2707–2716, 2013.
152. Catucci G, Bortolussi S, Rampolla S, et al. Flavin-containing monooxygenase 3 polymorphic variants significantly affect clearance of tamoxifen and clomiphene. Basic & Clin. Pharmacol. & Tox. 123, 687–669, 2018.
153. Ren L, Teng M, Zhang T, et al. Donors FMO3 polymorphisms affect tacrolimus elimination in Chinese liver transplant patients. Pharmacogenom. 18, 265–275, 2017.
154. Vanhove T, de Jonge H, de Loor H, et al. Comparative performance of oral midazolam clearance and plasma 4β-hydroxycholesterol to explain interindividual variability in tacrolimus clearance. Br. J. Clin. Pharmacol. 82, 1539–1549, 2016.
155. Park S, Lee NR, Lee KE, et al. Effects of Single-Nucleotide Polymorphisms of FMO3 and FMO6 Genes on Pharmacokinetic Characteristics of Sulindac Sulfide in Premature Labor. Drug. Metab. Disp. 42, 40–43, 2014.
156. Morandi A, Zusi C, Corradi M, et al. Minor diplotypes of FMO3 might protect children and adolescents from obesity and insulin resistance. Int. J. Obes. 42, 1243–1248, 2018.
157. Murphy C, Byrne S, Ahmed G, et al. Cost implications of reactive versus prospective testing for dihydropyrimidine dehydrogenase deficiency in patients with colorectal cancer: a single-institution experience. Dose Resp. 16, Article Number: 1559325818803042, 2018.
158. Amstutz U, Henricks LM, Offer SM, et al. Clinical Pharmacogenetics Implementation Consortium (CPIC) Guideline for dihydropyrimidine dehydrogenase genotype and fluoropyrimidine dosing: 2017 update. Clin. Pharmacol. & Ther. 103, 211–216, 2017.
159. Tong CC, Lam CW, Lam KA et al. A novel DPYD variant associated with severe toxicity of fluoropyrimidines: role of pre-emptive DPYD genotype screening. Front. Oncol. 8, 279, 2018.
160. Deenen MJ, Meulendijks D, Cats A, et al. Upfront Genotyping of DPYD*2A to Individualize Fluoropyrimidine Therapy: A Safety and Cost Analysis. J Clin. Oncol. 34, 227–235, 2016.
161. Gätke MR, Viby-Mogensen J, Østergaard D et al. Response to Mivacurium in Patients Carrying the K Variant in the Butyrylcholinesterase Gene. Anesthesiol. 102, 503–508, 2005.
162. Sokolowa S, Li X, Chena L, et al. Deleterious effect of butyrylcholinesterase K-variant in donepezil treatment of mild cognitive impairment. J. Alzheim. Dis. 56, 229–237, 2017.
163. Negrao AB, Pereira AC, Guindalini C, et al. Butyrylcholinesterase Genetic Variants: Association with Cocaine Dependence and Related Phenotypes. PLoS ONE 8(11), e80505. 2013.
164. Zhang P, Jain P, Tsao C, et al. Butyrylcholinesterase nanocapsule as a long circulating bioscavenger with reduced immune response. J. Cont. Rel. 230, 73–78, 2016.
165. Science News: Grant to develop therapy to protect against nerve agents https://www.sciencedaily.com/releases/2015/01/150128081849.htm.

166. Li X, Hu Z, Qu X, et al. Putative EPHX1 Enzyme Activity Is Related with Risk of Lung and Upper Aerodigestive Tract Cancers: A Comprehensive Meta-Analysis. PLoS One. 6(3), e14749, 2011.

167. Nock N, Tang D, Rundle A, et al. Associations between smoking, polymorphisms in polycyclic aromatic hydrocarbon (PAH) metabolism and conjugation genes and PAH-DNA adducts in prostate tumors differ by race. Canc. Epidemiol Biomarkers Prev. Jun; 16(6), 1236–1245, 2007.

168. Wadelius M, Chen LY, Eriksson N. Association of warfarin dose with genes involved in its action and metabolism Hum. Genet. 121, 23–34, 2007.

169. Daci A, Beretta G, Vilasaliu D, et al. Polymorphic variants of SCN1A and EPHX1 influence plasma carbamazepine concentration, metabolism and pharmacoresistance in a population of kosovar albanian epileptic patients. PLoS One 10 (11), e0142408, 2015.

170. Zhu X, Yun WT, Sun XF. Effects of major transporter and metabolizing enzyme gene polymorphisms on carbamazepine metabolism in Chinese patients with epilepsy. Pharmacogenet. 15, 1867–1879, 2014

171. Litvinov D, Mahini H, Garelnabi M. Antioxidant and anti-inflammatory role of paraoxonase 1: implication in arteriosclerosis diseases. N. Am. J. Med. Sci. 4, 523–532, 2012.

172. Mackness M, Mackness B, Human paraoxonase-1 (PON1): Gene structure and expression, promiscuous activities and multiple physiological roles. Gene 567, 12–21, 2015.

173. Kim DS, Marsillach J, Furlong CE, et al. Pharmacogenetics of paraoxonase activity: elucidating the role of high-density lipoprotein in disease. Pharmacogenomics 14(12), 1495–1515, 2013.

174. Dansette PM, Rosi J, Bertho G et al. Cytochromes P450 catalyze both steps of the major pathway of clopidogrel bioactivation, whereas paraoxonase catalyzes the formation of a minor thiol metabolite isomer. Chem. Res. Toxicol. 25(2), 348–356, 2012.

175. Wang D, Para MF, Koletar SL, et al. Human N-acetyltransferase 1 (NAT1) *10 and *11 alleles increase protein expression via distinct mechanisms and associate with sulfamethoxazole-induced hypersensitivity. Pharmacogenet Genom. 21(10), 652–664, 2011.

176. Zhang M, Wang S, Wilffert, B, et al. The association between the NAT2 genetic polymorphisms and risk of DILI during anti-TB treatment: a systematic review and meta-analysis. Brit. J. Clin. Pharmac. 84, 2747–2760, 2018.

177. N-acetyl transferase-2. https://www.snpedia.com/index.php/NAT2

178. Al-Ahmad MM, Amir N, Dhanasekaran S, et al. Studies on N-Acetyltransferase (NAT2) genotype relationships in emiratis: confirmation of the existence of phenotype variation among slow acetylators. Ann. Hum. Gen. 81, 190–196, 2017.

179. Ebeshi, BU, Bolaji, OO, Matimba A, et al. Genetic Polymorphisms of N-Acetyltransferase 2 (NAT2) in the Hausa, Ibo, and Yoruba Populations of Nigeria. Drug. Metab. Rev. 41, 75–76, Abs. 164, 2009.

180. Lotfi F, Bahrehmand F, Vaisi-Raygani A et al. Cytochrome P450 (CYP450,2D6*A), N-Acetyltransferase-2 (NAT2*7, A) and multidrug resistance 1 (MDR1 3435 T) alleles collectively increase risk of ulcerative colitis. Arch. Iran. Med. 21, 530–535, 2018.

181. Sacco J, Abouraya M, Motsinger-Reif A. Evaluation of polymorphisms in the sulphonamide detoxification genes NAT2, CYB5A, and CYB5R3 in patients with sulfonamide hypersensitivity. Pharmacogenet. Genom. 22(10), 733–740, 2012.

182. Alfirevic A, Stalford AC, Vilar J, et al. Slow acetylator phenotype and genotype in HIV-positive patients with sulphamethoxazole hypersensitivity. Br. J. Clin. Pharmacol. 55, 158–165, 2003.

183. Coleman MD, Scott AK, Breckenridge AM et al. The use of cimetidine as a selective inhibitor of dapsone N-hydroxylation in man. Brit. J. Clin. Pharmac. 30, 761–767, 1990.

184. Coleman MD, Holden LJ. The Methaemoglobin Forming and GSH depleting effects of dapsone and monoacetyl dapsone hydroxylamines in human diabetic and non-diabetic erythrocytes in vitro. Env. Tox. Pharmacol 17, 55–59, 2004.

185. Xiang Y, Ma L, Wu, WD, et al. More the incidence of liver injury in Uyghur patients treated for TB in Xinjiang Uyghur Autonomous Region, China, and Its association with hepatic enzyme polymorphisms NAT2, CYP2E1, GSTM1 and GSTT1. PLoS ONE 9 (1) e85905. 2014.

186. Katara P and Kuntal H. TPMT polymorphism: When shield becomes weakness. Interdiscip. Sci. Comput. Life. Sci. 8, 150–155, 2016.

187. Lennard L, Implementation of TPMT testing. Br. J. Clin. Pharmacol. 77, 704–714, 2014.

188. Taranu A, El Asmar K, Colle R, et al. The catechol-O-methyltransferase val(108/158) met genetic polymorphism cannot be recommended as a biomarker for the prediction of venlafaxine efficacy in patients treated in psychiatric settings. Basic. Clin Pharm. Toxicol. 121, 435–441, 2017.

189. Ji Y, Biernacka J, Snyder K, et al. Catechol O-methyltransferase pharmacogenomics and selective serotonin reuptake inhibitor response. Pharmacogenom. J. 12(1), 78–85, 2012.

190. Vasunilashorn SM Ngo LH, Jones RN, et al. The Association Between C-Reactive Protein and Postoperative Delirium Differs by Catechol-O-Methyltransferase Genotype. Am. J. Geriatr. Psych. 27, 1–8, 2019.

191. Hopkins SC, Reasner DS, Koblan KS. Catechol-O-methyltransferase genotype as modifier of superior responses to venlafaxine treatment in major depressive disorder. Psychiatry Res. 208, 285–287, 2013.

192. Benedetti F, Colombo C, Pirovano A, et al. The catechol-O-methyltransferase val(108/158)met polymorphism affects antidepressant response to paroxetine in a naturalistic setting. Psychopharmacol. 203, 5560, 2008.

193. Rajagopal VM, Rajkumar AP, Jacob KS, et al. Gene-gene interaction between DRD4 and COMT modulates clinical response to clozapine in treatment-resistant schizophrenia. Pharmacogenet. Genom. 28, 31–35, 2018.

194. Akay AP, Yazicioglu CE, Guney SA, et al. Allele frequencies of dopamine D4 receptor gene (DRD4) and catechol-O-methyltransferase (COMT) Val158Met polymorphism are associated with methylphenidate response in adolescents with attention deficit/hyperactivity disorder: a case control preliminary study. Psych. Clin Psychopharmacol. 28, 177–184, 2018.

195. Arseneault, L., Cannon, M., Poulton, R., et al. Cannabis use in adolescence and risk for adult psychosis: Longitudinal prospective study. BMJ. 325(7374), 1212–1213, 2002.

196. Lodhi RJ, Wang YB, Rossolatos D, Investigation of the COMT Val158Met variant association with age of onset of psychosis, adjusting for cannabis use. Brain. Behav. 7, 11, e00850, 2017.

197. Narendran R, Tumuluru D, May MA, et al. Cortical dopamine transmission as measured with the [C-11]FLB 457-Amphetamine PET Imaging Paradigm Is Not Influenced by COMT Genotype. PLoS ONE, 11(6), e0157867, 2016.

198. Hamidovic A Dlugos A, Palmer AA, et al. Catechol-O-methyltransferase val(158) met genotype modulates sustained attention in both the drug-free state and in response to amphetamine. Psychiatr. Genet. 20, 85–92, 2010.

199. Farheen S, Sengupta S, Santra A, et al. Gilbert's syndrome: High frequency of the (TA)7 TAA allele in India and its interaction with a novel CAT insertion in promoter of the gene for bilirubin UDP-glucuronosyltransferase 1 gene. World J. Gastroenterol. 12(14), 2269–2275, 2006.

200. Chiddarwar AS, D'Silva SZ, Colah RB, et al. Genetic lesions in the UGT1A1 genes among Gilbert's syndrome patients from India. Mol. Biol. Rep. 45, 2733–2739, 2018.

201. Gailite L, Rots D, Pukite L, et al. Case report: multiple UGT1A1 gene variants in a patient with Crigler-Najjar syndrome. BMC Pediatr. 18(317), 2018.

202. Bulmer AC, Bakrania B, Du Toit EF, et al. Bilirubin acts as a multipotent guardian of cardiovascular integrity: more than just a radical idea. Am J Physiol Heart Circ Physiol 315: H429–H447, 2018.

203. Iolascon A, Faienza MF, Centra M, et al. (TA)8 allele in the UGT1A1 gene promoter of a Caucasian with Gilbert's syndrome. Haematologica. 84, 106–109, 1999.

204. Heydari MR, Fardaei M, Kadivar MR, et al. Prevalence of 2 UGT1A1 Gene Variations Related to Gilbert's Syndrome in South of Iran: An Epidemiological, Clinical, and Genetic Study. Iran Red Cres. Med. J. 19, e44363, 2017.

205. Horsfall LJ, Zeitlyn D, Tarekegn A, et al. Prevalence of clinically relevant UGT1A alleles and haplotypes in African populations. Ann. Hum. Genet. 75, 236–246, 2011

206. Lin, R., Wang, X., Wang, Y., et al. Association of polymorphisms in four bilirubin metabolism genes with serum bilirubin in three Asian populations. Human Mutation 30(4), 609–615, 2009.

207. Kataoka R, Kimata A, Yamamoto K, et al. Association of UGT1A1 Gly71Arg with urine urobilinogen. Nagoya J. Med. Sci.73, 33–40, 2011.

208. Ehmer U, Kalthoff S, Fakundiny B, et al. Gilbert 's syndrome redefined: a complex genetic haplotype influences the regulation of glucuronidation. Hepatology 55(6), 2012.

209. Cho SK, Oh ES, Park K, et al. The UGT1A3*2 polymorphism affects atorvastatin lactonization and lipid-lowering effect in healthy volunteers. Pharmacogenet Genom. 22, 598–605, 2012.

210. Gupta N, Benjamin M, Kar A, et al. Identification of promotor and exonic variations, and functional characterisation of a splice site mutation in indian patients with unconjugated hyperbilirubinemia. PLoS ONE 10(12), e0145967, 2015.

211. Goey AKL, & Figg WD. UGT genotyping in belinostat dosing. Pharmacol. Res. 105, 22–27, 2016.

212. Turatti L, Sprinz E, Lazzaretti RK, et al. UGT1A1*28 variant allele is a predictor of severe hyperbilirubinemia in HIV-infected patients on HAART in southern Brazil. AIDS Res. Hum. Retroviruses. 28, 1015–1018, 2012.

213. Wang Y, Yi C, Wang, Y, et al. Distribution of uridine diphosphate glucuronosyltransferase 1A polymorphisms and their role in irinotecan-induced toxicity in patients with cancer Oncol. Lett. 14, 5743–5752, 2017.

214. Liu, D, Li, J, Gao, J, et al. Examination of multiple UGT1A and DPYD polymorphisms has limited ability to predict the toxicity and efficacy of metastatic colorectal cancer treated with irinotecan-based chemotherapy: a retrospective analysis. BMC Canc. 17, 437, 2017.

215. Milosheska D, Lorber B, Vovk T et al. Pharmacokinetics of lamotrigine and its metabolite N-2-glucuronide: Influence of polymorphism of UDP glucuronosyltransferases and drug transporters. Br. J. Clin. Pharmacol. 82, 399–411, 2016.

216. Li H, Hu B, Guo Z, et al. Correlation of UGT2B7 Polymorphism with Cardiotoxicity in Breast Cancer Patients Undergoing Epirubicin/Cyclophosphamide-Docetaxel Adjuvant Chemotherapy. Yonsei Med. J. 60, 30–37, 2019.

217. Divakaran K, Hines RN, McCarver DG. Human hepatic UGT2B15 developmental expression. Toxicolog. Sci. 141, 292–299, 2014.

218. Hildebrandt M, Adjei A, Weinshilboum R. Very important pharmacogene summary: sulfotransferase 1A1. Pharmacogenet. Genom. 19451861, 2009.

219. Motoi Y, Watanabe K, Honma H, et al. Digital PCR for determination of cytochrome P450 2D6 and sulfotransferase 1A1 gene copy number variations. Drug Disc. Ther. 11, 336–341, 2017.

220. Liu J, Zhao R, Zhan Y, et al. Relationship of SULT1A1 copy number variation with estrogen metabolism and human health. J. Steroid Biochem. Mol. Biol. 174, 169–175, 2017.

221. Klusek J, Nasierowska-Guttmejer A, Kowalik A, GSTM1, GSTT1, and GSTP1 polymorphisms and colorectal cancer risk in Polish non-smokers. Oncotarget. 9, 21224–21230, 2018.

222. Saadat M, Evaluation of Glutathione S-Transferase p1 (GSTP1) ile105val polymorphism and susceptibility to type 2 diabetes mellitus, a meta-analysis. EXCLI J. 16, 1188–1197, 2017.

223. Drozd E, Krzyszton-Russjan J, Marczewska J. et al. Up-regulation of glutathione-related genes, enzyme activities and transport proteins in human cervical cancer cells treated with doxorubicin. Biomed. & Pharmacother. 83, 397–406, 2016.

224. Tulsyan S, Chaturvedi P, Agarwal G, et al. Pharmacogenetic Influence of GST Polymorphisms on Anthracycline-Based Chemotherapy Responses and Toxicity in Breast Cancer Patients: A Multi-Analytical Approach. Mol. Diag. & Ther. 17, 371–379, 2013.

225. Lee HH, and Ho RH. Interindividual and interethnic variability in drug disposition: polymorphisms in organic anion transporting polypeptide 1B1 (OATP1B1; SLCO1B1). Brit. J. Clin. Pharmacol. 83, 1176–1184, 2017.

226. van de Steeg E, Stránecký V, Hartmannová H, et al. Complete OATP1B1 and OATP1B3 deficiency causes human rotor syndrome by interrupting conjugated bilirubin reuptake into the liver. J. Clin. Invest. 122, 519–528, 2012.

227. Tulsyan S, Mittal RD, Mittal B. The effect of ABCB1 polymorphisms on the outcome of breast cancer treatment. Pharmgenom. Pers. Med. 9, 47–58, 2016.

228. Sennesael AL, Panin N, Christelle LE, et al. Effect of ABCB1 genetic polymorphisms on the transport of rivaroxaban in HEK293 recombinant cell lines. Scient. Rep. 8(1), 41598-018-28622-4, 2018.

229. Hodges LM, Markova SM, Chinn LW, et al. Very important pharmacogene summary: ABCB1 (MDR1, P-glycoprotein). Pharmacogenet. Genom. 21(3), 152–161, 2011.

230. Jones LK, Rahm AR, Gionfriddo MR, et al. Developing Pharmacogenomic Reports: Insights from Patients and Clinicians. Clin. Transl. Sci. 11(3), 289–295, 2018.

231. Bush WS, & Moore JH Chapter 11: Genome-Wide Association Studies. PLoS Comput Biol 8(12), e1002822. 2012.

232. Han C, Wang SM, Bahk WM, et al. A pharmacogenomic-based antidepressant treatment for patients with major depressive disorder: results from an 8-week, randomized, single-blinded clinical trial. Clin. Psychopharm. Neurosci. 16, 469+, 2018.

233. Pérez V, Salavert A, Espadaler J, et al. Efficacy of prospective pharmacogenetic testing in the treatment of major depressive disorder: results of a randomized, double-blind clinical trial. BMC Psych. 17, 250, 2017.

234. Benitez J, Cool CL, Scotti, DJ, et al. Use of combinatorial pharmacogenomic guidance in treating psychiatric disorders. Personal. Med. 15, 481–494, 2018.

235. Mian, Paola; Allegaert, Karel; Spriet, Isabel; et al. Paracetamol in Older People: Towards Evidence-Based Dosing? Drugs & Ag. 35, 603–624, 2018.

236. Waring, RH, Harris RM, Mitchell SC. Drug metabolism in the elderly: A multifactorial problem? Maturitas 100, 27–32, 2017.

237. Tajiri K, Shimiz Y. Liver physiology and liver diseases in the elderly. World J. Gastroenterol. 19(46), 8459–8467, 2013.

238. McLachlan AJ, Pont LG. Drug metabolism in older people – a key consideration in achieving optimal outcomes with medicines. J. Gerontol. A. Biol. Sci. Med. Sci. 67(2), 175–180, 2012.

239. Doki K, Homma M, Kuga K, et al. CYP2D6 genotype affects age-related decline in flecainide clearance: a population pharmacokinetic analysis. Pharmacogen. & Genom. 22, 777–783, 2012.

240. Hatahira H, Hasegawa S, Sasaoka S. Analysis of fall-related adverse events among older adults using the Japanese adverse drug event report (JADER) database. *J. Pharm. Health Care & Sc. 4, UNSP* 32, 2018.

241. Khanassov V, Hu JY, Reeves D, et al. Selective serotonin reuptake inhibitor and selective serotonin and norepinephrine reuptake inhibitor use and risk of fractures in adults: A systematic review and meta-analysis. Int. J. Geriatr. Psych. 33, 1688–1708, 2018.

242. Leng Y, Diem SJ, Stone KL, et al. Antidepressant Use and Cognitive Outcomes in Very Old Women. J. Gerontol. Ser. A. Biol. Sci. Med. Sci. 73, 1390–1395, 2018.

243. Ku LC, Smith PB. Dosing in neonates: special considerations in physiology and trial design Pediatr. Pediatr. Res. 77, 2–9 Part: 1 2015.

244. Brussee JM, Vet NJ, Krekels EHJ, et al. Predicting CYP3A-mediated midazolam metabolism in critically ill neonates, infants, children and adults with inflammation and organ failure. Br. J. Clin. Pharmacol. 84, 358–368, 2018.

245. Neyro V, Elie V, Medard Y. mRNA expression of drug metabolism enzymes and transporter genes at birth using human umbilical cord blood. Fund. & Clin. Pharmacol. 32, 422–435, 2018.

246. Anderson BJ, Holford NHG. Negligible impact of birth on renal function and drug metabolism Pediatr. Anesth. 28, 1015–1021, 2018.

247. Linakis MW, Cook SF, Kumar SS, et al. Polymorphic Expression of UGT1A9 is Associated with Variable Acetaminophen Glucuronidation in Neonates: A Population Pharmacokinetic and Pharmacogenetic Study. Clin. Pharmacokinet. 57, 1325–1336, 2018.

248. Ng E. Taddio A, Ohlsson, A. Intravenous midazolam infusion for sedation of infants in the neonatal intensive care unit. Coch. Data. Syst. Rev. Article Number: CD002052, 2017.

249. Lu H, Rosenbaum S. Developmental pharmacokinetics in pediatric populations. J. Pediatr. Pharmacol. Ther. 19, 262–276, 2014.

250. Yueh MF, Chen S, Nghia, N, et al. Developmental, genetic, dietary, and xenobiotic influences on neonatal hyperbilirubinemia. Mol. Pharmacol. 91, 545–553, 2017.

251. Fernandez E, Perez R, Hernandez A, et al. Factors and mechanisms for pharmacokinetic differences between pediatric population and adults. Pharmaceut. 3, 53–72, 2011

252. Kelly LE, Chaudhry SA, Rieder MJ, et al. A clinical tool for reducing central nervous system depression among neonates exposed to codeine through breast milk. PLoS ONE, 8, e70073, 2013.

253. Autmizguine J, Guptill JT, Cohen-Wolkowiez M, et al. Pharmacokinetics and pharmacodynamics of antifungals in children: clinical implications. Drugs 74, 891–909, 2014.

254. Wallander KM, Ohman I, Dahlin M, et al. Zonisamide: pharmacokinetics, efficacy, and adverse events in children with epilepsy. Neuropediatr. 45, 362–U12, 2014.

255. Childress, A, Newcorn, J, Stark, JG, et al. A single-dose, single-period pharmacokinetic assessment of an extended-release orally disintegrating tablet of methylphenidate in children and adolescents with attention-deficit/hyperactivity disorder. J. Child Adol. Psychopharm. 26, 505–512, 2016.

256. Labellarte, M, Biederman, J, Emslie, G, et al. Multiple-dose pharmacokinetics of fluvoxamine in children and adolescents. J. Am. Acad. Child and Adol. Psychiatr. 43, 1497–1505, 2004.

257. Feuer AJ, Demmer RT, Thai A, et al. Use of selective serotonin reuptake inhibitors and bone mass in adolescents: an NHANES study. BONE 78, 28–33, 2015.

258. Jeong H, Kim S, Kim, MY, et al. Inhibitory and inductive effects of opuntia ficus indica extract and its flavonoid constituents on cytochrome P450s and UDP-glucuronosyltransferases. Int. J. Mol. Sci. 19, 3400, 2018.

259. Fontana RJ, Lown KS, Paine MF, et al. Effects of a chargrilled meat diet on expression of CYP3A, CYP1A, and P-glycoprotein levels in healthy volunteers. Gastrolenterol. 117, 89–98, 1999.

260. Walubo A, Coetsee C, Badenhorst AM, Effect of the South African traditional meat, biltong, on cancer-associated enzymes CYP2E1 and CYP1A2. South Afr. Med. J. 94(11), 903–905, 2004.

261. Peterson S, Schwarz Y, Li SS, et al. CYP1A2, GSTM1, and GSTT1 polymorphisms and diet effects on CYP1A2 activity in a crossover feeding trial. Canc. Epid. Biomark. Prev. 18(11), 3118–3125, 2009.

262. Tabart J, Pincemail J, Kevers C, et al. Processing effects on antioxidant, glucosinolate, and sulforaphane contents in broccoli and red cabbage. Eur. Food Res. Tech. 244, 12, 2085–2094, 2018.

263. Vanduchova A, Anzenbacher P, Anzenbacherova E, et al. Isothiocyanate from broccoli, sulforaphane, and its properties. J. Med. Food. jmf.2018.0024, 2018.

264. Palliyaguru DL, Yuan JM, Kensler TW, et al. Isothiocyanates: Translating the Power of Plants to People. Mol. Nutrit. & Food Res. 62(18) Special Issue, 1700965, 2018.

265. Loizou GD, Cocker J. The effects of alcohol and diallyl sulphide on CYP2E1 activity in humans: a phenotyping study using chlorzoxazone. Hum. Exper. Toxicol. 20, 7, 321–327. 2001.

266. Doepker C, Franke K, Myers E, et al. Key findings and implications of a recent systematic review of the potential adverse effects of caffeine consumption in healthy adults, pregnant women, adolescents, and children. Nutrients 10(1536), 2018.

267. Vitti FP, Grandi C, Cavalli RD, et al. Association between Caffeine Consumption in Pregnancy and Low Birth Weight and Preterm Birth in the birth Cohort of Ribeirao Preto. Revist. Brasilier. Ginecolog. Obstet. 40, 12, 749–756, 2018.

268. Mayerova, Michaela; Ustohal, Libor; Jarkovsky, Jiri; et al. Influence of dose, gender, and cigarette smoking on clozapine plasma concentrations. Neuropsychiatr. Dis. Treat. 14, 1535–1543, 2018.

269. Soldin OP, Mattison DR. Sex Differences in Pharmacokinetics and Pharmacodynamics Clin Pharmacokinet. 48(3), 143–157, 2009.

270. Islam M, Dubey NK, Poly TN, et al. Gender-based personalized pharmacotherapy: a systematic review. Arch. Gynecol. Obstet. 295, 1305–1310, 2017.

271. Damoiseaux VA, Proost JH, Jiawan VCR, et al. Sex differences in the pharmacokinetics of antidepressants: influence of female sex hormones and oral contraceptives. Clin. Pharmacokinet. 53, 509–519, 2014.

272. Uno Y, Takata R, Kito G, et al. Sex-and age-dependent gene expression in human liver: An implication for drug-metabolizing enzymes. Drug Metab. Pharmacokinet. 32, 1, 100–107, 2017.

273. Rościszewska D, Buntner B, Guz I, et al. Ovarian hormones, anticonvulsant drugs, and seizures during the menstrual cycle in women with epilepsy. J. Neurol. Neurosurg. Psychiatr. 49(1), 47–51, 1986.

274. Lavanya Y, Polasani N, Santh P. Influence of menstrual cycle on pharmacokinetic parameters of carbamazepine in epileptic patients. Int J. Pharm. Chem. Sci. 3(1), 28–35, 2014.

275. Jeong H. Altered drug metabolism during pregnancy: Hormonal regulation of drug-metabolizing enzymes. Exp. Opin. Drug Metab. Toxicol. 6(6), 689–699, 2010.

276. Zharikova OL, Fokina VM, Nanovskaya TN, et al. Identification of the major human hepatic and placental enzymes responsible for the biotransformation of glyburide. Biochem. Pharmacol. 78(12), 1483–1490, 2009.

277. Cunningham, FG. Williams Obstetrics. 21st ed. McGraw-Hill Medical Publishing Division; New York: 2001.

278. Dallmann A, Pfister M, van den Anker J, et al. Physiologically based pharmacokinetic modelling in pregnancy: a systematic review of published models. Clin. Pharm. & Ther. 104(6), 1111–1124, 2018.

279. Westin AA, Brekke M, Molden E, et al. Treatment with antipsychotics in pregnancy: changes in drug disposition. Clin. Pharmacol. & Ther. 103(3), 477–484, 2018.

280. Imaz ML, Oriolo G, Torra M. Clozapine use during pregnancy and lactation: a case-series report front. Pharmacol. 9(264), 2018.

281. Mor G, Cardenas I. The immune system in pregnancy: a unique complexity. Am. J. Reprod. Immunol. 63(6), 425–433, 2010.

282. Piedade R, Traub S, Bitter A, et al. Carboxymefloquine, the major metabolite of the antimalarial drug mefloquine, induces drug-metabolizing enzyme and transporter expression by activation of pregnane X receptor. Antimicrob Agents Chemother. 59(1), 96–104, 2015.

283. Haaland RE, Otieno K, Martin A, et al. Reduced nevirapine concentrations among HIV-positive women receiving mefloquine for intermittent preventive treatment for malaria control during pregnancy. AIDS Res. Hum. Retrovir. 34(11), 2018.

284. Kozlowski LT, O'Connor RJ. Cigarette filter ventilation is a defective design because of misleading taste, bigger puffs, and blocked vents. Tobacco Control 11(Suppl I), i40–i50, 2002.

285. Davies R. How big tobacco has survived death and taxes. The Guardian, 12th July 2017.https://www.theguardian.com/world/2017/jul/11/how-big-tobacco-has-survived-death-and-taxes.

286. Hendlin YH, Elias J, Ling PM. The Pharmaceuticalization of the Tobacco Industry. Ann Intern Med. Aug 15; 167(4), 278–280, 2017.

287. UK discussion paper on submission of notifications under article 20 of directive 2014/40/EU Chapter 6 – advice on ingredients in nicotine-containing liquids in electronic cigarettes and refill containers. https://assets.publishing.service.gov.uk/government/uploads/system/uploads/attachment_data/file/682739/Ingredient_guidance_final_draft_011116.pdf.

288. DIRECTIVE 2014/40/EU OF THE EUROPEAN PARLIAMENT AND OF THE COUNCIL of 3 April 2014 on the approximation of the laws, regulations and administrative provisions of the Member States concerning the manufacture, presentation and sale of tobacco and related products and repealing Directive 2001/37/EC https://ec.europa.eu/health/sites/health/files/tobacco/docs/dir_201440_en.pdf

289. Sachse C, Brockmöller J, Bauer S, et al. Functional significance of a C-->A polymorphism in intron 1 of the cytochrome P450 CYP1A2 gene tested with caffeine. Br. J. Clin. Pharmacol. 47(4), 445–449, 1999.

290. Laika B, Leucht S, Heres S, et al. Pharmacogenetics and olanzapine treatment: CYP1A2*1F and serotonergic polymorphisms influence therapeutic outcome. Pharmacogenom. J. 10, 20–22, 2010.

291. Kocar T, Freudenmann RW, Spitzer M et al. Switching from tobacco smoking to electronic cigarettes and the impact on clozapine levels. J. Clin. Psychopharmacol. 38, 528–529, 2018.

292. Fechter-Leggett ED, White SK, Fedan KB, et al. Burden of respiratory abnormalities in microwave popcorn and flavouring manufacturing workers. Occ. Env. Med. 75, 709–715, 2018.

293. Hendlin YH, Veffer JR, Lewis MJ et al. Beyond the brotherhood: Skoal Bandits' role in the evolution of marketing moist smokeless tobacco pouches. Tob. Induc. Dis. 15: 46, 2017.

294. Sanner T, Grimsrud TK. Nicotine: Carcinogenicity and Effects on Response to Cancer Treatment – A Review. Front. Oncol. 5: 196, 2015.

295. Niemelä S, Sourander A, Surcel HM, et al. Prenatal Nicotine Exposure and Risk of Schizophrenia Among Offspring in a National Birth Cohort. Am. J. Psychiat. 173:799–806, 2016.

296. Moran VE. Cotinine: beyond that expected, more than a biomarker of tobacco consumption Front. Pharmacol. 3, Article 173. 2012.

297. Šagud M, Cusa BV, Jakšić N, et al. Smoking in schizophrenia: an updated review. Psychiatria Danubina 30, Suppl. 5, pp S14–84, 2018.

298. Malleta J, Le Strat Y, Schürho F, et al. Cigarette smoking and schizophrenia: a specific clinical and therapeutic profile? Results from the FACE-Schizophrenia cohort. Prog. Neuro-Psychopharmacol. Biol. Psychiatr. 79, Part B, 332–339, 2017.

299. Gee SH, Taylor DM, Shergill SS, et al. Effects of a smoking ban on clozapine plasma concentrations in a nonsecure psychiatric unit. Ther. Adv. Psychopharmacol. 7(2) 79–80, 2017.

300. Williams R, Alexander G, Aspinall R, et al. Gathering momentum for the way ahead: fifth report of the Lancet Standing Commission on Liver Disease in the UK. Lancet 392, 2398–2441, 2018. https://assets.publishing.service.gov.uk/government/uploads/system/uploads/attachment_data/file/602132/Communicating_2016_CMO_guidelines_Mar_17.pdf

301. Attignon EA, Leblanc AF, Le-Grand B, et al. Novel roles for AhR and ARNT in the regulation of alcohol dehydrogenases in human hepatic cells. Arch. Toxicol. 91, 313–324, 2017.

302. Cederbaum AI. Alcohol metabolism. Clin. Liver Dis. 16(4):667–685, 2012.

303. Rusyn I, Bataller R, Alcohol and toxicity. J. Hepatol. 59(2): 387–388, 2014.

304. Tatsumi A, Ikegami Y, Morii R, et al. Effect of Ethanol on S-Warfarin and Diclofenac Metabolism by Recombinant Human CYP2C9.1 Biol. Pharmaceut. Bull. 32, 517–519, 2009.

305. Draganov P, Durrence H, Cox C, et al. Alcohol-acetaminophen syndrome. Even moderate social drinkers are at risk. Postgrad. Med. 107:189–195, 2000.

306. Watkins PB, Kaplowitz N, Slattery JT, et al. Aminotransferase elevations in healthy adults receiving 4 grams of acetaminophen daily: a randomized controlled trial. JAMA. 296(1):87–93, 2006.

307. Imani F, Motavaf M, Safari S, et al. The Therapeutic Use of Analgesics in Patients With Liver Cirrhosis: A Literature Review and Evidence-Based Recommendations. Hepat. Mon. 14(10): e23539, 2014.

308. Manchanda A, Cameron C, Robinson G. Beware of paracetamol use in alcohol abusers: a potential cause of acute liver injury. New Zeal. Med. J. 126, No 1383, 80–84, 2013.

309. Treating paracetamol overdose with intravenous acetylcysteine: new guidance. https://www.gov.uk/drug-safety-update/treating-paracetamol-overdose-with-intravenous-acetylcysteine-new-guidance

310. Macgregor S, Lind PA, Bucholz KK, et al. Associations of ADH and ALDH2 gene variation with self report alcohol reactions, consumption and dependence: an integrated analysis. Hum. Mol. Genet. 1;18(3):580–593, 2009.

311. Dasgupta A, Chapter 6-Genetic polymorphisms of alcohol metabolizing enzymes associated with protection from or increased risk of alcohol abuse. In 'Alcohol, Drugs, Genes and the Clinical Laboratory An Overview for Healthcare and Safety Professionals'. 107–116 Academic Press, Elsevier BV, 2017. https://doi.org/10.1016/B978-0-12-805455-0.00006-3

312. Fan Y, Chen ZY, Ye TT et al. Aldehyde dehydrogenase II rs671 polymorphism in essential hypertension. Clinica Chimica Acta 487, 153–160, 2018.

313. Li D, Zhao H, Gelernter J, Strong Protective Effect of The Aldehyde Dehydrogenase Gene (ALDH2) 504lys (*2) Allele Against Alcoholism And Alcohol-Induced Medical Diseases in Asians. Hum Genet. 2012 May; 131(5): 725–737, 2012.

314. Cui R1, Kamatani Y, Takahashi A, Usami M, et al. Functional variants in ADH1B and ALDH2 coupled with alcohol and smoking synergistically enhance esophageal cancer risk. Gastroenterol. 37(5), 1768–1775, 2009.

315. Breitkopf K, Nagy LE, Beier JI, et al. Current Experimental Perspectives on the Clinical Progression of Alcoholic Liver Disease. 33(10), 1647–1655, 2009.

316. Huang, CY, Kao CF, Chen CC, et al. No association of CYP2E1 genetic polymorphisms with alcohol dependence in Han Taiwanese population. J. Form. Med. Assoc. 117, Issue: 7, 646–649, 2018.

317. Motosugi U, Roldan-Alzate A, Bannas P, et al. Four-dimensional Flow MRI as a Marker for Risk Stratification of Gastroesophageal Varices in Patients with Liver Cirrhosis. Radiology 290, 101–107, 2019.

318. Pugh RN, Murray-Lyon IM, Dawson JL, et al. Transection of the oesophagus for bleeding oesophageal varices. Br. J. Surg. 60, 646–649, 1973.

319. Weersink RA, Bouma M, Burger DM et al. Evaluating the safety and dosing of drugs in patients with liver cirrhosis by literature review and expert opinion. BMJ OPEN 6, Issue: 10, e012991, 2016.

320. Artzner T, Michard B, Besch C, et al. Liver transplantation for critically ill cirrhotic patients: Overview and pragmatic proposals. World J. Gastroenterol. 24, 5203–5214, 2018.

321. Guy J, & Peters MG. Liver Disease in Women: The Influence of Gender on Epidemiology, Natural History, and Patient Outcomes. Gastroenterol. & Hepatol. 9, Issue 10, 633–639, 2013.

322. Stokkeland, K, Studies on alcoholic liver disease Karolinska University Press, ISBN-7140-853-3, 2006

323. Greenfield SF, Pettinati HM, O'Malley S et al. Gender Differences in Alcohol Treatment: An Analysis of Outcome from the COMBINE Study. Alcohol Clin. Exp. Res. 34(10): 1803–1812, 2011.

324. Chen H, Shen ZY, Xu W, et al. Expression of P450 and nuclear receptors in normal and end-stage Chinese livers. World J. Gastroenterol. 20(26): 8681–8690, 2014.

325. Delcò F, Tchambaz L, Schlienger R. Dose Adjustment in Patients with Liver Disease. Drug Saf. 28, Issue 6, pp 529–545, 2005.

326. Weersink, Rianne A.; Bouma, Margriet; Burger, David M.; et al. Evidence-Based Recommendations to Improve the Safe Use of Drugs in Patients with Liver Cirrhosis. Drug Saf. 41, 603–613, 2018.

327. Pentikainen PJ, Neuvonen PJ, Tarpila A et al. Effect of cirrhosis of the liver on the pharmacokinetics of chlormethiazole. Br. Med. J.2, 861–863, 1978.

328. Periáñez-Párraga L, Martínez-López I, Ventayol-Bosch P, et al. Drug dosage recommendations in patients with chronic liver disease. Rev. Esp. Enferm. Dig. 104, 165–184, 2012.

329. MacDonald AJ, Karvellas CJ. Emerging Role of Extracorporeal Support in Acute and Acute-on-Chronic Liver Failure: Recent Developments. Sem. Resp. Crit Care. Med. 39, Issue: 5, 625–633, 2018.

330. Gordon K, Stockwell J, Paul A, et al. 'Christine' dir. Carpenter, J. 110 min, Columbia Pictures, released December 9th 1983.

331. Bañares R1, Nevens F, Larsen FS Extracorporeal albumin dialysis with the molecular adsorbent recirculating system in acute-on-chronic liver failure: the RELIEF trial. Hepatology 57(3), 1153–1162, 2013.

332. Coutant DE, Hall SD, Disease-Drug Interactions in Inflammatory States via Effects on CYP-Mediated Drug Clearance. J. Clin Pharmacol. 58 (7), 849–863, 2018.

333. Zhong S, Han W, Hou C. Relation of transcriptional factors to the expression and activity of cytochrome P450 and UDP-glucuronosyltransferases 1A in human liver: co-expression. Network Analysis AAPS J. 19(1), 2017.

334. Zhuang YL, de Vries DE, Xu ZH. Evaluation of disease-mediated therapeutic protein-drug interactions between an anti-interleukin-6 monoclonal antibody (sirukumab) and cytochrome P450 activities in a phase 1 study in patients with rheumatoid arthritis using a cocktail approach. J. Clin. Pharmacol. 55(12), 1386–1394, 2015.

8 Role of Metabolism in Drug Toxicity

8.1 Adverse drug reactions: definitions

It is important to see the role of metabolism in drug toxicity as one component of the bigger picture of all adverse reactions suffered by patients during drug therapy. One view of this picture is on Figure 8.1. Different authors and textbooks classify these effects in various ways, but a very broad but convenient way to look at things can be to resolve all drug-adverse effects as either predictable (type A) or unpredictable (type B). This classification goes back broadly to the 1970s[1] and remains relevant[2]. This can be refined to accommodate whether a drug's effects are reversible or irreversible. Type A predictable effects can be reversible concerning pharmacological effects, but the effects from an overdose could well be irreversible as well as predictable. Type B unpredictable effects could be reversible or, if they are very severe, irreversible. At this point, it is worthwhile being more precise over the terminology of drug adverse reactions, in terms of what is reversible and what is irreversible and what is toxic. The term *toxicity* is a loose one and it has been used flexibly in this book so far. However, it is useful when looking at drug metabolism-mediated effects to define toxicity more accurately for the cell in particular:

Irreversible change in structure leads to irreversible change in function.

The key here is *irreversible;* it follows that a process that is not irreversible is not actually *toxicity* in the strict sense of the word. Reversible reactions are often either an intensification of the usual pharmacological response or, as has been mentioned previously, an 'off-target' pharmacological effect. There are, of course exceptions: this strict definition of toxicity is fine for the cell, but is not so clear for the relationship between the patient, their organs, tissues, and individual cells. Obviously, the effects of an anticancer alkylating agent are irreversible and toxic to a cell, but could ultimately save the life of the patient. Conversely, a heroin overdose reversibly inhibits central control of respiration and this leads to death, which is unarguably irreversible. In general, it is probably fair to say

Human Drug Metabolism, Third Edition. Michael D. Coleman.
© 2020 John Wiley & Sons, Inc. Published 2020 by John Wiley & Sons, Inc.

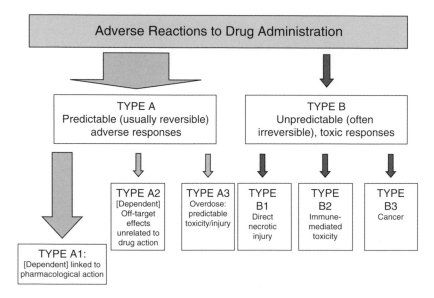

Figure 8.1   Summary of reversible (type A) and irreversible (type B) drug effects

that cumulative irreversible drug, pathological process, or pathogen-induced damage at the cellular level usually leads to patient morbidity and mortality.

8.2   Predictable drug adverse effects: type A

Type A describes the vast majority of adverse effects (Figure 8.1) and patients experience these in direct proportionality to the quantity of the drug and/or its metabolites in their tissues. These effects are sometimes described as *toxic,* but these are not usually irreversible effects. The exception is the damage caused by an overdose, which is often irreversible, even if the patient's life is saved. They can be resolved into three main causes, as follows.

8.2.1   Intensification of pharmacologic effect: type A1

This is proportional to drug concentration and happens if drug levels climb well above the therapeutic window. Anticonvulsants are membrane stabilisers, so at high concentrations they interfere with normal neuronal activity, causing sedation and confusion. Similarly, high concentrations of anticoagulants cause excessively long clotting times. Some drugs may display a series of known pharmacodynamic effects at therapeutic window levels, but exert other unwelcome effects at high concentrations. Some beta-blockers such as propranolol, at higher doses cause central effects such as nightmares. Naturally, when drug levels fall, the excessive pharmacodynamic effects subside also. These adverse effects, also known colloquially as *side effects,* are mostly an intensification of the main effects, and are thought to be the cause of more than 80% of patient problems with drug therapy. Many patients will experience an unpleasant concentration-dependent drug effect at some point in their lives. As we have seen in earlier chapters, drug metabolism-related changes in clearance caused by induction or inhibition of biotransformation systems can have profound and occasionally lethal effects on the patient. Other reasons for the intensification of drug effects include renal problems, overdosage, and problems with dosage calculations, or being too old, too young, or too sick.

8.2.2   Off-target reversible effects and methaemoglobin formation: type A2

As well as off-target pharmacodynamic effects caused by parent drugs, it is possible that at least one or other of the metabolites may disrupt cellular function in a way that is unrelated to the pharmacological effects of the drug, but the disruption is reversible and predictable. This situation is again dose related, in that the more drug that is absorbed, the more is cleared through the pathway that forms a 'problem' metabolite. This type of adverse effect, though reversible, has the potential to make a drug almost intolerable to take from the patient's viewpoint and can even be lethal. A good example of a reversible adverse effect is methaemoglobin formation, where the appearance of the patient's blood becomes a passable impression of chocolate milk.

To see the link between a product of drug metabolism and an adverse drug effect, it is first necessary to understand the endogenous system involved. Haemoglobin is a molecule that normally becomes reversibly oxygenated, rather than irreversibly oxidized. Our continued existence depends on the difference between those two terms. Haemoglobin, as you know, is oxygenated to transport the gas from lung to tissues and release it where required. It is a tetramer, which means it has four protein subunits, each of which contains an iron molecule as Fe^{2+}. The iron molecules bind the four oxygen molecules. The oxygen binding is dependent, like many enzymes, on the ability of the metal to gain and lose electrons, rather like charging and recharging a battery as mentioned previously. It is clear that a molecule like haemoglobin is potentially reactive and is going to need protection from oxidation so it can continue to function.

Thus, the erythrocyte has evolved to maintain haemoglobin so that it carries oxygen with minimal damage or change in structure for the duration of its lifespan. It accomplishes this with one of the highest intracellular GSH levels (Chapter 6.4) in the body after the liver and a series of protective enzymes, more of which later. Normally, the haemoglobin iron molecules (Fe^{2+}) bind the O_2 and form superoxoferrihaem complexes ($Fe^{3+}O_2^{-\cdot}$). These dissociate at the tissues and 99 times out of 100, the Fe^{2+} is restored and O_2 is delivered to the tissue. The 1 time out of 100, the oxygen retains the electron and becomes superoxide ($O_2^{-\cdot}$) and the Fe^{2+} becomes Fe^{3+}. This is not good, as Fe^{3+} will bind anything but oxygen, such as anions. So the Fe^{3+} means that this haemoglobin monomer is now methaemoglobin and useless for oxygen carriage[3]. Since this is a natural consequence of transporting a reactive gas with a reactive protein, the erythrocyte carries two systems for reducing the oxidized Fe^{3+} to Fe^{2+}, so restoring the function of haemoglobin. The systems are NADH diaphorase and NADPH diaphorase. The NADH diaphorase is similar to the cytochrome b_5 redox partner discussed in Chapter 3.4.9 and it operates for 95% of the time, as the NADPH system is incomplete for some reason and requires an artificial electron acceptor for full operation. If it were not for NADH diaphorase, methaemoglobin formation would be around 4–6% per day – more if you smoke. If you are a non-smoker and we measured your methaemoglobin levels now, they would be less than 1%, so the system is adequate to maintain haemoglobin in normal circumstances[3].

Several xenobiotics can react with haemoglobin to form methaemoglobin. Nitrites are moderately efficient at this process[5], but the aromatic hydroxylamines are particularly effective in forming methaemoglobin. They oxidize haemoglobin to methaemoglobin in two stages. Initially, the hydroxylamine directly reacts with oxyhaemoglobin to form methaemoglobin. This is known as a co-oxidation, as the hydroxylamine will be oxidized to a nitrosoarene[4]. This alone would not account for the speed at which methaemoglobin can form in patients with high levels of hydroxylamine metabolites in their blood (Figure 8.2). The nitrosoarene formed by the initial reaction is then rather unfortunately reduced to the hydroxylamine by all that erythrocytic GSH, so allowing the hydroxylamine to oxidize another oxyhaemoglobin molecule. The resultant nitrosoarene can then be re-reduced to the hydroxylamine and oxidize another oxyhaemoglobin and so on[4,5].

The initial oxidation is thus amplified by GSH, and the presence of millimolar levels of GSH is like pouring fuel on a fire. Each hydroxylamine molecule can oxidize at least four oxyhaemoglobin molecules. The result is a very rapid process, where as soon as the metabolite is produced, methaemoglobin starts forming in significant quantity, in direct proportion to the level of the metabolite released by the liver, which is of course in proportion to the drug dose[5]. In general, aromatic amine methaemoglobinaemia is more persistent and harder to treat than that caused by other drugs and toxins[6].

Clinical consequences of methaemoglobin

Methaemoglobin is measured as a percentage of haemoglobin. However, even modest levels affect oxygen release, due to the Darling–Roughton effect[7].

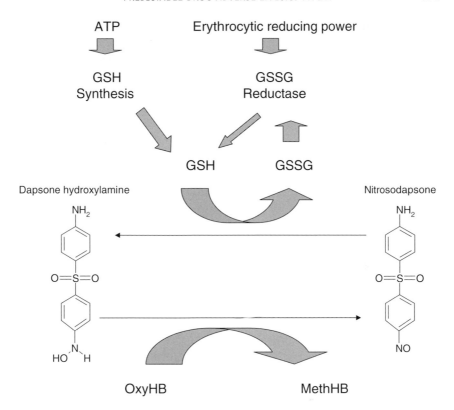

ATP   Erythrocytic reducing power

GSH   GSSG
Synthesis   Reductase

GSH   GSSG

Dapsone hydroxylamine   Nitrosodapsone

NH_2 NH_2

$O=S=O$ $O=S=O$

$HO\overset{N}{}H$ NO

OxyHB   MethHB

Figure 8.2   Basic process of methaemoglobin formation in the human erythrocyte. This is repeated many times per second in a futile cycle that is limited only by the levels of GSH in the erythrocyte

You might recall that haemoglobin is a tetramer and binds four oxygen molecules as oxyhaemoglobin. If a single Fe $^{2+}$ of one of the monomers is oxidized to an Fe^{3+}, then the remaining oxygen molecules are much more tightly bound and it is harder for them to escape and oxygenate tissues. If two monomers are oxidized then oxygen release is even more difficult. This means that a small percentage of methaemoglobin, say 5–10%, has a much greater effect than just removing 5–10% of haemoglobin[7]. The effects on the patient of methaemoglobin formation are a reflection of increasing inability of their tissues to receive enough oxygen. So at low levels, 4–6%, people of fair complexions may look slightly bluish and may have a mild headache. As levels are increased to 10–15%, this extends to headache, fatigue and sometimes nausea. At levels of 20–30%, hospitalisation is usually required, with the intensification of the previous symptoms plus tachycardia and breathing problems. Higher levels (50% plus) lead to stupor and loss of consciousness, with >70% leading to death[3]. Methaemoglobinaemia is a drug-dependent effect that may be either acute or chronic.

Acute methaemoglobinaemia

A particularly hazardous situation in terms of the impact of methaemo-globin formation can be as a result of benzocaine usage in anaesthetized patients. The local anaesthetic is sprayed on the oropharynx prior to intuba-tion, although not used much in the United Kingdom, it is still employed in some countries. Many formulations of the drug are around 14–20% and 3 one–second bursts of the spray could deliver perhaps 600 mg of the drug. A number of reports have shown that methaemoglobin levels of 30% or more can result in 30–40 minutes (Figure 8.3). This can be a serious prob-lem, as blood oxygenation is usually measured by pulse oximeters and these units overestimate blood oxygenation in the presence of increasing methaemoglobin[3,8].

This problem, coupled with the lack of recognition of the visual signs of methaemoglobin in a patient on the operating table, can lead to dangerously high methaemoglobin levels in half an hour or so, without the awareness of the staff. If in doubt, a CO oximeter must be used to reliably measure methaemoglobin formation. Benzocaine is unable to cause methaemoglobin itself, so it is thought that it is oxidized to a hydroxylamine, almost certainly by CYP2C9 and possibly 2C8, although CYP2E1 is thought not to be involved[8,9]. Other local anaesthetics can cause the effect also[10].

Rather counterintuitively, acute methaemoglobin formation has actually been a therapeutic goal: this would appear bizarre, but cyanide is an anion, so it binds very strongly to methaemoglobin, forming cyanomethaemoglobin[11]. The methaemoglobin has the effect of 'vacuuming' the cyanide out of the tissues and holding it in the blood, before a thiosulphate is given to finally facilitate the urinary excretion of the cyanide. Sodium nitrite has been employed as an antidote to cyanide poisoning; although it does not form large amounts of meth-aemoglobin quickly enough in cyanide overdoses where, as you can imagine, time is of the essence.

A series of phenones were designed in the 1950s and 1960s that were intended to be oxidized to potent methaemoglobin formers that would be used to protect military personnel prophylactically from cyanide toxicity[11]. How these personnel might cope with 15–20% methaemoglobinaemia and maintain their combat capabilities seems difficult to imagine. However, cyanides are one of the cheapest 'weapons of mass destruction' so they remain a long-term threat. The most extensively tested agent was 4-aminopropiophenone (4-PAPP)[11], which is also oxidized to a hydroxylamine and acts in a similar way to any aromatic hydroxylamine (Figure 8.3). Interestingly, 4-PAPP now lives on as part of elaborate devices designed to kill stoats in New Zealand[12], which since their introduction to control rabbits in the 1880s have been methodically annihilating the indigenous bird population.

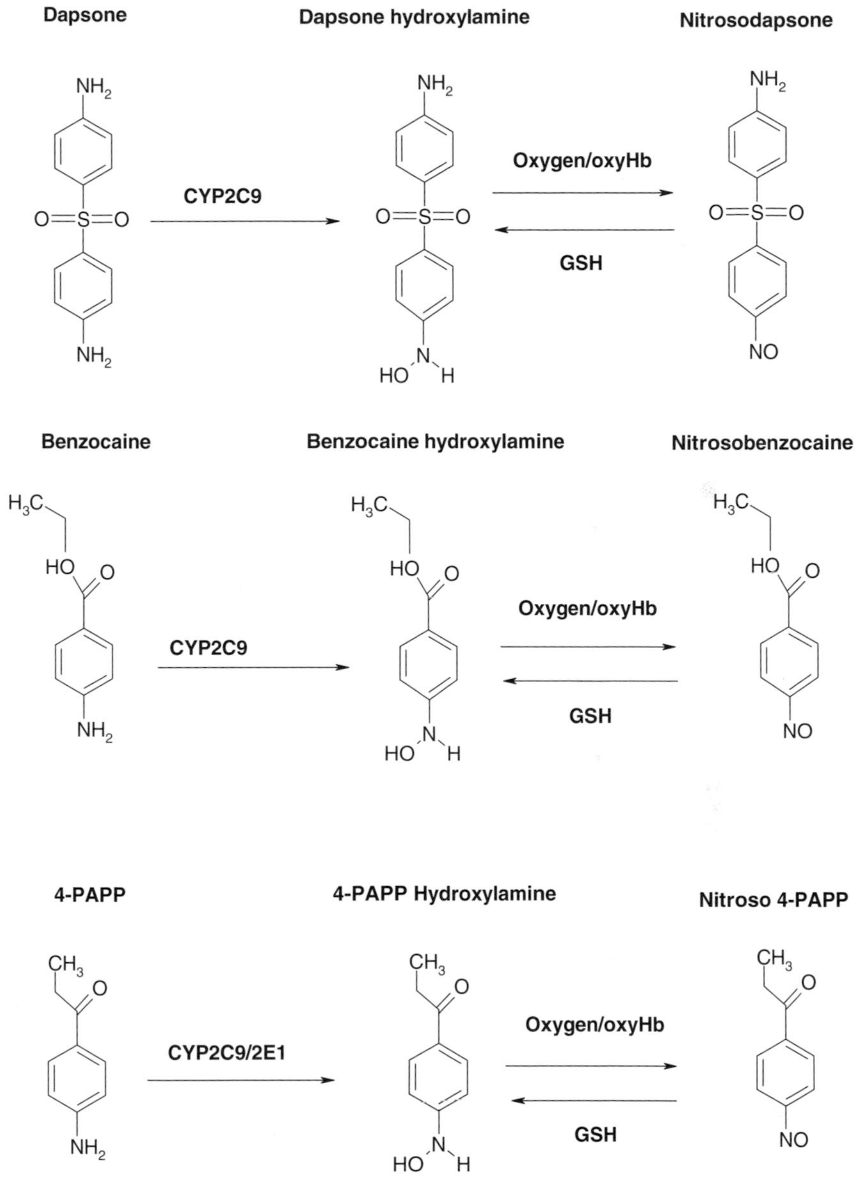

Figure 8.3   Production of various methaemoglobin-forming species from dapsone, benzocaine and 4-aminopropiophenone (4-PAPP). In each case, the hydroxylamine is formed from the parent compound by CYP-mediated oxidation and the presence of oxygen and/or oxyhaemoglobin (oxyHb) converts the agents to nitrosoarenes, whilst erythrocytic GSH re-reduces them to their hydroxylamines within the co-oxidation cycle

Chronic methaemoglobin formation

It is inconceivable that a new drug with an aromatic amine group would be approved for clinical usage today, on the grounds of toxicity. Of the few drugs that possess an aromatic amine, sulphonamides manage to retain their therapeutic place in HIV patient pneumonia treatment despite their cytotoxic and immuno-toxic hydroxylamine metabolites. However, fortunately, they form negligible levels of methaemoglobin. The sulphone, dapsone, is still the mainstay of the treatment of leprosy, dermatitis herpetiformis and conditions where neutrophil migration into tissues leads to inflammatory damage. Despite successful efforts to develop non-methaemoglobin forming and effective analogues of dapsone, the market is just not large enough to sustain their development[13]. This means that patients are essentially stuck with this amine and its dose-dependent methaemoglobin formation. Dapsone is usually given at 100 mg/day for leprosy and anything from 25 mg/week to 400 plus mg/day for dermatitis herpetiformis. Methaemoglobin peaks at around three hours post-dosage and the patients complain of a sort of permanent 'hangover' effect, although tolerance varies widely[14]. For an individual taking 100mg/day, methaemoglobin levels may peak at 5–8%, depending on their level of hydroxylamine production[15]. The half-life of methaemoglobin reduction is just under 1 hour, so after the initial pulse, the deleterious effects of the methaemoglobin wear off within three to five hours. Many patients have to take dapsone for several years, leading to significant impact on quality of life, although co-administration of cimetidine can ameliorate this situation (Chapter 5.8.6). Primaquine, the 8-aminoquinoline antimalarial also forms methaemoglobin, although this is used in much shorter courses for the elimination of *Plasmodium vivax, malariae,* and *ovale* (recurrent) malaria in those who are not returning to the area where they were infected. Fortunately, the drug can be effective in eliminating recurring malaria from a single 45 mg dose, so methaemoglobin formation is only a temporary problem. If longer courses are used, then methaemoglobin formation can be considerable, but usually better tolerated than that of aromatic amines[6,16].

Methaemoglobin as a protective process

Although potentially lethal, methaemoglobin formation is not really toxicity, as we have seen, the erythrocyte can reverse it. Once methaemoglobin is reduced to haemoglobin, it can resume its oxygen carriage function, the erythrocyte replaces its thiols and is physically undamaged. The co-oxidation process effectively 'ties up' the hydroxylamines in the cycle and eventually releases them (via glutathione conjugates) as the parent drug. The erythrocyte thus performs a temporary detoxification of the hydroxylamine. Although methaemoglobin formation seems like an own goal, effectively the structure of the erythrocyte has been protected from the protein reactive nature of the nitroso derivative as haemolysis does not occur[14]. The importance of GSH in the protective cycle is underlined by G-6-PD-deficient individuals (see below) suffering more haemolysis due to their inadequate GSH maintenance under high toxic pressure. Once the parent drug escapes the erythrocytes it could be either acetylated or re-oxidized by the liver, or it may undergo conjugation to an N-glucuronide. The hydroxylamine effectively cycles its way through erythrocytes for several weeks without inflicting significant damage, although eventually erythrocyte lifespan is shortened. However, the more fragile mononuclear leucocytes are easily destroyed by nitrosoarenes[14].

Glucose-phosphate dehydrogenase deficiency (G-6-PD)

This condition occurs in all races, but is most common in those of Afro-Caribbean descent, with about 100 million individuals affected worldwide. This genetic polymorphism of G-6-PD is normally not an issue, except when the individuals are exposed to methaemoglobin-forming drug metabolites. This enzyme supplies the majority of the reducing power of the erythrocyte and if it is poorly or nonfunctional, there is only sufficient reducing power available to supply GSH to protect erythrocytes from background levels of reactive species. As soon as the cells are exposed to high concentrations of hydroxylamines or primaquine metabolites, there is insufficient reducing power to tie up nitrosoarenes in the co-oxidation cycle and no detoxification occurs as little methaemoglobin formation happens. The nitrosoarene is then free to react with the structure of the erythrocyte. This causes the normal system of the erythrocytic 'sell by date' to be prematurely activated (they usually last 120 days) and the spleen automatically removes them from the circulation. This can happen so quickly that anaemia can result in days, with associated hepatic problems with the processing of large amounts of now waste haemoglobin. G-6-PD patients that must take either dapsone or primaquine must only receive half or less than the recommended dosage of the drugs. G-6-PD effectively converts what would be a reversible drug effect into true toxicity. If the individual is homozygous for Gilbert's syndrome (Chapters 6.2.4 and 7.2.7), they are already exquisitely sensitive, as their inadequate UGT1A1 expression will be rapidly overwhelmed by the large amounts of bilirubin formed during the erythrocytic destruction[17].

8.2.3   Predictable overdose toxicity: type A3

How drugs can cause irreversible effects

A drug overdose can often target an organ or tissue and whilst the patient could well survive, the damage may be irreversible which could cause permanent impairment. It is useful to consider from basic principles how drugs can cross the line from a pharmacological to a toxicological impact on cells, tissues and organs. All chemicals and of course drugs, interact with each other through three main routes. These range from relatively weak associations, such as van der Waals forces and hydrogen bonds, through to more potent ionic bonds and finally covalent bonds. Perhaps analogies for these interactions could be a child's toy magnet for van der Waals and hydrogen bonds, an electromagnet for ionic bonds and a spot-weld for a covalent bond. The first two interactions are reversible, varying in strength, but leading to no permanent changes in the participants. They are the basis for many pharmacological processes. A covalent bond is the product of a chemical *reaction*, rather than just an interaction and is not desirable unless the drug is intended to weld itself to its receptor to destroy its functional capacity. The covalent process is occasionally pharmacological but more often toxicological. Covalent binding is the process where anticancer drugs cross-link DNA or drugs that react with steroid receptors. However, the vast majority of drug action is propagated through reversible bonds, where a receptor is activated and the drug leaves to interact with other receptors, just as their endogenous counterparts usually do.

Unfortunately, drugs do cause unintended irreversible changes to organelles, cells and tissues, which could escalate towards organ damage. There are several ways that drugs might cause irreversible toxic effects, involving four interrelated pathways:

1. Drugs and/or their metabolites might chemically react directly with a variety of cellular structures, changing their structure and thus their function.

2. Drugs/metabolites may alter the expression of key genes in cellular homeostasis that may promote/cause irreversible damage.

3. The cumulative effects of 1 and 2 trigger a series of irreversible events in organelles such as mitochondria, which lead to cell death, either by apoptosis or worse, necrosis.

4. The cumulative effects of 1–3, may act to cause one group of cells to destroy another, such as by eliciting an immune response, which recruits cellular or antibody-mediated attacks on tissues.

These observations lead to several key questions: first, how would any given drug be responsible for these undesirable effects, what chemical interactions are involved and is it the parent drug and/or metabolites that are responsible? Starting with pathway 1, reaction with cell structures requires covalent binding and this would only occur in a highly unstable drug. With some exceptions (penicillin and alkylating agents), highly reactive entities are always weeded out in the drug discovery process, so pathway 1 is most likely to be caused by prior biotransformation to a reactive species either in or near to the tissue involved. It is not really possible to be precise about the processes occurring

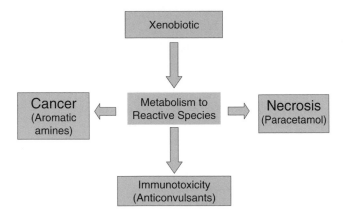

Figure 8.4   Main consequences of reactive species formation due to xenobiotic metabolism in different organs and tissues

in pathways 2 and 3, which may be caused by parent drug or by stable or reactive metabolites. So biotransformation is probably the major determinate of irreversible drug toxicity, but it is important to recognize that it is not the only factor involved (Figure 8.4). If processes 1–3 result in necrosis, this is more likely to lead to number 4, the attention of the immune system (see Section 8.3.4).

Role of biotransformation in causing drug tissue damage

There are many routes whereby reactive species may be formed. As you hopefully will have gathered, CYP enzymes can radically rearrange the structure of a molecule to make it more water-soluble and during this process molecular stability is usually reduced. Oxidation can transform some molecules from innocuous agents to highly reactive species that are lethal to cells if formed in sufficient quantity. In general, the less stable a compound is, the more likely it is going to react with cellular structures.

The CYP enzymes function in a similar way that machine tools do, where, say, a robot welds a piece of bodywork onto a car. The metal is subject to an intense, concentrated assault in a specific area that is designed to form a product. You can imagine what would happen if a live grenade was subject to this treatment. That would be the end of the robot. Obviously the robot is pre-programmed to weld anything of the appropriate dimensions that it is presented with, even something that could destroy it. You might feel that this is the Achilles' heel in drug metabolism. However, two billion years of research and development and an unlimited budget has indicated that there is really no other way to metabolise otherwise stable chemicals, and for the occasional reactive byproduct, there are sufficient repair and protection systems, as detailed in

Chapter 6 (nrf2, GSTs, GSH, etc.). Just as you are fully insulated inside your car from the engine's noise, heat, and emissions, the efficiency of the whole bio-transforming system is reflected in the relative rarity of organ damage in most therapeutic drug use.

On a molecular scale, some chemical structures that are subject to the rig-ours of CYP-mediated oxidation form reactive products because of unique inherent features of their structure. Examples include strained three-membered rings and epoxides. These structures are the chemical equivalent of the old explosive, nitroglycerine, which could be detonated by shock alone. Some reactive species, such as the mechanism-based inhibitors discussed in Chapter 5.4, can bind covalently to the active site and the enzyme is no longer functional. For biological function to continue, more enzymes must be syn-thesized. Drugs or chemicals that cause this effect are often termed suicide inhibitors, and as we saw in Chapter 5.4.3, grapefruit juice and norfluoxetine cause this effect.

The long period of inhibition of the CYPs that formed the reactive metabolites reflects the time taken for more enzymes to be synthesized. These metabolites are so reactive that they are paradoxically usually no problem, as they just destroy the enzyme that formed them and cannot reach the rest of the cell, although the immune system can sometimes detect damaged CYPs. Indeed, after all the available CYP has been inactivated by these metabolites, no more metabolite will be formed until more CYP is synthesized (Chapter 5.4.3). The rest of the parent drug may well leave the cell by other pathways. At the other extreme are metabolites such as hydroxylamines, which are no immediate threat if they are stabilised by cellular thiols or antioxidants such as ascorbate. These metabolites can be stable enough to travel through cells and leave the organ in which they were formed and enter the circulation, although this does depend on surrounding functional groups on the molecule[18]. Erythrocytes can thus detox-ify some of them through GSH cycling as described in Section 8.2.2. In certain conditions, they can spontaneously oxidize forming nitroso derivatives, which are tissue reactive and cytotoxic. Further metabolism is necessary before they can be detoxified. In between these extremes are metabolites that can react with any cell structures that are short of electrons and seek electron-rich structures. Potent electrophiles such as nitrenium ions (N^+) will react with nucleophiles, which are electron rich. If a CYP or any other enzyme forms a species that is missing electrons and seeks them, or is too electronegative, the net effect is a reactive species which has the potential to attack cellular proteins, DNA and membrane structures, forming covalent bonds, which can do sufficient damage to necessitate the resynthesis of that structure. It is important to consider also, that reductive metabolism can form equally reactive species that cause similar cellular damage to oxidatively produced species. Whatever process forms reactive species, the likely result will be some form of irreversible binding to cellular macromolecules.

Cellular consequences of CYP-mediated covalent binding

The obvious question is how covalent binding causes cell, then tissue, then organ damage. When a reactive species is formed in a cell, it can react with cellular organelles, enzymes, nuclear membranes, DNA, and the structure of the cell membrane as already stated. However, the hypothesis that a high rate of covalent binding to cellular structures leads directly to cell death is an oversimplification. When animals have been treated with antioxidants prior to exposure to a reactive species, the animals survive, despite their levels of covalent protein binding, which are as high as untreated animals that have developed fatal organ toxicity. So the fate of the cell is subject to a competition between various factors:

- The rate, quantity, and reactivity of the toxin formed

- The extent of reactive 'secondary' toxins (superoxide, various free radicals) formed from the initial reactive species

- The extent the cell can defend itself from the reactive toxin by rendering it harmless

- The period of time elapsed before cellular defensive resources are overcome

- Which specific molecules are damaged in the cell where irreversible damage occurs

- The extent of possible repair and restoration

- Whether the intra-and extracellular damage attracts the possibly malign attention of the immune system

These competing processes might lead to three main cellular outcomes: necrosis, where destructive forces overtake the cell; apoptosis, which might lead to an orderly dismantling of the cell and its contents; or finally the damage may be attenuated and repaired, resulting in cell survival. In the case of Type A3, drug overdose, necrosis is unfortunately the most likely outcome.

Sequences of events in necrosis

This process describes the effect where a cell sustains a great deal of damage to key cellular systems in a relatively short time (rather like the end of a James Bond film) that it cannot either protect itself from the toxic species or repair its systems fast enough to keep pace with the damage. In extreme cases, the cells may die before they can even initiate apoptotic processes. Cellular oxidative stress can cause such damage, and this process remains the focus of research in wide range of conditions, from aging through to disease pathology[19], as well as in drug toxicity[20]. This is usually defined as marked imbalance between reactive species production and detoxification. It is more likely that reactive drug metabolites, rather than the parent drug, will cause oxidative stress by themselves as well as by promoting redox cycles, which generate more reactive oxygen species as well as a host of destructive free radicals. This, in turn, accelerates GSH consumption, consuming reducing power and ATP[19,20]. Reactive species may derange other cellular processes, such as fatty acid metabolism and may also react with membrane lipids; this effect triggers a cascade where the lipids will generate their own radicals as they oxidize each other in a chain reaction. This process has a forest-fire effect on the membrane lipids, and the structure of the membrane will eventually break down, causing cell contents to escape[20]. Although cells such as hepatocytes are extremely robust, if the reactive species are produced in very large quantities over a short timescale, such as after an overdose, then the drug is oxidized in sufficient quantity to form enough toxic species to overwhelm even hepatic cellular defences, such as with paracetamol, as we will see. Other drugs may promote a more gradual but sustained oxidative stress that could be tolerated for some time, but may eventually defeat cell detoxification at lower dose levels over a longer period of time. In both cases, recent research has concentrated on the emerging and critical role of mitochondria in drug-induced acute and chronic organ toxicity[20].

Role of mitochondria in cell life and death

As you should know from basic biochemistry, mitochondria provide virtually all of our energy, and whilst a mononuclear leucocyte might only possess a few of these organelles, cells requiring high energy output, such as in the brain, the heart, or the skeletal muscle have thousands of them, and if more demands are made on the cell, more mitochondria duly appear. They have a smooth outer membrane that is permeable to many substances as well as nutrients, as well as a complex inner membrane, which is impermeable, and anything that enters must have a specific transporter. The inner membrane folds into cristae that are the site of the electron transport chains and the cristae reconfigure themselves according to energy demand[21].

How the cristae monitor and affect respiration involves extremely complex processes and are not fully understood[21]. The cristae have a negative membrane potential with respect to the rest of the cell, which is essential for them to make large quantities of ATP from the oxidation of our food components. In some high-energy output cells, mitochondria arrange themselves like large banks of batteries, rather like those in electric cars. They have their own DNA (mtDNA) with the means to express and control it, and mtDNA does not even replicate at the same time the nuclear DNA does. In essence, they look suspiciously like they might have been separate cells at one time, and it is now believed they were once indeed free-living bacteria[22]. Along with our CYPs and other biotransformational enzymes, we have appropriated them and their technology lock, stock, and barrel to power ourselves.

Whilst mitochondria are our passport to continued cellular life, providing they are not overwhelmed by necrosis, they also make the key decisions on whether a cell dies or not in an orderly fashion. Mitochondria are sensitive to a variety of different cellular stress factors that can drive them towards the decision to enter apoptosis, or programmed cell death. There are several reversible stages of apoptosis and right up to a certain stage, the process can be averted if the stress factor levels fall. Apoptosis has several very important functions. For example, in our uterine development, precisely sequenced programmed cell death is why we are not born with our fingers webbed together. Concerning a potentially toxic situation, if a cell is under high stress, its internal logic and various sensor systems will have reported that its DNA, both mitochondrial and cellular, is likely to be at risk or have already been damaged. In order to prevent the propagation of possibly dangerous genetic faults towards future cellular generations, the DNA is shredded during the latter stages of apoptosis, as would sensitive documents be, before a key installation is captured by the enemy. An apoptotic orderly dismantling of the cell and its contents prevents leakage of damaging enzymes such as lysozymes, but also ensures that the immune system does not detect any debris during the death process, which might cause some major and unwanted response (Section 8.3.4). So a completed apoptotic process is termed non-inflammatory cell death[23].

Many internal and external factors can damage mitochondrial inner membrane potentials, and this can lead to a fall in ATP output, which can be an apoptotic signal. Others include reactive oxygen species; various radicals and the oxidative stresses they encourage are some of the main causes of the inner membrane potential attrition, which is in itself apoptotic. These species may also derange electron transport and inhibit mitochondrial replication, as well as threaten the protein ultrastructure of the organelles.

If such apoptotic factors increase in quantity and effect, they start to promote the activity of what are termed *pro-apoptotic factors*[24]. These include factors known as BAX and BAK. If the toxic factor danger recedes, then the anti-apoptotic factors (the BCL-2 family of proteins)[23] can regain control and normal service is resumed. If the BAX-type factors gain the ascendency due to a resumption of serious oxidative stress, then they will promote a key stage of the apoptotic process where they link together and form pores in the outer mitochondrial membrane, increasing MOMP (mitochondrial outer membrane permeability)[23].

Some of the inner membrane contents, such as cytochrome c, then start to leak out into the cell through the BAX pores (mitochondrial permeability transition pores)[24]. Once cytochrome c escapes, it in turn promotes the activity of a family of proteins known as caspases that proceed to engage the full apoptotic process. Interestingly, it has been shown that if the caspases are blocked, then the mitochondrial inner membrane itself and the mtDNA is pretty much forced out of the pores, like toothpaste out of a tube, and this will trigger an immune response[23]. Hence, the caspases are essential for apoptosis to be orderly and noninflammatory.

As outlined in the previous section, during a drug overdose situation the scale of the oxidative damage overtakes apoptosis, leading to all the damaging consequences (inflammation, release of toxic cellular contents) that apoptosis is intended to avert[24]. Mitochondrial toxicity and apoptosis are under intensive study from a variety of perspectives, ranging from neoplastic disease (tumour cells circumvent apoptosis) to drug-induced cell death. Interestingly, the widespread toxicity of the first anti-HIV drugs, such as AZT, highlighted drug-induced toxicity of mitochondrial function, in terms of oxidative stress and its impact on gene expression[25].

Sites of biotransformational-mediated injury: the liver

Ironically, although the liver boasts extremely comprehensive detoxification defences and astonishing powers of recovery, hepatotoxicity is one of the most common reasons for the failure of a candidate drug in clinical trial, and even the withdrawal of a new drug from the market[26]. This is mainly because the liver contains the largest concentration of biotransformational enzymes in the body and bears the greatest burden in clearing drugs and other xenobiotics. This means that it is more likely to form reactive metabolites in quantities sufficient to cause cellular injury or to trigger an immune response. In addition, the central position of the liver in homeostasis, as we saw with the consequences of cirrhosis and MARS in Chapter 7.8, makes severe hepatic injury a life-threatening event that may only be remedied by a transplant.

Unlike toxicity in many other tissues, hepatotoxicity has been studied extensively in detail over the last few decades, and there are accurate diagnosis criteria for drug-induced liver injury. These validated criteria are applied to existing or newer drugs by regulatory organisations. They are termed *Hy's rule* by the USFDA, in honour of the work of the renowned hepatologist Prof. Hyman J. Zimmerman[27]. The rule states that if ALT levels equal or exceed three times the upper limit of normal (ULN) and serum bilirubin equal or exceed twice the ULN, then mortality can range from 10–50%, depending on the drug and patient pathology[26,28].

Hepatotoxicity may have many causes, but the commonest clinical manifestation of liver failure remains necrosis due to drugs. Often the term drug-induced liver injury (DILI) is used to describe liver damage in general, probably most often in relation to overdoses, as the term IDILI, or idiosyncratic drug induced liver injury describes liver damage with no specific cause which sometimes, but not always, involves the immune system (see Section 8.3.4). The following sections look at probably the best-known example of DILI, which is paracetamol.

Paracetamol

Although a number of compounds can cause liver necrosis and death in overdose, none of them is extremely cheap and available in virtually every retail outlet in the country. Every year in the United Kingdom, we get through around 1700 tonnes of paracetamol in various formulations and more than 40,000 people overdose on this drug annually. In 2017 the drug killed 218 people in the United Kingdom, which is less than half of those that died from antidepressant overdoses and just over a ninth of the opiate fatalities[29]. Paracetamol (also known as acetaminophen) is often the first and most convenient choice for those who wish to use an overdose to end their lives, although paradoxically, it is the safest way to commit suicide, as the chances of death are less than one in a hundred. It seems that intentional overdoses are less lethal than unintentional ones (very rare), and children are less susceptible to fatal overdose than adults.

The public is obviously aware that paracetamol can kill, hence the numbers taking it in overdose; however, despite more than 40 years of publicity, few are aware of the drawn-out and painful process that can lead to death with this drug. It is still used in suicide attempts because it works eventually (if the patient does not find medical help) and it is freely available in lethal amounts if you have a serviccable disguise and/or are prepared to walk to a few shops. It is possible that those who take overdoses feel that paracetamol is an ideal 'cry for help' drug, where there is time to reconsider, and this behaviour is associated with the aged 16–24 group, who are most likely to overdose. Ironically, you are much more likely to die of a paracetamol-related overdose if you are 40 and above. This is due to factors such as delays in seeking medical help and liver damage due to alcohol.

If the patient presents, or is carried to treatment, the rescue therapy usually works, but if it fails, then their relatives might hope for the fall-back position of a new liver. However, in the last chapter, it was underlined how difficult this can be with cirrhotic patients[30] and only around 20 transplants per year take place in paracetamol overdose patients[31]. Even in the light of restricted paracetamol pack sizes, which have made some impact in death rates[32], it remains incomprehensible to many health professionals that paracetamol is still available at all. Suicide is usually a result of the 'balance of the mind being disturbed', so it does not make sense to provide a readily usable method that is so easily obtainable. It is interesting that a number of much more therapeutically valuable drugs have been withdrawn from the market after accounting for a fraction of the deaths attributable to paracetamol.

Paracetamol: mechanism of toxicity

To say that paracetamol toxicity has been exhaustively investigated is perhaps an understatement, but the basic facts are now well established. It is often not considered that the drug can be highly nephrotoxic through deacetylation to 4-aminophenol, but the impact on the liver is usually the pressing problem. Around 95% of paracetamol is cleared hepatically to an O-glucuronide (UGT1A1/6/9 and UGT2B15) and sulphate (SULT1A1, 1A3/4 conjugates and about 2% escapes unchanged (Figure 8.5). A small fraction, maybe 1–3%, is oxidized mainly by CYP2E1 (with a fraction cleared by 1A2 and 3A4/5) to the reactive species every Toxicology student knows, N-acetyl-*para*-benzoquinone imine (NAPQI). Recent evidence suggests that CYP2D6 is also involved but probably only in overdose[33]. There are a number of other putative reactive species, such as dibenzoquinoneimines, but it is thought that NAPQI is the toxicologically important species. This compound is known to bind covalently to hepatocytes in various cellular areas, such as the cell membrane. Normally, however, there is negligible binding as the NAPQI is efficiently conjugated to GSH. It was believed that GST Pi was responsible for catalysing this reaction, although it may be a combination of direct reaction with GSH and catalysis through other antioxidant enzymes[33]. The net result is the excretion of NAPQI mercapturate in urine in small amounts[33]. In its therapeutic dosage range, it must be stressed that paracetamol is harmless. The most a 70 kg adult can take a day without problem is about 4 g. For children, this is adjusted to 50–60 mg/kg in children, depending on which authority is cited. Most agree that no child should receive more than 4 g/day in total.

Overdose of paracetamol is considered to be around 7–8 g (100–115 mg/kg) in an adult and more than 150 mg/kg in children. In overdose, the profile of paracetamol metabolism is similar, but sulphation is saturated first, then glucuronidation[33]. The proportion of the dose cleared to NAPQI in overdose does increase to around 15%, and of course 15% of a 20 g overdose is a very large amount of toxic metabolite compared with a recommended dose. Obviously, the demand for GSH will be high to detoxify the reactive species, and you might recall from an earlier chapter that GSH is effectively 'thermostatically' controlled (Chapter 6.4.2) in that if more is used, more will automatically be made to maintain organ concentrations at their normal preset level. Accelerated GSH consumption leads to an increase in the recycling of GSSG to GSH and the synthesis of more GSH from cysteine, glycine, and glutamate. Up until 10–12 hours after the overdose, the GSH system responds to the increased load and GSH levels are maintained through recycling and resynthesis. The one component of GSH manufacture that is in limited supply in a hepatocyte is unfortunately the key one, the thiol-containing cysteine. More can be transported in, and another sulphur-containing amino acid, methionine, can be converted to cysteine to meet the demand. Between 8 and 16 hours or so, GSH consumption by NAPQI gradually exceeds demand and cellular supplies of cysteine are exhausted[20,33]. Since GSH is acting as the cellular 'fire-extinguisher'

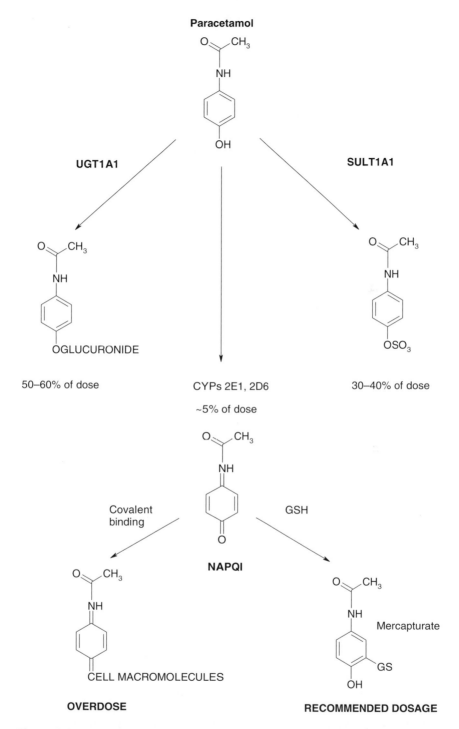

Figure 8.5 Main features of paracetamol metabolism. Aside from the toxic NAPQI (N–acetyl-*p*-benzoquinone imine), a 3-hydroxy derivative is cleared harmlessly by GSH to a catechol derivative

Table 8.1 Time frame and clinical markers of paracetamol toxicity

Post-overdose Time (h)	Hepatic GSH (%)	Transaminases (units/litre)	Symptoms
6	100	Normal range	None
16–24	40–70	300–500	Anorexia, nausea/vomiting
24–48	10% or less	> 1000–1500	Right upper quadrant pain, tenderness
48–72	Zero	> 3000	Jaundice, bleeding, organ failure/death

Other reactive species are formed as well as NAPQI. In this acute context, this is the cellular equivalent of running out of foam in the face of steadily encroaching flames. After 24 hours or so, there is nothing to prevent the NAPQI from binding to the hepatocytes and gradually killing them through necrosis. Interestingly, you might recall that with CYP2E1 Chapter 3.6.2, there is some attempt by microRNAs to rein in the toxic species production, but in most cases it does not appear to work in time. The key to the toxicity of NAPQI is which cellular macromolecules it binds. It is still not really clear (after 40 years of research) as to precisely which structures must be heavily bound before the cell dies. Eventually, such numbers of hepatocytes are killed that the central area of the liver starts to die also. It is termed *acute centrilobular hepatic necrosis*. This translates clinically to a series of symptoms shown in Table 8.1.

Each of the three phases of paracetamol hepatotoxicity is reckoned to last around 24 hours. The first phase, up to 24 hours after the overdose, involves nausea and vomiting and general reluctance to eat. Interestingly, some individuals do not show any symptoms at all during this time. Many do complain of a general sick feeling and start to look very pale and washed out. At this stage, the individual might be lulled into a false sense of security and might not think that he or she is in any real danger. In addition, many overdoses are accompanied by copious quantities of ethanol in various forms, and the devastating hangover that ensues can effectively mask the initial hepatic paracetamol-induced symptoms. It is also likely that in moderate to heavy drinkers, the ethanol will compete with the drug for CYP2E1, thus delaying NAPQI formation. The second phase, 24–48 hours, involves the appearance of some physical pain in the area of the liver. Heart rate can rise significantly and blood pressure falls. At this point, the liver is sustaining serious cellular damage and liver enzymes such as ALT and AST (Chapter 6.5.3) can be 30–50 times normal, which is an obvious sign of many disintegrating hepatocytes. In the third phase, the liver damage starts to become acute, although in less than 1 in 20 patients will this lead to organ failure. Symptoms of severe liver damage include jaundice, gut bleeding, general anticoagulation (INR>3) due to failure in clotting factor production, and even encephalopathy. Paracetamol liver failure can cause ALT levels to increase to over 10,000[31]. If organ failure does occur, then death follows within hours from cerebral oedema, renal failure, and even blood poisoning[20,34,35].

If the patient recovers and responds to rescue therapy (detailed below), they enter the fourth phase of sometimes-prolonged recovery from any liver damage that has been sustained. Over the next fortnight or so, gradually, the ALT, AST, and INR values head towards normal, and hopefully any renal damage that may have been sustained also improves, although the kidneys are much more fragile than the liver.

Paracetamol: rescue therapy

Clearly, the chances of survival depend on whether the patient undergoes treatment before too much liver damage occurs. If the individual does not seek medical help, then it is a matter of how much of the drug they consumed, how drunk they were, as well as their own liver's resistance to damage (and luck) as to whether they will survive. If the patient presents for treatment in less than an hour or so after the overdose, then it is likely to be worthwhile flushing out the stomach, as the entire drug dose may not be fully absorbed. If the patient appears within 3–4 hours of the overdose, then activated charcoal is sometimes given, which is controversial as to how effective it is, and it may interfere with subsequent antidotes. Upon admission to hospital, a blood sample will be taken and paracetamol levels measured. This sample is most informative as soon as possible 4 hours post overdose, as earlier samples are not relevant as drug absorption is still proceeding. Currently, we know that untreated paracetamol blood levels of below 150 mg/L or below can cause severe liver damage (alanine amino transferase >1000 U) in less than 10% of patients, whilst at 250–300 mg/L the figure rises to 40% and beyond 300 mg/L is uniformly fatal[36].

In most countries, the paracetamol plasma concentration measured in the patient after the overdose is compared with the values on a graph known as a nomogram, which reflects cumulative clinical experience of hundreds of patients over many years and it plots paracetamol blood levels with time elapsed since the overdose. Viewing the nomogram, the plotted curve helps predict whether these levels will cause liver damage if untreated. The best-known is the US 'Rumack–Matthew nomogram', which recommends treatment at 150 mg/L, but there are many others. In the United Kingdom, the current nomogram[37] is more conservative than the US version, as it indicates that for a blood level on or above 100 mg/L at 4 hours, the antidote, N-acetyl cysteine (NAC, Parvolex) should be immediately administered. In context, 100 mg/L of paracetamol is already six times higher than the standard therapeutic levels. Hence, following the curve, depending on when the patient presents, they should be treated if their levels are, say 70 mg/L at 6 hours, 40 mg/L at 9 hours and 15 mg/L at 15 hours post overdose.

Any nomogram is not very useful when the time of the overdose is unknown, so repeated blood samples will reveal if the drug levels are falling and how rapidly. Whenever there is any doubt of timing or frequency of the (staggered) overdose,

NAC is now given. Nomograms do not predict liver failure; rather they indicate the potential for liver damage. If the levels are above the upper line, then there is a greater than 60% chance of liver damage being sustained. The successive blood samples hopefully will track the elimination of the drug and as treatment progresses, should show the patient gradually moving out of the hepatotoxicity *danger zone*. The nomogram tracking starts 4 hours after the overdose and ends around 24 hours and some countries carry on dosing with NAC for 36 hours. Patient metabolic 'housekeeping' (INR, creatinine, and ALT levels) is checked before discharge to ensure hepatic and renal recovery is proceeding.

NAC acts to supply cysteine to GSH synthesis, so that hepatic GSH levels are maintained until all the paracetamol has been eliminated. Current advice is that the initial NAC dosing should be intravenous over one hour to minimize adverse reactions. These include headache, diarrhoea, skin-flushing, and lowered blood pressure in a small proportion of patients. This is typical of the intravenous administration of any thiol and happens with radioprotectant drugs also. Intravenously, in less than 1% of cases, anaphylactic shock can occur, with severe hypotension and bronchospasm, but the risks of all the adverse reactions can be minimized with accurate dosing. Currently, in the United Kingdom, the MHRA supplies weight tables to calculate the NAC dose as precisely as possible. Indeed, it is no longer accepted that hypersensitivity to NAC precludes treatment[37]. NAC does work just as well when given orally, and the argument is that large amounts are absorbed and enter the portal circulation and reach the liver in quantity and very rapidly. However, it tastes vile, and the fruit-juice idea does not help much, apparently, as it makes patients blow chunks repeatedly anyway. Antiemetics can help keep it down, but any vomiting within an hour of the dose means it has to be repeated. NAC must be given every four hours for three days to ensure full liver protection. In terms of guaranteeing the result, intravenous infusion seems the best plan, at a dose of 300 mg/kg for adults, over 20.25 hours.

NAC is virtually guaranteed to rescue the patient if they present for treatment at 8 hours or less after the overdose; no matter how much paracetamol he or she has taken, no hepatotoxicity will be sustained. Of course it is of particular importance to initiate the NAC therapy in pregnant women, as the NAPQI may be toxic to the foetus. If the NAC is started at 0–4 hours, there is no difference to the outcome; so early NAC treatment offers no advantage over later treatment up to the 8 hours. NAC is still effective up to 24 hours post-overdose, even though some hepatoxicity will have been sustained, and it can even be effective after this time if paracetamol is still in the plasma, and even if the liver is already damaged.

Paracetamol: methods for prevention of overdose

You might think that incorporation of a thiol with the paracetamol would solve the overdose problem instantly. The idea appeared in the early 1970s, soon after the mechanism of the drug's toxicity was discovered in the United States.

Unfortunately, nobody has ever determined the correct proportion of the thiol required to prevent liver damage in humans, nor are they likely to for ethical reasons. Therefore, to err on the side of caution, the only available product in the United Kingdom (Paradote) contains methionine in a 1:5 ratio with paracetamol[38]. It also means that everyone who takes this preparation would be ingesting methionine, and there are some long-term concerns about its possible toxicity, as well as some mild side effects, such as drowsiness, flatulence, and headache. Unless every paracetamol preparation (and there are hundreds of them) contains methionine, then the idea is unlikely to prevent lethal overdoses, due to the vast number of preparations that contain paracetamol without an antidote. The antidote/drug preparations are obviously much more expensive than plain paracetamol, which also means that their financial viability has not been sufficient for them to become popular. In the United States, rather than incorporate different agents within the dosage, measures were taken in 2011 to restrict table strengths to 325 mg on the grounds that there is no evidence that larger doses are any more effective[39].

Paracetamol: use in children

Opinion is divided as to the value of paracetamol for children in treating fevers and mild discomfort. Generally, paediatric fevers are overtreated and paediatric pain is undertreated[40]. Parents and many physicians are extremely wary of fevers in children due to perceived dangers of febrile seizure[41]. Not surprisingly, commercial pressures behind the marketing of paediatric preparations of paracetamol exploit these concerns extremely aggressively[41]. In hospitals, paracetamol is often used for what is termed minor discomfort in mild infections and for post-operative pain, so reducing opiate consumption. Whilst studies have shown that paracetamol is effective in pain relief, children can easily be given repeated doses that exceed the 4 g/day limit.

Although in general children are more resistant than adults to paracetamol toxicity[33], inadvertent overdosage in febrile children can lead to elevated transaminase enzyme levels and liver toxicity. This can occur where the doses of paracetamol have not been properly recorded, leading to hepatotoxicity from apparently low doses of the drug. Chronic paracetamol toxicity in febrile children is more likely to happen if they are less than three years old and are not eating or drinking enough fluid[40,41]. Paracetamol is useful in cases of febrile children in heart and respiratory failure, but has little value in those who do not have heart or lung problems, as the fever can be beneficial. This is because viruses and bacteria have specific temperatures where they multiply most efficiently and the febrile response ensures that the body temperature is higher than the infectious agents' optimum temperature. In addition, some components of the human immune system are more effective against viruses and bacteria at higher body temperatures[41]. This is borne out by observations that children do less well in cases of mild and severe infection when dosed with paracetamol[41]. Whilst I remember well my older son's great distress from one particular inoculation when he was a baby and I wished at the time I had given him paracetamol beforehand, evidence suggests that prophylactic use of the drug erodes antibody production, defeating the object of the exercise[42]. In general, paracetamol is effective in children when given as a scheduled dose over a period of time, rather just at need[40] and it is recommended that the maximum dose of 60mg/Kg/ should not be exceeded[40].

Paracetamol: conclusions

This drug will always be a best seller, despite its toxicity in overdose. This is partly due to the relatively narrow spectrum of drugs available for minor analgesia. Indeed, some of the mass-market alternative drugs, such as non-steroidal anti-inflammatories (NSAIDs), have significant issues, not least gastric irritation and even renal damage ranging from mild to severe, according to dose. Unfortunately, paracetamol is here to stay, even though to this day it is not even certain how it works. Its toxic effects are so obvious that if it had been submitted for testing even as late as the 1970s it would have been shown up as hepatotoxic and quietly dumped, like its ancestor phenacetin in 1983 and alongside thousands of other failed agents. Interestingly, co-proxamol (325 mg paracetamol, 32.5 mg dextropropoxyphene) was withdrawn in the United Kingdom in early 2005.

8.3　Unpredictable drug adverse effects: type B

8.3.1　Idiosyncratic and overdose toxicity: similarities and differences

With DILI caused by an overdose, such as with paracetamol, despite the life-threatening situation, if the patient seeks medical attention within a few hours they will be highly likely to recover. This is because even in the absence of a timeline of the overdose, the course of the process can be predicted to a significant extent, based on clinical experience and information from the patient's blood analysis. Hence, paracetamol, although it is toxic, survives because it is predictable. The chances of paracetamol causing major liver toxicity at 2 or 3 grammes per day are, as we have discussed, vanishingly small. With paracetamol overdose, the exogenous NAC 'refuels' the liver's GSH system so the organ essentially solves its own problem.

One of the nightmares of both the world's drug regulatory systems and pharmaceutical companies is a new drug that is either a variation of an existing theme, or could be the 'flagship' of a new and promising class of agents, which then causes an unpredicted aggressive hepatic failure in a substantial number of patients *at the prescribed dose*. That last phrase is important, as a catastrophic toxic effect involving a major organ normally would only be anticipated after an overdose. With IDILI, the drug concentrations are irrelevant to the process and tell the physician nothing useful, as he or she observes the patient's liver rapidly declining in health and capability. Precious time is lost whilst diagnosis and treatment decisions are improvised under pressure from the steadily worsening clinical situation. Not only is treatment of the IDILI patient problematic, the event often appears to be unrelated to a consistent time frame, dosage, or even specific patient risk factors.

Aside from the important issues of predictability, there are similarities between DILI and IDILI. Mitochondrial toxicity, macromolecule damage, GSH depletion, reduction in ATP formation, followed by apoptosis or necrosis, are all likely to be common events in the final stages of both situations. The key differences are what ignites and drives each process. With DILI, the high concentration of the drug, together with the formation of the toxic metabolite initiate and sustain the toxic process and drug concentration is effectively rate limiting. Once the drug has gone, the process should stop and the experience is not likely to impact future health for that patient, provided they fully recover. In IDILI, potential igniting events, like CYP-mediated reactive species generation and protective gene suppression start a toxic process whose momentum seems independent of the drug/metabolite trigger. What is difficult to understand is that reactive species generation and issues consequent to gene suppression are likely to occur to some degree in the majority of patients exposed to drugs that can cause IDILI, but in nothing like the scale required to inflict any obvious injury by themselves. The question is how such events on a cellular scale expand to eventually involve the whole tissue. The following examples of the anti-diabetic drug troglitazone and the antibiotic trovafloxacin illustrate the difficulties in understanding the sequences of events that in a relatively small number of individuals lead to liver failure. The sections also focus on the experimental methods used to try to reproduce and comprehend these processes.

8.3.2 Type B1 necrosis: troglitazone

Troglitazone's hepatotoxicity was all the more unexpected to practitioners and patients, as it was unconnected to any of its pharmacological actions and did not seem to be linked directly to the immune system. Troglitazone was the first of three new thiazolidinedione insulin sensitizers that were originally developed by a Japanese drug company for the treatment of the massive and still growing problem of type 2 diabetes, a condition that is linked strongly to affluence and obesity. The drug was interesting because it had new mechanisms of action, one of which was the stimulation of glucose uptake by skeletal muscle. The drug achieved this effect through activating AR γ and it also had many other beneficial anti-inflammatory actions, particularly in conditions related to type 2 diabetes[43]. Soon after it was introduced, reports were received of several fatal incidents involving hepatotoxicity in Japan and the United States. There were at least 35 deaths after approximately two million patients received the drug world-wide for just three months in late 1997. The FDA recommended that the drug should be monitored for hepatotoxicity; however, this was not carried out and has been cited as a tragic example of inertia in the regulatory system in the face of an obvious hazard to health[44]. Even worse, it emerged that the manufacturer went ahead with marketing in the full knowledge that the drug was toxic enough to elevate AST and ALT levels to 30-fold of the upper limit of normal in some patients in the clinical trials and was obviously a human hepatotoxin. An FDA employee (Dr John L. Gueriguian) recognized the problem, which perversely and unjustly cost him his position[45]. Toxicity could even occur within two days of starting therapy, although liver failure could occur long after the drug was discontinued. It was withdrawn after only two months in the United Kingdom, although it took until 2000 before worldwide withdrawal occurred[44].

Troglitazone did not 'survive' its toxicity despite its potential usefulness. This was partly due to the availability of other effective anti-type 2 diabetic agents, as well as the arrival of two sister compounds, rosiglitazone and pioglitazone, which were initially believed not to be subject to hepatotoxicity risks, due to structural differences. It is important to note that the tolerance of regulatory authorities to drug toxicity is strongly influenced by the intended disease target and the number and efficacy of alternatives available. Hence, the hideous toxicity profiles of some anti-cancer agents can be accepted, whilst much less damaging and more rare effects in other drugs for non-life-threatening conditions where there are also many alternatives lead to withdrawal. It is significant that troglitazone was administered to rats, mice and monkeys at doses of up to 1.2 g/kg daily for two years without registering any detectable hepatotoxicity during its preclinical testing[46], but turned out not only to be hepatotoxic, but particularly so to patients with the very clinical issues it was designed to treat.

Troglitazone mechanisms of toxicity

Troglitazone presents what is effectively an embarrassment of potential toxic pathways, and pinning down the cellular mechanisms that drive its effect has been challenging. Initially, it was shown to be cleared by CYP3A4, but human hepatocyte studies with CYP3A4 inhibitors did not abolish its cytotoxicity[47] and a CYP2C8 component was found[46]. Both CYPs form a quinone species that was assumed to be reactive, as several GSH troglitazone conjugates can be detected[48]. However, in a mouse model with a humanized liver in which the drug is hepatotoxic, the presence of a GSH-depleting agent (L-buthionine sulfoximine, BSO) paradoxically failed to cause serious cell damage, as would be expected if the quinone was forming reactive species through some form of REDOX cycling[46]. Alternatively, it was thought that cleavage of the thiazolidinedione ring might form reactive alpha-ketoisocyanate or sulfonic acid derivatives, but this is not necessarily thought to be a major toxic route either[48]. Troglitazone binds PXR and induces the main CYPs, as well as P-gp and several other transporters, although this is not necessarily connected to its toxicity. However, the drug downregulates the expression of a number of solute transporters and genes involved in oxidative stress handling[47]. Troglitazone also directly inhibits several transporters, including the bile salt export pump, causing accumulation of bile salts that can damage mitochondrial health due to their detergent qualities[49]. The drug is conjugated by SULT1A1, but in human hepatocytes it forms reactive sulpho-conjugates and it also causes a reduction in mitochondrial membrane potentials, thus promoting oxidative stress[43,49]. In other research, troglitazone's mitochondrial toxicity has been explored and found to be multifactorial[50] and the drug was most toxic in those with pre-existing hepatic insufficiency or fatty liver, which is seen in many type 2 diabetics, the target population for this drug[43]. The drug also inhibits much of fatty acid transport, so it would exacerbate issues related to lipid transport[51].

After troglitazone was withdrawn in 2000, rosiglitazone and pioglitazone fell under scrutiny, and for a few years continued in use, but eventually the mito-chondrial toxicity of this class of drugs surfaced in perhaps the organ which has the narrowest margin for error, regarding consistent energy demand and mito-chondrial health, the heart. Detailed clinical investigations revealed that rosigli-tazone increased the risk of congestive heart problems, and fatal infarctions, which sealed the drug's fate in 2010, when it was withdrawn[50]. Pioglitazone lives on, but is tarnished by an increased risk of bladder cancer[52,53] and has been with-drawn in many countries. Not surprisingly, for a particularly toxic drug, troglita-zone is under investigation as the basis for a series of anticancer agents[54]. The stark contrast between the glitazones' lack of impact on animal systems and multiple toxic interactions with human systems is echoed in the next failed drug, trovafloxacin.

8.3.3   Type B1 necrosis: trovafloxacin

In contrast to troglitazone, trovafloxacin was developed as part of a long and established line of fluoroquinolone antibiotics, such as ciprofloxacin, ofloxacin, and norfloxacin. These drugs had become firmly established in infection control and their use was increasing massively. From 1995–2002, prescriptions for these drugs increased from 7–22 million per year[55]. The fluoroquinolones were intended to be selective for bacterial DNA gyrase and topoisomerase IV. Trovafloxacin was seen as an advance, in that it had a broader spectrum of activ-ity and a better pharmacokinetic profile than the older drugs. In addition, it was better tolerated in terms of gastrointestinal effects than amoxicillin in patients, but was worse with central (headaches, dizziness) issues[56] and slightly more patients discontinued treatment due to side effects with trovafloxacin.

One of the clinical trials with the drug was conducted in children in Nigeria in 1996, which led to a number of deaths[57] and an out-of-court settlement 13 years later. The drug was introduced in early 1998 after trials involving more than 7000 patients. By the time it was withdrawn in 1999, more than 2.5 million people had used it and it had caused 140 cases of liver toxicity. Of these, 14 individuals went into liver failure and 4 received transplants, whilst 6 died[58]. The chances of the drug causing liver failure in a patient were estimated at 1:178,000[59].

Interestingly, several other new fluoroquinolones, such as grepafloxacin and temofloxacin were also toxic and went no further. In general, this class of drugs is under review[60] due to their various adverse reactions, which include significant impact on the CNS, the heart and in connective tissue[61]. In addition, the fluoroquinolones are more likely than other classes of antibiotics to cause hepatotoxicity[62]. However, adverse reactions are bound to be magnified by the sheer scale of the usage of fluoroquinolones[55] and antibiotics in general, which is an ongoing problem in itself.

With trovafloxacin, preclinical studies revealed no problems in the rat at doses of a gramme per kilo, but some reversible hepatotoxicity was seen in dogs[59]. Post-withdrawal work in rat hepatocytes at high doses also showed no problems[63]. In contrast to troglitazone, there were no clear indications in clinical trial that the drug could cause liver failure and you can see that even today no current Phase III trial structure would have the patient numbers capable of detecting this issue. Of course, the first concept to investigate is the possibility that the drug was metabolised to reactive species that then caused oxidative stress. The cyclopropylamine area of the molecule was a possibility as it can be oxidized to a reactive α, β-unsaturated aldehyde. It has been suggested that this can be achieved by CYPs or myeloperoxidases[64]. However, if reactive species generation were the main means of causation, then the effect would be to some extent dose, as well as species dependent and far more than 1:178000 people would be affected. Even if a polymorphic form of a toxic species generating biotransforming enzyme was responsible, then again, the frequency seems too rare for that. Studies in animals with trovafloxacin showed dramatic hepatotoxicity only in the presence of lipopolysaccharide (LPS), an immunogenic bacterial component. LPS stimulates interferon-γ and tumour necrosis factor α production, both powerful cytokines that are capable orchestrating cellular immune destruction[59,65]. The trovafloxacin/LPS/cytokine combination toxicity was not seen with nonhepatotoxic levofloxacin and could be stopped if the cytokine production was prevented. Indeed, in the animal models, trovafloxacin could actually increase cytokine production by itself[59]. Hence, trovafloxacin could theoretically recruit some form of immune response, but this could not be conclusively proved to operate in humans.

Unlike its sister fluoroquinolones, trovafloxacin is also a potent generator of oxidative stress at relatively low concentrations in HepG2 cell lines and human hepatocytes[66]. Its GSH depleting effect is more than 10-fold that of ciprofloxacin and nearly 15-fold greater than that of levofloxacin. DNA microarray studies in human hepatocytes were particularly damning, showing more than a thousand gene expression changes, including off-target pharmacodynamic effects, which showed that trovafloxacin-induced DNA gyrase and topoisomerase inhibitions are not actually selective for bacteria at all. The arrays also revealed widespread disruption of mitochondrial and signal transduction genes, as well as fatty acid and glucose metabolism[65,66]. The main problems appeared to centre, as with troglitazone, on mitochondrial function (loss of membrane potential and damage to electron transport), whilst GSTs and other nrf2-controlled oxidative stress response systems were also activated. The actual reactive species that cause these issues have not been conclusively identified, although trovafloxacin acyl-glucuronide has been suggested as a candidate in HepG2 cells[67]. Perhaps what is interesting is that trovafloxacin causes a significant series of toxic issues in a number of animal and human experimental platforms and it can directly, and/or through its acyl-glucuronide, influence gene expression and possibly the immune system. What is perhaps surprising is the severity of these issues in the experimental models does not seem to correlate with the much lower actual frequency of liver damage in patients.

Type B1 necrosis: role of the immune system

Whilst the preceding sections on troglitazone and trovafloxacin discussed many events that could conceivably ignite some form of cellular damage, there are clearly many possible mechanisms whereby this damage is fuelled so that it escalates to threaten to consume and then eventually kill the organ involved. In IDILI there is evidence that cellular events triggered either by the drug and/or its metabolites somehow attract the attention of the immune system by convincing it beyond doubt that a pathogen has established itself in the liver. IDILI can occur with the standard features of a system-wide immune response, such as fever, changes in white cell populations, and rashes. This is well documented with phenytoin, sulphamethoxazole, and halothane. This powerful and whole-body involvement with the liver reaction comes under the acronym DRESS, or drug reaction with eosinophilia and systemic symptoms[68,69]. Alternatively, it can be extremely focussed on the liver alone without systemic involvement, such as with the withdrawn anticoagulant ximelagatran[70,71]. The impact on the liver varies in intensity also, ranging from a relatively mild reaction which gradually subsides[72], all the way to a cytokine/cellular response that borders on the hysterical. This can mean that the clinician is faced with a situation that is in some ways analogous to trying to save a terrified, thrashing and panicking individual from drowning. The next section tries to explore what we know of role of the immune system in drug injury.

8.3.4   Type B2 reactions: immunotoxicity

The immune system: overview

If you listen to an everyday conversation about the brain, the words that crop up might be 'amazing', or 'remarkable' or 'awesome' or maybe the phrase 'mostly unused'. A similar and possibly less common conversation about the liver will probably include the words 'resilient' or 'tough' or the phrase, 'how is he still here after all that vodka'. With the immune system, it would be unusual if the words 'mysterious', 'unpredictable', and 'reaction' did not crop up. Most people are aware that one of the basic functions of the immune system is the detection and destruction of bacteria, viruses, prions, and infected cells. However, in the public consciousness, there is also growing awareness of the disorders of the immune system, such as the extent of the many unpleasant autoimmune conditions and their general refractoriness to treatment. Even on a small scale, there is a remorseless, almost 'Terminator-like' aspect to the immune system, which is capable of mushrooming into an all-consuming so-called *cytokine storm,* which seems to follow the 'better massively destructive than sorry' principle.

The immune system is a carefully honed product of evolution like the rest of our bodies, but the demands made on it are very different to other organs and systems. It needs to combine 24/7 vigilance in every organ, tissue, and cell in the body with the capability to summon the appropriate resources to respond very quickly to minor incursions of infectious agents, all the way to rapid, aggressive, and potentially overwhelming invasions. Consequently it has no 'base' or 'head-quarters', as the brain, the liver, and the gut do. Its mission is to be perpetually all-seeing and all-knowing, with the inexplicable ability to appear from apparently nowhere in overwhelming force immediately, if not sooner. These traits have been selected for ruthlessly by successive waves of pandemics, through history. We know of the Black Death of the Middle Ages, for example. This killed 40% of the European population and reappeared in England in the seventeenth century. In Henry VIII of England's reign, it was possible to die in just a few hours from a condition known as the 'sweating sickness', a terrifying prospect that could appear in the morning and kill before midnight[73]. Incidentally, Henry's second wife, Anne Boleyn, survived the sickness, only to be subsequently judicially murdered by her husband in 1536. In the early twentieth century, influenza caused several times more deaths than the entire Great War[74].

If you also consider the range of potentially infectious agents that we can encounter, you can see why the immune system's evolution has encouraged memory, flexibility, sensitivity, persistence, and savage ruthlessness. Foreign invaders appear in all shapes and sizes, combining incredible ingenuity and aggression with sheer weight of numbers. For example, intracellular pathogens are extremely adept at escaping and even manipulating the immune system, and even fairly basic organisms with modest genomes like mycobacteria can actually thrive within immune cells and resist drug intervention in very sophisticated ways, as we saw with tuberculosis in Section 4.4.3. At the other extreme, schistosomes can blatantly sit unmolested in the portal circulation causing havoc for 30 years or more. These particularly loathsome trematodal worms can be an *inch* long. Similarly, the filarial worms of onchocerciasis (river blindness) can also exist untroubled by the immune system as they too have a 'cloaking device' that is torn away by a drug called diethylcarbamazine. The resultant immune activity as the worms die is so intense that it can be fatal[75] and this drug has been largely outmoded by ivermectin[76]. Whilst this book has drawn attention to some of the unacceptable activities of the pharmaceutical industry, it should be noted here that Merck and GlaxoSmithKline's decisions to supply ivermectin and albendazole as open-ended donations guaranteed the success of treatment programmes with these drugs in Africa. Overall, we should forgive the human immune system some measure of paranoia, as everything 'out there' really is out to get us and sometimes does.

Given that infections can and do kill in hours rather than days, it is essential to find and control them right now this instant and thus be around to repair the damage at a later date. This is similar to the expectation that cancer patients will withstand severe toxicity during therapy, on the basis that survival should be attained at virtually any cost. With the immune system, this means in practice that it uses cells, antibodies, and cytokines to create a 'shock and awe' or 'storm' effect, against invaders, which can unfortunately destroy tissues and whole organs. This awesome destructive capability must be reined in by a strict gradation of response to avoid huge overreaction to minor amounts of infectious agent. The spearhead of the immune system, the activated neutrophil, creates a highly efficient bacterial death zone around it, through reactive species, degranulation, and even net formation[77]. Around 100 billion enter the circulation every 24 hours[78] and the severity of the lesions seen in diet-induced autoimmune diseases like dermatitis herpetiformis[79] show what several million activated neutrophils can do to noninfected tissue. It is a testament to the complexity of the immune system that a succession of scientific theories have appeared over the past decades that try to understand and predict immune function. Those who have devoted their lives to studying immune cells debate these theories endlessly and in many ways we are just beginning to understand its mysterious complexity.

Context of drug hypersensitivity

With IDILI linked to trovafloxacin, you would be exceedingly unlucky to develop this condition, but it has long been known that drugs can trigger immune assaults on any precinct virtually anywhere in the body. Indeed, drug-mediated severe hypersensitivity directed at many different tissues is much more common than that of trovafloxacin. As with IDILI, these situations are feared not just because they are relatively rare but due to their severity, they sometimes cause graphic tissue destruction accompanied by high fatality rates. Stevens–Johnson syndrome and its related condition Toxic Epidermal Necrolysis (SJS/TEN) are a good example (Section 8.4.3) where the patient's entire skin can be attacked and virtually destroyed. Anticonvulsant hypersensitivity syndrome can cause severe, widespread tissue and organ damage. Other reactions are less overtly destructive and more specific; agranulocytosis leads to the gradual and almost complete shutdown of neutrophil production and subsequent death from sepsis (Section 8.4.4).

As with IDILI, doctors treating patients suffering from these reactions are often faced with extremely limited treatment options and a rapidly deteriorating patient. In addition, withdrawal of the causative drug does not always lead to amelioration of disease progression. Obviously the immune system should not respond in this way, but predicting whether a new drug might cause a catastrophic immune reaction in a small number of patients is still a long way off. A serious drug-induced hypersensitivity reaction might affect maybe 1:5000, or even 1:10000 patients, so even though these reactions are around 15–20 times more common than trovafloxacin, it is still not realistic to organize clinical trials with numbers high enough to hope that these reactions can be detected.

However, progress has been made in the understanding of the immune system's normal operations and we now know enough to predict that certain groups of patients will be at higher risk from drug hypersensitivity. The difficulty remains in modelling such reactions experimentally so there is still much to learn about how the immune system operates, how drugs become immunogenic and how to prevent this happening to patients. As mentioned earlier, theories and hypotheses abound on immune function in general and specifically on drug immunotoxicity. Each of these ideas is supported by experiences with some drugs and undermined by others. Experimental platforms that are either human cellular based or chimeric (animal/human) systems are under continuous development and are our main source of information, as the option of rechallenging patients is not normally available, except in certain circumstances (Section 8,4.4). The following sections outline how the immune system is thought to operate and how it decides to see a drug as a danger that is just as life-threatening as an invading pathogen.

The immune system: aims and normal operation

Before we can study how drugs can cause immune-mediated toxicity, it is important to review what we expect the immune system to do so far:

- Detect non-self, foreign material at a specific site, or sites.

- Initiate an appropriate response that will lead to the destruction of that material.

- Retain the 'memory' of that material to expedite a response on rechallenge.

Tolerance and autoimmunity

Viewing the immune system mainly in terms of its ability to destroy pathogens, makes you think that it is engaged in a constant battle to detect non-self, such as invading organisms. The system identifies and responds strongly to certain structural features of these organisms, such as bacterial lipopolysaccharides (LPS) mentioned previously, which used to be termed *adjuvants*. However, these 'signature' molecules of infection are now grouped under the heading PAMPs, or pathogen-associated molecular patterns. The immune system learns about PAMPs from contact with the microbiome at mucosal areas such as the mouth, eye and particularly the gut. However, these PAMPs tell the immune system what other organisms look like, but do not provide any information as to whether they will be a problem[80]. Indeed, the vast majority of the microbiome, as we discussed in Chapter 6.2.12, is a positive benefit.

Hence, the initial view of the immune system was based on recognising the difference between self and non-self, sometimes termed the 'stranger hypothesis'. Indeed, we saw with trovafloxacin that the only way to trigger a reaction in the animal systems commensurate with the human response to the drug, was to use a key component of PAMPs, LPS, as the animals' immune systems would not normally attack their own tissue and the drug alone did not change this. Part of the assumptions behind the stranger hypothesis is that if it is self, then it is good and not a problem. Even if this hypothesis was expanded to interpret PAMPs to identify and distinguish dangerous from benign foreign material, it does not account for the sustained, damaging and apparently insane 'autoimmune' campaigns the immune system wages against our own tissues, ranging from diabetes to arthritis. Whilst it could be suggested that these processes might be been triggered by infection at some point, there is usually no trace or indication that bacterial or viral adjuvants were present and the obvious question is that immune system clearly does not respond to material solely according to source. It rather responds to the concept of 'threat' or, as the theory has it, 'danger'.

The 'danger hypothesis'[81] opens the door to other concepts that you may have thought of yourself already as you were reading this chapter. If we lay aside attrition on the cardiovascular system, the clearest threat to most people in the developed world is not infection, but 'self', or more precisely, 'malignant self'. The danger hypothesis thus allows for the concept that the immune system's vigilance includes a form of internal policing of self, leading to the recognition and destruction of malignant cells, thus protecting us from hundreds of potential cancers. If we take this further, we might ask why the immune system acts so ferociously against our own joints, tissues, and transplanted organs, yet fails to protect many of us from malignant cells and fails again to destroy or even contain most cancers. Indeed, to stretch the 'Terminator' analogy used earlier, when the cancer T-1000 appears, often our T-800 is nowhere in sight[82]. Returning to drug hypersensitivity, how does the immune system identify danger signals, and how can we predict and prevent this happening in the clinic during normal drug therapy?

The danger hypothesis suggests that the immune system controls its own state of readiness and sensitivity through a highly sophisticated listening system, where perhaps hundreds of different patterns of danger signals are processed every second to provide a picture of the health of a tissue. These DAMPs, or danger-associated molecular patterns have been the subject of much research[80,81]. What is very useful about this theory is that it can include virtually any molecule in the DAMPs and, indeed, PAMPs can be DAMPs and *vice-versa*. DAMPs can involve whole cellular states or processes, such as oxidative stress, or upregulation and/or downregulation of genes. It could also include key proteins or peptides that are not normally found in the circulation. Heat shock proteins, interleukins, as well as cytokines such as HMGB1 (high mobility group box 1 protein) may be among one of hundreds of signals ranging from large molecules such as DNA/RNA fragments to smaller agents like uric acid[83]. Other danger signals may be the products of damaged cellular plasma or nuclear membranes. This is logical, as pathogens can often thrash about in our tissues, wrecking cellular membranes and releasing their contents that include the danger signals. Again, if you have been paying attention, you will instantly contribute that tumours sometimes do exactly the same thing as they grow and expand. Alternatively, some 'silent' malignant or pathogenic cells can seem impossible to detect.

Silent malignancies resemble the hackneyed but still classic cinematic scenario of the police officer unexpectedly entering an apparently peaceful coffee shop full of outwardly calm but nervous hostages. The discovery of the desperate 'perps' relies on clues from the hostages' faintly abnormal and twitchy behaviour, as well as the police officer's acute nose for murderous scum, to detect the reality of the situation and mount an appropriate response, which rarely ends well, as we all know. The enormous scope of the DAMP concept means that in very vague terms, it suggests a means whereby our immune systems assess potential threat. Logically, the way to achieve this is to self-calibrate and adopt a complex series of thresholds and levels of response for different threats in or around different tissues. The success of self-calibration is probably the key to why most people do not develop or die from cancers or develop autoimmune disorders. The problems with self-calibration have been explored with the hygiene hypothesis, when lack of exposure to enough immunostimulation increases the likelihood of inappropriate immune activity[84].

In summary, even if we were all exposed to the same self-calibrating conditions, that is a large and healthy microbiome and a relatively wide but not too harrowing experience of live pathogens, some immune systems are more successful than others are. Possession of such a system means that the individual shrugs off infections, never develops cancer, or autoimmune problems. These systems do more than just passively 'listen out' for DAMPs. They use their experience, to actually create in advance the ability to respond to libraries of future DAMPs and their individual antigenic components[85] so they can 'get their retaliation in first'. This takes almost unimaginable cellular processing power. The successful immune system is more than just shock and awe – it is an intelligence so advanced it does not even require a nervous system.

Antigen processing and presentation

In practical terms, in order for the immune system to discover and assess threats, it must search everywhere for the enemy. So it must look inside the cells themselves as well as outside in the circulation and interstitial fluid. B and T lymphocytes usually detect antigens after they have been *processed* into forms where they can be *presented* to the B and T cells in such a way that they will recognize it as foreign and/or dangerous. The process can be resolved into intracellular and extracellular antigen detection; presentation and the sensitivity of the process is crucial in determining whether an individual develops an autoimmune disease or a severe drug hypersensitivity reaction.

Intracellular antigen detection and presentation

Normally within any given cell, a series of what are known as MHC I, or major histocompatibility complex I proteins (Figure 8.6) are expressed on the cell surface. In humans these MHC I proteins are known as HLAs (human leukocyte antigen). There are three human HLAs, A, B, and C. These HLAs are very polymorphic and there are hundreds of different alleles that code for slightly different versions of each HLA. The function of these complexes is to act as a

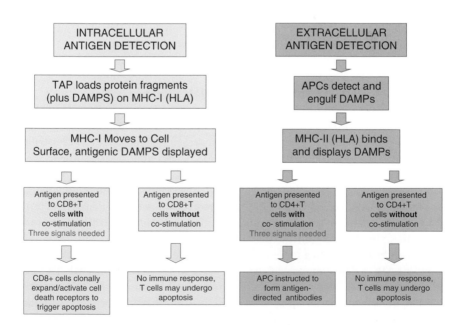

Figure 8.6   Extracellular and intracellular antigen processing

sort of 'presentation' or 'display case' for a wide variety of potentially immuno-genic material that has undergone an elaborate processing and packaging sequence of events that usually begins in the proteasome. Damaged or misfolded proteins as well as fragments of potential threats such as virus particles or even LPS, are moved into and through the rough endoplasmic reticulum by the peptide transporter TAP (transporter associated with antigen processing). Next they are moved onto the PLC or peptide loading complex, and they are trimmed into small representative peptides, which with the aid of a chaperone molecule called tapasin, they are loaded into the back of the HLA, which then displays them on the cell surface[86].

This is a form of constant quality control, a monitoring or sampling of the presence of self actually inside the cell, followed by a clear presentation of 'self' at the exterior of the cell. The HLA system is a vital link in the process that alerts the immune system to intracellular foreign antigens and abnormal expression profiles of endogenous proteins[86,87]. There is evidence that many other systems can present material to HLAs just as efficiently as TAP[86]. In the case of a viral infiltration, the established view is that viral proteins are picked up and shredded by the proteasome, loaded onto the HLA as above, then exhibited, rather like a 'hot dog in a bun', with the viral or foreign proteins acting as the hot dog. Cytotoxic T cells (CD8+) through their T-cell receptor systems (TCRs) recognise the HLA, so they will then accept the material bound to it as evidence of DAMP/non-self and then produce various cytokines such as tumour necrosis factor that together activate death receptors on the hepatocyte surfaces, which lead to apoptotic cell death[71]. This recognition by the T cell (using its TCR system) of the self of the HLA before antigen recognition prevents the T cell from indiscriminately attacking anything in sight and channels their cytotoxicity to where it is needed.

Extracellular antigen presentation and detection

Most bacteria and toxins are extracellular and viruses must be extracellular until they find a victim cell, so a series of antigen presenting cells (APCs) cruise the interstitial fluid around cells, as well as inside tissues hunting for antigens and 'listening' for DAMPs and PAMPs. These include B lymphocytes, macrophages, dendritic, and Langerhans cells in the skin. These cells carry immunoglobulin (Ig) molecules on their surfaces and they detect smaller entities than whole proteins, such as nucleic acid fragments, lipopolysaccharides, and lipids. At this point, the APC cells engulf the detected material, dismantle it, and tie up the protein of the complex to a series of HLAs from the class MHC II, which is only found in APCs. There are six of the APC HLAs, HLA DPA1, DPB1, DQA1, DQB1, DRA, and DRB1. They become functional by binding within various combinations, forming DPαβ heterodimers, which behave in the same way the other HLA A-C series do only this time the APC cell will present its processed protein fragments to helper T cells (CD4+), rather than the cytotoxic (CD8+)

ones (Figure 8.6). There are usually twice as many CD4+ than CD8+ cells, although during disease states the proportions can equalise[86,87].

Immunoactivation; the signal hypotheses

The vast majority of the protein fragments presented do not elicit an immune response, and the antigen processing steps outlined above are the first steps in the identification and assessment of the degree of risk that any given the antigen presents. Given the complexity of the immune system, we don't have a complete picture of how this is managed, but there are, as you can imagine, some hypotheses that attempt to describe the process. Logically, if you consider a potent weapons system, the prospect of the potential destruction if it operates out of control means that a very detailed sequence of command authorisations are required before full deployment occurs. We have all seen the various Bond films when our hero manages to frustrate the process of world domination by aborting some sort of fortuitously elongated countdown, which is followed by the rapid and explosive necrosis of yet another expensive Pinewood set.

The process of decision making in antigen processing was initially described by the '2–signal' hypothesis. The presentation of the HLAs and the fragments to the T cell surface receptors (TCR complex) triggers 'signal 1' that sets in motion a very involved sequence of events that includes the calcineurin pathway, which is effectively jammed up by cyclosporine and tacrolimus. At this stage, the T cells could be considered armed, but without further stimulation, the cells do not proliferate, and no further immune action is taken. Locking up the T cells in this signal 1 mode is the basis of successful organ transplantation. New drugs are being developed that will prevent the T cells from leaving this stage, such as voclosporin[88].

In the two-signal hypothesis, it was suggested that if the T cell receives signal 2, which is known as co-stimulation, full activation was thought to occur. Signal 2 comprises the binding of B7-1 and B7-2 receptors on the APC with a CD28 on the CD4+ T cell. T cells also express cytotoxic T-lymphocyte-associated antigen 4 (CTLA-4), which can bind the B7 molecules and stop the process of activation. Belatacept, a fusion protein, will mimic CTLA-4 and bind the B7's and lock up the system here and prevent activation proceeding further[88]. It was realised that a third signal was needed to fully activate the system. The presence of other immune cells, such as macrophages and dendritic cells is required to activate CD4+T cells to make lymphokines that instruct the APC to make antibodies[88]. These cells secrete cytokines and CD8+ T cells also require these third signals to become fully cytotoxic. The main cytokines involved are interleukins 1, 2, and 12, as well as several interferons, and probably many others. The third cytokine signal thus activates the B cell proliferation (memory cells), ensuring that the next time the antigen is detected there will be a much more rapid and intense response. As the CD4+ cells enable this aspect of the immune response, they are called T helper cells. The B cell amplification system is again designed to buy time. The immune system can respond to virtually any pathogen, provided it has the time to make enough cells and antibodies. This time period must not be slower than the proliferation rate of the infection, or death will result by weight of numbers and secreted toxins. The memory pool of cells provides the ability for rapid response to the pathogen when it is next encountered. All three signals are required to produce antibodies and to activate the CD+8 killer T cells[88].

This is a very simplified version of events. The full picture of antigen presentation, identification, and the subsequent decision on the element of risk made by the local cellular population is majestically complicated[89]. However, what does emerge to the nonexpert is that the immune system's decisions are far from impulsive and trigger happy. There are multiple decision stages for even quite obvious threats, such as PAMPs, as well as the more difficult to fathom DAMP interpretations. Even within the three-signal process, there are many opportunities for a default stand down, where the cells are receptive to multiple regulatory 'off' or delay interventions, such as CTLA-2. Other crucial stages, such as at signal 2, have windows when further activation is possible for only a few hours[88], before the system disarms. However, as we are well aware from apparently intractable autoimmune conditions, once the system is fully engaged, it is very difficult to switch off, but our understanding is improving.

How drugs might initiate immune responses

There are several current hypotheses that try to explain how drugs and other low-molecular-weight toxins manage to break through immune tolerance and elicit a major immune response. These hypotheses don't exclude each other, and there is experimental and clinical evidence to support all of them, so it is probable that to some extent they all occur, depending on the nature of the drug, individual, and their state of health (Figures 8.7 and 8.8).

If we consider the broad possibilities of how a drug, which is not a peptide, might trigger an immune response, we start with the idea that the drug circulates in the body as parent and metabolites. With some drugs, this is a considerable range of structures and stability. The molecular weight range of the drug or its metabolites will usually be no larger than around 900 Da. It is generally agreed that chemicals up to that size or smaller should not be directly antigenic[69], so the options might be that the drug and/or its metabolites either become attached to something that is antigenic, or that it somehow switches on the immune response pharmacologically.

Hapten hypothesis

In the 1930s, Landsteiner showed that the immune system would recognise very small molecules if they were to become covalently attached to much larger ones (40,000–50,000 plus). When a reactive small molecule does bind to a cell

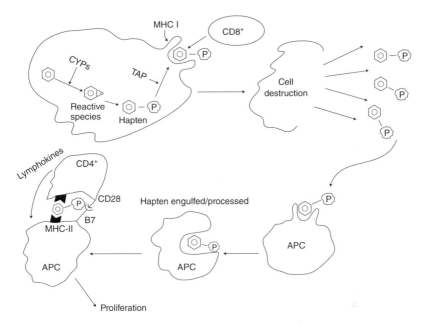

Figure 8.7 The hapten hypothesis as it might apply to CYP-mediated reactive species formation in drug hypersensitivity. The CD8+ HLA system may detect the hapten (reactive species bound to protein, P) and destroy the cell. The debris could then trigger the APC HLA system and signals 1–3 are supplied, followed by cell proliferation and antibody formation. Spontaneously reactive drugs like penicillin probably trigger both MHC systems at once

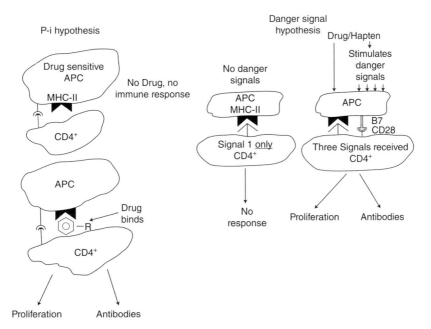

Figure 8.8 Aside from the hapten hypothesis, there are other competing hypotheses on the cellular mechanisms of drug hypersensitivity: the pharmacological interaction (p-i) hypothesis concept and the danger signal hypothesis. It is probable that no single hypothesis is sufficient to explain all the available clinical and experimental data and elements of all three may be involved in many drug reactions

macromolecule, the process of haptenation is said to occur. The process is not confined to drug reactions. It has been suggested to be responsible for the steep rise in general contact allergies over the past half-century, due to haptenation of proteins from various reactive chemicals in processed foodstuffs[90].

Injection of the hapten itself does not usually result in any reaction: the essential feature of this theory is that the hapten is bound to protein, thus making the hapten/protein complex immunogenic[59]. Another key feature of this process is that the hapten must be very chemically reactive in order to bind covalently. Penicillin spontaneously degrades and reacts with proteins and protein thiols both intracellularly and in the circulation and tissues, so the penicillin-related haptens could be detected through HLA pathways already described, intra and extracellularly. Studies from the 1960s onwards found antibodies directed at penicillin-bound protein material and the drug is certainly powerfully immunogenic in a substantial proportion of patients. A relatively high level of haptenation of proteins after penicillin use must occur in everyone who takes the drug and this you would think can be easily detected by the intra- and intercellular systems, but still only a minority, albeit a sizeable one, of individuals suffer severe reactions. Given that we need cytokines to provide our third signal for full immune activation of the T- and B-cell populations, it is not clear why and how cytokines would be released to trigger this process.

The hapten theory was extended to include reactive metabolites formed by CYPs (Figure 8.7). It is already known that suicide inhibition (Chapter 5) leads to CYP destruction and covalent binding and antibodies could be formed to the damaged CYP/hapten, or a slightly less reactive species may attack more distant macromolecules in the cell that could be equally antigenic. It seems more likely that this form of haptenation would be detected initially by MHC-I (HLA-A, B and C) intracellularly, rather than extracellularly as the CYPs are intracellular. Many drugs that cause hypersensitivity syndromes are extensively metabolised, sometimes to very short-lived reactive species, such as the AEDs carbamazepine and phenytoin. It could be speculated that once the CD8+s have destroyed the cell, debris that contained haptens could then escape the normal phagocytic 'clean up' and trigger the extracellular HLA/APC system (Figure 8.7). Alternatively, the TAP system might detect the debris in the cells that had phagocytosed the detritus. Assuming the haptenic antigens were processed, then if the three-signal hypothesis is true, cytokine release is also required to complete the activation of the T-cell populations. Again, it is not easy to explain where and why the cytokine release would happen. Overall, it is thought that hapten-peptide complexes are processed and 'read' by MHC-I and II, in order to elicit the full spectrum of T- and B-cell mediated immune reactions[69].

Whilst all this is plausible, it implies hapten formation must occur in every individual that is exposed to the drug. As we have discussed, IDILI reactions are extremely rare, unrelated to dose and have timelines bordering on the random. Hence, the relationship between hapten formation and immune response is unlikely to be a simple issue such as exceeding some form of threshold. In addition, antibodies related to diclofenac and halothane for example, have been detected in patients without hepatotoxicity and immune responses to drugs happen without serious liver damage[59]. So whilst haptenation is interesting, it does not explain how the immune system makes the decision to act after detecting it. Crucially, some drugs cause IDILI without haptenation or covalent binding occurring, such as the direct thrombin inhibitor ximelegatran and another hypothesis has appeared to try to account for the effect of this drug, which is describe below.

Pharmacological interaction hypothesis

Known as the pharmacological interaction with the immune system, or p-i hypothesis (Figure 8.8), this idea is based on experimental observations on the timeframe of activation of some populations of T cells that came from patients who had developed hypersensitivity reactions. This has been termed an off-target effect[69], although it could be argued that it is rather rare to be included in that category. Essentially, the problem drug was judged to have caused the hypersensitive response in the patients, as their T cells became activated too rapidly for antigen processing, haptenated or otherwise, to have occurred. The cells reacted in seconds, which suggests that the drug bypassed the MHC I & II antigen presentation process and acted pharmacologically, by either binding the MHCs themselves, and/or binding T-cell TCR complexes, which then activated the T cells. Ximelagatran, carbamazepine, and sulphamethoxazole are thought to bind directly to MHCs or TCRs in such a way that they promote T-cell activation[71]. Ximelagatran caused pro-inflammatory cytokine, chemokine release, and severe liver toxicity in a relatively high proportion of patients in trial and was withdrawn in 2006.

Some features of the p-i hypothesis distinguish it from haptenic reactions and have significantly reinforced the hypothesis over the past decade. The first is that the drug acts reversibly on the receptors of the immune cells without covalent binding involvement and indeed, the drug is unlinked with antigenic presentation or is antigenic itself. Instead, the drug binds directly to sites on the HLAs and/or the TCR complexes, which are almost infinitely variable; there are more than 10,000 MHC-1 and 3000 MHC-II alleles in humans, providing epic opportunities for pharmacological modulation of these receptor systems[69]. Further reinforcement for this theory is that drugs that cause immune responses through the p-i hypothesis only switch on T cells and not B cells, so the immune response differs from the complete T- and B-cell effect seen with haptenation, so no B-cell-derived IgE or IgGs are detected[69]. As the process is a pharmacological effect, the binding starts immediately, but the time delay before a response occurs coincides with the timeframe for the expansion of the activated T-cell populations, which is 5–7 days.

P-i reactions are now directly traceable to particular genetic variants of HLA, and the most investigated instances are with phenytoin and several other AEDs, with HLA-B*1502[91] being a very strong risk factor. This allele is very common in Thai and Han Chinese populations (8–9%) but more rare in Japanese and Caucasians (<2%). It is such a strong risk factor for SJS/TEN that the UK government has warned for some time against using AEDs that interact with this allele in the at-risk populations[92]. Likewise, the anti-HIV drug abacavir has a 50% chance of causing a hypersensitivity reaction in those with HLA-B * 5701[91] and up to 8% of Caucasians have this allele[93]. In fact, abacavir supports a refinement of the p-i hypothesis, in that its close association with HLA-B * 5701 is partly because the drug is believed to bind reversibly to the antigen binding cleft of 5701 and actually 'recalibrate' its peptide-binding activity[94].

Overall, it now appears that there is more evidence that drug-induced hypersensitivity is linked to the p-i hypothesis than haptenation, although some drugs are thought to actually use both pathways to cause their effects, such as flucloxacillin, which forms haptens and activates HLA B * 5701[69]. Overall, at least 25 drugs have now been associated with p-i related interactions with a host of different MHC-I and II HLAs, including anti-HIV (abacavir and nevirapine), antituberculosis drugs, and several antibiotics. Indeed, one of the most common causes of IDILI is amoxicillin/clavulanate, which is associated with the MHC-II DRB1*1501 and DQB1*0602 alleles[71].

Summary of causes of drug hypersensitivity reactions

Serious advances in understanding type B-drug hypersensitivity reactions have been made over the past decade. Indeed, some reactions, like those described in the previous section to various HLA alleles, can now be anticipated and tested for in appropriate populations. Polymorphisms in interleukins can be predictive of diclofenac IDILI[96]. Indeed, with diclofenac, integrin beta 3 (ITGB3) has been identified as a possible biomarker of IDILI induced by this NSAID, as its expression changes in peripheral blood after liver toxicity occurs[97]. However, the vast majority of drug hypersensitivity reactions remain rare, severe, and not predictable. Whilst much focus is placed on IDILI, as we have seen, drug hypersensitivity may or may not involve DRESS, or be focussed on other tissues such as the blood, bone marrow, or skin. Haptenation and the p-i hypotheses certainly between them can account for many instances of how a drug and/or its metabolites initially attract the attention of some or all of the immune system's cellular capability in different tissues. Hapten formation can occur anywhere in the body where immune reactions become manifest, from the liver to activated neutrophils[98] through enzymes such as myeloperoxidase. Phenytoin is more likely to engage a p-i type cutaneous reaction in an individual with slow CYP2C9*3*3 clearance, presumably because the higher concentrations of the drug promote its pharmacological interaction with the HLAs and TCRs[99].

However, these hypotheses can be accommodated in the overall view of the immune system as conforming to the danger hypothesis, but they do not necessarily cast light on how all the first stages of full immune response are progressed. For instance, the origin of the cytokines necessary for the completion of the three-signal hypothesis for T-cell activation prior to population expansion is not clear. Certainly once the condition has begun and the patient is seriously ill, cells that produce interferon γ are present in nearly half of patients, which presumably sustains the progression of the immune reaction[100]. The initial cytokine formation may be linked with complex cell attrition processes, such as those seen earlier with trovafloxacin and troglitazone, where a combination of gene expression modulation, mitochondrial impairment, and oxidative stress, or even viral infection, might provide the stimulus for the appropriate signals and satisfy the criteria that enough DAMPs are present to unleash the full immune response. Of course, if a drug caused such issues in substantial numbers of patients, it should be apparent in clinical trial, as we have seen with ximelagatran and troglitazone. It is still very difficult to account for the rarity of the reactions as well as their ferocity. Whilst future studies will uncover the mechanisms of how the immune system decides on the level of drug/metabolite-induced danger, the next section looks at the clinical impact of the immune response to drug therapy.

8.4 Nature of drug-mediated immune responses

The clinical presentation of drug hypersensitivity may depend on the time of onset, severity of symptoms, and particularly the site of the reaction. The first two types of reaction, anaphylaxis and anticonvulsant syndrome, lead to widespread reactions throughout the body, related to what is termed DRESS. The others involve organ direct reactions such as SJS/TEN, as well as effects on cell populations and on the tissues that form them, such as the gradual systemic loss of neutrophils (agranulocytosis).

8.4.1 Anaphylaxis

This can happen in response to a variety of foodstuffs (particularly nuts), food colourings, handling latex, contact with animals, insect bites and stings, as well as drugs such as penicillins, NSAIDs, and some intravenous preparations. The symptoms include fainting, swelling of the throat (laryngeal oedema), asthmatic symptoms, as well as difficulty in breathing, rashes, swelling, vomiting, and diarrhoea. If anaphylaxis progresses to anaphylactic shock, blood pressure plummets and tachycardia and arrhythmia occur alongside a racing pulse, coupled with extreme bronchospasm. Anaphylactic shock can be lethal in minutes.

This condition is caused by exposure to an antigen that provokes the formation (by B-lymphocytes) of a unique Immunoglobulin E (IgE) for that antigen. The IgE sits on the surface of mast cells and basophils, rather like a primed grenade. Another exposure to the antigen, or something similar in structure, and the IgE causes the cells to release massive amounts of vasoactive substances such as histamine. This process is known as degranulation[69]. Since mast cells are found in all the tissues, particularly the vasculature and bronchi, blood vessels are relaxed, causing the fall in blood pressure and the spaces between cells widen, leading to swelling in the tissues. The constricting effect on the bronchi is basically dependent on the amount of these agents released. Subsequent exposure to the antigen raises many more antibodies that cause more and more violent and widespread reactions through the body. This is the worst form of immunological 'own goal' as it is grotesquely out of proportion to the threat involved and can lead to death of the patient. The treatment is immediate injection of 300 micrograms (150 for a child) of adrenaline, which should control the metabolic mayhem, although recovery times can be considerable and a second reaction within 12 hours or so can happen in around 20% of victims[101,102].

Since haptenation is one possible route to a drug-related immune response, it would appear to be a bad idea to market drugs that were directly protein reactive, but penicillin derivatives as mentioned earlier, spontaneously degrade into a number of reactive derivatives in all patients that take them. The resulting penicilloyl haptenic derivatives act to sensitize the immune system, by activating B-cells, which form IgEs in response to the haptens and the next administration of the drug to the patient could be their last. This is essentially a disseminated reaction, i.e. a reaction that occurs in many areas of the body, including crucial ones such as the blood vessels and the lungs. About 10% of the population could be at risk from this type of reaction from penicillin or related antibiotics. Around 75% of fatal anaphylactic shock cases are linked to penicillin and its related drugs. Concurrent usage of beta-blockers makes anaphylaxis much worse as they prevent the beneficial effects of endogenous catecholamines released in response to the condition, as well as those administered to treat it[102]. In that case, isoprenaline or glucagon is effective[101].

Animal studies suggest that only a minuscule fraction of the penicillin dose needs to bind to provoke a reaction, so the degree of covalent binding itself is not the issue, rather it is the shape of what is bound and where it is bound. Penicillin originated from a eukaryotic organism (mould) and many fungally related organisms and their toxins are pathogenic to humans. Therefore it is not really surprising that the structure of the penicillin-related material can be so intensely antigenic. As mentioned earlier, the drug forms haptens in everyone without exception, but only a minority of patients develop anaphylaxis.

8.4.2   DRESS/Anticonvulsant hypersensitivity syndrome (AHS)

As you have seen so far, there are different terms that try to describe the various drug hypersensitivity reactions. It can be difficult to see consistency, mainly because some scientists/clinicians coin acronyms which can become widely used, but there always other authors who resist that term and either ignore it and/ or try to coin their own acronym which may or may not catch on. As mentioned previously, IDILI for example, can exist without systemic involvement, but when a generalised systemic immune response occurs, IDILI can become part of DRESS, a term coined in 1996[68]. The diagnostic criteria for DRESS involves multi-organ damage, skin eruptions, fever, and eosinophilia[103]. DRESS is linked with many drugs, including allopurinol and several sulphonamides, but the elements of the condition were first ascribed to the vintage anticonvulsants in the 1950s, starting with phenytoin and then carbamazepine, although lamotrigine and phenobarbitone also trigger it[103]. Hence, by the 1960s the condition became widely known as anticonvulsant hypersensitivity syndrome (AHS). This term AHS is still used, but as so many other drugs show the same pattern of response, it is now effectively under the DRESS umbrella. To confuse even further, the term AHS has been used to describe abacavir hypersensitivity syndrome[104]. Hence, rather than try to coin yet another new term, in this section I have used DRESS/AHS.

The chances of DRESS/AHS appearing can be from 1 in 1000, to 1 in 10,000[105] and the fatality rate can be up to 10%. Valproate is not usually associated with the syndrome, although it can promote the effect with lamotrigine, as valproate retards its clearance. DRESS/AHS is not as rapid in onset as anaphylaxis, but can begin within 2–12 weeks of starting therapy. Virtually all patients present with fever and this can persist for weeks after the drug has been withdrawn. A rash or more severe eruption affects the trunk and limbs; the face and lips may swell and blister appreciably. Around half the cases also have hepatitis; this is the most dangerous area of the reaction and up to a third of patients die from liver failure. When the condition is described as AHS, it has been recognised by the triad; of fever, rash, and internal organ involvement[105]. Kidney toxicity can also be apparent and general haematological abnormality results. Some patients go on to develop SJS/TEN (see next section) and DRESS/AHS can also cause a potentially fatal condition of the lungs known as BOOP (bronchiolitis obliterans organising pneumonia)[106], which is slightly different from bronchiolitis obliterans linked to popcorn lung (Chapter 7.6).

A serious clinical problem arises when the patient recovers, as the reaction may occur again with another anticonvulsant. This can happen accidentally[104,105] and the reaction will be much worse and appear more rapidly the second time, which is what you would expect with an immune reaction. Hence, patch testing is recommended before another anticonvulsant is started. Many authorities recommend that valproate or one of the newer AEDs be used as a safe substitute

AED, although it too can cause hepatotoxicity in rare cases. As so many of the symptoms resemble and thus present clinically as an infection, valuable time is often lost treating the patient with antibiotics whilst the offending drug is not always stopped immediately. Stopping the drug in time can prevent the condition from involving the organs and save the patient's life[105]. The triad should alert medical personnel to the possibility of DRESS/AHS in a presenting patient. Treatment usually involves high doses of steroids for as long as it is necessary to contain the condition, followed by gradual tapering of the dose with resolution of the symptoms and eventual recovery. Antihistamines are often used to suppress itching and irritation.

All the drugs responsible for this syndrome are extensively cleared by CYPs, often to reactive arene intermediates that are cytotoxic and tissue reactive *in vitro*. With carbamazepine, polymorphisms of microsomal epoxide hydrolases (EPHX1: Chapter 7, Section 7.2.4) are known to be responsible for altering the clearance of the 10, 11 epoxide, which may predispose to developing anticonvulsant syndrome[107]. This suggests that the DRESS/AHS could be triggered by haptenation, although it is now seen largely as a p-i mediated response[69]. DRESS/AHS is strongly associated with HLA-A*3101 (carbamazepine), as well as phenytoin with HLA-B*1502 and several other HLAs, as mentioned in Section 8.3.4. However it is triggered, the end-result is a powerful T-cell-mediated immune response that seems to require the presence of the drug to progress[69,105].

8.4.3   Stevens-Johnson syndrome (SJS) and toxic epidermal necrolysis (TEN)

These conditions are included in the acronym SCARS (severe cutaneous adverse reactions) and are extremely rare (1.5–8 cases per million person years), with only a few hundred diagnosed in the United States per year. They were thought to be two different conditions, but they are now seen as affecting the skin the same way, and the difference is mainly the level of skin destruction. Some publications term them as SJS/TEN, or EN, for epidermal necrolysis[108]. SJS is defined as less than 10% of detachment of body surface area and TEN is greater than 30%[109].

The word *detachment* clearly does not begin to describe what happens, but suffice to say that the skin dies and peels off and the treatment must take place in a hospital burns unit, which are often not sufficiently experienced in the nursing care required[108]. The patient must be protected from infection and dehydration, as well as medicated for the pain. More specialised problems such ocular and genital issues must also be managed, and cyclosporine is also beneficial[108].

Despite this, mortality from SJS is around 5% but more than 30% for TEN. The condition is most closely allied to AEDs as a class, which are nine times more likely to cause EN compared with other medicines and the highest risk drugs are phenytoin, lamotrigine, and carbamazepine[109,110]. The risk is reduced with a low starting dose and slow titration with lamotrigine, but otherwise, efforts to predict risk are based on the HLA alleles described earlier. These are the various HLA-B* alleles (1502/5801/1301), which predispose to risk for carbamazepine, allopurinol, and dapsone in different Asian groups and 5701 for abacavir in Caucasians[100,110]. Aside from the HLAs, TCRs play a key role, as one study found a particular TCR clonotype (VB-11–ISGSY) in 84% of SJS/TEN patients, but only 14% of otherwise healthy controls had it[111]. It is likely that all the EN conditions are likely to share very similar causes in terms of T-cell involvement, but they differ from other drug hypersensitivity syndromes, such as AHS and IDILI as they focus on the skin and they are many orders of magnitude more rare. There is clearly much progress to be made in understanding this group of conditions, as unfortunately, only a fifth of EN cases can be prevented even if HLA high-risk allele analysis is used to predict the risk of developing the condition as so many other factors are involved[100].

Since the late 1980s, there have been several reports that radiotherapy can trigger EN in patients taking AEDs such as carbamazepine. Even if the radiation is confined to the brain, an EN condition can appear and advance around the body. Radiation does not normally cause such lesions alone, and why it happens in the presence of AEDs is difficult to determine. Paradoxically, in some cases, the radiation has sometimes been immunosuppressive[112]. Radiation is widely used in cancer therapy because it is intensely cytotoxic through free radical and reactive species generation. Radiation requires the presence of oxygen to generate its maximum cytotoxicity. Unfortunately, bystander tissues contain much more oxygen than the tumour. Hence, it is likely that the widespread cellular stress induced by radiotherapy is capable of exerting influence on the co-activation process of the T cells.

Other cutaneous reactions

Aside from the devastating but extremely rare conditions like the ENs, cutaneous reactions of various types are also involved with other more common but still relatively rare DRESS/AHS conditions as mentioned previously. However, in the context of all drug-adverse reactions, skin reactions are the commonest reported events[113]. So there is a very wide spectrum of conditions, ranging from mild to intense maculopapular eruptions (MPEs), towards more serious conditions such as erythema multiforme. Their causes may be mixed, involving hapten formation as well as p-i mediated effects, which is reflected in the range of drugs that cause them. The commonest agents involved are the various penicillin-related agents, followed by NSAIDs and the drugs that cause DRESS/AHS, such as the AEDs, allopurinol, and the sulphonamides. Drugs such as dapsone can paradoxically both provoke cutaneous reactions and suppress them[14]. Clinically, there are so many causes of these conditions and they present in so many different ways, you can see that linking the reactions with a specific drug, particularly in polypharmacy, is problematic. If the conditions are caused by similar mechanisms, then the obvious question is, what processes govern the progression, or lack of progression, from mild MPE to intense, systemic and life-threatening conditions such as DRESS and the ENs?

8.4.4   Blood dyscrasias

These conditions describe the loss of a group of cells within the blood system, such as erythrocytes, platelets, or entire white cell populations. Although the various mechanisms of the reactions could fit within the current theories of drug hypersensitivity, the net effect can be either the targeting of the cells themselves, the organ or tissue that makes them, or a combination of the two.

Haemolytic anaemia

The premature intravascular destruction of erythrocytes can be caused by many circumstances, from physical damage to the cells (passage through surgical life-support systems) and accelerated wear and tear caused by drugs such as sulphones, through to immune-mediated causes related to drug therapy. Drug-induced haemolytic anaemia (DIHA), sometimes termed DIIHA (drug-induced immune haemolytic anaemia), is extremely rare, with estimates of the risk ranging from 1 in 10^6 to around 1:250,000[114,115]. A number of drugs are associated with DIHA, with diclofenac reported as the most frequent cause, but the major group responsible is antibiotics (penicillin, cephalosporins), as well as the platinum-based antineoplastic drugs and quinidine[114,116]. As far back as the 1960s and 1970s, high doses of penicillin tended to be used, which caused haptenation of erythrocytic membranes leading to recognition by B-cell-mediated immunoglobulins. This causes complement activation, although the other mechanisms besides haptenation cannot be ruled out. Drug linked or even auto-antibodies can be detected, and sometimes both[116]. Auto-antibodies were first found with the 1960s antihypertensive methyldopa, which is still used in pregnancy-induced hypertension[117].

As mentioned previously, red cells already have a 'sell-by date' system, where normal wear and tear (red cells can be reversibly deformed to an amazing degree) induces the appearance of various wear indicator molecules on their surface. As described early in this chapter (Section 2.2), the spleen 'reads' the number of these molecules and when the number corresponding to the 120-day lifespan appears, the spleen removes and destroys the cell. Clearly, a damaging and surface attritional process such as haptenation could accelerate the spleen's removal of the erythrocytes, as well as alerting the B cells. Whilst extremely rare as a therapeutic event, it is far more common in cystic fibrosis patients who comprise

40% of DIIHA victims, mainly linked with piperacillin/tazobactam treatment. The fatality rate is 7%, and it is not clear what the mechanism might be[116].

Clinical presentation could occur a day or so after the drug is started or even up to two weeks later. Haemoglobin levels fall to 3–5 g/dL, although it is recognised that less severe cases probably are not noticed, treated or reported. The condition when it is severe can involve jaundice and dark urine, due to the liver struggling with the disposal of so much haem in such a short time, leading to the formation of very high levels of bilirubin[114]. Symptoms appear that are related to tissue hypoxia, such as fatigue, shortness of breath, and tachycardia. As the condition progresses, renal and hepatic failure can ensue, which account for the high death rate, which can exceed 20%. Indeed, with one report, half the cases of DIHA triggered by ceftriaxone were fatal[116]. This outcome is not helped by misdiagnoses, where the condition is mistaken for a variety of problems, including biliary obstruction, uraemic syndrome, and even transient ischaemia[116].

Treatment is largely supportive after the key step of stopping the offending drug. However, with the advent of polypharmacy, the clinician might well ask, 'Stop which one?' In this case, it is recommended to withdraw all nonessentials, then amongst the 'essentials', start with the drug with the worst reputation for idiosyncratic reactions and work down[118]. Hopefully, the main organs will recover, erythropoiesis stimulation, folate, and transfusions can tide the patient over until normal red cell production eventually replaces the lost erythrocytes.

Aplastic anaemia

This condition is due to destruction of the bone marrow, which can be caused by radiation, infections or by an immune-mediated process. Drug-induced aplastic anaemia (DIAA), or again, drug-induced immune aplastic anaemia (DIIAA), involves damage to multipotent hematopoietic stem cells in the bone marrow and is a rare condition, with a risk range of around 1:500,000 with Caucasians to 1;150000 in Asian countries[117]. DIAA/DIIAA results in the gradual reduction of vascular levels of all blood cells, including white cell populations, erythrocytes, and platelets. The symptoms include those for haemolysis, alongside those of immune deficiency, such as oral thrush.

The condition is sustained by T cells and cyclosporine can be an effective treatment as it prevents T-cell activation. Several drugs cause aplastic anaemia, which include anticonvulsants, anti-cancer agents, chloramphenicol and phenothiazines and it is possible that it is another process that could be accounted for at least in part by the p-i hypothesis. Interestingly, the chronic use of illegal drugs, such as opiates and cocaine, increases the risk of aplastic anaemia. Indeed, severe aplastic anaemia can result on rare occasions from DILI caused by anabolic steroid abuse, which often has a very poor outcome; it is termed hepatitis-associated aplastic anaemia (HAAA)[119]. Aromatic organic solvents such as benzene and toluene are also linked with aplastic anaemia as well as a number of insecticides. The prognosis is generally poor, as even after a bone marrow transplant, survival over five years can be less than 60%[117].

Agranulocytosis

Neutropenia is diagnosed at less than 1500 cells per μLitre, as the normal range is 4500–10000 per μLitre. If levels fall below 500 per μLitre, this is agranulocytosis and there are not enough cells to prevent sepsis, which will happen within days unless the condition is treated[118]. This underlines the importance of neutrophils, as in spite of the existence of so many other immune cell types, life is only possible for a few days without neutrophils, which are the major form of polymorphonuclear leucocyte. Neutropenia can occur due to a variety of causes, although with drug-induced neutropenia, antineoplastic agents are the most obvious and predictable offenders. However, agranulocytosis due to noncytotoxic agents can become manifest with the symptoms of a serious systemic infection (unlike AHS/DRESS, it actually is an infection), with fever, malaise, and chills, leading to acute sepsis. The type of bacteria that infects the patients in this context includes *Staph. aureus* and various pseudomonads. In past decades, the death rate exceeded 20% and was probably due to failure to recognise the link between drugs and the condition and inadvertently continuing therapy with the trigger drug. As was learned with HIV, antibiotics are not really capable of resolving infections without immune system support, and this can only start to happen after removal of the trigger drug allows neutrophil numbers to gradually rise. Hence, better awareness amongst clinical staff means current death rates are now less than 5%[118]. Indeed, in the case of drugs such as clozapine, a robust clinical monitoring system ensures that patients cannot receive the drug without a normal white cell count. This has cut the odds from 1:100 to 3:10,000[118] and allowed a useful drug to continue in clinical practice.

Agranulocytosis was first seen in the 1930s with aminopyrine, a vintage drug introduced in 1897. By the 1950s various clinical studies demonstrated that aminopyrine stimulated the formation of antibodies to circulating neutrophils, which attacked in the presence of the drug. Just 10 mg of the drug would trigger the response in a sensitive patient[120]. If these studies were carried out today, it would lead to the arrest and subsequent imprisonment of all the participants. However, it was estimated in the 1960s that the frequency of agranulocytosis to aminopyrine could be approaching 1%[120]. More recently, drugs with the highest likelihood to cause agranulocytosis include the first-generation aromatase inhibitor aminoglutethimide, at around 2%, with clozapine at 1% and antithyroid drugs under 0.5%[95,118,121]. Sulphonamides, phenothiazines, and dapsone can also cause it, but their risk is much lower. With the anti-inflammatory sulphsalazine, the risk is 1:2400, initially, rising to 1:700 after about three months and then falling to >1:10000 after that[95]. The risk with dapsone seems to be greatest in those with existing autoimmune conditions[122]. Interestingly, in the 2000s, there were reports of heavy cocaine users developing agranulocytosis, and it was caused by the 'cutting' or adulteration of the drug with levamisole[118]. Overall, the major risk factors are being elderly, female, and a neutrophil count of 100 per μLitre or less, as well as other pathologies such as renal disease and immunosuppression[118]. As indicated with the sulphonamides, the condition often, but not always, follows the pattern of appearing 8–14 weeks after drug initiation. Treatment in sepsis is obviously intravenous antibiotics, along with granulocyte-macrophage colony-stimulating factor (GM-CSF) and granulocyte colony stimulating factor (G-CSF), which kick-start the bone marrow. These treatments do have risks such as immune reactions and the risk/benefits will be well explored before treatment[123].

Agranulocytosis: mechanisms

Neutrophils have a short lifespan of a day or so, depending on whether they are activated and they take about a week to produce from their precursors. Normally neutrophils appear in the circulation and die in their billions every day, but despite the vast numbers, neutrophil supply and demand is clearly well regulated. It seems that in agranulocytosis, disappearance of the cells from the circulation seems to coincide with the gradual slowing of production, rather than very significant accelerated cell destruction in the circulation. This is supported by observations that it is usually very difficult to find evidence of drug-dependent neutrophil antibodies, although autoantibodies may be found[118]. Of course it is very challenging in the circumstances of the patient's condition to try to carry out such tests and neutrophils are extremely unstable and problematic to assay. In any case, neutrophils are so short-lived and unstable that unless a fairly vigorous process such as seen with aminopyrine was happening, it is possible that immune destruction in the circulation is not the major issue in agranulocytosis with many drugs.

Hence, if we turn our attention to the effects of the condition on neutrophil production, in aminoglutethimide-mediated agranulocytosis, bone marrow is undamaged but has few cells. Once the drug is removed, production restarts[124], which does not happen in aplastic anaemia. As we know, the immune system can be extremely destructive, and whilst the damage seen in bone marrow in aplastic anaemia fits a typical immune T-cell mediated process, agranulocytosis, with its intact but dormant marrow, does not.

One way to further consider the role of an immune-mediated process is to rechallenge the patient, which should cause a faster and more aggressive effect. Obviously, we don't want to get arrested here, but in certain clinical circumstances rechallenge with a drug that caused neutropenia can be approved with drugs such as clozapine for valid therapeutic reasons. In one study, 70% of the patients did not develop a reaction, whilst in those that did, the effect was faster, but less severe[125]. In another study, 62% did not have a reaction on rechallenge, but in those that did, it was worse, lasted longer, and was faster in onset[126]. What is even more fascinating is that in the latter study, half the patients who had originally suffered the neutropenia continued on clozapine without further problems.

Hence, we could examine agranulocytosis in the light of the standard hypotheses of drug hypersensitivity. Considering the hapten process, many of the drugs that cause agranulocytosis are oxidised to reactive species and this can be achieved by activated neutrophils[118,124]. However, haptenation would occur in all patients to some degree and it could be expected that rechallenge would lead to a fulminant and aggressive T-cell response as seen with aminopyrine. Some of the patients on clozapine for example showed something of this pattern of response on rechallenge, but most did not. Haptenation can only occur where reactive species formed by enzyme systems such as myeloperoxidase in activated neutrophils can attack cell proteins. This could occur in the periphery, but as we noted earlier, there is little clinical evidence that a drug linked response to neutrophils appears. In addition, there is no evidence that haptenation could occur in the bone marrow, as the neutrophil precursor cells are not metabolically capable. Considering sulphasalazine, the active drug, sulphapyridine, is believed to form haptens and this drug has been associated with several distinct HLA alleles[95], such as HLA-B*08:01 and HLA-A*31:0, which incidentally are also linked with hypersensitivity to carbamazepine[95]. Sulphonamides are believed to be able to interact with HLAs either through haptens, or through p-i parent drug binding[127].

Agranulocytosis: summary

Taken together, it appears that the immune system stubbornly resists categorisation in our current theories and agranulocytosis seems to vary its mechanism according to the drug involved. What is striking is that there is evidence with clozapine and aminoglutethimide that progression towards drug-induced neutropenia is more common that is realised[124] and even in those that reach the stage of agranulocytosis, the system can counterintuitively correct itself. This has been described in IDILI with yet another hypothesis, called the *failure to adapt,* which tries to explain why most individuals that develop hypersensitivity manage to overcome it, but a minority do not and suffer the full force of the condition[59]. On balance, it could be speculated that agranulocytosis is mainly a p-i type effect, which might conceivably alter certain categories of HLAs so that the focus could be the gradual removal of key factors that regulate neutrophil production. This might occur in the manner abacavir impacts HLA binding. Agranulocytosis does not seem to elicit a classic T-cell response, does not damage the bone marrow and appears to be possible to override on rechallenge, perhaps through recruitment of other pathways of neutrophil regulation. What is entertaining and fascinating about this area, is that you might review all the papers I have read and come to a completely different conclusion.

8.4.5   Prediction of idiosyncratic reactions

In the past, perhaps our response to patient worries over potential drug hypersensitivity reactions may have been, 'Well, don't worry, it's unlikely to happen, but if it does, we will sort it out'. As we have seen from the previous sections, it is admittedly unlikely to happen, but if it does, it cannot always be sorted out. Perhaps hypersensitivity has, in our ignorance, been viewed passively as an inevitable price to pay in some unfortunate individuals. However, with recent advances, clozapine-style monitoring is being augmented and may even be left behind by prospective hypersensitivity risk testing. Certainly in some areas, clinical and scientific data accumulated so far allows us to try to estimate which drugs, classes of drugs, and crucially, which patient populations, are at risk from particular syndromes. Ultimately, as with polymorphic drug metabolism in Chapter 7, it could be argued that we should aim to test the patients and predict their individual risk to a range of hypersensitivity conditions, from the commonest mild rash to the rarest ENs. Then an informed clinical judgement can be made over the risk–benefit ratio for that drug choice for that patient.

Prediction of DRESS/SCARS

As with polymorphic drug clearance, genetic factors predispose to hypersensitivity in many cases, and certain alleles, haplotypes, and diplotypes have been linked to particular reactions and the technology to detect them is here, but is not necessarily cheap or robust enough for mass usage. HLA testing is already in clinical practice in some countries, particularly HLA-B*15:02 for carbamazepine and phenytoin in those of Asian ancestry and a strong case can be made for testing for HLA-A*31:01 with phenytoin in Caucasians[104,128]. It is now recommended by regulatory bodies that phenytoin should not be used in those with HLA-B*15:02[104]. In a key development, a more integrated assessment of risk can be made in the light of GWAS (Chapter 7.2.11, Genome-wide association studies). If you have a risk allele (s) which predisposes you to a reaction with a drug and another allele which delays the clearance of that drug, then this significantly multiplies your net risk of a reaction and this has been demonstrated with Japanese patients, with HLA-B*15:02 and CYP2C9*3 and phenytoin-linked SCARs[99]. Likewise, allopurinol should not be used with HLA-B*58:01 in Asian and European populations and HLA-B*57:01 status is now recommended by authorities such as the European Medicines Agency to be determined prior to abacavir therapy commencement.

Predicting agranulocytosis

The odds of developing this condition in some drugs are between 10–50 times shorter than developing DRESS, yet it is currently challenging to predict risk in this area. To test with reasonable certainty the odds of contracting sulphasalazine-linked agranulocytosis, not only HLA-B*08:01 and HLA-A*31:0, must be detected, but up to four other alleles must be found in the patient haplotype also, meaning that 1500 tests to carried out before one person is found to be at risk[95]. The authors of that study recommend that this is worthwhile with sulphasalazine and economically, it might be cheaper in the long run than the clozapine monitoring process, although this depends on the accuracy and predictive powers of the tests. As we saw in Chapter 7.2.11, this is not always a reliable process. For now, with high-risk drugs such as clozapine, all we can do is monitor, whilst with other agents clinical vigilance and swift recognition will minimize the risk until more reliable tests appear.

Predicting drug hypersensitivity: the future

As we have seen, many drugs in current clinical practice are associated with significant risks of a range of hypersensitivity reactions from the mild to the horribly and gruesomely fatal. However, if even current testing technology was fully applied, many of these reactions could be averted, and as we saw in Chapter 7.2.11, the economics are in many cases very much in favour of testing, given the exorbitant costs of hospital stays. If we factor in issues such as the loss of the patient's ability to work as well as the months or even years of care necessary due to the reaction, let alone the personal cost, this strengthens the case for prospective testing. With new drugs, as stated previously, it would never be possible to design trials large enough to detect a case of SCARs or EN. Fortunately, with our current knowledge, it is now even possible to build into the clinical trial process analysis of the risks a new entity might pose, given its structure and relationship to current drugs. Given the low frequency of the most serious reactions and the near universality of the clearance pathways and reactive species/haptenation risks of many drugs, with the most severe reactions, the causes lie with the unique combination of the individual's genetic and phenotypic responses. In the future, it should be possible from one blood sample to provide an analysis of the individual's complete risk profile for hypersensitivity that includes their polymorphic clearance capability. This could take a format that resembles that seen in Chapter 7.2.11, with the Commercial Pharmacogenetic Decision Supporting Kits (CPDSKs). The complete analysis must be backed up by an accurate risk assessment, which can then be considered by the physician. Such a process will augment rather than replace clinical monitoring, in order to save many lives and reduce the vast amount of suffering that drug hypersensitivity causes.

8.5   Type B3 reactions: role of metabolism in cancer

8.5.1   Sources of risks of malignancy

Oxidative and reductive metabolizing enzymes are responsible for the generation of some reactive species that interact with key cellular structures in a different manner and with a different result, in comparison with immunogenic or necrosis-inducing species. Carcinogenic metabolites must have the exquisitely precise structural and physicochemical characteristics necessary to react with individual DNA nucleotides. This can lead to cross-linking of the DNA, preventing transcription, or causing transcriptional errors. As stated earlier (Section 8.2.3) one of the chief defences against the propagation of damaged DNA is apoptosis. Whether specific or nonspecific binding is responsible for initiating the process, controlled and ordered cell death results, where reactive endogenous systems are shut down or made safe, DNA is destroyed and the cell dismantles itself. Apoptosis is a very complex process, but in terms of a response to reactive species, it is an attempt by the organism to contain damage and limit its consequences for other cells and, most importantly, the organism itself. The destruction of the DNA of the stricken cell prevents faulty genes from being propagated in cell division. The survival of the cell in this context could mean eventual death of the entire organism from a future neoplasm. If apoptosis is not triggered, the organism will rely on gene repair to prevent possible future malignancies.

Gene repair in dividing cells is an endless process, like roadworks on motorways. The balance between enzymatic reactive species generation, degree of detoxification, and actual DNA damage, as well as the efficiency of DNA repair ultimately dictates an individual's risk of malignancy. Anecdotally, some whales can live for over 150 years, partly because their DNA repair is so efficient. In terms of the consequences of gene damage, a rather old-fashioned analogy is one of those ancient films where somebody with a stethoscope is trying to open a safe by listening for the tumblers in the lock to drop. Nine down, ten required, it will not open and we will never know how close we came.

8.5.2   Risks of malignancy and drug development

The risks of carcinogenicity from drug therapy must be resolved into two clear categories. The first includes anticancer drugs, as many attack DNA such as cyclophosphamide. This and other agents, such as the anthracyclines and particularly the alkylating agents like procarbazine, carry the risk of causing a later malignancy themselves. This can occur sometimes many years after the treatment was apparently successfully concluded[129,130]. If a new anticancer drug is not an alkylating agent, then its risk of causing a malignancy is much reduced, but all small molecule antineoplastic agents are intended to powerfully disrupt malignant cell metabolism and there will always be the risk that they might in some way predispose bystander cells to some future malignancy. The second category, involves all other drugs and whilst many agents are only intended to be taken for a few days or weeks at most, many drug classes are now taken pretty much for decades. These include agents used in mental health (antidepressants and antipsychotics) as well as drugs used to treat cardiovascular function and type 2 diabetes. Whilst it is very rare that a prescription drug is linked with neoplasms, the recent case of the antidiabetic pioglitazone and bladder cancer[131] is a salutary warning, as nobody outside of the industry anticipated this problem and the subsequent fall-out generated was hugely damaging and costly (Appendix A.1). The drug industry is required to examine this possibility with carcinogenicity tests with many new drugs, particularly if they are likely to be taken for months and/or years[132]. After *in vitro* studies, such as the Ames test (Appendix A.6.6.) have shown a drug not to be mutagenic *in vitro*, the animals are exposed to the drug in carcinogenicity studies for their whole lifetimes with doses that greatly exceed the likely therapeutic level. These studies are usually still going on while the drug is in Phase I and II (human volunteers and patient) trials[132].

Although many drugs can interact with various nuclear receptor systems so causing profound changes in gene expression, the possibility that these changes could lead to malignancy can be detected to some extent through the use of DNA microarrays (A.6.9) in human cells as well as *in vivo* animal studies. The other more likely route is through DNA damage caused by reactive species, formed by biotransformation in the liver, lung or perhaps the kidney. Usually, but not always, reactive species may interact with human and animal DNA in a similar fashion. However, animal models may diverge from the human situation strongly as to which CYP enzymes actually form the reactive species. In the case of aromatic amines like sulphonamides, dapsone, aniline, β-naphthylamine, and 4-aminobiphenyl, different enzymes between the species form the same hydroxylamines. The recent examples of troglitazone, trovafloxacin, and ximelagatran underline the problem that humans are more susceptible to certain toxic events than animals. Hence, the Pharmaceutical Industry is aware of the danger that a drug might be metabolised to a reactive or potentially carcinogenic species in humans that does not happen in an animal model. The use of human liver CYP enzymes to activate the compounds prior to inclusion in the Ames test can address this issue, followed by the animal lifetime carcinogenicity studies. These studies can be backed up by the many DNA damage/repair assays (Appendix A.6.7/8) that can detect genotoxicity in human primary and cultured cell systems. Overall, it is likely that the chances of a new drug causing cancers in humans are low, but not as remote as we once believed. Outside of drug metabolism, the main causes of human cancers where drug-metabolizing enzymes are involved in the formation of DNA-reactive species include smoking, diet, occupation, and atmospheric pollution.

8.5.3   Environmental carcinogenicity risks

Although there is a vast literature on carcinogens in animal species, there are surprisingly few proven carcinogens in man. It is perceived that our greatest risks of exposure to carcinogens lie in what we eat, what we breathe (or inhale, if it is tobacco) and where we work. The earliest hard information about human carcinogenic risk came from occupational sources. The first documented example (eighteenth century) was the detection of scrotal cancers in young chimney sweeps, followed by polycyclic aromatic-mediated disease in the early twentieth century. Aromatic amine-mediated bladder cancer risks in the dye and rubber industries were established well before the 1950s[133]. This real-life experimental system is from the human standpoint tragic and costly, whilst from the scientific standpoint, it is relatively uncontrolled and incomplete. Exposure to carcinogens is confounded by factors such as genetic predisposition, age, smoking habits, changes in recommended exposure levels, and working environments. Individuals are often exposed to carcinogens for up to 40 years or, in some cases, their exposure is relatively brief and the latency period before the cancer appears can be several decades[133]. It is also increasingly clear that protective mechanisms that maintain DNA are also extremely sophisticated and vary in their effectiveness between individuals. So what is perhaps remarkable is that so many thousands of individuals smoked and drank heavily, ate all the wrong things, stayed up really late, worked in the rubber or dye industries and still died of old age.

8.5.4   Occupational carcinogens

Aromatic amines: introduction

As mentioned previously, the chances of any new drug being introduced containing, or had any chance of being metabolised to, an aromatic amine are vanishingly remote. This is because they are a 'red flag' in terms of ease of oxidation and reduction, which results in metabolites that attack cellular macromolecules and cause adverse reactions[98,124,134]. To understand why aromatic amines are so toxic and carcinogenic, it is first useful to look at some basic amine chemistry. It is easiest to see these compounds in terms of existing mostly in the environment as two main forms, both of which are essentially nontoxic by themselves, that is, they are stable and nontissue reactive. The first form is the amine itself and the second is the nitroderivative (Figure 8.9). From this figure you can see that the amine can be oxidized at stage 1 to the N-hydroxy, or hydroxylamine. This usually requires some serious oxidative energy. This is usually supplied by CYP isoforms, but also can result through other oxidising systems such as myeloperoxidase[98,134]. The product hydroxylamine varies in stability according to the rest of the structure of the molecule[14,134]. Stage 1 (Figure 8.9) essentially converts a stable entity into a relatively unstable one, in true CYP tradition, with two electrons being lost in the process. The hydroxylamine can now be conjugated, but this can lead to the formation of even more unstable products, which split, leaving a reactive nitrenium group (more later) that attacks DNA.

The other alternative is the spontaneous oxidation (stage 2) and loss of two more electrons to form a nitrosoarene, which can often be highly tissue reactive and rapidly cytotoxic, although some are much more stable. Nitrosoarene stability again depends on the structure of the rest of the molecule. The further spontaneous oxidation and loss of two more electrons (stage 3) forms a stable nitro

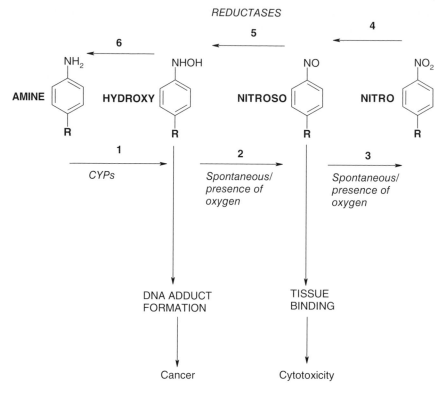

Figure 8.9 Stages of oxidation and reduction of aromatic amines

derivative, which is usually not a problem unless it meets a reductase enzyme, of which there are a large number, in the liver (NADPH reductases), erythrocytes (NADH reductases) and the cytosol of most cells. These return the electrons and form the (stage 4) nitroso and (stage 5) hydroxylamine and then the (stage 6) amine in turn. At each point, the same dangerous intermediates, the nitroso and the nitrenium ions, can be formed. So aromatic amines are a problem because they are relatively easy to oxidize or reduce, but essentially the same toxic intermediates are formed. So many different enzyme systems can activate them anywhere in the body, not just the liver. We have already discussed the role of activated neutrophils and their myeloperoxidase. Therefore, vulnerability to the different unstable aromatic amine metabolites is problematic as the vast majority of tissues do not have the in depth antioxidant protection of the liver and the lung. So aromatic amines can cause a range of problems from methaemoglobin formation, all the way to cancer in widely differing types of tissues[133]. As mentioned above, the pattern of toxicity caused by aromatic amines depends very much on their physicochemical characteristics and in particular, the stability of their hydroxylamines (Table 8.2). This is a generalisation, but usually the more

Table 8.2 The relationship between the stability of aromatic hydroxylamines and their mode of toxicity

Parent drug	Hydroxylamine half-life (pH 7)	Methaemoglobin formation	Immunogenicity	Carcinogenicity
S ' onamides	< 10 min	+	++++	No
Chlor ' col	< 20	++	+	No
Dapsone	< 60 min	++++	++	No
4-ABP	> 3 h	++++	No	++++
BNA	> 10 h	No	No	++++

S ' onamides   (sulphonamides);   BNA   (β-Naphthylamine);   4-ABP   (4-aminobiphenyl); Chlor ' col (chloramphenicol).

stable the hydroxylamine, it is less immunogenic, a poorer methaemoglobin former, but the more carcinogenic it might become.

Hopefully you can see how aromatic amines can be so toxic, so the following examples should be easier to understand. Chloramphenicol's nitro group is reduced to a hydroxylamine, which caused the methaemoglobin in babies in the 1950s (grey baby syndrome). This drug only survives today as a topical preparation for minor eye infections, so its systemic absorption is virtually zero. Interestingly, a number of other drugs in the last 20 years have failed in clinical trials because they were metabolised to amines that were then toxic. There are many aromatic amines consumed as foodstuffs (overcooked burgers), environmental pollutants and occupational agents. I have cited some examples below, but there are many more in the scientific literature. As mentioned previously, there are two main ways aromatic amines can be metabolised to carcinogens: through oxidative or reductive means.

Aromatic amine carcinogens: oxidative metabolism

β-Naphthylamine (BNA) was used in the rubber industry for many decades as an antioxidant to maintain pliability in rubber products, such as tyres. It could be found in levels that exceeded 1%. It was employed as an anti-rust additive in various oils used to coat metals between manufacturing stages. BNA was also used in the dye industry and it has been established as one of the few compounds that are accepted to be a human carcinogen even in a court of law, provided exposure can be documented. Benzidine, aniline and o-toluidine were all aromatic amines used along with BNA in the manufacture of various colourfast dyes. The pattern of aromatic amine carcinogenesis is interesting: it appears that in any given population of workers that were exposed to this agent, up to 10% may develop bladder cancers[133]. This may be from a relatively short exposure of as little as 1–6 years. Up to 30 years after exposure ends, the bladder tumours appear and often the first realisation that something is wrong for the patient is blood in the urine (haematuria), although many tumours are symptomless until quite late stages[133].

The following factors influence whether a worker might develop aromatic amine-related bladder cancer if they:

- Are slow acetylators (Chapter 7.2.5)

- Possess wild-type full-function CYPs and null detoxification genes (such as GSTs

- Are heavy smokers

- Have a diet low in vegetables and high in meat

- Work in a hot environment with poor fluid intake

Aromatic amines: mechanisms of toxicity

It seems that aromatic amine carcinogens all share a similar metabolic profile and it is possible to trace the metabolites formed and determine which are carcinogenic and which are not. It is also possible to explain each of the risk factors above. What is not easy to explain and is beyond the scope of this book, is why it takes 20–40 years before irreversible changes in the bladder lead to the appearance of a tumour.

From Figure 8.10, it is possible to see that there are several metabolic products of aromatic amine clearance. Usually low levels of parent amine are found in urine, but the main fate of these compounds is oxidative and conjugative metabolism. The hydroxylamine derivative is usually too unstable to be able to proceed through the bloodstream to be filtered by the kidneys to reach the bladder[14]. It is much more likely to form an O-or an N-hydroxy glucuronide. The parent drug can also form an N-glucuronide. Free parent amine in the bladder is probably not a problem, as the bladder is unlikely to be able to activate it to its hydroxylamine in quantity. The acetylated derivative is not a problem, provided it is not deacetylated to the parent drug that can then be oxidized by the liver to the hydroxylamine. The fact that slow acetylators are more likely to develop bladder cancers suggests that the acetylation process does 'hold up' the amine and protects it from oxidative clearance. Some aromatic amines are excreted in urine as acetylated derivatives, but these may also be glucuronidated and oxidized to form the N-hydroxyacetylated derivative, although this is usually a relatively minor pathway.

The majority of the dose of aromatic amines undergoes oxidative metabolism and the various glucuronides are stable in blood, but not in acid urine[14,133]. Most

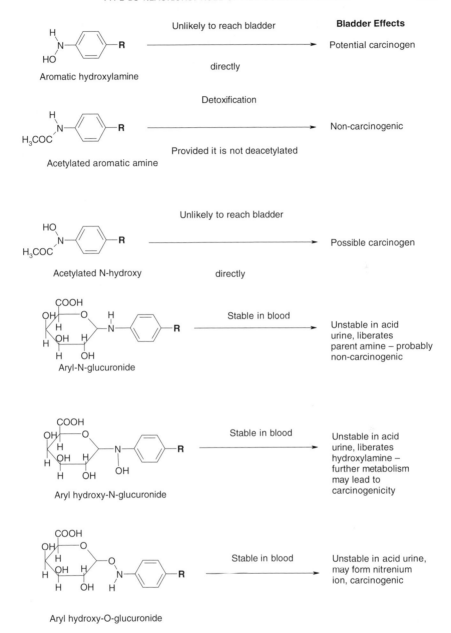

Figure 8.10 Some major metabolites of aromatic amines and their possible role in bladder carcinogenesis

people in the developed world eat too much meat and not enough vegetables. This leads to acid urine, which accelerates the decay of these metabolites to only a few minutes in some cases. In addition, men working in hot environments often do not drink enough fluid. They might have a flask of tea or coffee, but the caffeine in these drinks causes a net dehydration throughout the day. This, coupled with possible restricted access to water due to the work environment, means that often the workers' urine is very concentrated throughout the working day, leading to high levels of the various aromatic amine metabolites in the bladder. The decomposition of the glucuronides leads to the formation of parent amine, hydroxylamine and possibly nitrenium ions through a number of different pathways[128] (Figure 8.11).

Nitrenium ions (sometimes called aminylium ions) can be formed when a nitrogen/ oxygen bond is broken, which leads to the formation of a positively charged nitrogen with two unshared electrons. This makes for a very reactive species that seeks to find its missing electrons from a rich source of them, perhaps DNA or other cellular macromolecules. The first pathway where these ions could be formed is by the decomposition of the aryl hydroxy-O-glucuronide. There are other ways where nitrenium ions could also be formed (Figure 8.11), involving decomposition of the aryl-hydroxy N-glucuronide to yield the hydroxylamine. Nitrosoarenes would form in the presence of oxygen from the hydroxylamine and be protein reactive, although it is not clear whether they would be carcinogenic. It is suspected that nitrenium ions can be formed from reactions involving the hydroxylamine, but it is also possible that the bladder itself contributes to this by its acetylation of some of these metabolites, which may form more reactive species. There are thought to be many other reactive species formed by the oxidation of aromatic and alkyl amines, which may lead to carcinogenicity[135].

Aromatic amines: smoking

Smoking yields the same aromatic amines as the dye and rubber industries and as a result, the number one cause of bladder cancer is actually due to smoking, although it remains a relatively rare cancer; it is as of 2018 the tenth most common cancer, affecting three males to every female. A smoker would be unlucky to contract it – they would be nine times more likely to die of lung (the commonest cancer) rather than bladder cancer[133]. Many men who worked in the dye, rubber, automotive, and light engineering industries where they were in contact with aromatic amines in dyes, oils, and rubber also smoked, but it is likely that the greatest contribution of amines came from the workplace, as considerable numbers of nonsmokers who worked in these industries have developed bladder cancer. It is certainly true that smoking would add to the body burden of amines, but the chances of developing the disease are probably most strongly linked to the balance of carcinogenic metabolite formation, detoxification and DNA repair[133].

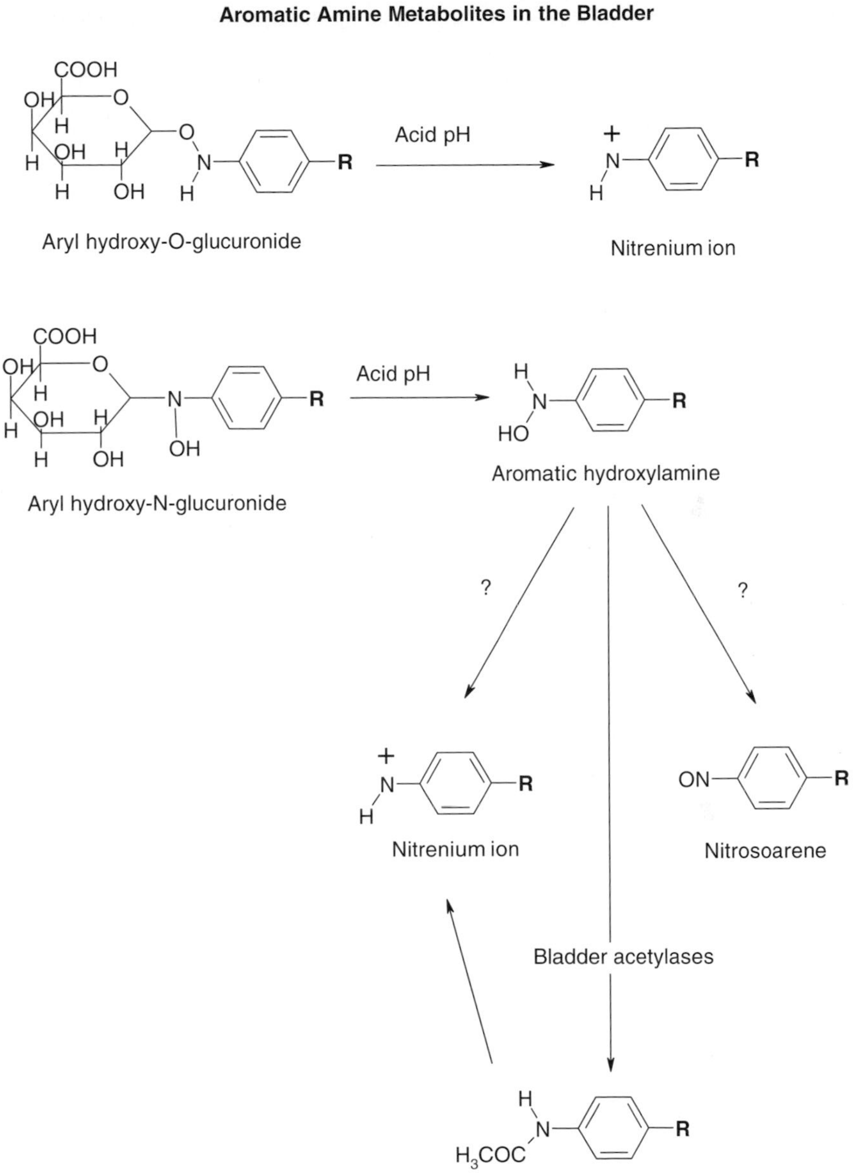

Aromatic Amine Metabolites in the Bladder

Figure 8.11   Final formation of aromatic amine-derived carcinogenic metabolites in the human bladder

Aromatic amines: human consequences of occupational exposure

Many men involved in these industries develop bladder cancer in their late 50s and early 60s, around the time when they would normally be looking forward to retirement. These individuals had worked since they were in their mid-teens, often for a number of long-since defunct organisations, so they struggle to obtain compensation through the UK legal system. They also face a long battle with the Department of Health and Social Care to establish that their condition is related to their occupations. The criteria laid down for this are strict and the onus is firmly on the individuals to prove they were exposed to some or all of a list of specific aromatic amines (BNA, benzidine, etc.); they may well die before they even receive a disability pension.

Since BNA was banned in the United Kingdom in 1949 and world manufacture was supposed to have been curtailed by 1971, it is increasingly hard for men to prove they were in contact with the extremely precisely described aromatic amines[133]. However, recent studies with rubber residues in various tyre factories in Europe have shown high mutagenesis in the Ames test, suggesting that the chemicals used as antioxidants to replace BNA and its equally toxic derivatives (phenyl BNA) may be just as carcinogenic as their lethal forbears. The current treatment for bladder cancer is removal of the bladder. Five-year survival for bladder cancer is currently improving, and with earlier diagnosis and awareness it can exceed 70%[133]. However, the pain and discomfort that these patients can suffer make their quality of remaining life relatively poor.

Aromatic amine carcinogens: reductive metabolism

As you know, most of atmospheric pollution is the result of burning fossil fuels. In general, the lighter the fraction of crude oil that is burned, the lower the polycyclic aromatic emissions become. Among the fractions used in transport, despite the obvious problems burning it cleanly, diesel was viewed as a 'greener' fuel than petrol, in that it contains more hydrocarbons so you can travel further on a given volume of fuel, with lower carbon dioxide emissions. However, it has facilitated the mass ownership of large, extremely heavy, aesthetically hideous four-wheel-drive Humvee-like vehicles that reflect the personalities of their owners without the single-figure urban fuel consumption (and fuel tax expense) that would result from using petrol. Early diesel engines put out seriously disgusting fumes, which contained large amounts of carcinogenic polycyclic aromatics alongside the particulates, and many studies over the last 25 years or so have shown the mutagenicity of diesel emissions. Although cleaner than they were, it seems that no matter which future emission standards they are said to comply with, diesels still smoke copiously on hard acceleration. Due to the high sulphur content of many European fuel blends, they also emit a harsh acidic stench alongside a mechanical clatter from which the manufacturers tried hard to isolate the owner, at the expense of the rest of us. The wasteful obsession of converting precious food resources into biofuels whilst so many starve make diesel emissions even worse[136].

Diesel car owners claim they bought their cars to save the planet, yet the primary reason was to save on anachronistic fuel taxes. The success of diesel cars means that their combined nitrogen oxide emissions have outstripped those of trucks and buses by several-fold[137]. This was compounded by the car manufacturers who deliberately designed software that could sense the circumstances of emissions testing so they could be cheated to improve performance and increase sales[138,139]. All this has meant that diesel emissions are the primary reason why urban air in the United Kingdom and other countries near busy roads is barely breathable; in fact, diesel emissions are one of the primary sources of mutagenic polycyclic aromatics and nitrogen oxides in the atmosphere–indeed, the most mutagenic substances yet measured are found in diesel emissions[140]. Although there are many mutagens in diesel, the nitroaromatics are the most potent and are discussed below.

Nitropolycyclic aromatics

These compounds are formed during the combustion of many fuels, but they appear to be produced in the greatest amounts in diesel exhausts, particularly when the engine is under high load and hard acceleration, such as in fully, or overloaded trucks. The previous record holder for the most mutagenic compound ever, 1,8 dinitropyrene, was supplanted in 1997 by 3-nitrobenzanthrone (3-NBA; Figure 8.12), which supplied 6 million mutations in the Ames test (A.6.6.) per nanomole, compared with the 4.8 million caused by 1,8 dinitropyrene[140]. The metabolic fate of nitroaromatics is a useful example of the conflicting roles that different enzymatic systems have in detoxifying, activating, and excreting these toxins. They also illustrate the effects of reduction, rather than oxidation, which is a minor route in drug metabolism, but an important route in carcinogenesis.

Reductase enzymes are found all over cells in the cytosol and you might (or not) recall that they are one of the 'fuel pumps' for CYP isoforms, as they supply electrons that power the enzymes (Chapter 3.4.8). Reductases are found in tissues that are most likely to meet significant levels of 3-NBA and other nitropolycyclics, such as the lungs and the gut. Since these compounds are thought to be lung carcinogens, it is likely that the reductases form the hydroxylamine which either hits DNA itself through oxidation to a nitrosoarene, or more likely, undergoes sulphation and GST conjugation reactions which lead to the formation of esters and sulphates which decompose, yielding DNA reactive nitrenium ions (Figure 8.12). Whatever reactive species is formed, it certainly binds DNA but does not seem to be formed by CYP enzymes, so it appears that most of the activation is through reduction, followed by conjugation. All the enzymes necessary

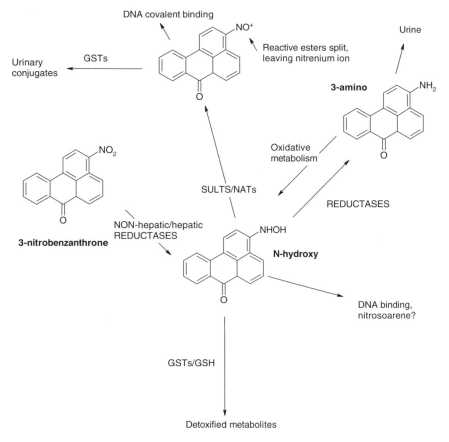

Figure 8.12 Processes of possible activation of 3-nitrobenzanthrone by reductive metabolism

to activate these compounds (reductases, SULTs, and GSTs) are found in the lung and the gut. Human reductases are particularly adept at this nitroreduction and many have suggested that nitropolycyclics are responsible for many cases of lung cancer in those exposed to diesel fumes, which, unfortunately, is all of us. The interesting feature of 3-NBA activation is that essentially the penultimate stage in carcinogenesis is the same as that for all aromatic amines, the formation of the hydroxylamine, which can then lead to nitrenium ion formation, via an unstable conjugated product. It is important to see the process in terms of varying stages of oxidation and reduction from aromatic amine to nitro derivative and back again. The nitrenium ions formed from 3-nitrobenzanthrone lead to transversion mutations, causing guanine/cytosine to change to thymine/adenine. Recent calculations involving nitrenium ion stability in the benzanthrone series and Ames test results indicate that the more stable the ion, the more mutagenic it is[140]. This is logical, as mentioned previously, extremely reactive species attack pretty much the first protein they encounter and are simply not stable enough to travel towards the nucleus and reach DNA. There is also some evidence that 3-aminobenzanthrone, one of 3-nitrobenzanthrone's human metabolites, is an inducer and substrate of the CYP1A series as well as NADPH reductases[141].

8.5.5 Dietary carcinogens: acrylamide

Acrylamide: background

Many potential carcinogens are thought to be present in foods, although they should be less likely to affect the developed world, due to the degree of processing which occurs in the justifiably maligned UK food industry. The exception is acrylamide, which is a highly reactive water-soluble agent that until the early 1990s was best known for its undoubted neurotoxicity in man and mutagenicity in test systems and animals. The acute neurotoxicity was well-documented in construction workers who by 1997 had been exposed to significant amounts of acrylamide in grouting processes during work on the highly problematic Hallandsås tunnel in Sweden[142]. Two years earlier, acrylamide was demonstrated to be a carcinogen in rats[143] and was already under suspicion as a possible human carcinogen. Whilst the spectrum of the toxicity of acrylamide was well understood in the scientific community, in the early 2000s it was discovered that during the thermal processing (>120 °C) of various cereal and potato products, significant, indeed alarming, amounts of acrylamide were formed[144]. This pushed acrylamide into the public domain, as it included pretty much anything we like to eat, from potato crisps (chips), potato chips (French fries), coffee, bread and cornflakes, to name just a few foods. Acrylamide formation is believed to occur through decomposition of asparagine and various sugars through complex Maillard reactions. Worryingly, many studies suggested that acrylamide exposure was up to three times that of adults in toddlers and young children[144].

Acrylamide: mechanisms of toxicity

Acrylamide can be spontaneously reactive, but is oxidized to glycidamide by CYP2E1, which is even more reactive than its parent. Glycidamide and its parent are electrophiles, which leads them to attack thiols such as free GSH, as well as cysteine residues in proteins. Both acrylamide and glycidamide trigger the nrf2 system and are also detoxified by the GSTs and epoxide hydrolases (Figure 8.13)[144,145]. The reactivity of acrylamide and glycidamide certainly accounts for the considerable mass of reported data in animal models which points to their carcinogenic, reproductive and cytoskeletal toxicity[144]. Studies in human models, such as A549 lung adenocarcinoma, revealed that both

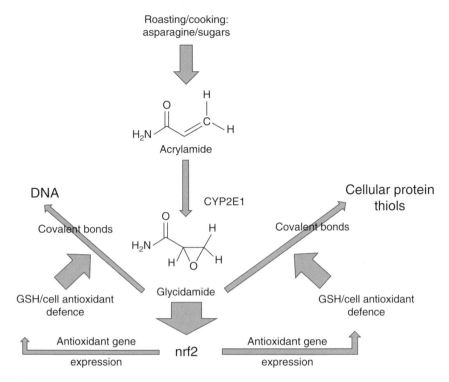

Figure 8.13   Acrylamide and its paths of human toxicity

acrylamide and its metabolite kill through mitochondrial damage and oxidative stress, which leads to apoptosis[145]. They also act as antiproliferative agents, blocking spindle fibre formation in cell division in the A549 cells[145]. All this toxic mayhem sounds terrifying, but interestingly, after more than a decade of clinical research, the relationship between dietary acrylamide and human carcinogenicity is not strong[144,146].

Acrylamide: a human carcinogen?

I am not fond of rhetorical questions in general, but this is a very important one to consider, given the prevalence of acrylamide in our diets. There are many reasons why acrylamide has not so far proved to be a human cancer risk. Some are due to the difficulties in designing studies that can detect the link between diet and specific cancers. It is hard for individuals to be precise as to their diets over previous decades and foodstuffs are prepared differently across the world, so estimating the degree of acrylamide exposure, let alone the relationship with effect (increase numbers of cancer cases) is not straightforward. Very heavy occupational exposure can be measured through haemoglobin adducts[147], but this is not likely to be practical for foodstuff-linked acrylamide. Another reason is the physicochemical nature of acrylamide and its metabolite, glycidamide. Both, as mentioned above, are electrophiles, which means that their electron deficiency means they will attack a wide variety of cell thiols, proteins, and DNA, but they are not very specific. They are likely to react with anything in their path, so thiols on the cell membrane would be first, followed by those in the cytoplasm and cellular organelles[145].

Most cells contain free thiols in millimolar concentrations such as GSH, as mentioned in Chapter 6.4.1. If we combine the thiols with the other detoxifying systems discussed in Chapter 6, then between the passive reaction with the cell structure and the active detoxifications, acrylamide and glycidamide would form adducts very rapidly and quite high doses would be necessary to saturate all the reaction sites in the cell. All the while, as noted in Chapter 6.4, GSH is being replenished through synthesis and recycling to 'fight the fire' so to speak. This would suggest that very high doses of acrylamide would be needed for enough of it and its metabolite to reach the nucleus to attack DNA. The dose of acrylamide needed to kill about half (inhibitory concentration for 50%; IC_{50}) of a culture of human transformed cells is in the 4–7 mM range[145], although primary cells, such as those isolated from the gut would be more sensitive. If we compare this concentration with the IC_{50} of cyanide in mammalian cells ($13\mu M$)[148], it looks like acrylamide is not a particularly specific or selective toxin. Indeed, the doses used in animal carcinogenicity studies are far in excess of our daily exposure[146].

It is useful to try to equate acrylamide's less than startling toxicity with what is estimated to be 'normal' human exposure. The highest levels of this toxin in some foodstuffs have been measured at roughly 2–5 mg/kg, whilst most other foodstuffs are around the 0.5mg/Kg level[144,146]. Immediately, you can see that most of us will hopefully not be consuming potato crisps by the kilo per day and our coffee intake is realistically only a few grammes. Indeed, quite significant strides have been made in the food industry to reduce acrylamide formation in foodstuffs anyway. The highest daily intakes of acrylamide are believed to be around 8 µg per kg body weight, so we are looking at around half a milligram for a 70 kg person. If we consider that the average person contains 55 litres of water and if we assumed that acrylamide distributed rapidly and uniformly in total body water like ethanol does (just as a very rough comparison), this gives us an exposure of 560 µg per day in 55 litres of water. This means around 10 µg acrylamide per litre in the body.

Given that the molecular weight of acrylamide is 71.8, this body exposure corresponds to a concentration of about 140 nanomolar, which is rather modest at best. Hence, even though acrylamide can deplete GSH in high-dose animal studies[144,145], it is difficult to see this happening in the average developed world consumer at realistic acrylamide exposure levels. It could be argued that the level of accumulation of acrylamide is not defined which might impact the risk. However, acrylamide and glycidamide are so reactive that the adducts do have a considerable half-life[147], but if they are not attached to DNA then this is not significant, with regard to mutagenicity. If you have been paying attention, you will immediately wonder if acrylamide is immunogenic, as it should be through haptenation. This again is not easy to define, but does seem to be the case, mainly through IgE-mediated mechanisms[149].

It could be speculated that a diet very high in acrylamide could multiply the risk of other more potent carcinogens reaching our DNA, most obviously from tobacco. Indeed, smoking itself is a significant source of acrylamide[150]. However, overall, it is possible to postulate that the carcinogenicity of acrylamide is difficult to prove in humans because the chemical is too reactive and nonselective, our dose is too low, and our antioxidant protection/DNA repair systems can attenuate the problem. Currently, it does not look as if acrylamide is a realistic threat to us if our diet is healthy and we avoid tobacco.

8.5.6   Dietary carcinogens: aflatoxins

Aflatoxins: introduction

In complete contrast to acrylamide, aflatoxins are formidable toxins and carcinogens in humans, and this is well-documented. This class of difuranocoumarins was unknown until 1960, when a bizarre condition wiped out over 100,000 turkeys in several countries, including the United Kingdom. Eventually the cause was narrowed to the contamination of the birdfeed by *Aspergillus* fungus[151,152]. These fungi grow on peanuts, corn, wheat, maize and many other oilseed crops. It is now believed that some form of mycotoxin contaminates perhaps a quarter of world grain production. *Aspergillus* grows best where conditions of storage are excessively damp and there is a lack of ventilation. The fungus uses at least 25 genes and an 18-step synthesis to form dozens of different aflatoxins[152], which range from merely extremely to ferociously toxic. They are unusual in that they are cytotoxic, mutagenic, immunogenic, and carcinogenic to anything living, from birds, animals, and fish to humans[153]. Their effects can be resolved into acute and chronic toxicity.

Aflatoxins: acute toxicity

Since the 1960s threat to the UK's Christmas lunch, there have been much more serious outbreaks of human acute aflatoxin effects, which is termed aflatoxicosis. The most severe outbreaks have been documented in India, where hundreds died of acute aflatoxicosis in the 1970s[154]. This has been due to poverty, where rural people had to eat mouldy cereals or nothing at all[154]. The effects of acute aflatoxicosis include rapid and massive liver damage, shown by severe jaundice, portal hypertension, abdominal ascites and a condition similar to the effects of cirrhosis, which is quickly fatal in 60–80% of cases[153,154].

Only a few milligrams per day of the aflatoxins is necessary for these toxins to induce liver failure in days or weeks. In some of the areas in India affected, all the domestic dogs died of the disease just before the humans developed it. In the early 2000s, similar outbreaks with mass fatalities were reported in Kenya[152]. These toxins can also cause a disease similar to Reye's syndrome, which is rooted in damage to mitochondria. This complaint is usually fatal and has been described widely. The risks of aflatoxins are strongly rooted in the developing world, but in developed countries the risk of eating imported food contaminated with aflatoxins and other mycotoxins is well recognized and monitoring of samples of potentially affected imports is carried out for mycotoxin contamination[152]. As well as in cereals and groundnuts, aflatoxins are found in human and cows' milk, as well as in various dairy products, so as well as the risk of acute exposure, there is effectively lifetime exposure to these agents at some level in human diets in many countries[152].

Aflatoxins: chronic toxicity

Looking at some of the many sources of cancer statistics, liver malignancies as of 2018 are around the sixth most common in the world, but are the fourth most lethal form of the disease after lung, colorectal, and stomach cancers. The disproportionate death rate in liver cancer is partly a reflection of the disease's aggression, but also its prevalence in developing countries with poor health infrastructure. Aflatoxins are a major driver of world hepatocellular cancers in many contexts, such as in workers in peanut-processing plants in the developed world, as well as food-processing workers in Third World areas[155]. Hepatitis B infection greatly increases the carcinogenic effects of aflatoxins[152]. What is particularly striking, is that liver cancers in developed countries are mostly linked with older individuals, but the influence of aflatoxins in developing countries cause the tumours to appear in the most productive years (30–45)[152]. Indeed, anything up to a third of world hepatocellular cancers are now attributable to aflatoxins[153]. This attribution prevalence is regional, closely tied into local foodstuff aflatoxin contamination. It ranges from more than 60% of cancers in parts of China and the Gambia, down to less than 20% in Brazil and Mexico[156]. This contrasts sharply with levels in developed countries such as the United States, where aflatoxin-linked cancers are mostly confined to immigrant populations who have been previously exposed to these toxins[156]. However, whilst aflatoxins are not considered to be significant in US liver cancer aetiology, aflatoxin adducts have been detected in US citizens in states such as Texas after periods of climate extremes presumably put pressure on food supplies[156].

Aflatoxins are reasonably chemically stable and difficult to eliminate with cooking processes, but dry roasting peanuts apparently decomposes them, so now you know. Unfortunately, eliminating them from milk is not possible with conventional pasteurizing techniques, so all that can be done is to set maximum limits of contamination[153]. Aflatoxins appearing in mother's milk is particularly problematic, as they affect growth in children and make them more susceptible to bacterial and other infections through immunosuppression[152,153].

Aflatoxins: activation

The most toxic of all these molecules is aflatoxin B1 (AFB1), followed by G1, B2, and G2 in terms of acute toxicity. The reason AFB1 is so dangerous is the presence of a double bond at the position in ring 1 (dihydrofuran; Figure 8.14) as well as other substituents of the coumarin (rings 4 and 5). These compounds are thought to rely on CYP activation for their toxicity and the major isoform involved appears to be CYP3A4, but in humans and other species, several other CYPs CYP1A1/1A2, CYP2A6, CYP3A5, and foetal CYP3A7 are also involved[157]. Many metabolites are formed, but the key carcinogenic and general macromolecule-reactive agent is the 8,9 exo-epoxide, formed mainly by CYP3A4. If you look at Figure 8.14, the best way to visualise the shape of the molecule is to imagine that rings 2–5 are basically planar (flat) while ring 1

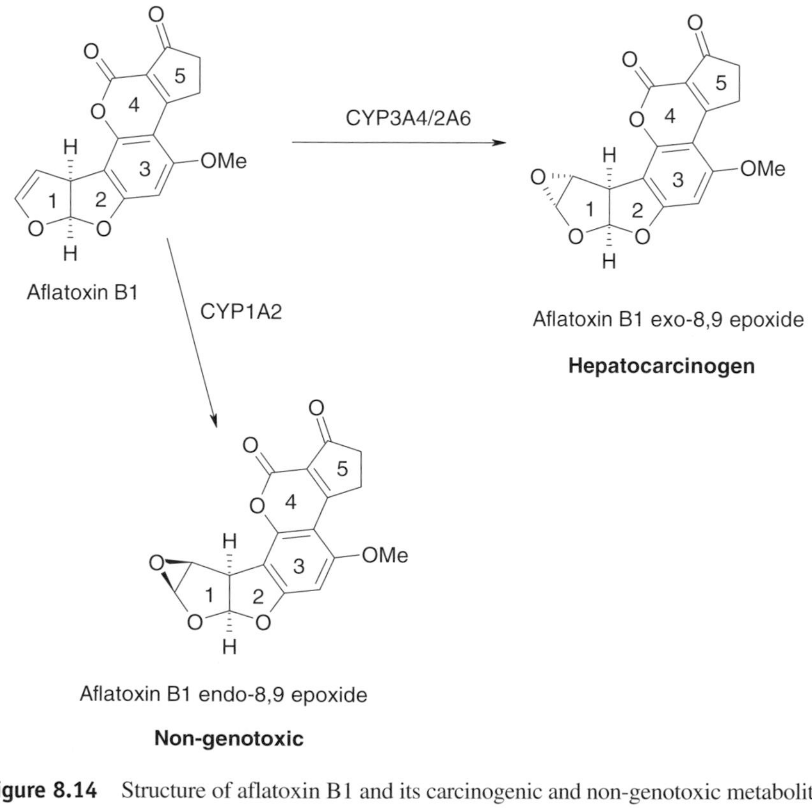

Figure 8.14   Structure of aflatoxin B1 and its carcinogenic and non-genotoxic metabolites

sticks up at an approximately 45° angle to the horizontal. The term *exo* for the epoxide means that the oxygen dips away from you as you look at the molecule on the page. There is an *endo* epoxide, where the oxygen would be oriented towards you as you looked at the molecule.

The reason the 8,9 exo-epoxide is so problematic is the precision where it interacts with DNA. It binds at the N7 position of guanine residues[158]. This is obviously a direct consequence of the three-dimensional shape of the molecule, as the 8,9 endo-epoxide and other aflatoxin metabolites do not intercalate with DNA in this position and are not as genotoxic. This intercalating effect of the epoxide causes a transversion of guanine and thymidine at codon 249 of the p53 tumour suppressor gene in the hepatocytes[156,158]. This changes the codon from AGG to AGT, switching the wild-type arginine to a serine. This is known as the R249S mutation, which is more generally termed TP53R249S.[156] Tp53's function is obviously key in preventing carcinogenesis and it is a commensurately tough gene to damage, as many SNPs have little or no impact on its function. However, TP53R249S is the direct cause of aflatoxin-mediated liver cancers, and it is in a position on the gene that is difficult to repair so once it appears, it looks like you are stuck with it[159]. TP53R249S is the marker that is used to detect long-term aflatoxin exposure and hepatocellular cancer risk[156,159]. It is likely that dietary inducers of CYP3A4 and probably CYP1B1 will increase aflatoxin activation, in both humans and animals[157].

One of the most sinister aspects of the aflatoxins, is the sheer depth and width of their toxic interaction with biological systems. As if their carcinogenicity was not enough, they also lung, gastrointestinal, and gall-bladder carcinogens. They distort at least 35 different metabolic pathways, ranging from steroid hormone to nuclear receptor activity, as well as mTOR and VEGF systems, as well as multiple microRNAs[153].

Aflatoxin: detoxification

It would be expected that processes would detoxify the 8,9 exo-epoxide and the most obvious candidates would be epoxide hydrolase, GSH and GST enzymes. The detoxification of this metabolite is not completely understood; although its half-life is so short (around 10 seconds), epoxide hydrolase may well not have a significant role in the formation of the diol, which apparently has an even shorter half-life. GSH does not react directly with the epoxide at pH 7, so an important route of clearance from the cell is likely to be via GST catalysed thiol conjugation. GST alpha, mu, and theta are thought to be involved with aflatoxin 8,9 epoxide clearance. Individuals with no expression of GST M1, (null individuals) have a much-increased risk of malignancy due to aflatoxin B1[158]. Even the hydroxylated metabolites of aflatoxins are problematic. The most investigated is AFM1, which not only is the main metabolite in maternal milk, but still manages to be mutagenic on the Ames test system, but only around a tenth as much as AFB1[153]. AFM1 is classed as an IARC group 2B carcinogen[153].

Aflatoxins: prevention of toxicity

Since these toxins make a massive impact in developing world health, the major concern is to educate ordinary food producers about the dangers of poor feed storage. In addition, the alleviation of poverty, although highly unlikely, would be invaluable in preventing the situation where there is no choice but to eat contaminated food. There are other measures, such as ammonia treatment, which very effectively destroys over 95% of these toxins in grains and feedstuffs[160]. Another effective measure involves dietary supplements, for animals and humans, such as clays and/or chlorophyllin, which bind the toxins and restrict absorption[152]. Other systems are being developed that use UV light to destroy the aflatoxins on the surface of nuts, although this is not effective against all the toxin's derivatives as yet and takes several hours.

If the aflatoxins cannot be prevented from forming or being absorbed, the next line of defence is some form of antioxidant-based protective measure. Studies with the chemopreventative dithiolthione oltipraz demonstrated that these agents have some promise in preventing the toxicity of dietary agents such as aflatoxins. The dithiolthiones operate through nrf2 (Chapter 6, Section 6.5.6) to increase GSH synthesis and they induce GST, thus stimulating detoxification of reactive species. Although clinical trials in China had mixed results, there was some reduction in urinary aflatoxin adducts, although there were issues with side effects as quite high doses were necessary to demonstrate efficacy[152]. Other newer members of the class of dithiolthiones, such as TBD, may be more effective in future in clinical trials. Vitamin C has been shown to be protective in cellular systems, with Vitamins E and A less so. Sulforaphane has been investigated clinically through preparing a drink made from 3-day-old sprouts. Whilst for many, this concoction might only rival N-acetyl cysteine solution as the worst beverage prospect ever, it was also not very startlingly effective, either, partly due to bioavailability issues[152].

The process of aflatoxin carcinogenicity is now reasonably well understood, and the main thrust of current research is to find a practical, yet inexpensive means to diminish the formation and effect of aflatoxin-related reactive species. In the real world, it is highly unlikely that aflatoxins can be removed from human diets. Therefore, interest remains in dietary activators of the KEAP1 nrf2-ARE system such as sulforaphane (Chapter 6.5.6) as well as the new dithiolthiones, which will attenuate the toxic species, even if their formation probably cannot be prevented.

8.6   Summary of biotransformational toxicity

Although drugs and toxins may occasionally act as parent agents to cause irreversible changes in cell structure and function, the major part of the process of necrosis, immunological damage and malignancy, is linked with

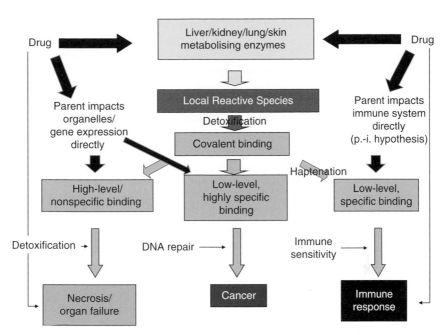

Figure 8.15 The links between drug metabolism, formation of active species, covalent binding, and irreversible damage to the organism. Conjugative detoxifying (UGTs, SULTs, and GST/GSH) defences are crucial to prevent organ failure from high levels of covalent binding, whilst detoxification mechanisms as well as DNA repair are essential to avoid carcinogenicity as a result of specific DNA binding of reactive species. For an immune response, the sensitivity of the individuals' immune system may be the major determinate of whether a reaction occurs. Although covalent binding is linked with most routes of toxicity, the effects of the parent drug may also be significant

the formation of unstable and damaging species from a xenobiotic compound (Figure 8.15), which leads to oxidative stress. Ultimately, virtually all of us form these species, but the combination of our net exposure to these toxins and our individual detoxification and repair mechanisms decide whether we will suffer significant toxicity or tolerate and clear the agent without ill effects.

References

1. Rawlins MD, Thompson JW. Pathogenesis of adverse drug reactions. In Davies DM, ed. Textbook of Adverse Drug Reactions. Oxford: Oxford University Press, 1977.
2. Iasella CJ, Johnson HJ, Dunn MA. Adverse drug reactions: Type A (intrinsic) or type B (idiosyncratic). Clin. Liver Dis. 21(1), 73–87, 2017.

3. Coleman MD and Coleman NA. Drug-induced methaemoglobinaemia. Drug Saf. 14, 394–405, 1996.

4. Coleman MD, Jacobus DP. Reduction of dapsone hydroxylamine to dapsone during methaemoglobin formation in human erythrocytes *in vitro*. Biochem. Pharmacol. 45, 1027–1033, 1993.

5. Coleman MD, Hayes PJ, Jacobus DG. Methaemoglobin formation due to nitrite, disulfiram, 4-aminophenol and monoacetyldapsone hydroxylamine in diabetic and non-diabetic human erythrocytes in vitro. Env. Tox. Pharmacol. 5, 61–67, 1998.

6. Kim YJ, Sohn CH, Ryoo SM, et al. Difference of the clinical course and outcome between dapsone-induced methemoglobinemia and other toxic-agent-induced methemoglobinemia. Clin. Tox. 54, 581–584, 2016.

7. Darling RC, and Roughton FJW. The effect of methemoglobin on the equilibrium between oxygen and haemoglobin. Am J. Physiol. 137, 56–68, 1942.

8. Coleman MD, and Taylor CH. Bioactivation of benzocaine in rat and human microsomal systems in vitro. Env. Tox. Pharmacol. 3, 47–52, 1997.

9. Hartman N, Zhou HF, Mao JZ, et al. Characterization of the methemoglobin forming metabolites of benzocaine and lidocaine. Xenobiotica 47, 431–438, 2017.

10. Erkuran MK, Duran A, Kurt BB, et al. Methemoglobinemia after local anesthesia with prilocaine: a case report. Am. J. Emerg. Med. 33, 602.e1 2015.

11. Coleman MD, Kuhns MK. Methaemoglobin formation by 4-aminopropriophenone in single and dual compartment systems. Env. Tox. Pharmacol. 7, 75–80, 1999.

12. Murphy E, Sjoberg T, Dilks P, et al. A new toxin delivery device for stoats-results from a pilot field trialNew Zeal. J. Zool. 45, 184–191, 2018.

13. Coleman MD, Thorpe S, Lewis S, et al. Preliminary evaluation of the toxicity and efficacy of novel 2,4 diamino 5-benzyl pyrimidine-sulphone derivatives using rat and human tissues in vitro. Env. Tox. Pharmacol. 2, 389–395, 1996.

14. Coleman MD, Dapsone: modes of action, toxicity and possible strategies for increasing patient tolerance. Brit. J. Dermatol. 129, 507–513, 1993

15. Coleman MD, Scott AK, Breckenridge AM et al. The use of cimetidine as a selective inhibitor of dapsone N-hydroxylation in man. Brit. J. Clin. Pharmac. 30, 761–767, 1990.

16. Veira JL, Ferreira MES, Ferreira MVD, et al. Primaquine in plasma and methemoglobinemia in patients with malaria due to *Plasmodium vivax* in the Brazilian Amazon Basin. Am. J. Trop. Med. Hyg. 96, 1171–1175, 2017.

17. Nicolaidou P, Kostaridou S, Mavri A, et al. Glucose-6-phosphate dehydrogenase deficiency and Gilbert syndrome: a gene interaction underlies severe jaundice without severe hemolysis. Pediatr. Hematol. Oncol. 22(7), 561–566, 2005.

18. Coleman MD, Tingle MD, Hussain F, et al. An investigation into the haematological toxicity of structural analogues of dapsone *in vivo* and *in vitro*. J. Pharm. Pharmacol. 43, 779–784, 1991.

19. Liguori I, Russo G, Curcio F, et al. Oxidative stress, aging, and diseases. Clin. Interv. Aging 13, 757–772, 2018.

20. Wang X, Wu Q, Liu A et al. Paracetamol: overdose-induced oxidative stress toxicity, metabolism, and protective effects of various compounds in vivo and in vitro. Drug Metab. Rev 49, 1–83, 2017.

21. Burke PJ, Mitochondria, bioenergetics and apoptosis in cancer. Trend. Canc. 3, 857–870, 2017.
22. van Esveld SL, Huynen MA, Does mitochondrial DNA evolution in metazoa drive the origin of new mitochondrial proteins? IUBMB LIFE 70, 1240–1250, 2018.
23. Riley JS, Quarato G, Cloix C, et al. Riley mitochondrial inner membrane permeabilisation enables mtDNA release during apoptosis. The EMBO J. 37, e99238, 1–16, 2018.
24. Armstrong JA, Cash NJ, Ouyang Y, et al. Oxidative stress alters mitochondrial bioenergetics and modifies pancreatic cell death independently of cyclophilin D, resulting in an apoptosis-to-necrosis shift. J. Biol. Chem. 293, 8032–8047, 2018.
25. Koczor CA, Jiao Z, Fields E, et al. AZT-induced mitochondrial toxicity: an epigenetic paradigm for dysregulation of gene expression through mitochondrial oxidative stress. Physiol. Genom. 47(10), 447–454, 2015.
26. Kullak-Ublick GA, Andrade RJ, Merz M, et al. Drug-induced liver injury: recent advances in diagnosis and risk assessment. Gut 66, 1154–1164, 2017.
27. Seeff LB, Obituary, Hyman J. Zimmerman, MD. JAMA. 283(6), 812, 2000
28. Zimmerman HJ. Hepatotoxicity: The Adverse Effects of Drugs and Other Chemicals on the Liver. 2nd edition. Philadelphia PA: Lippincott Williams and Wilkins, 1999.
29. Office of National Statistics; deaths related to drug poisoning by selected substances, 2018. https://www.ons.gov.uk/peoplepopulationandcommunity/birthsdeathsandmarriages/deaths/datasets/deathsrelatedtodrugpoisoningbyselectedsubstances
30. Artzner T, Michard B, Besch C, et al. Liver transplantation for critically ill cirrhotic patients: Overview and pragmatic proposals. World J. Gastroenterol. 24, 5203–5214, 2018.
31. Gulmez SE, Larrey D, Pageaux GP. Liver transplant associated with paracetamol overdose: results from the seven-country SALT study. Br. J. Clin. Pharmacol. 80(3), 599–606, 2015.
32. Buckley NA, Gunnell D, Does restricting pack size of paracetamol (acetaminophen) reduce suicides? PLoS Med 4(4), e152. 2007.
33. Mazaleuskaya LL, Sangkuhl K, Thorn CF, et al. PharmGKB summary: Pathways of acetaminophen metabolism at the therapeutic versus toxic doses. Pharmacogenet Genom. 25(8), 416–426, 2015.
34. Yoon E, Babar A, Choudhary M, et al. Acetaminophen-induced hepatotoxicity: a comprehensive update. J. Clin. Transl. Hepatol. 28; 4(2), 131–142, 2016.
35. O'Malley GF, Acetaminophen poisoning. https://www.msdmanuals.com/en-gb/professional/injuries-poisoning/poisoning/acetaminophen-poisoning
36. Prescott LF. The Chief Scientist Reports. Prevention of hepatic necrosis following paracetamol overdosage. Health Bull. (Edinburgh) 36, 204–212, 1978.
37. United Kingdom Medicines and Healthcare products Regulatory Agency (MHRA) Drug Safety Update. September, 6(2), A1, 2012.
38. Bebenista MJ, and Nowak JZ. Paracetamol: mechanism of action, applications and safety concern. Acta Polon. Pharmaceutica-Drug Res. 71, 11–23, 2014
39. FDA Drug Safety Communication: Prescription Acetaminophen Products to be Limited to 325 mg Per Dosage Unit; Boxed Warning Will Highlight Potential for Severe Liver Failure Acetaminophen Products to be Limited to 325 mg Per Dosage Unit; Boxed Warning Will Highlight Potential for Severe Liver Failure 13 January 2011, https://www.fda.gov/Drugs/DrugSafety/ucm239821.htm.

40. de Martino M, Chiarug A. Recent advances in pediatric use of oral paracetamol in fever and pain management. Pain Ther. 4, 149–168, 2015.

41. Sahib A, El-Radhi M, Fever management: Evidence vs current practice. World J. Clin. Pediatr. 8; 1(4), 29–33, 2012.

42. Prymula R, Siegrist CA, Chlibek R, et al. Effect of prophylactic paracetamol administration at time of vaccination on febrile reactions and antibody responses in children: two open-label, randomised controlled trials. Lancet. 374, 1339–1350, 2009.

43. Segawa M, Sekine S, Sato T, et al. Increased susceptibility to troglitazone-induced mitochondrial permeability transition in type 2 diabetes mellitus model rat. J. Toxicol. Sci. 43, 339–351, 2018.

44. Mayall SJ, Banerjee AK. The evolution of therapeutic risk management. In Therapeutic Risk Management of Medicines. 25–59. ISBN 978-1-907568-48-0 Woodhead Publishing, Elsevier LtD 448, 2014.

45. Patil D, Patwardhan B, Kumbhare K. Why and how drugs fail. In R. Chaguturu, B. Patwardhan (Eds.), Innovative Approaches in Drug Discovery: Ethnopharmacology, Systems Biology and Holistic Targeting (pp. 23–64). Netherlands: Elsevier. ISBN: 978-0128018224 / 9780128018149, 2016.

46. Kakunia M, Moritab M, Matsuo K, et al. Chimeric mice with a humanized liver as an animal model of troglitazone-induced liver injury. Tox. Lett. 214, 9– 18, 2012.

47. Lauer B, Tuschl G, Kling M, et al. Species-specific toxicity of diclofenac and troglitazone in primary human and rat hepatocytes. Chemico-Biol. Int. 179, 17–24, 2009.

48. Macherey C, Dansette PM, Chapter 25, Biotransformations leading to toxic metabolites: Chemical aspects. The Practice of Medicinal Chemistry (Fourth Edition) 585–614, Academic Press, 2015.

49. Saha S, New LS, Ho HK, et al. Direct toxicity effects of sulfo-conjugated troglitazone on human hepatocytes. Toxicology Letters 195, 135–141, 2010.

50. Varga ZV, Ferdinandy P, Liaudet L, et al. Drug-induced mitochondrial dysfunction and cardiotoxicity. Am. J. Physiol. Heart Circ. Physiol. 309: H1453–H1467, 2015.

51. Angrish MM, McQueen CA, Cohen-Hubal E, et al. Toxicity tests based on an adverse outcome pathway network for hepatic steatosis. Tox. Sci. 159(1), 159–169, 2017.

52. Ryder RE. Pioglitazone and bladder cancer. Lancet 378, 1544–1545, 2011.

53. FDA Drug Safety Communication: Updated FDA review concludes that use of type 2 diabetes medicine pioglitazone may be linked to an increased risk of bladder cancer. https://www.fda.gov/Drugs/DrugSafety/ucm519616.htm

54. Saha S, Chan DSZ, Lee CY et al. Pyrrolidinediones reduce the toxicity of thiazolidinediones and modify their anti-diabetic and anti-cancer properties. Eur. J. Pharmacol. 697, 13–23, 2012.

55. Strauchman FM, Morningstar MW. Fluoroquinolone toxicity symptoms in a patient presenting with low back pain. Clinics and Practice 2, e87, 2012.

56. Williams D, Hopkins S. Safety of trovafloxacin in treatment of lower respiratory tract infections. Eur. J. Clin. Microbiol. Infect. Dis. 17, 454–458, 1998.

57. Wise, J. Pfizer accused of testing new drug without ethical approval. BMJ. 322(7280), 194, 2001.

58. Poulsen KL, Albee RP, Ganey PE, et al. Trovafloxacin potentiation of lipopolysaccharide-induced tumor necrosis factor release from RAW 264.7 cells requires extracellular signal-regulated kinase and c-Jun N-terminal kinases. J. Pharmacol. Exp. Ther. 349, 185–191, 2014.

59. Shaw PJ, Ganey PE, Roth RA. Idiosyncratic drug-induced liver injury and the role of inflammatory stress with an emphasis on an animal model of trovafloxacin hepatotoxicity. Tox Sci. 118(1), 7–18, 2010.

60. FDA updates: warnings for fluoroquinolone antibiotics on risks of mental health and low blood sugar adverse reactions. https://www.fda.gov/NewsEvents/Newsroom/PressAnnouncements/ucm612995.htm.

61. Douros A, Grabowski K, Stahlmann R. Safety issues and drug–drug interactions with commonly used quinolones. Exp. Opin. Drug Metab. Tox. 11, 25–39, 2015.

62. Kleiner DE, Chapter 12, Drugs and toxins. Macsween's Pathology of the Liver (Seventh Edition). 673–779, Elsevier, 2018.

63. Liguori, MJ, Anderson MG, Bukofzer S, et al. Microarray analysis in human hepatocytes suggests a mechanism for hepatotoxicity induced by trovafloxacin. Hepatology 41, 177–186, 2005.

64. Zhu QS Ran, Foss FW, Macdonald TL. Mechanisms of trovafloxacin hepatotoxicity: Studies of a model cyclopropylamine-containing system. Bioorg. Med. Chem. Lett. 17, 6682–6686, 2007.

65. Granitzny A, Knebel J, Muller M, et al. Evaluation of a human *in vitro* hepatocyte-NPC co-culture model for the prediction of idiosyncratic drug-induced liver injury: A pilot study. Tox, Rep. 4, 89–103, 2017.

66. Liguori MJ, Blomme EAg, Waring JF. Trovafloxacin-induced gene expression changes in liver-derived *in vitro* systems: Comparison of primary human hepatocytes to HepG2 cells. Drug. Metab. Disp. 36, 223–233, 2008.

67. Mitsugi R, Sumida K, Fuji Y et al. Trovafloxacin acyl-glucuronide induces chemokine (C-X-C motif) ligand 2 in HepG2 cells. Biol Pharm Bull. 39, 1604–1610, 2016.

68. Bocquet H, Bagot M, Roujeau JC. Drug-induced pseudolymphoma and drug hypersensitivity syndrome (Drug Rash with Eosinophilia and Systemic Symptoms: DRESS). Semin. Cutan. Med. Surg. 15, 250–257, 1996.

69. Pichler WJ Hausmann O. Classification of Drug Hypersensitivity into Allergic, p-i, and Pseudo-Allergic Forms reactions Classification Drug hypersensitivity. Int. Arch. Allergy Immunol. 171, 166–179, 2016.

70. Keisu M, Andersson TB. Drug-induced liver injury in humans: the case of ximelagatran. In Adverse Drug Reactions, 407–418, Springer, Berlin, Heidelberg, 2010.

71. Iorga A, Dara L, Kaplowitz N. Drug-induced liver injury: cascade of events leading to cell death, apoptosis or necrosis. Int. J. Mol. Sci. 18, 1018 1–25, 2017.

72. Uetrecht J, Chapter 11, Role of the adaptive immune system in idiosyncratic drug-induced liver injury. Drug-Induced Liver Disease (Third Edition). 175–193, Academic Press, 2013.

73. Weir A. Henry VIII; King and Court, 207, Jonathan Cape, ISBN 0-22406022-8, 2001.

74. Porras-Gallo M, Davis RA, eds. The Spanish Influenza Pandemic of 1918–1919: Perspectives from the Iberian Peninsula and the Americas. Rochester Studies in Medical History. 30. Uni. Rochester Press. ISBN 978-1-58046-496-3, 2014.

75. Francis H, Awadzi K, Ottesen EA. The Mazzotti reaction following treatment of onchocerciasis with diethylcarbamazine: clinical severity as a function of infection intensity. Am J Trop Med Hyg. 34, 529–536, 1985.

76. Boussinesq M, A new powerful drug to combat river blindness. The Lancet, 10154, 1170–1172, 2018

77. Brinkmann V, Reichard U, Goosmann C, et al. Neutrophil extracellular traps kill bacteria. Science 303(5663), 1532–1535, 2004.

78. Teng TS, Ji AL, Ji XY, et al. Neutrophils and immunity: from bactericidal action to being conquered J. Immunol. Res. 9671604, 1–14, 2017.

79. Collin P, Salmi TT, Hervonen K, et al, Dermatitis herpetiformis: a cutaneous manifestation of coeliac disease. Ann. Med. 49, 23–31, 2017.

80. Cerboni S, Gentili, Manel N. Diversity of pathogen sensors in dendritic cells. In Advances in Immunology. 120, 211–237, Academic Press, 2013.

81. Matzinger, P. Tolerance, danger, and the extended family. Ann. Rev. Immunol. 12, 991–104, 1994.

82. Schwarzenegger A, Hamilton L, Patrick R, et al. Terminator 2: Judgement Day. Dir. Cameron J, 137 min, Carolco Pictures, Released July 1, 1991.

83. Emlet DR, Gomez H, Kellum JA. Chapter 21–pathogen-associated molecular patterns, damage-associated molecular patterns, and their receptors in acute kidney injury. Critical Care Nephrology (Third Edition). e3, 121–127, Elsevier, 2019.

84. Mutius E. Microbial influences on the development of atopy. Allergy, Immunity and Tolerance in Early Childhood, The First Steps of the Atopic March. 209–217, 2016.

85. Finn OJ, Immunosurveillance and Immunotherapy; A Believer's Overview of Cancer Immunotherapy; A Believer's Overview of Cancer. J. Immunol. 200, 385–391, 2018.

86. Oliveira CC, van Hall T. Alternative antigen processing for MHC class I: multiple roads lead to Rome. Front Immunol. 6, 298, 2015.

87. Hinz A, Jedamzick J, Herbring V, et al. Assembly and function of the major histo-compatibility. complex (MHC) I Peptide-loading complex are conserved across higher vertebrates. J. Biol. Chem. 289, 33109–33117, 2014.

88. Goral S. The three-signal hypothesis of lymphocyte activation/targets for immuno-suppression. Dialysis & Transplant. 40, 14–16, 2011.

89. Chen L & Flies DB. Molecular mechanisms of T cell co-stimulation and co-inhibition. Nat. Rev. Immunol. 13(4), 227–242, 2013.

90. McFadden JP, White ML, Basketter DA et al. Does hapten exposure predispose to atopic disease? The hapten-atopy hypothesis. Trend. Immunol. 30, 67–74, 2009.

91. Li X, Yu K, Mei S, et al. HLA-B*1502 Increases the risk of phenytoin or lamotrigine induced Stevens-Johnson syndrome/toxic epidermal necrolysis: evidence from a meta-analysis of nine case-control studies. Drug Res. (Stuttg). 65, 107–111, 2014.

92. Phenytoin: risk of Stevens-Johnson syndrome associated with HLA-B*1502 allele in patients of Thai or Han Chinese ethnic origin. https://www.gov.uk/drug-safety-update/phenytoin-risk-of-stevens-johnson-syndrome-associated-with-hla-b-1502-allele-in-patients-of-thai-or-han-chinese-ethnic-origin.

93. Mallal S, Phillips E, Carosi G, et al. PREDICT-1 Study Team: HLA-B * 5701 screening for hypersensitivity to abacavir. N Engl. J. Med. 2008.

94. Novoa SR, García-Gascó P, Blanco F, et al. Value of the HLA-B*5701 allele to predict abacavir hypersensitivity in Spaniards. AIDS Res.Hum. Retrovir. 23(11), 1374–1376, 2007.

95. Wadelius M, Eriksson N, Kreutz R, et al. Sulfasalazine-induced agranulocytosis is associated with the human leukocyte antigen locus. Clin. Pharm. Ther. 103, 843–853, 2018.

96. Aithal G P, Ramsay L, Daly A K, et al. Hepatic adducts, circulating antibodies, and cytokine polymorphisms in patients with diclofenac hepatotoxicity. Hepatology 39, 1430–1440, 2004.

97. Benesic A, Dragoi D, Pichler G, et al. Monocyte-derived hepatocyte-like cells in combination with proteomics identify a potential biomarker for drug-induced liver injury by diclofenac. EMJ Hepatol. 6(1), 61–62. Abs. Rev. No. AR5, 2018.

98. Ng W, & Uetrecht J, Changes in gene expression induced by aromatic amine drugs: Testing the danger hypothesis, J. Immunotoxicol. 10, 178–191, 2013.

99. Chung WH, Chang WC, Lee YS, et al. Taiwan Severe Cutaneous Adverse Reaction Consortium. Japan Pharmacogenomics Data Science Consortium: Genetic variants associated with phenytoin-related severe cutaneous adverse reactions. JAMA 312, 525–534, 2014.

100. Klaewsongkram J, Sukasem C, Thantiworasit P, et al. Analysis of HLA-B allelic variation and IFN-γ ELISpot responses in patients with severe cutaneous adverse reactions associated with drugs. The Journal of Allergy and Clinical Immunology: In Practice. 7, Issue 1, 219–227. e4, 2019.

101. Momeni M, Brui B, Baele P, Matta A. Anaphylactic shock in a beta-blocked child: use of isoproterenol. Paediatr. Anaesth. 17, 897–899, 2007.

102. Gandhi R, Sharma B, Sood J. Severe anaphylaxis during general anaesthesia in a beta-blocked cardiac patient: considerations. Acta Anaesthesiol. Scand. 52(4), 574, 2008.

103. Choudhary S, McLeod M, Torchia D, et al. Drug Reaction with Eosinophilia and Systemic Symptoms (DRESS) Syndrome. J Clin Aesthet Dermatol. 6(6), 31–37, 2013.

104. Gerogianni K, Tsezou A, Dimas K. Drug-induced skin adverse reactions: the role of pharmacogenomics in their prevention. Mol. Diagn. Ther. 22, 297–314, 2018.

105. Mehta M, Shah J, Khakhkhar T, et al. Anticonvulsant hypersensitivity syndrome associated with carbamazepine administration: Case series. J. Pharmacol. Pharmacother. 5(1), 59–62, 2014.

106. Ghandourah H, Bhandal S, Brundler M, et al. Bronchiolitis obliterans organising pneumonia associated with anticonvulsant hypersensitivity syndrome induced by lamotrigine. BMJ Case Reports bcr2014207182, 2016.

107. Hamm RL, Drug allergy: delayed cutaneous hypersensitivity reactions to drugs. EMJ Allergy Immunol. 1(1), 92–101, 2016.

108. Lerma V, Macías M, Toro R, et al. Care in patients with epidermal necrolysis in burn units. A nursing perspective. Burns 44, 1962–1972, 2018.

109. Borrelli EP, Lee EY, Descoteaux AM, et al. Stevens-Johnson syndrome and toxic epidermal necrolysis with. antiepileptic drugs: An analysis of the US Food and Drug Administration Adverse Event Reporting System. Epilepsia 59, 2318, 2324, 2018.

110. Phillips EJ. Defining regional differences in drug-induced Stevens– Johnson syndrome/toxic epidermal necrolysis: a tool to improve drug safety? Clin. Pharm. Ther. 105, 22–25, 2019

111. Ko TM, Chung WH, Wei CY, et al. Shared and restricted T-cell receptor use is crucial for carbamazepine-induced Stevens–Johnson syndrome. J. Allergy Clin. Immunol. 128(6), 1266–76.e11, 2011.

112. Waskiel A, Dyttus-Cebulok K Polak-Witka K, et al. Stevens-Johnson syndrome induced by combined treatment: carbamazepine and cranial radiation therapy. A case of EMDART? Przeglad Dermatologiczny 104(3), 331–336, 2017.

113. Farshchian M, Ansar A, Zamanian A et al. Drug-induced skin reactions: a 2-year study. Clin Cosmet Investig Dermatol. 8, 53–56, 2015.

114. Nicolini A, Perazzo A, Gatto P et al. A rare adverse reaction to ethambutol: drug-induced haemolytic anaemia. Int. J Tuberc. Lung Dis. 20(5), 704–770, 2016.

115. Carnovale C, Brusadelli T, Casini ML, et al. Drug-induced anaemia: a decade review of reporting to the Italian pharmacovigilance data-base. Int. J. Clin. Pharm. 37, 23–26, 2015.

116. Mayer B, Bartolmäs T, Yürek S, et al. Variability of findings in drug-induced immune haemolytic anaemia: experience over 20 years in a single centre. Transfus Med Hemother 42, 333–339, 2015.

117. Al Qahtani SA. Drug-induced megaloblastic, aplastic, and hemolytic anemias: current concepts of pathophysiology and treatment. Int. J. Clin. Exp. Med. 11(6), 5501–5512, 2018.

118. Curtis, BR, Non–chemotherapy drug–induced neutropenia: key points to manage the challenges. Hematology, 187–193, 2017.

119. Qureshi K, Sarwar U, Khallafi H. Severe aplastic anemia following acute hepatitis from toxic liver injury: literature review and case report of a successful outcome. Case Rep. Hepatol. 216570, 1–7, 2014.

120. Aminopyrine, Dipyrone and Agranulocytosis Can Med Assoc J. Dec 5; 91(23), 1229–1230, 1964.

121. Yang J, Lv Y, Zhang Y. Decreased miR-17-92 cluster expression level in serum and granulocytes preceding onset of antithyroid drug-induced agranulocytosis Endocrine 59, 218–225, 2018.

122. Coleman MD. Dapsone mediated agranulocytosis: risks, possible mechanisms and prevention. Toxicology 162, 53–60, 2001.

123. Ojong M, & Allen SN, Management and prevention of agranulocytosis in patients receiving clozapine. Ment. Health Clinician: 3(3), 139–143, 2013.

124. Ng W, Uetrecht JP. Effect of aminoglutethimide on neutrophils in rats: implications for idiosyncratic drug-induced blood dyscrasias. Chem. Res. Toxicol. 26, 1272–1281, 2013.

125. Prokopez CR, Armesto AR, Gil Aguer MF, et al. Clozapine Rechallenge After Neutropenia or Leucopenia J. Clin. Psychopharmacol. 36(4), 377–380, 2016.

126. Dunk LR, Annan LJ, Andrews CD, Rechallenge with clozapine following leucopenia or neutropenia during previous therapy. Br. J. Psych. 188, 255, 2006.

127. Weltzien, H.U., Dotze, A., Gamerdinger, K., Hellwig, S. & Thierse, H.J. Molecular recognition of haptens by T cells: More than one way to tickle the receptor. In: Madame Curie Bioscience Database, Vol. Accessed March 8, 2017 (Landes Bioscience, Austin (TX), 2000–2013).

128. Yip VL, Pirmohamed M. The HLA-A*31:01 allele: influence on carbamazepine treatment. Pharmacogenom. Pers. Med. 10, 29–38, 2017.

129. Schaapveld M, Aleman BMP, van Eggermond AM, et al. Second cancer risk up to 40 Years after treatment for Hodgkin's Lymphoma. N. Engl. J. Med. 373, 2499–2511, 2015.

130. Aidan JC, Priddee NR, McAleer JJ. Chemotherapy causes cancer! A case report of therapy related acute myeloid leukaemia in early stage breast cancer. Ulster Med. J. 82(2), 97–99, 2013.

131. Tuccori M, Filion KB, Yin H, et al. Pioglitazone use and risk of bladder cancer: population based cohort study. BMJ 352, i1541 1–8, 2016.

132. Colerangle JB. Chapter 25–Preclinical development of nononcogenic drugs (small and large molecules). In A Comprehensive Guide to Toxicology in Nonclinical Drug Development (Second Edition), 659–683, Academic Press, 2017.

133. Coleman MD. Chapter 4; Chronic and permanent injury; bladder cancer and occupation. 82–122, in Expert Report Writing in Toxicology: Forensic, Scientific and Legal Aspects: Wiley International; 224, ISBN: 978-1–118-43214-3, 2014.

134. Ng W, Metushi IG, Uetrecht JP. Hepatic effects of aminoglutethimide: A model aromatic amine. J. Immunotoxicol 12(1), 24–32, 2015.

135. Skipper PL, Kim MY, Sun HP et al. Monocyclic aromatic amines as potential human carcinogens: old is new again. Carcinogenesis 31(1), 50–58, 2010.

136. Project: Fuels Joint Research Group (FJRG). Strong mutagenic effects of diesel engine emissions using vegetable oil as fuel. Arch. Toxicol. 81(8), 599–603, 2007.

137. Bünger J, Krahl J, Munack A, NOx emissions from heavy-duty and light-duty diesel vehicles in the EU: Comparison of real-world performance and current type-approval requirements. ICCT Int. Council on Clean transportation https://www.theicct.org/publications/nox-emissions-heavy-duty-and-light-duty-diesel-vehicles-eu-comparison-real-world May 2017

138. Thompson GJ. In-use emissions testing of light-duty diesel vehicles in the United States. https://www.theicct.org/sites/default/files/publications/WVU_LDDV_in-use_ICCT_Report_Final_may2014.pdf

139. Neate R. Meet John German: the man who helped expose Volkswagen's emissions scandal. The Guardian, 26th Sept 2015. www.theguardian.com/business/2015/sep/26/volkswagen-scandal-emissions-tests-john-german-research

140. Arlt VM. 3-Nitrobenzanthrone, a potential human cancer hazard in diesel exhaust and urban air pollution: a review of the evidence. Mutagenesis 20, 399–410, 2005.

141. Stiborová M, Dračínská H, Martínková M, et al. 3–Aminobenzanthrone, a human metabolite of the carcinogenic environmental pollutant 3-nitrobenzanthrone, induces biotransformation enzymes in rat kidney and lung. Mutation Research/Genetic Toxicology and Envir. Mutagen. 676, 93–101, 2009.

142. Kjuus H. Acrylamide in tunnel construction-new (or old) lessons to be learned? Scand. J. Work Environ. Health 27, 217–218, 2001.

143. Friedman MA, Dulak, LH Stedham MA. A lifetime oncogenicity study in rats with acrylamide. Fundam. Appl. Toxicol. 27, 95–105, 1995.

144. Kumar J, Das S Teoh SL, Dietary acrylamide and the risks of developing cancer: facts to ponder. Front. Nutr. 5:14. doi: 10.3389/fnut.2018.00014, 2018.

145. Kacar S, Vejselova D, Kutlu HM, et al. Acrylamide-derived cytotoxic, anti-proliferative, and apoptotic effects on A549 cells. Hum, Exper. Toxicol. 37, 468–474, 2018.

146. Semla M, Goc Z, Martiniakova M, et al. Acrylamide: a common food toxin related to physiological functions and health. Physiol. Res. 66, 205–217, 2017.

147. Hagmar L, Törnqvist M, Nordander C, et al. Health effects of occupational exposure to acrylamide using hemoglobin adducts as biomarkers of internal dose. Scand. J. Work Environ. Health 27, 219–226, 2001.

148. Leavesley HB, Li L, Prabhakaran K et al. Interaction of cyanide and nitric oxide with cytochrome c oxidase: implications for acute cyanide toxicity. Toxicol. Sci. 101, 101–111, 2008.

149. Guo J, Yu DD, Lv N. et al. Relationships between acrylamide and glycidamide hemoglobin adduct levels and allergy-related outcomes in general US population, NHANES 2005–2006Environ. Pollut. 225, 506–513, 2017.

150. Vesper, HW, Caudill SP, Osterloh JD, et al. Exposure of the U.S. population to acrylamide in the National Health and Nutrition Examination Survey 2003–2004. Environ. Health Perspect. 118, 278e283, 2010.

151. Blount, W.P. Turkey "X" disease. Turkeys 9, 52–54. 1961.

152. Kensler TW, Roebuck WD, Wogan GN, et al. Aflatoxin: A 50–year odyssey of mechanistic and translational toxicology. Tox. Sci. 120(S1), S28–S48, 2011.

153. Marchese S, Polo A, Ariano A, et al. Aflatoxin B1 and M1: biological properties and their involvement in cancer development. Toxins 10(6), 214, 1–19, 2018.

154. Kumar P, Mahato DK, Kamle M, et al. Aflatoxins: A global concern for food safety, human health and their management. Front. Microbiol. 7, 2170, 2017.

155. Saad-Hussein A, Taha MM, Beshir S, et al. Carcinogenic effects of aflatoxin B1 among wheat handlers. Int. J. Occup. Environ. Health. 20(3), 215–219, 2014.

156. Jiao J, Niu W, Wang Y et al. Prevalence of aflatoxin-associated TP53R249S mutation in hepatocellular carcinoma in Hispanics in South Texas. Canc. Prev. Res. 11(2), 103–112, 2018.

157. Diaz GJ, Murcia HW, Cepeda SM. Cytochrome P450 enzymes involved in the metabolism of aflatoxin B1 in chickens and quail. Poult. Sci. 89(11), 2461–2469, 2010.

158. A review of human carcinogens. Part F: Chemical agents and related occupations / IARC Working Group on the Evaluation of Carcinogenic Risks to Humans (2009: Lyon, France). Aflatoxins 225–244 (IARC monographs on the evaluation of carcinogenic risks to humans; v. 100F), 2009.

159. Ortiz-Cuaran S, Cox D, Villar S, et al. Association between TP53 R249S mutation and polymorphisms in TP53 intron 1 in hepatocellular carcinoma. Genes Chromo. Canc. 52, 912–919, 2013.

160. Jalili M, A review on aflatoxins reduction in food. Iran. J. Health, Saf. & Environ. 3(1), 445–459, 2016.

Appendix A
Drug Metabolism in Drug Discovery

A.1 The pharmaceutical industry

Drug companies, often dubbed *Big Pharma,* have been much maligned, often with good reason. However, their business model is not an easy one, and it is unlike many other manufacturing concerns, such as the car industry, for example. Whilst there are plenty of examples of cars that were not very profitable, even the worst built and ugliest made something back for their manufacturers. The drug industry model is closer to what used to be known as the 'record' industry, which invested in thousands of acts, losing money on almost all, relying for their profits on the handful of artistes that became world stars. As you are probably aware, as soon as a new prospective therapeutic drug is patented, the pharmaceutical company usually has sole rights to the product for 20 years. In this time frame, it must invest hundreds of millions a year for nearly a decade in order to develop the drug to the point it is granted approval in European and US markets. If the drug does not perform well enough after its clinical trial, then it may be dropped and the company's investment is all lost. If the drug is approved, then the race is on over the remaining 8–10 years to recoup the billions invested and make a profit in order to finance all the other drugs in the company pipeline and, of course, the huge marketing and publicity machine needed to sell the new drug to the world.

When things go wrong

In some instances, development and manufacturing costs can run out of control and the market is just too small to recoup the costs. The advertising machine then has an impossible task to fulfil. Glybera, approved in Europe in 2012, was

Human Drug Metabolism, Third Edition. Michael D. Coleman.
© 2020 John Wiley & Sons, Inc. Published 2020 by John Wiley & Sons, Inc.

a gene therapy drug that was somewhat notorious as the first drug to cost US$1 million per treatment and its less than stunning efficacy and general poor performance led the manufacturer not to seek an extension of its licence in 2017[1]. Assuming a drug is licensed and health services around the world and insurance companies agree to fund treatment with it, marketing then propels the numbers of patients exposed to the drug from a few thousands to perhaps millions within months. In this case, any issues with adverse reactions may well appear with a relatively short period of time, but some drugs can be available for several years, or even decades, before they are withdrawn. The withdrawal process is usually a very significant financial problem for the manufacturer.

The largest event of this type was with the COX-2 inhibitor rofecoxib (Vioxx), introduced in 1999. The COX-2 inhibitors' blockbuster potential was predicated on their lack of gastrointestinal problems, which afflict most other NSAIDs, so they were immediately successful and as a class by 2004, they were worth US$5.5 billion in annual sales[2]. However, Vioxx increased the risks of myocardial infarction by 34%[3] compared with other arthritis medications, and a very conservative FDA estimate put the numbers of deaths caused by the drug as exceeding 27 000. More than 20 million in America alone used the drug and the manufacturer had to set aside just under US$5 billion to settle worldwide liability claims related to the drug. Evidence mounted that the manufacture knew much more than it admitted as to the risks of Vioxx before it was marketed, and this was compounded with an obvious reluctance to withdraw the drug before its final extinction in 2004.

After more than a decade on sale, the antidiabetic drug pioglitazone (Actos) was linked in 2011 with an increased incidence of bladder tumours. Whilst there is a predisposition for this cancer in type 2 diabetics[4], pioglitazone increased the risk of bladder malignancy by 63% compared with rosiglitazone[5]. Although not all studies showed this association[4], the drug was withdrawn in many countries and at one point the manufacturer faced over 8000 legal claims for damages, finally settling for US$2.7 billion.

Generally, it seems that the manufacturers are often aware of significant problems with drugs before they are revealed publically. They are also often very slow to withdraw their product and usually the profits made prior to withdrawal are an order of magnitude larger than the liabilities. Incidentally, the manufacturers of Glybera, Vioxx, and Actos are still in business and apparently thriving at the time of writing. Indeed, a lack of new 'blockbuster' drugs in the pipeline is a much more significant threat to a pharmaceutical company's existence than drug withdrawals in the long term.

Why things go wrong

Drugs are usually withdrawn as a result of the appearance of mounting numbers of published case reports, as well as prospective studies, where a drug is investigated to determine if there is a specific issue with safety. Meta-analyses are also used to compare and analyse all the published data about a drug to try to detect significant trends in toxicity[6]. In Europe, the major reason for drug withdrawal since 2002 has been cardiac toxicity in around half the cases, with hepatotoxicity the next most significant issue, with around a fifth of withdrawals.

Toxicity is usually directed at a specific organ or tissues, but as we have seen, biotransformation will have a significant bearing on any toxic events associated with a drug. For instance, if the drug accumulates in certain patients due to lack of clearance, this will exacerbate toxic issues, as we saw in Chapter 8.2. In drugs that have been withdrawn, in 70% of cases a reactive species was formed that contributed to the toxicity[6,7]. Although drug design and initial candidate selection usually begins with an overriding emphasis on pharmacodynamic performance, the drug industry's interest in biotransformation, in terms of its bearing on toxicity and how predictable a drug's disposition could and should be, is becoming an ever-stronger influence on drug design.

A.2   Drug design and biotransformation: strategies

Drug design and metabolism: the traditional approach

Drug design is informed by a variety of different influences. The new agent could be one of a series of entirely new structures drawn from basic laboratory research, or increasingly, a lead structure may result from virtual analysis of existing databases. Alternatively, a structure could be a variation on a well-worn theme that modulates the possibly problematic physicochemical properties of an already successful drug. This might be dismissed as a 'me too' agent, but it may have a very significant advantage over its predecessor. Whatever is driving the decision to pursue a specific structure or series, it is usually based on optimizing pharmacodynamic potency without creating agents that have properties that would undermine its future efficacy, such as very poor bioavailability or extremely rapid clearance, for example. Various *in-silico* predictions could be made regarding the possible biotransformation of a new structure, which would then be explored in various animal and human models. If the basic structure of the selected drug cannot be changed, then the metabolic fate of the drug is essentially a *fait a compli*. There are many examples of drugs with far from optimal pharmacokinetic performance that are tolerated simply because they are very old, or there is no prospect of a replacement appearing as the market is too small or too poor to fund the discovery process. Indeed, most drugs to date have been developed in this way, and the risks inherent in this approach are reflected in the rising numbers of withdrawals, as mentioned previously.

Certain key features of the biotransformational fate of the new drug in animal and human systems need to be ascertained. Poor results at this early stage before human clinical trials may mean that the drug is potentially too problematic to pursue, and whilst useful information is gathered, it is at vast and possibly unnecessary expense if the drug class is not judged to be viable. The following highlights some basic issues that should influence the final decision on a drug:

- *Rate of metabolism.* In comparison with similar drugs, is it very rapid, or very slow? Extremes of clearance can present problems in accurate control of plasma levels, and with very long half-life agents, the risk of toxicity can be magnified.

- *Multiple CYP metabolism.* This is a very positive attribute, as if many CYPs can clear the drug and one or two are inhibited by co-administered drugs, there will usually be a route of metabolism to prevent serious accumulation, so clearance is very robust.

- *Single CYP clearance.* This can be seen as vulnerability, as inhibition of the CYP will easily lead to accumulation and potential toxicity. Conversely, induction of this CYP could lead to treatment failure.

- *Likelihood of DDIs, inhibition.* If a drug is likely to inhibit a major CYP isoform, or its metabolism is blocked by a known series of inhibitors, then the likelihood of drug–drug interactions (DDIs) will be high.

- *Likelihood of DDIs, induction.* A powerful inducer will be problematic in complex regimens and will accelerate the clearance of other drugs, as inducers tend to be very broad in their effects on the main human CYPs. DDIs will be promoted through induction and treatment failures may result.

- *Polymorphisms.* If the metabolism of the drug is dependent on a CYP isoform that is unusually polymorphic, this erodes the mass-marketability and safety of the drug. Clinically, a drug that was subject to a single polymorphic CYP clearance may well be restricted in its usage or would require close medical supervision, as the risks of toxicity and drug failure would be considerable.

- *Linearity of metabolism.* Most drugs are cleared at a rate proportional to their intake (linear metabolism). Some (ethanol, phenytoin) are cleared at a constant rate, irrespective of intake (nonlinear metabolism). If the metabolism of the drug is nonlinear, then it will be hard to predict plasma levels with ascending dose, which will also make the drug difficult to use and subject to toxicity and potential failure.

- *Metabolism to reactive species/toxins.* Drug metabolism may lead to the formation of toxic or highly reactive products in some or all of the target patient group. This leads to risks ranging from organ-directed, to immunotoxicity.

- *Off-target gene expression effects.* The drug and/or its metabolites causes marked changes in gene expression in certain organs or specific patient groups with unique pathologies.

Once a clear picture of the drug's metabolic fate appears from the various experimental platforms employed, a decision must be taken on whether the agent has a biotransformational signature that is likely to be a practical proposition in terms of clinical use and will not present insuperable practical difficulties. This decision is made in light of the condition being treated, the alternatives (if any) that could be used, the nature of the patient cohort targeted, and which other drugs (perhaps for other ailments) are likely to be used alongside the new entity. Often, if the new drug is the first agent available to treat a life-threatening condition, then the bar, so to speak, is pretty low. Very undesirable properties, such as unusually long/short half-life, powerful inhibition and/or induction of CYP3A4, single route of clearance via CYP2D6 and nonlinear pharmacokinetics may well all be tolerated if the drug is likely to transform and/or save thousands of lives. Alternatively, if a drug is one of series of agents that are not much more effective than the existing treatment of a non-life-threatening condition, then it is likely to be discarded.

Drug design: integration of biotransformation

Rather than simply rely on the metabolism of a new drug to be 'hopefully something we can work with', it has been realised for some time that to maximize drug performance and minimize toxicity, a much more integrated approach must be taken in drug design. Innovative drug design now involves combining and optimizing pharmacokinetic, biotransformational, pharmacodynamic, and toxicodynamic/kinetic design. This means that a drug is carefully engineered to produce the bioavailability and half-life, which supports a once-daily dosage (so patients remember to take it), which is compatible with near ideal pharmacodynamics, along with a dose that minimizes local and systemic risks of toxicity. This means modifying drug structure perhaps quite extensively, in the light of initial human model testing data. This is not always possible, as it depends on identification and preservation of the pharmacophore so that rest of the drug can be engineered to carry it to the receptor in the most predictable way possible. This approach can be to either block all metabolism or modify it to manipulate and predict the progress of the drug through the patient.

Drug design: hard drugs

The term *hard drugs* is usually used in an entirely different context (see Appendix B), but in drug design, it means an agent that is resistant to biotransformation and its clearance is effectively renal, or it might be eliminated in bile.

This can make its pharmacokinetics very predictable and clinically manageable in many cases, except with the frail elderly and those with known renal issues. Some 'hard' drugs just happened to be resistant to biotransformation, such as atenolol. Others have been a product of structural modification of another agent that was intended to minimize or avoid biotransformation. The angiotensin converting enzyme inhibitor captopril was reasonably effective but its thiol group caused issues as it was reactive and formed haptens. Lisinopril and enaprilat (eventually esterified to enalopril to make it orally available) were designed without the thiol, undergo little metabolism, and are less likely to be problematic than captopril[8]. However, as we have seen in this book, the capacity of biotransforming enzyme systems to attack molecules of all types is formidable and, short of festooning your molecule with a dozen fluoride ions, it is very difficult to design a molecule that will not be metabolised by some system, somewhere.

In addition, the sheer scale of pharmaceutical pollution in the world's rivers and waterways is already extremely serious. Hence, to fill the environment with even more metabolically impregnable chemical entities would be irresponsible in the extreme. There is already a considerable list of synthetic endocrine disruptors and toxins (Chapter 2.4.1) in our rivers and seas that effectively defeat human and ecosystem biotransformation, ranging from the polybrominated diphenyl ether flame retardants to the ubiquitous dioxins.

Drug design: soft drugs

This strategy goes back to the 1970s and involves a compound design intended to be cleared in a controlled and predictable way. This is different from a prodrug, where an active agent is modified to improve its physiochemical characteristics and can only work when it has been metabolised. In contrast, a 'soft' drug is already pharmacologically active, but has been designed to facilitate clearance through pathways that do not create reactive and destructive species. This is best achieved by avoiding oxidative pathways in drug metabolism that produce these species. This is easier said than done, as three-quarters of drug metabolism is oxidative and linked to CYPs[9].

Another problem is that drugs that are cleared nonoxidatively, such as through conjugation, can form protein-reactive species. A good example of this is drugs with carboxylic groups, which are cleared to unstable acyl glucuronides (Section 6.2.7). Another issue is that compared with hydrolysis reactions, oxidative processes are relatively slow, so enzymes such as esterases have been spotlighted as preferred pathways for soft drug design.

An advantage of esterases is that because they are so ubiquitous, the drug should meet them and be cleared well before it encounters the oxidative systems. To facilitate this, the soft drug approach works best where a local effect is needed, such as the gut, lung, or eye. This also limits systemic exposure to the agent. The best examples of soft drugs include the β-blocker esmolol, which can be hydrolysed by esterases, as well as atracurium, which is cleared by Hofmann degradation as well as esterases and the opiate remifentanil, which is cleared also by ester hydrolysis[8,9].

To echo the issues mentioned earlier over drug pollution, 'soft' drugs are also less likely to pollute watercourses and safeguard our somewhat beleaguered environment.

Drug design: metabolic soft spots

Soft drugs are aimed at local targets with relatively low systemic exposure. For drugs that are intended to have a deeper systemic penetration, a modified approach is necessary. A vast library of information is now available on structure–metabolism relationships, biotransforming enzymatic active site preferences, and of course, structures to avoid.

In a series of prospective drugs, visual and software analysis will immediately predict that certain groups, such as benzyl carbons, O, S, and N methyl groups are metabolic soft spots[10]. Assuming there is no steric hindrance, possession of these groups as we saw in Chapter 3.12 usually means rapid clearance. There is also the issue that a very close structural analogue of an endogenous agent will probably share its metabolic fate.

For many years the drug industry has used bioisosteres[10], which are equivalent groups to soft spot groups, which are cleared more slowly rather than not at all, as we saw with the hard drugs. Other approaches are to block the metabolism of vulnerable groups with other substituents so CYPs cannot easily bind them. This is not easy, as CYP3A4 has almost limitless flexibility (Chapter 3.6.3). Another ingenious method to discover metabolic soft spots is to make the compound with deuterium, which is a stable isotope of hydrogen[10]. Biotransformation enzymes struggle to break into carbon-deuterium bonds, so an abrupt change in the half-life of the deuterated version in an experimental system compared with the starting structure indicates a soft spot has been located. Indeed, a deuterated version of tetrabenazine has been approved for Huntington's Chorea with less frequent dosing and less toxicity[11].

A.3   Animal and human experimental models: strategies

Platforms and predictions

It is clear that drugs fail in clinical trials usually because they either don't work or are toxic, or both. Drugs that survive clinical trials and are withdrawn after they enter the clinic usually work but are likely to be toxic[7]. The decision to allow the drug in the first place to progress towards human clinical trial is taken on the basis of data from pre-clinical (animal) and to some extent also on human *in vitro* data. The data from the whole animal biological system is very important to build a picture of the disposition of the drug[7] and in most countries, it remains a legal requirement for such animal studies to be carried out with a new drug. Clearly the hepatic metabolism of the drug will be determined *in vitro* using the same animal model.

Current animal models are problematic from many standpoints and they can be poor value scientifically and financially[12,13]. In reproductive toxicity, the predictive power of current animal studies is so poor there is no real scientific justification for even carrying them out. In addition, current animal models only have a 1 in 2 chance of predicting human hepatotoxicity[14], which as we have seen is one of the most common reasons for drug withdrawal, although animal studies are admittedly more predictive for gastrointestinal and cardiotoxicity. The species differences issue is practically insurmountable, as the best that can be hoped for in terms of agreement between whole animal and human clinical studies with regard to drug metabolism is 70%[15]. Given the industry's problems with drug failures, the missing 30% contributes to the increasing numbers of drugs being withdrawn due to systemic toxicity[2]. Chapter 8.3 provided drug examples that were safe in animals but toxic in man.

Alternatively, we will never know how many other agents that were toxic in animals, such as tamoxifen, might have turned out to be safe and valuable in humans. Aside from the obvious species differences, it has been recognised at the regulatory level that the key issue with adverse drug reactions is human variation, which is not present in in-bred animal models[16]. The challenge is how to model successfully in animals the full range of human variation. Some advances towards achieving this goal will be outlined later in this Appendix.

Platform strategies

The drug industry is addressing the issue of poor predictivity in its pre-clinical experimental platforms in two main ways. The first is to employ more human-based cell platforms, ranging from primary and stem-cell-derived models towards more complex systems as three-dimensional and organ-on-a-chip technologies advance. These can explore the impact of new therapeutic entities on human systems as early as possible in the drug development process. Using human models, it is possible to make reasonably accurate determinations of the essential features of a drug's biotransformation that will gradually minimize the current risks involved during Phase I initial human volunteer trials, as well as reducing unnecessary animal suffering[13]. Whilst *in silico* methods (see later on) can use these data to run simulations and make predictions, many of the more advanced human models lack validation and standardisation, as outlined earlier.

The second way forward is to humanize animal models with respect to pharmacodynamic, pharmacokinetic, and toxicokinetic studies. Such engineered humanized animal models are under intense development so that they can eventually explore the 'whole system' effect on a candidate drug. With the accumulation of data from a variety of sources, *in silico* systems are also actively assisting in the design of drugs by contributing to the design of pharmacodynamic/kinetic and toxicokinetic studies and integrating the data accumulated. Thus, the combination of more human models and humanization of animal platforms must eventually improve the predictivity of all the testing platforms and feed better real data into the *in silico* simulations. All this should reduce the current candidate drug attrition both before and after the approval processes.

In the next sections we will consider the current *in vitro* human models, which are based on human biotransforming enzyme systems and the various matrices used to support them. After that, we will examine the humanizing process of the whole animal models. Before we start to study the biotransformation of a candidate drug in any model, we need to select some practical means of detecting the drug and its various metabolites in biological fluids.

A.4   *In vitro* metabolism platforms and methods

A.4.1   Analytical techniques

When a candidate drug progresses to the point where it will be toxicologically and metabolically tested, it is necessary to choose a method or assay that will quantify the drug and its metabolites reliably in various *in vitro* and *in vivo* experimental models, which are usually aqueous and contain biological material such as enzymes and other proteins. The assay requirements are sensitivity, selectivity, and reproducibility, and some form of chromatography is usually employed. High-pressure liquid chromatography (HPLC) has always been one the most practical methods used to detect most low-molecular-weight (<400) drugs and related material.

Usually reversed phase HPLC is used where the column contains a nonpolar stationary phase (the packing) and the mobile phase is water-based and polar, involving solvents such as acetonitrile, methanol, and different additives. As most drugs are usually fairly nonpolar, they are repelled by the mobile phase and tend to associate with the nonpolar stationary phase for much longer than the more polar metabolites and the other cellular material from an experimental incubation. This leads to good chromatographic separation from unwanted material and facilitates accurate drug and metabolite quantification.

HPLC assay development usually starts with authentic standards of the drug; To 'see' the drug, some form of detection is necessary, which most often exploits UV or fluorescence absorption, as these are very common properties amongst therapeutic drugs as they often contain aromatic rings. Other agents that do not absorb UV or fluoresce can be measured using systems such as electrochemical detection, which is no longer the 'eye of newt, toe of frog' nightmare it was in the 1980s.

Once the drug can be visualised, the appropriate chromatographic column, solvent proportions and other conditions must be found through experimentation. These conditions must ensure that the drug is retained on the column long enough to be well clear of the more polar metabolites and that it should elute (move off the column) clear of any other drugs that might be in the same aqueous matrix. This can require significant trial and error to make the assay as efficient, versatile, and sensitive as possible[17]. At this point, another chemical entity that is very close in structure and physicochemical properties to the drug is selected and chromatographically tested to be an internal standard (IS). The IS becomes a yardstick with which to measure the drug and compensate for variation in the assay, and it should be retained on the column in a similar fashion to the drug. Once the chromatography is established, some form of extraction must be devised to remove the drug and the IS (added just before the assay) from the aqueous matrix. This can be through the use of organic solvents and/or disposable miniature solid phase columns. The solvents are intended to dissolve the drug and metabolites but not too much of the biological material, and they can be quickly evaporated after they have carried the drug out of the aqueous matrix. The dried sample is then resuspended in an alcohol. In some cases, the resuspended sample could be directly analysed after the solvent extraction if it is 'clean' enough, that is, there is virtually no protein material contaminating the sample. However, to extend column life, and make the HPLC run more reliably, the samples are subjected to a further clean-up stage using individual disposable solid phase columns. This starts with the priming of the solid phase columns with a suitable solvent; then the sample is introduced. The column holds the drug and metabolite and the residual biological material is flushed through. The drug and metabolites are then 'eluted' with methanol or a similar solvent. The main aim of this stage is to leave behind all of the proteins and cell contents and produce the cleanest uncontaminated sample that contains drug and hopefully some metabolites.

Once the extraction procedure is optimized, the suspected metabolites can be separated using the HPLC and then directly routed to more intensive analytical systems. These include mass spectrometry or more often NMR (nuclear magnetic resonance) machines that provide enough information to assign a structure to the metabolites, so authentic standards can then be made to study their stability and toxicity. Although there are other chromatographic methods, such as gas liquid chromatography, its advantages of high sensitivity can be outweighed by its operation mode, where the sample must be vaporized at high temperature to be analysed. Many unstable metabolites may not survive this process.

If it is suspected that a new drug is converted to a reactive species, clearly this can be very difficult to analyse and some species have been the subject of intense study, such as that of paracetamol and the precise structure is still debatable.

The isolation and assignment of structure to reactive species is challenging and often the best hope is to use a 'trap' molecule such as glutathione to react with the species to form a GS-adduct, which can be analysed with mass spectroscopy or NMR to determine which area of the molecule has attacked the thiol. You might recall in Chapter 5.4.3 the use of GSH in inhibition studies also. Other systems may use proteins to trap a species, such as haemoglobin adducts which are a useful clinical measure of acrylamide exposure[18].

Immunoassays can be used to quantify drugs of a variety of different structures and sizes, particularly peptides and high molecular weight antibiotics. These use the high specificity of fluorescently labelled antibodies to bind the drugs and these systems are often used for therapeutic monitoring of steroids and other agents and are capable of detecting sub-picomolar concentrations. Overall, most drug analysis systems can now be fully automated and are highly sensitive and selective. They are essential to determine the fate of the drug and its related metabolites in the different experimental systems. The following sections describe some of the main *in vitro* techniques employed in the evaluation of biotransformation and toxicity in drug development.

A.4.2   Human liver microsomes

The most basic tool for the determination of the metabolic fate of a candidate drug is the fractionation of human liver through differential centrifugation. The liver is homogenized on ice and centrifuged at 10000 *g* for 20 min to remove cellular and nuclear membranes, and larger cellular organelles. The supernatant is then centrifuged at 100000 *g* for around an hour to provide a pellet that contains the smooth endplasmic reticulum (Chapter 3.2) where the CYP isoforms and UGTs reside. This is termed a 'microsomal' pellet and it can be stored at least −70 °C until the content of the total CYP content is measured using a standard and vastly cited method that exploits the ability of carbon monoxide to bind the CYPs[19]. Every human liver also has a unique profile of individual CYP isoforms, and the livers are usually characterized in terms of their ability to metabolise certain substrates that are cleared by particular CYP isoforms. In general, CYP activity survives subcellular fractionation and cryopreservation, but some enzyme systems, such as microsomal epoxide hydrolase and GST activity, can be negatively impacted respectively, by those processes[20].

As there is so much variability in human liver microsomal systems, in trying to determine the metabolic profile of a candidate drug, it will be studied using many human livers to build up a picture as to which isoforms predominantly oxidize the drug. The products of the oxidation are usually extracted and purified, then analysed as described above. It is possible to confirm which CYP isoforms are oxidizing the drug by the use of specific CYP inhibitors as well as antibodies raised to CYP isoforms, which also block their activity. It is also possible by using specific substrates of various isoforms to determine if the new agent has any inhibitory effects on the main CYP isoforms.

Once it has been identified which CYP or CYPs are responsible for the oxidation of the drug, other microsomal studies can be carried out using human lung, kidney, and gut to determine what contribution these organs will make to the overall oxidative metabolism of the drug. The addition of UDGPA as a co-factor will activate the UGTs in the liver to provide a preliminary evaluation as to the degree of glucuronidation that may occur. Human liver cytosolic fractions (S9 fractions) can also be used to determine levels of sulphation and GST activity with the drugs and metabolites. In addition, cytosolic and microsomal epoxide hydrolase activity can be determined.

Supplies of human liver will always be problematic but happily, liver transplants are ever more common and successful than in previous years. The disadvantages of human liver homogenate preparations include the wide variation and the detailed observation of only some parts of a complex system of metabolism. If a very reactive but cytotoxic oxidative metabolite is detected, this suggests that the drug could be a hepatotoxin or may be reactive elsewhere. However, what cannot be determined with human microsomes is that detoxification and transport systems might just as well avidly clear the toxic metabolite and no systemic or organ-directed toxicity occurs. In microsomal studies, a drug may be a potent CYP inhibitor but *in vivo* it might not penetrate cells due to its particular physicochemical properties or lack of a suitable transporter. Although microsomal incubations are not a complete system, estimations of hepatic intrinsic clearance can be made (Chapter 1.3.4). Clearly microsomal studies are a useful but far from definitive stage in determining a candidate drug's metabolic profile.

A.4.3   Heterologous recombinant systems

Once the basic parameters of the metabolism of a new drug have been established, such as which CYPs and conjugative systems are involved, the specific components of the metabolic profile can be focused with heterologous systems. Although many of these data can be obtained from microsomes, the obvious advantage of heterologous systems is that the cell model can express just one isoform free from the influence of other pathways. In the late 1980s and early 1990s, the technology for the excision of complete human CYP and conjugative genes as well as their reductases and ancilliary support systems was developed and insertion into bacterial, eukaryotic, or viral vectors was perfected. With the models that expressed CYP3A4 in yeasts, disruption of the cells was necessary prior to experimental use, to free the CYP proteins. This was accomplished using a 'French Press'. This not only resembled something for jacking up trucks, but it took almost as much effort[21].

These recombinant systems allow the focus on specific human CYP isoforms, as well as many UGTs, SULTs, and GSTs. This facilitates extremely detailed studies of how a particular isoform metabolises a candidate drug, along with inhibition studies and detailed kinetic measurements, such as K_i measurements, rates of formation of metabolites and other determinations. This is a valuable foundation to the process of predicting possible drug–drug interactions (DDIs). The recombinant CYP (rCYPs) studies are particularly well suited to high-throughput assays that are necessary to generate sufficient data to make decisions in drug development as rapidly and economically as possible[22].

Recombinant CYPs can also be adapted to high-capacity nanolitre assays where the CYP is embedded in alginates and fluorescent visualization allows very high throughputs of test agents to screen for metabolite formation[22]. The burgeoning study of polymorphisms also allows detailed evaluation of the catalytic activities of various polymorphic isoforms, which might again inform on how a particular population may clear a drug. This is extremely relevant in the global marketing of drugs that occurs in the face of such radical human diversity, outlined in Chapter 7.2.3. Heterologous systems are also used to provide plentiful supplies of expression transporters such as the OATPs, P-gp and the MRPs, to determine whether the new entity is a substrate, inhibitor, (or both) on these systems[23]. There is an assumption that the gene cloned and expressed is identical to the CYPs in all patients, which may not necessarily be true. Of course, rCYPs are not applicable for enzyme induction studies, but for many years very sophisticated models have been developed to scale the data derived from heterologous systems to those expected from human liver microsomes in order to estimate intrinsic clearances[23].

A.4.4   Liver slices

This model has the advantage of retaining the architecture of the liver and specifically, the different zones of the liver, such as the centrilobular areas for example, can be studied, which of course cannot be done when the cells have been homogenized, fractionated, and/or separated. It is possible to carry out studies for up to 24 hours without too much loss of biotransformational capability. With suitable culture conditions, experiments can be carried out over days. The disadvantages of slices are that they are dying and CYP expression starts to decay within a few hours and that clearance is not as high as would be seen in hepatocytes and this is thought to be due to the drug not penetrating to all the available hepatocytes[24].

A.4.5   Human hepatocytes

Background

Although obtaining human liver suitable for the preparation of hepatocytes is always likely to be expensive and problematic, modern effective cryopreservation techniques mean that their use in drug discovery and academic study is now much more practical and prevalent. Hepatocyte preparations remain the key experimental technique for modelling drug biotransformation from a number of standpoints; indeed the phrase usually used rather monotonously in publications is 'gold standard'. Hepatocytes can demonstrate that a drug is physicochemically able to enter the liver, with or without influx transporter assistance. The complete oxidative and conjugative fate of a drug can be explored in a single model and their use in a suspension and similarly in two-dimensional (2-D) culture, models diffusion of drug into the cells quite well in comparison with *in vivo*, which as noted above is not so easy to show with liver slices. Hepatocyte preparations can allow for some exploration of induction processes[24]. From a drug company point of view, human hepatocytes are vital in determining the validity of their animal data on the same compound or series of compounds.

Hepatocyte suspensions

Hepatocytes are often sourced from livers unsuitable for transplants, as they are too old, fatty, young, diseased, damaged or unsterile, although if the liver is of very poor quality, data derived from it is only representative of the particular diseased state. Mostly the best researchers can hope for would be pieces of 'good' liver from resections, after the necessary histological procedures have been carried out on the tissue[15]. The best quality hepatocytes are obtained from livers that have been cold preserved for around 4 hours or less. The standard two-step method of hepatocyte suspension preparation involves a body temperature perfusion with calcium-free isotonic buffer containing a chelator to remove all divalent metals and to flush out any blood. This is followed by another warm flush, this time with collagenase to break up the tissue matrix[24,25]. Of course the collagenase digestion process causes oxidative stress in the cells and even if a whole liver is treated, only around 10% of the hepatocytes (there are ~190 billion in an adult healthy liver) will be harvested, although the yield is better from smaller liver fragments. However, despite the extreme process to which the cells are subjected, cryopreserved hepatocytes (stored with DMSO in liquid nitrogen) show, with one of two exceptions, broadly similar CYP, conjugative and OATP transporter viabilities to freshly isolated cells[20]. Suspensions of hepatocytes were employed for many years and remain useful for some studies. Whilst intrinsic clearances of various drugs can be estimated, they are highly variable and the cells start to lose CYP expression after just a few hours. Unfortunately, in any primary culture system of any cell type, the major disadvantage is that cells gradually lose their identity in a de-differentiation process. This means that as soon as they are separated from the 'mothership', so to speak, they are no longer instructed to retain a specific differentiated identity and they duly lose it in a day or so. If you recall from Chapter 2.2 the idea of biological systems' intense desire to control their environment, then in the light of the horrors of malignancy, the need to exert control over their own cells is even more important. This means many hepatocyte studies are only carried out over short timeframes of a few hours. One of the key challenges in drug metabolism research has been to circumvent this hepatocyte 'dead man's handle'. Transporter expression does not always survive any cryogenic processing and GSH levels can be low when the cells are revived. However, the donor variability problem is usually addressed by using cells from three to five livers to conduct studies.

Hepatocyte cultures

Human hepatocytes are now extensively used as primary cultures, which is a more demanding preparation, as the hepatocytes need to be isolated as describe above, but in a sterile environment, so that when the cells are freed from their tissue matrix, they can be seeded onto collagen coated plastic plates as sterile monolayers[25]. The cells retain their biotransformational capacity for longer than a day or so if a second 'sandwich' layer of collagen is supplied, which allows them to align themselves in apical and lateral membranes. This method is used extensively in the pharmaceutical industry to examine new drug metabolism[25]. Whilst the sandwich methods can preserve the functionality of the hepatocyte monolayer cultures for longer than single layer methods, full biotransformational capability in human hepatocytes is still a relatively temporary feature in such culture platforms.

Hepatocyte micropatterned co-cultures

It has been known for decades that human hepatocytes will retain their phenotypic expression of their biotransformational enzymes for days or even weeks if they are 'told to' by other cells, which don't necessarily have to be human. Models that involve the culturing of cryopreserved human hepatocytes with nonparenchymal cells (NPCs) go some way to restoring the situation *in vivo*, where the NPCs such as Kupffer cells, produce paracrine factors that maintain phenotype and functionality in the hepatocytes. A typical model can combine human hepatocytes with mouse-derived embryonic fibroblasts and this micropatterned culture approach (MPCC) extends full hepatocyte capability to at least a month[14]. This opens the door to productive studies with repeat dosages of drugs and even drug combinations and exploration into the molecular events linked with DILI and IDILI. The platform also provides the opportunity to use more advanced cell functionality techniques and biomarkers, which will be described in a later section to model the toxicity behind drugs such as troglitazone (8.3.2).

Hepatocyte immortalised cell lines

The idea of using immortalised hepatocyte-like cell lines looks as if the scarcity and de-differentiation issues can be addressed simultaneously. Immortalised cell lines are easy to buy, prepare, propagate, and maintain. However, they originate from tumours, and if you have either witnessed or even endured the remorseless effects of malignancy you will appreciate that these cells display rather too many 'Terminator'-like highly abnormal characteristics of their parent tumour, such as resistance to toxicity, overexpression of detoxification and transport systems, manic growth, and abnormal energy metabolism. One of the main reasons why hard tumours are resistant to radiotherapy is that this treatment acts by creating reactive oxygen species and oxygen is almost absent in anaerobic tumours, as they only build themselves a minimal blood supply while they grow. The cells derived from these tumours retain this anaerobic profile, using massive amounts of glucose in glycolysis instead of in oxidative phosphorylation. This is known as the 'Warburg effect' after Prof. Otto Warburg who first discovered it in tumours in the 1920s[26]. The net effect is to protect immortal cells from oxidative stress and various apoptotic pressures, although it is not clear why the cells do not override the Warburg state once they are freed from the anaerobic environment and the full significance of the process to tumour survival is not yet fully understood[26].

Unfortunately, in practical terms, immortal cells are seriously aberrant and they need a high carbon dioxide, reduced oxygen atmosphere to grow. Although their mitochondria are functional, they remain greatly underused, so metabolic toxins that target mitochondria are not very effective. Hence, many immortal cell models underestimate the effects of candidate drugs that cause oxidative stress through this route, so you can see that they would not be very informative about troglitazone and trovafloxacin. Nonetheless, the convenience and robustness of these lines make them a well-used screen for both toxicity and pharmacodynamic drug effects.

Amongst the many lines employed, HepG2 cells were isolated in the 1970s from a hepatoblastoma and they model some aspects of hepatocyte function, such as CYP induction. Unfortunately, their basal CYP expression is very low, relatively narrow in scope and variable between laboratories. Annoyingly, from the research perspective, even immortalised cells have a shelf-life of enzyme expression and after a certain number of 'passages' when they are re-seeded in a fresh vessel after they overgrow (over-confluent). Whilst they can grow well in three-dimensional culture systems, HepG2 mRNA levels are low and translation to actual CYP protein, with CYPs such as 1A1 and 1A2 is very much lower than intact animals, or human hepatocytes[24,27]. Hence, HepG2s are very limited in what they can do in context of drug metabolism studies and extrapolations to humans[28].

Currently the best plan in immortalised hepatocyte models seems to be the HepaRG™ cells, which are much superior to the HepG2s. HepaRGs™ are known to express a full complement of biotransforming and transporter systems, including the nuclear receptors such as PXR and CAR, as well as the ability to form bile caniculi in culture. They are stable and reproducible, but expression levels of CYPs are not yet on a par with human hepatocytes and they only represent a single donor phenotype, so they do have some limitations[24]. However, HepaRGs™ are a good platform for induction studies and they have very high expression of CYP3A4, possibly due to the presence of an inducer (dimethyl sulfoxide) in their stock medium[28]. Another issue is that you need to pre-culture them for a week to boost CYP activity, but they do compare favourably with human hepatocytes in terms of capability and performance[28].

The future: iPSC hepatocytes

To get beyond 'gold standard' to 'Holy Grail' in terms of hepatocytes would be a model that is human, easily accessible, reproducible worldwide, reflects human variability, and is exactly the same as hepatocytes in our livers. Hepatocytes formed from induced pluripotent stem cells (iPSCs) are believed by many to be the platform that can attain this very demanding standard. You start with a somatic cell, which is reprogrammed to a stem cell and then differentiated into a hepatocyte.

The early problems with this approach included very wide variation in different methods and level of functionality of the cells, along with many of the disadvantages of human hepatocytes. Biotransforming enzyme expression decayed rapidly and enzyme range and expression levels did not compare well with human hepatocytes[24]. Currently, iPSC-hepatocytes are commercially available, but CYP expression levels and profiles are still some way from human primary hepatocytes. Aside from CYP3A4, most of the major CYPs (2C, 2B6, and 2D6) as well as conjugating systems like sulphation and glucuronidation are poorly expressed in current models[28]. Even more importantly from a drug company perspective, despite the impressive nuclear receptor presence, their induction responses are suboptimal, and it has been suggested that several factors are missing in the cells from the induction control process[28]. Indeed, in one study with verapamil, the primary human hepatocytes produced an extra 10 metabolites that the iPSCs and other immortalised models could not provide[28].

For the future, high sensitivity of the iPSC-derived hepatocytes to preparation protocol in terms of the cells' enzyme profiles and expressions could be a strength, as it is then possible to produce several different phenotypes from the same cells using differing protocols. Alternatively, the same protocol could be used with different iPSC lines[28]. Either way, simulation of human variation in CYP expression and functionality is perhaps achievable.

The iPSC-derived cells are improving in their capability constantly and may in the future be able to duplicate primary capability and expression levels[29]. However, with all stem cell models, it is a rather like ordering a kit which when you build it, looks exactly like the product in many respects, but it does not quite work as there are always some key components missing. Realization dawns that you were supposed to supply those key components yourself. However, you have no idea what they do, where they are, how they work or how to make them.

This somewhat Kafkaesque[30] quality of stem-cell platforms is extremely frustrating. This is partly because other, supposedly more primitive human platforms allowed you and your colleagues to build amazing systems that ran like a Swiss watch and did everything that was asked of them[31]. Stem-cell experimental models in general will occupy scientists, rather like fiendishly complex toys, for decades to come until the desired *in vivo* replica systems eventually appear.

Hepatocytes: multidimensional microfluidic systems

Whilst the sources and types of cellular models are being intensively developed, experience in other fields, such as in cancer cell research, have indicated that it is not just what you are growing, but how it is grown. Growing cells in 2-D flat on various matrices restricts the ability of the cells to behave as they did in their original tissue, from which in the case of human hepatocytes, they were *untimely ripped*. Many authors have recommended that the various hepatocyte platforms would be more efficient and express their enzymes better in 3-D culture[28]. Several elaborate systems have been developed where the cells are grown into spheroids with the intention of recapitulating the tissue architecture of the liver. Spheroid culture has long been an area of study in cancer research but naturally without a blood supply their cores will run into necrosis once the spheroid exceeds around 100 µM in diameter[25]. Various methods are also used to increase oxygen supply to hepatocyte cultures, such as growing them in different bioreactor designs and various perfused culture vessels[25,27]. There are now many different two- and three-dimensional hepatocyte culture platforms in use. Whilst the 3-D systems in various multi-well microfluidic bioreactor platforms are very exciting, standardisation in their designs and protocols is some way away, as one particular method needs to 'catch on' so that it becomes the mainstream. There are many operational difficulties in optimizing these systems, such as defining oxygen and nutrient levels to provide a platform stable enough for reproducible data generation. In addition, in the drug company research and development environment, high throughput is essential due to the large numbers of different compounds that must be screened and the more advanced models are not yet capable of such up-scaling[24,27].

However, there are many miniature micro-scale organ-on-a-chip models currently available that show potential for adaptation to high-throughput analysis of candidate drugs. These platforms feature multiple cell types, plus built-in fluorescent biosensors, and can be adapted for any spheroid-like tissue, ranging from liver/gut interfaces, to many other tissues[13]. Such systems detect oxidative stress as well as signs of hepatoxicity and can feature iPSC-derived cellular systems. One of the most exciting concepts has been to string different bioreactors and organ-on-a-chip devices together to mimic systemic metabolism and the anatomical realities of some tissues (blood brain barrier and brain) can also be studied[31]. Whilst there are many different configurations of systems, eventually the most efficient, productive, and economic system platform will emerge and the models will become standardized. Once this happens, there is more chance of these platforms gaining acceptance as part of drug development data submissions to regulatory authorities. In advance of such acceptance, the application of allometric scaling approaches has already been developed that can use *in vitro* models of a variety of different designs and structures to scale data up to animal or human systemic levels[32].

Hepatocyte data

The initial use of hepatocytes in drug development is to calculate a more realistic intrinsic clearance of new agent compared with other models. To predict whether a drug might be a low or high extraction agent, the intrinsic clearance must be related to whether a drug is strongly protein bound or not *in vivo* as well as to blood flow. To compensate for the issue of free and bound drug some studies have included serum or various proteins in hepatocyte incubations, although this is controversial. Other authors have shown that hepatocyte data combined with mathematical models for protein binding and liver blood flow can yield convincing estimations of *in vivo* clearance. These scaling values can be applied to most aspects of biotransformation, including the activity of various transporters[33] through relative activity approaches. This is part of an allometric approach that scales up dosage ranges, so that drug development can progress towards preclinical animal and then in human trials. Hepatocytes are also used to determine metabolic pathways, such as those involving conjugative products. Indeed, if the major metabolites are cytosolic conjugates, then it is not really necessary to look at drug clearance from microsomal fractions.

Drug metabolism inhibition studies are well established in hepatocytes and they also demonstrate that both inhibitor and 'victim' drug can attain cellular concentrations that are relevant *in vivo*. Both reversible and time-dependent inhibition can be studied in hepatocyte cultures, often by the use of cocktails of CYP- specific substrates and various inhibitors in six-well plates, which maximize the data extracted from the valuable cultures. Some studies in hepatocytes demonstrate lower inhibitory drug effects compared with expressed systems (see next section) and some are the reverse. This may be linked to cellular binding concentrating effects and other metabolizing systems, which are again more representative of the real world.

The advent of the monolayer-plated and MPCC cultures described above are major advances in the modelling of enzyme induction, which is tracked by measuring increases in the appropriate mRNA as well the CYP itself. Human hepatocytes retain at least for a few hours the full functioning complement of nuclear and cytoplasmic receptors necessary to respond to any inducer and all *in vivo* inducers show their effects in hepatocytes and it is thought unlikely that a new agent shown to be a hepatocyte inducer would not act similarly *in vivo*. The ability to model induction and inhibition realistically is a powerful advantage of hepatocytes in the prediction of drug-drug interactions (DDIs) *in vivo*. The US FDA defines a new agent as an inducer in hepatocyte studies if it doubles control CYP expression or produces more than 40% of a positive control's inductive effect.

Hepatocytes in sandwich cultures also have immense value in the impact of the various influx and efflux transporters on drug metabolism (OATPs) and possible DDIs can be explored. Cultured human hepatocyte platforms are currently the mainstay of drug metabolism in drug development, but as discussed earlier, other immortalised and iPSC platforms will eventually augment and perhaps supplant them, as the issues of scarcity, flexibility, predictivity, reproducibility, robustness, and relative economy are all addressed.

A.5   Animal model developments in drug metabolism

A.5.1   Introduction

It is sometimes said that the ideal platform to test new drugs would be the intended recipients of the drugs – humans. However, we know that this cannot be carried out for ethical reasons at the very least, but even if we could find volunteers, the 'human' model would not be scientifically adequate, as we could not have total control over it. All we could do was observe what happened and make some modest and low-risk interventions to try to learn how the various cellular systems worked. Human cellular models such as hepatocytes are also very limited, as although they may provide a great deal of relevant information on the clearance of a drug as mentioned previously, they cannot yet model the complete living system of an intact human.

Current animal models at first glance look attractive, as unlike humans, we appear to have total power over every aspect of the experimental animal model. We can carry out any reasonable and humane investigation as long as it is ethically permissible and benefits humans. However, this 'total' control has been somewhat illusory, as we have had to accept serious compromises in terms of the extensive species differences in hepatic function, ranging from nuclear receptor-mediated induction, PXR, CAR, and the various detoxification and transport systems. Indeed, the attrition in drug development and the increase in drug withdrawals reflect this. However, in recent years, advances in gene editing technology have revolutionized the process by which rodent genomes can be manipulated. This technology over the next few decades will potentially give us total control of all aspects of the experimental process for the first time, so that humanized animal models will be aligned with the most advance stem cell models to create a drug evaluation process which will yield safe drugs that work in the vast majority of humans. In the next section you can gain some idea of the current state of the development of the rodent humanization process as a model for drug metabolism and toxicity in humans.

A.5.2   Genetic modification of animal models

The earliest significant genetic modifications to animal systems in drug metabolism were intended to tease out how the various systems operated. Since the mid-1990s, 'knockout' and 'knockin' animals have been produced where various genes, such as those that encode a complete enzyme system are been deleted or added. These studies involved a fair amount of trial and error, as some knockouts are not viable or are fatal (such as POR, Chapters 3.4.8). However, much can be learned of the homeostatic roles of various isoforms and also their activity in the presence of drug molecules. These systems are now known as genetically engineered mouse models (GEMMs) and they have progressed to the point where genes can be engineered to be switched on and off by dosing the animals with drugs such as tetracycline during the same experiment. These 'tet-on' or 'tet-off' systems allow a tetracycline concentration-dependent flexibility in expressing or not expressing different enzyme systems in the animals[34]. Other models such as RNA interference (RNAi) allow for the knockdown, rather than complete knockout of different genes[34].

A number of technologies have been developed that now allow intricate and complex multigene editing, where either human systems can be inserted, modified, or removed in cell lines and in experimental animals. The first systems to be used significantly were zinc finger nucleases (ZFNs) and the transcription activator-like effector nucleases (TALENs). ZFN has been used in many models to knock out transporters and was used in the HepaRG™ line to remove PXR. However both TALENs and ZFN require extensive protein engineering and are not easy to work with[35]. A major breakthrough in this area was the adaptation of the Clustered Regularly Interspaced Short Palindromic Repeats (CRISPR) gene editing system, such as CRISPR associated protein 9 (CRISPR-Cas9)[35]. This allows the flexibility to modify or eliminate a whole chromosome if necessary for a fraction of the cost and time involved with other systems. Multiple genes can be edited simultaneously, and there are many therapeutic applications under development. Interestingly the long tradition of 'theft' of bacteria systems, described in Chapter 2.2, continues here, as CRISPR originates from various primitive bacteria that use it as a sort of 'genome grooming' to hunt through their chromosomes to eliminate unwanted plasmid-derived DNA and viral genetic material[35]. This technology has produced rats with CYP2D6 and CYP3A expression and various transporter system knockouts and knockins that are much more efficient and have less background activity than with previous methods[36]. Whilst the technology needs more development as applied to biotransformation studies and is currently lacking in efficiency, it holds considerable promise for engineering human polymorphic biotransforming and transporter isoforms into animal systems.

A.5.3 'Humanized' mice

Background

As we have seen, the technology for humanizing animals, particularly mice, has progressed from the insertion and deletion of specific genes all the way to the creation of various types of 'transgenic' animal. This work has been a high priority over the past decade to try to improve the extrapolation of data derived from animals to the human situation. There are now two main approaches intended to humanize mice, and they have undergone considerable development, although neither is yet accepted as part of the regulatory drug development submission process, they are used increasingly to inform key decisions on compounds in drug development and try to bridge the considerable chasm between the preclinical and initial human phase of drug development.

Chimeric mice

This concept was developed as far back as the 1990s and there are three main types of chimeric mouse systems in use currently, the SCID, TK-NOG, and FRG systems[37]. They all use a similar approach, where mice that are immunodeficient have most of their hepatocytes removed and then replaced with human hepatocytes isolated, as described earlier, without any rejection issues. The liver blood flow does not change too radically after the transplantation and the hepatocytes function well and retain the characteristics of the donor, in terms of polymorphisms and responses to induction, inhibition, and clearance. CYP and conjugation systems are mostly well expressed, and many CYP protein levels were very broadly comparable with human liver expression levels[37]. A good feature of the chimeric mice is that they can be made according to the human hepatocyte characteristics, so cells from a liver with a particular polymorphic expression of biotransformational isoforms can be studied for a considerable period of time[38]. Not only can hepatocytes from livers used in earlier studies in a drug's progression through development be used in this model, but the superior width of metabolite formation in hepatocytes in comparison with the immortalised cells, can be reproduced in the chimeric animals. In addition, studies have shown that chimeric mice are good models for human hepatotoxicity. Diseased hepatocytes can also be used and excitingly, iPSC-generated hepatocyte chimeric mice are under development[37,38].

All these apparent advantages do come at a price. The hepatocytes need to be obtained or purchased, then prepared. The mice must be individually treated and operated on and the end result costs several thousand dollars *per mouse*[38]. If you add to this the often painfully expensive process of sustaining iPSC hepatocytes, you can see how development costs, even for a large corporation, can become very considerable in a relatively short-lived animal model that cannot reproduce itself.

Chimeric mice can be used to examine the whole system aspect of drug metabolism, including clearance, inhibition/induction, DDIs and other whole body issues such as volume of distribution. Of course the gut and the rest of the animal are obviously not human and there are residual mouse hepatocytes in some of the models, as 70–95% of the mouse cells are replaced. This can be an issue if, for example, a drug was cleared tenfold more rapidly in the mouse cells than the human, This would mean that even a small number of mouse cells would make a disproportionate impact on the clearance of the drug, so the model has some limitations.

'Humanized' mice

These mice have whole sections of their DNA removed and replaced with human single or whole groups of genes and even chromosomes. These insertions can range from the various major CYPs, several transporters, as well as the nuclear receptor systems such as PXR, CAR and the cytoplasmic AhR system[38]. Whilst some models contain just a single human CYP, others contain four different genetic insertions yielding a complete human inducible PXR/CAR CYP3A4/7 system[39]. The strength of this model is that often only two or three biotransforming isoforms might be involved in the clearance of a new drug and the other systems present in human liver are not involved. So the genetically humanized model expressing a small focussed number of human isoforms would be a useful model at the later stages of drug development, when such issues were already well defined from the enzymatic perspective[38] and information was needed on issues related to global clearance – like blood flow, for instance. Not only can the humanized mice be used for all the metabolic explorations detailed above, but also they are 'only' a few hundred dollars each and can be bred into a sustained colony. Other advantages include not being restricted to the liver – biotransforming human genes can be inserted into any organ, such as the lung or the kidney[38]. The possibilities for this model are very wide. For instance, specific conjugative systems such as GSTs, or other protective isoforms can be removed or modified and the impact on clearance and hepatotoxicity can be explored. There are some disadvantages, such as compensatory changes in the expression of other mouse isoforms that could distort the clearance and toxicity of a candidate drug[38,39].

A.6 Toxicological assays

A.6.1 Aims

For an experimental drug, much work will obviously focus on the particular metabolizing system that might form toxic reactive metabolites. However, it is also important to visualise what these species will actually do to a cell, either *in vitro* or in an experimental animal. This can be resolved in terms of the broadest to the narrowest focus. The broadest is the obvious; that is, are the cells alive or dead after some time in the presence of the drug? To answer that, with *in vitro* studies, viability assays are necessary. Of course it is not enough to know whether cells were dead or dying, but the mechanisms behind the death process, whether it was apoptosis or necrosis and key impacts on organelles such as mitochondria need to be uncovered. It is also necessary to design *in vitro* systems that model the movement and progress of reactive species through a matrix that resembles the *in vivo* situation. Whilst damage to cell energy maintenance processes are significant for the cell itself, damage to DNA is potentially significant to the tissue, the organ, and finally the animal or human, as the faulty and potentially malignant material will progress to future generations. Assays for DNA and chromosomal damage can be applied to *in vitro* cellular systems or cells removed from animal tissues.

A.6.2   Cell viability assays

There are several basic assays of cellular health that can be routinely applied to any cell- related system in toxicity screening. The simplest test of all is probably trypan blue exclusion, where recently dead cells cannot exclude the dye and swell as well as looking bloated and grotesque. Other assays measure a cell's functional capability, by reduction of a dye such as MTT. Although originally developed in the early 1980s, this assay is still used, although it is rather cumbersome and involves destruction of the cells to recover the blue dye for measurement on a fluorescent plate reader. The Alamar Blue assay is noninvasive and low in toxicity, so it is easier to use and the cells can be studied in real time for considerable periods, which is a good advantage over MTT[40]. ATP levels are also a useful measure of mitochondrial health and cellular GSH depletion assays reveal levels of oxidative stress. There are also several dyes, such as JC-1, that are selectively permeable to mitochondrial membranes and through changes in their fluorescence, mitochondrial membrane potential can be measured under toxic stress[40].

Flow cytometry is an extremely flexible technique that can be used to identify and count specific cells as well as to study the effects of a toxin on the cell cycle, both from the apoptotic and necrotic viewpoints. Different fluorescent antibodies can be used to count and separate populations of cells. In addition, various dyes are used, which either bind DNA or become activated within metabolically active cell areas, such as mitochondria. All these assays enable comparisons to be made between positive controls and a candidate drugs and a dose-toxicity relationship for the new agent can be produced in any given cell model.

The focus on examining the impact of toxins can be carried out in different experimental protocols and apparatus. Bioassays can show the effect of the toxins in terms of their ability to activate the immune system or kill human cells and the proximity of the biotransforming enzymes with victim cells can have a strong bearing on the toxicity of the species formed.

A.6.3 'One compartment' cell models

A number of systems incorporate human cells (from kind donors too slow to escape when you saw them pass the lab) *in vitro* to study or even predict possible idiosyncratic reactions, either with or without some source of biotransforming enzymes. The Basophil Activation Test (BAT) is intended to detect IgE-mediated reactions to various drugs, such as penicillins and NSAIDs. This test involves flow cytometry and is capable of predicting potential anaphylactic reactions[41]. The Lymphocyte Transformation Test involves radiolabelled thymidine and measures cell proliferation in the presence of the test drug and can be up to 80% predictive of drug hypersensitivity.

Another assay that is more directed at basic research is the lymphocyte cytotoxicity test, or sometimes the Spielberg test[41]. The simplest bioassay system is the mixture in one sample tube of victim cells such as human mononuclear leucocytes incubated with mouse microsomes and a source of reducing power[42]. This has been expanded to feature human liver microsomes,[43] hepatocytes, or even expressed human CYPs[21]. After a period of time, the human cells would be separated from the mix and assayed for toxicity, as detailed above. A more sophisticated version of this method is also used where human CYP systems are inserted heterologously into human cell lines, such as HepG2 cells. A reactive species could be made by the particular CYP and the degree of toxicity measured. These systems are very efficient at modelling short-lived species that cannot escape a cell or tissue before reacting with a macromolecule. Another facet to the test is that leucocytes from patients who have suffered idiosyncratic reactions to drugs such as phenytoin are more sensitive to the reactive species generated by the biotransforming enzymes, although the predictive capability of the test is controversial[41]. Whether or not the species would escape the liver, or even cause hepatotoxicity in a more complete model such as hepatocytes is debatable, and there is a considerable likelihood of 'false positives' appearing. This type of assay lends itself quite well to University or College practical courses to illustrate biotransformation-mediated cytotoxicity.

A.6.4 'Two compartment' models

Relatively stable toxins such as hydroxylamines are better modelled in multi-compartment systems that separate the source of the reactive species from the victim cells by a porous barrier, which keeps cellular material apart but allows small molecules to pass through. Dialysis systems comprising Teflon disks are useful for adapting to this purpose. The victim cells can be aseptically assayed, and the ability of a toxin to travel some distance and still inflict toxicity can be evaluated. Modular systems like those described above can be used with virtually any cell type and many different sources of reactive species[44].

A.6.5　DNA and chromosomal toxicity assays

It is essential to investigate much more subtle and early indications of drug and reactive species toxicity aside from the impact on cell organelles such as mitochondria. DNA-linked effects of drug toxicity as mentioned previously could have extremely serious long-term effects on a biological system. Indeed, the bladder cancer issues related to the antidiabetic drug Actos (pioglitazone) emphasize the importance of detecting the admittedly slight risk of malignancy from a therapeutic drug.

A.6.6　The Ames test

Invented in the early 1970s by Professor Bruce N. Ames, variants of this basic test are now a cornerstone of regulatory determinations worldwide and investigations for new drugs. It detects if a chemical is mutagenic and was subsequently adapted to include sources of biotransformation enzymes to model the formation of reactive species and their potential for DNA damage. Briefly, the test usually uses genetically altered *Salmonella typhimurium* bacteria that are dependent on supplied histidine in their media. If they are plated out onto media without histidine they will not grow. If a mutagenic agent is added, it causes reverse mutations and the bacteria grow on the deficient media as they can now make their own histidine. Essentially, there is a linear relationship between the colonies formed and the mutagenicity of the chemical[45]. Many different *S. typhimurium* strains are used and are particularly good at picking up frameshift mutations and base-pair substitutions, whilst *Escherichia coli* is more sensitive at detecting adenine/thymine mutations[45].

　　Ames himself quickly understood the necessity of incorporating biotransforming enzyme into the test[46] and any activating system will work, ranging from human microsomes and S9 fractions towards heterologous systems and cell lines. Indeed, major human biotransforming isoforms have been expressed in genetically altered strains of the Salmonella bacteria themselves for some time[47] to determine if the specific test agent/isoform combination has the potential to form a mutagen. It is important to carry out an Ames test before a large investment is made in a compound, as a positive result usually leads to abandonment of the compound. The reason why lies in the accuracy of the test, as an Ames test positive drug has a quite high chance (65–87%) of causing cancer in rat models[45]. However, there are cases of false positives. For instance, the bacteria obviously have different enzymes to us, and they can certainly convert an agent into a mutagen that we could never actually make. If a test compound contains histidine, it can cause growth of the bacteria, and impurities and degradation products also can lead to a false positive[48]. A positive Ames test does not always finish a drug, but a great deal of time and money must be spent on several *in vivo* transgenic mammalian mutagenicity/carcinogenicity assays and especially the 'Pig-a' rat mutation assay[49] – and they all have to be negative before the compound is 'forgiven', so to speak. Interestingly, several older drugs do fail it, such as isoniazid. Aflatoxins obviously fail, as, of course, do loathsome diesel emissions as noted in Chapter 8.5.4. A few years ago, I actually had the privilege of being sent an Ames research group manuscript to review for a journal and he was still publishing at the age of 90 at the time of writing. The manuscript? It was excellent, obviously.

A.6.7 Comet assay

This assay detects DNA damage in cells by subjecting the DNA to an electric current in a gel (electrophoresis) in solutions of different pH. The DNA moves through the gel forming a characteristic 'comet' shape and usually single DNA strand breaks (alkaline Comet assay) are visualised using fluorescent dyes. Double-stranded breaks (neutral Comet assay) are more usually evidence of exposure to cisplatin or radioactive material[50] and should be beyond the scope of what you might expect from exposure to a non-chemotherapy drug. The Comet assay can be used for cell line studies but from a regulatory standpoint it is used in rats to examine the impact of a drug on certain tissues, at certain periods post dose. Hence, most of the studies are carried out on areas of high drug contact, such as the gut and liver. The interpretation of Comet assay data can be controversial and which assays should be used to corroborate these data, as well as standardisation of protocols has not been fully decided. In addition, positive Comet assay data are not always reported to regulatory authorities[51]. This is concerning, as Comet assays are sensitive, as they detect early signs of drug-induced DNA instability and damage and data collected from them are of strong interest when set alongside other cytotoxicity tests conducted at the same time on the same experimental system.

A.6.8 Micronucleus test

This test is included in all regulatory submissions on a variety of different chemicals, not just drugs, but also pesticides, food additives and other environmental agents. The micronucleus test is used in conjunction with the Comet assay, *in vivo* or *in vitro,* and it detects micronuclei, which are formed during mitosis when a chromosome is lagging behind and may be lost (aneugenic event) or when a chromosome is broken (clastogenic event)[52]. These faults prevent the chromosomes from integrating into the daughter nuclei and can be counted. The measured frequency of these events is correlated with increasing toxicity. The micronucleus test can be carried out on any group of dividing cells and is often used to look at the impact of potential toxicity on bone marrow, as well as peripheral cell populations. One of the major focusses in recent years is to integrate it more humanely into animal studies to minimize usage and reduce suffering[52].

A.6.9   Toxicology in drug discovery

Background

As we have seen so far, a key feature of drug development is to employ many different experimental platforms coupled with specific biotransformational systems to produce a coherent mass of data that essentially convinces regulatory authorities that a lead structure in a series of candidates is likely to be safe in humans. With many drugs in current clinical use, this approach was adequate and no significant toxicity occurred. However, as we have seen, even if one out of hundreds of drugs marketed by a company turns out to be toxic and has to be withdrawn, the costs in terms of patient health and from the financial perspective are eye-watering. In future, it is likely that regulatory submissions will contain detailed reasons why a particular drug is unsuitable for specific populations and even groups of individuals, as the toxicological risk/benefit is tilted too far towards risk. Therefore, instead of hoping it will not occur, there is pressure to confront possible toxicity overtly so that new drugs are focussed towards some patients and away from others. To achieve this, it is not enough to refine existing techniques; a more radical approach is needed.

Systems toxicology: toxicogenomics

Of course, commercial drug discovery is not innately fascinated with toxicology for its own sake; rather, it must exploit toxicological concepts that can prevent drugs causing problems. To achieve this, it has been necessary to recognise the relatively 'reductionist' aspect of the techniques and approaches described in this Appendix, as they are focussed on specific isoforms, pathways, or metabolites. Toxicity is now approached from a 'systems biology' perspective that may allow us to augment our current 'local' towards a 'global' recognition of more potentially toxic drug-linked patterns in multiple cellular, tissue and organ systems[53]. This aspect of patterns and progressions in toxicity has been conceptualised in terms of adverse outcome pathways (AOPs)[54,55]. This is a recognition that all systemic toxicity begins at the level of molecular events in a cell that could be propagated through a series of steps that progress and amplify through to the level of the tissue, then towards the organ and finally they affect the individual in terms of symptoms. This recognition that toxicity is a result of pathways of a series of irreversible changes in structure and function gives us a framework we can use for predictions in toxicity. Hence, it is possible to see how events detected at the molecular level can be extrapolated towards their possible impact on the organism. If you look at the many examples in Chapter 8, you can string together AOPs for each chemical yourself. The pathways can thus be studied forwards, to develop predictive capability and in reverse, as a forensic 'why did this happen' approach.

From a commercial drug development standpoint, you can see the necessity for the development of AOP approaches 'forwards' to focus the next steps in drug development to ensure that you were not straying from pharmacological through to toxicology doses in a test platform, unless this is what you intended to do. In addition, studying AOPs assists key objectives in initial testing, that is to detect and predict organ and tissue directed toxicity in a given series of compounds. However, in the past, the perception was that when a company abandoned a lead or series of structures due to toxicity, they lost interest and moved on to other structures because they could not afford to engage with forensic toxicology. With AOP thinking, the opportunity exists to explore the pathways, identify the problematic area of a drug's structure as it interacts with the AOP, change it, and retain potency and virtually eliminate toxic risk.

To make AOP thinking a practical proposition, extremely sensitive techniques have matured over the past decade that shed light on the initial events in a toxicity pathway. These approaches echo the danger hypothesis (Chapter 8.3.4) and turn it around to detecting potentially injurious patterns caused by a drug in a biological system. How this can happen, is through the 'omics' sub-disciplines. These include genomics, proteomics, metabolomics, and transcriptomics, which are focussed on understanding different aspects of living system hierarchies. These disciplines are being gradually drawn together in *toxicogenomics,* which has differing definitions but mainly refers to changes in all those 'omics' areas in response to the presence of a drug or chemical[53]. This is a monumental challenge, as the myriad changes in gene expression, protein activity, and all the millions of different individual controlling molecular events that maintain human systems at first seem beyond analysis. However, several advanced techniques used in molecular biology have for some time been able to provide us with the means to start to 'listen' to these systems (Chapter 4.1.1) and software has appeared to analyse and help interpret the enormous mass of data that can be produced, to facilitate predictive and/or forensic approaches in drug development.

Microarrays

As we saw in Chapter 8.3.3, drugs and other toxic agents can change gene expression, either single genes or complex linked groups of genes, such as those concerned with cell oxidant defence. Gradually, it is becoming better understood that drugs sometimes do not just affect single gene expressions, but they down- or upregulate major cell-wide systems such as GSH maintenance, biotransforming enzyme expression of the CYPs, mitochondrial genes, as well as 'housekeeping' cell-cycle gene systems. DNA microarray systems have allowed us to explore these drug 'system impacts'. These assays depend on using sequences of nucleotides attached to a chip (usually called a probe) that binds reverse-transcribed cDNA in a sample. The imaging is provided by fluorescent tags. This detects the mRNA from cells or tissues exposed to toxic agents and compares it with control mRNA expression[53]. Single high-density microarrays are now available that contain groups of genes related to different functions or pathologies[56]. This can create a great deal of data that can require complex software systems to process. Of course, an array could show the down- or upregulation of a gene, but if we are unaware of the role of that gene, this information is not immediately relevant. Of course, it could be in the future. However, knowledge of genes specifically involved in responses to toxicity, as well as detoxification, are now much better documented, and as you will have read in Chapter 8.3.3, arrays now focussed in many different system areas.

The combination of array data and toxicogenomic databases of known toxicants are analysed in terms of the impact of certain predictive 'biomarker' genes whose 'fold-change' expression may be critical in the development of organ toxicity. Statistical analysis determines whether the fold-changes are significant with respect to changes in housekeeping genes[40]. Biomarker genes or 'tox' genes include inflammatory process genes, such as those expressing interleukins and other cytokines, as well as genes involved in the defence against oxidative stress, such as the GSH maintenance enzymes, superoxide dismutase, and catalase. DNA repair, cholestasis, immunotoxicity, and mitochondrial functional genes are amongst many that are radically changed in expression under toxic pressure.

Studies in the mid 2000s with microarrays in human hepatocytes[57] in the presence of known hepatotoxins troglitazone and trovafloxacin illustrated the power of the combination of these two highly relevant techniques to man. These drugs induced oxidative stress in human livers and this was reflected in their effects on gene expression. Troglitazone upregulated a large number of tox genes that its safer sister drugs, pioglitazone and rosiglitazone did not. Among the many genes affected by trovafloxacin, BAX and Mitofusin-1 were downregulated, which are important genes in the maintenance of mitochondrial health[57]. A decade or so later in MPCCs, microarrays[14] reiterated the scale and the detail of troglitazone's impact on the human genome, as many hundreds of individual gene's expressions were changed. The MPCC model allowed the incubation of the drug with the cells for several days, mimicking the *in vivo* situation. This revealed widespread stress responses throughout a range of cellular systems, from bile acid metabolism, all the way to 'wound healing' responses. Potent induction of both CYP3A4 and CYP2C8 was also evident, as was the sharp contrast between troglitazone and its much lower impact sister compound rosiglitazone[14]. Within the changes in gene expression revealed by microarrays, validation and a sharper focus can be brought about by the application of quantitative polymerase chain reaction (qPCR) and reverse transcription (RT-PCR) to explore the impact of a drug or chemical on specific sequences in detail[14,40]. Indeed, microarrays are a powerful technique, but often other confirmatory functional assays which are indicative of wider oxidative and cellular distress (GSH, ATP, and mitochondrial dyes) are necessary to support and help interpret array data and also to build a more complete 'systems' picture of the impact of the agent on the cells[40]. Presently, drug development programmes have access to combinations of focused microarrays, toxicogenomic databases of predictive gene biomarkers and 'learning' software that[58] processes these data and then is intended to make predictions on how a drug might impact possible susceptible patient populations[58].

Microarrays do have limitations; they are not always linear in their relationship between the amount of material being measured and the signal generated. In addition, they are not always very selective and specific in distinguishing between very similar genes. Finally, in chromatography, for example, you can detect an unknown peak and identify it in due course. In a microarray, you can only find what you are looking for; it cannot detect unknown genes as there is no probe for them in the array[56].

RNA-Seq

This technique has overtaken microarrays in many contexts, as it can find and quantify previously unknown RNA transcripts and is extremely sensitive. It involves mRNA extraction, conversion to cDNA, shredding into fragments, followed by processing, amplification, sequencing, and then identification[53,59]. The method is quantitative but is currently more expensive than microarrays and the derived data is more challenging to analyse, but these issues will resolve as usage becomes wider. This method is already being applied to drug-mediated cellular expression changes induced in immune systems *in vivo*[60].

A.7   *In silico* approaches

Drug design

The availability of specific software packages intended to assist in the design of new drugs combined with easy access around the world to extremely cheap and fast PCs have left older laborious methods of synthesizing thousands of probably useless compounds in the dust. These packages rely on algorithms, which are software-driven process instructions designed to accomplish the task of producing a viable drug structure with high efficacy and low toxicity. To achieve this, current software exploits huge databases accrued by either commercial or academic institutions to search and incorporate pharmacophores, i.e. specific areas of molecules that have high pharmacodynamic activity, and toxophores, i.e. areas that are likely to lead to either direct toxicity or the propensity to be metabolised to something unpleasant. The effectiveness of any package is reliant on the robustness of the algorithm, but it is mainly dependent on the quality, depth, and diversity of the data in the database. There are now many accessible world databases containing millions of 3-D crystalline structures of macromolecular therapeutic targets, such as various enzyme isoforms, nucleic acids, and proteins, as well as NMR/X-ray crystallographic data. Small molecule databases focus the design of agents to interact with these targets, and further packages of molecular dynamics and docking software study and fine-tune the series of agents to optimize binding, well before any molecule is synthesized in the real world[61].

This software design process is evolving extremely quickly and has been combined with combinatorial chemical synthesis (CCS). CCS began in the 1980s as a form of chemical 'mass production' to exploit many basic chemical structures, like modular platforms seen in the car industry. This meant that hundreds of analogues could be designed and then robotically synthesized in days to make small quantities of relatively impure compound. These were intended to be rapidly screened, and if there was a 'hit', i.e. pharmacodynamic activity, the effect could be fine-tuned in days rather than years. Now, CCS libraries of millions of structures can be manipulated and processed in silico[62] and combined with the docking and dynamics software to produce lead libraries, so carrying out the vast majority of the 'leg work' *in silico* before a single compound is actually made in the laboratory. This saves enormous amounts of money and time and when combined with *in vitro* pharmacodynamic and toxicodynamic testing high-throughput systems, lead structures have a high degree of refinement before they even reach experimental animals and human *in vitro* screens. Naturally, prospective drugs must be made in a real laboratory and evaluated using the *in vivo* and *in vitro* screens. As soon as these data are generated, they are 'invested' back into the databases and the *in silico* systems learning software builds up a picture of what does and does not work to optimize the algorithms and the learning software.

Drug metabolism

The relatively small number of CYPs relevant to human drug clearance at first glance makes the task of exploring and predicting biotransformation easier than perhaps locating new target macromolecules in drug discovery. Indeed, just designing *in silico* models for CYP3A4 and CYP2C9 comfortably has applications towards half of all current prescription drugs[63]. The combination of molecular descriptors, quantitative structure activity and software analysis allows very sophisticated predictions to be made on the CYPs and their major allelic variants involved with a candidate drug's clearance, all the basic enzymatic parameters of K_m and V_{max}, alongside all the possible exploration of inhibition. However, whilst hard and soft drug design that we touched on earlier suggests how far the drug industry has progressed from 'passive' drug development, there remains the occasional possibility that a novel agent may be cleared by a biotransforming system which does not have the huge database and *in silico* processing behind it that the CYPs do. Knowledge of many conjugation pathways is currently nowhere near that of the CYPs and some as we noted have not yet been crystallised, although progress may be driven forward when a very potent and useful agent is identified that happens to be cleared by a relatively obscure pathway.

Global in silico approaches

Of course, 'global' can have two meanings, in terms of the whole process of developing a drug, as well as developing it at multiple sites around the planet for the world market. Software platforms have now been refined and expanded to the extent that they are designed to actually assist in key decisions made at various stages in the candidate drug's progress through the pipeline. Packages such as Simcyp® Simulator originally conceived in the United Kingdom (University of Sheffield) are now part of larger worldwide commercial enterprises. The Simulator integrates pretty much all there is to find out about a drug, based on all the data accrued from the myriad different screens, to make predictions at whatever stage is required. The most sophisticated simulator software also incorporates allometric algorithms that process all the data from every *in vitro* and *in vivo* model and platform employed in the development of the drug to that point. The Simulator uses all this processing power, real multi-model data and key basic assumptions to run experiments, scenarios and trials virtually. This allows the Simulator to predict specific outcomes in animal models and eventually different patient groups (such as paediatrics) in terms of basic pharmacodynamics and pharmacokinetics. Such predictions are intended to assist with clinical trial design, but *in silico* simulators also aim to predict possible adverse reaction risks in terms of patients and clinical situations. Simulator data can also be used to track, understand and assemble AOPs, from both prediction and forensic standpoints. Indeed, ideally, certain patient groups would be excluded from the use of a particular drug based on these predictions, preventing them from adverse reactions before they even happen and saving untold suffering and costs. All the major drug manufacturers and developers are using this general approach, either in house, or through contracts with *in silico* suppliers and developers. Certainly once drugs reach the clinic, models such as Simcyp® Simulator are able to predict ranges of drug concentrations at a given time and dosing interval that are nearly 90% accurate[64].

Indeed, there are several *in silico*-based approaches that are part of the clinical trial design, assessment, and output interpretation. These include model-informed precision dosing (MIPD), which includes physiologically based pharmacokinetic (PBPK) modelling and simulation. These approaches can reduce the toxicity of drugs and improve efficacy[65-67]. Eventually, it will be possible for every patient in cancer chemotherapy or anti-HIV treatment to have their therapy managed through continuous software analysis to eliminate the risks of adverse reactions. Overall, the drug industry is still investing strongly in these systems from a variety of sources to minimize costs and allow for larger, more successful and safe drug pipelines.

What is also important, is that regulatory authorities such as the US FDA are content to receive *in silico* data simulations, provided they are given sufficient background information about the packages used, their key assumptions, and how real data are used to support these assumptions. In addition, they require detailed information on the simulations, how they are conducted, as well as their status and significance in the overall submission. Whilst careful not to endorse specific proprietary packages, the FDA is convinced that *in silico* simulations have a place in drug development, trial design, and prediction of possible adverse reactions[68].

A.8 Summary

Whilst there are so many hugely exciting biological drugs, peptides, and antibodies available, small molecules retain their importance because they are both our servants and our potential enemies, as we have seen. Since our bodies have a vast range of systems modulated by our own small molecules, low-molecular-weight drugs that often mimic them are also going to be around for some time to come. Even within the enormous complexity of the drug development machine, biotransformation remains the strongest predictor of whether a small-molecule drug will stay in its therapeutic window and treat the condition successfully without disappearing or accumulating. Hence, knowledge and understanding of drug metabolism will continue to be a vital component in drug development for the foreseeable future.

Figure A.1 tries to summarize all the major components of the early stages of drug development that involve the metabolism and toxicology of prospective therapeutic drugs.

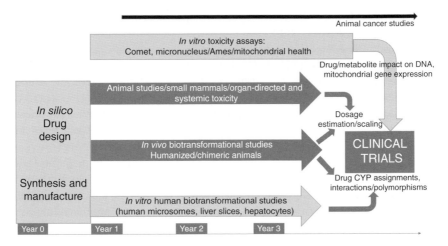

Figure A.1 Broad summary of the role of techniques based on drug metabolism and drug-mediated toxicity in the preclinical stages of drug development towards the commencement of clinical trials

References

1. Glybera (alipogene tiparvovec) European Medicines Agency October 2017. https://www.ema.europa.eu/en/medicines/human/EPAR/glybera
2. Collins JM, Simon KI, Tennyson S. Drug withdrawals and the utilization of therapeutic substitutes: The case of Vioxx. Journal of Economic Behavior & Organization 86, 148–168, 2013.
3. Graham DJ, Campen D, Hui R, et al. Risk of acute myocardial infarction and sudden cardiac death in patients treated with cyclo-oxygenase 2 selective and non-selective non-steroidal anti-inflammatory drugs: nested case-control study. Lancet 365, 9458, 475–481, 2005.
4. Filipova E, Uzunova K, Kalinov K, et al. Pioglitazone and the risk of bladder cancer: A Meta-. Diabetes Ther. 8(4), 705–726, 2017.
5. Filion KB, Yin H, Yu OH, et al. Pioglitazone use and risk of bladder cancer: population based cohort study. BMJ 352, i1541, 2016.
6. McNaughton R, Huet G, Shakir S. An investigation into drug products withdrawn from the EU market between 2002 and 2011 for safety reasons and the evidence used to support the decision-making. BMJ 4, e004221, 2014.
7. Denayer T, Stöhr T, Roy MV, Animal models in translational medicine: Validation and prediction. European Journal of Molecular & Clinical Medicine 2(1), 5–11. 2014.
8. Kebamo S, Tesema S, Geleta B. The Role of Biotransformation in Drug Discovery and Development J. Drug Metab. Toxicol. 6(5), 1–13, 2015.

9. Buchwald P, Bodor N. Recent advances in the design and development of soft drugs. Pharmazie 69, 403–413, 2014.

10. Zhang Z, Tang W. Drug metabolism in drug discovery and development Acta Pharmaceut. Sinica B 8(5), 721–732, 2018.

11. Cummings MA, Proctor GJ, Stahl SM. Deuterium Tetrabenazine for Tardive Dyskinesia. Clin. Schizophr. Relat. Psychoses. 11(4), 214–220, 2018.

12. Henderson VC, Kimmelman J, Fergusson D, et al. Threats to validity in the design and conduct of preclinical efficacy studies: a systematic review of guidelines for *in vivo* animal experiments. PLoS Med. 10, e1001489, 2013.

13. Esch EW, Bahinski A. Huh D. Organs-on-chips at the frontiers of drug discovery. Nat. Rev. Drug Discov. 14(4), 248–260, 2015.

14. Ware BR, McVay M, Sunada WY, et al. Exploring chronic drug effects on microengineered human liver cultures using global gene expression profiling. Tox. Sci. 157(2), 387–398, 2017.

15. Lee SML, Schelcher C, Laubender RP, et al. An Algorithm that Predicts the Viability and the Yield of Human Hepatocytes Isolated from Remnant Liver Pieces Obtained from Liver Resections. PLoS ONE 9(10): e107567, 2014.

16. Zuberi A, Lutz C. Mouse models for drug discovery. can new tools and technology improve translational power? ILAR J. 57(2), 178–185, 2016.

17. Coleman MD Russell RM, Tingle MD. et al. Inhibition of dapsone-induced methemoglobinemia by cimetidine in the presence of trimethoprim in the rat. J. Pharm. Pharmacol. 44, 114–118, 1992.

18. Hagmar L Törnqvist M, Nordander C, et al. Health effects of occupational exposure to acrylamide using hemoglobin adducts as biomarkers of internal dose. Scand. J. Work Environ. Health 27, 219–226, 2001.

19. Omura T, Sato R. The carbon monoxide-binding pigment of liver microsomes. I. evidence for its hemoprotein nature. J. Biol. Chem. 239, 2370–2378, 1964.

20. Wang Z, Fang Y, Teague J, et al, *In vitro* metabolism of oprozomib, an oral proteasome inhibitor: role of epoxide hydrolases and cytochrome P450s. DMD. 2017.

21. Coleman, MD; Smith, SN; Kelly, DE; et al. Studies on the toxicity of analogues of dapsone in-vitro using rat, human and heterologously expressed metabolizing systems. J Pharm. Pharmacol. 48, 945–950, 1996.

22. Ung YT, Ong CE, Pan Y. Current high-throughput approaches of screening modulatory effects of xenobiotics on cytochrome P450 (CYP) enzymes. High-Through. 7(29), 1–11, 2018.

23. Chen Y, Liu L, Nguyen, K. Utility of intersystem extrapolation factors in early reaction phenotyping and the quantitative extrapolation of human liver microsomal intrinsic clearance using recombinant cytochromes P450. Drug Metab. Disp. 39, 373–382, 2011.

24. Soldatow VY, LeCluyse EL, Griffith LG, et al. *In vitro* models for liver toxicity testing Toxicol Res (Camb). 2(1), 23–39, 2013.

25. Shulman M, Nahmias Y. Long-term culture and coculture of primary rat and human hepatocytes. Methods Mol Biol. 945, 287–302, 2013.

26. Liberti MV, Locasale JW, The Warburg effect: how does it benefit cancer cells? Trends Biochem Sci. 41(3), 211–218, 2016.
27. Bachmann A, Moll M, Gottwald E, et al. 3D cultivation techniques for primary human hepatocytes. Microarrays 4, 64–68, 2015.
28. Kvist AJ, Kanebratt KP, Walentinsson A, et al. Critical differences in drug metabolic properties of human hepatic cellular models, including primary human hepatocytes, stem cell derived hepatocytes, and hepatoma cell lines. Biochem. Pharmacol. 155, 124–140, 2018.
29. Sauer V, Roy-Chowdhury M, Guha C, et al. Induced pluripotent stem cells as a source of hepatocytes. Curr. Pathobiol. Rep. 2(1), 11–20, 2014.
30. Kafka, F. The Castle. Translated by JA Underwood. Originally printed 1926; reprinted by Penguin Twentieth Century Classics (ISBN 978-0-14-018504-1), 1997.
31. Woehrling EK, Parri HR, Tse EHY, Hill EJ, Maidment ID, Fox GC, Coleman MD. A predictive *in vitro* model of the impact of drugs with anticholinergic properties on human neuronal and astrocytic systems. PLoS ONE 10(3), e0118786, 2015.
32. Ucciferri N, Sbrana T, Ahluwalia A. Allometric scaling and cell ratios in multi-organ *in vitro* models of human metabolism. Front. Bioeng. Biotechnol. 2(74), 1–10, 2014.
33. Izumi S, Nozaki Y, Kusuhara H, et al. Relative activity factor (RAF)-based scaling of uptake clearance mediated by organic anion transporting polypeptide (OATP) 1B1 and OATP1B3 in human hepatocytes. Mol. Pharmaceut. 15(6), 2277–2288, 2018.
34. Lee, H. Genetically engineered mouse models for drug development and preclinical trials. Biomol. Ther. (Seoul). 22(4), 267–274, 2014.
35. Karlgren M, Simoff I, Keiser M, et al. CRISPR-Cas9: A new addition to the drug metabolism and disposition tool box. Drug Metab. Disp. 46, 1776–1786, 2018.
36. Lu J, Shao Y, Qin X et al. CRISPR knockout rat cytochrome P450 3A1/2 model for advancing drug metabolism and pharmacokinetics research. Scient. Rep. 7, 42922, 2017.
37. Naritomi Y, Sano S, Ohta S. Chimeric mice with humanized liver: Application in drug metabolism and pharmacokinetics studies for drug discovery. Drug Metabolism and Pharmacokinetics. 33, 31–39, 2018.
38. Scheer NM Wilson ID, A comparison between genetically humanized and chimeric liver humanized mouse models for studies in drug metabolism and toxicity. Drug Disc. Tod. 21, 250–263, 2016.
39. Hasegawa M, Kapelyukh Y, Tahara H, et al. Quantitative prediction of human pregnane x receptor and cytochrome p450 3a4 mediated drug-drug interaction in a novel multiple humanized mouse line. Mol. Pharm. 80, 518–528, 2011.
40. Nagel DA, Hill EJ, O'Neil J, Mireur A, Coleman MD, The effects of the fungicides fenhexamid and myclobutanil on SH-SY5Y and U-251 MG human cell lines. Env. Tox. Pharmacol. 38, 968–976, 2014.
41. Elzagallaai AA, Rieder MJ. *In vitro* testing for diagnosis of idiosyncratic adverse drug reactions: Implications for pathophysiology. Br. J. Clin. Pharmacol. 80(4), 889–900, 2015.

42. Spielberg SP. Acetaminophen toxicity in human lymphocytes *in vitro*. J Pharmacol. Exp. Ther. 213(2), 395–398, 1980.

43. Coleman MD, Brokenfridge, AM, Park BK. Bioactivation of dapsone to a cyto-toxic metabolite by human hepatic-microsomal enzymes. Brit. J. Clin. Pharmacol. 28, 389–395, 1989.

44. Riley RJ, Roberts P, Coleman MD et al. Bioactivation of dapsone to a cytotoxic metabolite – *in vitro* use of a novel 2 compartment system which contains human tissues. Brit. J. Clin. Pharm. 30, 417–426, 1990.

45. Beedanagari S, Vulimiri SV, Bhatia S, et al. Biomarkers in toxicology. Chapter 43, Genotoxicity Biomarkers: Molecular Basis of Genetic Variability and Susceptibility. 729–742, Academic Press, 2014.

46. Ames BN, Durston WE, Yamasaki E et al. Carcinogens are mutagens-simple test system combining liver homogenates for activation and bacteria for detection. Proc. Nat. Acad. Sci USA 70, 2281–2285, 1973.

47. Honda H, Minegawa K, Fujita Y, et al. Modified Ames test using a strain expressing human sulfotransferase 1C2 to assess the mutagenicity of methyleugenol. Genes Environ. 38(1), 1–5, 2016.

48. Dixit M, Kumar A, Chapter 4, In vitro gene genotoxicity test methods. In vitro Toxicology 67–89, Academic Press, 2018.

49. Bhalli JA, Shaddock JG, Pearce MG, et al. Sensitivity of the pig – an assay for detecting gene mutation in rats exposed acutely to strong clastogens. Mutagenesis 28(4), 447–455, 2013.

50. Calini V, Urani C, Camatini M. Comet assay evaluation of DNA single- and double-strand breaks induction and repair in C3H10T1/2 cells. Cell. Biol. Toxicol. 18, 369–379, 2002.

51. Vasquez M. Recommendations for safety testing with the *in vivo* comet assay. Mut. Res. 747, 142–156, 2012.

52. Hayashi M, The micronucleus test – most widely used in vivo genotoxicity test – Genes and Environ. 38, 18, 1–6, 2016.

53. Alexander-Dann B, Pruteanu LL, Oerton E, et al. Developments in toxicogenomics: understanding and predicting compound-induced toxicity from gene expression data. Mol Omics. 14(4), 218–236, 2018.

54. Ankley GT, Bennett RS, Erickson RJ, et al. Adverse outcome pathways: a conceptual framework to support ecotoxicology research and risk assessment. Environ. Toxicol. Chem. 29, 730–741, 2010.

55. Vinken M. The adverse outcome pathway concept: a pragmatic tool in toxicology. Toxicology 4(312), 158–165, 2013.

56. Bumgarner R. DNA microarrays: Types, applications and their future. Curr. Protoc. Mol. Biol. 22, 1, 2013.

57. Liguori MJ, Anderson MG, Bukofzer S, et al. Microarray analysis in human hepatocytes suggests a mechanism for hepatotoxicity induced by trovafloxacin. Hepatology 41(1), 177–186, 2005.

58. Shi L, Campbell G, Jones WD, et al. The MicroArray Quality Control (MAQC)-II study of common practices for the development and validation of microarray-based predictive models. Nat. Biotechnol. 28, 827–838, 2010.

59. Seesi SA Duan F, Mandoiu IA, et al. Chapter 10, Genomics-guided immunotherapy of human epithelial ovarian cancer. In Translational Cardiometabolic Genomic Medicine. 237–250, Academic Press, 2016.

60. Bowyer JF, Tranter KM, Hanig JP, et al. Evaluating the stability of RNA-Seq transcriptome profiles and drug-induced immune-related expression changes in whole blood. PLoS ONE 10(7), e0133315, 2015.

61. Bai Q, Li LL Liu S, et al. Drug design progress of in silico, *in vitro* and *in vivo* researches. In-vitro In-vivo In-silico J. 1, 1, 2018.

62. Liu R, Li XC Lam KS, Combinatorial chemistry in drug discovery. Curr. Op. Chem. Biol. 38, 117–126, 2017.

63. Nembri S, Grisoni F, Consonni V et al. In silico prediction of cytochrome P450-drug interaction: QSARs for CYP3A4 and CYP2C9. Int. J. Mol. Sci. 17, 914, 1–19, 2016.

64. Rowland A, van Dyk M, Hopkins AM, et al. Physiologically based pharmacokinetic modeling to identify physiological and molecular characteristics driving variability in drug exposure. Clin. Pharmacol. Ther. 104(6), 1219–1228, 2018.

65. Saechang T, Na-Bangchang K, and Karbwang J. Utility of physiologically based pharmacokinetics modeling (PBPK) in oncology drug development and its accuracy: A systematic review. J. Clin. Pharmacol, 74, 1365–1376, 2018.

66. Yoshida K, Budha N, and Jin JY. Impact of physiologically based pharmacokinetic models on regulatory reviews and product labels: Frequent utilization in the field of oncology. Cpt-journal.com, 101(5), 597–602, 2017.

67. Rowland A, van Dyk M, Mangoni A, et al. Kinase inhibitor pharmacokinetics: Comprehensive summary and roadmap for addressing inter-individual variability in exposure. Op. Drug Metab. Toxicol. 13(1), 31–39, 2017.

68. Physiologically based pharmacokinetic analyses – Format and content guidance for industry U.S. FDA (CDER), August 2018, Clinical Pharmacology https://www.fda. gov/downloads/Drugs/GuidanceComplianceRegulatoryInformation/Guidances/ UCM531207.pdf.

Appendix B
Metabolism of Major
Illicit Drugs

B.1 Introduction

It is certain that the use of various plant- and animal-sourced chemicals by humans for their pleasurable effects predates recorded time. These agents have been and continue to be manufactured, distributed, directly ingested, drunk, smoked, or inhaled, using a variety of processes. Some abused drugs have vital clinical roles, such as the opiates; others, such as cocaine, most amphetamines, and PCP, are of little or no current clinical use. Others show tantalizing glimpses into some therapeutic promise, whilst yet more are little more than a chemical death trap. Many of us are not content with what we see, how we feel, and what we perceive at any given time. Hence, we seek some form of licit or illicit chemical stimulation. We will pay any price, take any risk, and go to any length to obtain it, whilst ignoring the personal costs and the horrors of the drug traffickers' firm hand on the tiller.

Whilst in some areas cannabis has joined alcohol and tobacco in legality, judicial tolerance to other controlled substances shows no sign of weakening and governments continue to devote vast resources, usually fruitlessly, trying to stem the unquenchable thirst of their citizens for mind-altering substances. Whether or not a drug-mediated experience is right or wrong, good or bad, and indeed, whether it is real or unreal, is outside the scope of drug metabolism. However, we will discuss some of the pharmacological, toxicological, and biotransformation processes behind the user's personal chemical odyssey, however long it might last.

Human Drug Metabolism, Third Edition. Michael D. Coleman.
© 2020 John Wiley & Sons, Inc. Published 2020 by John Wiley & Sons, Inc.

B.2   Opiates

These agents include those extracted from opium (morphine) or are close structural analogues of morphine, such as diacetylamorphine (heroin), codeine, and dihydrocodeine, oxycodone, hydrocodone, and buprenorphine. Amongst the many synthetic opiates, these include methadone, fentanyl, and pethidine (meperidine). Most of the morphine derivatives are converted to morphine or morphine-like agents. Others, such as fentanyl, methadone, and buprenorphine, act directly on opiate receptors. Whilst opiates are the mainstay of serious pain relief and a great deal of effort is invested in trying to optimize their effects in patients, up to 70% of cancer sufferers have inadequate pain control[1]. Most opiates are quite crude and potentially dangerous drugs, as they are potent, addictive, and lack selectivity, causing many intolerable side effects. In addition, respiratory depression is an ever-present danger.

Morphine

The receptor classification of opiates can be confusing at first glance. Currently, you might see the main three receptors (G-protein coupled) referred to by the Greek letters μ, κ, and δ; also, MOP (mu-OPiate receptor), KOP (kappa-OPiate receptor), and DOP (delta-OPiate receptor)[2]. A number of endogenous ligands (enkephalins and endorphins) are intended to operate these receptors, and morphine, its relatives, and the synthetic opiates act with varying potencies on MOPs (μ-opioid receptors), but can also work on the other receptors[2]. Clinically, opiates are the most effective agents of relief available for most types of pain, although they are much better for short-term rather than longer-term chronic usage[2] and they can be inadequate for neuropathic pain. They have a wide range of unpleasant side effects and of course the euphoric aspects of morphine and its derivatives' effects have been the basis for their abuse and they induce tolerance and dependence. Addiction to morphine-like compounds is a worldwide problem, which is as acute in the countries that grow opium poppies as it is in the developed world. Morphine is poorly bioavailable orally, so addicts tend to use the intravenous route. This makes them vulnerable to the well-publicised, clichéd, but nonetheless real dangers associated this route. These include infections from shared needles and inadequate sterilisation, as well as the risk of fatal respiratory depression in overdose[2].

The pharmacological potency of morphine resides in several aspects of its structure (Figure B.1). The tertiary nitrogen and its methyl group are required for analgesic activity and changes to either moiety will reduce activity – indeed, substitution of the methyl group with larger groups yields antagonists. Substitutions to the 3-hydroxyl group on the phenolic group usually result in loss of activity, whilst changes to the 6-hydroxy group can increase activity. Aside from the poor oral absorption, it is not lipophilic enough to enter the brain very rapidly[2].

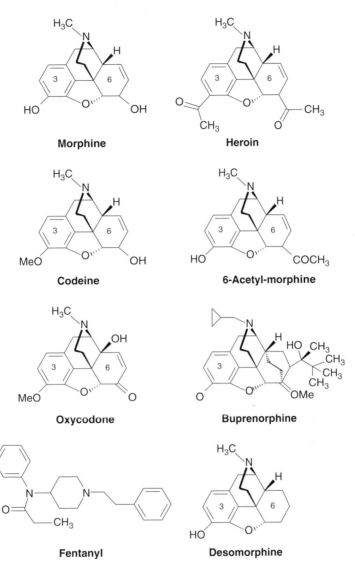

Figure B.1 Structures of some morphine-based and synthetic opiates

Morphine metabolism

Just over half of a morphine dose will be cleared to the 3-glucuronide, whilst 10% appears as the 6-glucuronide. Of the remainder, about 5% forms a 3-sulphate and small amounts of diglucuronides (3,6-diglucuronide) are also formed. Less than 1% of morphine is methylated to codeine (3-methoxymorphine), whilst around 10% of the dose appears in urine unchanged. In addition, N-dealkylation occurs at the tertiary nitrogen, leading to an N-oxide as well as the formation of normorphine[3]. The liver is probably the major site of morphine metabolism, as cirrhotic patients show a substantially increased half-life of the parent drug. Renal failure also increases morphine half-life in humans, and when renal function declines, the drug and its 6-glucuronide accumulate unless the dose is reduced. Although the 3-glucuronide has no analgesic activity it is thought to be rather toxic, having a stimulant effect as well as causing irritability and even hallucinations. By contrast, the 6-glucuronide is believed to be up to 50-fold more potent as an analgesic compared with the parent drug. Morphine 6-glucuronide penetration into the brain is poor, but the OATPs are believed to mediate its penetration and P-gp its efflux, although it is actually formed in the brain itself anyway[3]. For some years the 6-glucuronide has been promoted as a superior analgesic to morphine, and some authors regard morphine as a prodrug, as the vast majority of the drug's analgesic action is due to the 6-glucuronide[4]. The relationships between the clearance of morphine and its analgesic activities in different groups of patients remain to be fully explored, although UGT2B7 is responsible for the glucuronidation of the drug[5]. This UGT is polymorphic and subject to inhibition by NSAIDs such as mefenamic acid. It is thought that differences between patients' morphine responses may be linked with UGT2B7 expression, although the opposing effects of the glucuronides make this difficult to study. In addition, there is considerable variation in the *OPRM1* gene that codes for the opiate receptors, which has a major impact on sensitivity to the opiate effect[3]. Some studies have suggested that induction of UGT2B7 might reduce the effectiveness of morphine, as it increases the toxic 3-glucuronide at the expense of the parent drug, although more of the 6-glucuronide should also be formed. Several studies have shown rifampicin, an inducer of UGTs, to erode the analgesic effect of morphine, linked with an induction effect on P-gp[6]. If so, then rifampicin will diminish absorption, but also induce P-gp in the blood–brain barrier and reduce the 6-glucuronide's brain penetration also. Interactions with UGT metabolism through inhibition or competition can prolong morphine's residence in the body and increase its respiratory depression effects[7].

Heroin

Diacetylmorphine, or heroin, was first synthesized in 1874 with the aim of increasing the potency of morphine; it was even marketed in the early twentieth century as a 'nonaddictive' cough suppressant, a role in which it is fifty-percent effective. Unfortunately, it has become the choice of more than 70% of opiate abusers worldwide[8], as its onset or 'hit' is as rapid as it is potent. As we all know from fabulously gruesome cinematic depictions[9], heroin is injected, smoked, or snorted, according to the consistency and purity of the available drug. It can range in appearance from pharmaceutical-grade white powder all the way to the black-tar version manufactured in Mexico and popular in the United States. With good-quality pure heroin, obviously, intravenous dosing is the most efficient method of administration, but the dragon-chasing inhalation method allows very impure drug to be used, despite poor bioavailability. The half-life of heroin in humans is around the 5–7 minute mark, indicating rapid clearance. In addition, the two acetyl groups make the drug much more lipophilic than morphine so it reaches the CNS more rapidly and is better orally absorbed than morphine[3].

Once in the CNS after rapid penetration of the blood–brain barrier, heroin can be hydrolysed quickly by various hepatic, plasma, and CNS esterases. These include butyrylcholinesterase (BChE), human carboxylesterase-1 (hCE1) and acetylcholinesterases. Heroin is a prodrug, as it shows little binding to opiate receptors, but if the 3-acetyl group is hydrolysed away, the 6-acetylmorphine does bind, and it is probably more potent than morphine. The 6-acetyl derivative is then hydrolysed to morphine. The 6-acetyl derivative of morphine has a half-life of 6–25 minutes in blood and is stable in urine and has been used for many years as conclusive proof of relapsed heroin abuse in those enrolled on methadone programmes[10]. This is because no other opiate is metabolised to the 6-acetyl derivative.

The problem with 6-acetylmorphine as a marker of heroin abuse is its short half-life, and when addicts finally expire, it is necessary to determine the source of the morphine metabolites in their systems. There might not be enough 6-acetyl derivative to detect, depending on the time elapsed since death and the state of decomposition. An ingenious way around this is to exploit the fact that most illicit heroin contains codeine that is not a heroin metabolite. In a postmortem blood sample, if the morphine codeine ratio is greater than one, this is taken in the absence of 6-acetylmorphine to be evidence of a death from heroin intoxication[11].

The metabolic profile of heroin use is the same as morphine, with the exception of the 6-acetyl derivative. Although heroin addicts are exposed to many different substances and are subject to various systemic abnormalities (e.g. effects on glucose tolerance), there is evidence that heroin addicts form much more of the pharmacologically active morphine 6-glucuronide than the inactive 3-glucuronide, in comparison with non-addicts receiving heroin[12]. Interestingly, aside from the usual risks of hepatitis and HIV, intravenous users of black tar heroin rapidly incapacitate their exposed veins and are reduced to subcutaneous administration, which can protect from HIV[13], but can lead to the development of botulism, from spores introduced when the heroin is cut with whatever is at hand, such as boot polish or soil.

Desomorphine 'krokodil'

In Russian, крокодил, or in Ukrainian, крокодила, has become in a relatively short space of time the tabloid and media 'zombie shock horror' drug, Such hysterical reports are often not taken very seriously, but it seems with krokodil/krokodyla, as rendered phonetically in English, they may be understating the truth. As with many illicit drugs, its popularity originates in the darkest of desperation, in that addicts turn to it when heroin is unavailable or prohibitively expensive.

I think we can all agree that 'home-made' is terrific for cakes, chocolates, and various fruit preserves, but from a pharmaceutical standpoint there are some serious issues of purity and contamination that are often somewhat inadequately addressed. Krokodil is relatively easy to make at home and the instructions are readily available on the internet. Indeed, experience with illicit methamphetamine manufacture is helpful, but not essential. To make some krokodil, it is necessary to visit several local pharmacies (with appropriate multiple disguises) to buy codeine tablets. Next, it is important to assemble some other ingredients that would normally be found in a well-stocked kitchen, larder, or garage. You need hydriodic acid, red phosphorus, petrol/gasoline (leaded for extra 'body' or unleaded will do at a pinch) or paint thinners, some serious drain cleaner, and battery acid[14]. Now, codeine must be extracted from the tablets with the solvent, then the two-step synthesis is begun, and less than an hour later, the product is ready to inject. The product is very acidic, but if there is any baking soda available, or failing that, cigarette ash[14], it can be buffered somewhat.

The major component of the product is desomorphine (Figure B.1), an opiate that appeared before World War II and was available as Permonid in Switzerland, but it is 10-fold more potent than morphine, so it was withdrawn in 1952 due to the usual respiratory depression issues, but perhaps the Swiss might find other uses for it today. The krokodil product contains desomorphine and considerable number of its derivatives, as well as various codeine analogues[15]. The main issue with injecting krokodil is its acidity and the catastrophically toxic impurities, such as the phosphorus and iodine derivatives. Filtering it with tampons is apparently not sufficient to address the problem[15]. The result is that the user's skin gradually turns into a lumpy, loathsome and passable impression of, well, a crocodile. The user generally progresses towards gangrene, loss of limbs, and jaw necrosis, the latter problem linked to the phosphorus (phossy jaw)[14,16]. There is a very long list of other devastating neurological effects, and any online image search is likely to be a short and unpleasant one. This all seems baffling to outsiders, as even the first time it is used, the local pain at the site of injection is apparently very significant. However, aside from the pleasure aspect, the whole krokodil experience appears, even by the standards of opiate addiction in general, to be one of industrial strength self-harm. Perhaps the condition is a hideous distortion of reward-seeking behaviour, so it makes the user view their wounds in the same way an artisan might view an item they just handcrafted. The fascination with injecting krokodil is probably because smoking it may not be an option due to the asphyxiatingly corrosive nature of the fumes. However, with attention to detail and care taken in the synthesis, some individuals can apparently inject krokodil without turning green and lumpy for many years. Despite the horror stories, it seems unlikely that this agent will sweep the world in the way crack cocaine did from the 1990s. This is a local drug for local people – nothing for you here!

Buprenorphine

Introduced in the late 1970s (Figure B.1), this semi-synthetic opiate is of interest pharmacologically, as it can act as a partial as well as full agonist on μ-opioid receptors (MOPs), but as antagonist on κ-receptors (KOPs). This means that it shows the usual complement of opiate effects and is 30 to 100-fold more potent than morphine, but at higher doses it is less likely to depress respiration and opiate effects are then diminished[2]. However, this is negated by the concurrent administration of benzodiazepines and other CNS depressants that potentiate its effects leading to fatalities. The drug still causes euphoric effects and is an efficient, long lasting (1–3 days) analgesic, although in around a third of patients it has a sedating effect. As it is subject to such high first-pass clearance, it is given sublingually[8] and has found extensive use in skin patches, although the cachexia due to terminal cancer may influence absorption, as a good friend of mine experienced some rather disturbing hallucinations due to a buprenorphine patch.

It has found increasing use in managing opiate withdrawal instead of methadone and some studies suggest it could be superior[17]. Interestingly, its partial agonist status means that it is not as physiologically dependence inducing as heroin or morphine and that the withdrawal is not quite as severe as from the major narcotics. Unfortunately, naloxone is not very effective at reversing severe respiratory depression if it does occur. This is believed to be due to more potent binding and reduced dissociation from the opiate receptors, which is an effect that also contributes to its long-lasting action. Where buprenorphine has been marketed in 4:1 preparations (Suboxone) with naloxone to abolish the euphoric effect when injected, relatively little abuse of that preparation occurs. This is in contrast to countries where buprenorphine was obtainable alone where abuse is considerable, sometimes with benzodiazepines[8]. The best the drug abuse fraternity can do for a street name appears to be 'bupe', although there are others. Buprenorphine is cleared predominantly by N-dealkylation to norbuprenorphine and the hydroxylation of the hydroxytrimethylpropyl sidechain also occurs prior to glucuronidation. There is a buprenorphine and norbuprenorphine 3-glucuronide formed and buprenorphine-3-glucuronide is the main urinary metabolite, by UGT2B7 and UGT1A1[18]. There are at least five hydroxylated derivatives formed by CYP3A4[18]. Concurrent efavirenz induced buprenorphine's clearance[19] and the drug will be subject to all the standard CYP3A4 inducers and inhibitors.

Oxycodone and hydrocodone

Oxycodone (Figure B.1) has been available for nearly a century and is a semi-synthetic opiate originally designed to be a less dangerous alternative to heroin, although it has nearly double the opiate effect of morphine. The less potent hydrocodone was synthesized a few years later and is structurally related to codeine, but is a much stronger analgesic. Both drugs are part of several popular analgesic combinations; oxycodone is found in Percocet, Percodan and OxyContin, a potent sustained release version known as 'hillbilly heroin' as it was a popular choice in high-poverty areas of the United States. Hydrocodone is found in Vicodin, which is often linked with more 'white collar' addiction, as well as media personalities, both real and fictional. Vicodin abuse (15–75 tablets daily) and unintentional overuse can cause sensorineural irreversible hearing loss[20]. Since the preparation contains 500 mg acetaminophen (paracetamol) as well as the 5 mg of hydrocodone, the opiate addition drives many into serious hepatic damage[21]. It is unknown why the hearing loss does not occur with all abusers with this agent and not with other opiates, but the only remedy is the somewhat drastic one of a cochlear implant[20]. Over 47,000 Americans died from opiate overdoses in 2016[22], which is actually 9,000 more than were shot dead in the same year[23]. A very high proportion of these opiate-related deaths are due to prescription synthetic opiates such as oxycodone and hydrocodone and the role of the pharmaceutical companies in this issue is as shameful as it is well-documented[24].

Oxycodone metabolism was covered in detail in Chapter 7.2.3. The drug is cleared by CYP3A4 and CYP2D6 forms oxymorphone, which is 14-fold more potent than the parent, but it is only formed from 10% of the dose[25], so the parent drug is seen as the main pharmacologically effective agent[26]. Indeed, CYP2D6 metaboliser status has pharmacokinetic effects in terms of the levels of oxymorphone, but this does not translate into any significant differences in terms of pain relief[27]. Interestingly, adjusted for weight, its plasma levels can be up to a quarter higher in women than in men, and this should be taken into account during prescribing[26]. The drug can accumulate significantly in patients with either severe liver disease or renal impairment, possibly reducing clearance by more than half.[26] Hydrocodone, unlike oxycodone, is cleared almost entirely by CYP2D6 and is subject to the usual issues related to this route. Interestingly, a minor fraction (<10%) of codeine is cleared to hydrocodone[26]. Inducers can almost wipe out oxycodone potency through CYP3A4 clearance to low-potency metabolites and UGT-mediated clearance of the active metabolites[25], although they will have less impact on hydrocodone. CYP inhibitors will cause the parent drugs to accumulate, just as in patients with liver and kidney problems.

Fentanyl and designer opiates

Although krokodil is a more recent and rather extreme example, there is at least a 40-year history of using various disguises, a basic knowledge of synthetic chemistry and structure–activity relationships, to home-make opiates. It does seem difficult to believe, but something far more horrifying than krokodil has been made and distributed already.

In 1976, Barry Kidston began home manufacture of 'synthetic heroin', which was a pethidine (meperidine; Demerol) analogue known as MPPP (1-methyl

4-phenyl-4-propionpiperidine). Mr Kidston unfortunately was a rather sloppy organic chemist; in retrospect, he would have benefitted from advice from Dr Richard Kemp, who received his Doctorate in Chemistry from Liverpool University, on how to make a very pure illicit drug (LSD) in industrial quantities. The resulting errors during the synthesis managed by Mr Kidston made MPTP (1-methyl 4-phenyl-1,2,5,6-tetrahydropyridine) appear, which is a protoxin than causes a permanent Parkinsonian condition in its users, as it is metabolised by MAO-B to MPP+, a potent and selective mitochondrially acting neurotoxin[28]. The memorable and indeed constructive advice from the motion picture *Scarface*[29] (Don't get high on your own supply) unfortunately was several years too late for Mr Kidston, who developed the Parkinsonian condition, which is a hallmark of the drug, and ended his life with cocaine less than two years later.

Home opiate design, synthesis and illegal distribution has tended to centre on the synthetic opiate fentanyl (Figure B.1) introduced in the 1960s. This agent is already more than a 100-fold more potent than morphine and is used in clinical practice in a variety of forms, including lozenges, various tablets, and transdermal patches, as well as many parenteral forms. Its rapid onset and potency makes it highly effective in chronic pain management as well as in surgical pain control[30]. It is abused due to the intensity of its euphoric effects. Fentanyl derivatives in general are cleared by CYP3A4 by N-dealkylation mainly to norfentanyl-like derivatives and hydroxylated metabolites[26]. Azoles markedly inhibit the drug's clearance, as will any CYP3A4 inhibitor. If fentanyl accumulates, respiratory depression is a real danger. Since the 1970s, methylated fentanyl derivatives have been made illicitly and are thousands of times more potent than heroin and their metabolism/clearance is virtually irrelevant as they may stop the addict breathing in a few minutes. Since 2005, of the 14 different unregulated opiates that have appeared in Europe, many are fentanyl derivatives[31].

One might question the wisdom of developing and marketing a legitimate new fentanyl derivative that was 5- to 10-fold more potent than fentanyl, and many apparently did. However, in November 2018, sufentanil[32] gained FDA approval and is delivered sublingually[33]. Sufentanyl is an order of magnitude less potent than carfentanyl; this agent is 10,000 times the potency of morphine[31] and is useful if you need to tranquillise inmates of Jurassic Park. Carfentanyl is intended for veterinary use, involving some sort of tranquilliser dart, but of course now figures in human deaths from overdoses[31]. Carfentanyl and remifentanil in a halothane aerosol were used in the Moscow Theatre Hostage Crisis of 2002[31]. The failure of the Russian authorities to communicate the nature of the agent to the medical personnel involved massively increased the number of deaths in that incident. The strongest opiate obtainable currently is cis-fluoro analogue of ohmefentanyl, which is nearly twice the potency of carfentanyl and about 18,000 times more potent than morphine[34] – because we need that.

Methadone

This synthetic opiate (Figure B.2) was first synthesized in Germany during World War II in response to opiate shortages. Although it does not resemble other opiates chemically, it does act on opiate receptors and is cleared relatively slowly. Methadone has been unfairly stigmatized as its best-known use clinically is as a substitute for heroin, so aiding gradual withdrawal from addiction; this is because its half-life is much longer than other opiates. The drug is used in a variety of formulations worldwide, but in the United Kingdom, it is dispensed to addicts as the ubiquitous sweet green methadone 'mixture' (1 mg/mL), which is distinct from the less potent (2 mg/5 mL) methadone linctus[35]. The latter preparation is intended for severe coughing in terminally ill patients. The green methadone mixture was so coloured to prevent confusion with other medicines,

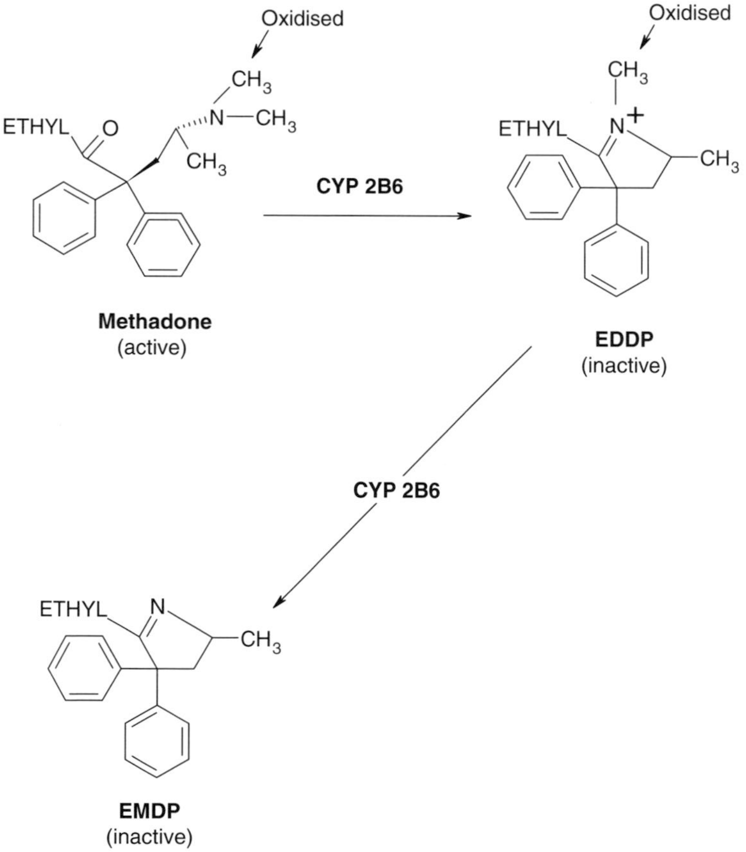

Figure B.2 Dual, sequential N-demethylation of methadone to inactive products by variants of CYP2B6 in the gut and liver

but tragically it has been irresistible to children on several occasions and now recovering opiate addicts are required to consume their methadone mixture whilst supervised in a pharmacy.

However, methadone can be very effective in cancer patients who are unable to control their pain with morphine or who have experienced severe morphine side effects. Although a Cochrane analysis[36] did not consider methadone any better than morphine, other studies show much less dose escalation with methadone compared with morphine in terminal cancer[37]. Interestingly, the drug has anticancer properties and can act as a chemosensitizer[38]. However, clinically, it is recommended that if there is no medical practitioner available who is well acquainted with its use, a different opiate should be used. There are several reasons for this; methadone clearance is astonishingly variable, with reported half-lives ranging from a few hours to more than six days[38]. Consequently, steady state may take more than a week to achieve and the analgesic effect can be dissociated from its half-life, which means that the patient may take the drug more frequently to maintain analgesia, but the drug quietly accumulates. Indeed, a quarter of fatal outcomes with methadone occur in the first two weeks of therapy[39].

As detailed in Chapter 7.2.3, methadone is pharmacologically active through its R-isomer but toxic through its S-isomer. Although it can cause *torsades des pointes* rarely (Chapter 5.7.2) the real danger, as with all opiates, is respiratory depression. However, it is believed that low initial doses, conversion of the dose to reflect the previous opiate tolerance, and gradual dosage escalation to achieve adequate analgesia can exploit the capabilities of this drug without the attendant toxicity. This approach has been summarized as 'start low, go slow'[40].

The biotransformation of methadone was outlined in Chapter 7.2.3) and the drug undergoes two sequential N-demethylations (Figure B.2); the first methyl group is removed from the tertiary nitrogen and the molecule essentially reacts with itself and immediately cyclizes, forming EDDP (2-ethylidene-1,5-dimethyl-3,3-diphenylpyrrolidine), which is then demethylated again to form EMDP (2-ethyl-5-methyl-3,3-diphenylpyraline). Both metabolites are inactive. Methadone clearance is a CYP2B6 issue primarily and it is highly vulnerable to change caused by the major various CYP inhibitors and inducers[41]. Clinically, it has long been known that inducers such as phenytoin and rifampicin can accelerate the clearance to the point that not only is analgesia lost, but also opiate withdrawal is precipitated[42]. The azoles (fluconazole, ketoconazole), and cimetidine and fluvoxamine all retard clearance-causing accumulation of methadone[43].

Methadone is a substrate for P-gp that governs its entry into the CNS as well as many other tissues. From Chapters 4.4.7 and 5.6 you will recall that P-gp can be induced or inhibited to a very significant degree. Considering that CYP2B6 is a very polymorphic human CYP, this has a strong influence over the wide variability in the clearance and pharmacological effects of methadone[44].

B.3   Cocaine

Cocaine originated in Latin America, and most of the world's supply still comes from this continent. The drug is a potent inhibitor of presynaptic dopamine and noradrenaline reuptake by transporter systems and the pharmacological effect is through the persistence of the neurotransmitters. Judging from its popularity and the cravings that it causes in addiction, to classify the drug merely as a stimulant is a spectacularly inadequate description. Cocaine also acts to block sodium channels and inhibit action potentials in peripheral nerves. During its use for its local anaesthetic effects, its vasoconstrictive actions are useful to restrict bleeding. However, many other safer analogues can be used that do not have the attendant cardiotoxic effects. Although its stimulant effects when smoked as the base (crack) or when injected intravenously occur within 10–15 seconds and are intense, this does not last more than 5–10 minutes. This is due to the extremely rapid clearance of the drug[45].

As with heroin, butyrylcholinesterase (BChE; Chapter 7.2.4) and human carboxyesterases (mainly hCE-1 and hCE-2) metabolise most of the drug (Figure B.3) as they are found in the plasma as well as most other organs, particularly the liver. The hCE enzymes are high capacity, low affinity and are found in the endoplasmic reticulum in the liver, although their presence in plasma appears variable. Drugs like cocaine are subject to widespread and rapid attack by several esterases, hence their short half-lives. About half the dose is de-esterified to ecgonine methyl ester, which is low in toxicity, as it has little vasoconstrictor effects or cardiotoxicity. Benzoylecgonine, unfortunately, is responsible for most of cocaine's acute toxicity, as it a more powerful vasoconstrictor and its half-life lasts days as it is a zwitterion, is resistant to hydrolysis and its presence in biological fluids betrays cocaine use[46].

Cocaine is optically active and the (−) isomer is the natural version, whilst the (+) isomer is found when it is synthesized illegally. The (+) is poorly cleared by esterases but 200 times less potent than the natural version.

CYP3A4 mediated demethylation to the pharmacologically active norcocaine[47] also leads to toxicity due to vasoconstriction and oxidative stress, as norcocaine is hydroxylated, eventually forming a reactive electrophilic nitrosonium ion (triple-bonded Nitrogen/oxygen group), which is one of the routes whereby cocaine use leads to hepatotoxicity[48]. Hence, if the cocaine user also consumes CYP3A inducers (rifamycins, St John's Wort and barbiturates) then the possibility for toxicity due to excessive norcocaine effects depends on how quickly the hCEs and serum cholinesterases can clear the parent drug (Figure B.3). If the individual has poor plasma cholinesterase performance (BChE polymorphisms, Chapter 7, Section 7.2.4), then the chances of them exhibiting cocaine toxicity (tremors, agitation, paranoia, high blood pressure, weak pulse) are much higher.

As we know, serious recreational drug users are always interested in new pharmacological challenges, and some have taken to self-administering insecticides, such as organophosphates or carbamates, which will block the esterases to

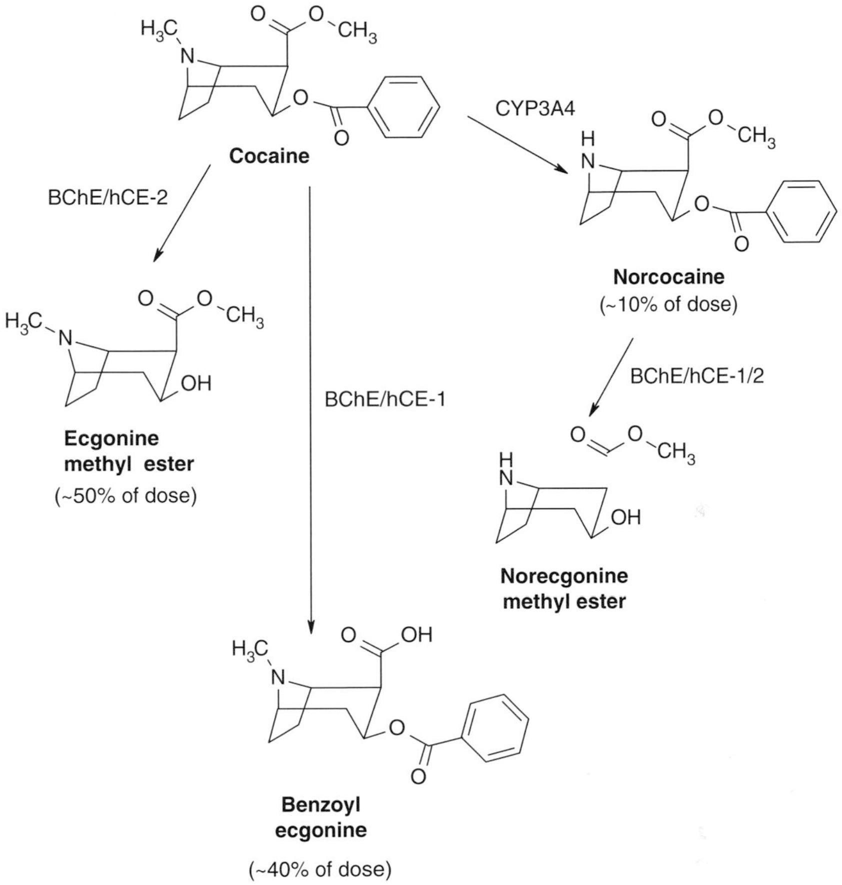

Figure B.3 Cocaine metabolism, showing the role of CYP3A4 in demethylation (toxic pathway) and human carboxylesterases (hCE-1 and 2) and butyrylcholinesterase (BChE). Esterases are found in virtually every tissue, clearing cocaine extremely rapidly

prolong cocaine's survival in the plasma[49]. Clearly, this could increase or prolong the high but equally go horribly wrong and induce severe toxicity, such as convulsions and arrhythmia, leading to fatal cardiotoxicity[49,50].

As recreational drug users often like to party with several drugs, it was noticed that cocaine's effects were 'improved' when the user had consumed considerable quantities of ethanol. This is because, rather unusually, the use of cocaine with ethanol leads to the formation of a separate and unique 'hybrid' metabolite of both agents, known as cocaethylene[51]. It is thought that human carboxylesterase 1 (hCE-1) transesterifies cocaine with ethanol to form cocaethylene. This is followed by clearance by either hCE-1 or 2 to the more usual cocaine ester metabolites (Figure B.4). Cocaethylene is thought to induce a more potent state

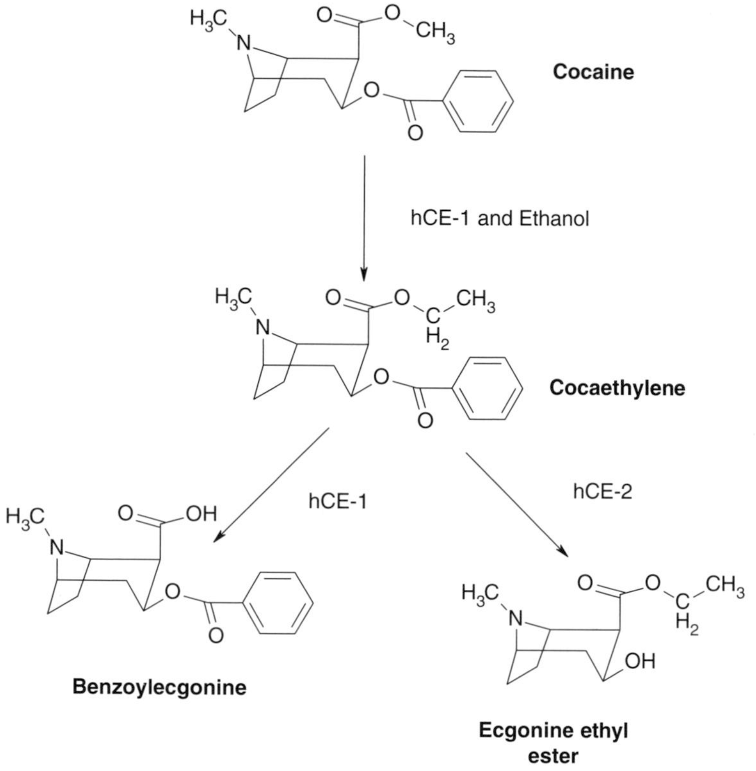

Figure B.4   Formation of cocaethylene by esterification with ethanol and the metabolism of the product to other cocaine esters and metabolites by esterases (hCE-1/2)

of euphoria than cocaine and the dysphoria (down) is much less unpleasant. Usefully for those who indulge, cocaethylene addresses the brevity of the parent drug's effects as its half-life is up to five-fold longer, so giving participants more bang for buck, so to speak. The downside is that it is more cardiotoxic and can induce convulsions at lower concentrations and predisposes to strokes[52].

After decades of research, there is no reliable treatment for chronic cocaine abuse and acute poisoning is difficult to manage. This has led to a variety of different approaches tested in animal models to try to attenuate the drug's effect with antibodies (made from haptens)[52] or by injecting esterases to accelerate drug removal and this latter idea has at the time of writing reached clinical trial[46]. The public perception of cocaine is markedly different from the public-toilet sleaze of heroin and the blue-collar desperation of OxyContin. Cocaine retains a sort of glossy, raffish air of decadence that is at odds with the medical issues in treating and rescuing users from its consequences. Although all addictions are destructive, cocaine in all its forms is particularly efficient at consuming its victims remorselessly, both physically and mentally.

B.4   Hallucinogens

LSD

Discovery

For those of us who have worked in laboratories, it is almost unimaginable to have made a compound that ignited a whole counterculture movement and remains influential worldwide to this day. In a book chapter in 1970 Dr Albert Hofmann clearly wished, as cliché runs, to 'set the record straight'[53] to emphasize that his synthesis of lysergic acid diethylamide in 1938 was not just a search for mind-altering substances, although ironically he did spend subsequent decades doing exactly that. He worked for the pharmaceutical concern Sandoz AG in Basle, Switzerland, and they were engaged in a systematic pharmaceutical development programme to make respiratory stimulants. LSD was similar in structure to a drug called nikethamide, a vintage and superseded stimulant[53]. Although the chemistry involved was a new procedure, the twenty-fifth analogue made (LSD-25) did not impress immediately, although some preliminary animal data indicated that it cause 'marked excitation', as well as some other not-unexpected effects[53].

In 1943, he thought it would be a good idea to revisit LSD-25 and whilst he was an excellent chemist, his health and safety awareness was rather of his time, and one day he suffered a series of what we now know to be psychedelic experiences that passed after a couple of hours. He realised he had basically contaminated and poisoned himself accidentally, but he had diligently recorded what he had been working with and narrowed the agent down to the d-isomer of LSD-25. Therefore, he decided at 4.20 p.m. on 16th of April, to ingest what he believed would be 'the lowest dose that might be expected to have any effect, i.e., 0.25 mg LSD.'[53] By 5.00 p.m., he had ceased to make much sense and he could not continue to make notes in his laboratory book. Because it was wartime, there were no cars available, so he had to ride home on his bicycle accompanied by a colleague and spent the next six hours or so inventing the first acid 'trip'. He did not fully recover until 14 hours had elapsed[54]. He admitted that what he took was effectively the first overdose of LSD, approximately five-fold more than was necessary[53].

Dr Hofmann suffered no obvious ill effects, living on to the exceptionally impressive age of 102. Sandoz organized trials in volunteers and wanted to market LSD to try to get it established in psychotherapy. The US military flirted briefly with the drug as a nonlethal weapon and entertainment figures such as Cary Grant used it under supervision as part of various experimental psychiatric treatments in the 1950s and 1960s. Increasing abuse and flying accidents ('Look, I can fly!!') led to its proscription in the late 1960s. LSD is thousands of times more potent than psilocybin in magic mushrooms (first synthesized in the laboratory by Dr Hofmann) and 5000 times stronger than mescaline. The d-isomer of LSD is active and the rest of the usually racemic mixture that is synthesized illicitly (iso-LSD) is inactive.

The seeds of the 'Morning Glory' plant contain structural relatives of LSD (lysergic acid amide and isolysergic acid amine), which have milder hallucinogenic effects, although these agents cause nausea and vomiting in most people, as well as occasional panic attacks and even a hangover. Of course, the good Doctor Hofmann was also the first to officially discover this effect also, as his life-long interest in psychedelics involved spending time in Mexico to research and synthesize a considerable number of different fungal hallucinogens[53].

Effects

LSD's action on the brain begins within 1 or 2 hours and may last for up to 12 hours. There are many descriptions available of its effects[55], and suffice to say, they range from radiant visual hallucinations of colours and sounds, overwhelming monomania (I just solved everything! Wow!), towards rather more disturbing effects such as reliving your own birth and total loss of self and loss of reality. Unfortunately, if it is not going well, like the 'mental rollercoaster', you cannot get off when you want to. The drug is usually taken in very low doses (25–100 µg) and is extensively metabolised, so is difficult to detect and is not usually included on standard laboratory drug screening systems. Excessive use quickly leads tolerance within only a few doses, although the chief danger from the drug is not really from direct toxicity, but from accidents caused by loss of fear and general craziness during intoxication[55].

After effects

Assuming an individual feels they have just touched the face of God, they might be forgiven for feeling that such an experience is beyond earthly, temporal and neurological consequences. However, sometimes it is and sometimes it isn't. The majority of users recover from its use, but some may not and LSD does have the tendency in some users to unmask psychotic reactions from which recovery is very slow if it occurs at all (acid casualties). None of us can know in advance if we are vulnerable to this risk, and significantly many hardened chemical abusers refuse to take it. Indeed, it has long been observed acid casualties present clinically in a very similar way to newly diagnosed schizophrenics, more of which later[56].

Assuming you do not become a casualty, there are other issues. Flashbacks are a celebrated badge of psychonautical voyaging, but it seems they can be rather difficult to live with if you are in your 30s and they have been going on since the last of your 30 trips was experienced when you were 18[57]. The prevalence of flashbacks, or 'hallucinogen-persisting perception disorder', is difficult

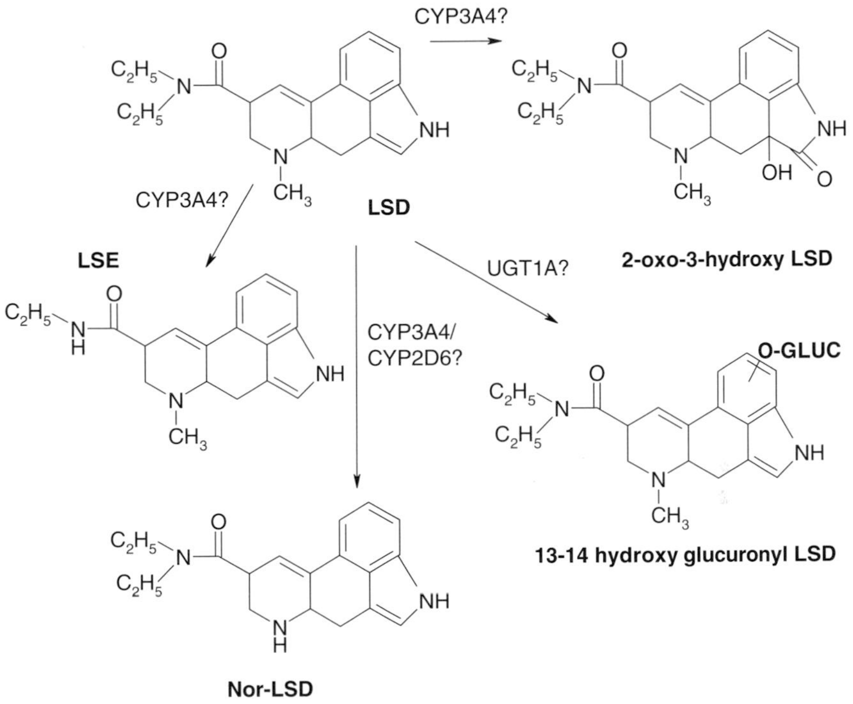

Figure B.5 Metabolism of LSD: formation of lysergic acid ethylamide (LSE), nor (N-demethylated) LSD and its hydroxylated derivatives

to determine, but might be in the 5–10% range of users, mainly in LSD veterans[57]. This can manifest itself in visual distortions of perception, like haloes, points of light and sensations of movements in the corner of the eye. Other issues include micropsia (things look much smaller than they are) as well as a sense of levitation. This amounts to a permanent deficit in visual processing and so is very difficult to treat. Indeed, in one patient, antipsychotics over a decade of treatment were to no avail, and SSRIs made the condition worse. Finally, lamotrigine resolved the symptoms[57]. It seems that with hallucinogens, you need to have some expertise and select a safer and less potent alternative to LSD, such as mescaline, which does not cause flashbacks in the Navajo[58].

Metabolism

Despite its long period of effect, it is rapidly metabolised, primarily by dealkylation, or de-ethylation, to be precise (Figure B.5). The main metabolite in the urine of individuals who have used the drug is 2-oxo-3-hydroxy LSD, which is not found in the plasma, presumably as it is rapidly filtered by the kidneys.

This metabolite is found in much greater quantity than the parent drug in urine. Aside from the de-ethylated derivative (lysergic acid ethylamide) a demethylated metabolite (nor-LSD), the side chain and the top phenyl ring (13/14 position on the molecule) can be hydroxylated and at least one glucuronide is formed from the 13/14 hydroxylated derivative. For some reason, there is still no confirmed CYP assigned to these metabolites, although the molecule has some lipophilicity (log P of 1.3) and vaguely resembles a steroid shape, so CYP3A4 is a likely candidate, and/or CYP2C9 and CYP2D6 might be involved in the hydroxylation and dealkylation of LSD. The resemblance of LSD to the ergot drugs that are CYP3A4 substrates[54] and the general preference of CYP2D6 for alkaloids and 5-HT agonists strengthens the idea that these CYPs are involved. The drug may also be a substrate of P-gp.

The doors of perception

In recent years, LSD has been used as a 'window' to research brain function, particularly regarding consciousness and information processing. The cerebral cortex controls consciousness and reasoning and it has long been a research focus as to how the cortex communicates with the rest of the world through the other organs of the brain such as the thalamus. The current theory involves the idea of cortico-striatal-thalamo-cortical (CSTC) feedback loops[59]. These allow the thalamus to control or 'gate' information to and from the cortex. This gating process is modulated by the striatum, through 5-HT2A receptor activation. Information processing in the brain is difficult to comprehend, but perhaps a tentative analogy might be if the driver in a World Rally Championship car powering through a forest stage is the cortex, the navigator in the passenger seat constantly yelling detailed coded instructions on hazards, tightness of corners, and sudden dips is the thalamus. Indeed, when we are driving ourselves through a busy city being directed by a passenger, we all know the feeling of slight panic when we approach a roundabout in fast traffic when our passenger is silent and basically lost: 'WHICH WAY??'. In this state, we are not rational and we are not certain ourselves what we will actually do.

In the brain, the data overload caused by the failure of the thalamo-striatal gate seems to lead to a sort of cortical panic. This results in frightening vortex of cognitive disruption, sensory confusion and eventually psychosis. This is very similar situation to the major symptoms of schizophrenia and it is now believed that this condition is also related to a form of sensory overloading due to the failure of the thalamic gate[59]. The action of LSD and its related psychedelics on striatal 5HT-2A receptors downregulate the striatum, which temporarily causes cortical overload[59]. This excessive thalamic-cortical activity is also seen in major depression, and it is likely that LSD has really opened new directions in drug targeting of depression and other conditions[59,60].

Interestingly, Dr Hofmann always believed passionately that LSD did have potential medical benefit despite its inherent dangers and he would have been genuinely thrilled by this research[59]. Hofmann was not in favour of the attitude of 1960s counterculture luminaries such as Dr Timothy Leary, who championed the drug's use as a psychedelic recreational agent. However, the increasing value of these drugs in terms of understanding brain function and opening new treatment options suggests that the outwardly staid research chemist creator of LSD was in his own way just as much a visionary as Dr Leary.

Serotonin-based hallucinogens

The Sonoran desert toad, or Colorado River toad (*Bufo alvarius*) found in Arizona, as well as many other species such as the Australian cane toad, are really massive by toad standards; *B. alvarius* can grow to 8 inches long and weight a couple of pounds. They can eat practically anything, including other toads and even scorpions. However, some hardy seekers of alternative realities still brave the wrath of these killer toads of doom for the hallucinogenic properties of their skin secretions. These are exuded by the toads when frightened or alarmed, to incapacitate predators. Their sheer numbers point to the success of this strategy, and until recently it was thought their only known enemies were Australians with cricket bats. Apparently, the toads are so arrogant that they just sit there and refuse to move when predators attack them, confident in the power of their chemical defence. However, some species of ant are immune to the toxin and can kill the younger toads.

It is now better understood amongst the public that toad venoms contain many exceedingly toxic agents. These include bufagins (bufendienolides) that have cardiac glycoside-like effects and catecholamines[61]. The net result is heart failure, seizures and vasoconstriction leading to death in animals and at least some recorded cases of seizures in children and a man in Korea died in 2017 as a result of eating toads that he thought were bullfrogs. The collective effect from an entire toad skin could be lethal to the average weight adult. The hallucinogenic fraction of the venom contains a series of serotonin-like compounds, the indolealkylamines, which are commonly named bufotenines. Bufotenine itself (5-hydroxy-N, N-dimethyltryptamine) is a weak hallucinogen, which has potent cardiodepressive effects, sufficiently severe to induce circulatory crises. This derivative is actually an endogenous metabolite of tryptamine metabolism in man. Individuals with autistic disorders and schizophrenia eliminate more of it than normal in their urine[62] and its levels correlate with hyperactivity.

The potent hallucinogen in the toad venom is 5-Methoxy-N′N′ dimethyltryptamine (known also as 5-MeO-DMT, O-methylbufotenine and 5-methoxy bufotenine; Figure B.6)[63]. This agent has a number of 'street', or should that be

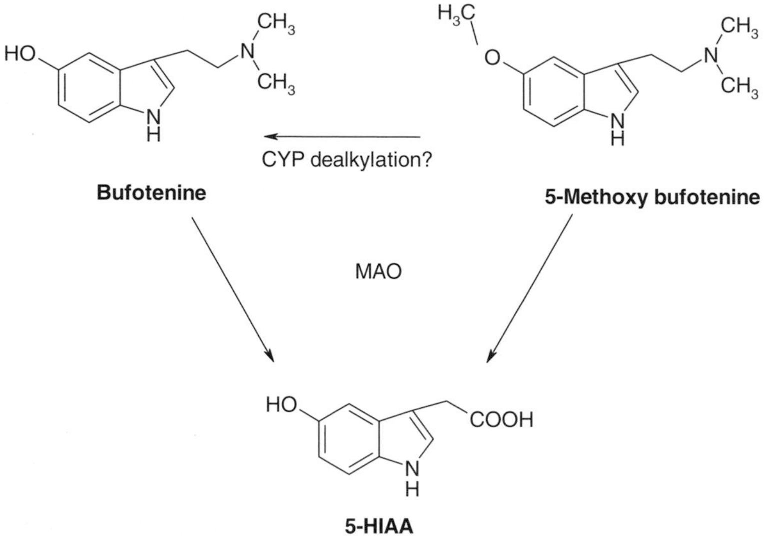

Figure B.6 Some routes of dimethyltryptamine derivative clearance leading to 5-hydroxyindole acetic acid (5-HIAA)

'desert track', names, such as five-methoxy, 5-MEO, or my favourite, 'The Power'. Orally, these derivatives have little effect, as MAO-A clears dimethyltryptamines to 5-OH indole acetic acid in the gut, although the 4-hydroxy substitution of the close structural analogue psilocybin (O-phosphoryl-4-hydroxydimethyl-tryptamine) makes it orally active[63,64]. 5-MeO DMT was first identified in the 1930s but it was not until the late 1960s that it was found in the toad exudates. Inhalation of 5-methoxybufotenine as an aerosol results in significant effects through smoking the dried toad exudate. Due to the variability of the active 5-MeO DMT content of the venom exudates, the effects have been described as 'from bliss to horror', which sounds somewhat perturbing. Chemically synthesized bufotenines are now available illicitly, although they are Class A drugs in the United Kingdom. In the United States, opinion is predictably divided as to whether licking the toads is harmful to the toads or unpleasant for the user, as upsetting the animals causes them to produce more exudate. The toads are an ecological disaster in Australia, but it does not look like they can be eliminated anytime soon. It is now established that 5-MeO DMT is O-demethylated to bufotenine and both are deaminated by MAO-A[64]. Hence, many of the effects of 5-MeO DMT will be subject to CYP2D6 and MAO-A polymorphisms. Interestingly, MAO-A inhibitors such as harmaline will increase The Power's persistence and effects, although harmaline is also cleared by CYP2D6 so the net effect of this could lead to serotonin syndrome[63] (Chapter 5.5.2), which is extremely bad.

B.5 Amphetamine derivatives

Introduction

Amphetamines are heavily substituted analogues of phenethylamine, a naturally occurring neuromodulating constituent of the CNS. The structural changes affect the major amine neurotransmitter systems causing a wide range of effects, from mild stimulation to psychosis. They are not difficult to make, with the starting materials such as ephedrine and pseudoephedrine for methamphetamine, or benzyl methyl ketone for amphetamine and piperonyl methyl ketone (to make MDMA) relatively easy to find[65]. New variants can be designed and manufactured with a decent working knowledge of chemical synthesis techniques. When the authorities raid amphetamine 'factories', they never find any starting materials left over, so the manufacturing is tightly organized and batches are made to order, so as not to attract too much unwelcome attention from chemical suppliers[65]. Not surprisingly, the manufacturers dump the toxic byproducts anywhere they can, occasionally injuring bystanders as well as the environment[65].

Amphetamines themselves have been popular drugs of abuse for more than 70 years. Their first major application was during World War II, to address combat fatigue. Whilst all sides used the drugs, the German Armed Forces were particularly enthusiastic in issuing methamphetamine (Pervitin) to their service members in vast amounts in the initial stages of WWII[66,67]. Pervitin fuelled Blitzkrieg and barbarous slaughter extremely effectively, but its toll on the men's health and performance, lead to its eventual withdrawal[66,67]. MDMA may well be the most popular amphetamine manifestation in current usage (see below). Amphetamines are known by the usual litany of tedious street-names, and the most popular variant of the more serious forms at the time of writing is methamphetamine, known mainly as 'ice' or 'crystal meth', and users are termed methheads. By 2011, mostly on the strength of MDMA, but also on crystal meth, this drug class had reached number two behind cannabis in the rankings for the world's favourite illicit chemical recreation[68].

Amphetamine pharmacology

Amphetamines and their derivatives are used intravenously, orally or smoked; this depends on the physical form of the drug, as well as rapidity of onset desired. They potentiate CNS and peripheral biogenic amine effects extremely strongly by causing amine release and preventing their re-uptake and destruction[69]. These sustained elevated amine levels, (particularly dopamine) lead to a sort of CNS turbocharging. The user experiences boundless physical energy, intense euphoria, happiness, and loss of inhibitions, with some possible hallucinations thrown in also[69]. There is also evidence that methamphetamine addicts show impairment in seratonergic systems as well as dopaminergic areas.

The general effects of amphetamines are similar to that of cocaine, but with differences. Cocaine is a transient but very intense 'high', perhaps only a few minutes, whilst amphetamines can maintain their potent effects over more than half a day, but the onset is slower and the high is less powerful than cocaine[69]. In general, serious drug users expect highs to be as intense as possible, immediately, if not sooner (again), so the fastest way to do this is through smoking or intravenous use[69,70].

The dysphoric effects of amphetamines once they have been metabolised are notoriously bad, due to severe synaptic biogenic amine depletion. Of course, the tight neuronal regulation of biogenic amine adapts rapidly through mechanisms such as receptor down-regulation. These erode the pharmacological effects over time, leading to tolerance and dependence with repeated usage[69,71]. The drugs cause significant neurotoxicity over time, through oxidative and mitochondrial stress, as well as excitotoxicity, particularly with methamphetamine; these effects are most significant in dopamine and 5-HT neuronal terminals[71]. The drugs also cause neuroimflammation and astrogliosis[71], and these effects are magnified with HIV infection[71].

Addicts escalate their doses and try to beat the tolerance and the dysphoria by taking the drug for days alongside depressants such as ethanol or heroin. 'Tweaking', as it is known, leads to continuously wakeful states that may exceed two weeks and can make these individuals exceedingly dangerous to themselves and others; indeed, they should definitely avoid operating machinery (Chapter 5.7.3).

Amphetamines can cause a long list of toxic effects, from hypertensive crises to strokes and seizures[70]. They can even induce paranoid schizophrenia in some individuals, as well as repetitive stereotypic effects. The impact on the physical appearance of serious addicts over time is genuinely shocking, and part of this problem are the difficulties in treating addicts and completing successful rehabilitation[70]. Unlike opiates, with their lethal respiratory 'Russian roulette' dangers, crystal meth addiction is sustainable for years. The drug acts like a sort of predatory insect, gradually digesting the user inwardly and outwardly at its leisure, whilst helpfully providing endless images of personal catastrophe for the internet (look online for 'Faces of Meth')

The variety in amphetamine clearance routes reflects their closeness in structure to endogenous biogenic amines. Their main route of metabolism seems to be ring–hydroxylation mostly by CYP2D6. Methamphetamine is 4-hydroxylated and demethylated by the same CYP[68]. Potent inhibitors of 3A4 and 2D6 such as some of the HIV protease inhibitors are thought to be capable of causing fatal amphetamine accumulation from normally safe dosages as a consequence of inhibition[68]. Amphetamines can also be N-oxidised by flavin monooxygenases (FMO-3) and deaminated (Chapter 3.12.2) by various enzyme systems such as the MAOs. MDMA is a good example of how the metabolic fate of an amphetamine has been gradually unravelled and how this might relate to the long-term sequences of its usage.

Ecstasy (MDMA) and its congeners: mode of action and acute toxicity

Relative lethality

Since the mid-1990s, MDMA has established itself as the illicit drug of choice for between 3–7% of the 16–24 and around 2% of the 16–59 age group[72]. Considering the scale of MDMA usage, it is perhaps surprising that it manages to kill 'only' roughly one user a week in the United Kingdom. Indeed, there were 56 confirmed deaths due to MDMA in the United Kingdom in 2017, which should be set against recorded deaths with paracetamol (218), cocaine (432), and heroin and morphine (1164) overdoses[73]. Surprisingly, more UK people in 2017 died from gabapentin overdose (60) than MDMA[73]. Of course, the perception amongst the 16–24 age group of the scale of the hypocrisy behind media lectures on the dangers of illicit MDMA is fuelled by the number of UK deaths related to licit alcohol consumption in the same year (7697)[74].

Acute effects

MDMA (Figure B.7) is the best-known representative of a group of N-substituted methylenedioxyamphetamine derivatives, which also include the N-ethyl derivative ('Eve') and MDA, the primary unsubstituted amine derivative. These

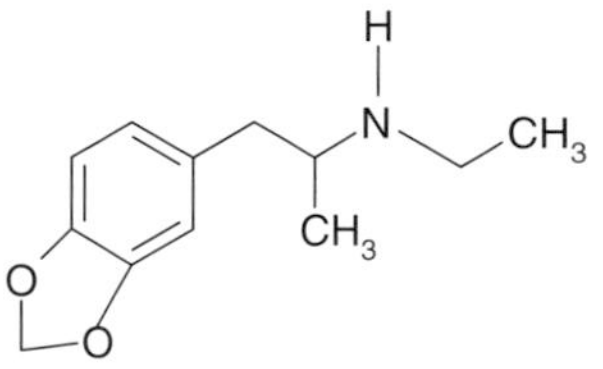

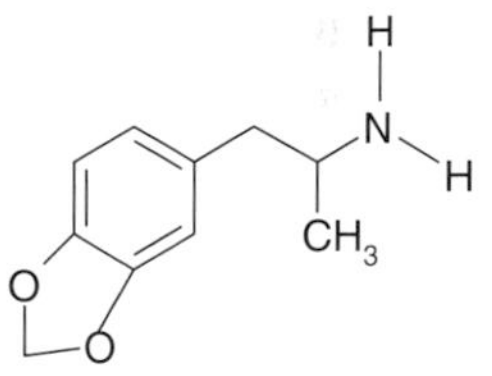

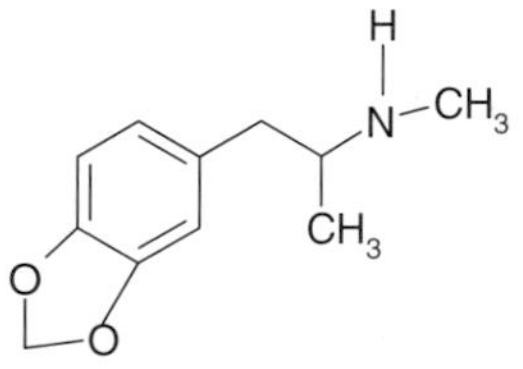

Figure B.7 Methylenedioxy amphetamine derivatives: MDMA (methylenedioxymethyl amphetamine: ecstasy), methylenedioxy amphetamine (MDA) and methylenedioxyethyl amphetamine (MDEA: Eve)

drugs differ from other amphetamine derivatives, in that they much more selective for 5-HT transporters, compared with dopamine and glutamate transport[71]. This means they retain some of the characteristic amphetamine stimulant effects, but they also induce feelings of empathy and warmth towards oneself and others; the word that was coined in the 1980s to describe them is 'entactogen'[75]. MDMA analogue toxicity can be resolved into acute and chronic toxicity – the acute toxicity is well understood and described. Overdose of these agents can lead to strokes, hyperthermia, high blood pressure, rhabdomyolysis and multiple organ failure[71,76]. Deaths attributed to these drugs are sometimes exacerbated by repetitive violent physical activity ('dancing') and its attendant dehydration.

Unfortunately, whilst entactogen users usually know which agent they are looking to buy and ingest, there is no guarantee they will be successful. This is partly because alongside most illicit drug users, those seeking entactogens are curiously selectively trusting. They refuse to believe authority figures that there are significant dangers inherent in these drugs. Yet the affable nematode-like representative of the criminal classes they have just met in a night-club will unquestionably sell them high-quality MDMA free from unpleasant impurities. As part of a general *caveat emptor* vibe, it is useful to consider the MDMA analogue 4-methylthioamphetamine (4-MTA), which was created as part of an experimental programme researching MDMA brain effects[77] and was 'pirated'. It was synthesised in quantity and distributed illicitly, often selling as ecstasy. A clue to its toxicity lies in its street name (flatliners)[77] and it is indeed far more likely to damage and/or end it all for you than MDMA, according to the statistics.

In the light of Dr Hofmann's attitude to LSD, it is perhaps instructive to read the personal reaction of Professor David Nichols, who designed and synthesised 4-MTA, to the news that his creation has taken a number of lives[78]. In response to these problems of purity and adulteration, in some European countries MDMA testing facilities are available near where the drug sells to guarantee a reasonable standard of purity.

Chronic toxicity and metabolism

After the drug became popular in the md-1990s, a considerable number of studies were carried out in rodents and non-human primates. As with many laboratory-designed animal studies, they used high pure drug doses, acute and/or chronic, to find organ-directed toxicity, rather as described in Appendix A and this they achieved. In the rodent studies, MDMA damaged 5-HT neuronal systems, causing deficits in transmission, although it did not cause the neurotoxicity seen in other amphetamines[71]. MDMA also impaired neuronal antioxidant status and mitochondrial oxidative stress and mDNA damage was the result[71]. Long-term motor impairment occurred in the animal models[79]. The results were publicised and were at first clearly at odds with the experience of the hundreds of thousands of users partying and empathizing furiously up and down the country. It is probably fair to say that animal testing protocols model human use of prescription drugs more diligently than human illicit drug use. Indeed, the inability of many rodent models to reproduce MPTP toxicity at first sight undermined their credibility and predictive power[80]. MDMA users vary dosage and often take several other substances (particularly cocaine) at the same time, intermittently and/or chronically over different periods of their lives.

It is difficult to be sure of the individual's drug history and so relating neuroimaging and motor testing observations to specific drug use is not straightforward. Human prospective investigations on illicit drugs assume that the individual had normal spatio-motor activity before their drug use, which may not necessarily be the case[79]. In addition, it has to be considered whether the deficits recorded in an individual are because of the drug – were they instead previously psychologically impaired, which somehow predisposes them to take the drug?[75,79]

However, a mixture of controlled and compensated neuroimaging, spatial/motor and other testing protocols has been developed for humans and it appears that chronic use of MDMA does downregulate 5-HT pathway activity[81], and the drug also impacts motor activity. This translates as a deficit in thalamo-corticostriatal connectivity and behavioural control and the region tries to compensate with increased activity, but to little gain. This is rather like a thalamic 'slipping clutch' effect and means the chronic MDMA users need to concentrate more to complete motor tasks compared with nonusers. The thalamic deficit is also seen in depression[79]. Taken together, it seems that using the drug once or twice is not likely to be a significant issue, but chronic use does damage seratonergic transmission in such a way that thalamic processing is less productive. Although not generally realised, it is recognised that at least 15% of MDMA users are dependent on the drug[82] and to what extent these users will recover once they eventually stop using it remains to be determined.

MDMA metabolism

The metabolism of MDMA (Figure B.8) is quite complex. MDMA derivatives in general are cleared by CYP2D6, and MDMA itself is demethylenated by this CYP to the catechol HHMA. Roughly a third of the dose is cleared by CYP2D6[68] and MDMA is metabolised about a hundred-fold faster than methamphetamine[68]. However, this process comes to a grinding halt after 1–2 hours or so, when the drug inhibits CYP2D6 mechanistically, to the point that all CYP2D6 activity is eliminated for several days[68,83]. This essentially phenocopies all females and just under 70% of males towards CYP2D6 poor metaboliser status. This seems to suggest that the more you take, the less is cleared, but there is evidence that CYPs 2C19, 2B6, and 3A4 also oxidise MDMA and it even increases CYP1A2 activity[83,84]. The drug is also cleared by MAO and COMT, so CYP2D6 phenotype has only a minor impact on the clearance of the drug[83].

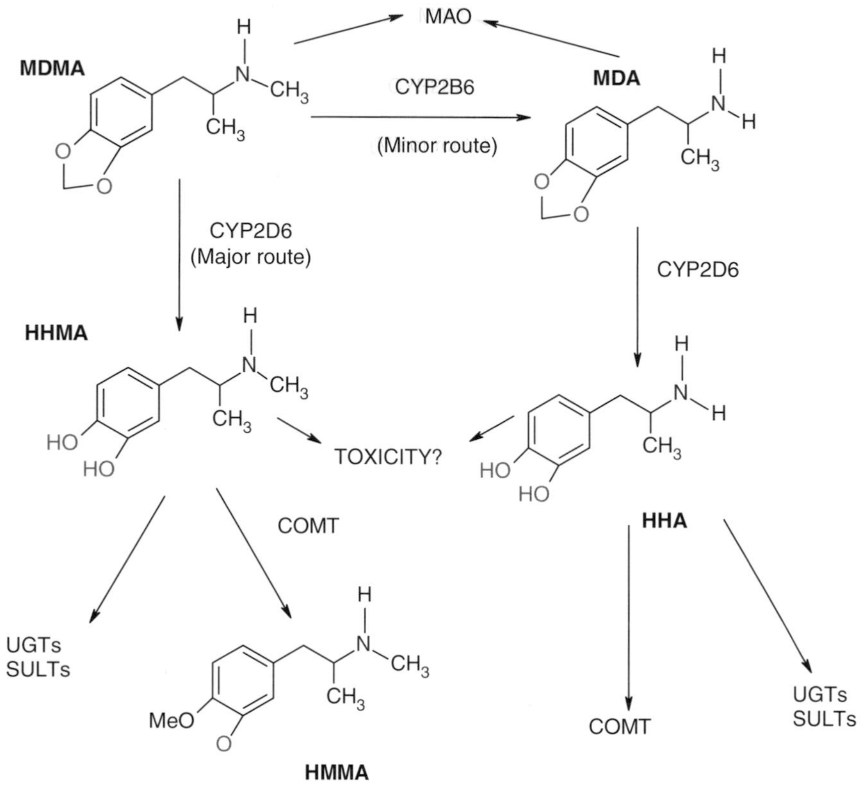

Figure B.8 Metabolism of MDMA: the main route of clearance is initially via CYP2D6 to HHMA (3,4,dihydroxymethamphetamine), which is cleared either by conjugation or may be methylated by catechol-O-methyl transferase (COMT) to HMMA (4-hydroxy-3-methoxymethyl amphetamine). The demethylated MDA may then be demethylenated to HHA (3,4 dihydroxyamphetamine), which can also be methylated and undergo conjugation. MAO may also oxidize MDA, or MDMA

The ethyl side-chain in 'Eve' and the methyl group in MDMA can also be dealkylated by CYPs 2B6, 3A4 and 2C19, although this route is a minor one in humans, as the active metabolite MDA comprises only about 5% of parent drug in plasma [84].

Whilst formation of the catechol metabolites due to CYP2D6 stops fairly rapidly, as mentioned above, other CYPs can still make the potentially toxic HHMA[85] despite the CYP2D6 inactivation by other drugs such as paroxetine or MDMA itself. Hence, the parent drug can act to cause potent pharmacological effects, whilst the catechol metabolites HHMA and HHA can enter redox cycles intracellularly, such as in mitochondria and this is thought to be how MDMA mediates seratoninergic pathway attrition[68]. The other pathways such as COMT, CYP2B6 and CYP2C19 are, as we know, subject to polymorphisms, and slow versions of CYP2C19 (*2/*2) show increased cardiovascular responses to MDMA[84]. In addition, as CYP2D6 capacity is inactivated, CYP2B6 participation increases in terms of clearance, and CYP1A2 also can increase MDA levels, but this is not a major issue[84].

Essentially, the acute (and worrying) cardiovascular impact of the drug may well be dependent on its non-CYP2D6 clearance, so the risk of acute toxicity is probably magnified by poor metaboliser status with CYP2C19 (*2) and CYP2B6 (*6)[84]. The cumulative neuronal toxicity of MDMA is likely to be dependent on a fine balance of CYP activation alongside efficient cellular detoxification in terms of intracellular thiol status for instance, as well as clearance systemically and of course how much of the catechols are made locally in the brain. This means that acute and chronic MDMA toxicity is something of a lottery, but from the chronic perspective it will almost certainly cause deficits in all heavy users eventually.

MDMA CYP2D6 inhibition will significantly affect the clearance of any prescription drugs that are cleared by this route that the user is currently taking, such as antipsychotics, antidepressants (SSRIs and TCAs) and opiates. In addition, MDMA effects will be blunted by the SSRIs and SNRIs, due to 5-HT pathway competition[68].

Other amphetamine derivatives: 2C-P and 4-FA

As with University experience, dissatisfaction can extend to illicit drugs that do not meet user expectations in terms of effects, as well as price and availability. We have seen with Mr Kidston where such dissatisfaction can lead, and problems with MDMA, such as confusion and disorientation, have made would-be psychonauts seek alternatives. Many have found other amphetamine derivatives. One is 2,5-dimethoxy-4-n-propylphenethylamine, known as 2C-P (Figure B.9). This agent was synthesized and then first overdosed on, by the prolific psychedelic synthesizing chemist Dr Alexander Shulgin, who was himself wary of the potency and toxicity of 2C-P[86] as well as its narrow 'therapeutic index', or should that be 'psychedelic index'[86]. He was proved right in the case for one unfortunate young lady who was said to be 2C-P's first UK fatality at the time of writing. There appears to be little or no scientific literature on this drug, but it is described as 'powerful' 'overwhelming' and it causes agitation and delirium. Since it has a long duration of action and even its creator was not enthusiastic about it, taking 2C-P is not recommended at all.

A more controllable and less toxic phenethylamine derivative is 4-fluoroamphetamine (4-FA, or 'flux', Figure B.9) on the grounds that it provides some of the entactogen MDMA experience, along with an amphetamine-like stimulation later on during the 4–6 hour duration of the effect. The drug is said to have a more dopaminergic effect than 5-HT[87] and anecdotally, the comedown has been

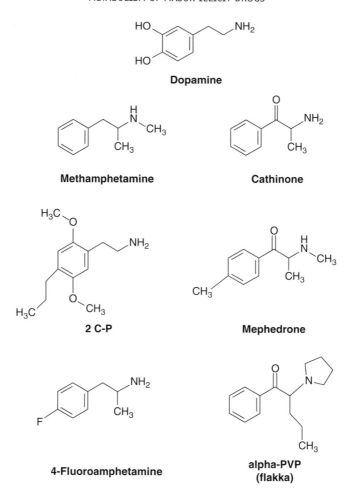

Figure B.9 Various amphetamine-like derivatives, alongside dopamine

believed to be less brutal than with other amphetamines, but clinical data contra-
dicts this perception[87] and 4-FA clearly has some issues. Users have again
reported anecdotally that 4-FA is more amphetamine-like at lower doses (<100
mg) and more MDMA/entactogenic at higher doses. If this is a true perception
amongst users, then this is actually quite concerning. This is because when 4-FA
was studied clinically, even at 100 mg doses it drives up blood pressure to around
130–150 mmHg and heart rates up to 165 bpm[87]. Indeed, the Dutch regulatory
authorities advised clinicians running the study that they not continue with the
original plan to dose the required number of subjects at 150 mg on the grounds
of safety. Those cardiovascular effects can be intermittent for the duration of the
trip and might account for the drug's tendency to cause strokes[88].

The clinical study showed that 4-FA makes its users more vigorous and friendly in the first hour or so, improves reaction times, and may erode memory, much in the same way MDMA does[87]. In my own personal experience, I have achieved a transient increase in friendliness and vigour, without any negative impact on memory or cardiovascular performance, by being offered a large bag of Reese's Peanut Butter Cups. Afterwards, the negative issues include mild annoyance that there were none left alongside some minor dehydration due to the admittedly shocking salt content.

Amphetamine natural highs: the cathinones

Khat

The best-known incarnation of these drugs is the favourite chemical diversion of the Horn of Africa, *khat*, which comprises the leaves and stems of *Catha edulis*, a flowering shrub that grows in arid and semi-arid areas. If it finds good soil, it can apparently grow to over 30 feet tall, but you wonder if it ever gets the chance given the effect of chewing it. The place of khat culturally is perhaps similar to that of coca leaves in Bolivia, although a difference is that khat chewing is a male activity and not culturally acceptable for women[89]. Khat is regarded as a complex mix of medicine, support and entertainment across societies in Ethiopia and Somalia. These three essentials were supplied by alcohol in developed countries in the past, although the balance has shifted precipitously over the last century towards alcoholic entertainment, with the support and medicine angles now thoroughly discouraged and the gaps latterly filled by antidepressants. The idea of a 'medicinal brandy' is now confined to appropriately lavish period TV drama.

Khat contains over 70 different known pharmacologically active agents, including at least three groups of various cathinone derivatives with S(-) cathinone (Figure B.9) as the most potent[90]. Cathinones are potent analogues of amphetamines, with the attendant dopamine release and inhibition of reuptake, but the various cathinone derivatives all have subtle differences that probably vary as to how the plant is cultivated. Concerns about khat centre on its many peripheral pharmacological effects (cardiovascular, gastrointestinal, and urinary tract) as well as the usual amphetamine-like central stimulant effects, which are accompanied by significant craving[90]. This latter issue, along with an unpleasant withdrawal experience, makes it very difficult to stop using it, and this is something of a common feature amongst the cathinone derivatives. This is combined with its LSD-like propensity for unmasking psychiatric problems in its chronic users[91]. Khat is such a complex and variable mixture of different agents, it is of course problematic to be precise about its impact on prescription medicines, but to date it is clear that the cathinones in khat are both cleared by CYP2D6 and can inhibit it, although this thought at the moment to be a competitive rather than a mechanism-based effect[89]. Cathinone itself is metabolised to phenylpropanolamine by

CYP2D6, but the drug is also converted to norephedrine and cathine by other routes. Other inhibitory effects on CYP3A4 and CYP2C19 are minor but may have some therapeutic consequences[89]. The disposition and pharmacological effects of khat have to be seen in the context of the relatively common CYP2D6 multiple gene-copying effect seen in indigenous peoples in the Horn of Africa. The preference of this isoform for alkaloids suggests an adaptation to plants such as khat in the local diet[89]. Unfortunately, khat's CYP2D6 inhibition can cause DDIs with the antimalarial chloroquine and primaquine. The latter drug, in order to effect a radical cure (elimination of the hepatic forms of the malarial parasite) requires activation by CYP2D6[92] and khat will probably diminish efficacy[89], but the user may well not be particularly concerned.

Unnatural highs: synthetic cathinones

Several synthetic cathinone derivatives are now available in quantity. After appearing as legal highs in the mid-2000s, they were marketed vigorously as 'plant fertiliser' and even at one point as a vacuum cleaner fragrance, with the clear instruction that they were not for human consumption. Mephedrone (Figure B.9) was initially the most popular ('meow meow', MCAT) and a decade on is still widely available and there are plenty of other synthetic cathinones being designed and manufactured to satisfy the demand for these drugs[93]. These agents are almost always used with other drugs, usually through insufflation (snorting) but orally as well. What makes these drugs dangerous is partly the mixing, but also how users binge on them, consuming up to a gramme in a session to maintain the effects. Purity is variable (50–88%) and users tend to be rather vague on which drug they are actually taking[94]. This chaotic, almost freefall approach has led to numerous fatalities with the synthetic cathinones after hypertensive crises led to strokes[93].

Mephedrone is cleared by CYP2D6 to N-demethylated (normephedrone), keto-reduced and hydroxylated metabolites, as well as several glucuronides. Normephendrone is also conjugated rather unusually with succinic acid and the significance of this pathway in biogenic amine metabolism is not fully understood[95]. The question of whether normephedrone has pharmacological activity posed by some authors[95] was answered emphatically by its manufacture and successful distribution alongside all the other synthetic cathinones. Close relatives of mephendrone include ephedrone and metaphedrone (3-MMC)[94]. These drugs are, like mephedrone, N-demethylated, and ketoreduced again probably by CYP2D6. Interestingly, metaphedrone, amongst all the usual amphetamine-like stimulant effects, causes its users to become extremely verbose and stutter at the same time[94].

Another very successful synthetic cathinone is flakka (α-pyrrolidinovalerophenone; α-PVP; Figure B.9). The term *flakka* is said to be from Spanish slang for a beautiful woman[96], although literally translated it apparently means 'very skinny girl' (La Flaca). Various mixes of α-PVP, mephedrone and another agent

3,4-methylenedioxypyrovalerone (MDPV), are said to be the active ingredients in 'bath salts'. The α-pyrrolidinophenones are more lipophilic than many of the other cathinones, as well as those found in khat and absorption and the onset of the effect is much quicker than that of the plant agents[96].

The problem with the synthetic cathinones in general is their tendency to cause psychotic aggression, hallucinations, and dangerous behaviour[96]. In this area, these drugs contrive to mimic elements of cocaine, amphetamines and LSD quite significantly. Interestingly, the main metabolites of α-PVP are a hydroxylated derivative and a lactam, and analysis by gas chromatography causes conversion in the heat of the injector of the hydroxyl to the lactam, so HPLC is preferred (see Appendix A.4.1). Most of these agents, including α-PVP and its derivatives, are believed to be CYP2D6 substrates. It is thought that their similarity to the khat cathinones means some are likely to be inhibitors also[96].

Synthetic cathinones have soared in popularity in recent years and now comprise about a third of the new psychoactive drug market[65,96]. This is a problem, because although the various 'bath salts' concoctions have been widely proscribed around world, it is not easy to treat users in overdose. With α-PVP for example, it is more potent than cocaine and other amphetamines in inhibiting dopamine, noradrenaline, and vesicular monoamine transporters, with negligible 5-HT interactions[96]. Hence, their effects are not like that of the entactogens, and they can cause some fairly deranged and aggressive behaviour, more akin to PCP (see later) than amphetamines.

Piperazine derivatives (BZP, TFMPP)

As we have seen, amphetamines of various types are close to the heart of many an illicit drug user. Interestingly, in the early 2000s in New Zealand there was actually a culture of tolerance to amphetamine-type 'legal highs' so they could be sold commercially locally and exported, as their prohibition was not complete internationally. The process whereby the New Zealand government first attempted to regulate these agents through legislation and then decided to effectively ban them is an interesting sequence of events that is very revealing of governmental struggles with psychoactive substances[97].

The legal highs or 'party pills' were mainly a series of piperazine derivatives (Figure B.10), mainly N-benzylpiperazine (BZP) and trifluormethyl phenylpiperazine (TFMPP). These agents are either used singly or in combination, when they are intended to recreate the effects of ecstasy. Other popular derivatives include methylenedioxybenzyl-piperazine (MDBP) methoxyphenylpiperazine (MeOPP) as well as methylbenzylpiperazine (MPZP)[97,98]. Piperazines are often sold in pill form under a variety of jolly names, most usually, 'party pills'. Whilst piperazines are still used to de-worm animals, the idea that BZP was developed for the same reason apparently is not true; you can't whack all your intestinal parasites and get high at the same time.

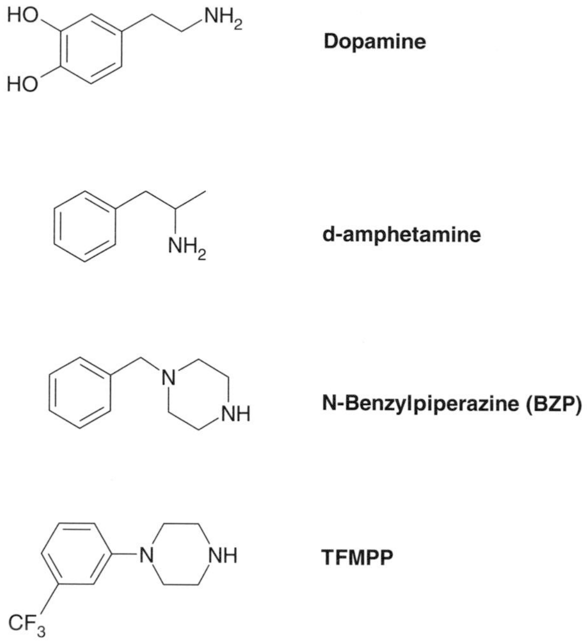

Figure B.10 Structures of dopamine, d-amphetamine, N-benzylpiperazine (BZP) and trifluoromethylphenylpiperazine (TFMPP)

However, whilst BZP was not much use at stunning helminths, it was found to be a terrific racehorse and athlete stimulant until it was banned in 2010[98]. BZP was intended to be an antidepressant, but was rather too amphetamine-like and the drug company suspected after early 1970s clinical trials that it could become a drug of abuse[98]. BZP and its relatives do have many amphetamine-like actions, including the usual stimulant effects (increased blood pressure and heart rate). BZP itself is a sympathomimetic like amphetamine, as it releases dopamine in the CNS through its effects on the dopamine transporter. The drug has shown some addictive potential in non-human primate studies, whilst TFMPP, which is a seratinonergic agonist, did not. The combination of BZP and TFMPP does appear to promote an MDMA-like 'entactogen' effect[98], where a feeling of general well-being is accompanied by the occasional hallucination at no extra charge. BZP has been responsible for deaths and multi-organ failures, but often with other agents[99]. In the United Kingdom, it is not to be found on Home Office lists of drug related deaths for 2017[73] and for the moment, appears to be out of fashion.

Piperazine metabolism

The possibility that BZP and its congeners may be problematic when taken with prescription drugs is borne out by preliminary studies on their human biotransformation and metabolism. *In vivo*, relatively little is known of BZP 's clearance, although one study suggested that its bioavailability is fairly low. The half-life was estimated at 5.5 hours and it is cleared much more slowly than methamphetamine. Hydroxylation of the aromatic ring was noted at the 3 and 4 positions, which was followed by sulphation rather than glucuronidation. There is also sulphation at one of the nitrogens in the piperazine group[100]. Using human liver microsomes, it was found that BZP and TFMPP were metabolised mainly by CYP2D6, CYP1A2 and CYP3A4, but not by CYP2C19 and CYP2C9, although BZP does not appear to block CYP2C19, it does inhibit the rest of the major CYPs[100,101]. Indeed, these drugs as a class appear to be major CYP inhibitors. Interestingly, BZP and TFMPP also inhibit each others' metabolism[100,101], which may partly explain their mutually enhanced pharmacodynamic effects on the CNS. Although more studies on the biotransformation of these drugs needs to be carried out, it seems certain that the piperazines will cause drug-drug interactions with CYP2D6 substrates such as antipsychotics and SSRIs as well as CYP3A4 substrates. The toxic interaction between MDMA and BZP is probably at least partly caused by irreversible CYP2D6 inhibition leading to BZP accumulation[100]. It seems that the general propensity of drug users to take combinations of various agents may be particularly risky when this involves the benzylpiperazines.

B.6   Cannabis

Introduction

The hemp plants *Cannabis sativa, indica* and *ruderalis* are the main source of world cannabinoids, a group of around 100 terpene psychoactive agents, which are the most commonly used illicit substances for recreational purposes. Cannabis reigns supreme in terms of illicit drug use, involving around 2.5% of the world's population and 1 in 12 Americans[102]. The richest source of the cannabinoids is the resin removed from the leaves of the female plant, although there are more than 500 chemicals discovered in these plants to date[103]. The most potent of the cannabinoids is Δ^9-tetrahydrocannabinol (Δ^9-THC; Figure B.11), although Δ^8-THC is as psychoactive and is chemically more stable. Cannabinoids are usually prepared from the dried flowering tops and leaves (marijuana, up to 5% Δ^9-THC), the resin itself combined with the flowers (hashish, up to 25% Δ^9-THC) and the industrial strength version, hash oil. This is resin that has been extracted and purified with solvents and concentrated and it can exceed 70% in Δ^9-THC content. Cannabis is rather bulky to smuggle and is not economical enough to ship in the average Drug-Lord's quite impressive fleet of homemade submarines and law-enforcement interception rates have improved also. Hence, the growers and dealers have exploited efforts made since the 1970s to develop higher THC yielding strains (so-called 'skunk') alongside further breeding to produce plants that could be grown under lights near their actual market. This means there was no need to ship comical yet surprisingly carefully wrapped

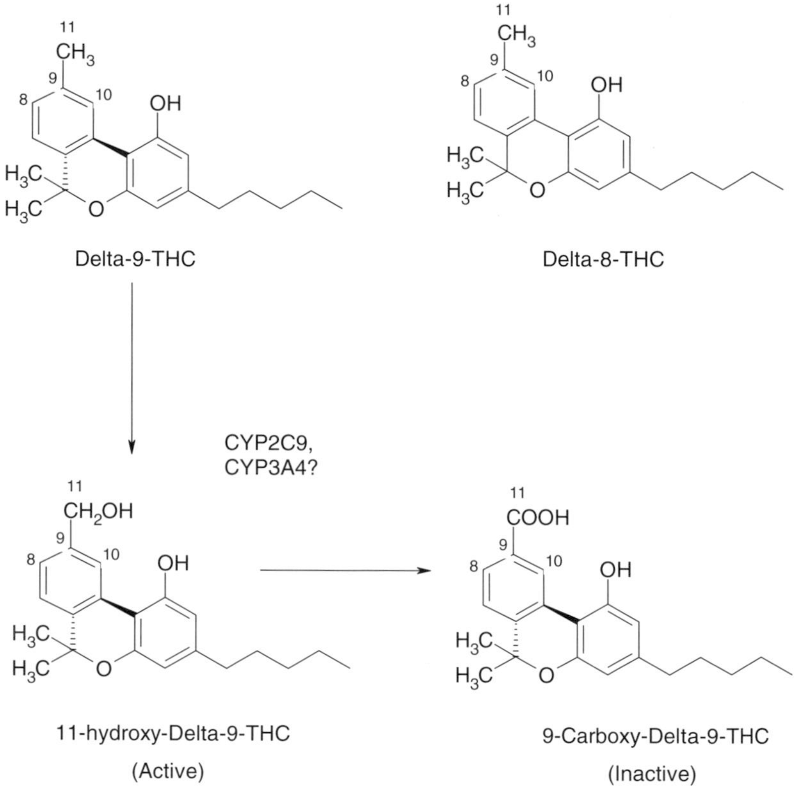

Figure B.11 The main active constituents of *Cannabis* variants, Δ^8- and Δ^9-THC and the clearance of Δ^9-THC to its hydroxyl carboxy products by CYPs 2C9 and possibly CYP3A4

massive bales of dope around the world whilst being chased by assorted angry uniformed people in powerboats.

The results are hardy, small, potent, and highly profitable plants, grown unobtrusively in quantity in a suburban house next door to you. High THC content cannabis can have three times the Δ^9-THC content compared with the old stuff shipped over in the bales from somewhere remote and hot. A great deal of illicit research and development has led to the use of high intensity lighting, which induces the plants to form more resin that is part of the effort to produce greater Δ^9-THC content. Originally, the indoor growing process was a race against time before the power company detected the stadium-sized electricity bill. However, improvements in general stealth standards, alongside cunning abstraction of power from multiple streetlights, made the process more dependable for the growers. This can change after a heavy snowfall, when all the snow melts off the roof of one particular house very quickly – you know, where the very quiet people live, that nobody seems to see or hear.

Cannabinoid pharmacology

The cannabinoids are quite potent, and a dose of less than 10 mg is required for the standard 'high' effects, although they also have stimulant, sedative, and even hallucinogenic properties. It is now clear that these agents have far more in common with opiates than other drugs of abuse such as the amphetamines and cocaine, which just crudely disrupt biogenic amine pathways. The cannabinoids can activate a wide range of endogenous receptors as well as participate in what has been termed a 'psychoneuroimmunology' axis. This is a complex network of the main G-protein cannabinoid receptors known as CB_1 (CNS) and CB_2 (immune system and CNS), as well as other G-protein coupled receptors GPR55 and 119, the vanilloid TRPV1 receptor, as well as the PPAR family[103,104]. These are acted on by the endogenous agonists (endocannabinoids), such as anandamide and 2-AG, which are made on demand and then hydrolyzed by an enzyme known as FAAH, which is of interest as a possible drug target to potentiate the effects of the endogenous molecules[104]. The endocannabinoids also operate through a complex network of 'entourage' small molecules, which, in turn, modulate their action as they control vast multi-tissue organ and systemic regulatory process[102], the very least of which is the classic 'high' gained from smoking the plant.

Administration is as efficient through smoking or vaporisation as it is when given intravenously (experimentally), but orally, absorption is slow and erratic[102]. When smoked the drug's effects last around 3 hours, whilst you can trip on an orally dosed drug for more than 12 hours[102]. It seems that cannabinoid receptors act by inhibiting the release of neurotransmitters responsible for anxiety, like glutamate and GABA. THCs are agonists on CB_1 receptors, whilst other agents in cannabis, such as cannabidiol (CBD) act as antagonists and are not psychoactive. Earlier strains of cannabis had a less extreme THC/cannabidiol ratio, but growers have been determined not just to increase the THC level but to reduce the CBD level, skewing the ratio towards maximizing the potency of the drug[105].

Essentially, the mixture of psychotropic (mainly THC mediated) and anxiolytic (CBD) effects will vary according to the contents of the respective agents, route of administration and dose. Hence, cannabis consumption leads to a complex mixture of effects targeted on areas of the brain rich in the CB_1 receptors, such as those associated with cognition, memory and movement coordination. CB_1 receptor activation by higher doses is noticeably different from lower doses and this can cause the intense paranoia and psychosis[105], linked with heavy use, which can lead to users being 'sectioned' under the Mental Health Act. This process can be exacerbated by the accumulation of active THC derivatives in heavy usage, as well as the low CBD levels in current plants. The CB_1 receptors are also found in the periphery alongside CB_2 receptors, and their endogenous functions are wide-ranging and not entirely understood. Several synthetic agonists and antagonists have been synthesized. The first CB_1 antagonist, rimonabant, was briefly introduced for treatment of obesity, but was withdrawn in 2008 due to severe disorders in mood[106], which suggests that we have some way to go to understand the full profile of the cannabinoid receptor function. In contrast, Sativex, the first of several medicinal cannabis preparations, is a mouth spray that consists of equal proportions of CBD and Δ^9-THC. This continues to be effective for the pain and spasticity associated with a variety of conditions ranging from rheumatoid arthritis to multiple sclerosis and cannabis oil preparations enriched with a cannabidiol/THC ratio of 20:1 have shown promise in refractory childhood epilepsy[107].

It is possible to overdose and die from cannabis-related products and in 2017 in the UK 29 individuals did just that[73]. However, given the scale and quantity of the drug that is used, it is much safer than many prescription drugs and it seems the main reason that cannabinoids are not comparable in terms of toxicity with opiate receptors, for example, is that they do not exert such control of respiration or heart rate. It is possible that more subtypes of cannabinoid receptors will be found and the full medicinal potential of these receptors will emerge in years to come. The anticancer effects of these agents are under significant investigation[103].

Metabolism

THC-derivative clearance is extensive, with dozens of metabolites formed; indeed, only small amounts of parent Δ^9-THC appear in urine. The half-life of Δ^9 THC in blood is around 20 hours, although the effects from smoking appear within 5 – 10 minutes. When cannabis products are smoked, Δ^9-THC is cleared within a few minutes to 9-carboxy Δ^9-THC (Figure B.11), which is pharmacologically inactive and is the major urinary metabolite. This metabolite is used to monitor cannabis usage in drug-testing protocols. CYP3A4 and CYP2C9[108,109] form hydroxy derivatives of Δ^9-THC, such as the 11-hydroxy derivative that retains activity, although CYP1A2 is also involved, probably through the pyrolysis hydrocarbons[110]. Cannabidiol is cleared by CYP3A4 and CYP2C19[102]. Ketaconazole has a significant impact on the clearance of Δ^9-THC and cannabidiol[102]. Various UGTs are also thought to clear the hydroxylated metabolites[110]. Generally there is little in terms of significant drug interactions likely to be problematic from heavy cannabis consumption, aside from those on rifampicin will have to buy and smoke about twice as much to get the same effect[108-110]. It is interesting that school friends of mine who used the drug in the late 1970s insisted that it was far better consumed orally than smoked and that the effects were more pleasant and longer in duration. Indeed, despite the poor bioavailability of around 8–10%,[111] oral dosage promotes the formation of 11-hydroxy Δ^9-THC, which is not only more potent than Δ^9-THC, but enters the brain more quickly[111,112]. It is possible that chronic use leads to some induction of THC metabolism and absorption is better with heavy smokers[112]; however, the metabolites are not easy to clear due to their lipophilicity and accumulate in fatty areas. After around a week, more than one-third of the dose is still in the user. Elimination of the various metabolites usually occurs in faeces and can take months after high doses. Hence, the consumption of small quantities can lead to the failure of a random drug test several weeks later[112].

Cannabis and the lung

As we know, the impact of tobacco on the lungs has been exhaustively documented. However, the effects of chronic cannabis smoking has been much more difficult to determine, obviously because so many smoke cannabis and tobacco at the same time or at different times. What has emerged, is where it has been possible to compare the two substances, cannabis has some obvious similarities, but some surprising differences on the lungs in comparison with tobacco. The similarities are the usual wheezing, gloopy sputum and smoker's cough. The differences seem to be that tobacco is far more destructive to lung function, causing COPD (chronic obstructive airway disease) where cannabis does not[113]. It has been suggested that this is linked with the way cannabis and tobacco users smoke differently, with the cannabis-style breath holding somehow improving lung function, but this is not thought to be convincing[113]. It seems most likely that the pharmacological actions (bronchodilation and anti-inflammation) of the cannabinoids act to preserve lung function, whilst tobacco is wholly irritant and damaging.

Regarding the possibility of cannabis causing lung cancer, this remains controversial after perhaps 25 years of studies, many of which are inconclusive due to various flaws[113]. Indeed, different authors view the same literature and come to different conclusions[113,114]. Some feel that smoking cannabis is not carcinogenic[113], whereas others cite several studies that demonstrate the carcinogenicity, mutagenicity and mitochondrial toxicity of the cannabinoids[114]. Whether vaping cannabis products might further reduce what risks that exist will take some time to emerge, but in the meantime, users can decide for themselves whether to smoke, eat, or vape the plant and its extracts – or pass altogether.

Cannabis and the young

The impact on health due to mild to moderate use of cannabis might be fairly marginal regarding most consenting adults, but there is no doubt that it is toxic to the young from the earliest age possible. Smoking of any kind is teratogenic, but smoking cannabis seems to be particularly problematic. It is now clear that cannabinoids act in the same way antineoplastic drugs such as taxols, in that they disrupt mitotic spindle formation in dividing cells, which eventually leads to severe damage to chromosomes, leading to breaks and deletions[114]. This happens at attainable concentrations from smoking the plant and many published studies have demonstrated that parental significant cannabis use leads to a lifelong impairment in their offspring, in terms of development, intelligence and their ability to concentrate, as well as promoting hyperactivity[114]. As the offspring grow in a warm family atmosphere of cannabis fumes, there is plenty of evidence that mental health issues such as depression, anxiety, and psychosis are strongly linked with the use of the plant in the 15–27 age groups[115]. People of this age group are especially vulnerable to stress-related mental illness, and they do not comprehend the established vulnerability of their developing brains to agents such as cannabis and its ability to cause long-term epigenetic changes in their personalities and behaviour[116]. Ultimately, using cannabis for pleasure, escape, or as a form of self-medication is not likely to be of any assistance to adolescents and younger individuals and ultimately will make their lives considerably more difficult.

Synthetic cannabis: 'Spice' & 'K2'

In the early 2000s, a veritable smorgasbord of different synthetic cannabinoids appeared, perhaps from the same stable as the other legal highs. These ranged from several cyclohexylphenols (CP-47, 497, and derivatives) and naphthoylindoles (AM-678/JWH-18, Figure B.12) and indol-3-ylcycloalkyl derivatives, plus at least a dozen other compounds. They were usually sold in predetermined mixes of various drug combinations that had been dissolved in a volatile solvent and then sprayed onto tobacco or some other plant material, ready to be smoked. The mixes were sold with different names, some rather reminiscent of 1970s fireworks. These products were termed 'K2' in the United States or 'Spice' in the United Kingdom. The dose and the effects appear to be pretty much completely random, which presumably is part of the attraction. All these agents have been intended as CB1 receptor agonists, often up to 100-fold the potency of the natural Δ9 high[117,118]. Part of this was to evade military or company urinary drug detection tests, but also just to find something thrilling and exciting that the users would be prepared to buy. As fast as they are made illegal, more are synthesized and tested around the world and then marketed in a sort of 'slash and burn' approach to chemistry, pharmacology, and ultimately to the health of the users.

The potency of these agents leads to a wide range of symptoms, from cardiovascular toxicity through to psychosis and organ damage. This contrasts with the perception amongst the users that synthetic cannabis agents were safer than other drugs, such as cocaine and opiates[118]. The method of administration, the somewhat chaotic preparation process and general lack of quality control, alongside the potency and unpredictability, means that acute intoxication might be as much as surprise to the user as to the emergency services. The onset can be in a few minutes and followed by several hours of a much more exciting rollercoaster compared with that of LSD[118]. The drugs are well known for impairing coordination and judgment, causing users to fail roadside sobriety tests; a wide range of behaviour was seen from aggression and panic to relaxed[118]. These drugs are also rather habit forming and can induce craving and intermittent psychosis,

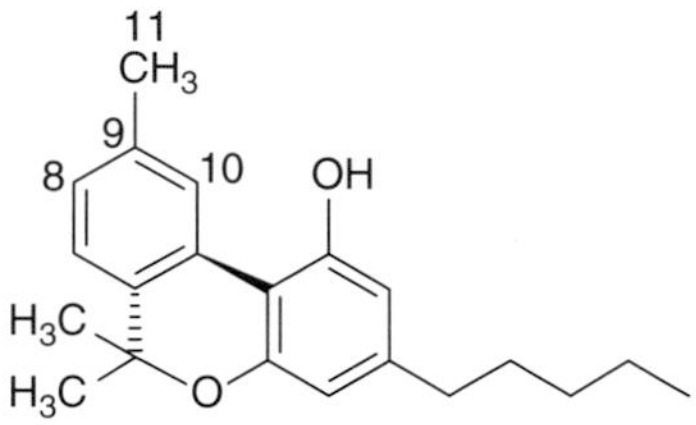

Delta-9-THC

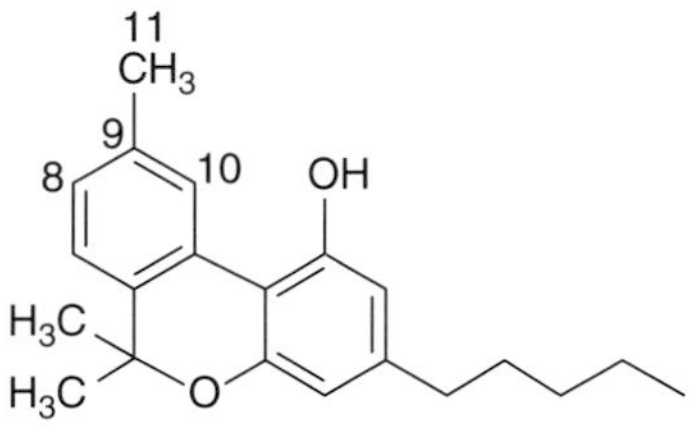

Delta-8-THC

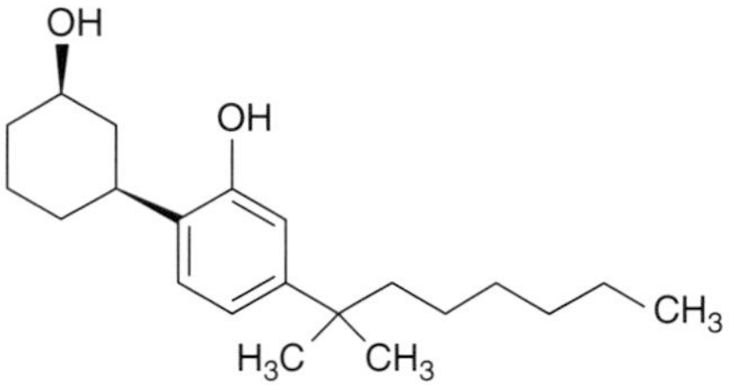

CP 47,497

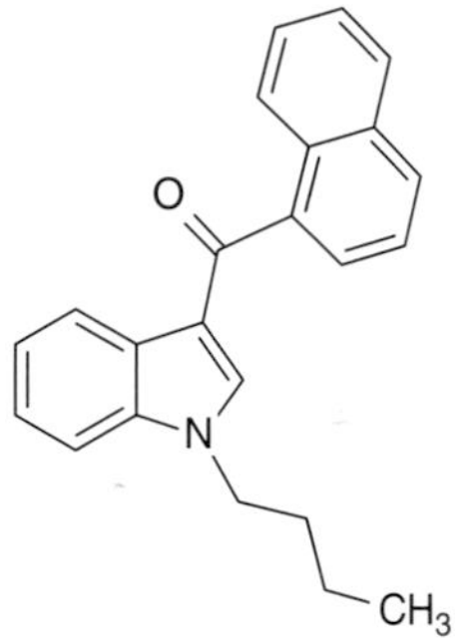

AM-678/JWH-18

Figure B.12 Synthetic cannabinoid derivatives found in 'Spice' shown alongside Δ^8 and Δ^9 THC structures

including paranoia and suicidal tendencies; in one instance an individual could only gain control of his symptoms by stealing his roommate's quetiapine and this was eventually prescribed to him for treatment[118].

The synthetic cannabis agents generally appear to be cleared extensively by UGT isoforms (UGT1A1, 1A9, and 2B7), as the glucuronides are found in considerable amounts in the urine of users[117]. The drugs are also thought to be oxidised by CYPs 2C9 and 1A2. In the brain CYP2D6 is thought to regulate their effects in the vicinity of CB1 receptors[117]. Whether these agents interfere with prescription drugs one would guess might be the least of the user's worries. These drugs seem to be only a moderate step below the 'real deal' for those who demand the most extreme and unpredictable illicit experiences, the dissociative anaesthetics.

B.7 Dissociative anaesthetics

PCP

Phencyclidine (PCP; Figure B.13), an arylcyclohexylamine, was synthesized in 1956, and soon after patented by Parke-Davis as a dissociative anaesthetic, where the patient feels detached from themselves and their surroundings. PCP acts at the

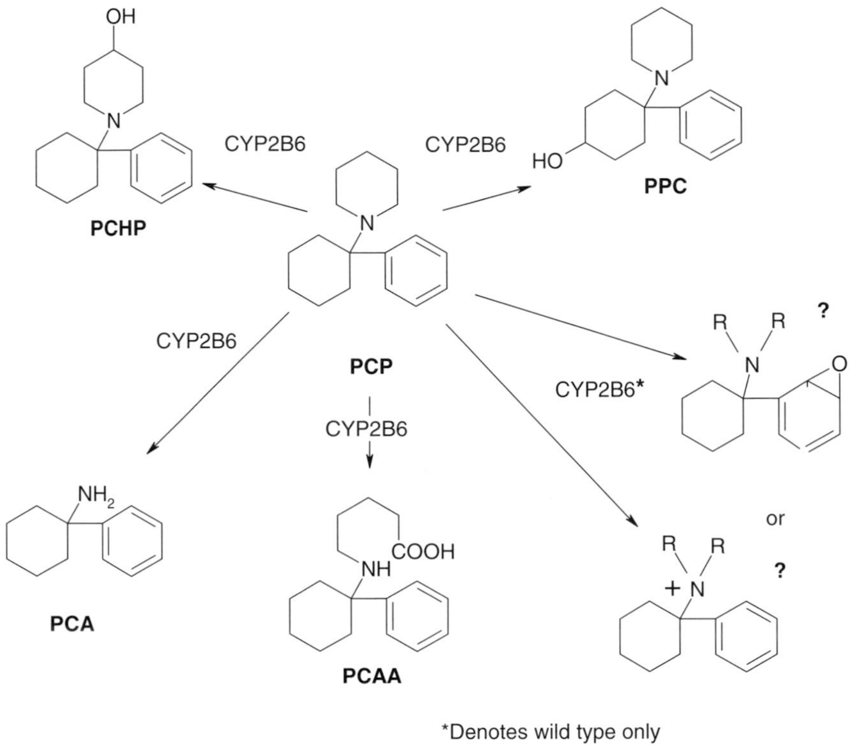

*Denotes wild type only

Figure B.13   Major metabolites of PCP in man; CYP2B6 K262R cannot form the reactive species that inactivates the enzyme

glutamate-binding NMDA receptor as a noncompetitive antagonist. It had a potent analgesic effect, but at clinically effective doses it did not depress respiration or the cardiovascular system, which is highly desirable clinically[119]. Although initially promising, it being the 1960s, the drug was actually marketed even though it caused aggression, hallucinations, general crazed behaviour, and even catatonia in clinical trials. The drug was only withdrawn by the manufacturers as they had developed a suitable replacement, ketamine (see later). By the late 1960s, PCP began to be abused, with unpredictable and often horrible effects that were seriously scary, so mass popularity eluded it. It turned its users into what presented clinically as schizophrenics[119]. Once it was discovered it could be sprayed onto tobacco or cannabis and smoked to exert some rudimentary control of its effects, it gained more adherents, although it remains yet another local drug, confined to some large cities in the United States, notably Los Angeles and Philadelphia.

The drug is effective orally as well as through smoking PCP oil-soaked tobacco. The 'oil' is an ether extract of the drug when illicitly manufactured, which in the right circumstances and quantity is explosive and can reveal the whereabouts of the former illicit factory by the position of the crater and wreckage. On the plus side, it is not difficult to make and the starting materials such as PCC are cheap so profits are very high. When purchasing the drug on the street, if the product has a fishy, metallic, and bitter taste, then this is good. However, PCP with excessive starting material contamination (20% or more) is not good, as the PCC forms cyanide when smoked[119].

In the pure form, PCP is a white crystalline powder, but the street drug can be anything from a powder or pill to a brownish syrupy gum. The powdered form of the drug (angel dust) is virtually pure PCP and is so lipophilic it can be absorbed through the skin in pharmacologically effective quantities. The drug's use declined in the 1980s due to its general scariness, and the relative safety of crack cocaine may well have driven PCP to its nadir in popularity in the mid-1990s. However, interest has revived, perhaps helped along by some more arresting street names, including 'hog', 'wack', and 'embalming fluid'. It is now apparently available with MDMA in a combination called 'elephant flipping'[119].

Physicochemically PCP base is so lipophilic it crosses the blood–brain barrier easily and it exhibits a very wide range of central stimulant and depressant effects, including paranoid delusions, depression, and hallucinations. Some scientists greeted the horrors of PCP enthusiastically, as it could mimic schizophrenia more accurately than MPTP mimics Parkinson's disease. Hence, PCP's NMDA receptor antagonism grew into the basis for a serious challenge to the reigning dopamine theory of schizophrenia, known as the NMDA hypofunction hypothesis. If PCP could cause a schizoid condition by blocking NMDA receptors, then the condition could be due to low stimulation and/or activity linked to the same receptors. Stimulating them with an agonist such as glutamate should then improve the condition[120]. However, NMDA receptor pharmacology is extremely complex, relying on multiple sequential and parallel binding events by numerous ions and amino acids[121]. This might be why 30 years of investigation has failed to bring a new drug to the market that can exploit the NMDA receptor hypothesis.

Those suffering from acute PCP intoxication can be diagnosed by their generally bizarre and violent behaviour, nystagmus (eye oscillations and visual impairment), and a positive urine test for the parent drug. It causes seizures and can be lethal at only 1 mg/kg, often due to a variety of severe reactions, from strokes, respiratory arrest, status epilepticus, and hyperthermia. The drug is spectacularly addictive on repeated use and it is said that users often die violently or commit suicide under its influence. Fortunately, for those who consider PCP as wholly inadequate for their recreational requirements, there are several designer analogues available, such as PCEEA and PCMEA[122], so good luck with those.

PCP metabolism

PCP is extensively metabolised to several main hydroxylated derivatives, including PCHP (1-(1-phenylcyclohexyl)-4-hydroxypiperidine), PPC (4-phenyl-4-(1-piperidinyl) cyclohexanol) and PCAA (5-[N-(1-phenylcyclohexyl)]-aminopentanoic acid;

Figure B.11). Another metabolite has also been found, known as 1-phenylcyclohex-ylamine (PCA). The drug has a long half-life in humans and the effects of one dose can last several days in chronic users. The metabolites are difficult to clear, but are eventually eliminated in the urine and faeces, where they can be detected up to 28 days after drug use. Interestingly, several reactive species were formed by human CYPs and PCP can form an iminium ion catalysed by several CYPs including CYP3A4[123,124].

CYP2B6 is the main isoform that metabolises PCP, although some of its designer equivalents are CYP2D6 substrates[122]. The drug is also capable of forming a mechanism-dependent reactive species that inactivates wild-type CYP2B6[124] but not the K262R variant[125,126] (Chapter 7.2.3), which is probably why PCP is so persistent in heavy users. Interestingly, the K262R variant can form all the dealkylated metabolites the wild-type can. The wild-type lysine is not near the active site, but it promotes substrate binding that facilitates reactive species formation[124,126]. The substitution of the polymorphic arginine prevents this process[124]. The reactive species can be trapped with GSH and is thought to be either a dihydroxylated iminium, or a monohydroxylated epoxide deriva-tive[126]. Hence, the effects of CYP2B6 inhibitors and inducers may depend on the drug user's CYP2B6 expression status. Overall, PCP is something of a 'Nantucket Sleigh ride' and is not recommended.

Ketamine

Effects and clinical usage

Ketamine, as mentioned previously, was an attempt to refine the nightmare out of PCP, whilst retaining the useful aspects (Figure B.14). The drug was first used widely during the Vietnam War from 1970 and has a number of desirable proper-ties aside from its dissociative anaesthetic action and the drug retains a valuable place in clinical practice[121]. This is because it is a good analgesic and hypnotic, with the key benefit of not causing respiratory depression. It does not affect car-diac output, so it is very useful in hypotensive patients those with who have blood pressure and organ perfusion instabilities[121]. Many centres use it for sedat-ing artificially ventilated asthmatics as it is said to have bronchodilatory proper-ties, although this is controversial. Ketamine also improves the effects of opiates, thus reducing doses and, if used during surgery, can lessen pain, both acute and chronic postoperatively. It is generally dosed intravenously or intramuscularly, as oral bioavailability is poor and its effects are less predictable, with more of a sedative rather than an anaesthetic effect, unless the doses are high. The drug is usually given as a racemate, although the S-isomer is four-fold more potent as an analgesic and anaesthetic than the R-isomer[121]. The drug is highly effective at chemically restraining animals; just 350 mg of ketamine with the same dose of

PCP   **Ketamine**

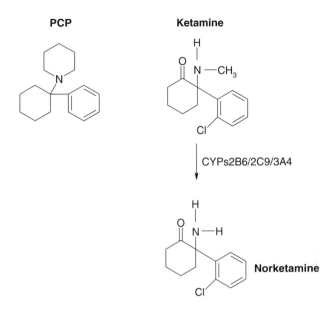

Figure B.14   Structure of ketamine and its major demethylated derivative norketamine, alongside the structure of PCP

xylazine will immobilise a three-ton Asian cow. Indeed, it is very effective if you need any veterinary work done on one of your elephants[127].

Ketamine and depression

This drug continues to be investigated for its benefits in intractable depression and recent studies have focussed on unravelling its CNS actions, which involve repairing glutamine deficits with respect to NDMA receptor activity. Ketamine also can indirectly improve the structural connectivity between synapses[128]. The drug has been used in up to six infusions to cause a sustained increase in mood; this happens relatively rapidly. Indeed, ketamine's effects are reinforced by lithium[121]. Ketamine's success in treating intractable depression[129] has led to the approval by the US FDA of an intranasal formulation of the S-enantiomer (esketamine; Spravato) in March 2019, although clinical reaction has been cautious.[130]

Ketamine abuse

Ketamine is abused for its hallucinogenic, dissociative, and narcotic properties, and has been scheduled as a controlled drug in most countries. It appears to detach users successfully from who they are, why they are here, and what they are doing, so it is not surprising that it has been said that is an 'ideal' date-rape drug[131]. It can be sniffed or snorted and is known by a number of street names, such as 'Special K', 'Kit-kat', 'wonk' and 'boing'. OK, so the last name I made up. Using ketamine admits an individual to 'k-land' or into the 'k-hole', apparently.

Although ketamine is not conventionally physiologically addictive, tolerance occurs that promotes a compulsive desire to use it[131]. Whilst the veracity of online accounts of behaviour linked to the drug might be suspect, one blog entry I read concerned a frustrated would-be ketamine tourist, who was convinced that Vietnam was a wonderland of gullible and friendly vendors of cheap ketamine. Insisting they knew what they were doing, this young man and his girlfriend had toured over 50 pharmacies and several hospitals vainly trying 'believable' cover stories on the not-as-it-turned-out very gullible locals.

On a more significant level, abuse of this drug is now recognised to cause ketamine cystitis in around a third of chronic users[129]. This progressive and dose/frequency of use condition is characterized by severe bladder mucosa damage and pain, increased urinary urge and sometimes-attendant renal problems[129]. Relatively heavy use (several sessions a week for months) can cause the syndrome and the only resolution is to stop using the drug. If the user cannot stop, the condition will progress towards surgery and possible ileostomy, a very high rate of surgical complications due to the inflammation[131] and possible lifelong infirmity. The mechanism has been proposed to be direct cytotoxicity by the drug and/or its metabolites alongside an inflammatory cascade of events[131]. It does not inspire confidence that one of its leading advocates and several people since have drowned in their baths while under its influence. That might well be because it is a dissociative and paralysing anaesthetic. Users often combine ketamine with crystal meth, cocaine, cannabis, and alcohol – perhaps at the same time. Even when used medically, rapid intravenous injection can stop a patient breathing, so it is more than unwise to disrespect ketamine, unless your friends are particularly loyal, sober, and adept at CPR.

Ketamine metabolism

The poor (8–24%) bioavailability of ketamine is due to its high hepatic extraction that is equal to blood flow. Hence, the design of the intranasal formulation of Spravato, which circumvents this problem[130]. Around 80% of the dose is rapidly N-demethylated to norketamine (N-desmethylketamine), which is thought to have at best a third of the analgesic action of the parent. Norketamine, as well as the parent is also hydroxylated. Ketamine metabolism has some stereoselective aspects, as the S-isomer when given alone is cleared faster than the R-isomer and it is thought that the latter may inhibit the former when the drug is given as a racemate[121]. CYP2B6 is believed to be the main high affinity low capacity route of clearance, as well as CYP2C9, with some CYP3A4 involvement as the low affinity, high capacity isoform[121]. This is supported by *in vitro* data where methadone and diclofenac (CYP2B6 and CYP2C9 substrates) are known to inhibit ketamine conversion to norketamine in human microsomes, probably by competition[132]. The clearance of ketamine is likely to be affected by many factors, not least the high variability of CYP2B6 and to a lesser extent, CYP2C9, as well as the effects of inducers and inhibitors of these CYPs. The CYP2B6 inhibitor ticlopidine can inhibit S-ketamine clearance[133], whilst inducers such as rifampicin accelerate ketamine clearance. From its emergence in the various dance/rave scenes in the 1990s, ketamine abuse is increasing and the problems it causes are becoming more significant in terms of its impact on the health of its users.

Methoxetamine (MXE)

This NMDA antagonist began as a legal addition to the class of dissociative anaesthetics, selling in 'head' (drug paraphernalia) shops before it attracted the attention of the forces of law and order. It is known on the street as 'MEX', 'm-ket', and 'Special M', which underscores the link between ketamine users. It is closer structurally to ketamine than to PCP, fortunately. MXE became popular as it resembles ketamine, but with longer duration[134]. Aside from the basic dissociative floating effects, the drug has the antidepressant and some of analgesic properties of ketamine. The analgesic properties are less pronounced in comparison with ketamine, because MXE has a 3-methoxy group where ketamine has a 2-chloro group[134]. What makes MXE a 'crossover' drug is that it blocks all the main biogenic amine transporters, as well as that of serotonin, giving it both amphetamine-like and entactogen qualities. These include cardiovascular as well as CNS stimulation, but along with the usual euphoria, hallucinations and sensory distortions, empathy and friendliness is a welcome bonus. The negative effects are pretty much as you might expect, along with what was listed rather worryingly as 'a near death experience'[134], which presumably is not good. It can cause cerebellar toxicity, which manifests as ataxia, coordination problems, inability to keep balance, and reduced state of consciousness[135]. The patients recovered after a few days, so no problem there.

MXE can be sniffed, snorted, or used orally, as well as injected, and the effects last for up to seven hours, so it is cleared more slowly than ketamine, but through similar pathways with the same CYPs. It is mostly O and N-demethylated and hydroxylated by CYP2B6 and CYP3A4[134]. Those that make and sell MXE have promoted it as 'bladder friendly'[136] on the basis that it is more potent than ketamine so will not cause so much damage. Time and the world's medical case literature will explore this claim.

B.8   Charlie Don't Surf![137]

To paraphrase John Gillespie Magee's (1922–1941) poem 'High Flight'[138], most us at some point like to slip the surly bonds of earth and touch the face of whatever God we favour. This could be in a Spitfire, or through music, sport, or even work. As we have seen, many others wish to 'cut to the chase' towards transcendence and use drugs, licit and illicit. If legal highs are not enough, users will choose from the range of illegal agents available, and they will eventually meet the brick wall of the law. Governments have rarely tried to depart from the standard blanket bans of what are perceived as dangerous substances, and when they do, the fear of the electoral consequences and a genuine concern over the potential casualties makes them return to the proscribing fold. Indeed, most societies conduct ongoing debates for and against legalisation of drugs, seemly without resolution. Most would agree that prohibiting psychoactive substances to minors is the right path, and some feel that those that contravene the law in this respect should be 'corrected' in the harshest possible way. However, as soon as an individual attains their legal majority, whatever the law might state and whatever law enforcement might desire, the key choices for the individual remain – which substance (s), how much and how often?

Whilst we are all well aware of the somewhat impulsive nature of most psychoactive substance use, ideally, if choice is inevitable, it should be informed. To achieve this, it makes sense to gather as much information as possible. Hence, I hope this Appendix has provided the reader with some basic knowledge that can be built on, in terms of the direct pharmacological, toxicological, and mental health consequences of using various drugs, including alcohol from Chapter 7.7. It could be said that this is as far as a text on drug metabolism should or needs to go – indeed, it could also be said that this tome has gone rather further into this area than is necessary. In addition, it does not take any in-depth analysis of this Appendix to discern that the author feels that using mind-altering substances is definitely 'Charlie's Point'[137]. However, this book was intended to be more than a basic guide to drug metabolism, so in that spirit, there is one more perspective that needs to be conveyed that will assist the would-be illicit drug user in their choices.

Each psychoactive substance, including alcohol and some prescription medicines, such as the opiates, benzodiazepines, and antidepressants, as well as all the illicit ones, is its own vast enterprise. The design, synthesis, manufacture, distribution, selling, consumption, effects, and consequences of each drug could be said to comprise an independent 'living' organism – an 'entity'. The drug entity may be local, national, or international, it may be thriving or waning, but it is usually available to be found and joined by the user on demand. In joining the drug entity, the users might know what they want from the relationship – but perhaps they should pause and ask, what might the drug entity want from them?

One way to address this question is to watch a Dutch motion picture called *Spoorloos*[139], rendered in English as *The Vanishing*. A better translation is

Traceless. Rather than check it out on the net or attempt any prior investigation, I would invite you to just watch the film and ignore the grossly inferior Hollywood remake. You can never know everything you need to know, but you can know more than you do now.

> Were such things here that we do speak about, or have we eaten on the insane root that takes the reason prisoner?

Act 1, Scene 3, *Macbeth*, by William Shakespeare

References

1. Davis MP, Walsh D. Epidemiology of cancer pain and factors influencing poor pain control. Am. J. Hosp. Palliat. Care 21(2), 137–142 2004.
2. Pathan H & Williams J. Basic opioid pharmacology: an update. Brit. J. Pain 6, 11–16, 2012.
3. De Gregori S, De Gregori M,Ranzani GN, et al. Morphine metabolism, transport and brain disposition. Metab. Brain Dis. 27(1), 1–5, 2012.
4. Klimas R, & Mikus G. Morphine-6-glucuronide is responsible for the analgesic effect after morphine administration: a quantitative review of morphine, morphine-6-glucuronide, and morphine-3-glucuronide. Brit. J. Anaesth. 113 (6), 935–44, 2014.
5. Chau N, Elliot DJ, Lewis BC, et al. Morphine glucuronidation and glucosidation represent complementary metabolic pathways that are both catalyzed by UDP-Glucuronosyltransferase 2B7: Kinetic, inhibition, and molecular modeling studies. J. Pharmacol. Exp. Ther. 349, 126–137, 2014.
6. Fudin J, Fontenelle DV, Payne A. Rifampin reduces oral morphine absorption: a case of transdermal buprenorphine selection based on morphine pharmacokinetics. J Pain Palliat. Care Pharmacother. 26(4), 362–367, 2012.
7. Kotlinska-Lemieszek A, Klepstad P, Faksvåg Haugen D. Clinically significant drug–drug interactions involving opioid analgesics used for pain treatment in patients with cancer: a systematic review. Drug Des. Dev.Ther. 9, 5255–5267, 2015.
8. Yokell MA, Zaller ND, Green TC, et al. Buprenorphine and buprenorphine/naloxone diversion, misuse, and illicit use. Curr. Drug Abuse Rev. 4(1), 28–41, 2011.
9. McGregor E, Bremner E, Lee Miller J, et al. In 'Trainspotting' Dir. Boyle D. 93 min, Channel 4 Films Released 23 February, 1996.
10. Cone EJ, Welch P, Mitchell JM, et al. Forensic drug testing for opiates: I. Detection of 6-acetylmorphine in urine as an indicator of recent heroin exposure; drug and assay considerations and detection times. J. Anal. Toxicol. 15(1), 1–7, 1991.
11. Ellis AD, McGwin G, Davis GG, et al. Identifying cases of heroin toxicity where 6-acetylmorphine (6-AM) is not detected by toxicological analyses. Forensic Sci. Med. Pathol. 12, 243–247, 2016.
12. Antonilli L, Semeraro F, Suriano C, et al. High levels of morphine-6-glucuronide in street heroin addicts. Psychopharmacol. (Berl). 170(2), 200–204. Epub 2003.

13. Ciccarone D, Bourgois P. Explaining the geographical variation of HIV among injection drug users in the United States. Subst. Use Misuse. 38(14), 2049–2063, 2003.

14. Alves EA, Cornelis Grund JP, Afonso CM, et al. The harmful chemistry behind krokodil (desomorphine) synthesis and mechanisms of toxicity. Forens. Sci. Int. 249, 207–213, 2015.

15. Savchuk SA, Barsegyan SS, Barsegyan IB et al. Chromatographic study of expert and biological samples containing desomorphine, J. Anal. Chem. 63, 361–370, 2008.

16. Marx RE. Uncovering the cause of "phossy jaw" Circa 1858 to 1906: oral and maxillofacial surgery closed case files-case closed. J. Oral Maxillofac. Surg. 66(11), 2356–63. 2008.

17. Gowing L, Ali R, White JM, Mbewe D. Buprenorphine for managing opioid withdrawal. Coch. Database Syst. Rev. CD002025, 2017.

18. Khanna IK, Pillarisetti S. Buprenorphine – an attractive opioid with underutilised potential in treatment of chronic pain. J. Pain Res. 8, 859–870, 2015.

19. McCance-Katz, EF, Moody, DE, Morse, GD, et al. Interactions between buprenorphine and antiretrovirals. I. The nonnucleoside reverse-transcriptase inhibitors efavirenz and delavirdine. Int. Rev. Clin. Infect. Dis. 43, SUPPL. 4, S224–S234, 2006.

20. Novac A, Iosif AM Groysman R et al. Implications of sensorineural hearing loss with hydrocodone/acetaminophen abuse. Prim Care Companion CNS Disord. 17(5), 2015.

21. Csete M, Sullivan JB. Vicodin-induced fulminant hepatic failure. Anesthesiology. 79(4), 857–860, 1993.

22. Drug and Opioid-Involved Overdose Deaths – United States, 2013–2017. Morbidity and Mortality Weekly Report 67(5152), 1419–1427, 2019. https://www.cdc.gov/mmwr/volumes/67/wr/mm675152e1.htm?s_cid=mm675152e1_w.

23. Rhodan M. November 6, 2017 Gun related deaths keep going up http://time.com/5011599/gun-deaths-rate-america-cdc-data/.

24. McGreal C. Don't blame addicts for America's opioid crisis. Here are the real culprits. The Guardian August 13, 2017. https://www.theguardian.com/commentisfree/2017/aug/13/dont-blame-addicts-for-americas-opioid-crisis-real-culprits

25. Samer CF, Daali Y, Wagner M et al. Genetic polymorphisms and drug interactions modulating CYP2D6 and CYP3A activities have a major effect on oxycodone analgesic efficacy and safety. Br J Pharmacol. 160(4), 919–930, 2010.

26. Smith HE. Opioid metabolism. Mayo Clin. Proc. 84(7), 613–624, 2009.

27. Andreassen TN, Eftedal I, Klepstad P, et al. Do CYP2D6 genotypes reflect oxycodone requirements for cancer patients treated for cancer pain? A cross-sectional multicentre study. Eur. J. Clin. Pharmacol. 68, 55–64, 2012.

28. Langston JW, Palfreman J. The Case of the Frozen Addicts: How the Solution of a Medical Mystery Revolutionized the Understanding of Parkinson's Disease. 264, Revised ed. edition (15 Jan. 2014) ISBN-10: 9781614993315, IOS Press, 2014.

29. Pacino A, Pfeiffer M, Bauer S, et al. In Scarface, dir. De Palma, B, 170 min, Universal Pictures, Released US-wide Dec 9th 1983.

30. Roxburgh A, Burns L, Drummer OH, et al. Trends in fentanyl prescriptions and fentanyl-related mortality in Australia. Drug and Alcoh. Rev. 32(3), 269, 275. 2013.

31. Fomin D, Baranauskaite V, Usaviciene E, et al. Human deaths from drug overdoses with carfentanyl involvement – new rising problem in forensic medicine. Medicine (Baltimore). 97(48), e13449, 2018.

32. Deshpande, CM, Mohite, SN, and Kamdi, P. Sufentanil vs fentanyl for fast-track cardiac anaesth. Ind.J. Anaesth. 53(4), 455–462, 2009.

33. Gottlieb S. US FDA Approval; Dsuvia (sufentanyl). https://www.fda.gov/NewsEvents/Newsroom/PressAnnouncements/ucm624968.htm.

34. Yong Z, Hao W, Weifang Y, et al. Synthesis and analgesic activity of stereoisomers of cis-fluoro-ohmefentanyl. Pharmazie. 58(5), 300–302, 2003.

35. British National Formulary; Methadone hydrochloride https://bnf.nice.org.uk/drug/methadone-hydrochloride.html

36. Nicholson AB, Watson GR, Derry S, et al. Cochrane database of systematic reviews. Methadone for cancer pain (Review) 2, CD003971, 2017.

37. Mercadante S, Casuccio A, Agnello A, et al. Morphine versus methadone in the pain treatment of advanced-cancer patients followed up at home. J. Clin. Oncol. 16(11), 3656–3661, 1998.

38. Michalska M, Katzenwadel A, Wolf P. Methadone as a "tumor theralgesic" against cancer. Front Pharmacol. 8, 733, 1–5 2017.

39. Taheri Fl, Yaraghi A, Sabzghabaee AM, et al. Methadone toxicity in a poisoning referral center. J Res Pharm Pract. 2(3), 130–134, 2013.

40. Hawley P, Chow L, Fyles G, et al. Clinical Outcomes of Start-Low, Go-Slow Methadone Initiation for Cancer-Related Pain: What's the Hurry? J. Palliat. Med. 20(11), 1244–1251, 2017.

41. Kharasch, ED, Bedynek PS, Park S, et al. Mechanism of ritonavir changes in methadone pharmacokinetics and pharmacodynamics I. Evidence against CYP3A mediation of methadone clearance. Clin. Pharmacol. Ther. 84(4), 497–505, 2008.

42. Kreek MJ, Garfield JW, Gutjahr CL, et al. Rifampin induced methadone withdrawal. New Engl. J. Med. 294(20), 1104–1106, 1976.

43. Tarumi Y, Pereira J, Watanabe S. Methadone and fluconazole: respiratory depression by drug interaction. J. Pain and Symp. Man. 23, 148–153, 2002.

44. Ahmad T, Valentovic MA, Rankin GO. Effects of cytochrome P450 single nucleotide polymorphisms on methadone metabolism and pharmacodynamics. Biochem Pharmacol. 153:196–204, 2018.

45. Pomara C, Cassano T, D'Errico S, et al. Data available on the extent of cocaine use and dependence: biochemistry, pharmacologic effects and global burden of disease of cocaine abusers. Curr. Med. Chem. 19, 5647–5655, 2012

46. Chen X, Zheng X, Zhan M, et al. Metabolic enzymes of cocaine metabolite benzoylecgonine. ACS Chem. Biol. 19; 11(8), 2186–2194, 2016.

47. Leduc B, Sinclair PR, Shuster L, et al. Norcocaine and N-Hydroxynorcocaine Formation in Human Liver Microsomes: Role of Cytochrome P-450 3A4 Pharmacology 46(5), 294–300, 1993.

48. Larrey D, Ripault MP. Chapter 25-Hepatotoxicity of Psychotropic Drugs and Drugs of Abuse, in Drug-Induced Liver Disease (Third Edition), 443–462 Academic Press, 2013.

49. Herschman Z, Aaron C. Prolongation of cocaine effect. Anesthesiology 74(3), 631–632, 1991.

50. Schindler CW, Goldberg SR, Accelerating cocaine metabolism as an approach to the treatment of cocaine abuse and toxicity. Fut. Med. Chem. 4(2), 163–175, 2012.

51. Andrews P. Cocaethylene toxicity. J Addict Dis. 16(3), 75–84, 1997.

52. Treadwell SD, Robinson TG. Cocaine use and stroke. Postgrad. Med. J. 83(980), 389–394, 2007.

53. Hofmann, A. The Discovery of LSD and Subsequent Investigations on Naturally Occurring Hallucinogens Chapter 7 of Discoveries in Biological Psychiatry, eds Ayd FJ & Blackwell B, eds., Lippincott Company, 1970.

54. Schiff PL, Ergot and Its Alkaloids. Am. J. Pharm. Educ. 70(5), 98, 2006.

55. Masters R & Houston J. The Varieties Of Psychedelic Experience, 336, Park Street Press, ISBN-10: 0892818972, 2000.

56. Vardy MM, Kay SR. LSD psychosis or LSD-induced schizophrenia? A multimethod inquiry. Arch Gen Psych. 40(8), 877–883, 1983.

57. Hermle L, Simon M, Ruchsow M, and Geppert, M. Hallucinogen-persisting perception disorder. Ther Adv Psychopharmacol. 2(5), 199–205, 2012

58. Halpern J. Hallucinogens: an update. Curr. Psych. Rep. 5, 347–354, 2003

59. Preller KH, Razib A, Zeidman P, et al. Effective connectivity changes in LSD-induced altered states of consciousness in humans PNAS 116 (7) 2743–2748, 2019.

60. Carhart-Harris RL, Bolstridge M, Rucker J, et al. (2016) Psilocybin with psychological support for treatment resistant depression: An open-label feasibility study. Lancet Psychiatry 3:619–627, 2016.

61. Yoo WS, Kim YS, We Y, et al. Secretio Bufonis: A Traditional Supplement in Asia J. Acupunc. Merid. Stud. 2, Issue 2, 159–164, 2009.

62. Colombo EE, Martinelli R, Brondino V, et al. Elevated urine levels of bufotenine in patients with autistic spectrum disorders and schizophrenia. Neuro. Endocrinol. Lett. 31, 117–21, 2010.

63. Tittarelli R, Mannocchi G, Pantano F, et al. Recreational Use, Analysis and Toxicity of Tryptamines. Current Neuropharmacology, 13(1), 26–46, 2015.

64. Shen H, Jiang X, Winter JC, et al. Psychedelic 5-Methoxy-N,N-dimethyltryptamine: Metabolism, Pharmacokinetics, Drug Interactions, and Pharmacological Actions. Curr. Drug Met. 11, 659–666, 2010.

65. European Monitoring Centre for Drugs and Drug Addiction and Europol., EU Drug Markets Report: Strategic Overview, EMCDDA – Europol Joint publications. Publications Office of the European Union, Luxembourg, 2016, http://www.emcdda.europa.eu/start/2016/drug-markets#panel

66. Ulrich A. Hitler's drugged soldiers. Der Spiegel 6.5.2005. http://www.spiegel.de/international/the-nazi-death-machine-hitler-s-drugged-soldiers-a-354606.html.

67. Kamienski L, Shooting Up: A Short History of Drugs and War. 111–113. Oxford University Press. ISBN 9780190263478, 2016.

68. de la Torre R, Yubero-Lahoz S, Pardo-Lozano R, et al. MDMA, methamphetamine, and CYP2D6 pharmacogenetics: what is clinically relevant? Front. Genet Article 235, 1–8, 2012.

69. Heal DJ, Smith SL, Gosden J, et al. Amphetamine, past and present – a pharmacological and clinical perspective. J. Psychopharmacol. 27, 479–496, 2013.

70. Buxton JA, Dove NA, The burden and management of crystal meth use. CMAJ 178, 1537–1539, 2008.

71. Yamamoto BK, Moszczynska A Gudelsky, GA. Amphetamine toxicities Classical and emerging mechanism. Ann N. Y. Acad. Sci. 1187: 101–120, 2010.

72. UK Home Office; National Statistics; Drug misuse: findings from the 2017 to 2018 CSEW. https://www.gov.uk/government/statistics/drug-misuse-findings-from-the-2017-to-2018-csew.

73. UK Office for National Statistics; Deaths related to drug poisoning by selected substances https://www.ons.gov.uk/peoplepopulationandcommunity/birthsdeathsandmarriages/deaths/datasets/deathsrelatedtodrugpoisoningbyselectedsubstances.

74. UK Office for National Statistics. Alcohol-specific deaths in the UK: registered in 2017. https://www.ons.gov.uk/peoplepopulationandcommunity/healthandsocialcare/causesofdeath/bulletins/alcoholrelateddeathsintheunitedkingdom/registeredin2017.

75. Jungaberle H, Thal S, Zeuch A, et al. Positive psychology in the investigation of psychedelics and entactogens: a critical review. Neuropharmacol. 142, 179–199, 2018.

76. Di Trapani L, Eiden C, Mathieu O, et al. Life-threatening intoxications related to persistent MDMA (3,4-methylenedioxymethamphetamine) concentrations. Toxicol. Analyt. Clin. 30, 80–83, 2018

77. Winstock AR, Wolff K, Ramsey J. 4-MTA: a new synthetic drug on the dance scene Drug and Alc. Dep. 67, 111–115, 2002.

78. Nichols D. Legal highs: the dark side of medicinal chemistry. Nature. 469 (7328), 7, 2011.

79. Salomon RM, Karageorgiou J, Dietrich MS, et al. MDMA (Ecstasy) association with impaired fMRI BOLD thalamic coherence and functional connectivity. Drug Alc. Depend. 1; 120(1–3), 41–4, 2012.

80. Boyce S, Kelly E, Reavill C, et al. Repeated administration of N-methyl-4-phenyl 1,2,5,6-tetrahydropyridine to rats is not toxic to striatal dopamine neurones. Biochem. Pharmacol. 33, 1747–1752, 1984.

81. McCann UD, Szabo Z, Seckin E, et al. Quantitative PET studies of the serotonin transporter in MDMA users and controls using [11C]McN5652 and [11C]DASB. Neuropsychopharm. 30, 1741–1750, 2005.

82. Steinkellner T, Freissmuth M, Sitte HH, et al. The ugly side of amphetamines: short-and long-term toxicity of 3,4-methylenedioxymethamphetamine (MDMA, 'ecstasy'), methamphetamine and D-amphetamine. Biol. Chem. 392, 103–115, 2011.

83. Schmid Y, Vizeli P, Hysek CM, et al. CYP2D6 function moderates the pharmacokinetics and pharmacodynamics of 3,4-methylene-dioxymethamphetamine in a controlled study in healthy individuals. Pharmacogen. Genom. 26, 397–401 2016.

84. Vizeli P, Schmid Y, Prestin K, et al,.Pharmacogenetics of ecstasy: CYP1A2, CYP2C19, and CYP2B6 polymorphisms moderate pharmacokinetics of MDMA in healthy subjects. Eur. Neuropsychopharmacol. 27, 232–238, 2017.

85. Segura M, Farré M, Pichini S, et al. Contribution of cytochrome P4502D6 to 3,4-methylene dioxymethamphetamine disposition in humans: use of paroxetine as a metabolic inhibitor probe. Clin. Pharmacokinet. 44, 649–660, 2005.

86. Shulgin, A and Shulgin A. PiHKAL: A Chemical Love Story. Transform Press, pp 978 1991.

87. de Sousa Fernandes Perna EB, Theunissen EL, Dolder PC, et al. Safety profile and neurocognitive function following acute 4-fluoroamphetamine (4-fa) administration in humans. Front. Pharmacol. 9, 713, 2018.

88. Wijers CH, Van Litsenburg RT, Hondebrink L, et al. Acute toxic effects related to 4-fluoroamphetamine. Lancet. 389(10069), 600. 2017.

89. Bedada W, de Andrés F, Engidawork E, et al. Effects of khat (catha edulis) use on catalytic activities of major drug-metabolizing cytochrome P450 enzymes and implication of pharmacogenetic variations. Sci. Rep. 8, 12726, 2018.

90. Feng LY, Battulga A, Han E, et al. New psychoactive substances of natural origin: a brief review. J. Food & Drug Anal. 25, 461, e471, 2017.

91. Hoffman R, Al'Absi M. Khat use and neurobehavioral functions: suggestions for future studies. J. Ethnopharmacol. 132, 554e63, 2010.

92. Bennett JW, Pybus BS, Yadava A, et al. Primaquine failure and cytochrome P-450 2D6 in Plasmodium vivax malaria. N. Engl. J. Med. 369, 1381–1382, 2013.

93. Ordak, M, Nasierowski T. The problem of mephedrone in Europe: causes and suggested solutions. Eur. Psychiatr. 55, 43–44, 2019.

94. Ferreira B, Dias da Silva D, Carvalho F, et al. The novel psychoactive substance 3-methylmethcathinone (3-MMC or metaphedrone), a review. For. Sci. Int. 295, 54–63, 2019.

95. Pozo OJ, Ibáñez M, Sancho JV, et al. Mass spectrometric evaluation of mephedrone in vivo human metabolism: identification of Phase I and Phase II metabolites, including a novel succinyl conjugate Drug Metab. Dispos. 43, 248–257, 2015.

96. Kolesnikova TO, Khatsko SL, Demin KA, et al. DARK classics in chemical neuroscience: α-pyrrolidinovalerophenone ("flakka") ACS Chem. Neurosci. 10 (1), 168–174, 2019.

97. Hutton F. BZP-'Party pills', populism and prohibition: Exploring global debates in a New Zealand context. Aust. New. Zeal. J. Crim. 50, 282–306, 2017.

98. Kerr JR, Davis LS. Benzylpiperazine in New Zealand: brief history and current implications, J.Roy. Soc. New Zeal. 41:1, 155–164, 2011.

99. Gee P, Jerram T, Bowie D. Multi-organ failure from 1-benzylpiperazine ingestion legal high or lethal high. Clin. Tox. 48, 230233, 2010.

100. Antia U, Tingle MD, Russell BR. Metabolic interactions with piperazine-based 'party pill' drugs. J. Pharm. Pharmacol. 61, 877–882, 2009.

101. Antia U, Lee HS, Kydd RR, et al. Pharmacokinetics of 'party pill' drug N-benzylpiperazine (BZP) in healthy human participants. For. Sci. Int. 186, 63–67, 2009.

102. Bridgeman MB, Abazia DT. Medicinal cannabis: history, pharmacology, and implications for the acute care setting. PT. 2017 Mar 42(3), 180–188.

103. Maurya N, Velmurugan BK. Therapeutic applications of cannabinoids. Chem. Biol. Interact. 293:77–88, 2018.
104. Uranga JA, Vera G, Abaloa R. Cannabinoid pharmacology and therapy in gut disorders. Biochem. Pharmacol. 157, 134–147, 2018.
105. Wilson J, Freeman TP, Mackie CJ et al. Effects of increasing cannabis potency on adolescent health. Lan. Child. Adol. Health 3, 121–128, 2018.
106. Sam AH, Salem V, Ghatei MA. Rimonabant: From RIO to ban. J. Obes. 432607, 1–4 2011.
107. Hausman-Kedem M, Menascu S, Kramer U, Efficacy of CBD-enriched medical cannabis for treatment of refractory epilepsy in children and adolescents-An observational, longitudinal study. Brain & Dev. 40, 544–551, 2018.
108. Watanabe K, Yamaori S, Funahashi T, et al. Cytochrome P450 enzymes involved in the metabolism of tetrahydrocannabinols and cannabinol by human hepatic microsomes. Life Sci. 80, 1415–1419, 2007.
109. Sachse-Seeboth C, Pfeil J, Sehrt D, et al. Interindividual variation in the pharmacokinetics of delta-9-tetrahydrocannabinol as related to genetic polymorphisms in CYP2C9. Clin Pharmacol Ther. 85:273–276, 2009.
110. Stout SMI, Cimino NM. Exogenous cannabinoids as substrates, inhibitors, and inducers of human drug metabolizing enzymes: a systematic review. Drug Metab Rev. 46(1), 86–95, 2014.
111. Schwilke EW, Schwope DM, Karschner EL. Δ^9-Tetrahydrocannabinol (THC), 11-hydroxy-THC, and 11-nor-9-carboxy-THC plasma pharmacokinetics during and after continuous high-dose oral THC. Clin. Chem. 55(12), 2180–2189, 2009.
112. Sharma P, Murthy P, Bharath S. Chemistry, metabolism, and toxicology of cannabis: clinical implications. Iran J Psych. 7(4), 149–156, 2012.
113. Ribeiro L, Ind PW. Marijuana and the lung: hysteria or cause for concern? Breathe 14, 196–205, 2018.
114. Reece AS, Hulse GK. Chromothripsis and epigenomics complete causality criteria for cannabis-and addiction-connected carcinogenicity, congenital toxicity and heritable genotoxicity. Mut. Res. 789, 15–25, 2016.
115. Leadbeater, Bonnie J.; Ames, Megan E.; Linden-Carmichael, Ashley N. Age-varying effects of cannabis use frequency and disorder on symptoms of psychosis, depression and anxiety in adolescents and adults. Addiction 114, 278–293, 2019.
116. Szutorisza H, Hurdab YL. High times for cannabis: Epigenetic imprint and its legacy on brain and behavior Neurosc. & Biobehav. Rev. 85, 93–101, 2018.
117. Fantegrossi WE, Moran JH, Radominska-Pandya A, et al. Distinct pharmacology and metabolism of K2 synthetic cannabinoids compared to Δ9-THC: Mechanism underlying greater toxicity? Life Sci. 27, 97(1), 45–54, 2014.
118. Castaneto MS, Gorelick DA, Desrosiers NA, et al. Synthetic cannabinoids: epidemiology, pharmacodynamics, and clinical implications. Drug Alcohol Depend. 1, 0: 12–41, 2014.
119. Bertron JL, Seto M, Lindsley CW, et al. DARK classics in chemical neuroscience: phencyclidine (PCP). ACS Chem. Neuro. 9, 2459–2474, 2018.

120. Moghaddam B, Javitt D. From revolution to evolution: the glutamate hypothesis of schizophrenia and its implication for treatment. Neuropsychopharmacol. 2012 Jan; 37(1), 4–15, 2012.

121. Peltoniemi MA, Hagelberg NM, Olkkola, KT, et al. Ketamine: a review of clinical pharmacokinetics and pharmacodynamics in anesthesia and pain. Therapy. Clin. Pharmacokinet. 55, 1059–1077, 2016.

122. Sauer C, Peters FT, Schwaninger AE, et al. Investigations on the cytochrome P450 (CYP) isoenzymes involved in the metabolism of the designer drugs N-(1-phenyl cyclohexyl)-2-ethoxyethanamine and N-(1-phenylcyclohexyl)-2-methoxyethan-amine. Biochem. Pharmacol. 77, 444–450, 2009.

123. Jushchyshyn MI, Kent UM, Hollenberg PF. The mechanism-based inactivation of human cytochrome P4502B6 by phencyclidine. Drug Metab. Disp. 31, 46–52, 2003.

124. Jushchyshyn MI, Wahlstrom JL, Hollenberg P, et al. Mechanism of inactivation of human cytochrome P450 2B6 by phencyclidine. Drug Metab. Disp. 34(9), 1523–1529, 2006.

125. Talakad JC, Kumar S, Halper JR. Decreased susceptibility of the cytochrome P450 2B6 variant K262R to inhibition by several clinically important drugs. Drug Metab. Disp. 37, 644–650, 2009.

126. Shebley M, Hollenberg PF. Mutation of a single residue (K262R) in P450 2B6 leads to loss of mechanism-based inactivation by phencyclidine. Drug Metab. Dispos. 35(8), 1365–1371, 2007.

127. Jacobson, ER. 1988. Chemical restraint and anesthesia of elephants. Proc. Ann. Elephant Workshop 9. Pages: 112–119.

128. Colic L, McDonnell C, Li M, et al. Neuronal glutamatergic changes and peripheral markers of cytoskeleton dynamics change synchronically 24 h after sub-anaesthetic dose of ketamine in healthy subjects. Behav. Brain Res. 359, 312–319, 2019.

129. Zheng W, Zhou YL, Liu WJ, et al. Neurocognitive performance and repeated-dose intravenous ketamine in major depressive disorder. J Affect. Dis. 246, 241–247, 2019.

130. Perez-Esparza R, Kobayashi-Romero LF, Garcia-Mendoza AM, et al. Promises and concerns regarding the use of ketamine and esketamine in the treatment of depression. Acta Psychia. Scand. 140, 182–183, 2019.

131. Misra S, Ketamine-associated bladder dysfunction – a review of the literature. Curr. Bladder Dys. Rep. 13, 145–152, 2018.

132. Capponi L, Schmitz A, Thormann W, et al. *In vitro* evaluation of differences in phase 1 metabolism of ketamine and other analgesics among humans, horses, and dogs. American J. Veterin. Res. 70, 777–786, 2009.

133. Peltoniemi MA, Saari TI, Hagelberg NM, et al. Exposure to oral S-ketamine is unaffected by itraconazole but greatly increased by ticlopidine. Clin Pharmacol Ther. 90, 296–302, 2011.

134. Botana CJ, de la Peña B, Kim HJ, et al. Methoxetamine: A foe or friend? Neurochem. Int. 122, 1–7, 2019.

135. Shields JE, Dargan PI, Wood DM, et al. Methoxetamine associated reversible cerebellar toxicity: three cases with analytical confirmation. Clin. Toxicol. 50, 438–440, 2012.

136. Lawn W, Borschmann R, Cottrell A. Methoxetamine: prevalence of use in the USA and UK and associated urinary problems. J. Subst. Use 21,115–120, 2016.

137. Brando M, Duvall R, Sheen M, et al. 'Apocalypse Now' dir. Coppola FF, 153 min, Omni Zoetrope Productions, first US release, August 15th 1979. 'Apocalypse Now: Final Cut', 183 min. released August 15th, 2019.

138. BBC 'Inside Out' 23rd February, 2007. http://www.bbc.co.uk/insideout/yorkslincs/series11/week7_poem_flying.shtml.

139. Donnadieu BP, Bervoets C, ter Steege, J, et al. 'Spoorlos', Dir. Sluizer G. 107 min, Golden Egg Films, Released 27 October 1988.

Appendix C
Examination Techniques

C.1   Introduction

This section does not just apply to drug metabolism but to almost any life-science subject. Obviously, the type of examination you might be sitting for this subject could be anything from continuous assessment, through to multiple choice or essay-style questions. It is true that the vast majority of students extract from the university system more or less exactly what they put in. Every student, if they are honest, knows that they will get the degree they deserve, and most put in enough effort to achieve this. If all you want is a modest degree, then you will not exert yourself, no matter what your lecturers say. However, there will be a fair number of students who would like to obtain the very best degree they can, but are unsure as to how to achieve this. Often, it is possible to emerge from 13 years of school examinations and still be none the wiser as to how to produce a first-class performance, *even though you are capable of it*. So how is it done?

C.2   A first-class answer

This does not consist of simply reproducing, like an elaborate living photocopier, the notes your lecturer gave you; much more is needed than this. The detail of the lectures is a starting point, a platform, if you like, for building a first-class answer. If you have learned the course, then you can understand and exploit the opportunity for extra reading and graft the extra knowledge onto your existing knowledge. This means you can show that you know and understand more than you were given.

Human Drug Metabolism, Third Edition. Michael D. Coleman.
© 2020 John Wiley & Sons, Inc. Published 2020 by John Wiley & Sons, Inc.

The initial source of this extra reading could be a textbook, but at the highest level it is better to use primary knowledge from journals. You may not be able to understand all of a scientific paper, but the introduction and discussion will provide an overview of the subject and if you read a few papers on the subject, you will see that they basically say the same thing.

Another essential component in the construction of a first-class answer is the integration of your knowledge, perhaps with different courses and particularly in how you answer the question. At this point, your answer will be clearly different from the run-the-mill student effort, but it still lacks a vital component for top marks. This final component can only appear once you have mastered the previous ones. You need to be creative. Your answer must show depth of thought about a subject that means you have evaluated the available knowledge and come up with your own view and even your own interpretation on it. At the stage of a final degree, the subject is still evolving, and 5, 10, or 20 years later, you might look back and see how wrong the prevailing wisdom of the time was, in terms of understanding of a particular phenomenon. So it is right to question the ideas and theories of current scientific literature, providing it is done through logical argument. So the five components are:

1. Learn and understand the lecture notes.

2. Do extra reading of primary literature.

3. Demonstrate integration of knowledge.

4. Answer the specific question.

5. Show originality of thought and analysis.

If you can manage all this, your answer will practically leap off the page and stun your examiner, as the majority of students are either not prepared or not intellectually able to go to these lengths to succeed, and this shows in their answers. It is unlikely that most of you can write first-class answers on every subject in your course.

Can a perfect answer be written? I thought not, until 2017, when I finally awarded 100% to an essay written in a Toxicology examination, after twenty-six years of teaching that course – so it can be done.

Perhaps you might not manage 100%, but you should be able to absorb enough information to retain, discuss, and analyse the subject so you can hit a first-class (70% plus) mark on courses that you find particularly interesting. This is because your interest makes the work easier to absorb and analyse. On courses that you really do not enjoy or have always had trouble with, work on them to the point where you can get a good second-class answer no matter what the question.

C.3   Preparation

From the above, you can see what is *required* to achieve a first-class answer. You may be wondering exactly *how* you might achieve this. Up to university level, most students have evolved a method of learning their work that has served them reasonably well, well enough indeed, to actually get to university. This may not be enough to prosper at the highest level. The usual problems encountered are:

- lack of sufficient time devoted to revision;

- poor productivity from time spent;

- inability to recall what was learned;

- examination terror/horror/panic (or worse).

Lack of sufficient time

This is obviously related to commitment and determination. Starting revision two or three weeks before a major examination is startlingly inadequate, even with an eidetic (photographic) memory. To a very great extent, efforts must be made throughout the academic year to understand and process information accrued through lectures/practicals, etc.; certainly, several months before major examinations, revision should have been started.

Many courses tend to communicate the bulk of the work required early in the year, so the majority of the course has usually been covered several months before the finals. It is a case of 'how committed are you?' You may be capable of first-class work, but do not have the inclination to fulfil your potential. Regarding the future outside university, perhaps it is worth considering how much a good degree distinguishes you from 'the pack'. This gives you the widest possible choice in future directions, and in the context of near 50% participation in higher education in the United Kingdom, you will need every edge you can get to find a specific job or position that you covet.

Poor productivity

It is very easy to sit in front of notes or a book for several hours and then adjourn to the nearest café or drinking establishment while transported in a haze of self-congratulation at your academic prowess. It is quite another thing to actively test how much you actually learned, by using past exam papers and working on written tests. This can be a little demoralizing at first, but the truth must be found in terms of how much you have retained. Once you have been through your courses, you must start again and again and as many times as you can to commit them to long- term memory. After the second run-through, this is the time to incorporate extra reading and original thought, as you have worked out most of the basics and have something to build on.

Inability to recall

You may have been annoyed by a friend while watching an old film on the TV, who then announces who did it and why, having seen it once before in a drunken haze at party seven years previously. We appear to have almost unlimited space in our brains for storing information; the problem arises when we try to recall it.

Obviously, you will remember where you went on holiday last year and have problems remembering what you did at 4.25 pm last Tuesday. The key appears to be some form of indexing system that provides a focus for recall. One way around this is to condense each lecture or parcel of a particular subject to a few lines, memorize them (word perfect), then memorize a list of these short condensates for each course. It is time consuming, but it means that you will be able to recall the entire course in serious detail.

One of the keys to good performance in examinations is mastery of detail. This impresses examiners and can be attained with enough commitment. You may find your own way to 'reach' the knowledge you have learned. It is important that you use some method, as otherwise there is little point in learning your work if it cannot be recalled.

Exams: 'The horror ...'

Attitude

The actual examination process can account for a considerable proportion of any lack of fulfilment of potential and often has dogged a student's academic career. There is no easy answer, other than building confidence by using the techniques described above.

Often, poor examination performance in otherwise good candidates is due to 'rabbit in the headlights' – inexplicably strange choices made in the exam. Consider a crude and not entirely appropriate analogy: you will be aware that members of the armed forces, firefighters, and the like, develop fears about what they have to face like anyone else, but they can overcome these fears and function during otherwise terrifying situations by sheer repetition in their training. How often do you read about some heroic individual who usually says something like

'I was staring certain death in the face, then my training kicked in and I saved the day. I didn't have time to be scared, we had trained for this for months, etc.'. If they can do it, you can, by training yourself to face the exam horror by focusing on preparation, building confidence, and looking forward to and embracing the concept of the exam.

Preparation

Before the exam you will have:

- learned the basic notes from the lectures;

- supplemented with primary literature;

- worked on your integration and understanding of the courses;

- where you can, applied original thought to the work;

- ensured that even for courses which are less than thrilling, you can answer any question to high second-class standard.

Another important task prior to the big day is to become completely familiar with the examination layout, question type, number of marks per question and above all, question timing. It is your right as a student to be given full disclosure of the examination structure. Students often ask about past papers; in some cases, a new structure may be introduced and there are no past papers. Again, the university must explain to you the new format fully so there are no surprises.

Past papers

Regarding past papers, where they are the same format as the examination for which you will sit, some institutions have misgivings with providing them as then they feel they also have to provide model answers, which indicate a very detailed level of direction. Students sometimes also write out their own model answers and ask for academics to scrutinize them. This puts the academics in a difficult position, not least because of the time involved, but also because this again borders on micromanaging the students' answers – almost dictating the material to the student in a way. In addition, such dictation can leave the academic open to the accusation that their guidance, however detailed, was not precise enough – not exactly what will be asked in the examination. Indeed, the students can sometimes feel driven to crave and actually demand a disproportionate and unsustainable level of guidance to gain what they perceive as the best university degree.

Examinations: their purpose

Hence, there is a clear problem with student expectation, but crucially comprehension issues arise as well, as to what the examinations are actually really for and in what context. For example, if we are examining a process, such as one in Pharmacy where a medicine is to be dispensed legally, accurately, and safely, then all the stages of the process will be marked and it is straightforward to show that students have, or have not met the appropriate standard. In a life-science course, we assume some detailed knowledge and skills, but what is required at the highest level is to demonstrate student creativity, not the creativity of the academic staff. Guidance must be given on how to analyse, discuss, and consider an issue, but that essential level of creative input must come from the student.

When a student is at school or college up to the age of 18, it is not unreasonable for much of their interaction with their teachers to be 'corner shop'. You know, you have known the proprietor since you were child - 'Yes, I can get that for you next week, how are your parents? OK, pay me tomorrow'. University should be like a supermarket – the staff are there to stock the shelves and guide you to where the coffee and Reese's Peanut Butter Cups are, but nobody – nobody at all – puts any groceries in your basket or trolley. That is your job.

At a good university, the staff do not make you into a scientist, they help you to lay the foundations of turning yourself into one. Moreover, the staff and the students must come to terms with the wide range of student creativity, from nonengagement/zero to very insightful and impressive ideas. In addition, students might lack confidence: What do I know? What can I contribute? Whilst it is true that the academic staff know more information, they are not invariably any more clever or always as insightful as some of the students they teach. What is important is that higher education at this level is not a competition – it should be nearer a partnership, where mostly the staff lead the students towards the truth of the subject, but the students can sometimes take a turn leading also, and together they make a real university in the fullest sense.

C.4   The day of reckoning

Hopefully you have trained yourself to turn a situation you dread into an occasion you actually want to attend so you can shine. The exam slowly changes from a potential nemesis to your personal showcase. Interestingly, no matter how well you got along with a lecturer during a course, as soon as you see the questions, you might immediately but momentarily think rather dark thoughts about your instructor – including an irrational sense of betrayal. This should fire you up to set about the task at hand – the gauntlet has been thrown down.

For those who might not be so bullish, you can avoid the 'rabbit in the headlights' feeling by using some important tips that should be burned into the consciousness so no amount of panic will erase them:

- Read the question: every word will have been scrutinized by maybe a dozen external and internal academics, so clearly every word, punctuation mark, etc. is essential and cannot be ignored.

- If there is a choice, *do the easiest question first*: you gain confidence with 'money in the bank'. You will save time also, which you can use in tackling harder questions. The questions should be done in order of difficulty.

- Do not write things that you know are not relevant to the question: this is double jeopardy because you are getting no marks and are losing time that you could have used to answer a question where you might have excelled.

Finally, you have done all you can do. A university/college is the last opportunity in your life where you will receive an absolutely fair deal, whether or not you might have extenuating circumstances or whatever might apply. The staff and the external examiners will give you every consideration so that the degree you received is a fair and accurate reflection of your commitment and aptitude. You will never again encounter a more 'level playing field', so make the most of it.

C.5   Foreign students

The advice given in this Appendix pertains to some aspects of how academic courses and examination systems operate in the United Kingdom. However, the university system is changing constantly and there are many courses now where there is little or no opportunity to write examination essays, partly because of the perception that they are excessively subjective. This can mean difficulties in defending a given mark in the light of subsequent challenge, or even litigation, perhaps. So many institutions worldwide as well as in the United Kingdom have moved towards multiple choice-style or oral examinations which are evaluated in a different manner from the system that I participated in as a student, postgraduate and then academic. Indeed, many students come to the United Kingdom having had a totally different formal examination experience to the one described in this Appendix.

As an individual from outside the United Kingdom who might wish to study at the undergraduate or postgraduate (Master's) level in the United Kingdom, you probably have considered or are in the process of considering which course or university you wish to attend. You will be looking at the huge issue of living away from home in another country, culture, and language. You will be considering the course structure and content, as well as a host of other specific issues related to the university or the city or town in which it is situated. How many prospective applicants to UK universities consider the examination system used by that university/college in their choices? Do the universities all provide a clear and logical exposition of their system for you to assist in your application? Do car manufacturers ever advertise cars in the crushing reality of giant crawling traffic jams, or do they show their product gliding elegantly and alone, down a ramrod straight, but starkly beautiful desert highway at sunrise?

Therefore, I would recommend that you research this area yourself – ask questions about how the course is examined. You need to consider this information in the light of your own experience from your national college or university system. Whether you are an undergraduate or postgraduate, if you take care to explore this aspect of your possible academic experience in the United Kingdom, then as mentioned in Appendix B, you can't know everything, but you can know more than you know now.

Appendix D
Summary of Major CYP Isoforms and Their Substrates, Inhibitors, and Inducers

This is not an exhaustive list, and many drugs are metabolized by several CYP isoforms to varying degrees.

CYP	Substrates	Inhibitors	Inducers
1A1	Polycyclic aromatic hydrocarbons (PAHs), organochlorine insecticides	α-naphthoflavone	PAHs, smoking, omeprazole organochlorines
1A2	Amitriptyline, imipramine, tizanidine caffeine, fluvoxamine, clozapine, haloperidol, mexiletine, ondansetron, propranolol, tacrine, theophylline, verapamil, R-warfarin, zolmitriptan, PAHs	Amiodarone, cimetidine, furafylline, fluvoximine, ticlopidine ciprofloxacin	PAH amines in barbecued/flame-broiled meat, Brussels sprouts, broccoli, insulin, tobacco, omeprazole, phenytoin, modafinil, smoking
1B1	Tamoxifen, polycyclic aromatic, aromatic amines, aflatoxins, docetaxel, oestrogens	PAHs (e.g. pyrenes) stilbenes, some flavonoids and coumarins, resveratrol, mitoxantrone, flutamide, taxols	PCBs(Aroclor 1254), β-naphthoflavone, PAHs from smoking

(Continued)

Human Drug Metabolism, Third Edition. Michael D. Coleman.
© 2020 John Wiley & Sons, Inc. Published 2020 by John Wiley & Sons, Inc.

CYP	Substrates	Inhibitors	Inducers
2A6	Coumarins, aflatoxins, nicotine, tegafur, efavirenz, valproate, halothane, 1,3 butadiene, letrozole, valproate	Tranylcypramine, methoxsalen, grapefruit juice	Phenobarbitone, rifampicin
2B6	Bupropion, coumarins, cyclophosphamide, mephenytoin, metha-done, methoxychor (pesticide), PCP, ketamine, efavirenz	Tranylcypramine, thiotepa, ticlopidine thiotepa	Phenobarbitone, rifampicin, efavirenz modafinil, methoxychlor
2C8	Amodiaquine, cerivastatin, paclitaxel, tolbutamide montelu-kast, enzalutamide, repaglinide	Quercetin, glitazone drugs, gemfibrozil efavirenz, gemfibrozil (1-O-β-glucuronide), quercetin, saquinavir	Phenobarbitone Rifampicin
2C9	Amitriptyline, dapsone, fluoxetine, fluvastatin, NSAIDs, phenytoin, sulphonyl ureas, tamoxifen, S-warfarin, losartan, flurbiprufen.	Isoniazid, fluvastatin, fluvoxamine, lovastatin, sulphafenazole diclofenac	Secobarbitone, rifampicin enzaluta-mide, elvitegravir
2C19	Barbiturates, citalo-pram, mephenytoin, proton pump inhibitors e.g. omeprazole, phenytoin, proguanil, R-warfarin, clopi-dogrel, voriconazole	Tranylcypramine, cimetidine, fluox-etine, ketoconazole, ticlopidine fluvoxamine	Carbamazepine, norethindrone, prednisone, rifampicin enzalutamide
2D6	TCAs, antipsychotics (haloperidol, etc.), anti-arrhythmics: flecainide, mexiletine, beta-blockers (e.g. timolol, S-metoprolol, bufuralol), MDMA, SSRIs, opiates (e.g. tramadol, codeine), venlafaxine	Amiodarone, cimetidine, ranitidine, histamine H-1 receptor antagonists (e.g. chlorphe-niramine), quinidine, SSRIs (e.g. fluox-etine, paroxetine), St John's Wort	Pregnancy, rifamy-cins, corticosteroids

CYP	Substrates	Inhibitors	Inducers
2E1	Benzene, chlorzoxazone, ethanol, flurane anaesthetics (e.g. halothane), paracetamol	Sulphides (e.g. DDC, diallyl sulphide, disulfiram)	Ethanol, acetone isoniazid starvation, diabetes
3A4/5	Aflatoxin B1, antihistamines (terfenadine, astemizole), calcium channel blockers (e.g. felodipine), cannabinoids, cyclosporine, macrolides (e.g. erythromycin), opiates (buprenorphine), fentanyl, sufentanyl, protease inhibitors (e.g. ritonavir), statins (except pravastatin), tacrolimus, paclitaxel midazolam, nefazodone, dasatinib paclitaxel	Amiodarone, azoles (ketoconazole fluconazole, voriconazole), cimetidine, grapefruit juice, macrolides, endogenous steroids, protease inhibitors	Barbiturates, carbamazepine, glucocorticoids, glitazones, nevirapine, phenytoin, rifampicin, St John's Wort, modafinil, mitotane, nevirapine, efavirenz, clotrimazole, enzalutamide, dicloxacillin, aprepitant

Index

Human Drug Metabolism, Third Edition. Michael D. Coleman.
© 2020 John Wiley & Sons, Inc. Published 2020 by John Wiley & Sons, Inc.